Computational Neuroscience
Trends in Research, 1998

Computational Neuroscience
Trends in Research, 1998

Edited by

James M. Bower
California Institute of Technology
Pasadena, California

Springer Science+Business Media, LLC

Library of Congress Cataloging-in-Publication Data

Computational neuroscience : trends in research, 1998 / edited by
James M. Bower.
       p.   cm.
    "Proceedings of the [Sixth] Annual Computational Neuroscience
Conference, held July 6-10, 1997, in Big Sky, Montana"--CIP t.p.
verso.
    Includes bibliographical references and index.
    ISBN 978-0-306-45919-1   ISBN 978-1-4615-4831-7 (eBook)

    DOI 10.1007/978-1-4615-4831-7
    1. Computational neuroscience--Congresses.   I. Bower, James M.
II. Computational Neuroscience Conference (6th : 1997 : Big Sky,
Montana)
QP357.5.C642   1998
573.8'01'1--dc21   98-25793
                                                              CIP

Proceedings of the Annual Computational Neuroscience Conference,
held July 6 – 10, 1997, in Big Sky, Montana

ISBN 978-0-306-45919-1

PREFACE

This volume includes papers presented at the Sixth Annual Computational Neuroscience meeting (CNS*97) held in Big Sky, Montana, July 6–10, 1997. This collection includes 103 of the 196 papers presented at the meeting. Acceptance for meeting presentation was based on the peer review of preliminary papers originally submitted in January of 1997. The papers in this volume represent final versions of this work submitted in January of 1998. Taken together they provide a cross section of computational neuroscience and represent well the continued vitality and growth of this field.

The meeting in Montana was unusual in several respects. First, to our knowledge it was the first international scientific meeting with opening ceremonies on horseback. Second, after five days of rigorous scientific discussion and debate, meeting participants were able to resolve all remaining conflicts in barrel race competitions. Otherwise the magnificence of Montana and the Big Sky Ski Resort assured that the meeting will not soon be forgotten.

Scientifically, this volume once again represents the remarkable breadth of subjects that can be approached with computational tools. This volume and the continuing CNS meetings make it clear that there is almost no subject or area of modern neuroscience research that is not appropriate for computational studies.

In order to emphasize the interrelated nature of computational neuroscience, the papers in this volume are grouped into very general levels of investigation and analysis. The papers found in each category represent research undertaken with a wide range of experimental preparations, analysis techniques, and technical approaches. The range of subjects presented here is unusual in modern biology, and one of the strengths of our field and of this meeting. This volume represents work focused on figuring out how brains compute rather than on a particular animal, brain structure, or technique.

For a student or someone new to the field, this book provides an overview of some of the best work currently being done in this field. For a library, this book is the best available representation of the current state of computational brain studies. For those that participated in the meeting in Montana, it is my hope that this book reminds you both of the exciting science we heard about AND what it was like to spend five days under Montana's big sky.

Jim Bower

REVIEWERS FOR CNS*97

The papers presented in this volume were submitted in January of 1997. Each submitted paper was peer reviewed prior to its acceptance at the meeting under the supervision of the program committee. The meeting organizers are particularly thankful for the efforts of the reviewers in assuring acceptance of the highest quality papers

CNS*97 ORGANIZING AND PROGRAM COMMITTEE

- Jim Bower (California Institute of Technology)
- John Miller (University of California, Berkeley)
- Charlie Anderson (Washington University)
- Axel Borst (Max-Planck Institute, Tuebingen, Germany)
- Leif Finkel (University of Pennsylvania)
- Anders Lansner (Royal Institute of Technology, Sweden)
- Linda Larson-Prior (Pennsylvania State University Medical College)
- Christiane Linster (Harvard University)
- Maureen Rush (California State University, Bakersfield)
- Karen Sigvardt (University of California, Davis)

CNS*97 REVIEWERS

Larry F. Abbott, Brandeis University; Charles H. Anderson, Washington University School of Medicine; Upinder S. Bhalla, National Centre for Biological Sciences; Alexander Borst, Max-Planck-Society; Ron Calabrese, Emory University; Erik De Schutter, University of Antwerp–UIA; Bard G. Ermentrout, University of Pittsburgh; Leif H. Finkel, University of Pennsylvania; Michael E. Hasselmo, Harvard University; William R. Holmes, Ohio University; Gwen Jacobs, Montana State University; Leslie M. Kay, Caltech; Nancy Kopell, Boston University; Anders Lansner, Royal Institute of Technology; Linda J. Larson-Prior, Pennsylvania State Univ. Med. College; Gilles Laurent, Caltech; Christiane Linster, Harvard University; William W. Lytton, University of Wisconsin; Bartlett W. Mel, University of Southern California; Kenneth D. Miller, University of California at San Francisco; John Miller, Montana State University; Mark E. Nelson, University of Illinois; Bruno A. Olshausen, University of California at Davis; Michael Paulin, University of Otago; Klaus Pawelzik, Max-Planck-Institut; John Rinzel, MRB/NIDDK/NIH; Maureen E. Rush, California State University at Bakersfield; Idan Segev, Hebrew University of Jerusalem; Shihab Shamma, University of Maryland; Gordon Shepherd, Yale University School of Medicine; Karen A. Sigvardt, University of California at Davis; Nelson Spruston, Northwestern University; Michael Stiber, University of Washington at Bothell; Greg Stuart, Australian National University; David S. Touretzky, Carnegie Mellon University; Philip S. Ulinski, University of Chicago; Gene V. Wallenstein, Harvard University; Charles Wilson, University of Tennessee

CNS*97 CONFERENCE SUPPORT

Judy G. Macias (California Institute of Technology)
Monica Oller (California Institute of Technology)

SUPPORTING AGENCIES

National Institute of Mental Health and National Science Foundation

CONTENTS

SECTION II: CELLULAR

SECTION III: NETWORK

SECTION IV: SYSTEMS

SECTION V: METHODOLOGY

Computational Neuroscience
Trends in Research, 1998

RESPONSE-FIELD DYNAMICS IN THE AUDITORY PATHWAY

D.A. Depireux, Powen Ru, S.A. Shamma and J.Z. Simon

Center for Auditory and Acoustic Research
Institute for Systems Research
University of Maryland
College Park MD 20742 U.S.A.

I. INTRODUCTION

Natural Sounds are characterized by loudness, pitch and timbre (i.e. the dynamic envelope of the spectrum).

Our Question: how is timbre encoded in primary auditory cortex (AI)?

Our Approach: beyond the sensory epithelium, principles used by neural systems are universal. So we view the basilar membrane as a 1-D retina, and use the method of gratings to study single units in AI.

Important Concepts

• *Response Field* (RF): range of frequencies that influence a neuron (as a function of time).

• *Ripple*: broadband sound of sinusoidally modulated spectral envelope ("auditory grating").

• Data analysis based on **linear systems theory** to characterize response field. By varying ripple frequency and velocity, we measure the transfer function. The inverse Fourier transform gives the *spectro-temporal* RF (STRF).

We Find:

• Cells can be characterized by an STRF, separable or not.

• Cells behave like a linear system: when presented with a sound made of up the sum of several profiles, the response of the cell is the sum of the responses to the individual profiles.

• Response fields in AI tend to have characteristic shapes both spectrally and temporally.

• Cortical cells with all center frequencies, all spectral symmetries, bandwidths, latencies and temporal impulse response symmetries.

We Show predictions of single-unit responses in AI to complex spectra, verifying:

• **Linearity** of AI responses to all types of dynamic ripples: responses to up and down moving ripples can be superimposed linearly to predict responses to arbitrary combinations of these ripples.

• **Separability** of spectral and temporal measurements of the responses: spectral properties can be measured independently of the temporal properties.

We Conclude: Because of linearity of cortical responses with respect to spectral envelope, we can use the ripple method to characterize auditory cortical cell responses to dynamic, broadband sounds. AI decomposes the input spectrum into different spectrally and temporally tuned channels. Another view is that a population of such cells effectively represents the input spectrum at multiple scales. AI performs a multi-dimensional, multi-scale wavelet transform of the auditory spectrum. The combined spectro-temporal decomposition in AI can be described by an affine wavelet transformation of the input, in concert with a similar temporal decomposition.

II. THEORY

A. Spectro-Temporal Fourier Transform

Since the cochlea performs (to first order) a Fourier transform along the log frequency axis, we measure spectral distance in log(frequency). Since the Fourier transform is time-windowed, we also require a time axis. For this reason we will focus attention on two-dimensional functions of log(frequency) and time. For linear systems, the spectro-temporal domain and its Fourier domain are equivalent. Analysis is often conceptually simpler in the Fourier domain. Real functions in the spectro-temporal domain give rise to complex conjugate symmetric functions in Fourier space.

The next figure illustrates the envelope of a speech fragment (*"Water all year"*), in both its spectro-temporal and Fourier representations. In the Fourier representation, the function is highly concentrated near zero.

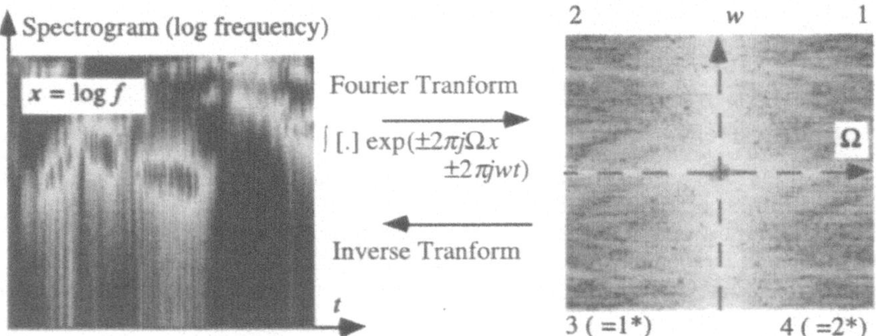

Figure 1: w = ripple velocity, Ω = ripple frequency

B. Spectro-Temporal Response and the Fourier Transform (Transfer Function)

Properties of AI cells are typically derived using pure tones or clicks akin to using dots of light or flashes to study cells in the visual pathway. We use the auditory version of drifting gratings[1] to characterize response properties of cells to dynamic broadband sounds, so as to gain insight to how timbre is encoded. The method presented here allows us to simulta-neously determine temporal and spectral properties, using the same set of stimuli for a vari-ety of cells. We use the Response Field (RF), a function measured using broadband sounds. It is given in the form of a function, with positive values describing excitation) and negative values inhibition.

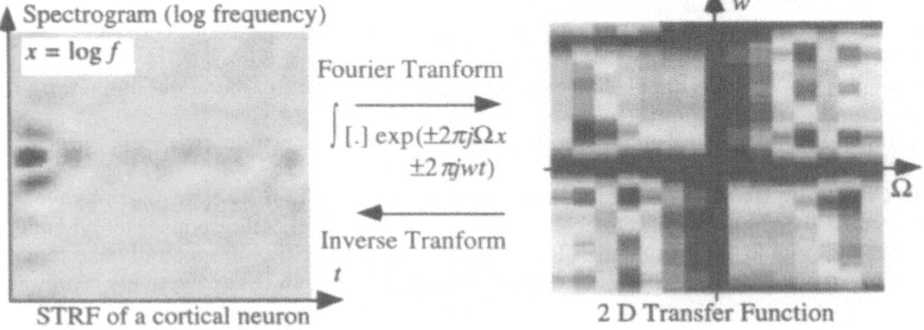

Figure 2: Spectro-temporal RF of a neuron, and its Fourier dual, the transfer function.

C. Spectro-Temporal Stimulus and the Fourier Transform

Natural sounds, such as environmental sounds and speech, are classified along several perceptual axes: loudness, pitch and timbre. Pitch is what changes when we pronounce the same vowel with different tonal heights. Timbre is what changes when, keeping the same tonal height, we pronounce different vowels. In this work we address timber only. Figure 3 illustrates the spectral envelope of a sound, i.e. its timbre. It can be viewed as a low-order polynomial fit of the (time-windowed) spectrum of the sound. A common method for the extraction of the envelope is the Linear Predictive Method (LPC).[2]

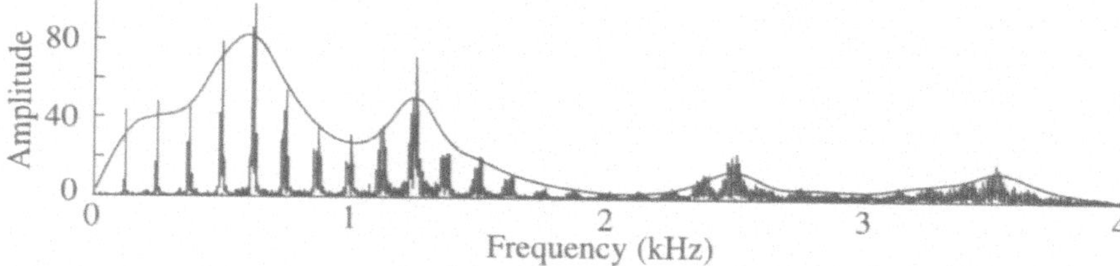

Figure 3: The spectrum of /aa/ spoken by one author, with the spectral envelope superimposed.

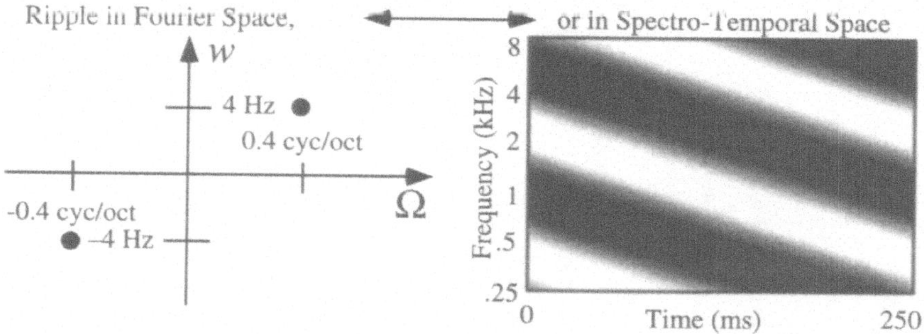

Figure 4: Points in the Fourier space correspond to broadband sounds with a sinusoidally modulated spectral and temporal envelope. The Fourier transform of a ripple has support only on a single point (and its conjugate).

D. Quadrant Separability

An STRF can fall into one of three categories:
• **Non-separable**: The transfer function is an arbitrary function of ripple frequency and ripple velocity.
• **Quadrant separable**: The transfer function within each quadrant is a product of a function of ripple frequency and a function of ripple velocity. The *envelope* of the STRF is the product of a function of spectrum and a function of time.
• **Fully separable**: The transfer function is the product of a function of ripple frequency and ripple velocity everywhere. The resulting STRF is a product of a function of spectrum and a function of time.

E. Linearity

The guiding principle behind our research program is that cells behave like a linear system with respect to the spectral envelope. The proof of linearity is that when cells are presented with a sound made of up the sum of several spectral envelopes, the response, as measured assuming a rate code, is the sum of the responses to the individual envelopes. A response linear in frequency and time is characterized by a two-dimensional impulse response (or time-dependent response field) or equivalently, its two-dimensional Fourier transform.

As indicated for a 4 Hz ripple in Figure 5, the response of a cell as a function of time is modulated at the same (temporal) frequency as that of the stimulus. Therefore, we just have to extract the phase and the amplitude of the response.

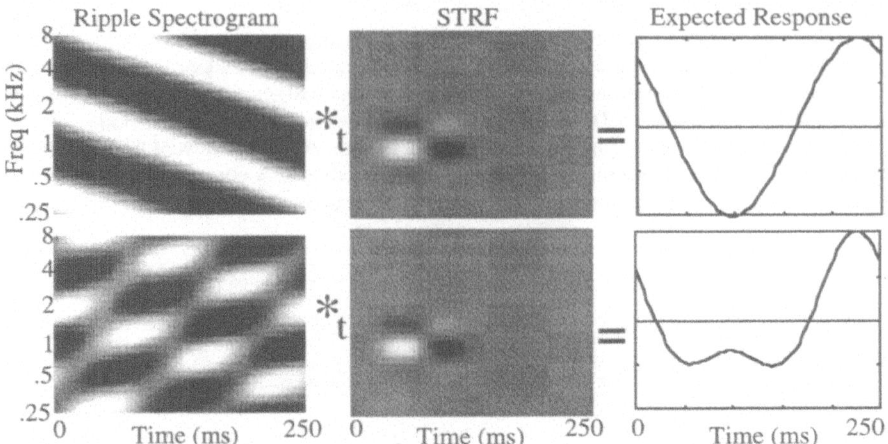

Figure 5: Assuming linearity, the STRF predicts the response to any broadband dynamic stimulus, including single ripples moving in either direction (first row) and combinations of upward and downward moving ripples.

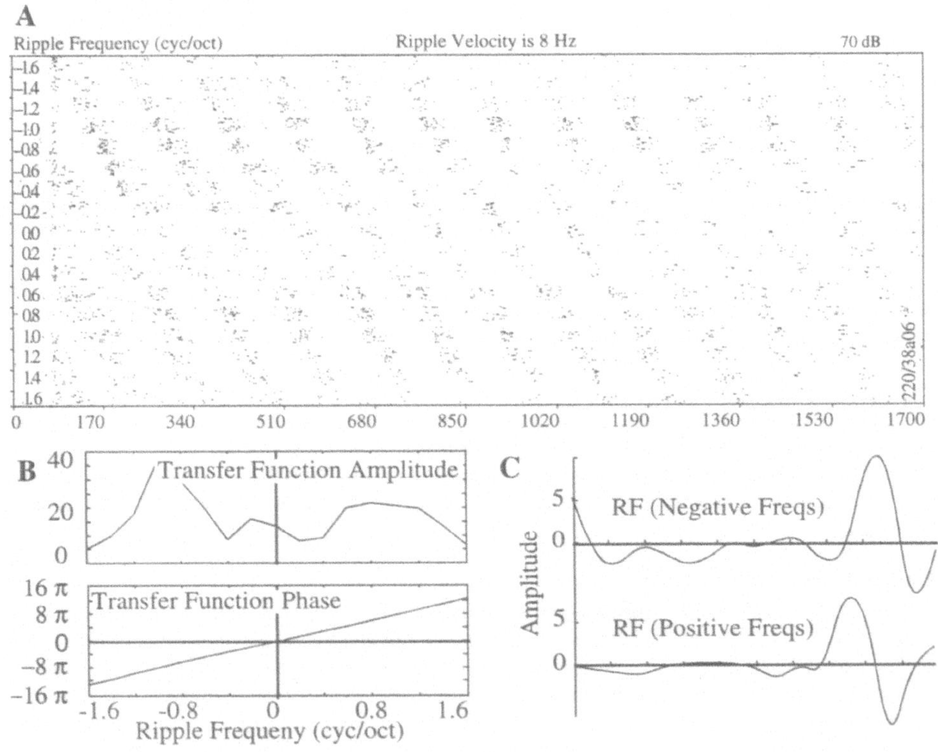

Figure 6: Data analysis using ripples of fixed velocity and varying frequencies. **A**: Raster plot of responses. Each point represents an action potential, and each paradigm is presented 15 times. **B**: Magnitude and phase of the period histogram fits. **C**: Separate inverse Fourier transforms for positive and negative ripple frequencies of **B**, obtaining a slice of the RF.

4

III. EXPERIMENT AND RESULTS

Data were collected from the auditory cortex of domestic ferrets anesthetized with ketamine and xylazine; with sounds presented in the contralateral ear. AI cells were typically isolated in cortical layers III and IV[3]. For details see Shamma et al.[3]

A. Obtaining the Transfer Functions

We measure cells' transfer functions by presenting, at a fixed ripple frequency, ripples of varying velocities; then, for a fixed velocity, we present ripples of varying ripple frequencies.

A typical example of the analysis is shown in Figure 6. Ripples were presented at 8 Hz, for ripples frequencies from −1.6 cyc/oct to 1.6 cyc/oct in steps of 0.2 cyc/oct, with the ripple starting to move at t = 0 ms, and being acoustically turned on starting at 50 ms. Each ripple is presented 15 times. Once the onset activity has died away, the cell goes into a steady-state response. For each ripple frequency, we compute a period histogram excluding the onset response. To assess the strength and phase of the phase-locked response, we compute the phase and the strength of the response of the cell by Fourier transforming a 16 bin period histogram of the response, extracting the phase and amplitude of $T(\Omega, w = 8\ \text{Hz})$ from the first component of the transform

The magnitude and phase of the transfer function is shown in panel **B**. In **C**, we have inverse Fourier transformed separately the transfer function in quadrant 1 and 2, or equivalently for down- and up-moving ripples, after removing the constant (temporal) phase factor $2\pi w \tau_d + \theta$, where $w = 8\ \text{Hz}$.

The extraction of the temporal cross-section of the transfer function as in Figure 6 would proceed the same way. Ripples are presented at 0.4 cyc/oct, for ripple velocities from −24 Hz to 24 Hz in steps of 4 Hz. For each ripple frequency, we compute a period histogram to assess the strength and phase of the phase-locked response. The amplitude and phase of the response is then evaluated by performing a Fourier transform of the data, and extracting the phase and the amplitude of $T(\Omega = 0.4\ \text{cyc/oct}, w)$ from the first component of the Fourier transform. We inverse Fourier transformed separately the transfer function for down- and up-

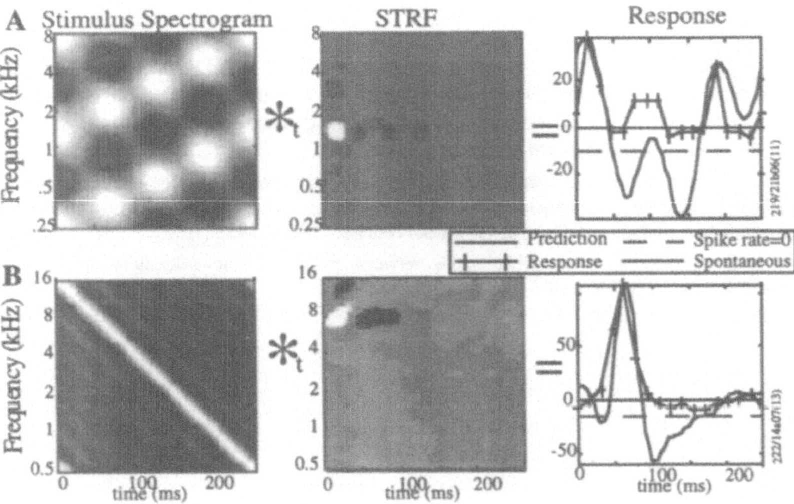

Figure 7: Predictions of response to complex dynamic spectra using the STRF. **A** A prediction is computed by convolution (along t) of the STRF with the spectrogram The stimulus shown consists of 2 ripples (0.4 cyc/oct at 12 Hz and −4 Hz). The prediction is shown juxtaposed with the actual response (crosses) over one stimulus period. **B** Another example: the stimulus consists of a combination of ripples with ripple frequencies 0.2 cyc/oct at 4 Hz, 0.4 cyc/oct at 8 Hz, … 1.2 cycles/octave at 24 Hz, in cosine phase, resulting in an FM-like stimulus.

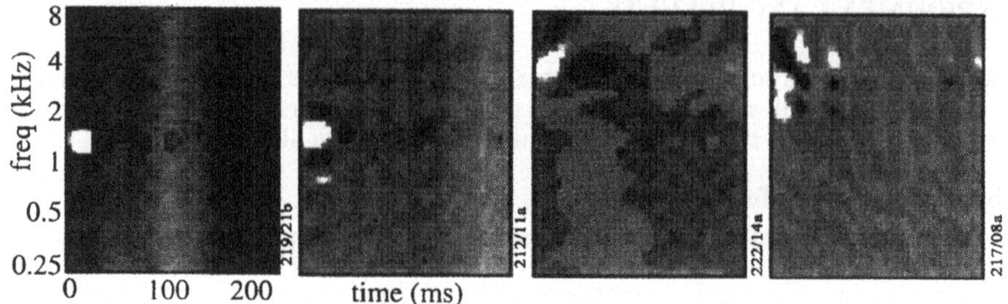

Figure 8: Examples of Spectro-Temporal Response Fields.

moving ripples, after removing the constant (spectral) phase factor $2\pi\Omega x_m + \phi$, where $\Omega = 0.4$ cyc/oct.

B. Separability and Linearity

In vision, some cortical simple cells are fully separable,[4] but all are at least quadrant separable.[5] We have found both types in AI as well; Figure 8 shows examples of each. A fully separable cell has an STRF that is a simple product of an RF and an IR, as in the left two examples. A quadrant separable cell, as in the right two examples, does not, since it has different responses for upward and downward moving ripples: the STRF is not symmetric about x_m. The separability of a cell does not affect the linearity of responses to ripple combinations.

IV. ACKNOWLEDGEMENTS

Work supported by grants from the Office of Naval Research (MURI grant N00014-97-1-0501), from the NIDCD (T32 DC00046-01), and the National Science Foundation (NSFD CD8803012).

V. REFERENCES

1. R.L. De Valois and K.K. De Valois, Spatial Vision, Oxford University Press, New-York (1988).

2. L.R. Rabiner and R.W. Schafer, Digital processing of speech signals, Prentice-Hall, New-Jersey (1978).

3. S.A. Shamma, J.W. Fleshman, P.R. Wiser and H. Versnel, Organization of response areas in ferret primary auditory cortex, J. Neurophys. **69**, 367-383 (1993).

4. J. McLean and L.A. Palmer, Organization of simple cell responses in the three-dimensional frequency domain. Vis. Neurosc. **11**, 295-306 (1994). G.C. DeAngelis, I. Ohzawa and R.D. Freeman, Receptive-field dynamics in the central visual pathways. Trends Neurosc. **18**, 451–458 (1995).

5. B.W. Andrews and D.A. Pollen, Relationship between spatial frequency selectivity and receptive field profile of simple cells, J. Physiol. (London) **287**, 163–176 (1979). S.M. Friend and C.L. Baker, Spatio-temporal frequency separability in area 18 neurons of the cat, Vision Res. **33**, 1765–1771 (1993).

RAPID CATEGORIZATION OF EXTRAFOVEAL NATURAL IMAGES : IMPLICATIONS FOR BIOLOGICAL MODELS

Michèle Fabre-Thorpe, Denis Fize, Ghislaine Richard and Simon Thorpe

Centre de Recherche Cerveau et Cognition (UMR 5549)
Faculté de Médecine de Rangueil
133 route de Narbonne
31062 Toulouse, France

INTRODUCTION

Despite intensive research over the past 30 years, the performance of artificial visual systems in object recognition is still poor when compared with humans. For humans, the identification of objects in natural scenes appears both effortless and fast. Just how fast was a question we addressed recently in a study that associated behavioral measurements and event related potential (ERP) recordings[1]. The task used was a go/no-go visual categorization task in which human subjects had to respond when a photograph of a natural scene contained an animal. The photographs had never been seen before and were flashed centrally on a screen for only 20 ms. Humans scored 94% correct, moreover, their ERPs recorded on animal and non-animal trials showed a clear difference on all frontal electrodes that started around 150 ms after stimuli onset. Thus, it appears that the human visual system can process such previously unseen complex natural scenes in less than 150 ms, a level of performance well above that of any currently available artificial system. This is despite the fact that the neurons that constitute the human visual system are relatively slow - firing rates rarely exceed 200 Hz, a value far slower than the transistors of a modern microprocessor, which can change state over a million times faster.

This original experiment provided strong data against two commonly used "reasons" for the superiority of biological systems : the ability to use intelligent eye movements and the ability to make use of contextual information. The first of these is ruled out by our use of a very brief 20 ms presentation which effectively blocked the use of exploratory eye movements, while the second is ruled out by the use of a very wide range of stimuli and the use of trial unique presentations. However, in this original study, the images were presented in central vision, and since they were obtained from commercial sources, one could argue that targets will have been centered by the photographer. Thus the image of the target would naturally fall near to the fovea, making unnecessary the use of a saccade to center the target. The fovea is a very special zone on the retina in terms of both the density of cells and the size of its representation in the cortex. We therefore wondered how performance would be affected by presenting the images in extrafoveal vision. To investigate this question, subjects were required to perform the same task, but the images were presented at random at one of three positions, either centrally, on to the left or right of the fixation point. Moreover the original task was proposed alternatively with the three position task in order to determine the cost of widening the attention across the visual field.

MATERIAL AND METHODS

Photographs of natural scenes were presented on a computer monitor, and subjects had to perform a go/no-go categorization task, releasing a button if the image contained an animal. All the pictures were natural scenes taken from a vast CD-ROM data bank (Corel) allowing access to several thousands of stimuli. Targets included fish, birds, mammals, reptiles and insects presented in their natural environments. Distractors included landscapes, trees, buildings, flowers and fruits. Subjects had no a priori information concerning the type of animal to look for, its size, its position in the image or even the number of animals present. In the original 1-position task (1-P), the images (384 by 256 pixels, viewed from about 1 m corresponding to 6.7° by 4.5° of visual angle) were presented centrally around a fixation point; in the new 3-position task (3-P) the images were presented at random at one of three locations on the screen, either centrally, or at an eccentricity of 3.6° to the left or right of the central fixation point. Because of the unpredictable location of the stimulus in the 3-P task, the subjects were compelled to widen their attention on a visual field that was just over twice the size than in the 1-P position task. Each visual stimulus was only presented once in order to avoid learning, and it was flashed for a very short duration (20 ms). This very short stimulus duration had two main advantages : (i) the subjects had no time to perform exploratory eye movements, and (ii) the brief presentation of stimuli in the left or right hemifield allowed the lateralisation of the visual inputs to the controlateral visual cortical areas and provided the data to compare the performance of both hemispheres on such a task.

A group of 13 human subjects (7 males and 6 females, aged 22-55 years) were tested on both tasks using blocks of 100 trials. During one recording session, 1-P and 3-P task blocks were randomly presented to the subjects who were always aware of the task they were involved in. Behavioral data and associated evoked potentials were recorded. Behavioral performance (proportion of correct trials and reaction time distributions) were statistically compared for different task conditions (using chi-square test and two tailed t-test for each subject, paired Wilcoxon or paired t-test for the whole group of subjects).

Event-Related Potentials were recorded using a 32-electrodes Electrocap bonnet connected to a Neuroscan SynAmps system sampling at 1 000 Hz. The results reported here concern only 11 of the 13 human subjects as the recordings were not satisfactory for the two other ones. ERPs on target and non-target trials were averaged for each given condition. The differential brain activity between the animal and non-animal trials was computed as in the previous study [1] and for all different conditions, with particular attention being paid to its onset latency and magnitude. The onset latency of the differential brain activity was evaluated using the statistical criteria proposed by Rugg et al.[2]; at least 15 consecutive t-test values had to exceed the 0.05 level of significance.

RESULTS

This study provided three main results. (i) Widening of the attentional window to a much larger part of the visual field had essentially no effect on performance or brain processing; (ii) presenting the images with an eccentricity of 3.6° had very little effect on the categorization performance and associated brain activity; and (iii) no hemispheric superiority could be detected when comparing categorization performance and brain activity for left and right extrafoveally presented images.

Categorization performance in the 1-P task

In the present study, the group of 13 subjects averaged 94.5% of correct responses with a median reaction time (RT) at 410 ms (mean RT : 421 ms) in the 1-P task.

The ERPs recorded during the task performance were analyzed and the figure 1 shows the averaged responses of all seven frontal electrodes computed separately for animal and non-animal trials. The difference between target and non target brain activities clearly shows the sharp onset of this differential activity at 150 ms. These results are a straight replication of the recently published study[1] in which human subjects averaged 94% correct with a median RT at 445 ms, and a brain differential activity between animal and non-animal trials at 150 ms. Another finding is that although the present group of subjects was on average about

20 ms faster than the original group of subjects, this rapidity was apparently not the result of faster visual processing since the latency of the differential brain activity was very consistent. The subjects were probably just faster to trigger their motor responses.

Categorization performance for the centrally presented stimulus in both tasks

We were particularly interested in the effects that would be induced by the increased number of possible spatial locations in which the stimuli could appear, as the subjects would have to share their attention between different parts of the visual field. We thus analyzed separately the behavioral performance obtained for the centrally presented stimuli in the 3-P task, considering that its apparition was equiprobable with the 2 other lateral locations. The behavioral results were identical to those obtained in the 1-P task condition : the 13 subjects averaged 94.6% of correct responses with a median RT at 407 ms (mean RT : 419 ms). These data clearly indicate that there was no time penalty despite the very high demand made on the visual system by the task itself and by the increase of the attentional demand.

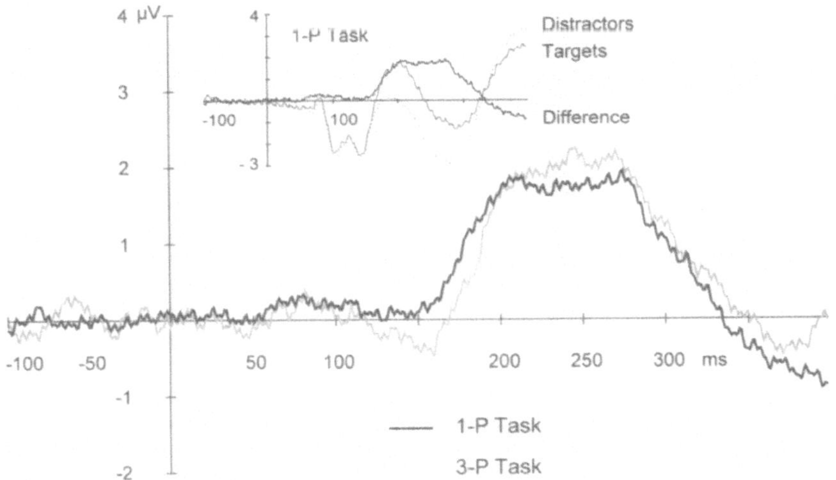

Figure 1. The differential activities calculated in the 1-P task (black line) and in the 3-P task (gray line) for the centrally presented stimuli are compared. The insert illustrates, for the 1-P task, how the difference curve was calculated. The evoked potentials were averaged on all seven frontal electrodes of 11 human subjects separately for target trials (thin gray line), and non-target trials (dotted gray line), the result is the same difference curve (black line) that is presented in the main picture compared with the 3-P task difference curve. Note that there is no difference between the two task conditions for the onset of the differential activity.

The ERPs recorded to centrally presented stimuli during the 3-P task were compared with those recorded in the 1-P task. The brain activity difference curves between the target and non target trials recorded in both tasks (figure 1) superimposed well before stimulus onset and for some time after stimulus onset. They both diverge very sharply from the baseline reaching roughly the same amplitude. However, the latencies of both curves are remarkably similar. No statistical difference could be found when comparing the two curves averaged for the whole group of subjects or when considering individually the latencies of both curves for each subject and running a paired t-test.

The behavioral results in terms of accuracy and reaction time, together with the onset of the differential activity between target and non target conditions were both very consistent regardless of the 1-P or 3-P task conditions. Thus increasing the attentional demand by multiplying the number of spatial locations where the stimulus could randomly appear did not induce any significant cost in terms of information processing required to reach decision.

Categorization performance in extrafoveal vision

The other question that was addressed in this study concerned the performance impairment that would be induced by presenting the natural image extrafoveally. In the 3-P task the performance of our group of 13 subjects[3] was analyzed separately for left, right and central stimuli, each of which were intermixed at random and equiprobable. The behavioral performance for laterally and centrally presented stimuli were compared and showed that the expected performance impairment for laterally presented stimuli was much smaller than one could have expected given the fact that acuity drops rapidly as retinal eccentricity is increased. For all extrafoveally presented stimuli, the average performance of the 13 subjects was 91.4% correct with a median RT of 418 ms. The behavioral time penalty for processing extrafoveally presented stimuli could thus be evaluated at about 10 ms.

The comparison of the behavioral results for the left (L) and the right (R) presentations of the natural images showed that the level of performance was affected similarly (figure 2), for the percentage of correct responses (L : 91.7%, R : 91.1%) and for the RT recorded for correct responses (median RT : 418 ms on the left and 417 ms on the right).

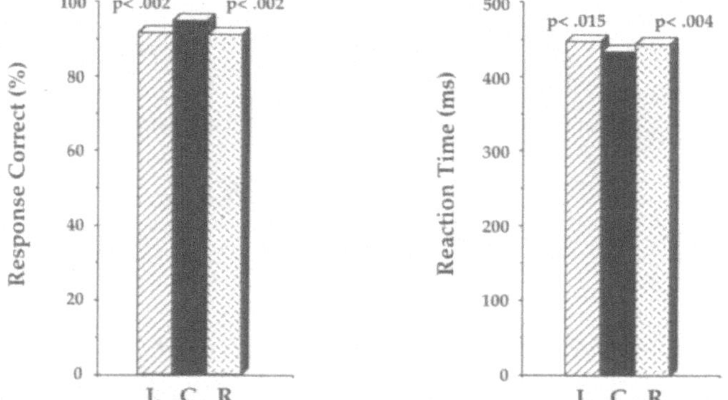

Figure 2. Averaged performance of 13 human subjects on the 3-P task. Percentage of correct responses (left) and median reaction time of correct go responses (right). Solid Bar : centrally presented stimuli (C), hatched pattern : stimuli presented in the left hemifield (L), stippled pattern : stimuli presented in the right hemifield (R).

The performances impairments for L or R presented stimuli when compared to centrally presented stimuli (C) were statistically significant both for the decrease of correct responses (paired Wilcoxon test: p<.002 for L vs. C and R vs. C) and for the increase of go responses RT (paired t-test: p<.004 for L vs. C, p<.015 for R vs. C). When considering these four criteria (decrease of correct responses on the L or on the R side and increase of reaction time for L or R presented stimuli) individually for each of the 13 subjects, only 2 subjects showed a deficit on all of them whereas they were all unaffected for 3 other subjects.

The associated ERPs were analyzed separately for animal and non-animal trials and for each of the 3 possible locations of the stimuli on the screen. The onset of the differential activity was less sharp for laterally presented stimuli (L, and R) than for centrally presented stimuli, the slope of the two difference curves was less pronounced, and the maximal amplitude of the differential activity was lower. Thus, although this was not statistically significant it may be that processing of a laterally presented stimulus induces a time penalty of about 10-20 ms. The lower amplitude may just be due to the larger number of cells recruited when the stimulus image is formed on the fovea.

This study showed that this rapid visual categorization task can be performed with very little performance deterioration when the natural images are presented extrafoveally (at 3.6° eccentricity) rather than foveally.

As in the study by Biederman and Cooper[3] we found no evidence for any hemispheric specialization in this task, since there were no consistent differences between the behavioral

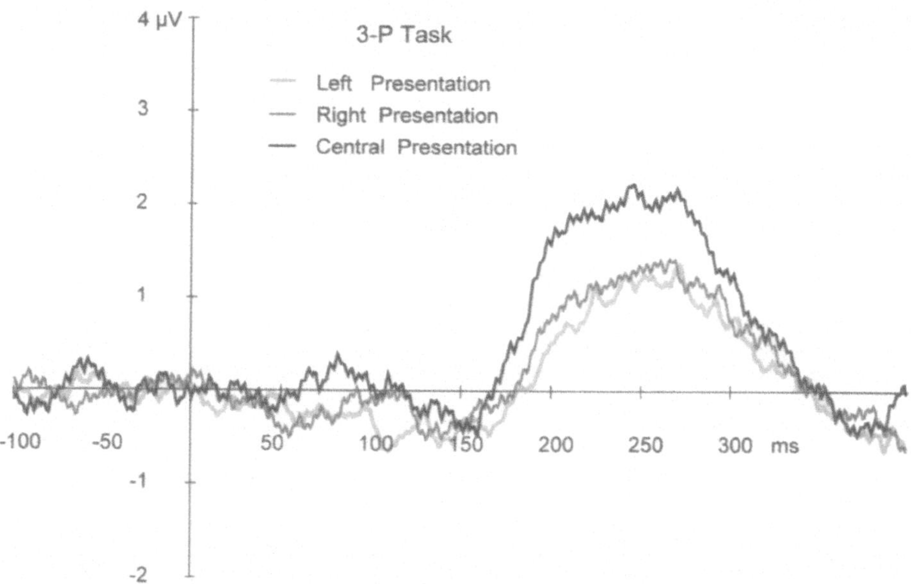

Figure 3. Averaged difference curves computed from evoked potentials recorded on target and non-target trials on all seven frontal electrodes of 11 human subjects performing the 3-P task. Black line : centrally presented stimuli, thick gray line : stimuli presented on the left, thin gray line : stimuli presented on the right.

or ERPs results for images presented in the left and right hemifields. We thus failed to replicate the left hemisphere superiority in accessing stored representation of known objects reported by Vitkovitch and Underwood[4].

DISCUSSION

The study provides two main types of results. First, by alternating between blocks of trials where the images were always presented in the same central position, and blocks where the position was varied at random, we were able to show that the unpredictability of stimulus position had essentially no effect on central performance, either in terms of accuracy and behavioral reaction time or when considering the minimal processing time necessary to reach a decision determined by the ERPs. Secondly, we showed that performance was only slightly less good with laterally presented stimuli, and that time penalty for behaviorally responding to extrafoveally presented stimuli did not exceed 10-20 ms, even when the location of the stimulus presentation was totally unpredictable.

Increasing the attentional dedication to a spatial location also increases the amount of stimulus information extracted from that given location[5, 6]. The attentional resources can be all allocated to a small area of the visual field with a very high perceptual resolution, alternatively they may be allocated to a larger field but with a cost on the perceptual resolution[7]. In such conditions one may have reasonably expected that performance on the centrally presented stimulus in the 3-P condition would have been altered either in RT or/and in correct response rate when compared to the 1-P condition. The absence of performance impairment showed that the cost of widening the attentional window could just have been filled by increasing the attentional resources allocated to the task. In that case this means that the task can be done with broadly distributed attentional resources of relatively low perceptual resolution.

Presenting the stimuli laterally had little effect on performance since accuracy was only 3-3.5 % lower than at the central position and the RT increase was about 10 ms. Although statistically significant, these relatively small effects on performance when using extrafoveal presentations imply that high acuity vision is not essential for object recognition as already suggested[8]. Thus the widely-held view that foveation is required for recognition may be less true for natural scenes than it is for certain types of artificial stimuli such as letters and words that have traditionally been used to investigate the performance of the visual system. On the other hand this decrease of performance may be due to an attentional gradient. It has been shown that the attentional facilitation decreased progressively from an attended location that could be the fixation point in our task; in this case at 3.6° of eccentricity, our results could well be explained by the behavioral cost of this attentional gradient[9].

Such data have important implications both for the understanding of the visual processing required for object recognition and for the development of biologically inspired artificial vision systems. First, the human visual system is clearly able to process natural scenes very efficiently, even in the absence of contextual help, without eye movements or even foveation, and without *a priori* information on the exact nature of the object to look for. This is especially striking in the case of animals as there are few (if any) features in common between a snake and a giraffe! Second, the remarkable rapidity with which information can be processed by all the processing stages between the retina and high level cortical areas argues that much of it must be possible on the basis of essentially feed-forward mechanism[10]. Indeed, the fact that there is essentially no serious time penalty for processing uncentered images implies that the position invariance that characterizes human vision is achieved mainly by the use of massively parallel processing. This parallel processing could be seen both in terms of parallel computation of different features and of simultaneous involvement of different hierarchical visual areas (pipe-line architecture). It has been suggested recently[11] that this sort of position invariance might depend upon dynamically determined rerouting of visual information. If the visual system had to dynamically reroute information in response to stimuli presented to the left and right of the fixation point, one ought to predict some cost in terms of processing speed, at least in the case where the position of the stimulus could not be predicted in advance (as in the present experiments). However, the present data indicate that this shifting mechanism would have at most 10-20 ms in which to operate - a temporal constraint which would be difficult to meet with relatively slow neuronal circuits. In fact, such a special mechanism for rapid visual object recognition may have evolved because it does not require too much attention, it relies on especially fast information processing and, at least in the case of familiar objects, allows identification with a remarkably low rate of errors.

1. S.J. Thorpe, D. Fize, and C. Marlot, Speed of processing in the human visual system. *Nature* 381: 520-522 (1996).
2. M.D. Rugg, M.C. Doyle, and T. Wells, Word and nonword repetition within-modality and across-modality. An event-related potential study. *Journal of Cognitive Neuroscience* 7: 209-227 (1995).
3. I. Biederman and E.E. Cooper, Object recognition and laterality : null effects. *Neuropsychologia* **29**: 685-694 (1991).
4. M. Vitkovitch and G. Underwood, Visual field differences in an object-recognition task. *Brain and Cognition* 19: 195-207 (1992).
5. C.J. Downing, Expectancy and visuo-spatial attention : Effects on perceptual quality. *Journal of Experimental Psychology : Human Perception and Performance* 13: 228-241 (1988).
6. N. Lavie and Y. Tsal, Perceptual load as a major determinant of the locus of selection in visual attention. *Perception and Psychophysics* 56: 183-197 (1994).
7. C.W. Eriksen and Y.Y. Yeh, Allocation of attention in the visual field. *Journal of Experimental Psychology : Human Perception and Performance* 11: 583-597 (1985).
8. J.M. Henderson, K.K. McClure, S. Pierce, and G. Schrock, Object identification without foveal vision : Evidence from an artificial scotoma paradigm. *Perception and Psychophysics* 59: 323-346 (1997).
9. T.C. Handy, A. Kingstone, and G.R. Mangun, Spatial distribution of visual attention: Perceptual sensitivity and response latency. *Perception and Psychophysics* 58: 613-627 (1996).
10. S.J. Thorpe and M. Imbert, Biological constraints on connectionist models., in: *Connectionism in Perspective.*, R. Pfeifer, Z. Schreter, F. Fogelman-Soulié, & L.Steels, eds., Elsevier: Amsterdam. 63-92. (1989).
11. B.A. Olshausen, C.H. Anderson, and D.C. Van Essen, A multiscale dynamic routing circuit for forming size- and position-invariant object representations. *Journal of Computational Neuroscience* 2: 45-62 (1995).

CORTICAL ACTIVITY PATTERN IN COMPLEX TASKS

F. Frisone,[1] P. Vitali,[2] and P. Morasso[1]

[1]DIST - Department of Informatics, Systems and Telecommunication
University of Genova
Via Opera Pia 13, 16145 Genova, Italy
e-mail:friso@dist.dist.unige.it
[2]Institute of Neurophysiopathology
DISM - Department of Motor Sciences
University of Genova - Ospedale S. Martino, L.go R. Bensi 10, 16132
Genova, Italy
e-mail:vita@dism.unige.it

INTRODUCTION

The goal of this study is to investigate the organization of the human cortical activity in complex tasks like verbal production, taking into account some recent experimental results in the fields of neuroanatomy/neurophysiology obtained applying the *functional magnetic resonance imaging* (FMRI) system.

In recent years FMRI has become one of the most promising tools in cerebral neurophysiology. In particular, the BOLD technique (*Blood oxygen level dependent*) is sensitive to increased blood oxygenation related to neural activation. FMRI is a modern, non-invasive tool for brain mapping which adds good temporal resolution to excellent spatial resolution even in deep structures. This new technique is sensitive to increased capillary and venous blood oxygenation related to neural activation: the subsequent decrease in deoxihemoglobin concentration can be measured as an increase in the local $T2^*$-weighted magnetic resonance signal [1]. FMRI has dramatically increased our understanding of localization of the cerebral functions in the last years: previous reports described brain mapping in motor and perceptual (somatosensory, visual, auditory, olfactory) paradigms, but the most common application is in cognitive tasks; for this purpose it is necessary to use two similar stimuli which are presented to the subject in alternating phases, called activation and control phases, differing only for the investigated process. Language is the main human cognitive process, and its neurophysiology typically reflects the dichotomous categorization of disturbances in linguistic output and input (i.e. expressive/receptive, fluent/non fluent, motor/sensory), which is related to two classical areas located in the inferior frontal gyrus and in the posterior superior temporal gyrus, respectively the Broca's and Wernicke's areas in the dominant hemisphere. Recent studies by Position Emission Tomography and FMRI (review in [2]) have provided strong evidence for more diffuse integration of multiples areas in language tasks: Friston described that as "functional connectivity"[3].

The aim of the performed FMRI experiments was to evaluate multiple cortical areas involved in verbal output and input during two different covert language tasks, namely "*verbal fluency*" and "*verbal understanding*". We discuss the relevance of such data for a theory of cortex dynamics which emphasizes the role of long-range lateral connections somehow contradicting the conventional view of the cortex as a continuous computational medium characterized by a localization of tasks and an associated locality of lateral connections. In fact, a 2-dimensional somatotopic organization is found in many cortical areas. Moreover, long-range lateral connections for linking non-local area in the cortex are not in agreement with a general underlying ecological pressure on brain formation towards minimizing the total length of cortical wiring[4] and reducing the communication load[5]. Short-range lateral connections stress the fact that nearby columns of cells in the cortex tend to prefer stimuli with similar features. Cells with similar properties communicate more often than cells with different properties. Consequently, these cells might be expected to interconnect more densely through their axons and dendrites. Thus, there are at least two good reasons for using only short-range lateral connections. Nevertheless, the observed cortex activity patterns, obtained applying the FMRI technique for complex tasks appears to support the opposite view.

EXPERIMENTAL METHOD AND OBSERVATIONS

Let us now define more precisely the two covert language tasks, called *verbal fluency* and *verbal understanding* which are used for evaluating multiple cortical areas involved in verbal output and input. *Verbal fluency*: during the activation phase the subject thinks as many words as possible beginning with a letter said by the examiner. During the control phase the subject silently repeats nonsense word ("bla"). *Verbal understanding*: the subject mentally finds the words corresponding to the definitions given by the examiner. During the control phase, the subject silently answers "calm" when the examiner says "you must stay".

Eighteen transaxial anatomic slices (T1 weighted, encompassing the whole brain) were acquired by means of a 1.5 T Siemens scanner equipped with head coil and EPI software. Forty T2* weighted functional sequences were acquired during each task, which was composed by 4 cycles. Each cycle alternated 5 acquisitions in the task activation phase and 5 in the control phase. The 720 (40x18) functional images were statistically evaluated on a Siemens workstation by a special software ("Stimulate") which performed cross-correlation analysis and cluster filtering: the significantly activated pixels were superimposed on anatomical images.

As shown by the figure, a widespread activation was present in all subjects in both hemispheres. During verbal fluency the majority of activated pixels was not only in the left inferior frontal gyrus (Broca's area) but also in the left dorsolateral prefrontal cortex, which has been shown [6] to be related to "willed action" tasks; other activation patterns are in premotor and motor areas, in anterior gyrus cinguli, left hippocampus, thalamus, insula and parieto-occipital areas. Premotor and motor areas, seen in other tasks of word production, are probably involved in planning of speech, even if not executed. Limbic areas seem to be related to attentional and memory effort, whereas parieto-occipital areas to visual recall of things corresponding to words[2]. Broca's area (and somewhere the controlateral inferior frontal gyrus) are seen active even during verbal understanding task, with multiple parietotemporal areas, i.e. superior and middle temporal, angular and supramarginal gyrus. Indeed recent acquisitions show primary

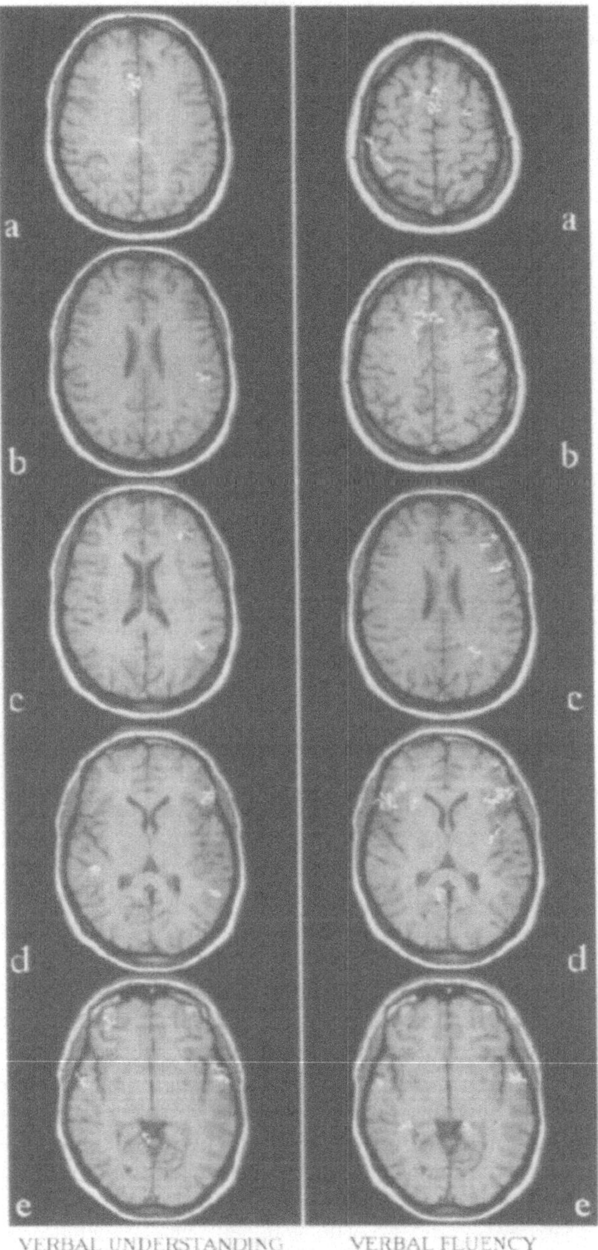

Figure 1. Cortex activity during verbal understanding and verbal fluency tasks. During verbal understanding task we can observe activation in: anterior and posterior cingulus (a), supramarginal gyrus (b), angular gyrus and BA 44 or Broca's area (c), BA 45 of the inferior frontal gyrus and posterior medium temporal gyrus or Wernicke's area (d), bilateral superior temporal gyrus (e). During verbal fluency task we can observe activation in: primary and supplementary motor areas (a), anterior cingulus and dorsolateral prefrontal cortex (b), BA 44 or Broca's area (c), bilateral BA 45 of the inferior frontal gyrus (d), bilateral superior temporal gyrus and hippocampus (e).

activations to Wernicke's area in verbal understanding, but this function seems split into semantic (prefrontal) and phonological (parietotemporal) areas. This second task produced an even more diffuse activation with respect to verbal fluency.

COMPLEX TASKS AND LONG RANGE CONNECTIONS

The described cortical activation patterns are in agreement with a new computational point of view about the cortex which is considered as a continuously adapting dynamical system, shaped by competitive and cooperative lateral connections[7]; moreover, such organization is not static but changes with ontogenetic development[8]. Shortly, it has been suggested that cortical areas can be seen as a massively interconnected set of elementary processing elements, which constitute a computational map[9]. From the modeling point of view, the most common misconceptions about cortical functionality can be reduced to: *flatness* of cortical maps, related to the locality of lateral connections; *fixed* lateral connections, i. e. lateral connections are determinated a priori and not in the learning phase; *Mexican-hat function* of lateral interactions (it implies a significant amount of recurrent inhibition for the formation of localized responses by lateral feedback). The flatness assumption which characterizes the classic map models[10, 11] is contradicted by the fact that the structure of lateral connections is not genetically determined but depends mostly on electrical activity during development. More precisely, they have been observed to grow exuberantly after birth and reach their full extent within a short period; during the subsequent development, a pruning process takes place so that the mature cortex is characterized by a well defined pattern of connectivity, which includes a large amount of non-local connections: this rules out all the models limited to a purely 2-D circuitry. Moreover, the superficial connections to non-neighboring columns are organized into characteristic patterns: a collateral of a pyramidal axon typically travels a characteristic lateral distance without giving off terminal branches and then it produces tightly terminal clusters (possibly repeating the process several times over a total distance of several millimeters). Such characteristic distance is not a universal cortical parameter and is not distributed in a purely random fashion but is different in different cortical areas[12, 13, 14]. Thus, the development of lateral and afferent connections depends on the cortical activity caused by the external inflow, in such a way to capture and represent the (hidden) correlations in the input channels. Each individual lateral connection is "weak" enough to go virtually unnoticed while mapping the receptive fields of cortical neurons but the total effect on the overall dynamics of cortical maps can be substantial, as is revealed by cross-correlation studies[15]. Lateral connections from superficial pyramids tend to be recurrent (and excitatory) because 80% of synapses are with other pyramids and only 20% with inhibitory interneurons, most of them acting intra-columnarly[16]. Recurrent excitation is likely to be the underlying mechanism which produces the synchronized firing which has been observed in distant columns. The existence (and preponderance) of massive recurrent excitation in the cortex is in contrast with what could be expected, at least in primary sensory areas, considering the ubiquitous presence of peristimulus competition (or "Mexican-hat pattern") which has been observed time ago in many pathways, as the primary somatosensory cortex, and has been confirmed by direct excitation of cortical areas as well as correlation studies; in other words, in the cortex there is a significantly larger amount of long-range inhibition than expected from the density of inhibitory synapses. In general, "recurrent competition" has been assumed to be the same as "recurrent inhibition", for providing an antagonistic organization that sharpens responsiveness to an area smaller than would be predicted from the anatomical funneling of inputs. Thus, an intriguing question is in which manner long-range competition can arise without long-range inhibition and a possible solution is the mechanism of *gating inhibition* based on a competitive distribution of activation, proposed by[17] and further investigated by[18, 20, 19] In particular, we have investigated the following kind of model:

$$\frac{dV_i^x}{dt} = -\gamma_i V_i f(h_i^{lat} + h_i^{ext}) \tag{1}$$

in which V_i corresponds to the global "activation level" of a cortical neural module, say a micro-column, and the equation simply says that this variable evolves under the action of three competing influences:

1. a self-inhibition (weighted by $\gamma_i > 0$);

2. a net input h_i^{lat} coming from the set of lateral connections;

3. a net external input h_i^{ext} (related to thalamo-cortical connections or cortico-cortical connections across different cortical areas).

The h_i^{lat} term is a recurrent, excitatory input, intended to express the massive lateral excitatory connections which are assumed to satisfy a criterion of topological organization. The external input provides a feedforward pattern of excitation which is rather diffused and can be approximated by a broad Gaussian function. These two excitatory influences are balanced by non-linearities in the network dynamics: (i) a mechanism of gating inhibition of the recurrent input, which means that the activity level of each neural module is normalized (as in Reggia's model) according to the average activity of its immediate neighbors; (ii) a shunting effect on the external input, which tends to focus the input excitation on the active population of neurons. In such a model when a stimulus occurs a non-linear *diffusion* process takes place: we can observe that initially the population code is flattened (spreading the activity pattern over a large part of the network) and then a re-sharpening process makes the activation higher around the interested areas.

CONCLUSION

According to the illustrated experimental evidence, speech-related tasks involve the simultaneous and concurrent activation not only of the specifically speech-related areas (Broca's and Wernicke's) but also of a number of other cortical and sub-cortical formations. The proposed computational model is consistent with this view. However, this only highlights the static part of the cortical machinery, related to the geometric organization of the activation patterns. The timing structure is not observable with the standard FMRI techniques but the proposed model provides a stimulus and an explanatory tool for integrating the FMRI technique, which has good spatial resolution but coarse timing resolution, with EEG recordings, which are characterized by complementary properties. The future experimental investigation will be aimed in this direction, i.e. the analysis of the spatio-temporal structure of cortical activity in normal subjects and pathological conditions, such as epilepsy. The underlying hypothesis is that the observed patterns of cortex activity, which mix local and non-local areas, have a specific computational function: representing complex tasks onto the cortex. In this framework, different local areas are used for mapping onto the cortex high-dimensional input spaces (e.g. multimodal stimuli) thus setting up a number of inter-related multi-dimensional lattices which cooperate in carrying out complex tasks such as speech. The

stability of the mechanism is determined by the delicate balance between the massive
recurrent excitation and the non-linear stabilizing effects described above and it is not
surprising that such stability can be broken even by small local parametric changes.
The theory emphasizes the importance of the global network dynamics and provides a
starting point for analyzing normal and pathological cortical activity.

REFERENCES

1. J. Kwong, J.W. Belliveau, and D.A. Chessler. Dynamic magnetic resonance imaging of human
    brain activity during primary sensory stimulation. *Proceedings of National Academy of Sciences
    USA*, 89:5675–5679, 1992.
2. M. Habib, J.F. Demonet, and R.S.J. Frackowiak. Neuroanatomie cognitive du language: con-
    tribution de l'imagerie fonctionelle cerebrale. *Revue Neurologique*, 152:249–260, 1996.
3. K.J. Friston, C.D. Frith, P.F. Liddle, and R.S.J. Frackowiak. Functional connectivity: the prin-
    cipal component analysis of large (pet) data sets. *Journal of cerebral blood flow and metabolism*,
    13:5–14, 1993.
4. R. Durbin and G. Mitchison. A dimension reduction framework for understanding cortical
    maps. *Nature*, 343:644–647, 1990.
5. M. E. Nelson and J. M. Bower. Brain maps and parallel computers. *Trends in Neuroscience*,
    13:403–408, 1990.
6. C.D. Frith, K.J. Friston, P.F. Liddle, and R.S.J. Frackowiak. Willed action and the prefrontal
    cortex in man: a study with pet. *Proceedings Royal Society of London*, 244:241–246, 1991.
7. J. Sirosh, R. Mikkulainen, and Y. Choe. *Lateral interactions in the cortex*. Hypertext Book,
    1989. URL:www.cs.utexas.edu/users/nn/web-pubs/htmlbook96.
8. L.C. Katz and E.M. Callaway. Development of local circuits in mammalian visual cortex.
    Annual Review of Neuroscience, 15:31–56, 1992.
9. E. I. Knudsen, S. du Lac, and S.D. Esterly. Computational maps in the brain. *Annual Review
    of Neuroscience*, 10:41–65, 1987.
10. S. Amari. Dynamics of pattern formation in lateral-inhibition type neural fields. *Biological
    Cybernetics*, 27:77–87, 1977.
11. T. Kohonen. Self organizing formation of topologically correct feature maps. *Biological
    Cybernetics*, 43:59–69, 1982.
12. C.D. Gilbert and T.N. Wiesel. Morphology and intracortical projections of functionally identified
    neurons in cat visual cortex. *Nature*, 280:120–125, 1979.
13. H.D. Schwark and E.G. Jones. The distribution of intrinsic cortical axons in area 3b of cat
    primary somatosensory cortex. *Experimental Brain Research*, 78:501–513, 1989.
14. W. Calvin. Cortical columns, modules and hebbian cell assemblies. In M. Arbib, editor, *The
    handbook of brain theory and neural networks*, pages 269–272. MIT Press, Cambridge, MA, 1995.
15. W. Singer. Development and plasticity of cortical processing architectures. *Science*, 270:758–
    764, 1995.
16. A. Nicoll and C. Blakemore. Patterns of local connectivity in the neocortex. *Neural Computa-
    tion*, 5:665–680, 1993.
17. J.A. Reggia, C.L. D'Autrechy, G.G. Sutton III, and M. Weinrich. A competitive distribution
    theory of neocortical dynamics. *Neural Computation*, 4:287–317, 1992.
18. P.G. Morasso and V. Sanguineti. How the brain can discover the existence of external egocentric
    space. *Neurocomputing*, 12:289–310, 1996.
19. F. Frisone and P. G. Morasso. Extending the TRN model in a biologically plausible way. In
    ICANN97- Int. Conf. on Artificial Neural Networks, Lousanne, October 1997. In Press.
20. F. Frisone and P. Morasso. Representing multidimensional stimuli on the cortex. In C. von der
    Malsburg, W. von Seelen, and J. C. Vorbrüggen, editors, *Artificial Neural Networks - ICANN96*,
    pages 649–654, Bochum, Germany, July 1996. Springer-Verlag.
21. C. Von der Malsburg and W. Singer. Principles of cortical network organization. In P. Rakic
    and W. Singer, editors, *Neurobiology of Neocortex*, pages 69–99, New York, 1988. Wiley.
22. F. Frisone, V. Sanguineti, and P. Morasso. A novel hypothesis on cortical map: topological
    continuity. In *WIRN96 - VIII Italian Workshop on Neural Nets.*, pages 194–198, Vietri sul Mare
    (SA), Italy, May 1996. Springer-Verlag.

RELATIONS AMONG EEGS FROM ENTORHINAL CORTEX, OLFACTORY BULB, SOMATOMOTOR, AUDITORY AND VISUAL CORTICES IN TRAINED CATS

G. Gaál[1]* and W.J. Freeman[2]

[1]*Emory University
School of Medicine
Department of Neurology
WMB 6000
Atlanta GA 30322

[2]129 Life Sciences Addition,
Department of Molecular and Cell Biology,
University of California,
Berkeley, CA 94720

INTRODUCTION

We are testing the hypothesis that a rapid exchange of corollary discharges and perceptual signals occurs between the limbic system and the primary sensory cortices (PSCs), in the 1 s time segments preceding a discriminated CS and between a CS+ and a CR in a fixed ITI appetitive bar release paradigm. We have focused on the entorhinal cortex (EN) owing to the bi-directional paths linking it to the PSCs. We also postulate that signals lasting on the order of 50 - 100 ms, which constitute a basis for implementing rapid shifts in attention, can be detected in EEGs simultaneously recorded from electrode arrays on the surfaces of the EN and PSCs.

It has been described earlier that in awake subjects the olfactory bulb[1] (OB), the somatomotor[2] (SM), the auditory, the visual cortices[3] (AI; VI) and also the entorhinal cortex[4] (EN) can generate repeated bursts of EEG oscillations. There is no general agreement about the role of such oscillations, although they have been attributed to attention, feature association, visual awareness and other forms of higher level cortical representation[5]. It has been proposed that synchronization plays an important role in the integration of distributed neuronal activity[6] which can synchronize over long distances both in primary sensory input[3] and motor output cortical areas[7,8]. Visual areas can also synchronize with motor cortical areas in awake cats[9] or primates[10].

Very little is known about the synchronization between PSCs and EN. The role of the EN in the processing of olfactory information and its relationship with olfactory areas as well as the hippocampus during different types of odor sampling behavior have been thoroughly analyzed in cats[4]. The objective of the present study was to find whether behavior-related changes in EEG activity may occur in PSCs and also at the level of the EN. If this could be demonstrated it would provide physiological evidence of a functional relationship between the activity of the entorhinal cortex and sensory, namely visual and auditory information processing.

METHODS, ANALYSIS AND RESULTS

Our aim was to evaluate the role of the EN in multisensory integration and forming the unity of perception. We were also interested in exploring the presence and timing of beta- gamma band (~10-90 Hz) oscillations in perception and attention tasks in awake subjects and finding out whether multiregional cortical coherence occurs at various frequencies. We explored the timing of feedforward and feedback communication delays between pairs of brain areas (Figure 1).

Computational Neuroscience
edited by Bower, Plenum Press, New York, 1998

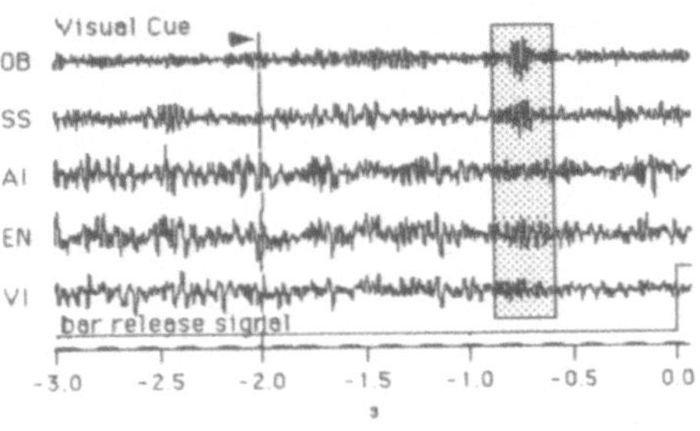

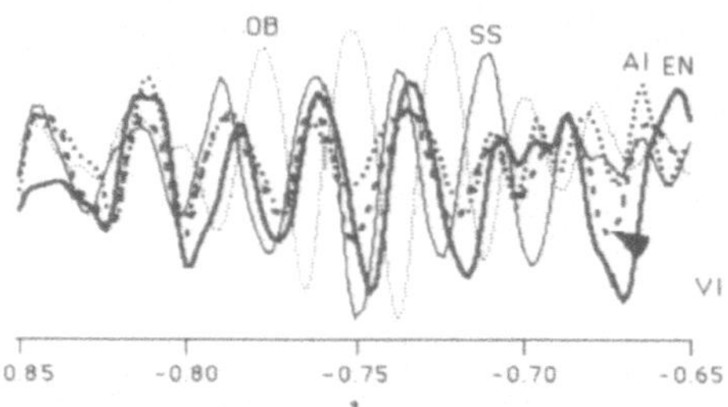

Figure 1. Lead and lag relationships during sensory stimulus identification and preparation for bar release in S3. Upper panel shows five simultaneous recordings filtered between 10-100 Hz. Lower panel shows an expanded time view of the period when oscillatory bursts synchronize across several brain areas. EN is synchronized with VI and AI at first, then lags behind these sensory cortices. SM appears to lead the other four areas for several cycles in this spindle and briefly synchronizes with the EN.

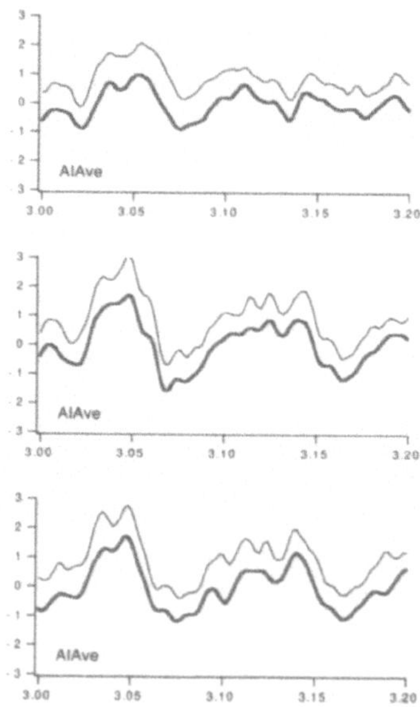

Figure 2. Time series averages of responses evoked by 500 Hz 0.1 s sound at rest and in two consecutive days in the auditory task in S1. The subject already anticipated and recognized the sound as a CS+ stimulus. Intertrial intervals at rest were not fixed, so the subject could not anticipate the precise timings of the sounds. Filtering: 10-100 Hz. The rise time of the auditory evoked response is faster and the amplitude is more prominent in the task than at rest. Average evoked responses on two consecutive task days are fairly reproducible. Recordings in 6 s were normalized, averaged for 16 electrodes per trial then time averages of 20 trials of spatial averages were taken. Upper curves are added standard deviations.

We implanted two depth electrodes in OB, placed 4x4 square epidural arrays (with 0.8 mm spacing) over SM, AI and VI and slid an 8+8 flat epidural array into an opening on the occipital groove along the boni tentorium to record from the surface of the EN.

We monitored neural activity during (A) rest , (B) appetitive operant conditioning task in a go-no-go paradigm with CS+ (rewarded after appropriate response) and CS- (nonrewarded) visual cues and (C) appetitive operant conditioning task using auditory cues for sensory discrimination. DC lamp provided CS+ light stimulus in the upper left corner of the operant conditioning panel for 3 s. DC lamp provided CS- light stimulus in the upper right corner of the operant conditioning panel. 500 Hz 0.1 s tone was used as CS+ auditory stimulus. 5000 Hz 0.1 s sound was used as CS- auditory stimulus; at a later stage of the learning task, the contingencies of these two sounds were reversed. Auditory mapping with sinusoidal sounds of various frequencies (in steps of 200 Hz) lasting for 0.1 s at rest was done during several sessions before the auditory operant conditioning task began and was repeated for two more sessions after the animals were trained in the auditory learning task. Visual array placement was tested in a single session for each subject at rest with bright or dim light flashes delivered via a PS22 photostimulator (same intensities as used in Barrie et al., 1996 for aversive conditioning experiments in rabbits; but note that the flash is accompanied by an audible click, so it is unsuitable for rigorous multisensory or cross modality studies) as well as with the same DC lights which were used in the task.

The basic characteristics of the EN EEGs were compared to those of the OB and PSCs using methods that take into account linear ((A) spatial and time ensemble averages, (B) auto, cross power and phase spectra as well as the corresponding auto and cross correlograms calculated in 64-128 ms sliding windows) and nonlinear relationships (estimating the Jacobian matrix of the nonlinear neural activity[11,12].

Evoked sensory responses could be seen also in individual trials. 5000 Hz 0.1 s stimuli of the same amplitude gave more prominent responses than 500 Hz 0.1 s responses. Off-responses at the end of the 0.1 s sound stimuli were often present. On average, more prominent auditory response could be observed after the animals were trained to anticipate and identify the sound stimulus in a CS+ trial in sessions with fixed 6 s intertrial interval and fixed 3.0 s waiting period (Figure 2).

Spatial ensemble averages computed for up to 16 traces of the simultaneously recorded waves over the various cortices revealed a common waveform across each array, but different waveforms for different arrays. The spatial amplitude modulation of the waveform varied within trials after presentation of identical stimuli, but recurrence of amplitude (root-mean-square - rectified or original) spatial patterns could also be noted both in rest and task data. Auditory spatial pattern recurrence matrices calculated for rest data are shown in Figure 3.

OB recordings monitored on a dorsal and ventral electrode pair had a classic appearance; they consisted of a slow wave with a period of 0.5-2 s that corresponds to the respiration cycle of the animal. Superimposed on this slow wave, bursts of high frequency sinusoidal activity could be seen after almost every inhalation. Bursts of somewhat lower frequency could also occur in the troughs of the respiratory slow waves. The pair of waves were approximately in phase in three cats and in antiphase in the fourth cat, S1. Time ensemble averaging of rectified waves or Fourier amplitudes revealed that in three cats the amplitude of OB bursts tended to increase after the onset of the sensory stimuli, while in a third cat, S2, the amplitude decreased. Auditory evoked potentials could be identified in individual trials both at rest and in the task; visual evoked potentials in the operant

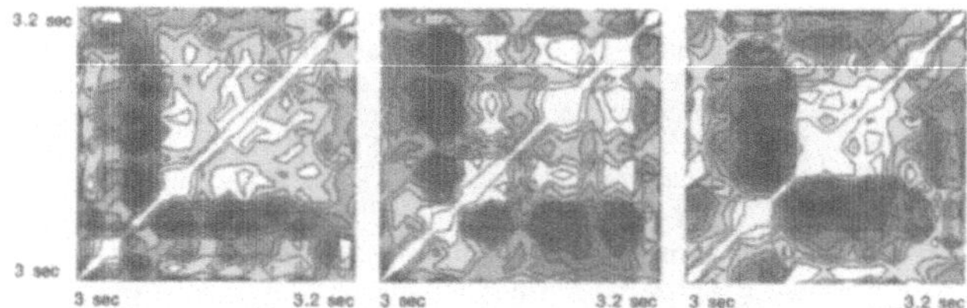

Figure 3. Recurrence matrices for three consecutive auditory responses evoked in individual trials at rest in S1. 500 Hz and 5000 Hz sound stimuli were randomly mixed with control trials (no sound). Auditory amplitude spatial patterns were cross-correlated with each other at every 20 msec between 3.0 and 3.2 s Auditory stimulus started at 3.0 s and lasted for 0.1 s - same rest data as in Fig. 2. White patches indicate that auditory evoked patterns tended to persist and even recur for tens of milliseconds. Density scale is -1.0, +1.0 (dark-to-light).

conditioning task were revealed only after time ensemble averaging (the light sources were protruding above the shoulder of the subjects). No slow waves could be seen at rest in PSC and EN EEGs, but slow waves, that might have reflected periodic licking, head movement or paw movement were not infrequent in task conditions. Such slow waves did not reveal strong correlates with the OB respiratory rhythm. The OB bursts centered on 40 Hz in the task were often noted within 500 msec preceding the sensory stimuli (either CS+ or CS-) and before bar release (CR), but were not strictly time locked to either stimulus delivery or bar release. They were regularly preceded or accompanied by bursts of similar frequency in SM and occasionally also in EN but rarely in AI or VI. Bursts of lower frequency residing in the troughs of the respiratory waves tended to synchronize across more than two brain areas also at rest. Gamma bursts in AI became more apparent when the animals were retrained in the auditory task following visual discrimination.

Phase patterns derived from FFT calculations in sliding windows showed longer continuous segments at rest, in which EN activity precedes the activity in PSCs in individual trials. Figure 4. shows a matrix of spatial density plots of coherence (left corner), FFT amplitude (main diagonal) and phase (right corner) values averaged for 20 CS+ trials in a visual task for S2.

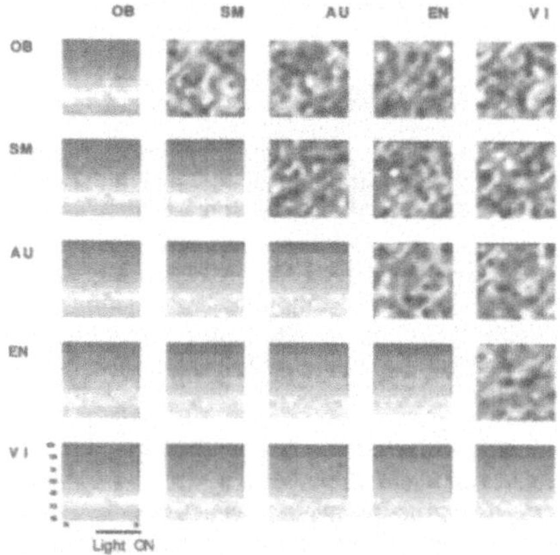

Figure 4. FFT amplitude (main diagonal), coherence (left corner) and phase (right corner) patterns for pairs of brain areas for 20 CS+ trials calculated in 128 msec sliding windows stepped at 100 msec. Filtering: 10-100 Hz. The OB gamma activity is characterized by a narrow band centered on 40 Hz. 40 Hz patches also appear in the SM spectrographs; the activity in AI, EN and VI is more broadband and the dominant frequencies shift towards the beta range. Vertical axes are frequencies between 15-110 Hz in steps of 7 Hz; horizontal axes are time between 2.4 s and 3.6 s. Light stimulus onset is at 3.0 s, the center of the squares. Ave. bar release at: 4.3+-0.8 s.

Note that there are subtle changes in the strength of coherence pre- and post stimulus, but these proved to be less reliable than the presence/absence of center stripes (spots) in sliding cross correlographs. Taking the values of coherence as a measure of linear coupling between populations of neurons, this coupling is strong between OB and the SM, the OB and the EN, and is weaker between AI and VI and the OB, but strongest between AI-EN, AI-VI and VI-EN. The coherence, FFT and cross correlogram plots are very different from similar plots calculated for Gaussian noise signals. Gaussian noise is more broadband even after digital filtering in the beta-gamma range and has a higher cut-off frequency than the 55-60 Hz found in cats after filtering in the 10-100 Hz range. In our control studies, one of the waves was shifted relative to the other by one trial period as described in Murthy and Fetz (1996) and as used for extracellular unit shift predictor controls (Gerstein et al., 1989). Center stripes were entirely absent control cross correlogram density graphs.

Mutual delays between pairs of brain areas were first compared by locating EEG minima (which usually corresponded to minima of gamma oscillations in waves filtered between 10-100 Hz and minima of oscillations located in troughs of slow waves in unfiltered waves) and calculating auto and cross-correlograms for such events similarly to extracellular unit action potential correlograms[13]. We found side peaks characteristic of the presence of oscillations in the gamma range in auto and cross correlograms (Figure 5).

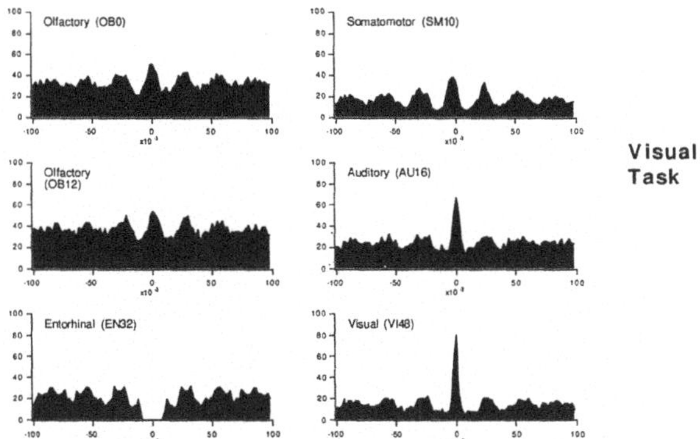

Figure 5. Cross correlograms for EEG minima. Oscillations were most apparent and synchronization among the five brain areas was highest in S3, but the pattern of differences in the strength of the center peaks of the cross correlograms was similar also in the other subjects. The strength of the cross correlogram center peaks was lower at rest than in either visual or auditory task in all subjects and the functions tended to be less well centered on zero. Ave. bar release at: 4.4+-0.8 s.

Figure 5 shows cross-correlograms calculated for EEG minima with reference to EN activity in S3 in a visual task with 3 s CS+ light stimulus. There is a stronger tendency for the locations of the peaks to average out to zero at the center during task versus rest conditions. The cross correlogram strength markedly increased in task versus rest in all four cats, especially for EN-VI, EN-AI and AI-VI pairs (which were highest to start with also at rest). This could be demonstrated also in sliding cross correlogram spatial plots in the form of a central stripe which was often absent or offset at rest compared to task.

Figure 6A. shows sliding cross-correlogram functions calculated between AI and EN neural activity for 500 Hz and 5000 Hz auditory stimuli at rest. C and D panels of Figure 6B. show sliding cross-correlograms calculated for VI and EN activity during visual mapping with bright and dim light flashes. Figure 7. shows results of similar calculations for auditory and visual tasks. Note that the increase in correlation strength between the PSCs and the EN manifests itself in the brighter central stripe

Lead and lag relationships were apparent in the presence of evoked sensory responses when the PSCs were leading other areas also in task. In some cases, it could be demonstrated also on average, that the EN was leading the VI and AI before the onset of the sensory stimuli or SM before

bar release (Figure 8C). Generally, however, such lead and lag relationships, although present in individual trials, but without precise time locking to external events, averaged out to zero.

CONCLUSIONS AND IMPLICATIONS FOR FUTURE STUDIES

We conducted a series of experiments in cats in which EEGs were recorded from primary sensory areas simultaneously with EEGs in EN during sensory discrimination in an appetitive operant conditioning task and at rest[14]. Earlier studies in this laboratory found spatial patterns in rabbits in respect to sensory stimuli first in olfactory[15] then later also in other primary sensory

areas[16]. Spatial patterns in the gamma band are believed to reflect perceptual categorization, which could also be found in one primate. The patterns took the form of amplitude modulation of a spatially coherent oscillation in the gamma frequency (20-80 Hz) range. The spatial patterns lacked invariance with respect to conditioned stimuli, showing a dependence on brain state, behavioral context and

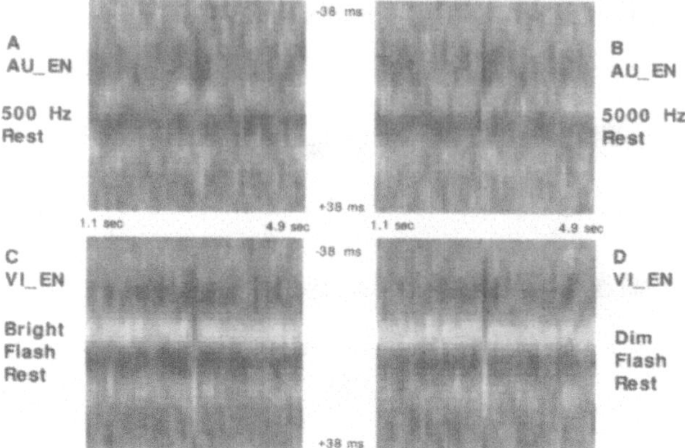

Figure 6. Averaged sliding cross correlograms calculated with reference to EN EEG recorded in S2 during auditory mapping with 500 Hz (A) and 5000 Hz (B) 0.1 s sound stimuli at rest or during visual mapping with bright (C) or dim (D) light flashes. Note that the PSCs lead the EN at the time of the evoked response.

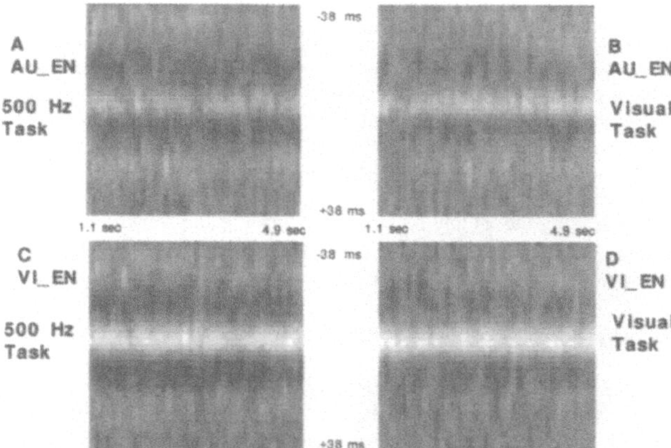

Figure 7. Averaged sliding cross correlograms calculated with reference to EN EEG recorded during auditory task with 0.1 s 500 Hz CS+ (A) or visual task with CS+ light stimulus for S2 (B). (C) and (D) show sliding cross correlograms calculated for VI and EN activity for the same tasks and trials as (A) and (B). Density scale is set between -0.7 and +0.7 (dark to light). Vertical axes show time from -38.0 to +38.0 msec for center cross correlograms; horizontal axes show time along trials from 1.1 to 4.9 s; stimulus onset is at 3.0, the center of the squares; 128 msec windows were stepped at 100 msec. Ave. bar release at 4.5+-0.85 s and 4.3+-0.8 s (visual).

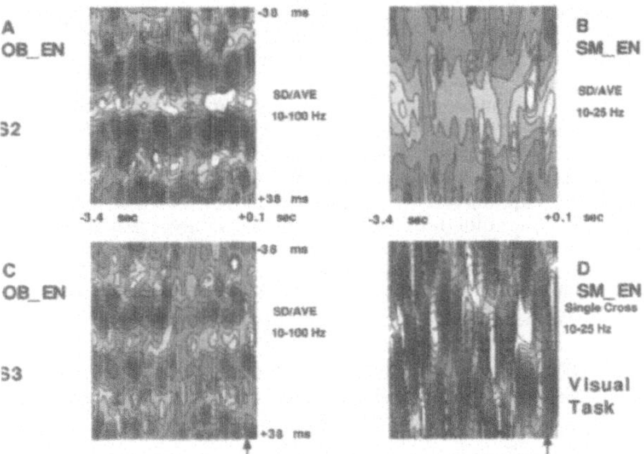

Figure 8A, B. SD divided by mean of cross correlograms for 20 CS+ trials in a visual task for S2. Trials were aligned for movement onset (1.4 s before and 0.2 s after the onset of movement was plotted; movement onset is labeled by arrows at the bottom spectrographs). Note the increase in strength between OB-EN in 10-100 Hz and beta band for SM-EN before movement onset. **Figure 8C, D.** SD/Mean (C) for S3 in a visual task and single trial cross correlogram in the beta band for SM-EN. Note increased strength before bar release. EN appears to lead OB also on average in Figure 8C (white spot in the upper right corner). Density scales: -5.0 - +5.0 in Figures 8 A,B,C and - 0.7 - +0.7 in D. Ave. bar release at 4.5+-0.7 (S2) or 4.4+-0.36 s (S3).

training history . The claim was that the context reflected the meaning of the stimuli for the subject, but no recordings were done from multiple cortical areas simultaneously, nor has it been explored how spatial patterns would change in an operant conditioning task where the behavior of the subject can be monitored more precisely than in aversive classical conditioning. Therefore, epidural arrays of up to 16 electrodes were used in the present studies in a multisensory recording arrangement.

Delay and phase plots between pairs of traces indicate that the EN can lead the PSCs by 6-20 ms preceding the onset of the CS; it can also lead the OB or SM before movement onset. The PSCs lead after the onset of the CSs. Such leads are most apparent in the beta band. Coherence between SM and other areas is strong just before CS onset, then immediately before and during bar release. Such periods of strong coherence often coincided with valleys in errors between estimates of Jacobian matrices of neural activity but were preceded or followed by peaks in the error estimates[12]. The strength of cross correlation between pairs of brain areas is weaker at rest than in visual or auditory tasks in all four subjects and the cross correlograms in task often showed zero time lag synchrony. Cross correlogram strength for VI-EN, AI-EN and AI-VI pairs are often strongest after the onset of CS+ stimuli, weaker after CS-, especially in beta band (although not for all sessions investigated), and weakest at rest.

The aperiodic waveforms, the rapid global state changes, the context dependence of coherence and phase patterns between pairs of brain areas namely the EN and PSCs and the possible phase transitions in neural activity at various stages of the operant conditioning tasks suggest that the neural events formed during perception in awake, behaving animals, are constructed by cooperative nonlinear dynamics of the cortical neuronal populations. The results imply that the EN is actively engaged in multisensory integration and in creating the unity of perception instead of simply providing a relay station between sensory input areas and the memory related limbic system. The interareal changes in the patterns of delays/interactions in the scale of tens of milliseconds indicate that the interactions between neuronal groups are highly flexible. The findings suggest that dynamically changing synchronization may play an important role in multisensory somatomotor integration. The EN plays a central role in creating the unity of visual and auditory perception by orchestrating active exploration, such as orienting towards the source of expected sensory input, anticipating and integrating input, then helping to compare with what is stored in memory and gating out irrelevant information. Our studies might lead to a theory of sensory or perceptual coding.

In order to test whether the characteristics of the coherent electrical activity were related to the sensory stimuli themselves or to their behavioral context, extensive auditory mapping was carried out before, during and after the auditory operant conditioning learning experiment. The learning paradigm was done in an operant conditioning task to minimize the possibility of habituation, that might occur if the same sensory stimulus is repetitively administered in aversive tasks. More detailed analysis of the spatiotemporal neuronal activity monitored over the course of several months will be presented later. The auditory mapping and was done to get preliminary data and results for a revised version of a grant proposal (unsubmitted) related to tonotopic, visuotopic and perceptual mapping in awake subjects in sensorimotor tasks and at rest. It was proposed that combined depth and surface recordings should be done from behaving subjects in a similar paradigm to allow us to explore how patterns of extracellular spike activity and their synchronization give rise to spatiotemporal EEG patterns observable on the surface of the cortex. The results could reveal how plasticity observed in auditory tuning functions and dynamical receptive field changes of individual units in sensory discrimination studies[17,18]. relate to drifts in surface spatiotemporal patterns in awake subjects. Such experiments should also reveal how the auditory maps characterized in anesthetized and awake animals[19] in unit studies or with optical recording techniques manifest themselves also at the given electrode spacing in this particular task and how tonotopic patterns and isoamplitude maps transform into patterns which reflect the meaning of the stimuli as revealed in RMS or FFT amplitude patterns used as components of Euclidean vectors.

ACKNOWLEDGMENTS

This work was funded by NIMH 06686 and ONR N63373 N00014-93-1-0938 grants.

REFERENCES

1. W.J. Freeman, Mass Action in the Nervous System. Examination of the Neurophysiological Basis of Adaptive Behavior through the EEG. Academic Press, New York. (1975).
2. J.J. Bouyer, M.F. Montaron, J.M. Vahnee, M.P. Albert, and A. Rougeul, Anatomical localization of cortical beta rhythms in cat. *Neuroscience* 22: 863. (1987).
3. A.K. Kreiter and W. Singer, Oscillatory neuronal responses in the visual cortex of the awake macaque monkey. *Eur. J. Neurosci.* 4: 369. (1992).
4. P.H. Boeijinga and F.H.L. da Silva, Differential distribution of beta and theta EEG activity in the entorhinal cortex of the cat. *Brain Res.,* 448: 272. (1988).
5. F. Crick and C. Koch, Towards a neurobiological theory of consciousness. *Sem. Neurosci.* 2: 263. (1990).
6. A. Riehle, S. Grun, M. Diesmann, and A. Aertsen, Spike synchronization and rate modulation differentially involved in motor cortical function. *Science*, 278: 1950. (1998).
7. V.N. Murthy and E.E. Fetz, Oscillatory activity in sensorimotor cortex of awake monkeys: Synchronization of local field potentials and relation to behavior. *J. Neurophys.* 76: 3949. (1996).
8. J.P. Donoghue, J.N. Sanes, N.G. Hatsopoulos, and G. Gaál, Neural discharge and local field potential oscillations in primate motor cortex during voluntary movement. *J. Neurophys*, in press. (1998).
9. P.R. Roelfsema, A.K. Engel, P Konig, and W. Singer, Visuomotor integration is associated with zero time-lag synchronization among cortical areas. *Nature*, 3385: 157. (1997).
10. S.L. Bressler, R. Coppola, and R. Nakamura, Episodic multiregional cortical coherence at multiple frequencies during visual task performance. *Nature* 366: 153. (1993).
11. G. Gaál, Relationship of calculating the Jacobian matrices of nonlinear systems and population coding algorithms in neurobiology. *Physica D* 84: 582. (1995).
12. G. Gaál and W.J. Freeman, Nonlinear functions interrelating neural activity recorded simultaneously from olfactory bulb, somatomotor, auditory, visual and entorhinal cortices of awake, behaving cats, Trends in Research, Jim Bower (ed.), Plenum Press New York, pp. 653-659 (1997).
13. G.L. Gerstein, P. Bedenbaugh, and A.M.H.J. Aertsen, Neuronal assemblies. *IEEE Trans. Biomed. Eng.* 36: 4. (1989).
14. G. Gaál and W.J. Freeman, Relations between EEGs recorded from olfactory bulb, somatomor, auditory, visual and entorhinal cortices in trained cats. submitted to *J. Neurophys.* (1997).
15. W.J. Freeman and G. Viana Di Prisco, Relation of olfactory EEG to behavior: time series analysis. *Behav. Neurosci.* 100: 753. (1986).
16. J.M., Barrie, W.J. Freeman, and M.D. Lenhart, Spatiotemporal analysis of prepyriform, visual, auditory and somesthetic surface EEGs in trained rabbits. *J. Neurophys.* 76: 520. (1996).

17. G.H. Recanzone, C.E. Schreiner, and M.M. Merzenich, Plasticity in the frequency representation of primary auditory cortex following discrimination training in adult owl monkeys. *J. Neurosci.* 13: 87. (1993).
18. C.E. Schreiner, Order and disorder in auditory cortical maps. *Current Opinion in Neurobiology.* 5: 489. (1995).
19. N.M. Weinberger, N.M., Dynamic regulation of receptive fields and maps in the adult sensory cortex. *Ann. Rev. Neurosci.*, 18: 129. (1995).

Naive Preference and Filial Imprinting in the Domestic Chick: A Neural Network Model

Lucy Hadden

Department of Cognitive Science
University of California, San Diego
9500 Gilman Drive, MC 0515
La Jolla, CA 92093-0515
hadden@cogsci.ucsd.edu

Background

Filial imprinting in domestic chicks is of interest in psychology, biology, and computational modeling[1,2] because it exemplifies simple, innately programmed learning. Horn et al.[3] have recently found a naive tendency to approach heads and necks, which develops over the course of the first three days of life. It is not species-specific; chicks respond to heads and necks of ducks and chickens equally[4]. This preference interacts interestingly with filial imprinting, or learning to recognize a parent. Chicks can still learn about (and imprint on) other objects even in the presence of this predisposition, and the predisposition can override previously learned preferences[5]. These interactions are reminiscent of other systems which rely on naive preferences and learning, not least human language acquisition, as well as human babies' preference for human faces[6].

While the neurological basis of imprinting learning is fairly well understood, that of this predisposition is only beginning to be investigated. Imprinting learning is known to take place in IMHV (intermediate and medial portions of the hyperstriatum ventrale)[3], and to rely on noradrenaline[7]. (See Fig. 1.) IMHV in chicks is the target of afferents from a number of visual areas, including the visual Wulst, the avian analogue of V1[3]. The predisposition's location is currently unknown, but its strength correllates with plasma testosterone levels[8].

No previous models of imprinting have tried to account for the naive preference for heads and necks and its interactions with learning in filial imprinting. In this project, I develop a model which incorporates both the predisposition and imprinting, and which is as consistent as possible with the known neurobiology. The goals are to clarify how this predisposition might be implemented, and to examine more generally the kinds of representations that underlie naive preferences that interact with and facilitate, rather than replace, learning.

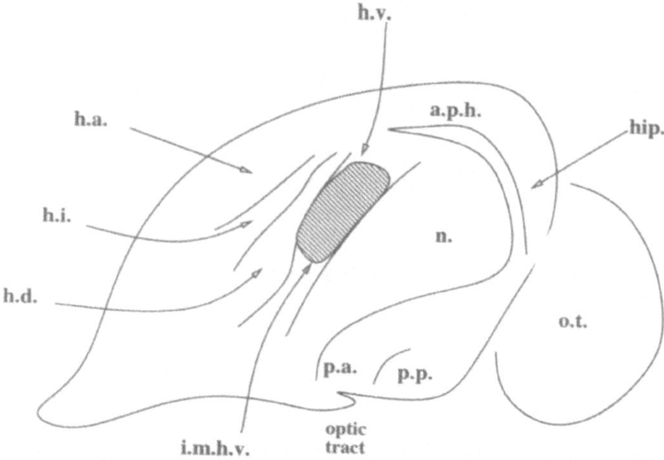

Figure 1. A longitudinal view of the chick brain, adapted from Horn 1985[3]. Structures shown include the hippocampus (hip), optic tectum (o.t.), and various portions of the hyperstriatum (h.a., h.i., h.d., and h.v. (of which the i.m.h.v. is a part).

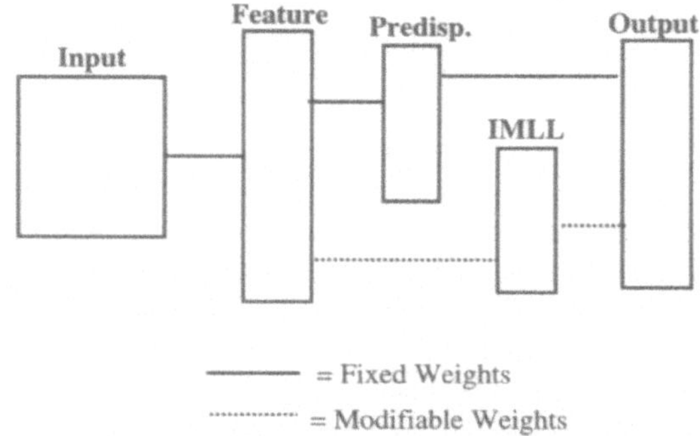

Figure 2. The network architecture sketched

Methods

The model is a neural network, and consists of 4 subparts (each a single layer): a feature detector preprocessor, a "predisposition" network, an "imhv" network, and an output layer. The output units are taken to represent different action patterns; in particular (following Bateson and Horn[2]), the output layer consists of an "approach" unit and a "withdraw" unit. These are the two chick behaviors which researchers use to assess a chick's degree of preference for a particular stimulus. The feature detector provides input to the "predisposition" and "imhv" layers; the output layer gets input from those two layers. All units in the network represent groups of neurons rather than single ones, and have a hyperbolic tangent activation function. Where there are connections, layers are fully interconnected.

The lowest level of my network is a feature-detecting preprocessor. This network takes (for now) crude 6x6 pictures, and spits out a 9-place floating-point vector. The feature detector was trained on random inputs using a Hebbian learning rule, and its weights were then frozen.

The "predisposition" layer was trained using the outputs of the feature detector. The pattern produced by the "head-neck" picture in the feature detector was trained to excite the "approach" output unit and inhibit the "withdraw" unit; other patterns were trained to a neutral value on both output units. These weights are stored, and treated as fixed in the larger network. (In fact, these weights are scaled down by a constant factor before being used in the larger network.) Since this is a naive preference, these weights are assumed to be fixed evolutionarily. Thus, the method of setting them is irrelevant; they could also have been found by a genetic algorithm.

The onset of the predisposition is modeled by connecting the "predisposition" subnetwork to the outputs only after the network receives an "experience" signal. In chicks which are reared in darkness, the onset of visual experience (such as training), or even being carried across the room in a dark box can trigger the onset of the preference for heads and necks[7].

The "imhv" layer is a winner-take-all network which modifies its connections with the feature detector's outputs by a Hebb rule. The output layer also learns via a Hebb rule. Its connections with both "imhv" and the "predisposition" layer are learned. The overall behavior of the network is a function of the values of the two output units.

Training

In the chick experiments on which this model is based, chicks are kept in the dark (and in isolation) except for training and testing periods. Training periods involve visual exposure to an object (usually a red box); testing involves allowing the chick to choose between approaching the training object and some other object (usually either a stuffed hen or a blue cylinder)[3]. So in these simulations, the network gets the training pattern during training periods, and random input patterns (simulating the random firing of retinal neurons) otherwise. While real chicks can be tested only once because of the danger of one-trial learning, the network's weights can be kept constant during testing, and the networks are tested many times.

Results and Discussion

At this time, the predisposition and imprinting portions of the model are functional in isolation, and their interactions have not been tested. For easier comparison to the behavioral results in chicks, the results are reported as changes in preference scores. Preference scores are the percentage of total activation of the "approach" unit (in networks) or time (in chicks) over two inputs allocated to the input of interest. Thus, for the networks, pref. score $= 100 \times a_t/(a_t + a_c)$, where a_t is the activation of the approach unit when the network is presented with the training (or target) picture, and a_c is the activation of the approach unit given the comparison picture. It is assumed that both values are positive; otherwise, the approach unit is taken to be off.

Given only random inputs the network over time turns on the approach unit when exposed to the "head/neck" pattern and no others (see Fig. 3-b). In particular, it does so more after the "predisposition" has been allowed to develop than it did initially, as is true in chicks. Since the preference score changes over time, the mere existence of the "predisposition" weights is not sufficient to explain these results; instead, the imll layer is also building up a representation of the head/neck.

The network also "approaches" the training pattern more strongly at the end of training than at the beginning, and indeed "prefers" the box picture, on which it was trained, to the other pictures more after training is done than before it started

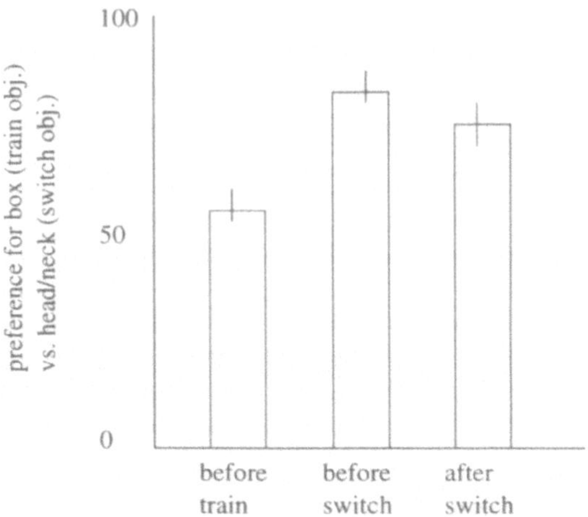

Figure 3. A summary of the results of the model. All bars are differences in preference scores between conditions for chicks (open bars) and the model (striped bars). a: Imprinting (change in preference for training object):
trained − untrained. b: Predisposition (change in preference for fowl):
experience − no experience (predisposition − no predisposition). (Chick data adapted from Horn[3], Bolhuis et al [9].)

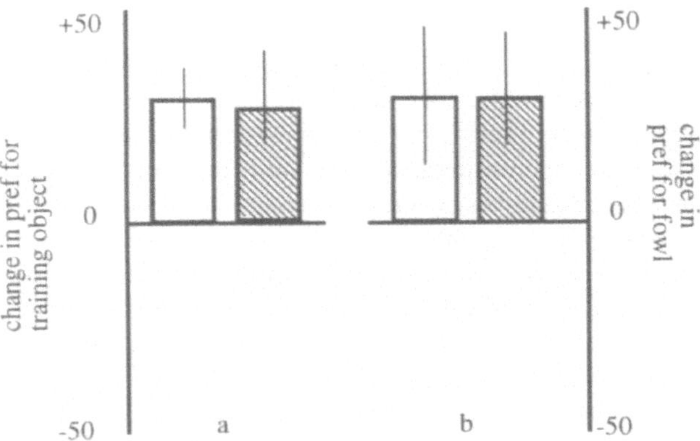

Figure 4. Like chicks, after a certain amount of training the networks cannot switch imprinting targets easily. This indicates a "sensitive period" in imprinting.

(see Fig. 3-a). Furthermore, as in chicks, the networks exhibit a "sensitive period" in imprinting, characterized by a difficulty in swiching imprinting targets once imprinting has occurred, and greatly decreased learning about a second object once they have "imprinted" on a first (see Fig. 4). As in other sensitive period simulations, the sensitive period is due to the imll units' activations being pushed very far into the nonlinear portion of the activation function.

Thus, the networks both imprint and show a predisposition. The predisposition in this model is not a bias on particular features, but instead exists as a set of connections in a higher-level visual area. Whether the predisposition in chicks has the same form has yet to be discovered. Thus far, the model is largely incomplete, and does not account for many of the relevant interactions between the predisposition and imprinting learning, nor does it account for the details of timing seen in chicks. It does, however, provide an instance of a predisposition which expresses itself more and more fully through learned connections, and which requires only random inputs to do so.

REFERENCES

1

1. R. C. O'Reilly and M. H. Johnson. Object recognition and sensitive periods: A computational analysis of visual imprinting. *Neural Computation*, 6(3):357–389, May 1994.
2. P. Bateson and G. Horn. Imprinting and recognition memory: A neural net model. *Animal Behaviour*, 48(3):695–715, 1994.
3. G. Horn. *Memory, Imprinting, and the Brain: An inquiry into mechanisms*. Clarendon Press, Oxford, 1985.
4. M. H. Johnson and G. Horn. Development of filial preferences in dark-reared chicks. *Animal Behaviour*, 36:675–683, 1988.
5. M. H. Johnson, J. J. Bolhuis, and G. Horn. Interaction between acquired preferences and developing predispositions during imprinting. *Animal Behaviour*, 33(3):1000–1006, Aug 1985.
6. J. Morton and M. H. Johnson. Conspec and conlern: a two-process theory of infant face recognition. *Psychological Review*, 98(2):164–181, 1991.
7. D. C. Davies, M. H. Johnson, and G. Horn. The effect of the neurotoxin dsp4 on the development of a predisposition in the domestic chick. *Developmental Psychobiology*, 25(2):251–259, May 1992.
8. J. J. Bolhuis, B. J. McCabe, and G. Horn. Androgens and imprinting: Differential effects of testosterone on filial preference in the domestic chick. *Behavioral Neuroscience*, 100(1):51–56, Feb 1986.
9. J. J. Bolhuis, M. H. Johnson, and G. Horn. Interacting mechanisms during the formation of filial preferences: The development of a predisposition does not prevent learning. *Journal of Experimental Psychology: Animal Behavior Processes*, 15(4):376–382, 1989.

A COMPUTATIONAL MODEL OF RETINOGENICULATE DEVELOPMENT

Gary L. Haith[1] and David Heeger[1]

[1]Department of Psychology
Stanford University
Stanford, CA, 94305-2130

BACKGROUND

The retinal projection to the lateral geniculate nucleus (LGN) of the thalamus (the "retinogeniculate projection") has become an ideal system for studying activity-dependent neural development. As the primary visual pathway to the cortex it is well studied, and it develops complex structures under the influence of spontaneous activity. Three models of development in the retinogeniculate system have been suggested.[1,2,3] None of these models encompasses the full range of structural refinement in the system and none considers the physiological development of LGN neurons. This paper attempts to provided a cohesive, realistic, and biologically predictive framework in which to investigate the development of retinotopy, eye-specific layers and on/off sub-layers in the retinogeniculate projection. A brief introduction to the biological system will help situate the model.

Waves

In neonatal ferrets, retinal ganglion cells (RGCs) fire spontaneous bursts of action potentials that propagate in "waves" across the retina. The waves involve both on- and off-center ("on" and "off") RGCs, and generally occur one at a time, separated by quiet periods of $\sim$60 seconds. Because they propagate synaptically and occur fairly rarely, the waves in the two eyes are assumed to be uncorrelated.[4,5]

In ferrets, the retinal waves are present at least from birth (post-natal day 0, P0) and they disappear just prior to eye opening (P31). In the third week (P14-P21) the on and off RGCs develop distinct patterns of activity. Although both cell types still participate in the waves, the off RGCs start to fire many additional bursts of action potentials that are uncorrelated with the firing of neighboring on and off RGCs.[6]

Retinogeniculate Refinement

RGC axons in the ferret reach the LGN 2 weeks before birth. The axons are initially intermixed, with all the axons innervating and synapsing in all the future layers. Although initially intermingled, RGC arbors are not exuberantly wide at any time from birth to eye-opening.[7]

During the first two weeks (P0-P14) the retinal afferents and LGN cells segregate into eye-specific layers. In the third week (P14-P21) the on and off afferents segregate into separate sub-layers. Interference with either the retinal or LGN activity at least partially disrupts layer formation.[8,9]

It has been established that there is a coarse retinotopic bias in the retinogeniculate projection at birth[10], but the development of precise retinotopy in the LGN has not been investigated. By maturity each LGN cell receives inputs from only a few neighboring on or off RGCs from one eye, and each mature RGC projects to only a few of the functionally appropriate cells within its axonal arbor.[11] The projection thus forms a precise retinotopic map within each sub-layer.

Neither mechanical nor trophic/chemical mechanisms seem able to account for this degree of precision. The fibers in the optic tract are relatively disorganized[12], and trophic/chemical factors would be unlikely to differentiate between adjacent cells of the same type. These limitations suggest that activity-dependent processes substantially contribute to the refinement of retinotopy in the LGN.

Immature LGN Physiology and Anatomy

During development, three factors potentiate activity-dependent self-organization in the LGN. First, developing LGN neurons are especially responsive to retinal activity due to both more efficacious NMDA-driven excitation and attenuated resting potentials.[13] Second, because of a lack of GABAa inhibition, immature LGN neurons respond with longer EPSP's ($\sim$2 orders of magnitude longer) than do mature cells.[14] Third, immature LGN cells exhibit long-term potentiation (LTP) when driven with wave-like stimuli.[15]

Most models of self-organizing map formation rely a center-surround arrangement of long-range inhibitory and short-range excitatory lateral connectivity (a "neighborhood function") that, in many cases, shrinks over the course of development.[16] There is no evidence for such a pattern of connectivity in the developing LGN (and little evidence for this in the cortex). Indeed, both inhibition[14] and LGN interneuron arborization[17] are initially small and progressively grow over the course of development.

THE MODEL

Our goal was to produce a model with biologically realistic inputs, LGN physiology and RGC axonal arbor development.

Inputs and Initialization

Matching input parameters to in vitro anatomical and physiological measurements is relatively straightforward. A wave is modeled as a bar-like section of a spreading ring with Gaussian cross-section. The width, frequency, height, and speed of the waves matches the reported retinal measurements.[4] Both on and off RGCs are indistinguishable based on activity for the first "2 weeks" of model development, after this point bursting noise is added to the off cell population. This change brings the bursting rates and correlational structure of the input into agreement with in vitro measurements of retinal activity during on/off sub-layer segregation.[6]

The model is initialized to mimic the projection at birth from a small retinal strip (1x20 cells x 2 cell types, on/off) from each eye to the same small slice (1x80 cells x 4 layers) of LGN tissue. Thus, there are 4 strips of distinct RGCs projecting to a 4 neuron deep LGN slice. In addition there are 4 LGN cells for every RGC, roughly matching ferret anatomy. Each RGC arbor is initialized as a Gaussian ($\sim 100um$ wide) that is centered on a random, though slightly retinotopically biased position. Each arbor initially innervates all layers — with a slight bias toward the appropriate layer and sub-layer (see Fig. 2A).

Model Dynamics

Although the developmental processes occur concurrently, we iterate through a four step algorithm for computational convenience:

1. Compute LGN activity in response to a "snapshot" presentation of a retinal wave
2. Hebbian weight update
3. Synaptic sprouting
4. Weight normalization and synaptic retraction

LGN Update. Geniculate potentials are computed as a (weighted) sum of their inputs from the retina and their neighbors in the geniculate. We assume that the LGN cells reach steady state for a given retinal input because the speed of wave propagation is much slower than synaptic transmission and membrane depolarization. The steady state is reached by iterating the following update equation until the geniculate potentials stabilize:

$$\frac{\delta l_{[x]}}{\delta t} = \frac{\lambda}{1 + 2\lambda}\left(\frac{1}{\sqrt{\lambda}}\sum_y W_{[xy]}r_{[y]} + \sum_{x \in \eta} l_{[x]}\right) - l_{[x]} \tag{1}$$

where $l_{[x]}$ is the potential of LGN cell x, λ is a constant, $W_{[xy]}$ is the synaptic weight from RGC y to LGN cell x, $r_{[y]}$ is the firing rate of RGC y, and η includes LGN cell x's 2 nearest intra-layer neurons and the 1 nearest inter-layer neuron. The LGN cell potential is calculated as a scaled product of its synaptic inputs and the RGC firing rates in addition to the potentials of the neighboring LGN cells and self-inhibition.

In signal processing terms the intra-LGN connections act as a feedback linear system that blurs the retinal input. The amount of blurring and the amplification of the LGN response are both scaled by λ, which decreases over development (see Fig. 1). The range of λ (20 at birth and 1 at eye opening) as well as the scale factor ($\frac{1}{\sqrt{\lambda}}$) on the effective inputs to the LGN cells were chosen to match the roughly three-fold decrease in geniculate responsivity observed over the course of development.[14]

Because it supports global order early in development and local precision later in development, this decrease in blurring plays an analogous role to more traditional neighborhood functions that shrink over development.[18]

A biophysical interpretation of Eq. 1 can be stated in terms of the membrane equation:

$$C\frac{\delta l_{[x]}}{\delta t} = -(g_{l[x]}l_{[x]} + g_{e[x]}(l_{[x]} - E_e) + g_{i[x]}(l_{[x]} - E_i)) \tag{2}$$

which describes a geniculate cell's membrane potential ($l_{[x]}$) as a function of its excitatory ($g_{e[x]}$), inhibitory ($g_{i[x]}$), and leak conductances ($g_{l[x]}$), their respective reversal potentials ($E_e, E_i,$ and 0), and the capacitance (C) of the membrane.

Eq. 2 becomes algebraically equivalent to Eq. 1 by eliminating the arbitrary scale factor $\frac{1}{\sqrt{\lambda}}$ from Eq. 1 and equating the following terms:

$$C = \frac{1 + 2\lambda}{\lambda}$$

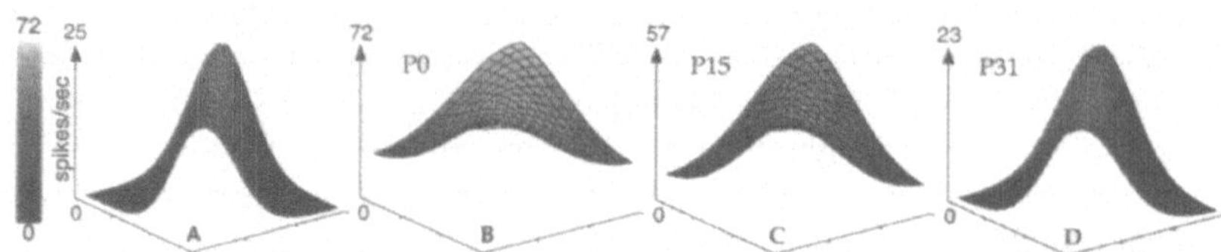

Figure 1. (A) Retinal activity during a wave. (B-D) LGN response at different stages in development. The retinal signal evokes a large, blurred response in the neonatal LGN, a less blurred response later in development, and a virtual copy of the retinal input at eye opening. For the purpose of this figure, the retinogeniculate projection was assumed to be 2-dimensional and perfectly retinotopic in all three cases.

$$g_{e[x]} = \frac{\sum_y w_{[xy]} r_{[y]} + \sum_{x \in \eta} l_{[x]}}{E_e}$$

$$g_{i[x]} = \frac{(g_{l[x]} + g_{e[x]} - \frac{1+2\lambda}{\lambda}) l_{[x]}}{E_i - l_{[x]}}$$

Simulations using either Eq. 2 or Eq. 1 display the appropriate changes over development. The decrease in λ over the course of development (see above), indicates that the capacitance C will increase by a factor of 1.5, and the inhibitory conductances $g_{i[x]}$ increase by a factor dependent on the magnitude of leak conductance $g_{l[x]}$ assumed. These dynamics are consistent with physiological measurements of developing LGN relay cells.[13,14] Note that although algebraically identical, the two equations behave slightly differently because the inhibitory conductances are rectified such that they cannot go below 0, and because we have eliminated the factor $\frac{1}{\sqrt{\lambda}}$. These two differences tend to compensate for each other.

Hebbian Weight Update. Second, the retina-to-LGN weights are updated using a modified Hebbian rule:

$$\delta w_{[xy]} = h \, w_{[xy]} \, l_{[x]} \, r_{[y]} \tag{3}$$

The weight change ($\delta w_{[xy]}$) is the product of the learning rate (h), the current weight ($w_{[xy]}$), the steady-state geniculate activity ($l_{[x]}$), and the retinal activity ($r_{[y]}$).

Scaling the Hebbian update by the previous weight generates a feed-back cycle in which strong weights tend to become stronger. This tendency allows for the development of precise connectivity , even in the case of multiplicative normalization (see Eq. 4 below) which would otherwise tend to favor graded synaptic connectivity.[19]

That LTP would be weighted by the previous weight is intuitively plausible (e.g. a cell with many synapses to another cell can potentiate all the synapses simultaneously), but empirically untested.

Synaptic Sprouting. Third, the weights are blurred (i.e. convolved with a Gaussian) to mimic a branching factor. This approach yields RGC axonal arbors that can shift their geniculate position appreciably, yet change their width only slightly over development, consistent with studies of the developing LGN.[7]

Notably, because sprouting via this mechanism is sensitive to the previous concentration of weights in the area, it biases neighboring LGN cells toward correlated firing by encouraging them to have the same afferent connections. This action plays a similar role to the direct excitation provided by neighborhood functions. Indeed, a projection with sprouting can topologically organize, and in some cases segregate into layers, in the absence of lateral interactions in the target tissue.[1,3]

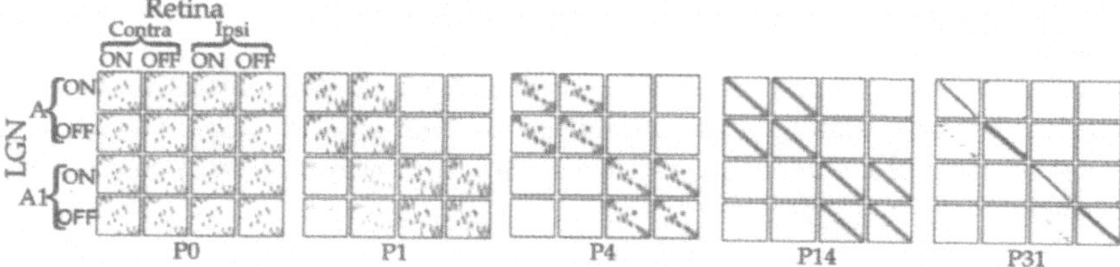

Figure 2. The figures depict retinotopic refinement and layer segregation during development. Each dot indicates a synaptic connection from an RGC (columns) to an LGN relay cell (rows) — darker dots indicate higher synaptic strengths. The 16 smaller boxes in each figure correspond to the projections from each RGC subpopulation to each LGN sub-layer. "Contra" and "Ipsi" specify which RGCs are from the eye contralateral or ipsilateral to the simulated LGN nucleus. A perfectly segregated and retinotopic projection with the appropriate orientation would resemble the identity matrix (P31).

Weight Normalization. Finally, the weights are normalized divisively such that the total connection strength of each RGC and LGN neuron stays roughly constant.

$$w_{[xy]} = \frac{2w_{[xy]}}{\sum_x w_{[xy]} + \sum_y w_{[xy]}} \tag{4}$$

Because all of the weight changes are initially positive, this step is crucial to prevent the weights from increasing without limit. The assumption here is that neurons will compete for synaptic sites and that each neuron can only support a limited number of synapses.

Alternative Formulation

We have implemented an alternate version of the model in which the normalization is subtractive and the Hebbian update is unscaled by the previous weight. This alternative version yields similar results, as predicted by previous analytic work.[19,20]

RESULTS AND DISCUSSION

The mechanisms in the model can support the development of eye-specific layers, on/off sub-layers and precise retinotopy from an initially disordered projection (see Fig. 2, P0 and P30). The time-course of refinement is qualitatively similar to refinement in the actual projection: eye-specific layers segregate early in development, on/off sub-layers segregate during differential on/off activity in the retina, and the refinement of retinotopy is relatively prolonged (see Fig. 2).

The realism of the model will make it a useful guide for future investigations into this system. In particular, the model makes two concrete predictions, neither of which has been tested.

First, in order for layer segregation to develop properly in the model, the inter-layer physiological interactions must be much weaker than the intra-layer interactions in the LGN. The model thus predicts that the spread of excitation should be wider within a layer than between layers. In addition, the spread should attenuate as development progresses. To test this prediction one could excite an LGN cell with an intracellular electrode while simultaneously recording from other LGN cells at various distances within the same layer and across layers. Failure to find this strongly asymmetric spread of excitation would suggest that additional mechanisms (e.g. a stronger bias in initial ingrowth and/or trophic factors) contribute more substantially to layer segregation than this model assumes.

Second, the biophysical interpretation of the LGN update equation implies a push-pull relation between excitatory and inhibitory conductances in LGN relay cells. This interpretation thus predicts an inverse relationship between inhibitory and excitatory conductances.

REFERENCES

1. R. Keesing, D. G. Stork, and C. J. Shatz, "Retinogeniculate development: The role of competition and correlated retinal activity," in *Advances in Neural Information Processing Systems 4*, (San Mateo, CA), pp. 91–97, Morgan Kaufmann, (1992).

2. C. W. Lee and R. O. L. Wong, "Developmental patterns of on-off retinal ganglion cell activity lead to segregation of their afferents under a Hebbian synaptic rule," in *Society For Neuroscience Abstracts*, vol. 22, p. 1202, Society for Neuroscience, (1996).

3. S. Eglen, *Modelling the development of the retinogeniculate pathway.* PhD thesis, University of Sussex, (June 1997).

4. R. O. L. Wong, M. Meister, and C. J. Shatz, "Transient period of correlated bursting activity during development of the mammalian retina," *Neuron*, vol. 11, pp. 923–938, (November 1993).

5. M. B. Feller, D. P. Wellis, D. Stellwagen, F. S. Werblin, and C. J. Shatz, "Requirement for cholinergic synaptic transmission in the propagation of spontaneous retinal waves," *Science*, vol. 272, pp. 1182–1187, (May 24 1996).

6. R. O. Wong and D. M. Oakley, "Changing patterns of spontaneous bursting activity of on and off retinal ganglion cells during development," *Neuron*, vol. 16, pp. 1087–1095, (June 1996).

7. D. W. Sretavan and C. J. Shatz, "Prenatal development of retinal ganglion cell axons: Segregation into eye-specific layers within the cat's lateral geniculate nucleus," *Journal of Neuroscience*, vol. 6, pp. 234–251, (January 1986).

8. C. J. Shatz and M. P. Stryker, "Prenatal tetrodotoxin infusion blocks segregation of retinogeniculate afferents," *Science*, vol. 242(4875), pp. 87–89, (1988).

9. J.-O. Hahm, R. B. Langdon, and M. Sur, "Disruption of retinogeniculate afferent segregation by antagonists to NMDA receptors," *Nature*, vol. 351, pp. 568–570, (June 13 1991).

10. G. Jeffery, "The topographic relationship between shifting binocular maps in the developing dorsal lateral geniculate nucleus," *Experimental Brain Research*, vol. 82, pp. 408–416, (1990).

11. D. N. Mastronarde, "Non-lagged relay cells and interneurons in the cat lateral geniculate nucleus: Receptive-field properties and retinal input.," *Visual Neuroscience*, vol. 8, no. 5, pp. 407–441, (1992).

12. T. Voigt, J. Naito, and H. Wassle, "Retinotopic scatter of optic tract fibers in the cat," *Experimental Brain Research*, vol. 52, pp. 25–33, (1983).

13. A. S. Ramoa and D. A. McCormick, "Developmental changes in electrophysiological properties of LGNd neurons during reorganization of retinogeniculate connections," *Journal of Neuroscience*, vol. 14, pp. 2089–2097, (April 1994).

14. A. S. Ramoa and D. A. McCormick, "Enhanced activation of NMDA receptor responses at the immature retinogeniculate synapse," *Journal of Neuroscience*, vol. 14, pp. 2098–2105, (April 1994).

15. R. Mooney, D. V. Madison, and C. J. Shatz, "Enhancement of transmission at the developing retinogeniculate synapse," *Neuron*, vol. 10, pp. 815–25, (May 1993).

16. N. V. Swindale, "The development of topography in visual cortex: A review of models," *Network: Computation in Neural Systems*, vol. 7, pp. 161–247, (May 1996).

17. J. K. Sutton and J. K. Brunso-Bechtol, "Dendritic development in the dorsal lateral geniculate nucleus of ferrets in the postnatal absence of retinal input: A golgi study," *Journal of Neurobiology*, vol. 24, no. 3, pp. 317–334, (1993).

18. T. Kohonen, "Physiological interpretation of the self-organizing map algorithm," *Neural Networks*, vol. 6, pp. 895–905, (1993).

19. K. D. Miller and D. J. C. MacKay, "The role of constraints in Hebbian learning," *Neural Computation*, vol. 6, pp. 100–126, (1994).

20. L. Wiskott and T. Sejnowski, "Constrained optimization for neural map formation: A unifying framework for weight growth and normalization," *Neural Computation*, (In Press).

CLUSTER STRUCTURE OF CORTICAL SYSTEMS IN MAMMALIAN BRAINS

Claus C. Hilgetag[1], Gully A.P.C. Burns[2], Mark A. O'Neill[1], and Malcolm P. Young[1]

claus.hilgetag@.ncl.ac.uk, gully@usc.edu, mao@crunch.ncl.ac.uk, m.p.young@ncl.ac.uk

[1]Neural Systems Group
University of Newcastle upon Tyne, Department of Psychology
Ridley Building, Newcastle upon Tyne NE1 7RU, UK
[2]USC Brain Project
University of Southern California, Hedco Neuroscience Building
Los Angeles, CA 90007, USA

ABSTRACT

Information about the numerous connections among the many cortical areas in mammalian brains has been published extensively. The organization of the cortical systems that these connections define, however, is still poorly understood. We analyzed this complex connectivity structure with the help of mathematical and computational tools in order to better understand cortical processing. Here we present the results of a number of different analyses that we used to decide whether cortical areas in the brains of the cat and the Macaque monkey can be grouped into separate clusters, based on the organization and strength of connections between the areas. We employed non-parametric cluster analysis together with multi-dimensional scaling, as well as a new computational tool based on the global stochastic optimization of interconnected structures. Our results revealed a well-defined clustered organization, which was particularly clear in the primate visual system. The identified clusters largely agreed with suspected functional sub-divisions of cortical regions. Our results provide a basis for modeling studies, for the interpretation of functional studies of the brain; as well as for predicting future anatomical experiments.

INTRODUCTION

The complex network of anatomical fibers which interconnects areas in the brain determines each area's inputs and its outputs, and so defines the significance of any one area for the functions of the whole nervous system, and hence for behaviour. A wealth of information exists in the anatomical literature about the existence or absence, and also about some finer details, of connections linking certain areas with others. Information on how neural systems are organized, however, is unlikely to be gained simply by inspection of these complex data. We have, therefore, pursued an approach using formal analysis of neuroanatomical data, applying mathematical and computational tools. In this framework, data have been assembled in a way that acknowledges both experimental detail and the requirements of formal representation (see Burns *et al.*, 1997), and various methods of analysis have be applied. In previous studies (e.g., Young, 1992, Young *et al.*, 1995; Scannell *et al.*, 1995) we have used multi-dimensional scaling, matrix transforms and seriation algorithms to investigate the general topology of neural systems in primates and cats, and have used hierarchical analysis (Hilgetag *et al.*, 1996) to study other aspects of neural systems' organization.

An influential idea in cortical neuroscience has been the concept of clusters or 'streams' of areas that are related functionally and anatomically. Although such concepts have been long established for the primate visual system (e.g., Ungerleider and Mishkin, 1982), and have recently been invoked also for the cat cortical system (Lomber *et al.*, 1996), the existence of streams in cortical systems has been debated repeatedly (e.g., Simmen *et al.*,1994). We decided to use two independent approaches, non-parametric cluster analysis and optimal set analysis, to elucidate these issues further.

METHODS

We analyzed 4 different sets of connectivity data for cortical systems in two species: the Macaque monkey visual system (interconnections of 32 areas: 321 existing connections and 373 potential connections that have been found absent), the somato-sensory system in the same species (63 connections between 14 areas), a global data set of connections in the Macaque cortex (834 connections between 73 areas), and the whole cat cortical system (892 interconnections of 55 areas). The first two data sets were taken from Felleman and Van Essen's (1991) compilation, and contain information simply about the existence or absence of a connection. The second data set, compiled in Young (1993), also states the existence or absence of a cortical connection. The latter cat data set is updated (Scannell, personal communication) from Scannell *et al.* (1995), and also includes information on whether a particular existing connection is of weak, moderate or of high density.

We used two analytical strategies to delineate the organization of connectivity clusters in the mammalian cortex.

(I) We employed statistical procedures that detect significant spatial relationships among distributions of points in metric coordinate systems. Therefore, the ordinal data of connections between cortical areas had to be converted into metric distances. We used non-metric multi-dimensional scaling (e.g., Kruskal, 1964) for this transformation. The method provides optimal, low-dimensional (for the sake of visualization, typically 2D or 3D) metric representations of high-dimensional sets of simple connectivity rankings.

We used a weighted dissimilarity transformation to pre-process the connectivity matrices (see 'wdsm1' in Young *et al.*, 1995) to generate proximities that could be analyzed with non-metric multidimensional scaling, in two, five, ten and fifteen dimensions. We analyzed the structure of these high dimensional configurations with non-parametric cluster analysis, which is capable of detecting irregular clusters with non-uniform shapes (using the MODECLUS procedure in SAS, SAS User's guide, 1990). We then used clustering kernels based both on fixed radius and nearest neighbor paradigms, and repeated the analysis using different density smoothing parameters to show which clustering features are found most consistently. The above procedures were automated with Perl 5 scripts, and results were pooled into summary diagrams showing the relative frequencies of any two cortical areas appearing in the same clusters, averaged over all employed methods and parameters (see Figure 2).

(II) We developed a new optimization tool which previously has been employed successfully in the analysis of hierarchical relationships in the cortex (Hilgetag *et al.*, 1995, Hilgetag *et al.*, 1996). We used a relational network processor that was designed to represent the anatomical connectivity structure in the real brain based on a data structure of interconnected objects. The approach relied on the specification of a cost function evaluating the overall arrangement of objects, which cost was then optimized using a modified simulated annealing algorithm (Van Laarhoven and Aarts, 1987; Hilgetag *et al.*, 1997a). Here we defined a two-component cost in which an optimal cluster arrangement of the system has as few connections running *between* the clusters as possible (this cost component we termed 'attraction'), and as few as possible non-connections *within* the clusters ('repulsion'). If it was possible on the basis of the available data, we distinguished between explored absent and hitherto unstudied connections; the latter would be excluded from the analysis. The system determined automatically the number of clusters for the

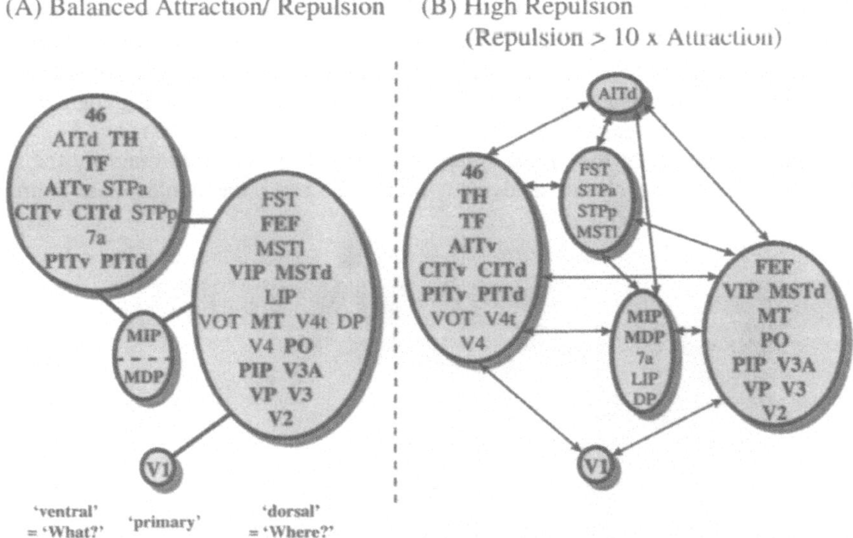

Figure 1. Optimal cluster arrangement of primate cortical visual areas obtained with evolutionary optimization. While the relative positions of the clusters themselves are not relevant, the sequences of the areas *within* the clusters follow a statistical summary of 150,000 optimal hierarchical arrangements for the visual system (Hilgetag *et al.*, 1996; 1997a). These diagrams, therefore, reflect two important features of primate visual anatomy, hierarchy and clustered connectivity.

Areas printed in bold face remained in their respective clusters for a wide range of different attraction and repulsion weights. Four relatively stable set of areas could be distinguished: a 'ventral', a 'dorsal' cluster and a cluster solely composed of primary visual area V1. The two main clusters agreed with suspected subdivisions of the primate visual system into main components thought to be responsible for object and for spatial recognition. The cluster containing areas MIP and MDP can be considered as an artifact, because the connectivity of these two areas is hardly known. For abbreviations of area names and definition of the areas see Felleman and Van Essen (1991).

1(A). Two optimal solutions were obtained for balanced attraction and repulsion weights. The solutions only differed in MIP and MDP being in one shared or in two separate clusters, respectively. The optimal arrangements had 91 connections running between the four clusters, and 42 non-existing connections were still contained within the clusters, yielding a total cost of 133 for the configurations.

1(B). A unique optimal arrangement of six clusters existed for weighting the repulsion component of the cost function more than ten times values larger than the attraction component. All of the non-existing connections were located outside of the clusters, while 188 area connections ran between them.

optimal solutions. Changes in the number and composition of the optimal clusters could be achieved by varying the weights for attraction and repulsion in the component cost, from high attraction weights (yielding fewer clusters, or only one in the limit case of overwhelming attraction) to high repulsion (yielding a larger number of clusters, until all non-existing connections were located outside of the clusters).

RESULTS

In the case of the primate visual system, we obtained between one and six clusters for different degrees of attraction and repulsion in the network optimization approach. With balanced attraction and repulsion the optimization yielded two main clusters, which broadly corresponded to the suspected sub-division of the visual system into the ventral and dorsal streams identified by previous studies (Figure 1).

These two main clusters remained largely stable over a wide range of different attraction and repulsion values. Another cluster appearing in computations with higher repulsion than attraction was made up by primary visual area, V1. For repulsion weights at least ten times higher than the attraction weights, all non-existing connections were located outside the clusters, so that any further increase of the repulsion weight could not further decompose the six clusters (Figure 1b). These clusters, therefore, represent 'building blocks' of the visual system's connectional organization, and their place in the cortical geography gives important clues about principles of cortical organization (Scannell, 1997).

The general picture of the visual system's organization was confirmed by the results from the non-parametric cluster analyses. Figure 2 shows the pooled data from five-

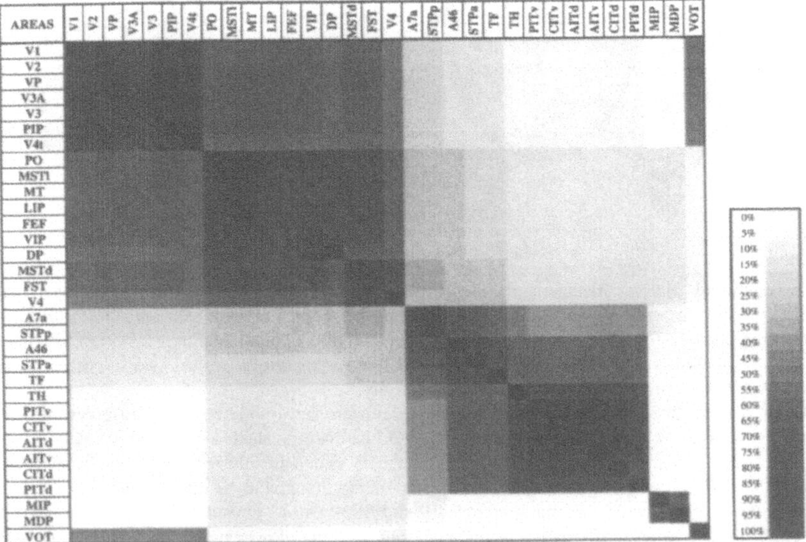

Figure 2. Pooled results of high-dimensional (D=5), non-parametric cluster analyses. The graph shows the relative frequency with which any two areas shared a cluster throughout four approaches using different cluster radius and smoothing parameters. Dark shading indicates a high, lighter shading a low frequency of cluster sharing, see palette bar. The list of areas was sorted along the main dimension as to bring together linked clusters. Three main clusters can be easily distinguished ('dorsal', 'ventral' and MDP/MIP) which agree almost perfectly with the arrangement of areas found in the network optimization approach (compare Figure 1A), the only mismatches being the assignments of V1 and VOT.

dimensional clustering computations for the primate visual system using a wide range of parameters for describing the radius and shape of potential clusters.

For cluster searches in lower-dimensional metric representations of the system, even fewer and more clear-cut clusters resulted. Three main clusters can be distinguished again, a group composed of occipital-dorsal areas, another main cluster of inferotemporal-ventral areas, and a probably spurious MDP/MIP cluster.

A number of stable clusters were also found in the computations for the Macaque somato-sensory connectivity and the global data sets for the cat and Macaque cortical systems. Figure 3, for instance, shows a summary of 2069 optimal solutions for the cat cortex for a wide variety of different attraction and repulsion weights. Despite the changes in cluster sizes and composition, which led to the many weakly-linked sets in the plot, a number of stable clusters can be easily recognized. Particularly striking were the very stable clusters of somato-motor areas apparent in both cat and Macaque data sets, which mirrored a similar result for the cluster analysis of semi-functional data (Hilgetag *et al.*, 1997b). Generally, however, these data sets show a larger variety in number and composition of clusters than for the primate visual system.

CONCLUSIONS

Our study showed that a well-defined cluster structure of mammalian cortical systems can be detected reliably by independent statistical and optimization approaches. The cluster arrangements that we found were clearest for the primate visual system, confirming a sub-division into three main groups of primary, 'ventral' and 'dorsal' areas. The larger variety

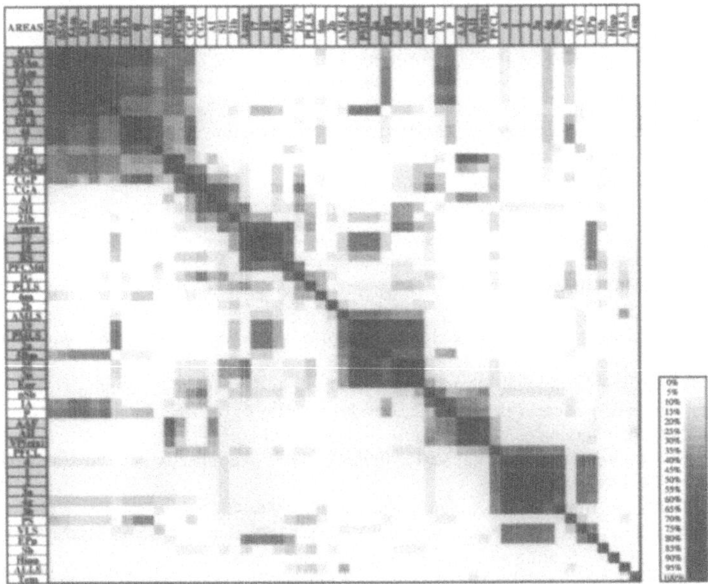

Figure 3. Optimally connected sets of cortical areas in the cat brain. The diagram which is in the same format as figure 2 shows a summary of all 2069 optimal configurations, obtained for a range of cost component weights from 20 times repulsion (46 optimal solutions with 14 to 17 clusters) to 10 times attraction (one optimal solution with one cluster). Groups of more than three areas which were strongly associated and shared clusters in more than 50% of all solutions are highlighted. Such groups largely followed suspected sensory modalities, in particular for the somato-motor system. For abbreviations of area names see Scannell *et al.* (1995).

in the results for the other systems might be due to the structure of the respective data, which did not explicitly distinguish between known absent and so far unstudied connections, distorting the analysis since all the connections not studied were assumed not to exist. This emphasizes the importance of explicitly reporting absent connections in anatomical experiments.

The results allow predictions for further experiments: if two areas (whose interconnections have not yet been studied) are consistently found in one cluster, there is a high likelihood that the areas are connected. This probability might be further specified through the relative association of these areas as computed by non-parametric cluster analysis.

Acknowledgments

Our work has been supported by the Royal Society, the MRC, the University of Newcastle upon Tyne, and the Wellcome Trust. We thank Jack W. Scannell for kindly letting us use his updated compilation of cat connectivity data.

REFERENCES

Burns, G. A. P. C., O'Neill, M. A., and Young, M. P., 1997, Calculating finely-graded ordinal weights for neural connections from neuroanatomical data from different anatomical studies, in: *Computational Neuroscience: Trends in Research, 1997*, J. M. Bower, ed., pp. 53-57, Plenum, New York and London.

Felleman, D. J., and Van Essen, D. C., 1991, Distributed hierarchical processing in the primate cerebral cortex, *Cerebral Cortex* **1**: 1-47.

Hilgetag, C. C., O'Neill, M. A., Scannell, J. W., and Young, M. P., 1995, A novel network classifier and its application: Optimal hierarchical orderings of the Cat visual system from anatomical data, in: *Genetic Algorithms in Engineering Systems: Innovations and Applications*, IEE Publication No. 414, Sheffield.

Hilgetag, C. C., O'Neill, M. A., and Young, M. P., 1996, Indeterminate organization of the visual system, *Science* **271**: 776-777.

Hilgetag, C. C., O'Neill, M. A., and Young, M. P., 1997a, Optimization analysis of complex neuroanatomical data, in: *Computational Neuroscience: Trends in Research, 1997*, J. M. Bower, ed., pp 925-930, Plenum, New York and London.

Hilgetag, C. C., Stephan, K. E., Burns, G. A. P. C., O'Neill, M. A., Young, M. P.., Zilles, K., and Kötter, R., 1997b, Systems of primate cortical areas defined by functional connectivity, *Soc. Neurosci. Abstr.* **23**, 514.13., New Orleans.

Kruskal, J. B., 1964, Multidimensional scaling by optimizing goodness of fit to a nonmetric hypothesis, *Psychometrika* **29**: 1-27.

Van Laarhoven, P. J. M., and Aarts, E. H. L., 1987, *Simulated Annealing: Theory and Applications*, Kluwer, Dordrecht.

Lomber, S. G., Payne, B. R., Cornwell, P., and Long, K. D., 1996, Perceptual and cognitive visual functions of parietal and temporal cortices in the Cat, *Cerebral Cortex* **6**: 673-695.

SAS/STAT User's Guide (version 6), 1990, 4th ed. SAS Institute Inc.

Scannell, J. W., Blakemore, C., and Young, M. P., 1995, Analysis of connectivity in the Cat cerebral cortex, *J. Neurosci.* **15**: 1463-1483.

Scannell, J. W., 1997, Determining cortical landscapes, *Nature* **386**: 452.

Simmen, M. W., Goodhill, G. J., Willshaw, D. J., and Young, M. P., Scannell, J. W., Burns, G. A. P. C., and Blakemore, C., 1994, Scaling and brain connectivity, *Nature* **369**: 448-450.

Ungerleider, L. G., and Mishkin, M., 1982, Two cortical visual systems, in: *Analysis of Visual Behaviour*, D. G. Ingle, M. A. Goodale, and R. J. Q. Mansfield, eds., pp. 549-586, MIT Cambrigde MA.

Young, M. P., 1992, Objective analysis of the topological organization of the primate cortical visual system, *Nature* **358**: 152-155.

Young, M. P., Scannell, J. W., O'Neill, M. A., Hilgetag, C. C., Burns, G., Blakemore, C., 1995, Non-metric multidimensional scaling in the analysis of neuroanatomical connection data and the organization of the primate cortical visual system, *Phil. Trans. R. Soc. Lond. B* **348**: 281-308.

DYNAMIC MEMORY MAINTENANCE

David Horn, [1] Nir Levy, [1] and Eytan Ruppin [2]

[1] School of Physics and Astronomy
[2] Departments of Computer Science & Physiology
Tel Aviv University, Tel Aviv 69978, Israel

INTRODUCTION

Memories can be maintained for very long periods of time, even during our whole lifetime. A fundamental dogma in the Neurosciences is that memories are engraved in the brain via specific, long-term, alterations in synaptic efficacies. However, synaptic turnover is relatively widespread in the mature nervous system[1,2,3]. How then are memories maintained for very long periods? Clearly memories can be maintained if synaptic weights can be kept fixed, which is the purpose of several mechanisms that were suggested in the literature. An interesting alternative, that we will explore below, is maintaining memories with altered synaptic values, i.e., synapses change dynamically and still encode the original memories[4].

In our model[5] memory maintenance is carried out on the neuronal level and compensates for synaptic degradation. It has the interesting property of normalizing basins of attraction, and prevents the formation of pathologic neural assemblies. To perform memory maintenance, the neurons in our model regulate their overall level of synaptic inputs (i.e., average post-synaptic potential) by activating *neuronal regulatory* (NR) processes that jointly modify all the incoming synapses of the neuron by a common factor.

Our proposal is biologically motivated by the extensive experimental evidence of homeostasis mechanisms that act to maintain neuronal activity (see [6] for a comprehensive review). The main new feature introduced in this work is the view of NR as a common change in the synaptic efficacies of a neuron that, depending on the neuron's activity, keeps the relative weights of different synapses unchanged. This key feature has recently received direct experimental support from the work of Turrigiano *et al.*[7], showing that neocortical pyramidal neurons regulate their firing rates by scaling the strength of their synaptic connections up or down as a function of activity. This is a slow process, affecting AMPA-mediated synaptic transmission. Just as in the model[5], it produces long-lasting regulation in the desired multiplicative post-synaptic fashion. Our work focuses on studying the functional significance of neuronal-regulated common synaptic changes, on the network level. As will be shown, such multiplicative

NR mechanisms are potentially highly important, as they can counteract the effects of synaptic turnover and enable life-long memory maintenance.

SYNAPTIC DEGRADATION AND NEURONAL REGULATION

We study NR in the framework of an excitatory-inhibitory associative memory network[8], having M memory patterns, N excitatory neurons, and sparse coding level $p << 1$. The initial synaptic efficacy $J_{ij}(t = 0)$ between the jth (presynaptic) neuron and the ith (postsynaptic) neuron is chosen in the Hebbian manner

$$J_{ij}(t = 0) = \frac{1}{Np} \sum_{\mu=1}^{M} \eta^{\mu}{}_{i} \eta^{\mu}{}_{j} \qquad (1)$$

where η^{μ} are the stored memory patterns. Synaptic weakening due to metabolic turnover, or synaptic degradation, is modeled by

$$J_{ij}(t + \Delta t) \to (1 - \epsilon_{ij}) J_{ij}(t) , \qquad (2)$$

where the time t denotes the number of degradation and maintenance steps, or epochs. The deterioration factor ϵ_{ij} is newly chosen at every deterioration cycle in a random fashion (with mean ϵ and variance σ^2). Synaptic strengthening resulting from NR is represented by

$$J_{ij}(t + \Delta t) \to c_i J_{ij}(t) , \qquad (3)$$

in which the regulation factors c_i correct the values of all excitatory synaptic connections projecting on neuron i,

$$c_i = 1 + \tau \tanh \left[\kappa \left(1 - \frac{\langle h^e{}_i(t) \rangle}{\langle h^e{}_i(t = 0) \rangle} \right) \right] \qquad (4)$$

where κ and τ are rate constants. This choice of c_i maintains the excitatory neuronal input field $h^e{}_i$, since it counterbalances the effect of any shift, as can be easily seen from the linear approximation which is valid for small changes in the field. In our numerical simulations we use $\kappa = 10$ and $\tau = 0.01$.

MAINTENANCE AND NORMALIZATION

This mechanism can counter-balance the average deterioration of the system, i.e. the accumulated ϵ effect. It works nicely as long as the accumulated variance is small. Interestingly, it can also counteract the formation of pathologic attractors. The latter are strongly embedded patterns, that dominate all other memory patterns. Suppose that at some point of time such pathologic attractors are formed, and the system finds itself with a synaptic efficacy matrix

$$J_{ij}(t) = \frac{1}{Np} \sum_{\mu=1}^{M} g^{\mu} \eta^{\mu}{}_{i} \eta^{\mu}{}_{j} \qquad (5)$$

where some of the memories are encoded with weights g^{μ} larger than 1. We find that if at this point the NR mechanism is applied, allowing the system to evolve through degradation and maintenance cycles, such attractors are trimmed down, as demonstrated in Figure 1. We display here the basins of attraction of our model, as measured by a retrieval process which is initiated by random inputs. Whereas at the beginning the strong memories dominate the scene, their weights are gradually reduced by the maintenance method, until an almost homogeneous embedding is achieved.

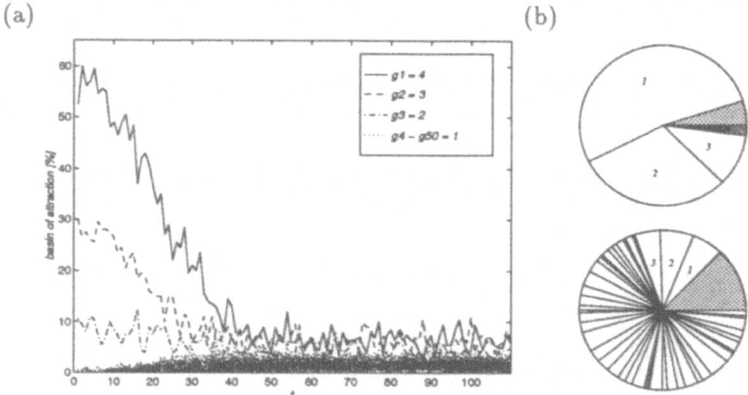

Figure 1. (a) Size of basins of attraction as measured by the percentage of retrievals of specific memories. This simulation of an $N = 1000$ network has 50 memories stored such that three have strengths of $g = 4$, 3 and 2, and all the rest have $g = 1$. (b) Shares of memory space (relative sizes of basins of attraction) at the beginning (upper figure) and the end (lower figure) of the simulation. Random inputs lead either to encoded memories or to the null attractor (gray shading) in which all activity stops.

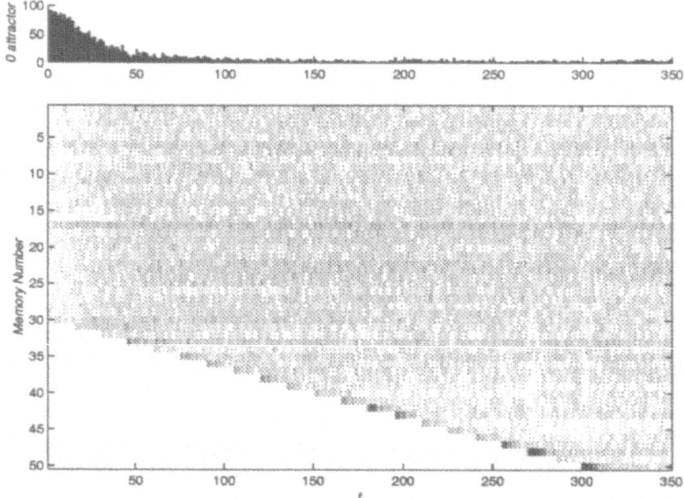

Figure 2. Alternating synaptic learning and maintenance. We start out with a system of $N = 1000$ neurons holding 30 memories. Every 15 epochs a new pattern is stored during 5 epochs, followed by 10 epochs of regular synaptic degradation and maintenance. The top figure shows how the null attractor gradually vanishes. The lower figure portrays the basins of attraction of the different memories (larger basins are darker) at subsequent epochs. As evident, homogeneous memory retrieval is maintained throughout the simulation.

Neuronal regulation works well also when it is combined with ongoing learning of new, unfamiliar, memory patterns. This is demonstrated in Figure 2. Here every few epochs the network acquires another memory in an activity dependent manner. A new memory is presented to the network via an external input and the synaptic efficacies of co-active neurons are allowed to change in a Hebbian fashion.

Even when optimal synaptic compensation is successfully realized via NR, the memory system eventually collapses because of the increasing accumulated variance in the synaptic degradation process. However, we find that if synapses are appropriately bounded, it leads to an asymptotic stable memory system. The example shown in Fig. 3 corresponds to the case of strong variance in the synaptic degradation process. During the NR process some synapses die while others approach the upper synaptic bound and remain in its vicinity, realizing long-term memory maintenance. Memory maintenance may therefore be achieved even though the synapses are not maintained at their original values.

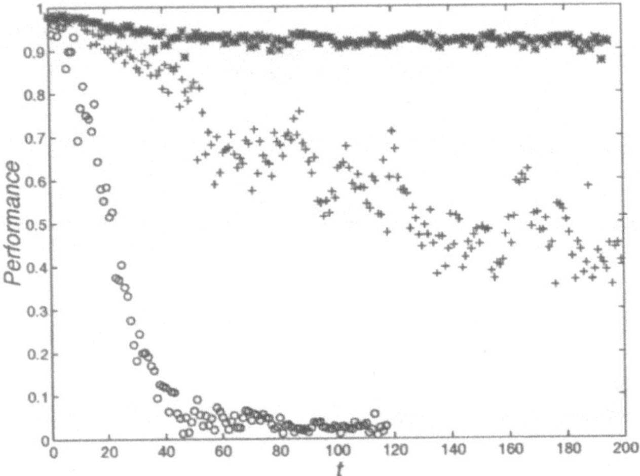

Figure 3. The effect of synaptic bounds. The small circles denote the performance of the network without synaptic bounds. It declines fast in this example because of high variance in the degradation process, $\sigma_\epsilon = 0.2$ while $< \epsilon_{ij} > = 0.005$. The '+' symbols denote the performance of a network with synaptic bounds that are equal to $r = 8$ times the size of a synapse that stores one memory at $t = 0$, while the '*' symbols correspond to the case of $r = 3$. The parameters of this simulation were $N = 500$, $M = 25$, $p = 0.075$.

DISCUSSION

The NR mechanism relies on activation of the memory system by random inputs, thus testing all basins of attraction without requiring the explicit knowledge of the memory patterns themselves. As such, it is reminiscent of previous suggestions[9, 10] that utilize random activity to unlearn spurious attractors in the network. Such attractors are rare in the Tsodyks model, and therefore were irrelevant in our study. Notice, though, that NR does weaken the memories that are frequently retrieved through random activation, without the need of assuming anti-Hebbian learning. Random activation of cortical memory systems may be triggered by PGO waves[11] during REM sleep. It is however still unclear whether this is indeed the appropriate and the only period in which NR-driven synaptic maintenance occurs. In any case, it seems preferable to have a clear separation between the processes of memory storage and memory maintenance since they probably require activation of different (and complementary) mechanisms.

NR can be viewed as a particular realization of 'dynamic stabilization', a term that describes the idea that during sleep there exist dynamic processes that maintain synaptic efficacies[4]. The instrumental potential of NR in obtaining memory maintenance shown in this paper, coupled with morphometric evidence showing that the average total synaptic area per unit volume is maintained throughout normal aging[12, 13], make it highly likely that NR plays an important functional role in adulthood too. Moreover, previous computational studies have shown that disturbances of NR mechanisms may underlie some of the clinical pathological manifestations of both Alzheimer's disease[14] and schizophrenia[15] (see [16] for a review).

We envisage memory maintenance as taking part during sleep, when memory systems are processing random inputs. Under such conditions the neuron can compare its activity to some baseline and, accordingly, activate its NR mechanism to maintain its synaptic values. However, as time goes by, the random synaptic degradation caused by synaptic turnover may destroy the encoded memories. Yet, memory demise can still be overcome if synapses are bounded. In that case, over long time periods, NR leads to synaptic variation, either growing up to the bound or decreasing and disappearing. Nonetheless it can still maintain memory systems intact. In fact, over long time periods we witness a transition from continuous synaptic encoding to an effective binary encoding, guaranteeing efficient storage of memories.

REFERENCES

1. P. Goelet, V. F. Castellucci, S. Schacher, and E. R. Kandel. The long and the short of long-term memory - a molecular framwork. *Nature*, 322:419–422, (1986).

2. J. Lisman. The CAM kinase hypothesis for the storage of synaptic memory. *Trends In Neural Science*, 17(10):406–412, (1994).

3. J. R. Wolff, R. Laskawi, W. B. Spatz, and M. Missler. Structural dynamics of synapses and synaptic components. *Behavioural Brain Research*, 66:13–20, (1995).

4. J. L. Kavanau. Sleep and dynamic stabilization of neural circuitry: a review and synthesis. *Behavioural Brain Research*, 63:111–126, (1994).

5. D. Horn, N. Levy, and E. Ruppin. Memory maintenance via neuronal regulation. Neural Computation, 10(1), (1998).

6. A. van Ooyen. Activity-dependent neural network development. *Network*, 5:401–423, 1994.

7. G. G. Turrigiano, K. Leslie, N. Desai, and S. B. Nelson. A biological mechanism for synaptic stability in developing neocortical circuits, these proceedings.

8. M. V. Tsodyks. Associative memory in neural networks with the hebbian learning rule. *Modern Physics Letters B*, 3(7):555–560, (1989).

9. F. Crick, and G. Mitchison. The function of dream sleep. Nature 304, 111-114 (1983).

10. J. J. Hopfield, D. I. Feldman, and R. G. Palmer. 'Unlearning' has a stabilizing effect in collective memories. Nature 304, 158-159 (1983).

11. J. A. Hobson, and R. W. McCarley. The brain as a dream state generator: an activation-synthesis hypothesis of the dream process. Am. Jour. of Psychiatry 134, 1335-1368 (1977).

12. C. Bertoni-Freddari, W. Meier-Ruge, and J. Ulrich. Quantitative morphology of synaptic plasticity in the aging brain. Scanning Microsc., 2, 1027-1034 (1988).

13. C. Bertoni-Freddari, P. Fattoretti, T. Casoli, W. Meier-Ruge, and J. Ulrich. Morphological adaptive response of the synaptic junctional zones in the human dentate gyrus during aging and Alzheimer's disease. Brain Research, 517, 69-75 (1990).

14. D. Horn, N. Levy, and E. Ruppin. Neuronal Based Synaptic Compensation: A Computational Study in Alzheimer's Disease. Neural Computation, 8, 1227-1243 (1996).

15. D. Horn, and E. Ruppin. Compensatory Mechanisms in Attractor Neural Network Model of Schizophrenia. Neural Computation, 7(1), 182-205 (1995).

16. E. Ruppin. Neural modeling of psychiatric disorders. Network, 6, 636-656 (1995).

ENCODING CONTEXT IN SPATIAL NAVIGATION: ONE ROLE OF DENTATE GYRUS

Karl Kilborn, Gary Lynch, and Richard Granger

Center for the Neurobiology of Learning and Memory
University of California
Irvine, CA 92697-3800

{kkilborn,glynch,granger}@ics.uci.edu

ABSTRACT

A simulation of hippocampal fields CA3 and CA1 that encodes information necessary for spatial navigation is extended to incorporate dentate gyrus. Dentate gyrus' projection of mossy fibers to field CA3 is presumed to carry information that takes advantage of its unique anatomy and physiology. It is hypothesized that these projections, based on context-specific activity in dentate, radically influence the specific recurrent dynamics in CA3 by governing its principal cells' excitability. The simulation reveals an ability of the hippocampus to differentially encode associations between familiar stimuli depending on context, and demonstrates this ability in its navigation of two distinct three-dimensional "worlds."

INTRODUCTION

A simulation of hippocampal CA3 and CA1, presented elsewhere, demonstrates the influential role that specifics of LTP physiology play in encoding associations between cues (Kilborn and Granger, 1995; Kilborn, 1997; Kilborn and Granger 1997). In this work, we wish to extend this model and demonstrate a hypothesized purpose of the mossy fiber projections from dentate gyrus to field CA3.

The role of context in hippocampal spatial navigation has long been of interest (see, for example, Nadel and Willner (1980)). We adopt the hypothesis that dentate gyrus is responsible for separating associations learned in one context from those learned in another. Given that CA3 principal cells are innervated by two orders of magnitude fewer mossy fibers than perforant path and recurrent afferents (Brown and Zador, 1990), it is unlikely that the mossy inputs carry the same quantity or nature of information as the other pathways, but rather modify the dynamics of CA3 in some other fashion.

Computational Neuroscience
edited by Bower, Plenum Press, New York, 1998

We will first summarize the model of hippocampal fields CA3 and CA1 and proceed to describe the incorporation of dentate gyrus.

THE HIPPOCAMPAL MODEL

The model of hippocampus used in this investigation tries to capture several important aspects of known long-term potentiation (LTP) physiology in area CA1 as well as hypothesized dynamics in area CA3. Of primary interest is determining the kinds of computations suggested by the LTP physiology operating within the unique architecture of the hippocampus.

In particular, feedforward inhibitory neurons (interneurons) in CA1 enforce LTP *priming*, the finding that eliciting LTP in principal cells using endogenous stimulation patterns requires an initial pulse to send feedforward interneurons to a refractory state so that subsequent stimulation will potentiate (Larson and Lynch, 1986; Diamond *et al.*, 1988).

In the simulation, area CA3 operates as a short-term memory, a function whose dynamics emerge from the area's densely recurrent collaterals (Taketani *et al.*, 1992; Granger *et al.*, 1996). Computer modeling of this area suggests that, under certain conditions, which subset of cells represents the short-term memory will slowly vary over time.

Area CA1 is hypothesized to be functionally divided into *patches*, delineated by the targets of feedback interneurons, where only the most depolarized of the cells within a patch will fire before feedback inhibition prevents further firing (Coultrip *et al.*, 1992). We further hypothesize that CA1 is functionally divided into patches delineated by feedforward interneurons. We also make the simplifying assumption that patches defined by feedback and feedforward interneurons coincide, although we have shown that this is not a critical assumption for the behavior of the CA1 simulation (Kilborn and Granger, 1997).

Another LTP finding, that recently elicited LTP can be erased by a different endogenous stimulation pattern (Staubli and Lynch, 1990; Larson *et al.*, 1993) has also been incorporated in the model. LTP erasure has the effect of retaining only associations between different inputs (e.g., $A \to B$) while selectively erasing associations between the same inputs (e.g., $A \to A$) and therefore increases storage (Kilborn, 1997; Kilborn and Granger, 1997).

Figure 1 illustrates the operation of the model, which operates in discrete time steps of 200 ms, during the presentation of two different inputs. Quantitative analyses of priming and erasure are presented elsewhere (Kilborn, 1997; Kilborn and Granger, 1997).

SIMULATION OPERATION IN TWO SPATIAL CONTEXTS

The simulation operation is demonstrated by presenting it with inputs derived from exploration in a 3D "world." It is presumed that significant regularization of the visual input takes place before it reaches the hippocampus; for the demonstration, an orthographic projection of the current view is encoded as perforant path input activity. In other words, it is assumed that scale and illumination invariance has already been applied in earlier processing. Inputs to the simulation, therefore, consist of identified objects along the periphery of each world, or context.

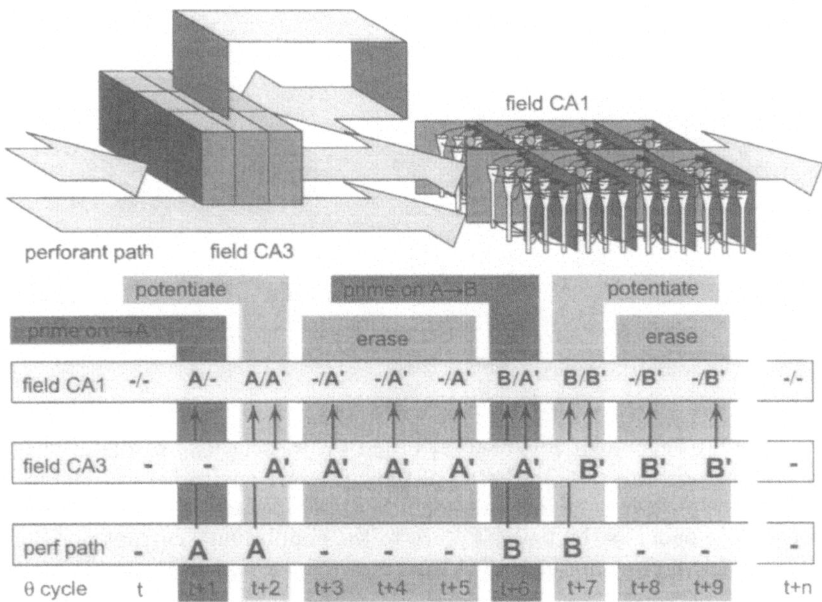

Figure 1 Schematic of the CA1 priming and erasure model. At theta cycle t, both fields CA3 and CA1 are relatively quiescent. At theta cycle $t+1$, input A is presented at the perforant path, which sets up recurrent excitatory activity in CA3 and fires feedforward interneurons in CA1 (and therefore primes a subset of patches in CA1 as the interneurons enter a refractory period). At theta cycle $t+2$, input A is still present at perforant path while A-driven activity (A') continues in CA3. The combination of Schaffer collateral input from CA3 to CA1 and direct perforant path innervation of CA1 (denoted as A/A' in the figure) potentiates the most depolarized of the primed (disinhibited) cells in CA1. During theta cycles $t+2$ through $t+5$, we hypothesize that principal cell firing activity, due to the lack of perforant path input, will be of a temporal pattern more conducive to LTP erasure than potentiation. Much of the potentiation elicited during cycle $t+2$, therefore, will not consolidate. Priming during cycle $t+6$ will select patches of cells in CA1 for potentiation during the next cycle according to the directional association of input A (whose earlier encounter set up CA3 recurrent activity) and input B (now present at the perforant path). This association will therefore be potentiated during cycle $t+7$. Subsequent cycles will prevent many of the synapses which are not unique to the $A \rightarrow B$ association from consolidating their potentiation.

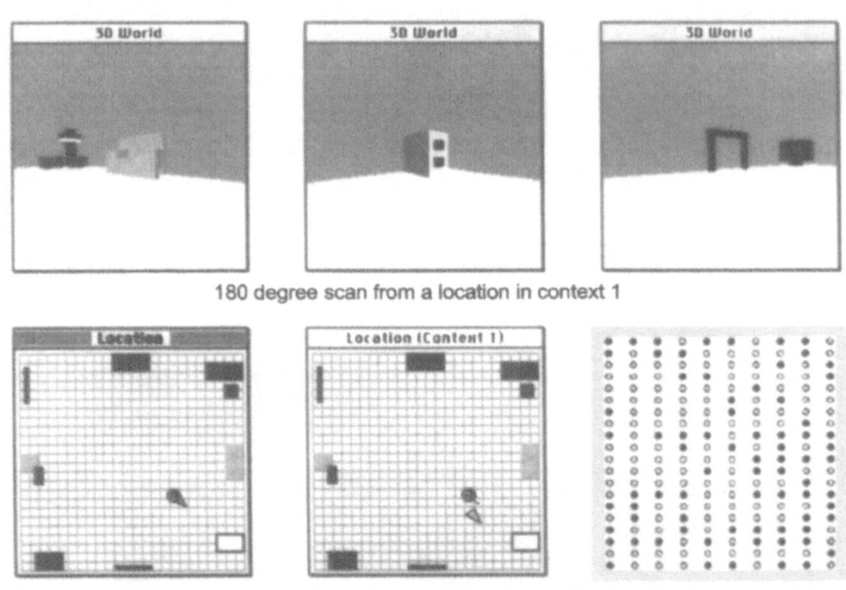

Figure 2 Context-dependent localization based on associations of distal cues. The top panels show portions of a 180° scan of the context 1 environment. The lower left panel shows a top-down view of a remembered place which is correctly recalled in the lower middle panel as the nearest place to the simulation's current position. The lower right panel shows cumulative CA1 output activity at the current position.

Figures 2 and 3 illustrate the simulation's operation in two distinct contexts. While many individual objects are shared between the two contexts, their relative configuration differs.

The simulation is first familiarized with the individual inputs and then trained during unsupervised exploration through the world. When the simulation sees a familiar input, corresponding activity is presented via the perforant path to hippocampus. A pattern of activity in CA3 is initiated and evolves as exploration in the world continues. When another familiar input is encountered, CA1 cells will fire based on the new input in conjunction with the activity in CA3 set up by the previous input. Therefore, CA1 output activity will depend on the number of theta cycles that elapsed between the two inputs as well as the inputs themselves. Later processing, presumed to be carried out in neocortex, can use successive CA1 outputs of recognized associations to estimate position, as the accumulation of these outputs establishes a partial ordering of all encountered inputs as well as a spatially-coded sense of the time that transpired between individual inputs. The principal navigation operations of the hippocampus, then, are to store the associations as well as perform the necessary time dilation to encode the temporal associations in a single spatial pattern.

As the simulation enters a new context, a new pattern of activity is set up in dentate gyrus. This activity, in turn, determines the relative excitability of CA3 cells. Therefore, depending on context, an entirely different pattern of recurrent activity will be present in CA3 for the same individual input. This ensures that identical associations between inputs will be encoded differently for separate contexts.

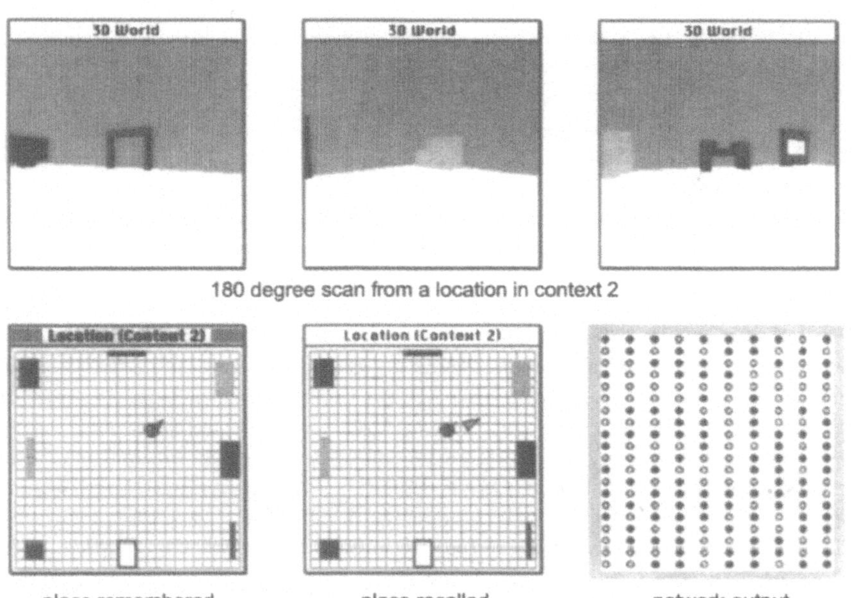

Figure 3 Context-dependent localization based on associations of distal cues in context 2.

In the current simulation, dentate gyrus and field CA3 are modeled using 100 threshold and fire units each. In addition to perforant path input and CA3 recurrent connections, CA3 units also receive input from one corresponding unit in dentate gyrus. Presence in a particular context is modeled as an orthogonal set of activated units in dentate gyrus which in turn comprise approximately 30% of the synaptic weight of those CA3 units that they contact.

For instance, the contexts illustrated in figures 2 and 3 share an arch, a couch, and a television set (viewed in the upper panels in figures 2 and 3). Estimation of location based on stored cumulative CA1 output activity (during a 180° scan) is correctly performed when dentate influences CA3 activity. The lower left panels in figures 2 and 3 illustrate a top-down view of the environment and a remembered place within that environment. The lower middle panels illustrate the best match of remembered locations given the simulation's current location (denoted by a triangle). The lower right panels illustrate the cumulative CA1 output for the simulation's current location in each environment. Despite the fact that three out of five identified objects are shared between the two scans, network output easily identifies the remembered place within the correct context. Without the modeled dentate gyrus influence, the simulation is more prone to erroneously recalling places stored in a different context.

CONCLUSION

The simulation illustrates how altering excitability of CA3 principal cells via mossy fiber projections from dentate gyrus may be instrumental in correctly resolving the environmental context in which input associations are identified.

The model presented in Treves and Rolls (1992) adopts a similar hypothesis of dentate gyrus function, although through a different mechanism. In their model, dentate gyrus serves to "pre-orthogonalize" inputs through a process of competitive learning. This permits sparser activity in CA3 as well as the ability to communicate different configurations of the same cues (e.g., AB vs. A or B) to field CA3. In the current model, dentate gyrus instead directly alters CA3 recurrent dynamics.

Another potentially important role of dentate is that of avoiding catastrophic forgetting. As reviewed recently in McClelland *et al.* (1995), it is widely observed in neural models that repeated learning of one set of inputs may compromise network performance on a set of inputs learned previously. By influencing which cells are firing in CA3, however, dentate ensures further orthogonalization of inputs to CA1, and thereby reduces the effect that associations learned in a new context can have on ones learned earlier in a different context.

A remaining question is how does dentate determine context in the first place. One possibility is that CA1 output, through an indirect feedback to dentate, helps set this up. Principal cell firing patterns in CA1 have been observed to be correlated with either recognizing or failing to recognize a particular input (i.e., "match" and "mismatch" cells described in Ranck (1973)). The simulation mirrors this observation in that distinct firing patterns emerge depending on whether a familiar or unfamiliar association has been observed. It is possible, then, that sufficient "mismatch" responses from CA1 will trigger a context switch and may ultimately call for exploratory behavior necessary to construct a representation of a new context.

REFERENCES

Brown, T. H. and Zador, A. M. (1990). Hippocampus. In Shepherd, G. M., editor, *Synaptic Organization of the Brain*, pp 346-388. Oxford Press.

Coultrip, R., Granger, R., and Lynch, G. (1992). A cortical model of winner-take-all competition via lateral inhibition. *Neural Networks*, 5:47-54.

Diamond, D. M., Dunwiddie, T. V., and Rose, G. M. (1988). Characteristics of hippocampal primed burst potentiation in vitro and in the awake rat. *Journal of Neuroscience*, 8:4079-4088.

Granger, R., Wiebe, S. P., Taketani, M., and Lynch, G. (1996). Distinct memory circuits composing the hippocampal region. *Hippocampus*, 6:567-578.

Kilborn, K. and Granger, R. (1995). Influence of LTP priming and decrement rules on temporal association. In *Proceedings of the World Congress on Neural Networks, Volume I*, pp 12-16. Lawrence Erlbaum Associates.

Kilborn, K. (1997). *The effects LTP induction rules have on memory*. PhD Thesis, UC Irvine.

Kilborn, K. and Granger, R. (1997). *In preparation*.

Larson, J. and Lynch, G. (1986). Induction of synaptic potentiation by patterned stimulation involves two events. *Science*, **232**:985-988.

Larson, J., Xiao, P., and Lynch, G. (1993). Reversal of LTP by theta frequency stimulation. *Brain Research*, **600**:97-102.

McClelland, J. L., McNaughton, B. L., and O'Reilly, R. C. (1995). Why there are complementary learning systems in the hippocampus and neocortex: insights from the successes and failures of connectionist models of learning and memory. *Psychological Review*, **102**:419-457.

Nadel, L. and Willner, J. (1980). Context and Conditioning: A place for space. *Physiological Psychology*, **8**:218-228.

Ranck, Jr., J. B. (1973). Studies on single neurons in dorsal hippocampal formation and septum in unrestrained rats. I. Behavioral correlates and firing repertoires. *Experimental Neurology*, **41**:461-531.

Staubli, U. and Lynch, G. (1990). Stable depression of potentiated synaptic responses in the hippocampus with 1-5 Hz stimulation. *Brain Research*, **513**:113-118.

Taketani, M., Ambros-Ingerson, J., Myers, R., Granger, R., and Lynch, G. (1992). Is field CA3 a reverberation short term memory system? *Society for Neuroscience Abstracts*, 18:1211.

Treves, A. and Rolls, E. T. (1992). Computational constraints suggest the need for two distinct input systems to the hipppocampal CA3 network. *Hippocampus*, 2:189-199.

LARGE SCALE SIMULATIONS OF HIPPOCAMPAL-NEOCORTICAL INTERACTIONS IN A PARALLEL VERSION OF GENESIS

J. C. Klopp [1,4] P. Johnston [1,3] V. I. Nenov [1,2] N. Goddard [6] G. Hood [6] &
E. Halgren [2,5,7]

1 Brain Monitoring and Modeling Laboratory, Div. Neurosurgery, UCLA
2 Brain Research Institute, UCLA
3 Div. Psychology, University of Northumbria, Newcastle, England
4 Interdepartmental Neuroscience Ph.D. Program, UCLA
5 INSERM, Rennes and Marseilles, France
6 Pittsburgh Supercomputing Center, Pittsburgh
7 VAMC, West Los Angeles, CA

ABSTRACT

A hippocampal-CA3 memory model was constructed with PGENESIS, a recently developed version of GENESIS that allows for distributed processing of a neural network simulation. A number of neural models of the human memory system have identified the CA3 region of the hippocampus as storing the declarative memory trace. However, computational models designed to assess the viability of the putative mechanisms of storage and retrieval have generally been too abstract to allow comparison with empirical data. Recent experimental evidence has shown that selective knock-out of NMDA receptors in the CA1 of mice leads to reduced stability of firing specificity in place cells. Here a similar reduction of stability of input specificity is demonstrated in a biologically plausible neural network model of the CA3 region, under conditions of Hebbian synaptic plasticity versus an absence of plasticity.

The CA3 region is also commonly associated with seizure activity. Further simulations of the same model tested the response to continuously repeating versus randomized non-repeating input patterns. Each paradigm delivered input of equal intensity and duration. Non-repeating input patterns elicited a greater pyramidal cell spike count. This suggests that repetitive versus non-repeating neocortical inpus has a quantitatively different effect on the hippocampus. This may be relevant to the production of independent epileptogenic zones and the process of encoding new memories.

INTRODUCTION

Much is known regarding the phenomenology of memory at the behavioral or psychological level. Evidence from cognitive neuropsychology has identified brain regions that are critical to various memory systems through lesion studies in animals and "natural experiments" in humans who have suffered head trauma. Similarly, at a lower level, much is known regarding the neuroanatomy and cellular function of those brain regions important for declarative memory. What is lacking from our current conception of the human memory system is a clear understanding of how the "strings and sealing wax" of neuroanatomy and cellular electrophysiology give rise to the more fancy stuff of encoding, storage, retrieval, consolidation and forgetting. The emerging discipline of computational neuroscience attempts to address such issues through numerical simulations of complex systems.

Basics of Memory

In the spirit of Marr (1971), we employ a methodology that draws heavily on the observations of cognitive psychology in outlining the fundamental properties of human memory. In this approach our modeling efforts proceed in a top-down manner. However, it is recognized that in a complex and dynamic system, it may not be possible to predict emergent properties that might arise through interactions of low level constraints. For this reason, we model the HippoCampal Formation (**HCF**) with a level of biological detail that exceeds previous large scale modeling endeavors (Read et al 1994).

The hippocampus has direct or indirect reciprocal connections with all neocortical multi-modal association areas (**figure 1**), as well as many sensory areas, and the synapses within the hippocampus are capable of rapidly changing the efficacy of their post synaptic response through activity dependent changes. These characteristics are considered essential for the HCF's role in normal brain function as a temporary repository of memory traces.

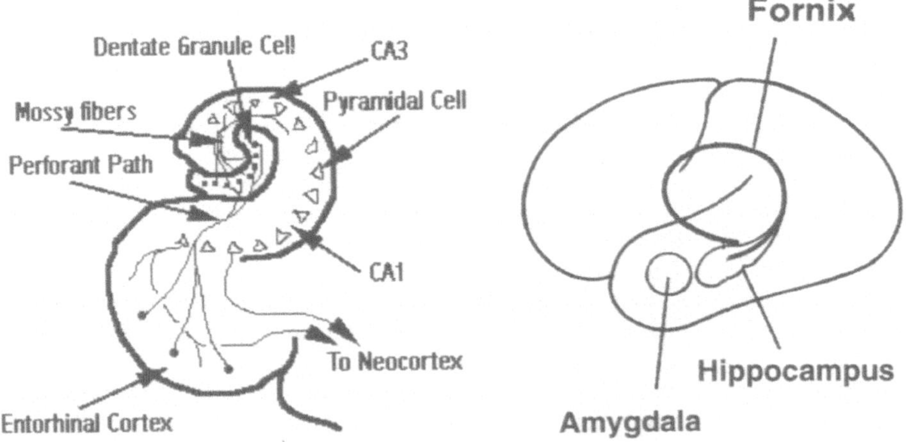

Figure 1
Structural subdivisions of the HCF and output to NC

Figure 2
Location of the HCF within the brain

A number of recent neural models have identified the HCF as a temporary storage site for the declarative memory trace (Halgren 1984; Alvarez et al 1994; Treves et al. 1994; McLelland et al. 1995). Generally, these models consider the HCF to constitute an associative network in bi-directional communication with the NeoCortex (**NC**). Temporary traces are set up in the HCF that assist the NC in reconstructing recent memories, and / or in consolidating neocortical traces of long term memory. A number of these models have specifically identified the auto-associative network of pyramidal cells within the CA3 field as a prime candidate for the substrate of rapid short-term declarative memory formation and storage.

Basics of Epilepsy

The most prevalent form of epilepsy is complex partial epilepsy. Complex Partial Seizures (**CPS**), also known as psychomotor seizures, account for 40% of all epilepsy cases (McNamara 1994) and are often characterized by complicated illusory phenomena and semi-purposeful complicated motor acts. CPS reflect hyper-synchronous activation of cortico-limbic neuronal networks that often start in, or spread through, limbic structures of the Medial Temporal Lobe (**MTL**) including the hippocampus proper, dentate gyrus and parahippocampal gyrus. In many cases it seems that a neocortical scar in post-Rolandic cortex results in an epileptogenic zone that involves not only the tissue immediately surrounding the scar, but also the HCF. The process of epileptogenesis is generally regarded as an interplay between intrinsic properties of individual neurons (Wong and Prince 1981) and properties of the synaptic connections of the network (Traub and Miles 1991). Ammon's horn sclerosis, a degeneration of principal cells in the HCF, is commonly associated with seizure activity and its relation to epilepsy is controversial as either a cause or a consequence.

In-between seizures, epileptogenic foci produce hyper synchronous bursts. These bursts seem to fulfill the criteria of convergent co-activation that are necessary for inducing plastic changes in the hippocampal synapses. A possible connection between learning and epileptogenesis is found in the phenomenon of kindling, first described by Graham V. Goddard in the 1960s. Kindling is the process whereby daily sub-convulsive electrical shocks delivered to specific regions of a rat's brain become capable of generating full tonic-clonic seizure-like convulsions. The HCF, located in the MTL (**figure 2**), has a low threshold for electrically induced seizure activity. For this reason the HCF is a frequent kindling study target. Once kindled, an animal exhibits a long lasting susceptibility to seizure activity. This was interpreted as a crude form of learning, or perhaps more accurately an undesirable side effect of normal memory mechanisms (Goddard et al. 1975). A subtle form of kindling may occur in humans through HCF-NC reciprocal connections. In a sense, mechanisms that establish memories under normal conditions may be usurped by hyper-excitable tissue for epileptogenesis (Mayank et al. 1993).

Although epileptogenic spikes have been well characterized, the fundamental mechanisms of neural plasticity (such as long term potentiation and long term depression) remain largely unknown. On one hand this introduces a critical weakness in the model that will be improved as the state of knowledge in this area improves. On the other hand the issue of how fundamentally low level processes combine to perform HCF-related functions is in essence the problem we wish to address.

We hypothesize that the spikes generated in the neocortical focus produce an epileptogenic zone in the HCF using the mechanisms of plasticity and long-range recurrent connections that are also essential for normal hippocampal function. A potential key to the spread of seizure activity and memory consolidation resides in the HCF's wide connections with associational neocortex (Van Hoesen 1982; Amaral 1990; Suzuki 1996). Due to the difficulty of testing all aspects of this hypothesis in vivo, we have turned to a moderately realistic computational model of the hippocampus.

COMPUTATIONAL METHODOLOGY

Modeling Platform

Computational models implemented at the Pittsburgh Supercomputing Center were written in **PGENESIS** (Goddard & Hood 1997), a parallel version of **GENESIS** (**GE**neral **NE**ural **SI**mulation System) (Bower et al 1995). PGENESIS allows a computer simulation to be distributed across multiple processors and has been optimized and compiled for the CRAY T3E. The standard GENESIS simulation package allows easy integration of simulation objects (such as cellular compartments and cell-membrane channel conductances) written by other neural modelers.

The Single Cell Model

The model includes compartmental representations of excitatory pyramidal neurons, inhibitory interneurons and a large input layer of dentate granule cell spike generators. Pyramidal neurons consist of multiple compartments with a minimally branched dendritic morphology and incorporate fast Na^+, K^+(dr), K^+(ahp) and Ca^{++} conductances (**figure 3**). These neurons are capable of producing a variety of firing patterns including complex bursts and single spikes. Interneurons are modeled as a single compartment and include fast Na^{++} and K^+(dr) currents. Synaptic interactions are modeled using a generalized alpha-function.

The Network Model

The model consists of three regions (**figure 4**). A population of 20736 spike-producing elements represent the dentate gyrus. Compartmental representations of 5184 pyramidal neurons and 576 interneurons of an intermediate degree of complexity represent the CA3 region. Principle connections include sparse projections from the granule cells to CA3 pyramidal and interneurons, sparse, fast, recurrent excitatory connections within the CA3 region, diffuse fast feedback inhibitory and recurrent inhibitory connections with a slower time-course representing $GABA_B$ inhibition. Both types of cells receive feed forward excitation via sparse connections from the dentate gyrus spiking input elements. However, this 'mossy fiber' input does not yet account for non-associative LTP seen in these synapses. A Hebbian algorithm updates synaptic efficacy of the associative pyramidal connections. Contact probability diminishes with distance and the processor workload is distributed in a design that biases output targets to remain on the originating processor (**figure 4**). This reduces inter-processor communication and optimizes the speed of simulation computation.

RESULTS

Under all simulation conditions performed thus far the pyramidal cells of the model reliably produce a significantly greater number of spikes (P=1.4e-5) during non-repeating input patterns. As would be expected, repeating input patterns produced a more periodic and regular response that, over time, activated a smaller percentage of the total cell population. After training the model on a repeating pattern (20 repetitions of the same pattern) it appeared more reactive (produced about 100% more spike events in the pyramidal cell population) to new patterns as compared to an identical network trained under a non-repeating pattern condition (20 repetitions of unique patterns). Both conditions displayed strong initial sensitivity and slow habituation to input.

The model's activity in response to input patterns and partial input patterns was examined in terms of individual unit responses, and gross activation levels. A large number of cells displayed specificity to a given input pattern, or preferentially showed

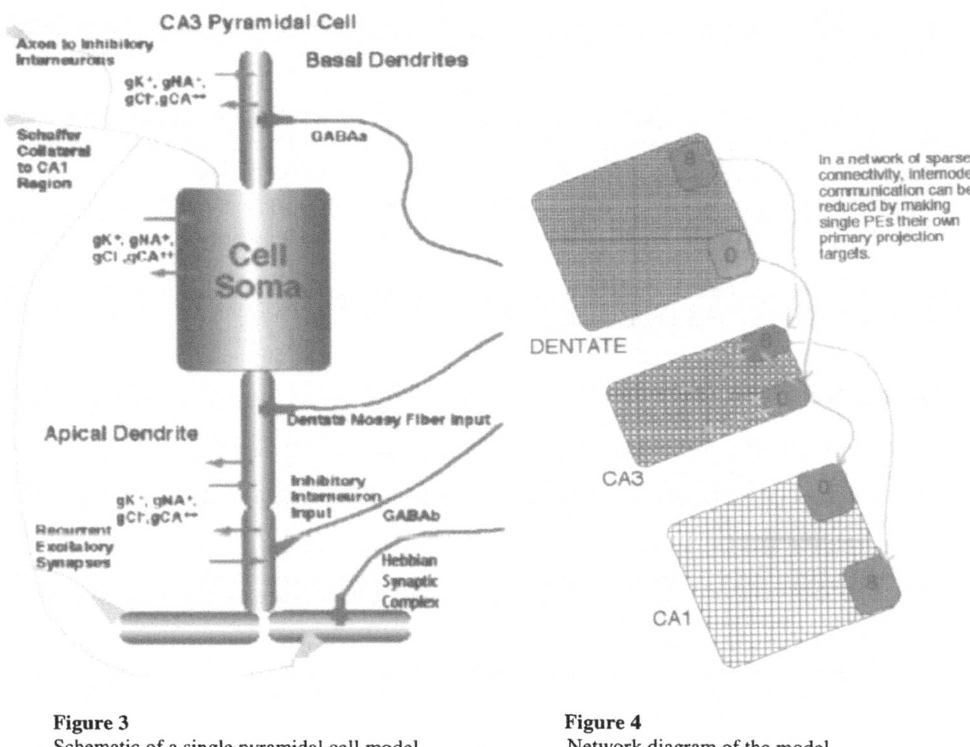

Figure 3	**Figure 4**
Schematic of a single pyramidal cell model	Network diagram of the model

higher firing to specific patterns. Correlation values of spikes per-cell during repetitions of patterns in the plasticity condition (r=0.38) was significantly greater than the same measure for repetitions of patterns without plasticity (r=0.0005).

DISCUSSION

These simulations show a decrease in total cell activity during repeating input (epileptogenic) versus non-repeating (normal) input patterns. This finding may be correlated with an observed decrease in PET metabolism during interictal periods (Engle et al 1982). The habituation to continuous input in the model could have a variety of possible causes. Most likely is that slow inhibition built up as a result of over-stimulation from feed back and feed forward activation of the inhibitory interneuron population.

In response to input patterns the simulated CA3 pyramidal neurons exhibited a range of behaviors broadly comparable to those observed in the real CA3 through unit recordings. At the level of individual units, greater input specificity is seen when plasticity is present, than when it is absent. Correlation measures show that 38% of the variance of the model's response during recognition of a previously presented pattern is described by the model's response to the learning phase of the same pattern only when plasticity is activated. Under conditions lacking plasticity (similar to the NMDAR1 knockout mice) virtually none of the model's response during recognition of a previously seen pattern is described by the model's response to the learning phase of the same pattern.

Results of the model memory tests resemble results from recent gene knock-out experiments (McHughs et al. 1996). Removing LTP from the model did not prevent neurons from firing selectively, or preferentially to given input patterns. However, the stability of this selectivity was decreased if synapses were not plastic. These findings are particularly impressive as a test of the model's plausibility, as they demonstrate the selective loss of a subtle but crucial emergent property due to the loss of a particular molecular property. Future modeling studies will concentrate on the relationship between this property and the contribution of the hippocampus to the recall of neocortical patterns. This will require a larger model including further hippocampal fields as well as neocortical areas.

Future simulations will be of a larger scale that can attain a sparseness of connectivity that is often considered crucial for effective memory recall. If epileptogenesis can occur as a result of dysfunctional memory mechanisms then a properly scaled network model may reveal differences in the response to repeating versus non-repeating stimulation that would otherwise not appear in the current model.

REFERENCES

Alvarez P, Squire LR: Memory consolidation and the medial temporal lobe: A simple network model. Proc Natl Acad Sci USA 91:7041, 1994

Bower J & Beeman D: The Book of Genesis: exploring realistic neural models with the GEneral NEural SImulation System. TELOS, New York, Inc. 1995.

Amaral D. G. and R. Insausti. The hippocampal formation. In The Human Nervous System, G. Paxinos, eds., Academic Press, New York, pp. 711-755, 1990.

Engle J. Jr, Kuhl DE, Phelps ME, Maziotta JC: Interictal cerebral glucose metabolism in partial epilepsy and its relation to EEG changes. Annals of Neurology, 12:510-517, 1982.

Goddard GV & Douglas RM: Does the engram of kindling model the engram of normal long term memory? The Canadian Journal of Neurological Sciences, pp385-394, 1975.

Goddard N, Hood G: Parallel genesis for large scale modeling; in Bower J Ed., Computational Neuroscience: Trends in Research, Plenum Press, 1997.

Halgren E: Human hippocampal and amygdala recordings and stimulation: Evidence for a neural model of recent memory. In: The Neuropsychology of Memory, Squire L, and Butters N (eds) New York: Guilford, pp. 165-181, 1984.

Marr D: Simple memory: A theory for archicortex. Philos Trans R Soc Lond [Biol] 262:23, 1971.

Mayank RM, Dasgupta C, Ullal GR. A neural network model for kindling of focal epilepsy: basic mechanism. Biological Cybernetics, 68(335-340), 1993.

McClelland J, McNaughton B & O'Reilly R: Why there are complementary learning systems in the hippocampus and neocortex. Psychological Review 102(3):419-57, 1995.

Mcnamara J: Cellular and molecular basis of epilepsy, The Journal of Neuroscience 14(6):3413-3425, 1994.

McHughs T, Blum K, Tsien J, Tonegawa S & Wilson M: Impaired hippocampal representation of space in CA1-specific NMDAR1 knockout mice. Cell, Vol. 87 pp. 1339-1349, 1996.

Read W, Nenov V, Halgren E: Role of inhibition in memory retrieval by hippocampal area CA3. Neuroscience and Biobehavioral Reviews 18(1):55-68, 1994.

Suzuki W.A. The anatomy, physiology and functions of the perirhinal cortex. Curr. Opin. Neurobiol. 6: 179-186, 1996.

Treves A & Rolls E: A computational analysis of the role of the hippocampus in memory. Hippocampus, Vol. 4, pp. 374-92, 1994.

Traub RD, Miles R. Neuronal Networks of the Hippocampus. Cambridge University Press, 1991.

Van Hoesen G.W. The parahippocampal gyrus: New observations regarding its cortical connections in the monkey. Trends Neurosci. 5:345-350, 1982.

Wong RKS, Prince DAJ. After-potential generation in hippocampal pyramidal cells. Neurophysiology 45:86-97, 1981.

PRODUCTION OF PHASE LAG IN CHAINS OF NEURAL NETWORKS OSCILLATING THROUGH AN ESCAPE MECHANISM

Jeanette Hellgren Kotaleski,[1] Anders Lansner,[2] and Sten Grillner [1]

[1]Dept. of Neuroscience
Karolinska Institutet
S–171 77 Stockholm, Sweden
[2]Dept. of Numerical Analysis and Computing Science
Kungliga Tekniska Högskolan
S-100 44 Stockholm, Sweden

INTRODUCTION

Understanding the behavior of oscillators consisting of neural elements and how they couple together to synchronize or display a phase delay has become an important issue. Factors influencing the phase lag along chains of reciprocally coupled inhibitory half–center oscillators, oscillating through an escape mechanism, are analyzed by means of simulations. A simplified network is used consisting of the basic building blocks of the spinal segmental CPG generating swimming in the lamprey. This network is well characterized using neurophysiological and computer techniques[1]. An important characteristic of the whole spinal cord network is the head-to-tail propagation of the activity with a time lag between the activation of successive segments constituting a constant fraction of the cycle duration independent on the cycle duration (i.e. a constant phase lag). To explain a constant phase lag the spinal cord has been considered as a chain of coupled limit–cycle oscillators and the general theory of such systems[2, 3] has been applied. A critical feature of this theory is that the effect of one oscillator on another can be described by an H-function that depends only on the phase differences between the sending and receiving oscillator and this function gives the change in frequency of the receiving oscillator caused by signals from the sending one. Simulations using simplified connectionist models have provided results of a constant phase lag and are compatible with this interpretation, while more biophysically detailed models have failed to show a constant phase lag although several other experimental characteristics have been captured[4, 5, 6]. Recently other types of models for coupling between oscillators of a relaxation type, that does not follow the theory of phase–coupled oscillators, have been defined. They can couple through a fast threshold modulation mechanism (FTM)[7]. In chains of such oscillators a phase lag has also been reported[8] although the interest have often laid on prerequisites for synchronization.

The subject of this paper is therefore to further compare qualitatively the lamprey segmental network model with oscillators of relaxation type and also discuss this in relation to phase oscillator models. Questions that are addressed are phase lag versus frequency behavior, buffering against frequency differences and perturbations, and also estimations of the intersegmental coupling functions (C–function) are performed[9].

METHODS

Compartmentalized Hodgkin-Huxley type model neurons equipped with Na^+, K^+, Ca^{2+}, Ca^{2+} dependent K^+ ion channels are used[4, 10, 11, 12]. Synaptic interactions are modeled as conductance increases in the different compartments. Three types of cells are modeled: E, C and M cells[4, 12]. The excitatory synaptic inputs to the interneurons come from ipsilateral excitatory interneurons (E), while inhibition is produced by contralaterally projecting inhibitory interneurons (C). Motoneurons (M) are modeled as passive output integrators used as probes to measure the local activity and the phase lag. In this network model the C neurons are given an important burst termination role, the alternating left–right activity is produced through the escape mechanism[13]. This means that factors deciding the frequency, i.e. when an inhibited cell escapes the inhibition from the active side, are strength of inhibition in relation of the excitatory drive. Effective inhibition is controlled by actual synaptic strength as well as the spiking frequency. Presence of low voltage activated calcium channels underlying rebound phenomena[14] will lead to earlier escape. Simulations were run using the SWIM simulator[15]. Magnitude of Ca^{2+} dependent K^+ conductance as well as accumulation and removal rates of the underlying Ca^{2+} have both effects on the active and inhibited side by controlling spike frequency and hyperpolarizing currents, respectively[12].

In some of the simulations the hemisegments are replaced by Morris–Lecar relaxation oscillators with parameters set in order for the left and right oscillator to work in the escape mode when reciprocally coupled, as described in [13]. The oscillation of an uncoupled Morris–Lecar oscillator is shown in Fig. 1 and is the one used here. One way to assure that two such oscillators, when reciprocally coupled with a fast synaptic coupling function, oscillate through an escape mechanism is that the duty cycle of an oscillator receiving inhibition is more that 50%. The parametric values used for the oscillator are the same as in [7], but with v_k=-0.9, v_l=-0.15, I_{ext}=0.2 and lambda is here 0.02.

SIMULATION RESULTS AND DISCUSSION

Phase lag increases with frequency and decreases with length

In Fig. 2 the phase lag is measured in a network corresponding to 10–20 segments, when each hemisegment consists of one C, E and M cell. Fig. 2A shows activity of C cells at different levels when the C cells inhibit the contralateral cells at the same level, as well as the next contralateral hemisegment one segment caudally. In Fig. 2B the phase lag measured on one side and expressed in % of cycle duration per segment is plotted for different frequencies when the intersegmental synaptic strength is 100%, 10% or 2% of the intrasegmental coupling strength. Weaker intersegmental C coupling than 2% in this model often results in decoupling between the segments. This would imply that a network of this type is not well described by a phase oscillator model[2] in practice. In all cases the phase lag increases with frequency. Longer intersegmental projections reduce the phase lag (Fig. 2C) which, however, is a general result also with phase–coupled oscillators[16].

Buffering against frequency differences and perturbations; comparison with relaxation oscillators

In Fig. 2D, M cell activity is shown in a network without well–defined segmental boundaries, as described in [4]. The cell and synapse parameters are randomly distributed[12] which would give the different "segments", if cut apart, different local frequencies[5]. In this case C cells project locally corresponding to half a segment rostrally and caudally, as well as weaker (10%) only one segment further caudally. Following a perturbation (at the arrow) by a short

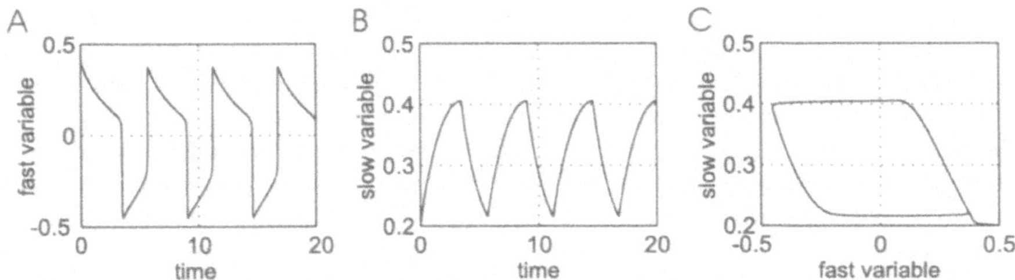

Figure 1. The fast (voltage) variable of the Morris–Lecar relaxation oscillator used is plotted against time (as arbitrary units) **A**, and in **B** the slow variable is shown. In **C** ordinate represents slow and abscissa fast over the same time.

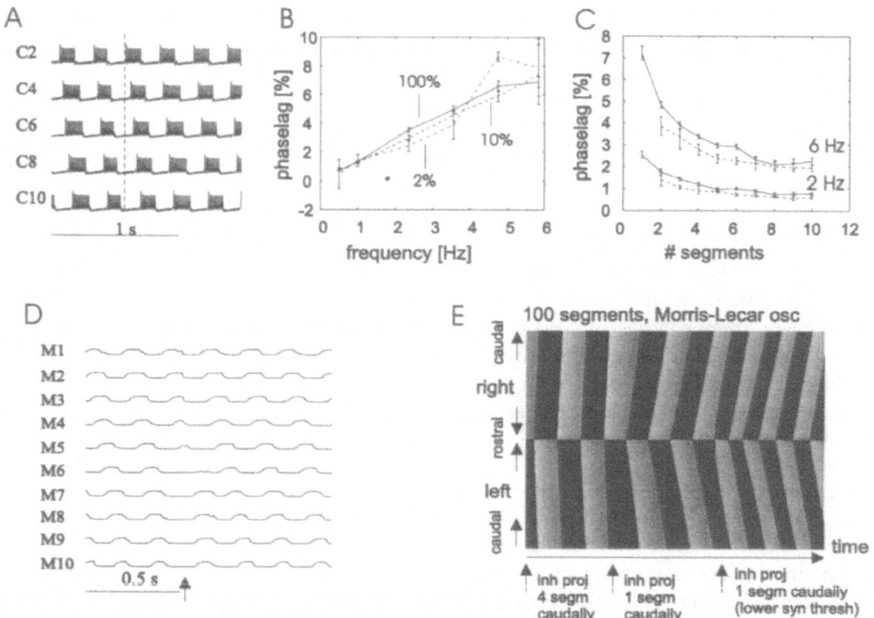

Figure 2. **A** Example of activity in C cells on one side of a 10 segmental network when the inhibition extends one segment caudally. **B** Phase lag versus frequency for the lamprey CPG model, plotted for 3 different intersegmental coupling strengths expressed as % of the intrasegmental coupling strength. **C** Phase lag versus how many segments caudally the intersegmental coupling reaches in a 20 segmental network (relative intersegmental strength 100%). **D** M cell activity in a network without well–defined segmental boundaries following a perturbation at the arrow. The coordination between the levels reestablishes within one or two cycles. **E** Phase lag dependency on frequency and length of projections when a Morris–Lecar relaxation oscillator is replacing the hemisegment. When the inhibitory coupling is longer the phase lag along such a chain decreases. If the frequency is increased by lowering the presynaptic threshold, thus reducing the time it takes for the inhibited cell to escape, the phase lag increases with frequency since the absolute slope is not much changed.

hyperpolarizing current pulse the phase lag is reestablished within one or two cycles. Buffering against frequency differences along chains and fast phase locking is a characteristic feature of relaxation oscillators also when coupled with relatively weak and short projections[7, 8]. This is in contrast to phase models that either need strong or long intersegmental connections to buffer against perturbations or frequency differences[17, 18] (compare, however, [19]).

A qualitative comparison with relaxation oscillators coupled together with an FTM mechanism[7] is done in Fig. 2E. Here the C cells are replaced by Morris-Lecar relaxation oscillators with parameters set to produce left–right oscillations through an escape mechanism[13]. A simulation is shown when the length of the caudal projections is decreased from four to one segments and when the frequency is increased. The phase lag decreases if the projection length is increased and the absolute phase delay does not scale with frequency leading to an increasing phase lag with frequency also here. Note that the relaxation oscillators show fast phase locking following the changes.

Estimation of intersegmental coupling functions

Intersegmental coordination in the lamprey spinal cord has been investigated by considering it as a chain of coupled limit–cycle oscillators, in which the effect of the coupling is described by an H-function. To bridge the gap between the theory and a real system Williams[9] developed a method for obtaining an approximation (C-function) of the stable region of the H-function by making measurement of the phase lag between two oscillators when their inherent frequencies are varied. Qualitatively similar results are achieved if estimations of the C–functions are done in the detailed model Fig. 3A and B, or if the contralateral hemisegments are modeled using Morris–Lecar relaxation oscillators that are set to work in the escape mode, using an FTM coupling mechanism (Fig. 3C). The C–functions behave in a rigid way when the frequency is varied, and stable values around zero or with reversed sign are hard to achieve in all cases. Note that these C–functions would be frequency dependent here.

If the C–function could be approximated by a linear expansion around their zero crossing values and a chain of oscillators are coupled symmetrically in the rostral and caudal directions the phase lags along such a chain would change gradually between consecutive oscillators[9]. This is not the case here (Fig. 3D) when the intersegmental coupling is produced by C cells projecting one segment rostrally as well as caudally. A characteristic and abrupt change of the sign occurs somewhere along the chain. Qualitatively similar results are achieved when Morris–Lecar relaxation oscillators are replacing the hemisegments (Fig. 3E).

Burst proportions in a chain of oscillators

One side in a local network of a right and left relaxation oscillator interacting through an escape mechanism would show bursting activity during 50% of the cycle. This is also the result if the intersegmental projections are short (Fig. 4A). However, if such networks are coupled with inhibitory projections extending over several segments a burst proportion of less than 50% can be achieved (Fig. 4B). This has also been observed in more biophysically detailed simulations[4].

CONCLUSIONS

The modelling approach using phase coupled oscillators has generated a number of experimental predictions some of which have been investigated experimentally[20, 21, 22] and by using connectionist unit models[9, 23, 24]. It can not be excluded that a local CPG of the lamprey behave according to this theory. However, using populations of neurons and simulating the different types thought to underly locomotion and with reasonable synaptic projections, a constant phase lag was not achieved[4], similar to the results above using a simplified model. The qualitative resemblance between the simplified network representing the spinal CPG in

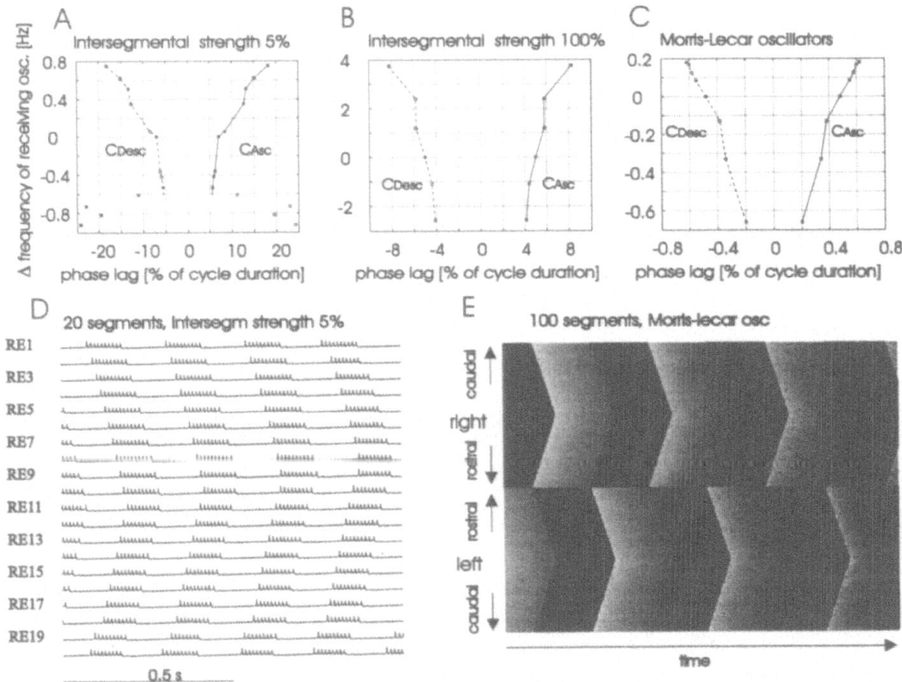

Figure 3. An estimation of intersegmental coupling function, assuming the theory for phase–coupled oscillators is valid, is shown for a weaker **A** (5%) or stronger **B** (100%) intersegmental coupling strength when the intersegmental coupling consists of inhibitory coupling contralaterally one segment down. **C** The C–functions when the Morris–Lecar oscillators replaces the hemisegments (intersegmental strength 100%). **D** Activity of C cells at every second level in a 20 segment chain when the intersegmental coupling is one segment in both the rostral and caudal directions. **E** As in D but Morris–Lecar relaxation oscillators are replacing the hemisegments. In both cases the phase lag, but not necessarily the absolute values, changes sign abruptly somewhere along the chain.

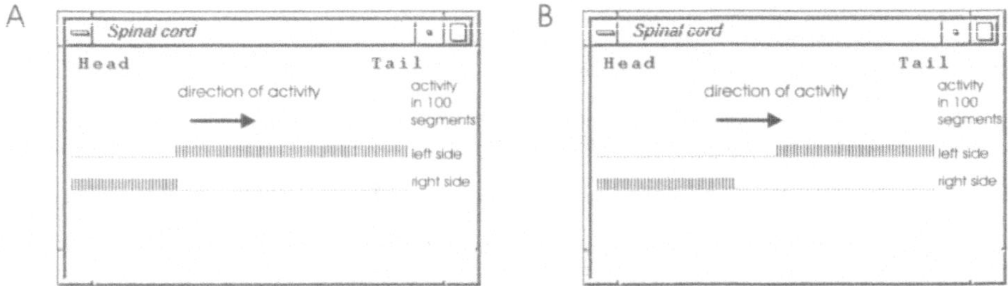

Figure 4. The activity in 100 segments of Morris–Lecar oscillators is shown. In **A** the intersegmental inhibition is only one level caudally leading to around 50% burst proportions. Burst proportions can be less than 50% if the intersegmental projection length is increased **B**.

the lamprey and a chain of relaxation oscillators suggests that further exploration of how production of a phase lag can occur in chains of relaxation oscillators coupled through an FTM mechanism would give valuable insight, and predictions from such studies would complement those generated from models building on phase coupled oscillators.

REFERENCES

1. S. Grillner, T. Deliagina, Ö. Ekeberg, A. El Manira, R. Hill, A. Lansner, G. Orlovsky and P. Wallén, Neural networks that co–ordinate locomotion and body orientation in lamprey, *Trends in Neurosci.* 18:270 (1995).

2. N. Kopell and B. Ermentrout, Coupled oscillators and the design of central pattern generators, *Math. Biosci.* 90:87 (1988).

3. N. Kopell, Chains of coupled oscillators, *in*: "The Handbook of Brain Theory and Neural Networks", A. Arbib, ed., MIT Press, Cambridge (1995).

4. T. Wadden, J. Hellgren, A. Lansner and S. Grillner, Intersegmental coordination in the lamprey: simulations using a network model without segmental boundaries, *Biol. Cybern.* 76:1 (1997).

5. J. Hellgren, S. Grillner and A. Lansner, Production of phase lag along chains of simplified models of the lamprey locomotor network. I: Oscillations occur through an escape mechanism, *in preparation.*

6. J. Hellgren, A. Lansner and S. Grillner, Production of phase lag along chains of simplified models of the lamprey locomotor network. II: Hemisegmental oscillations are produced by mutually coupled excitatory neurons, *in preparation.*

7. D. Somers and N. Kopell, Rapid synchronization through fast threshold modulation, *Biol. Cybern.* 68:393 (1993).

8. D. Somers and N. Kopell, Waves and synchrony in networks of oscillators of relaxation and non–relaxation type, *Physica D* 89:169 (1995).

9. T. Williams, Phase coupling in simulated chains of coupled oscillators representing the lamprey spinal cord, *Neural Comp.* 4:546 (1992).

10. L. Brodin, H. Tråvén, A. Lansner, P. Wallén, Ö. Ekeberg and S. Grillner, Computer simulations of N-methyl-D-aspartate (NMDA) receptor induced membrane properties in a neuron model, *J. Neurophysiol.* 66:473 (1991).

11. Ö. Ekeberg, P. Wallén, A. Lansner, H. Tråvén, L. Brodin and S. Grillner, A computer based model for realistic simulations of neural networks. I: The single neuron and synaptic interaction, *Biol. Cybern.* 65:81 (1991).

12. J. Hellgren, S. Grillner and A. Lansner, Computer simulation of the segmental neural network generating locomotion in lamprey by using populations of network interneurons, *Biol. Cybern.* 68:1 (1992).

13. F. Skinner, N. Kopell and E. Marder, Mechanisms for oscillation and frequency control in reciprocally inhibitory model neural networks, *J. Comp. Neurosci.* 1:69 (1994).

14. J. Tegnér, J. Hellgren-Kotaleski, A. Lansner and S. Grillner, Low voltage activated calcium channels in the lamprey locomotor network - simulation and experiment, *J. Neurophysiol* 77:1795 (1997).

15. Ö. Ekeberg, P. Hammarlund, B. Levin and A. Lansner, SWIM — A simulation environment for realistic neural network modeling, *in*: "Neural Network Simulation Environments," J. Skrzypek, ed., Kluwer, Hingham, (1994).

16. N. Kopell, W. Zhang and B. Ermentrout, Multiple coupling in chains of oscillators, *SIAM J. Math. Anal.* 21:935 (1990).

17. D. Kammen, P. Holmes and C. Koch, Cortical architecture and oscillations in neural network: feedback versus local coupling, *in*: "Models of Brain Funtion" R. Cotterill, ed., Cambridge University Press, Cambridge (1989).

18. N. Mellen, T. Kiemel and A. Cohen, Correlational analysis of fictive swimming in the lamprey reveals strong functional intersegmental coupling, *J. Neurophysiol.* 73:1020 (1995).

19. T. Williams and G. Bowtell, The calculation of frequency-shift functions for chains of coupled oscillators, with application to a network model of the lamprey locomotor pattern generator, *J. Comp. Neurosci.* 4:47 (1997).

20. K. Sigvardt, Intersegmental coordination in the lamprey central pattern generator for locomotion, *Semin. Neurosci.* 5:3 (1993).

21. T. Williams and K. Sigvardt, Intersegmental phase lags in the lamprey spinal cord: Experimental confirmation of the existence of a boundary region, *J. Comp. Neurosci.* 1:61 (1994).

22. K. Sigvardt and T. Williams, Effects of local oscillator frequency on intersegmental coordination in the lamprey locomotor CPG: Theory and experiment, *J. Neurophysiol.* 6:4094 (1996).

23. J. Buchanan, Neural network simulations of coupled locomotor oscillators in the lamprey spinal cord, *Biol. Cybern.* 66:367 (1992).

24. T. Williams, Phase coupling by synaptic spread in chains of coupled neural oscillators, *Science* 258:662 (1992).

A DYNAMIC NEIGHBOURHOOD FUNCTION
IN VOLUME LEARNING

Bart Krekelberg[1,2] and John G. Taylor[2]

[1]Allg. Zoologie und Neurobiologie
Ruhr University Bochum
44780 Bochum, Germany
Email: bart@neurobiologie.ruhr-uni-bochum.de
[2]Centre for Neural Networks
King's College London
London WC2R 2LS, England

INTRODUCTION

Nitric oxide (NO) is produced in both somata and post-synaptic dendritic specialisations in response to depolarisation. One of the many effects of NO at a synapse is to induce LTP or LTD depending on the presynaptic firing rate[1]. Volume learning takes into account the fact that NO can spread over a considerable distance in tissue and thus lead to *non-local, non-synapse-specific* LTP and LTD.

We have shown that this form of volume learning can lead to the development of cortical maps[2]. During the development of such a map, the spread of the nitric oxide defines which synaptic connections will be modified. The region of neurons with suprathreshold NO therefore seems analogous to Kohonen's neighbourhood function[3, 4]. There are, nevertheless, important differences between the two concepts which we explore here.

We first recapitulate the volume learning model[2] in the next section. Then, in section 'Map Formation', we present simulations of the *dynamics of map formation*. In these simulations the behaviour of the neighbourhood function as defined in volume learning can be observed and contrasted with the Kohonen neighbourhood. Whereas the Kohonen neighbourhood function is defined independently of the cortical map, the volume learning neighbourhood evolves with the cortical map. This difference is crucial in adult plasticity which is investigated in section 'Map Reorganization'.

THE VOLUME LEARNING MODEL

We modelled the effects of a non-local messenger on the processes of LTP and LTD in a framework based on the early work of Amari and colleagues[5, 6]. In this neural field theory, self-organization of the synaptic efficacies connecting neurons in a sensory layer S with neurons in a cortical layer C is studied. The stochastic dynamics of this system can be summarised in a set of three equations governing the membrane depolarisation at position x in the cortical layer C, $u(x, t)$, the concentration of NO at position x, $n(x, t)$ and the synaptic efficacy connecting the neuron at position y in the sensory layer S with the neuron at position x in cortex, $s(x, y, t)$. The essential difference with competitive models of map formation is that there is no (electrical) lateral interaction between the neurons in the cortical layer. As a consequence, there is no winner-take-all mechanism: the (soft) competition for cortical space takes place in the learning dynamics and is mediated by the diffusing messenger NO.

$$\tau_u \dot{u}(x, t) = -\gamma_u u(x, t) + \int_S dy'\, s(x, y', t) a(y', t)$$

$$n(x, t) = \int_0^t \int_C dt'\, dx'\, G^*(x - x', t - t') \Theta\big[u(x, t) - u_n\big] \qquad (1)$$

$$\tau_s \dot{s}(x, y, t) = Q\big[n(x, t) - n_s\big]\, R\big[a(y, t)\big].$$

The membrane is considered to be a simple linear leaky integrator, and NO is produced by parts of the membrane that are above threshold u_n. ($\Theta[x] = 1$ if $x > 0$, 0 otherwise.) G^* represents the effecitve Green's function for diffusion. Its precise form depends on the process of NO production and uptake, but it will generally be bell-shaped[2, 7]. The learning dynamics depend on the pre-synaptic depolarisation $a(y, t)$ and the post-synaptic NO concentration. The functional dependence $R[\ ]$ has been measured[1] and involves a negative (i.e. LTD) part for small a. Furthermore, the function $Q[\ ]$ is such that learning takes place only when the NO concentration is above some threshold n_s. This threshold defines the NO neighbourhood.

The dynamics (1) can be solved and lead to a (transcendental) equation for the size of the cortical point image in a (linear) cortical map[2]. This fixed point equation determines relations between the cortical point image and the properties of the diffusing messenger. These dependencies have been extracted and await testing[2].

Here our focus is not on the fixed point of this equation but on the dynamical evolution from an initially disorderd state to a cortical map. To study these, simulations were set up in MATLAB.

METHODS

A network with two layers consisting of 100 equidistant neurons was simulated in MATLAB. For computational efficiency the adiabatic approximation is assumed to hold for the membrane depolarisation. The Greens function for diffusion is modelled with a Gaussian (width ρ^*). The dependence of the weight change on the pre-synaptic firing rate a is modelled after the function measured by Zhuo et al[1]. The minimum pre-synaptic firing rate for LTD to occur is a_{ltd}, the minimum firing rate for LTP a_{ltp}. Weight change is linearly proportional to the amount of NO present above the threshold n_s. The learning rate is ϵ.

Activity on the sensory layer during development is modelled as Gaussian bubbles whose centres are drawn independently from a homogeneous probability distribution. In the simulation a stimulus is drawn, presented to the network and the corresponding membrane depolarisations in the network are determined. From this, the NO production and concentration are determined and the learning rule is applied to update the weights of those neurons that have an above-threshold NO concentration. This collection of neurons is defined to be the *volume learning-* or NO *neighbourhood*. The average neighbourhood is defined as the mean of the neighbourhood size over the whole set of stimuli. The weight update process is repeated until the change in weight averaged over all stimuli is less than 1%.

As is the case for the neural field cortical map formation model[6], this algorithm is not expected to converge to a cortical map from arbitrary initial conditions. Rather, we assume that some, presumably acitivity independent and chemoaffinity related mechanism has led to a bias in the weights[8] before activity dependent mechanism starts to operate . The task of activity dependent volume learning is to mould the initial bias into an accurate map that reflects the statistics of the environment. In the simulations we modelled the bias by giving a (small) percentage of the weights values that correspond to a cortical map. The other weights were randomly set to one (60%) or zero. This initial configuration leads to high NO concentrations due to the fact that each stimulus will lead to NO production in multiple locations in the network.

After the convergence to a cortical map, the networks were perturbed in two ways. First cortical perturbation was implemented by randomly resetting the weights of a region of neurons. The total sum of all weights before and after the lesion is kept constant, only their distribution is changed from an ordered configuration to a disorderd one. In the second experiment, the sensory surface was "lesioned" by removing a number of cells.

MAP FORMATION

Figure 1 shows a snapshot of the development of a cortical map in volume learning. At this stage of the learning process, the mean firing rate of the post-synaptic neurons as well as the NO concentration is still unstructered and high in large parts of the network (see for instance the response to the stimulus centred 50). On the other hand, some stimuli have already been allocated unique positions in the map; see the stimuli centred around position 75. After many iterations the network converges to a map in which a single region of firing neurons (the'winners') is found for each stimulus (not shown). In this final stage neighbouring stimuli lead to neighbouring cortical 'winners'; a map has been formed.

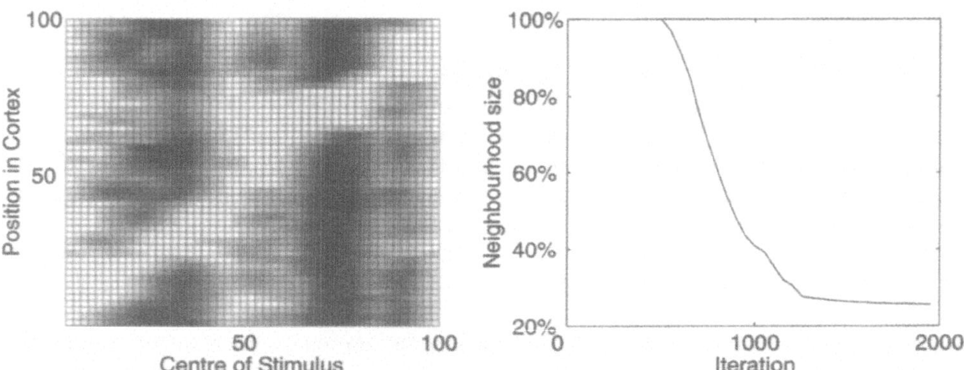

Figure 1. Left) A snapshot of the NO concentration during the learning process. The concentration of NO at a particular neuron in the network (y-axis) given the presentation of a particular stimulus (x-axis) is indicated by grey scale: low concentrations are represented by black. (Simulation parameters: $\rho^* = 5, u_n = 0.3, n_s = 0.1, a_{ltd} = 0, a_{ltp} = 0.2, \epsilon = 0.01$, Random weights: 90%.) **Right)** The dynamical evolution of the average neighbourhood size in volume learning. Initially the whole network is involved in weight change during the presentation of an arbitrary stimulus (Neighbourhood size ~ 100%). In the final converged cortical map, only a small region of neurons still refines its weights.

Due to the initially disordered state of the network, the neighbourhood encompasses nearly all of cortex. In figure 1 this can still be seen in the response to the stimulus centred at position 50. Interestingly, this is a phenomenon that has been found to be beneficial to self-organization in the Kohonen algorithm. There, the neighbourhood function is initially chosen to be large in order to allow learning to take place in as many neurons as possible and thus speed up the ordering process[9]. During the ordering phase the SOM-neighbourhood is gradually reduced. In our volume learning model the neighbourhood is not a free parameter but emerges from the dynamics of the diffusing messenger. The assumption of initial disorder, however, automatically leads to a large initial neighbourhood function. Figure 1 shows how the average size of the neighbourhood evolves. Clearly, there is a sharp reduction in the size of the neighbourhood during convergence.

Another important difference between the two views of a neighbourhood is that while the Kohonen algorithm restricts learning to a single connected region around the winner neuron, the neighbourhood function in this model can be seen to consist of multiple disconnected regions. In fact, the volume learning neighbourhood always encompasses those regions in which the disorder is large, thus allowing those neurons to reorganize their weights and speed up the convergence to a cortical map.

Finally, note that in the converged cortical map, the neighbourhood does consist of a single connected region surrounding the 'winner' neuron, thus coinciding conceptually with the Kohonen neighbourhood.

The dynamics of the neighbourhood function emerge naturally in the volume learning framework. Note also that, in contrast to many implementations of the Kohonen algorithm, the neighbourhood size in volume learning never reaches zero. This implies that the neurons always change their weights, even after a cortical map has been formed. In the cortical map stage, however, the overall change induced by the set of all stimuli averages out to zero, hence the map is stable *on average*. Although this may seem to be a weak kind of stability it is actually beneficial in a changing environment. Due to a non-zero neighbourhood size, any change in the statistics of the environment will be noticed by the network and the map will adapt to the new situation. Two examples of such plasticity are considered in the next section.

MAP REORGANIZATION

Whereas any fixed decrease of the neighbourhood (as in the SOM algorithm, or the interpretation of the shrinking as age-dependent loss of NO synthase[4]) would severely limit adult plasticity, the dynamic neighbourhood size implemented by NO can maintain plasticity in adult life. This section will show that the volume learning neighbourhood function is an efficient way to deal with changes in the environment.

Cortical perturbations

In this experiment (see 'Methods') we perturbed a converged cortical map by randomising some of the weights. The consequence of such a perturbation is that the neighbourhood size increases rapidly in the perturbed area. This allows the weights to reorganize to a cortical map (See figure 2).

The NO in this situation acts as a detector of disorder: a change in the organization leads to an increase in learning (more NO) as well as a larger region of neurons that change their synapses in response to a particular stimulus (wider spread of NO). Note that this increase in neighbourhood size is *local*, i.e. only around the affected region.

Sensory Perturbations

In a second experiment we test whether the adaptivity of the NO neighbourhood allows the network to adapt to changes in its sensory environment. We investigate the effect of an "amputation" of the sensory surface (see 'Methods').

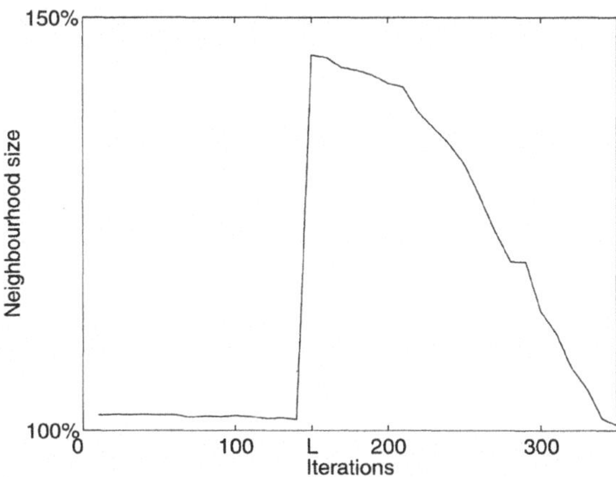

Figure 2. Evolution of the average neighbourhood size after a cortical perturbation. Immediately after the lesion ('L') the average neighbourhood size increases to 150% of its size in the converged map. Due to the re-organization, the NO neighbourhood size decreases, reaching its pre-perturbation level after the map has been restored (iteration 350).

The inevitable reduction of cortical activity that results from the removal of part of the sensory input *reduces* the NO production in the deafferented area. Due to the diffusion of NO from still active areas, however, these deafferented areas can still modify their weights. The deafferented neurons shift their receptive fields towards a region of the sensory surface that provides them with input. Figure 3 shows how a particular neuron's receptive field moves away from the amputated region.

The total shift of the receptive field centre depends on the distance from the deafferented area. Some neurons get a new lease of life through this RF shift while others, in the centre of the deafferented area, fail to find new sensory neurons to connnect to (See figure 3). This has been observed in experiments: deafferentation leads to reorganization of receptive fields in which the centres move away from the deafferented area[10]. The increase in the *size* of the RF which has been observed in experiments, is found in the model as well (not shown). In animal experiments, there is a limit to reorganization: large amputations usually leave some parts of the cortical tissue permanently without input. The volume learning model predicts that this limit can be manipulated by increasing the neighbourhood size. Experimentally this could be achieved by stimulating the NO synthase or, alternatively, comparing reorganization in normal mice to mice in which the NOS gene has been knocked out.

DISCUSSION AND CONCLUSION

It has been suggested before that the neighburhood function as conceived by Kohonen may be implemented by a diffusing messenger[4]. Our work shows this in more detail and thus provides a biological interpretation of a computational mechanism. Volume learning, however, adds functionality that is not present in the original propositions for (the implementation of) a learning neighbourhood.

The volume learning neighbourhood function is a dynamic quantity that is coupled to the membrane dynamics and the overall order in the map. The dynamics that emerge from this coupling turn out to mimic a shrinking neighbourhood that has been used to improve convergence in SOMs. Secondly, volume learning leads to multiple, disconnected neighbourhoods. This allows learning to take place in multiple regions in which disorder is large. Thirdly, the volume learning neighbourhood remains dynamic in the self-organizing process, even when

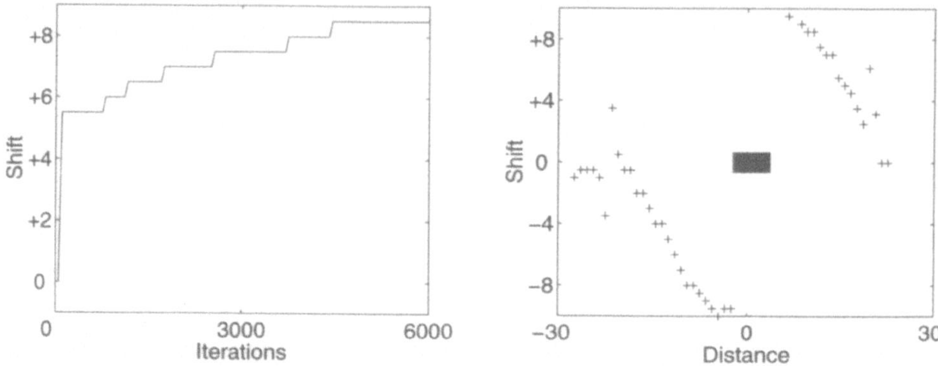

Figure 3. Left) The centre of the receptive field (RF) of a neuron just to the right of the deafferented cortical area. The first large shift in the RF centre is an artefact of the deafferentation. The following slow shift to the right, i.e. *away* from the amputation, is due to reorganization of the weights. **Right)** The total shift of the RF centre as a function of the distance from the centre of the deafferented cortical region. Neurons near the deafferented area move their receptive fields in order to receive input. The shaded area shows the neurons that remain without input even after reorganization.

a map-state has been reached. This leads to the concept of an adaptive neighbourhood, which increases *locally* in response to the amount of disorder in map. An interesting application of reverse engineering would be to extend the Kohonen neighbourhood with such an adaptation mechanism. This could be beneficial to a SOM operating in a changing environment.

The dynamic neighbourhood allows the cortical map to respond to changes in the environment. We have shown how this can lead to reorganization of receptive fields in deafferented cortical areas. The rough features of this reorganization agree with the experimental data. The feature that distinguishes this model from competitive learning models, the improved reallocation of sensory surface to deafferented neurons in the presence of an increased NO production, however, remains to be tested.

REFERENCES

1. M. Zhuo, E. Kandel, and R. Hawkins. Nitric oxide and cGMP can produce either synaptic depression or potentiation depending on the frequency of presynaptic stimulation in the hippocampus, *Neurorep.* 5:1033 (1994).
2. B. Krekelberg and J. G. Taylor. Nitric oxide: what can it compute?, *Network* 8:1 (1997).
3. T. Kohonen. Self-organized formation of topologically correct feature maps, *Biol. Cyb.* 43:59 (1982).
4. T. Kohonen. Physiological interpretation of the self-organizing map. *Neural Networks* 6:895 (1993).
5. A. Takeuchi and S.-I. Amari. Formation of topographic maps and columnar microstructures in nerve fields, *Biol. Cyb.* 35:63 (1979).
6. S.-I. Amari. Topographic organization of nerve fields, *Bull. Math. Biol.* 42:339 (1980).
7. B. Krekelberg. "Modelling Cortical Self-Organization by Volume Learning," PhD thesis, London University, King's College London (1997).
8. S. B. Udin and J. W. Fawcett, Formation of topographic maps, *Ann. Rev. Neurosci.* 11:289 (1988).
9. T. Kohonen. "Self-Organizing Maps," Springer Verlag (1995).
10. J. H. Kaas. The reorganization of sensory and motor maps in adult mammals, *in:* "The Cognitive Neurosciences," M. S. Gazzaniga, ed., MIT press, Cambridge, MA (1995).

BASAL GANGLIA PERFORM DIFFERENCING BETWEEN 'DESIRED' AND 'EXPERIENCED' PARAMETERS

András Lőrincz

Associative Computing, Inc.
Budapest, Konkoly-Thege út 29-33
Hungary H-1121

INTRODUCTION

The recently introduced first order SDS feedback control scheme[1] has been suggested as a suitable model of the basal ganglia - thalamocortical (BTC) loops[2]. The first order SDS model of the BTC loops can be extended to plants of any order[3] by slight modification of the original scheme. This SDS model identifies the direct and the indirect pathways of basal ganglia (BG) that respectively tend to enhance and suppress the reentrant thalamocortical excitation[4] with the desired (planned) and experienced channels of the SDS scheme. The model predicts that a relatively high contribution from the desired channel results in Huntington's disease whereas a relatively high contribution from the experienced channel results in Parkinson's disease. The model can be considered as a mathematical framework of the phenomenological thalamic disinhibition (TDI) model of DeLong[4]. The TDI model suggests that if the BG output is increased (decreased) it shifts the motion towards hypokinetic (hyperkinetic) states. Despite explaining several features of the BTC loops, the TDI model also predicts that lesion to the BG output recipient zones of the thalamus results in akinesia, which is not the case. This pitfall is avoided in the SDS model. The features of the general new scheme utilized to model the BTC loops are described in this paper.

THE SDS SCHEME

The SDS feedback control scheme deals with speed field tracking: we have an estimate or model of the true inverse dynamics Φ and we "know" the state of our plant, this is our observed state (q, $q \in \mathbf{R}^n$). Our aim is to modify that state with the 'desired speed vector' v. With the estimate of the inverse dynamics we are armed with the 'desired control vector' $u_d = \hat{\Phi}(q, v)$, where $\hat{\Phi}$ denotes the model of the inverse dynamics and $u_d \in \mathbf{R}^N$, with $n \leq N$. The following form is assumed:

$$u_d = \hat{A}(q)v(q) + \hat{b}(q) \tag{1}$$

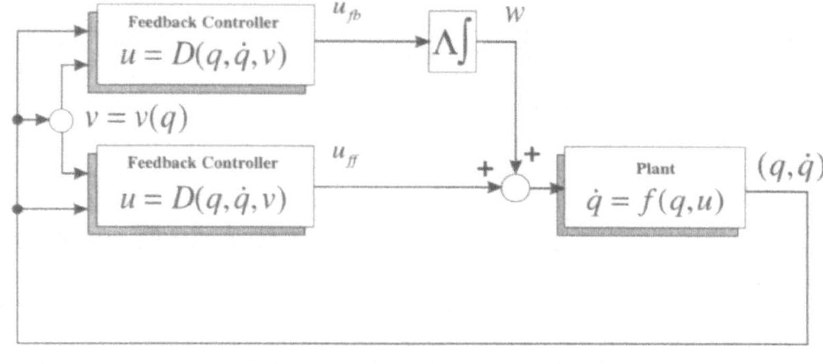

$$D(q,\dot{q},v) = \hat{\Phi}(q,v) - \hat{\Phi}(q,\dot{q})$$

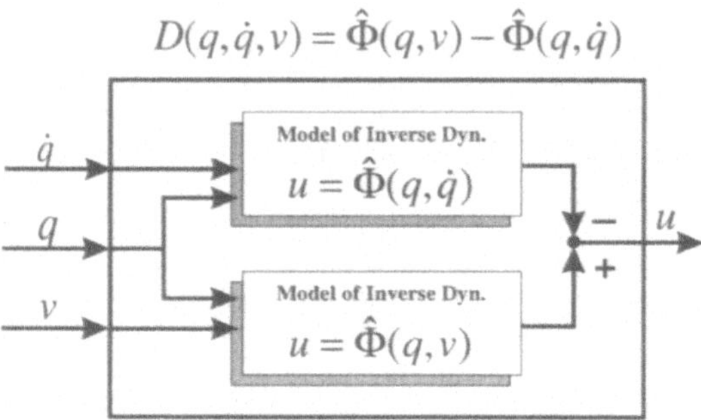

Figure 1. Compensatory control by using the inverse dynamics controller in differencing mode. The scheme utilizes identical feedforward and feedback controllers (ffc and fbc) built from estimates of the approximate inverse dynamics. The ffc and fbc provide control vectors $\mathbf{u}_{ff}$ and $\mathbf{u}_{fb}$, respectively. The feedback control vector is time integrated and the resulting control vector is added to the feedforward control vector. The state equation of the plant gives rise to changes in the state of the plant subject to its actual state, the control vector, and the dynamics of the plant. The identical ffc and fbc are detailed in the lower part of the figure. According to the scheme each controller comprises two copies of the model of the inverse dynamics. These models are inputted by the 'desired' and the 'experienced' speeds, respectively. The outputs of these models undergo differencing.

However, it is known that imprecise inverse dynamics can lead to instabilities and thus $\mathbf{u_d}$ cannot be applied directly. During motion we experience the speed vector $\dot{\mathbf{q}}$. The SDS model makes use of an identical copy of the very same controller and computes the 'experienced control vector' $\mathbf{u_e} = \hat{\boldsymbol{\Phi}}(\mathbf{q},\dot{\mathbf{q}})$ using the experienced speed, where the dot denotes temporal derivation. This control vector can be used as a correcting means. The SDS scheme of plants of any order is equipped with a feedforward controller (ffc) and a feedback controller (fbc). The ffc utilizes the control vector $\mathbf{u}_{ff} = \mathbf{u_d} - \mathbf{u_e}$; the fbc control vector is the time integrated and amplified ffc vector: $\dot{\mathbf{u}}_{fb} = \lambda \mathbf{u}_{ff}$, where λ is the gain of the feedback channel (Fig. 1).

The observed state of a second order plant $\mathbf{q}$ is given by its position vector and momentum vector. The temporal derivative of that state, i.e. the 'experienced speed vector' $\dot{\mathbf{q}}$, comprises the ('experienced') momentum vector and the ('experienced') acceleration vector. The 'desired speed vector' also has two components, viz. the 'desired momentum vector' and the 'desired acceleration vector'. Similar constructs can be developed for plants of any order. The resulting scheme is globally stable provided that the perturbation is 'sign-proper'. 'Sign-properness' means that for stable control the sign of the speed vector to control vector association should be properly set breaking up the state space into sign-proper domains. Note that there is no condition on the magnitude of the strength of this association making the first learning step reinforced sign-selection instead of tuning. Tuning then can be on-line, provided that the condition on sign-properness is met. Details of speed field tracking (Eq. 1) and the Liapunov-function approach based proof of stability can be found elsewhere[3].

THE SDS MODEL OF THE BASAL GANGLIA

The putamen and the caudate nucleus of BG are of modular structure with two types of modules, the striosomes and the matrisomes[5,6]. The striosomes are neurochemically specialized patchy input–output zones that tend to collect inputs related to the limbic system and to project to the dopamine–containing substantia nigra pars compacta. In contrast, the matrisomes receive sensorimotor and associative inputs and project to the output nuclei of the BG.

The SDS model deals with the matrisomes. First, we note that limb/body configurations and/or tasks (such as mirror writing) require the recognition and the selection of the appropriate sign-proper feedback channels which is conform with the parallel organization of the BTC loops[7].

The SDS model of the BTC loops assumes that the desired and experienced momenta are formulated in the 'where pathway' and all the differencings that follow are computed by the BTC loops modelling that the BG has direct and nonreciprocal projections from parietal area 7b, part of the 'where pathway'[8,9], whereas most of the cortical mantle projects to the BG in a reciprocal manner. The 'desired acceleration' can be approximated by taking the difference between the 'desired momentum' and the 'experienced momentum'. It is assumed that this difference – like other differences of similar kind – is computed by the BG.

We can thus characterize computational units (model neurons) suggested by the SDS model follows. (1) Model neurons representing the desired acceleration will have higher firing rates in the so called preparatory phase, i.e., before motion is initiated. (2) The firing of model neurons representing experienced momentum is motion related. (3) Model neurons that represent the desired momentum have mixed characteristics. (4) Model neurons that represent the output of the approximate inverse dynamics, i.e., neurons that express the components of the control vector, behave as 'muscle like' neurons and exhibit motion related firing. (5) Information about the actual task should reach model neurons in order to select the sign-proper feedback channels, e.g., by competition. Population coded approximation of the inverse dynamics can be built from such model neurons[3,10,11] The classification of model neurons is in general agreement with experimental findings and classification[12]. Other classes may be discovered according to the SDS model for perturbed motions. For this case the SDS model predicts that 'muscle like' neurons related to the desired and experienced control components can be dissociated. In this respect, prediction of the appearance of the experienced control components – control components that would appear provided that the experienced motion was the desired one – acts as a test of the validity of the model.

The model has been tried beyond the conditions of the theorem for a three joint robotic arm benchmark problem[13]. In particular, the differencing function has been tested[14]. It was found that the model is fairly robust against imprecisions of the differencings. Precision of the differencing was parametrized. If the parameter is 1.0 then the differencing is perfect. If the parameter is 0.0 then only the desired components contribute. It was found that the most sensitive term from the various differencings is the difference between the desired and the experienced momenta needed to compute the desired acceleration. Still, with a parameter value of 0.5, to the naked eye there was no difference between the trajectories. A parameter value of 0.1 resulted in damped oscillation *and* faster control; a further decrease of the parameter produced undamped and irregular oscillations, and these may be identified with some forms of Huntington's disease.

If we consider the observed appearance of dysmetria and unsmooth movement in cerebellar disorders when BG are the means of control then the model may indicate that BG differencing gives a higher contribution to the desired channel, the other option being that processing in BG is subject to considerable delay. Since the smaller the differecing parameter the faster is the control, the model suggests that processing delays of BG are compensated by smaller differencing parameter values (i.e., by higher contributions from the desired channel) and that the case of perfect differencing where the theorem holds, already corresponds to Parkinson's disease. The main underlying reason of Parkinson's disease, according to present model of BG, is that the contribution of the experienced momentum in the computation of the desired acceleration is large. Large experienced momentum signals overcompensate the desired momentum signals and that, in turn, counteracts the desired motion up to the point that the contribution from the experienced momentum signals becomes smaller than the contribution from the desired momentum signals. Motion is slowed down under these conditions.

The change of the feedforward controller to another external controller can be smooth if the feedforward channel of the BG is weak and if the external feedforward controller is precise. In this respect it is important that computations conducted *without* the feedforward channel show minor differences and that differences appear at the start of the motion[14] indicating the viability of this idea. Considering this type of robustness of the scheme and that integration and differencing are linear operations that can be interchanged we conclude that the two channels, viz. the feedforward and the feedback channels, are unified into a *single* channel that undergoes integration with a weighted memory function, or temporal kernel. Such temporal kernels can be represented by reentrant loops. This concept of the BG that unifies the two channels by means of a weighted memory function seems to accord well with the findings that (1) activities in the supplementary motor area, the motor cortex and the putamen strongly overlap, and (2) the activities in the cortical regions precede the activity in the putamen[12]. The requirement of simple sign-properness of the SDS model suggests that the homogeneous responses of dopamine neurons[15] can form a homogeneous reinforcement signal within the putamen and the caudate nucleus that serve to select sign-proper matrisome channels since selective type training suffices at this early stage of learning.

The peculiar feature that lesions to the thalamic recipient zones of BG outputs do not result in akinesia can be explained by (i) the presence of direct corticocortical connections between the supplementary motor area and the motor cortex; possibly these connections represent a direct channel for the approximate inverse dynamics without any kind of differencing. Such a direct channel represents another source of unbalancing in the differencing that – judging from the numerical studies – the SDS scheme is robust against. (ii) If the motion is controlled by an external motor programme generator and if it goes as planned then the BG provides little (or no) contribution (since the desired and the experienced parameters are close to each other). The 'no output case', however, corresponds to the lesioned case and thus akinesia is not a necessary consequence of the said lesion.

It has been suggested that the striosomes of the BG and the dopaminergic neurons together form a temporal differencing reinforcement learning (TDRL) scheme[16,17]. We note that the TDRL scheme can be rephrased as the computation of temporal differences between the 'expected reward' and the 'experienced reward' values. It is thus tempting to consider the basal ganglia as the substrates that take part in the computation of the differences between 'desired' ('expected', 'planned') and 'experienced' ('actual') parameters of motion and cognition.

CONCLUSIONS

The SDS model of the BTC loops assumes that the control actions are formulated by means of an approximate inverse dynamics controller. The control actions are made precise by extensive differencings all of which can be formulated in terms of desired and experienced parameters. The desired acceleration is the difference between the desired momentum and the experienced momentum whereas the output of one stage of the controller is the difference between the desired control vector (given by the approximate inverse dynamics inputted by the desired speed) and the experienced control vector (given by a copy of the approximate inverse dynamics inputted by the experienced speed). The desired and experienced channels of the SDS scheme are identified with the direct and indirect pathways of BG. Numerical simulations indicated that exact balancing of the two pathways is not required and the scheme can tolerate fairly large differences. According to the model, when the relative contribution of the direct pathway is large then moves 'out of measure' appear; in contrast, a relatively large contribution from the indirect pathway gives rise to slow motion with apparent friction. These extremes were identified with Huntington's disease and Parkinson's disease, respectively.

It has been shown that the resulting neuronal classification precisely corresponds to the experimental findings[12].

We have argued that according to the SDS model the BG can form part of a training hierarchy that starts with selective type homogeneous learning at the level of the striosomes, tunes the matrisomes, and upon tuning can provide an error signal to an external motor programme generator owing to the supplementary nature of the SDS control scheme. The strong feedback loop utilizing temporal integration will not build up a strong control signal if a well tuned external controller is in charge. This provides further support why lesions to the thalamic recipient zones of BG outputs do not give rise to akinesia. A further benefit is that apart from sign-properness, other requirements are not necessary to reach global stability. Thus the SDS scheme allows the smooth and symbiotic development of an external feedforward motor programme generator, a puzzling feature of the biological system.

The SDS model of the matrisomes and the temporal differencing reinforcement learning model of the striosomes can be unified into a common framework by saying that the basal ganglia take part in the computation of the differences between desired and experienced parameters of motion and cognition.

Acknowledgments

This work was partially funded by OTKA-Hungary grant T017110 and the US-Hungarian Joint Fund grant 519/95-A

REFERENCES

1. Cs. Szepesvári, Sz. Cimmer, and A. Lőrincz, Neurocontroller using dynamic state feedback for compensatory control, *Neural Networks* 10:1691 (1997).
2. A. Lőrincz, Static and dynamic state feedback control model of basal ganglia – thalamocortical loops, *J. Int. Neural Systems* (In press).
3. Cs. Szepesvári and A. Lőrincz, Approximate inverse-dynamics based robust control using static and dynamic state feedback, *in*: "Applications of Neural Adaptive Control Thechnology" J. Kalkkuhl, K. J. Hunt, R. Zbikowski, and A. Dzielinski, eds. World Scientific, Singapore (1997).
4. M. R. DeLong, Primate models of movement disorders of basal ganglia origin, *TINS* 13:281 (1990).
5. A. M. Graybiel and C. W. Ragsdale, Jr., Histochemically distinct compartments in the striatum of human, monkeys, and cat demonstrated by acetylthiocholinesterase staining, *PNAS* 75:5723 (1978).
6. R. Malach and A. M. Graybiel, Mosaic architecture of the somatic sensory-recipient sector of the cat's striatum, *J. Neurosci.* 12:3436 (1986).
7. J. E. Hoover and P. L. Strick, Multiple output channels in the basal ganglia, *Science* 259:819 (1993).
8. J. T. Weber and T. C. T. Yin, Subcortical projections of the inferior parietal cortex (area 7) in the stump-tailed monkey, *J. Comp. Neurol.* 224:206 (1984).
9: L. D. Selemon and P. S. Goldman-Rakic, Longitudinal topography and interdigitation of corticostriatal projections in the rhesus monkey, *J. Neurosci.* 5:776 (1985).
10. Cs. Szepesvári and A. Lőrincz, An integrated architecture for motion-control and planning, *J. Robotic Systems* (In press).
11. T. Fomin, T. Rozgonyi, Cs. Szepesvári, and A. Lőrincz, Self-organizing multi-resolution grid for motion planning and control, *J. Int. Neural Systems* 7:757 (1996).
12. G. E. Alexander G. E. and M. D. Crutcher, Control of goal-directed limb movements in primates: Neurobiological evidence for parallel, distributed motor processing, *in*: "Neurobiology of Motor Programme Selection," J. Kien, C. R. McCrohan, W. Winlow, eds., Pergamon, Oxford (1992).
13. C. W. Anderson and W. T. Miller III, Challenging control problems, it in: "Neural Networks for Control," W. T. Miller III, R. S. Sutton, P. J. Werbos, eds. MIT Press, Cambridge MA (1990).
14. Cs. Szepesvári and A. Lőrincz, (Unpublished observations).
15. W. Schultz, P. Apicella, R. Romo, and E. Scarnati, Context–dependent activity in primate striatum reflecting past and future behavioral events, *in*: "Models of Information Processing in the Basal Ganglia", J. C. Houk, ed., MIT Press, Cambridge (1995).
16. J. C. Houk, J. L. Adams, and A. G. Barto, A model how the basal ganglia generate and use neural signals that predict reinforcement, *in*: "Models of Information Processing in the Basal Ganglia", J. C. Houk, ed., MIT Press, Cambridge (1995).
17. G. S. Berns and T. J. Sejnowski, How the basal ganglia makes decisions, *in*: "Neurobiology of Decision Making", A. Damasio, ed., Springer-Verlag, Berlin (1995).

NEURAL MODEL OF TRANSFER-OF-BINDING
IN VISUAL RELATIVE MOTION PERCEPTION

Jonathan A. Marshall, Charles P. Schmitt,
George J. Kalarickal, and Richard K. Alley

Department of Computer Science, CB 3175, Sitterson Hall
University of North Carolina, Chapel Hill, NC 27599-3175, U.S.A.

INTRODUCTION

Human visual systems are much more sensitive to relative motion than to absolute motion. For example, the relative motion of two dots on a blank background is more easily detected than the motion of a single dot on a blank background. If both dots are also moved relative to the background, then their motion relative to each other remains more easily detected than their motion relative to the background. Each dot thus seems to provide a *reference frame* for the other's motion.

It is challenging to construct a theory of relative motion perception because such a theory must answer a difficult question: Which objects belong to which reference frames? We have designed and simulated a neural circuit model that provides an answer to this question. The model uses a simple competitive neural circuit to bind visual elements into a representation of an object. The model also uses information about the spiking pattern of neurons to *transfer* the bindings of an object representation from location to location in the neural circuit as the object moves. The model exhibits many characteristics of human perceptual relativity and solves several key neural circuit design problems: (1) the receptive fields that an object activates are selected dynamically within the context of the reference frame to which the object belongs; (2) a neural mechanism for "binding" or grouping operates to specify which parts of a scene form the reference frame for each object; (3) each object's bindings persist despite the object's motion; and (4) the groupings operate hierarchically, so that reference frames are represented relative to other reference frames.

Grouping example: Visual relative motion perception

As a conceptual example to help us design the model, we used the simple dot motion displays of Figure 1, in which correct grouping leads to improved predictions. When several dots have the same motion (Figure 1A), they are seen as constituting a group (i.e., an object, Figure 1B).[1,2,3,4] When one dot in such a grouping becomes invisible (Figure 1C), it is perceived as moving *behind* an occluder.[5] If it is represented as *continuing to exist* (Figure 1D) behind the occluder, then this representation can help the system predict correctly that it will reappear when it passes the occluder and will do so while maintaining the same relationships (e.g., distance, colinearity) to the other dots.[6] In Figure 1E, the visible moving dots change direction while one dot is occluded. If the dots all really do belong to the same object, then the best representation is one in which the whole object changes direction (Figure 1F). What is the best prediction that the system can generate for the motion of the invisible dot? If grouping mechanisms did not exist, then the system could generate predictions based only on the motion of the dot itself, extrapolating from the dot's motion when it was visible, predicting a linear trajectory (Figure 1G). But if a grouping mechanism does exist and the dots are grouped together, then the system can predict a trajectory that follows the group motion, as indicated by the visible dots (Figure 1H). If the dots are genuinely part of the same object, then the predicted reappearance of the dot will match its actual reappearance, and this prediction will be more accurate than the predictions not based on grouping.

Figure 1I depicts a three-dot display in which the middle dot is seen as having a horizontal motion relative to the motion of the group of dots (Figure 1J), and in which the group motion changes direction while the middle dot is occluded. A linear prediction would fail to match the reappearance of the occluded dot (Figure 1K). By what neural circuits can a human visual system produce a prediction for the case where the middle dot belongs to the group (Figure 1L)? This case is especially difficult because the group-based prediction for the middle dot (that its motion should be vertical) does not equal the motion of the group itself (diagonal).

Value of grouping: Improved predictions

Figure 1 suggests that the predictive performance of visual motion circuits is improved if the circuits incorporate information about grouping. This conclusion is based on the assumption that visual elements (like the dots) that behave coherently (e.g., move with the same velocity) will most often continue to behave coherently (at least over the short term). This Gestalt "common fate" assumption is reasonable in normal visual environments.

The transfer-of-binding problem

Even if two visual elements have common motion, they might belong to different objects. For example, suppose that two dots first move in different directions and later move with a common direction and speed. In that case, there is ecologically less reason to believe that they will continue having a common motion (which might just be coincidental) than for two dots whose motion had never been observed to differ. A visual system must be able to represent both cases: in the absence of contrary information, common motion should enhance grouping; but evidence of prior non-common motion should weaken grouping. This implies that visual systems should base their groupings on history as well as common motion. It also implies that a given dot has the potential to belong to many different groups (one at a time), consistent with its motion. Thus, a dot's actual belongingness, or bindings, must be represented *dynamically*, in a way that can be altered according to the dictates of the scene and the motion history.

How can a visual system use past bindings to predict and influence future bindings? This is the *transfer of bindings* problem. It may arise in many aspects of perception[7]

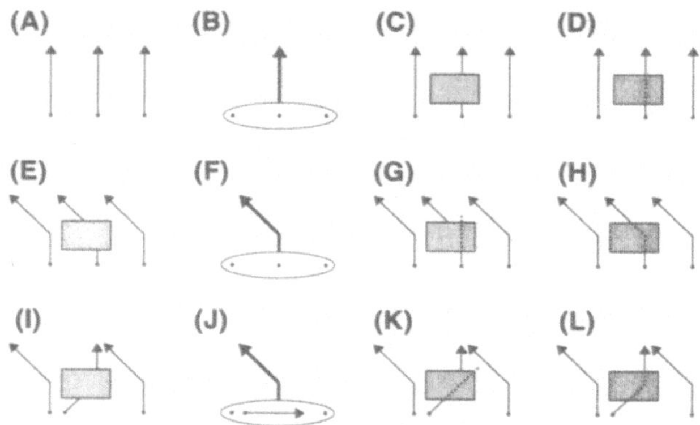

Figure 1. Grouping improves predictions in 3-dot motion displays. (A) Three dots with common motion are perceived as being grouped together (B) as parts of the same object. When one dot is temporarily occluded (C), it is perceived as continuing to exist (D) "behind" the occluder. How should the object's motion be represented if it changes while one dot is occluded (E)? If the three dots are indeed part of the same object (F), then a linear representation of the middle dot's motion (G) "behind" the occluder would fail. The best representation of the object's motion would be for the whole object, including the middle dot, to be represented as changing direction (H), on the basis of the two visible end dots' motion. If the middle dot has a horizontal motion relative to the group motion (J), then its reappearance from behind an occluder (I) is harder to predict. Again, a linear prediction (K) would fail. Howevever, assigning the group motion to the occluded dot would not work; instead, the group motion must be *added* to a remembered representation of the middle dot's motion, to predict the correct group-based change in the middle dot's motion (L).

and cognition.[8] It is exemplified by the three-dot examples of Figure 1, but it is an inherent problem that arises because neurons and synapses cannot physically move to keep up with their moving or transforming inputs. Our model neural circuit for dynamic binding correctly performs transfer of binding for moving objects.

METHODS: NEURAL CIRCUIT SIMULATION

The model comprises five neural processing layers with a highly specific pattern of excitatory and inhibitory connection pathways (symbolized in Figure 2). The pattern of connections can be established by simple activity-dependent learning rules, in response to ordinary moving visual stimuli during a developmental period.[9,10,11,12,13] Binding is effected by a model inhibitory interneuron gating circuit that causes feedforward excitatory afferent axonal pathways emanating from a common source neuron to compete for the "right" to carry signals, based on feedback signal strength. When a neuron e, representing a visual element, is bound to a neuron g, representing a group (or object), the $e{\to}g$ and $g{\to}e$ pathways are open, and other pathways $e{\to}g'$ and $g'{\to}e$ are gated shut by competition. This allows the spike pattern of neuron e to become synchronized to the spike pattern of neuron g.

Transfer of binding is effected by a mutual facilitation arising from synchrony of input spike firing patterns. When several visual elements are identified as belonging together to a group (or object), the neurons representing those elements become dynamically bound to a neuron representing the group; this "group" neuron then provides synchronizing feedback to the "element" neurons. Subsequently, after additional motion, the synchrony of the spike patterns from the element neurons enhances the likelihood that they will be represented together by a common group neuron at their new spatial positions.

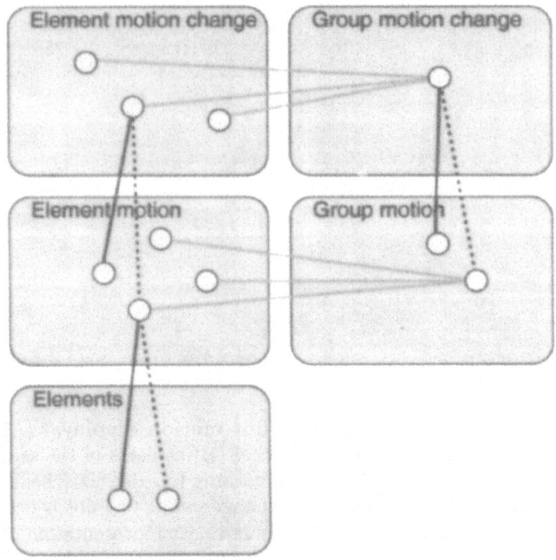

Figure 2. Neural architecture for relative motion perception, with transfer-of-binding.
Five processing areas are shown. Input arrives in the Elements area. The Element Motion area
contains simple bilocal motion-sensitive neurons, using time-delay pathways (dotted lines). (Other
architectures for motion-sensitive neurons would work as well.) The Element Motion neurons can be
dynamically bound to neurons in the Group Motion area, via binding pathways (gray lines).
Second-order changes (e.g., curves or accelerations/decelerations) in the motion trajectories across
the Element Motion and Group Motion areas activate neurons in the Element Motion Change and
Group Motion Change areas. When a group motion change occurs, predictions for the new motion of
each element are transmitted via reciprocal feedback pathways to the Element Motion Change
neurons, further back to the Element Motion neurons, and further back to the Element neurons.
Interneuronal binding circuits (not shown) implement competitive decision-making about temporal
correspondence matches and motion assignments. Neuron spiking patterns (not shown) are
transmitted between the areas and influence the decisions about which Group Motion neurons
become active, so that elements that were grouped together at one motion step are more likely to
remain grouped together at the next motion step.

RESULTS: PREDICTIVE TRACKING BASED ON GROUP COHERENCE

Figure 3 shows the results of a computational simulation of the model's behavior in
response to the visual relative motion stimulus of Figure 1I. The simulation illustrates
how the model uses information about the behavior of some elements of a group to
predict the behavior of other elements of the group. The simulation's state is shown
at the moment that the motion of the two outer dots has changed direction. The
simulation generates a feedback signal that predicts the middle dot's location, based on
group motion, rather than on element motion of the middle dot alone. The simulation,
which used 2341 simulated neurons, was run in approximately 20 minutes on a Hewlett-
Packard J210 workstation computer.

SIGNIFICANCE

This study has identified a computational problem, *transfer*-of-binding, that arises
in visual relative motion perception, as well as in other areas. The distinction between
binding and transfer-of-binding helps clarify the demands on biological visual systems
in interpreting relative motion displays.

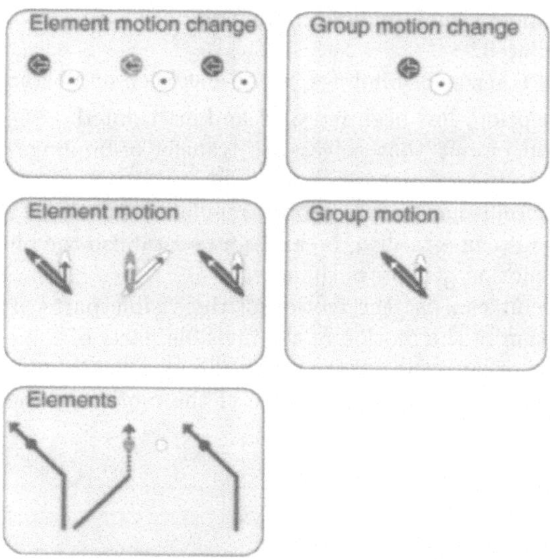

Figure 3. Simulation results: Time=3. The state of the simulated network at time=3 is shown. From time=1 to time=2, the two outer dots moved vertically upward, and the middle dot moved diagonally upward (heavy arrows in Element area). At time=3, the two outer dots (black dots in Element area) move diagonally upward, and the middle dot disappears, as if behind an occluder. Delayed excitation begins to activate two Element neurons at candidate locations for the occluded middle dot (two circles near center of Element area). Two Element Motion neurons are activated, and they bind to a common Group Motion neuron (black ellipses). The directional change of the group (from vertical to diagonal) activates a Group Motion Change neuron (black circle with leftward arrow). This sends feedback to the Element Motion Change neurons, which combines with candidate feedforward excitations to fully activate the two outer Element Motion Change neurons and to partially activate the middle Element Motion Change neuron (hatched circle). In turn, this partial activation feeds back excitation to the middle Element Motion neuron (hatched ellipse) and to the middle Element neuron (hatched circle). This feedback represents the network's prediction (hatched circle) that the motion of the (invisible) middle dot changes in accordance with the group motion, not as a mere ballistic extrapolation (open circle) of the middle dot's previous motion. If the firing pattern of the neurons representing the middle dot had not been synchronized with the firing pattern of the neurons representing the other two dots, then the prediction would *not* be influenced by the motion change of the two outer dots, and the prediction *would* be a linear extrapolation of the middle dot's prior motion.

The model suggests a plausible neural mechanism for binding and a plausible functional role for spike synchrony in transfer-of-binding. These aspects of the model are neurophysiologically plausible for several reasons:

- the competitive interneuron circuit for binding uses only well-known and widely-accepted components;
- a consideration of specific spike patterns at individual neurons is not required; only synchrony *between* spike patterns is examined; thus, only a weak model of spiking is needed;
- spike synchrony provides only an amplification factor to the strength of neural signals; the activation of neurons depends only partially on the synchrony; thus, again, only a weak model of spiking is needed.

Our research has yielded several insights into the principles that determine how transfer-of-binding works and how visual relative motion perception works:

- Highly specific grouping/binding circuits are needed for precise relative motion perception.

- A new conception of visual binding, combining both gating *and* spiking synchrony, has been formulated.
- The problem of transfer-of-bindings, which has been an obstacle for prior models of motion perception, has been revealed and articulated.
- The first neural circuit that solves the transfer-of-bindings problem has been designed.
- Such a neural circuit can have a simple, regular, hierarchical structure.
- The neural circuit can establish, break, and re-establish the bindings for visual elements whose motion groupings change.
- The neural circuit can use the motion of the visible parts of an object to *steer* the representation of the motion of the invisible parts of the object.
- The hierarchical organization of the neural circuit allows the motion of an element to be represented in the reference frame of the motion of the group to which it belongs.

ACKNOWLEDGMENTS

Supported in part by ONR grants N00014-93-1-0208 and N00014-93-1-1296. We thank Vinay S. Gupta and Viswanath Srikanth for useful discussions.

REFERENCES

1. K. Koffka. "Principles of Gestalt Psychology," Harcourt, Brace and Company, New York (1935).
2. J. Ternus, Experimentelle Untersuchungen über phänomenale Identität, *Psych. Forsch.* 7:81-136 (1926).
3. G. Johansson. "Configurations in Event Perception," Uppsala: Almquist and Wiksells Boktryckeri AB (1950).
4. G. Johansson, About visual event perception, *in*: "Persistence and Change: Proceedings of the First International Conference on Event Perception," W. H. Warren and R. E. Shaw, eds., LEA, Hillsdale, NJ (1985).
5. V. S. Ramachandran, V. Inada, and G. Kiama, Perception of illusory occlusion in apparent motion, *Vision Research* 26:1741-1749 (1986).
6. J. A. Marshall, R. K. Alley, and R. S. Hubbard, Learning to predict visibility and invisibility from occlusion events, *in*: "Advances in Neural Information Processing Systems 8," D. S. Touretzky, M. C. Mozer, and M. E. Hasselmo, eds., Morgan Kaufmann Publishers, San Mateo, CA (1996).
7. J. K. Kruschke and M. M. Fragassi, The perception of causality: Feature binding in interacting objects, *in*: "Proceedings of the Eighteenth Annual Conference of the Cognitive Science Society," Lawrence Erlbaum Associates, Hillsdale, NJ (1996).
8. D. Kahneman, A. Treisman and B. J. Gibbs, The reviewing of object files: Object-specific integration of information, *Cognitive Psychology* 24:175-219 (1992).
9. R. S. Hubbard and J. A. Marshall, Self-organizing neural network model of the visual inertia phenomenon in motion perception, Technical Report 94-001, Department of Computer Science, University of North Carolina at Chapel Hill 26 pp., (1994).
10. J. A. Marshall, Self-organizing neural networks for perception of visual motion, *Neural Networks* 3:45-74 (1990).
11. J. A. Marshall, Unsupervised learning of contextual constraints in neural networks for simultaneous visual processing of multiple objects, *in*: "Neural and Stochastic Methods in Image and Signal Processing," S. S. Chen, ed., Proceedings of the SPIE, San Diego, CA, 1766:84-93 (1992).
12. J. A. Marshall, Adaptive perceptual pattern recognition by self-organizing neural networks: Context, uncertainty, multiplicity, and scale, *Neural Networks* 8:335-362 (1995).
13. C. P. Schmitt and J. A. Marshall, Development of a neural circuit for motion grouping and disambiguation, *Investigative Ophthalmology & Visual Science* 36(4):377 (1995).

ANALYSIS OF COUPLED CHAOSCILLATORS EMBEDDED WITHIN THALAMOCORTICAL AND CORTICOCORTICAL REENTRANT LOOPS ENCOMPASSING DYNAMICS ON MULTIPLE TIME SCALES

James M. E. Patterson, Mark E. Jackson, and Lawrence J. Cauller

Neuroscience Program, University of Texas at Dallas
P.O. Box 830688, GR 41
Richardson, TX 75083-0688

ABSTRACT

The dynamical behavior of reciprocal thalamocortical and corticocortical interactions was analyzed using a large-scale GENESIS simulation consisting of interconnected networks of reconstructed neuron models with active kinetics and firing modes. Connections between cortical areas were based upon neuroanatomical tracing and intracellular neurophysiological experiments. Over a wide range of connection parameters, this system generated chaotic activity including regions of fractal activity identified by fractal exponents, an event-counting statistic derived from Fano- and Allan-factor analyses. This neural system can be described in terms of coupled "chaoscillators" within an extensively reentrant architecture with multiple time scales.

INTRODUCTION

We are interested in the dynamics of reciprocal corticocortical interactions in the presence of sensory inputs. For this study we have constructed biologically realistic models of the asymmetric reciprocal corticocortical connections between primary and secondary somatosensory areas (SI and SII) which have been thoroughly characterized in rats[1]. In previous investigations, interactions between simulated pairs of reconstructed and simplified pyramidal and stellate neurons demonstrated a dependency of the fractal nature of the spike train patterns on connection parameters including synaptic weight and conduction delay[2]. Specifically, in the presence of subthreshold random background activity, a random synaptic input elicited a fractal firing pattern from the SI pyramidal cell. In isolation, this behavior was most sensitive to the top-down synaptic weight. As the influence of the reciprocal synapse increases over a given range, the fractal dimension of the spike train increased. We have coined the term "chaoscillators" to describe these elementary circuits that are capable of generating chaotic activity. Figure 1 shows a pair of chaoscillator circuits representing a pyramidal-stellate cell pair from SI interconnected to a

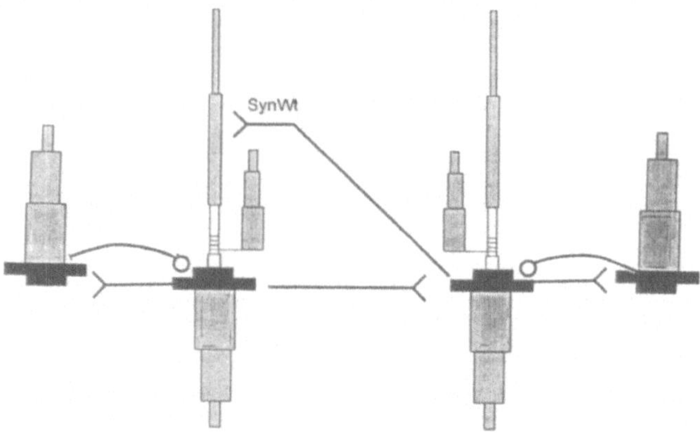

Figure 1. The elementary unit within a corticocortical reentrant loop. The center two cells are reduced compartment models of reconstructed dye-filled pyramidal neurons from the rat somatosensory cortex. The outer two cells are similarly generated stellate neurons. The leftmost chaoscillator circuit represents a functional unit in SI whereas the rightmost pair are in SII. The anatomical separation between the two areas is represented by a conduction delay. Inhibitory synapses are represented as open circle connections and the "Y" shaped projections are excitatory. The bottom-up projections are point-to-point synapses onto the soma of both the SII pyramidal and stellate cell types. The backward top-down connection is reciprocal but more wide spread across the distal dendrites of SI pyramidal neurons only.

pyramidal-stellate pair in SII, highlighting the asymmetry in the connections. Subsequent work has demonstrated that reciprocal connections between thalamocortical and thalamic reticular neurons likewise generate chaotic activity during the transition between subthreshold oscillations and burst-firing modes[3]. Therefore, chaoscillators appear to be distributed throughout local circuits within both the cortex and thalamus. As the next step toward characterizing the complex interactions between bottom-up and top-down cortical processes, the goal of this project was to construct a realistic network of neural chaoscillators based upon thalamocortical and corticocortical connections, and analyze its behavior during sensory pattern activation using the Fano-factor to describe the fractal nature of spike trains in conjunction with other measures developed for the analysis of dynamical systems.

METHODS AND RESULTS

To incorporate the reentrant dynamics found in reciprocal connectivity patterns, primary and secondary somatosensory areas were modeled using GENESIS by constructing two cortical areas, SI and SII, each consisting of 100 pairs of simplified pyramidal excitatory and stellate inhibitory cells. The component cell models were reduced reconstructions of intracellularly dye-filled neurons from rat somatosensory cortex[4] containing calcium dynamics and the currents I_{Na}, I_{Kdr}, I_{KA}, $I_{K(Ca)}$, and I_L sufficient to produce adaptive firing behaviors. The reciprocal connections between areas are asymmetrical in the sense that bottom-up projections of the pyramidal cells in SI terminate topographically on the proximal dendrites of SII pyramidal and stellate cells, while the backward top-down projections from SII pyramidal cells are widely distributed only to the distal dendrites of the SI pyramidal cells without directly activating the inhibitory cells[5]. Reciprocal thalamocortical interactions were incorporated with a network of 100 relay-reticular cell pairs with topographic projections between SI and thalamus. The thalamocortical projections excited focal clusters of cortical chaoscillators and the corticothalamic projections excited focal clusters of thalamic chaoscillators (*i.e.* both pathways engaged both the excitatory and inhibitory cells in the elementary chaoscillator circuit). The only input to the system was a constant current injection to a single thalamic neuron just sufficient to elicit action potentials.

Spike trains from the SI pyramidal cell corresponding to the thalamic neuron receiving the input were recorded for a series of trial conditions. The independent variable under investigation was the relative strength of the top-down synaptic weight (SynWt) between SII and SI. SynWt is scaled such that a value of one implies that when a pyramidal cell in SII fires an action potential, all of its target neurons also fire. A SynWt of zero reflects the no backwards connections condition. For an individual spike train fractal properties were measured using the Fano-factor. For a given time window, T, the spike train is divided into bins of length T and the number of spike events per bin is counted. The Fano-factor is then defined to be the variance in the number of counts divided by the mean of the counts[6]. As T increase a Fano-factor time curve (FTC) is generated. If the signal is fractal, the FTC grows in a power-law fashion for larger values of T. The exponential value of this power-law growth can be used to compare different spike trains and is referred to as the fractal exponent. However, the fractal exponent derived from the FTC is limited to values less than one. Since all of the data sets under consideration yielded fractal exponents close to one, there was the possibility that the true value was greater than one. To investigate spike trains whose fractal nature imply a more intricate structure, the Allan-factor is used. To calculate the Allan-factor and the corresponding Allan-factor time curve (ATC), first the Allan variance must be computed. The Allan variance is defined as the mean of the square of the difference between successive counts for a given bin size, T. The Allan-factor is then the Allan variance divided by twice the mean of the individual bin counts[7]. Fractal exponents derived from the ATC are limited to values less than three and yield values equivalent to those derived from the FTC when the exponent is less than one. Figure 2 shows the FTC and the ATC for a typical trial as a log-log plot of the respective curves as a function of the bin size, T. By plotting the FTC and ATC on a log-log scale the fractal exponent can readily be estimated by linear regression on the rightmost portion of the plot. Values of SynWt were chosen to sample the range of possible firing behaviors. With no top-down inputs (the zero SynWt condition) the activity in the SI pyramidal is periodical and the exact pattern of activity is governed by the firing rate of the input-driven thalamic

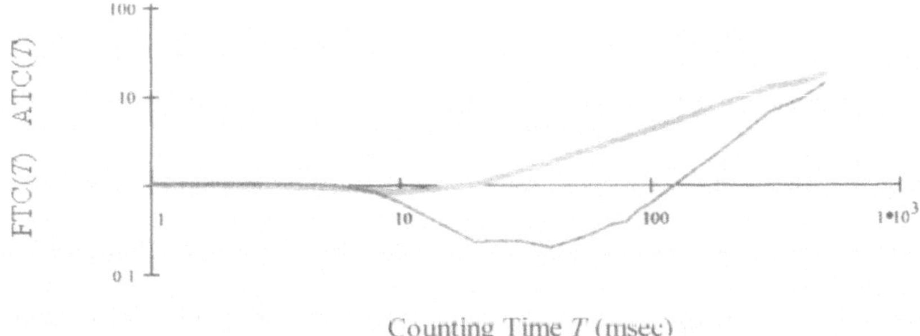

Figure 2. Log-log plot of the Fano-factor time curve (thick line) and the Allan-factor time curve (thinner line) from the spike train of a SI pyramidal neuron on a typical trial. Both curves display the characteristic power-law growth for larger counting times, indicative of fractal behavior. Notice however, that the slope of the increase in the Fano-factor is approximately unity, the saturation value for this analysis. Therefore, the Allan-factor gives a truer estimate of the fractal exponent.

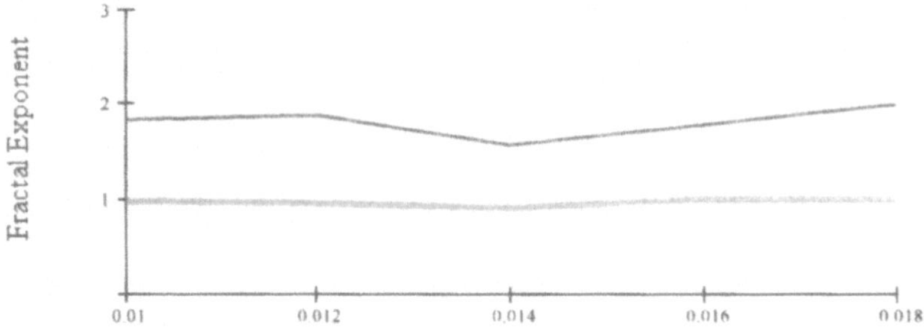

Figure 3. Plot of the fractal exponents derived from the Fano-factor time curve (thick line) and the Allan-factor time curve (thinner line) as a function of SynWt.

cell. For values of SynWt near one, all of the cells are driven at their maximum firing rate. However, for the range of tested SynWt values, fractal structure appears uniformly evident. Figure 3 depicts the range of Fano- and Allan-factor derived fractal exponent values over the tested range of SynWt.

CONCLUSION

An isolated chaoscillator pair is capable of a range of dynamic behavior. For the range 0 < SynWt < 0.2, the spike trains of the SI pyramidal cell is random when driven by random noise. For values of SynWt greater than approximately 0.5 the pyramidal cell is being driven so strongly by the SII pyramidal cell that the resulting spike trains are simply tonic firing. However, for SynWt values in-between these extremes, spike trains exhibit statistics, namely the fractal exponent derived from Fano-factor time curves, which indicate the presence of pattern correlations across time scales. That is, the spike trains are fractal in nature. Moreover, as SynWt increases over this range the magnitude of the fractal exponent increases. In contrast, a chaoscillator embedded within a network of chaoscillators appear to be more limited in its firing activity. Random activity is not seen in this model. More noticeably, the onset fractal activity occurs at much smaller values of SynWt (< 0.01) and results in fractal exponents which are more uniform in distribution and greater in magnitude (the average value is 1.8).

The principle goal of this project was to expose any differences between the isolated model and a chaoscillator circuit embedded within a realistic network. The inclusion of reciprocal connections between SI and SII and between SI and the thalamus introduced a correlated component to the ongoing background activity not seen when activity is driven by purely random inputs in the isolated model. This suggests that the SI output fractal dimension is apparently dependent upon the dynamics associates with time-scale differences defined by three chaoscillator networks: 1. Thalamocortical relay-reticular cell pairs; 2. Local cortical pyramidal-stellate cell pairs; 3. And the interareal (thalamus-SI and SI-SII) reciprocity which distributed the dynamics across the whole system. The fractal exponent indicates that the time scales relating to the local dynamics of the individual chaoscillator circuits were intricately interwoven with the slower time-scales of the coupling projections between areas.

REFERENCES

1. Cauller, L. J., Clancy, B., and Connors, B. W. (in press). Cortical origins of long horizontal axons in layer I of primary somatosensory cortex in rats. *J. Comp. Neurol.,*.
2. Jackson, M. E., Patterson, J. M. E., and Cauller, J. L. (1996). Dynamical analysis of spike trains in a simulation of reciprocally connected "chaoscillators": Dependence of spike train fractal dimension on the strength of feedback connections. In *Computational Neuroscience*, Ed. J. M. Bower, Academic Press, San Diego, 209-214.
3. Paul, K., Jackson, M. E., and Cauller, L. J. (1998). Presence of a chaotic region between subthreshold oscillations and rhythmic bursting in a simulation of interconnected thalamocortical relay and reticular neurons: Dependence of chaos on inhibitory synaptic conductance from reticular neurons. *ibid.*
4. Jackson, M. E. and Cauller, L. J. (in press). Evaluation of simplified compartmental models of reconstructed neocortical neurons for use in large-scale simulations of biological neural networks. *Brain Res. Bull.*.
5. Cauller, L. J. and Connors, B. W. (1992). Functions of very distal synapses: Experimental and computational studies of layer I synapses on neocortical pyramidal cells. In *Single Neuron Computation*, Eds. T. McKenna, J. Davis, and S. F. Zornetzer, Academic Press, San Diego, 199-229.
6. Teich, M. C. (1989). Fractal character of the auditory neural spike train. *IEEE Trans. Biomed. Engin.*, **36**(1):150-160.
7. Teich, M. C., Heneghan, C., Lowen, S. B., Ozaki, T., and Kaplan, E. (1997). Fractal character of the neural spike train in the visual system of the cat. *J. Opt. Soc. Am. A*, **14**(3):529-546.

PRESENCE OF A CHAOTIC REGION BETWEEN SUBTHRESHOLD OSCILLATIONS AND RHYTHMIC BURSTING IN A SIMULATION OF THALAMOCORTICAL RELAY AND RETICULAR NEURONS

Kush Paul, Mark Jackson and Larry J. Cauller

Cognition and Neuroscience Program
University of Texas at Dallas
Richardson, TX 75083.

INTRODUCTION

The complex interplay of ionic conductances within the thalamo-reticular network give rise to multiple modes of activity. They include the spindle oscillations in the 7-14 Hz range observed during drowsiness and the slow 0.5-4 Hz oscillations that is observed during deep sleep. This activity is a function of both the circuitry as well as the intrinsic properties of the neurons. The functional connectivity of the thalamic cells (TC) and reticular neurons (RE) are of particular importance in the synchrony and other network properties. The thalamus and the cortex have reciprocal excitatory connections and also send excitatory collaterals to the RE. The RE have inhibitory projections to TC[1,2]. The TC cells show a post-inhibitory rebound bursting property that is mediated by a low-threshold Ca^{2+} current (I_T)[3]. The voltage range of activation and inactivation of the low - threshold Ca^{2+} current (I_T) is between -80 and -50 mV. A hyperpolarizing current results in the deinactivation of the low-threshold Ca^{2+} conductance and a rebound burst is produced at the end of the hyperpolarizing pulse due to the activation of the low-threshold Ca^{2+} conductance that depolarizes the membrane potential to the threshold for the fast Na^+ spikes. The interplay of I_T and a mixed Na^+/K^+ current I_H which is activated by hyperpolarization in a subthreshold range has been shown to be essential for the intrinsic oscillations of TC neurons[2]. A similar low-threshold Ca^{2+} current (I_{TS}) is also observed in RE cells with slower kinetics. The RE cells also show intrinsic rhythmic firing properties which are mediated by a set of three currents, I_{TS}, a Ca^{2+} activated K^+ current $I_{K[Ca2+]}$ and a non-specific cation current I_{CAN}[2].

Wang,[4] showed in a model of thalamic relay neurons that the transition from subthreshold to bursting oscillation is characterized by an apparently chaotic state near the onset of rhythmic bursting. In this project we have studied the regions of apparent chaotic behavior in the genesis of rhythmic oscillations in the frequency range 0.5 - 3 Hz and 6 - 9 Hz. This analysis of the chaotic behavior was done qualitatively using phase portraits and the observations were further confirmed by quantifying the results using Lyapunov exponents. The Lyapunov exponent is a measure of how sensitive a system is to slightly different initial conditions[5]. Consider a one-dimensional system with initial condition x_0. The system is allowed to evolve over time along a particular orbit in state space. Now, an identical system is started with initial condition $x_0 + E_0$ where E_0 is very small (i.e., a small perturbation is introduced into the system). The orbits evolving from both systems are followed for a short period of time. The Lyapunov exponent is the average growth of the relative difference between the systems caused by the perturbation. Therefore, it characterizes the divergence or convergence of orbits traced by the system starting from very slightly different initial conditions[5]. Formally, the Lyapunov exponent is given by

$$\lambda = \frac{\lim}{n \to \infty} \frac{\lim}{E_0 \to 0} \frac{1}{n} \sum_{k=1}^{n} \log \left| \frac{E_k}{E_{k-1}} \right|$$

n is the number of time steps, E_0 is the is the initial "error" or perturbation, E_k and E_{k-1} are the errors at the kth and k-1th iteration.

METHODS

We have used GENESIS to simulate a network consisting of one RE and one TC cell which are mutually connected. The schematic of the model is shown in Fig.1. The existing TC cell model in the GENESIS environment was used and this model included the ionic currents I_T, I_H, I_{Na}, I_K and I_{GABA}. The RE model was created and consisted of I_{TS}, $I_{K[Ca2+]}$, I_{CAN}, I_{Na}, I_K and I_{AMPA}. All the currents were modeled by the generic Hodgkin - Huxley equation

$$I_j = \overline{g_j} m^M h^N (V - E_j)$$

where $\overline{g_j}$ is the maximal conductance, m and h are the activation and inactivation variables, V is the membrane potential and E_j is the reversal potential.

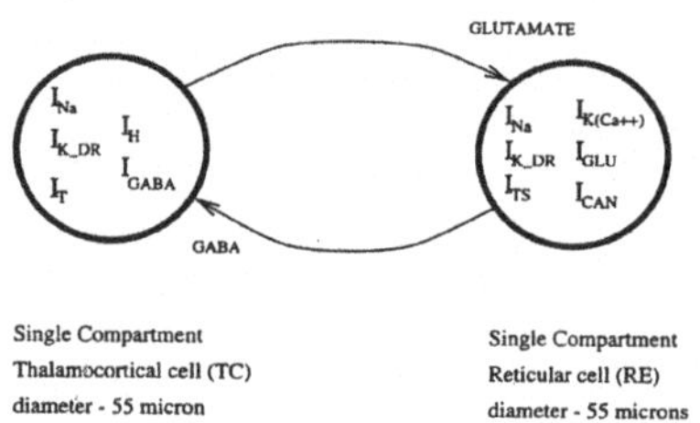

Figure.1 The reciprocal lumped soma model of the TC-RE circuit.

The expressions for the activation and inactivation variables were either directly taken or modified from published literature[6,7,8,9]. Numerous ionic currents have been observed and modeled in TC and RE cells but our objective in this project was to have a working subset of these currents such that oscillatory activity is observed and then to characterize the narrow region of transition between subthreshold oscillations and rhythmic bursting. As such, the waxing and waning nature of the spindle oscillations[6] was not simulated.

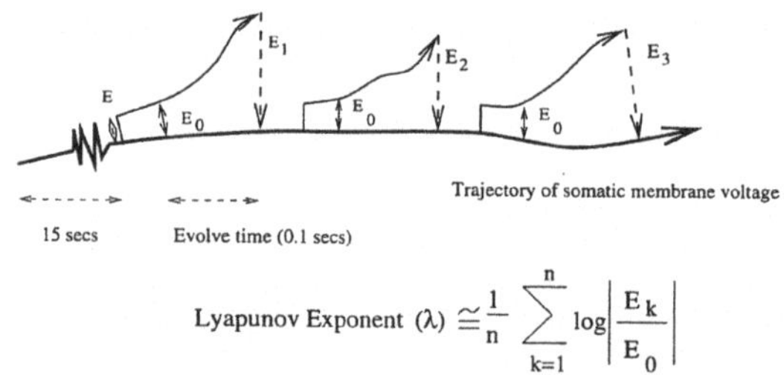

Figure.2 Algorithm for computing Lyapunov Exponents.

The method for computing Lyapunov exponents involved a direct approach which was possible within the GENESIS environment. This algorithm is shown in Fig.2. Two identical networks of the TC-RE reciprocal model was created. Both systems were run for a period of 15 secs simulation time in order to eliminate any transients. An iteration time step(dt) of 10^{-5} secs was used. After 15 secs, a perturbation of 0.005 mV was introduced into the copy of the original system. The two systems were now run for 5 time steps in order for the initial error introduced into the membrane voltage to filter through to all other parameters. The error E_0 was measured at this time and the systems were now run for an evolve time of 0.1 secs during which time the "error" was allowed to amplify. The parameter of interest was the soma membrane voltage of the TC cell. The data was collected from both the original system and the perturbed copy and the log of the absolute difference("error") was computed. All parameters of the perturbed system were reset to the values of the original system and the procedure was repeated for 500 and 1000 iterations. The average rate of divergence or convergence (i.e. the Lyapunov exponent) was then computed by the formula

$$\lambda \cong \frac{1}{n}\sum_{k=1}^{n}\log\left|\frac{E_k}{E_0}\right|$$

Care was taken to avoid the divide by zero situation. This procedure of perturb, run, collect data, reset and perturb again prevented the errors from becoming unbounded thus enabling the algorithm to detect any local "stretches" while avoiding the global "fold" . The main parameter to be explored was the GABAergic synaptic strength (InhWt) from the RE cell to the TC cell. InhWt was adjusted to take values from 0 to 1 and Lyapunov exponents were computed for each synaptic strength.

RESULTS AND DISCUSSION

Adjustment of the maximal conductance parameters along with the strength of the synaptic conductance was able to vary the mode of activity of the neurons. Fig.3c shows an example of oscillations at 6-9 Hz. The simulation was set such that the TC neuron was initially at rest at -56 mV whereas the RE neuron was intrinsically oscillating at 2-3Hz from a resting potential of -60 mV. The RE neuron entrained the TC neuron to oscillate due to post-inhibitory rebound and thereafter the reciprocal connectivity induced spindle oscillations at 6-9 Hz. A particular set of maximal conductance parameters produced 2-3Hz oscillations whereas another set of parameters produced 6-9 Hz oscillations.

Fig.3 shows the voltage time course and corresponding phase portraits of the TC cell for three different values of inhibitory connection strength parameter (InhWt) projecting from RE to TC. For InhWt = 0.01 (Fig.3a), no spiking activity is observed since with such weak inhibition from RE the low-threshold Ca^{2+} conductance is not deinactivated enough in order to give rise to a substantial low-threshold spike that will depolarize the membrane voltage to the threshold for fast Na^+ activation. The corresponding phaseplot of Ca^{2+} concentration $vs.$ Ca^{2+} inactivation variable (h) on the right hand side shows 3 periodic cycles for 25 seconds of data. In Fig.3b, the InhWt = 0.08, and the corresponding phaseplot shows infinite

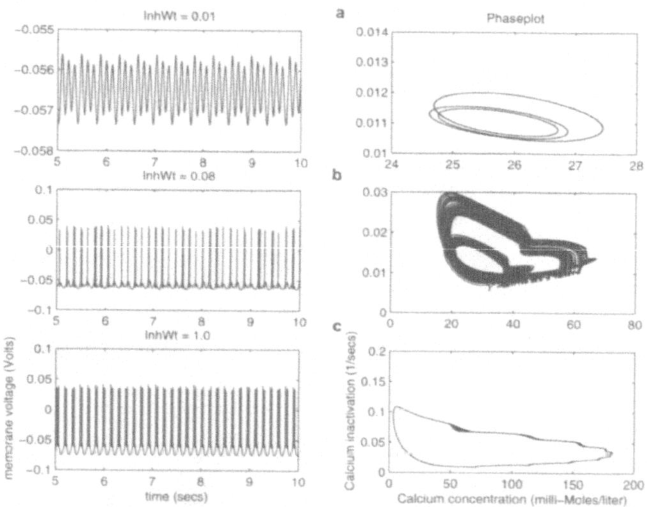

Figure.3 The three figures on the left show the membrane voltage time course of the TC cell for 5 seconds of data at three different values of the inhibitory synaptic conductance from RE cell. The topmost graph is at InhWt = 0.01, which corresponds to subthreshold oscillations. The corresponding phaseplot of the Ca^{++} concentration $vs.$ Ca^{++} inactivation (h_∞) for 20 seconds of data is shown on the right hand graph. The middle graph at InhWt = 0.08 shows an irregular aperiodic behavior suggesting a chaotic regime since this is a completely deterministic system. The corresponding phaseplot shows a rich structure of complex behavior. The bottom graph corresponds to InhWt = 1.0, at which point the strong inhibition causes regular rhythmic bursting behavior.

cycles suggesting the presence of a chaotic regime in this completely deterministic system. A definite band structure is also apparent in the phase portrait. With very strong inhibitory connection strength of InhWt = 1.0, the system moves completely into a rhythmic bursting mode of 6 - 8 Hz and the corresponding phase portrait once again reduces to a single periodic cycle while losing the richness of the intermediate connection strength value. In certain areas of the phase plot in Fig.3c some amount of spread is observed which correspond to variability in the intra-burst periods. These plots strongly suggest that in between subthreshold oscillations where I_T is not activated and the rhythmic bursting oscillations at about 7 Hz where the I_T is almost completely deinactivated first, due to strong inhibition from RE and then activated after the RE stops spiking, there exists a narrow region in the parameter space where aperiodic cycles of bursts occur in some cycles. In the Fig.3b, some cycles show single spikes, whereas others show two or more spikes. Furthermore, in some cycles no spikes occurred at all and only subthreshold oscillation was seen (not shown in Fig.3b). This is probably due to a critical deinactivation of I_T that occur in some cycles and not in others. Fig.4 shows the voltage time course and corresponding phaseplots for the TC cell when the synaptic inhibition from RE is replaced by random synaptic inhibition to TC. In this case, the spiking pattern is visually very similar to the "chaotic" condition (i.e., Fig.3b). However, the phaseplots of both Figs.4a and 4b lack the band structure observed in Fig.3b. The strength of the synaptic inhibition is identical in all three plots (i.e., InhWt = 0.08). Fig.5 shows the Lyapunov exponents computed on the membrane voltage of the TC neuron as a function of synaptic inhibition from the RE cell (InhWt). Negative Lyapunov Exponents (LE) were obtained at very small values of InhWt. At a critical point between InhWt = 0.015 and InhWt = 0.016, the LE suddenly became robustly positive. This strength of inhibition also corresponded with the occurrence of the first spikes from the TC cell. Therefore, this amount of inhibition was just sufficient under the working set of parameters to deinactivate the I_T enough in order to give rise to a LTS that would activate a single Na^+ spike. Robust positive LEs > 1 were obtained consistently for both 500 and 1000 iterations until InhWt = 0.25. For greater InhWt values, a gradual decrease occurred in LE values and negative LEs were obtained for InhWt $\geq$ 0.32. In contrast to the negative to positive LE transition which occurred suddenly at InhWt > 0.015, the positive to negative transition coinciding with greater rhythmic activity in the spindle range was gradual. It is to be noted that transition coinciding with greater rhythmic activity in the spindle range was gradual. It is to be noted that the abscissa in Fig.5 has not been drawn to scale. This is because more simulations were carried out in the regions of interest (i.e., regions were the transitions occurred).

In Figs. 6a and 6b, the inhibition from RE to TC was replaced by random synaptic inhibition at rates of 100 Hz and 8 Hz respectively. In both cases, for 0.03 < InhWt < 0.3, LEs were obtained that were close or equal to zero. This was in dramatic contrast to Fig.5, where for the same region of interest, robust positive LEs were obtained. From this we can conclude that from the point of view of the technique used to compute the LEs, our method of computing the LEs is clearly discriminating between random activity and deterministic chaos. Also, from the viewpoint of the phenomenon being observed, even though the visual observation of voltage - time plots show very similar activity, all of which appear to be random in the regions of large positive LEs, there is obviously a very different phenomenon occurring when there is weak inhibition from RE to TC as opposed to weak random inhibition to TC.

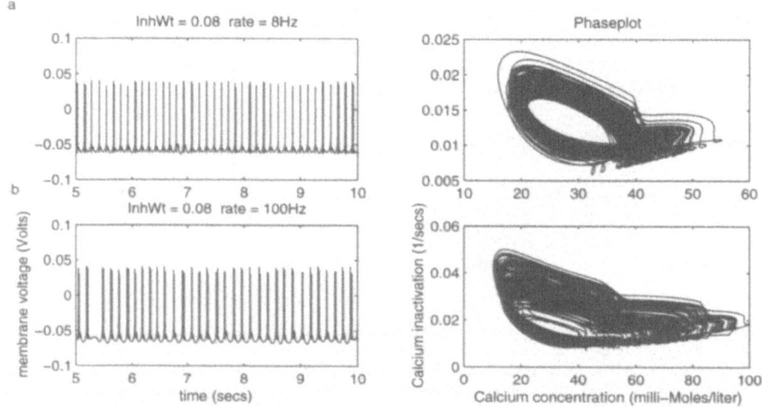

Figure.4 Inhibition to TC from RE is replaced by random inhibition at rates 8 Hz and 100 Hz. Spiking pattern is similar to "chaotic" condition (Fig.3b) however, the phaseplots do not show the band structure of Fig.3b.

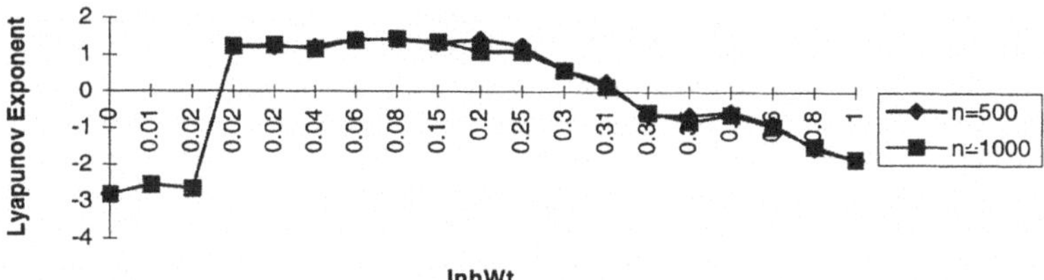

Figure.5 This graph shows the Lyapunov exponents as a function of InhWt. In the subthreshold oscillation range robust negative Lyapunov exponents were obtained. At the onset of the first spiking activity at InhWt = 0.016, the Lyapunov exponents changed drastically to positive values showing sensitivity to initial conditions and suggesting chaos. As the InhWt is increased the Lyapunov exponents gradually fall off as the behavior of the network becomes increasingly rhythmic. Beyond InhWt = 0.3, the spiking pattern is rhythmic oscillations and the Lyapunov exponents are negative.

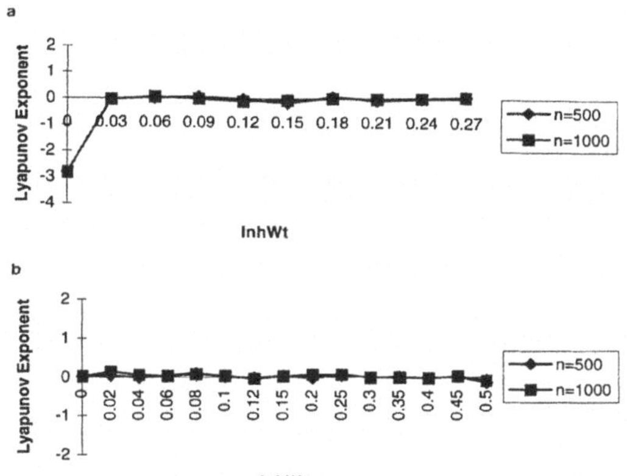

Figure.6 Lyapunov exponents are shown as a function of InhWt with a random inhibitory input of average frequency 100 Hz (Fig.6a) and 8 Hz (Fig.6b) replacing the inhibition from RE. As can be seen, the Lyapunovs average around zero for all values of InhWt in which the network showed chaotic behavior in the upper graph.

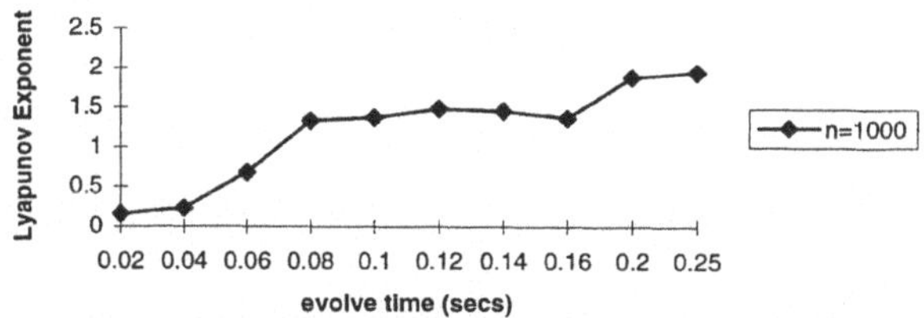

Figure.7 This graph shows the Lyapunov exponents as a function of evolve time. For InhWt = 0.08, that showed strong chaotic behavior in the evolve time = 0.1 secs, varying the evolve time causes the Lyapunov exponent to become larger. This is because in the chaotic regions with measurements taken at larger evolve times causes the copy system to diverge farther away from the original system.

Fig.7 shows the LEs as a function of the evolve time. The strength of the synaptic inhibition is fixed at InhWt = 0.08, and the evolve time is varied. Increasing the evolve time causes the LE values to become larger. This is because within the chaotic regions, measurements taken at larger evolve times causes the copy system to diverge farther away from the original system. It is expected however, that as the evolve time is increased indefinitely, the dynamical system will "fold" over in the attractor space thus giving variable (positive or negative) LEs. This simulation however, could not be carried out presently since larger evolve times take much longer times to complete the simulation.

Lyapunov Exponent analysis in a biological system is extremely rare. The most difficult condition in such a study is that the biological system remain stationary during the data collection period. One major objective of this project was to develop an alternative to the method of analyzing time - series data generated from biological systems. This alternative was to design a computational model that captures the essential properties of the biological system under study, and perform the Lyapunov Exponent analysis on this computational model that may be kept stationary indefinitely. It must be remembered that quantitive comparison of the LE values between the very few studies that have been done on biological computational models is fraught with difficulties. This is because of the various parameters in the analysis (i.e., evolve time, initial error), the choice of which is extremely subjective and depend on the specific system being studied. For e.g., in this study, the choice of the initial error of 0.005mV was due to the fact that simulation runs had indicated this value to be a good discriminator between various types of signals in our biological model. Thus, the LE values in different studies have varied widely. Canavier[9] computed Lyapunov Exponents in the range of 0.0001 - 0.0002 between the beating and bursting modes in an Aplysia cell model. In our study large LEs were obtained in the range -3 to + 2. What is important here is that the relative values of LE in this study showed strong effect of the mode of the system activity (i.e., periodic, chaotic, rhythmic bursting and random).

REFERENCES

1. G. Avanzini, M. De Curtis, F. Panzica, R. Spreafico, Intrinsic properties of nucleus reticularis thalami neurones of the rat studied *in vitro. J. Physiol.* 416:111 - 122 (1989).
2. T. Bal, D.A. McCormick, Mechanisms of oscillatory actvity in guinea-pig nucleus reticularis thalami *in vitro*: A mammalian pacemaker. *J Physiol.* 468: 669-691. (1993)
3. H. Jahnsen, R.R. Llinas, Ionic basis for the electroresponsiveness and oscillatory properties of guinea-pig thalamic neurones *in vitro. J Physiol.* 349: 227-247 (1984).
4. X.J. Wang, Multiple dynamic modes of thalamic relay neurons: rhythmic bursting and intermittent phase-locking. In *Neuroscience* 59(1): 21-31 (1994).
5. H. Peitgen, H. Jurgens, D. Saupe. *Chaos and Fractals - New Frontiers in Science.* Published by Springer- Verlag, New York, (1992).
6. A. Destexhe, D.A. McCormick, T.J. Sejnowski, A model for 8-10 Hz spindling in interconnected thalamic relay and reticularis neurons. In *Biophysical Journal.* 65 p.2473-2477 (1993).
7. J.R. Huguenard, D.A. McCormick, Simulation of currents involved in rhythmic oscillations in thalamic relay neurons. In *Journal of Neurophysiology* 68(4) p.1373 - 1383 (1992).
8. D.A. McCormick, J.R. Huguenard, A model of the electrophysiological properties of the thalamocortical relay neurons. In *Journal of Neurophysiology* 68(4) p.1384-1400 (1992).
9. G.V. Wallenstein, A model of the electrophysiological properties of Nucleus Reticularis Thalami. In *Biophysical Journal.* 66, p.978 - 988 (1994).
10. C.C. Canavier, J.W. Clark, J.H. Byrne, Routes to chaos in a model of bursting neuron. *Journal of Biophysics.* 57 : 1245-1251 (1990).

THE ROLE OF THE HIPPOCAMPUS IN THE MORRIS WATER MAZE

A. David Redish[1] and David S. Touretztky

Computer Science Department &
Center for the Neural Basis of Cognition
Carnegie Mellon University
Pittsburgh, PA 15213-3891
dredish@cs.cmu.edu, dst@cs.cmu.edu

[1] Current affiliation:
Neural Systems, Memory and Aging
Life Sciences North Bldg., Room 384, P.O. Box 24-5115
University of Arizona Tucson AZ 85724

INTRODUCTION

The Morris water maze consists of a submerged platform placed somewhere within a pool of water made opaque with milk or chalk. Normal rats quickly learn the location of the platform: if the platform is removed, the rats search at the place where the platform had been [6]. We suggest that the hippocampus plays two roles allowing rats trained on the Morris water maze to find the hidden platform: Upon re-entry into the environment, the animal must determine its location relative to the platform (*self-localization*). From this information, it can determine the direction it needs to swim in order to reach the platform. When the animal travels along a specific path, routes are stored in the recurrent connections of CA3. During sleep, the recent routes are replayed because, when primed with noise, the hippocampal formation settles to a stable representation of a location which then drifts along the routes stored in the CA3 recurrent connections (*route-replay*).

Our previous work laid out a theory of the role of the hippocampus in navigation [10]. The key components here are the following.

Path integration* occurs via a loop including the superficial layers of the entorhinal cortex (ECs), the subiculum (Sub), and the parasubiculum (PaS). Sensory cues (called the *local view*, but not solely visual) enter the hippocampal formation from high-level sensory association areas (e.g. parietal cortex). The path integration and local view representations are first combined in ECs but any conflicts are resolved by the recurrent connections in CA3.

---

*Path integration is the ability to return to a starting point, even after a long, circuitous path, using only idiothetic cues [5].

On re-entry into a familiar environment, competitive dynamics in hippocampus allows the system to settle to a coherent place code[t] even with ambiguous sensory cues. This coherent code resets the path integrator so that multiple experiences of the same environment are compatible with each other.

During sleep, recurrent connections within CA3 force a coherent code to form from noise, but due to asymmetric connection strengths produced during training, the represented location precesses along the learned routes, effectively replaying routes traversed during the task. Slowly-learning cortical networks (long term memory or LTM) can be trained by these "dreams" of routes traveled so that eventually the animal can perform the task without a hippocampus.

The theory requires the hippocampus to show two major modes of operation: a *storage* mode and a *recall* mode. The hippocampus does show two modes of activity, referred to by the very different EEG traces recorded during the two modes [2].

SELF-LOCALIZATION

In order to navigate within a familiar environment, an animal must use a consistent representation of position from session to session. Although visual cues can serve to inform the animal of its position, ambiguous cues require a mechanism to settle on a consistent representation of location. We believe intrinsic competitive dynamics force the hippocampus to settle on a coherent code.

We suggest that the competitive dynamics realized in the rodent proceeds thusly: both path integrator and hipppocampus are initially noisy; sensory cues passed through ECs into the hippocampus bias the random firing rates with candidate locations. Recurrent connections in CA3 allow one candidate location to win out by forming a coherent place code. Connections between CA1 and subiculum reset the path integrator to the correct representation of the animal's location. This happens in the course of a single sharp wave. In our simulations, the place code is coherent within 50-100 ms. Figure 1 shows the first 70 ms of a simulated sharp wave.

Simulation details are given in [11]. Two 2D neural sheets were used, one excitatory and one inhibitory, connected so as to produce local excitation and global inhibition. Figure 1 shows an example of an ambiguous local view representation resolved in the CA3 representation into a coherent representation of position.

ROUTE-LEARNING

There are four neurophysiological effects that allow the hippocampus to store routes as the animal travels: (1) **LTP is preferentially asymmetric** (see [1] for review). If neuron a fires shortly before neuron b then it is the $a \rightarrow b$ connection that is potentiated, not the $b \rightarrow a$ connection. (2) **The actual timing of spikes fired by hippocampal place cells precesses along the theta cycle** [7, 13]. Effectively, the position represented in the place code sweeps across the actual position from back to front with each theta cycle. (3) **LTP is correlational** (see [3, 4] for reviews). The increase in connection strength between two neurons a and b is proportional to the product of their firing rates F_a and F_b. (4) **Cells near the route will have higher-firing neighbors closer to the route.** A cell with a place field not centered at the location of the animal will be more strongly connected to cells that have place fields

---

[t]The place code is *coherent* if all neural activities are consistent with a representation of the same location in space.

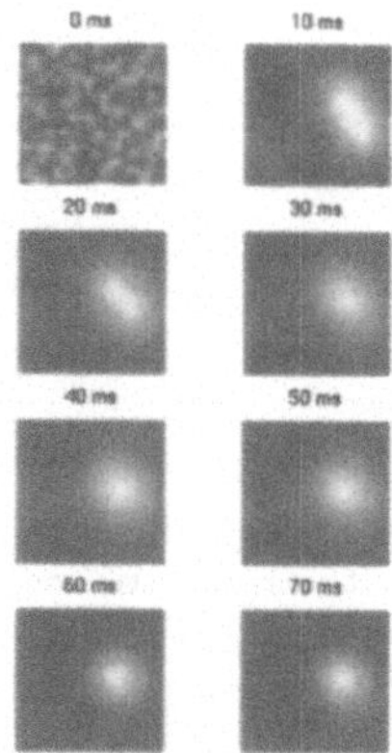

Figure 1. Starting from random noise, a coherent place code forms in less than 50 ms. Plot shows firing rates of simulated place cells. Cells are laid out in a 2D sheet with their locations in the sheet corresponding to the centers of their place fields in the environment. Intensity values have been interpolated for clarity. White indicates high firing rate, black low.

centered at the location of the animal than to cells with distant place fields. Effects 1 and 2 combine to produce asymmetries along the routes traveled; effects 3 and 4 combine to produce asymmetries leading towards those routes.

The route-learning simulation consisted of the same network as used in the previous section, with the addition of a new hippocampal mode. As before, simulation details are available in [11]. Figure 2a shows the paths traveled to reach the goal from the four cardinal points. These are the four routes that will be stored in the hippocampal simulation.

We model the asymmetric nature of LTP by making the learning rule dependent on the *synaptic drive* of the presynaptic neuron and the firing rate of the postsynaptic neuron, see [11]. Synaptic drive is the effect of a neuron on all the neurons on which it synapses divided by the synaptic weight over each synapse [9]. We do not model phase precession as an emergent result of a complex process; instead we assume that phase precession exists and show that, when combined with the asymmetric temporal nature of LTP, routes are stored in the recurrent connections of the hippocampus. These effects combine to store routes in the recurrent connections of HC. They produce a vector field pointing toward the path and then leading to the goal. Figure 2b shows the routes stored by an animal traversing the four paths.

REPLAY OF ROUTES DURING SLEEP

When there is sensory input into the hippocampus and the hippocampus is in LIA mode (i.e. the animal is awake, but not moving), sensory cues enter the system via ECs, and those CA3 cells that are consistent with the current local view will be more active than those that are not. This biases CA3 to settle to a place code that is consistent with the local view, and thus can serve as a self-localization procedure.

On the other hand, when there is no sensory input, this bias will be absent, but due to the recurrent connections in CA3, the hippocampus will still settle into a coherent activity pattern. Due to the asymmetric connections that were stored when the animal traversed the routes to the goal, the place code will precess along one of these

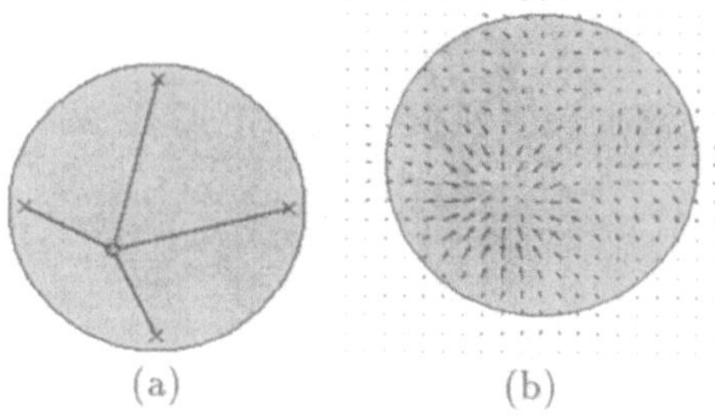

Figure 2. (a) Four routes to be stored in the asymmetric connections of the hippocampus. (b) Vector field of routes to a goal stored in the recurrent connections. For each cell j, we calculated the center of mass of the output connection weights, and plotted an arrow from the place field center towards the center of mass. Length of arrow corresponds to linearly scaled distance to center of mass of the output connection weights.

remembered routes. The bias provided by the sensory input should be enough to keep the system from precessing when awake, but in the absense of sensory input (during sleep), there is nothing to stop the precession.

During sleep, when sharp waves occur without sensory input, we expect to see replay of routes. This is shown in Figure 3. Given an initial noisy state, a coherent code forms within half a second and then over the next few seconds, it drifts along a remembered route. Data supporting a replay of recent experience in hippocampus during sleep exists [8, 12, 14].

The route replay simulations used for Figure 3 were identical to the self-localization simulation used for Figure 1, except that the weight matrix used was the asymmetric one produced by LTP in the previous section. In order to simulate "sleep," the sensory input representation was set to zero.

THE DUAL-ROLE PLAYED BY THE HIPPOCAMPUS

We have suggested here that the hippocampus plays two roles: *self-localization* and *route-replay*. These two roles can co-exist because when there are sensory signals providing candidate biases to CA3, they are enough to counter the drift provided by the asymmetric connections. In the absence of sensory input, there is nothing to stop the drift and the representation replays a recent route.

These two processes require what at first seem to be incompatible weight matrices. Self-localization requires that the recurrent connection weights in CA3 be mostly symmetric, while the precession of the replayed route occurs because of asymmetric connections. These two roles are not incompatible because the sensory input during self-localization is sufficient in our simulations to counteract the asymmetry in the weights.

Figure 4 shows the x-coordinate of the CA3/1 representation after a simulated sharp wave occured under under three conditions: **Light bars:** When the simulated animal was at five different locations in an environment, but the recurrent connection matrix between CA3 excitatory cells consisted of symmetric weights only. **Medium bars:** When the simulated animal was at the same five locations, but the CA3 recurrent connection matrix consisted of both symmetric and asymmetric weights. **Dark bars:** In the absense of sensory input, but when the representation began at each of the five locations above.

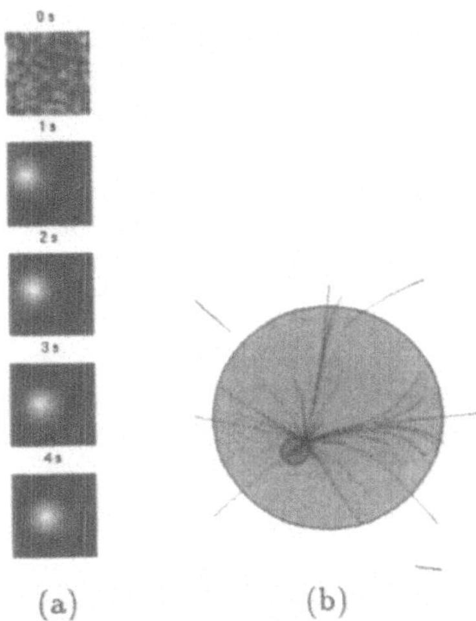

Figure 3. (a) Replay of routes during LIA without sensory input. A coherent code forms quickly and then slowly drifts to the goal over the subsequent few seconds. (b) Estimated position of the simulation during 20 5-second dreams, sampled at 100 ms intervals. Each point indicates the position represented in the model HC at each sampled time.

There is no difference at all between the results shown in the light and medium bars, indicating that the local view input is sufficient to hold the representation in place. On the other hand, when there is no sensory input to hold the representation in place, it drifts (Figure 3). Even though the representation started out at the five initial locations, in the absence of sensory input, by the end of the process it had drifted to the goal location. This is why the dark bars do not match the medium and light bars. We conclude, therefore, that the two modes are in fact compatible (at least in simulation.)

PREDICTIONS

Coherency of the place code during a sharp wave. The hypothesis that self-localization must occur on re-entry into a familiar environment predicts that rodents re-entering a familiar environment will show a sharp wave before movement, and that the place code will begin in an incoherent state and become coherent through the course

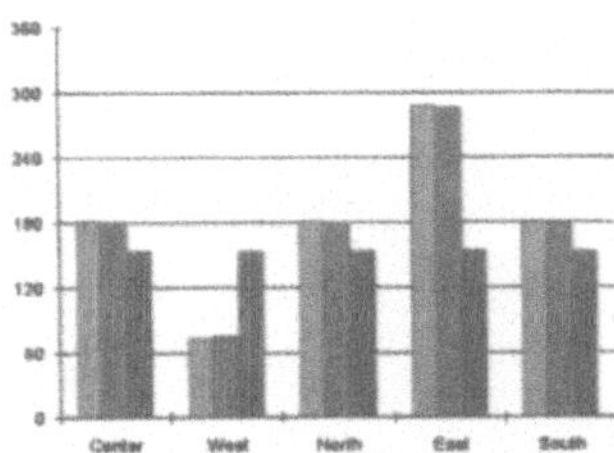

Figure 4. X-coordinate (on a toroidal map) of the final representation of the parallel relaxation process with (light bars) local view input and symmetric weights only, (medium bars) local view input and both symmetric and asymmetric weights, and (dark bars) both symmetric and asymmetric weights without local view input. Y-coordinates (not shown) are similar.

of that sharp wave. Coherency can be rigorously defined as the inverse width of the confidence interval of the location represented by the place code.

ACKNOWLEDGEMENTS

This work was supported by National Science Foundation grant IBN-9631336.

REFERENCES

1. L. F. Abbott and K. I. Blum. Functional significance of long-term potentiation for sequence learning and prediction. *Cerebral Cortex*, 6(3):406–416, 1996.

2. G. Buzsáki. Two-stage model of memory trace formation: A role for "noisy" brain states. *Neuroscience*, 31(3):551–570, 1989.

3. R. C. Malenka. LTP and LTD: Dynamic and interactive processes of synaptic plasticity. *The Neuroscientist*, 1(1):35–42, 1995.

4. B. L. McNaughton. The mechanism of expression of long-term enhancement of hippocampal synapses: Current issues and theoretical implications. *Annual Review of Physiology*, 55:375–396, 1993.

5. M. L. Mittelstaedt and H. Mittelstaedt. Homing by path integration in a mammal. *Naturwissenschaften*, 67:566–567, 1980.

6. R. G. M. Morris, P. Garrud, J. N. P. Rawlins, and J. O'Keefe. Place navigation impaired in rats with hippocampal lesions. *Nature*, 297:681–683, 1982.

7. J. O'Keefe and M. Recce. Phase relationship between hippocampal place units and the EEG theta rhythm. *Hippocampus*, 3:317–330, 1993.

8. C. Pavlides and J. Winson. Influences of hippocampal place cell firing in the awake state on the activity of these cells during subsequent sleep episodes. *Journal of Neuroscience*, 9(8):2907–2918, 1989.

9. D. J. Pinto, J. C. Brumberg, D. J. Simons, and G. B. Ermentrout. A quantitative population model of whisker barrels: Re-examining the Wilson-Cowan equations. *Journal of Computational Neuroscience*, 3(3):247–264, 1996.

10. A. D. Redish. *Beyond the Cognitive Map: Contributions to a Computational Neuroscience Theory of Rodent Navigation*. PhD thesis, Carnegie Mellon University, Pittsburgh PA, 1997.

11. A. D. Redish and D. S. Touretzky. The role of the hippocampus in solving the Morris water maze. *Neural Computation*, 10(1), 1998. In press.

12. W. E. Skaggs and B. L. McNaughton. Replay of neuronal firing sequences in rat hippocampus during sleep following spatial experience. *Science*, 271:1870–1873, 1996.

13. W. E. Skaggs, B. L. McNaughton, M. A. Wilson, and C. A. Barnes. Theta phase precession in hippocampal neuronal populations and the compression of temporal sequences. *Hippocampus*, 6(2):149–173, 1996.

14. M. A. Wilson and B. L. McNaughton. Reactivation of hippocampal ensemble memories during sleep. *Science*, 265:676–679, 1994.

A STATE SPACE MODEL OF GERBIL COCHLEA

Bilin Zhang Stiber,[*,1] Edwin R. Lewis,[1] and Kenneth R. Henry[2]

[1]Dept of Electrical Engineering & Computer Science
145 Cory Hall, University of California
Berkeley, CA 94720
{*bilin,lewis*} *@eecs.berkeley.edu*
[2]Dept. of Psychology, Univ. of California
Davis, CA 95616

INTRODUCTION

A mammal's cochlea appears to be an array of band-pass filters with steep high-frequency band edges and overlapping pass bands, each covering a slightly different range of frequencies and each serving as a separate signal-processing channel. With the channels ordered according to their pass band peaks, the system's response to a stimulus can be visualized as a three-dimensional spectrograph (the response versus time over an ordered array of filters), which contains the information passed to the nuclei of the auditory brainstem.

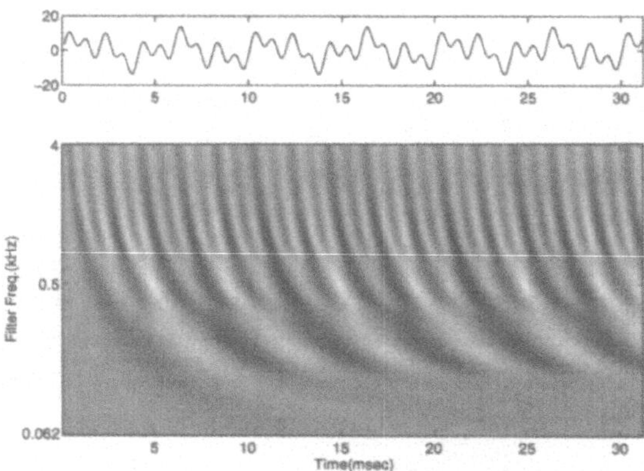

Figure 1. Low pass filter based artificial cochlea response to a three-tone signal (200, 500, 1000Hz).

An ordered array of low-pass filters with steep band edges occasionally has been taken to be an idealized model of such a system[1]. The lower frame in Figure 1 shows the spectrograph

---

[*]To whom correspondence should be addressed: c/o Dr. Michael Stiber, Computing and Software Systems, Univ. of Washington, Bothell, 22011 26th Ave. SE, Bothell, WA 98021-4900.

created by such a model in response to the three-tone signal (200, 500, 1000 Hz) in the upper frame. Each horizontal line in the spectrograph corresponds to one filter in the array, and shows the result (as gray level) of convolving the waveform in the upper frame with the impulse response of that filter. The vertical axis in the spectrograph is the center frequency of the corresponding filter.

In this paper we present a procedure for producing a more realistic cochlear model— replacing the impulse responses in the model of Figure 1 with impulse responses estimated by reverse correlation applied to gerbil cochlear axons.

GERBIL COCHLEA MODEL
REVCOR Data Collection

The data collection methods were described by Lewis and Henry[2]. Briefly, band-limited Gaussian white noise was presented and spike trains were recorded from individual fibers in the auditory nerve. A fixed (rms) noise amplitude was set for all data. Over 600 different datasets were obtained. The impulse response of each fiber was estimated by reverse-correlation between its spike train and the noise stimulus that generated it[3]. Because reverse correlation is a noisy process, the resulting estimate (often called the REVCOR function) is noisy.

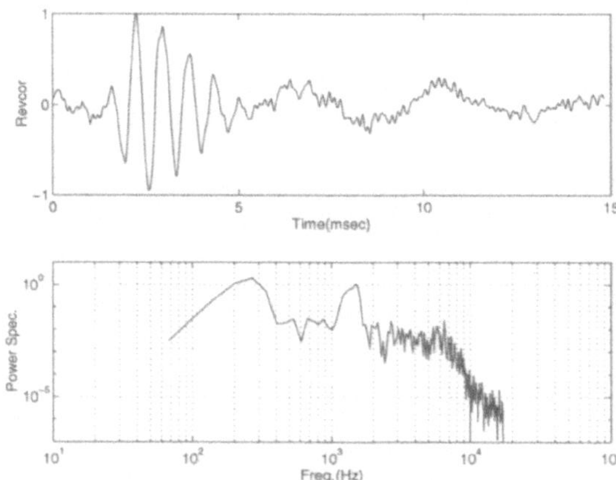

Figure 2. A Gerbil REVCOR function (top) and its spectrum with CF around 1.5kHz (Bottom).

Figure 2 shows one example of a REVCOR function (top) and its amplitude tuning curve or spectrum with a characteristic frequency (CF) of 1.5kHz (bottom). The noise in this waveform includes considerable energy above 2kHz and a slow wave attributable to an audible heartbeat in the animal's ear canal.

Noise Elimination and State Space Model

Ideally, a bank of clean filters is desirable for the model. To eliminate major noise, we first applied a Butterworth band-pass filter to the REVCOR function. The filtered REVCOR function of Figure 2, can be seen in Figure 3. Note the low and high noise frequency ranges mentioned previously; the result is a good primary output for further smoothing.

In the next step, a filtered REVCOR function is converted to a state-space model (SSM) by a minimum-error fitting routine[4]. The purpose of this procedure is further smoothing. A

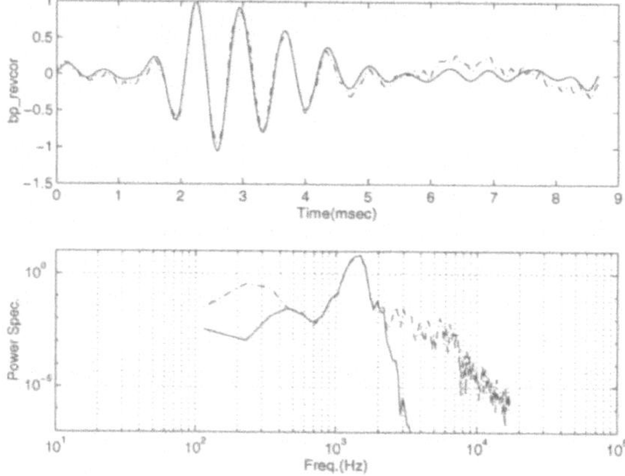

Figure 3. REVCOR function of Figure 2 (dashed line) along with the filtered version of the same function (solid line).

SSM is a dynamical model of a linear time-invariant system. Briefly, its representation is,

$$\begin{cases} x(t+1) = Ax(t) + Bu(t) \\ \quad y(t) = Cx(t) + Du(t) \end{cases} \tag{1}$$

where $x(t)$ is state vector, $u(t)$ is input, $y(t)$ is output and A, B, C, D are matrices.

Continuing with the same data as in Figures 2 and 3, a SSM-fitted REVCOR function is shown in Figure 4. The dash-dotted line is the original data, the dashed line is the filtered REVCOR function from in Figure 3, and the solid line is the result of SSM fitting. We can see that the SSM captures the major features of the fiber's response, while eliminating fluctuations.

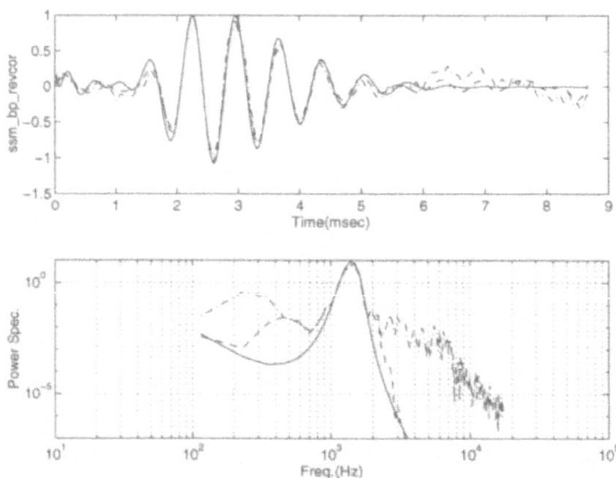

Figure 4. The original Gerbil REVCOR function (dash-dotted line), the filtered version (dashed line) and the result of SSM fitting (solid line). Top: REVCOR functions, Bottom: their amplitude spectra.

Generating a Filter Bank

By repeating this process of bandpass and SSM for REVCOR functions from many fibers, we generated a filter bank to serve as a model of cochlear signal processing. This was done for over 600 fibers with CFs ranging from 450Hz to 5.3kHz

In the bank, we make an array of processed filters with center frequencies distributed logarithmically, based on experimental evidence from living preparations. In other words, we generated log_2 based frequency bins (*fbins*) for a histogram

$$fbin_i = 2^{\{[log_2(ef)-log_2(sf)]/nbin\}*i+log_2(sf)} \qquad (2)$$

where $i = 1, 2, ..., nbin$, $sf=$ 450Hz and $ef=$ 5.3kHz are the start and end frequency of the entire filter bank and $nbin=$ 10 is the number of bins in the histogram. We equalize counts in each $fbin_i$ by randomly choosing filters to meet the requirement of smoothly and logarithmically rising frequencies. The histogram is plotted in Figure 5 and the path of center frequencies along filter number is shown in Figure 6, where the theoretical result (solid line) accords to equation (2) and the marks '*' and 'o' respectively indicate *fbin* and actual data. The experimental data, which formed the filter bank or gerbil cochlea model, followed well a logarithmic arrangement.

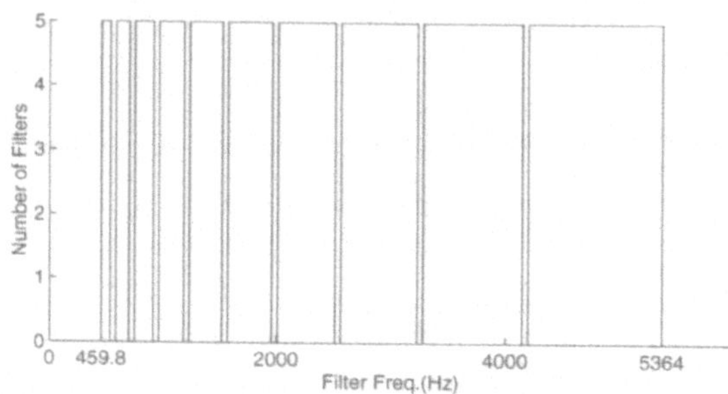

Figure 5. Histogram of filters; bin widths increase logarithmically with center frequency.

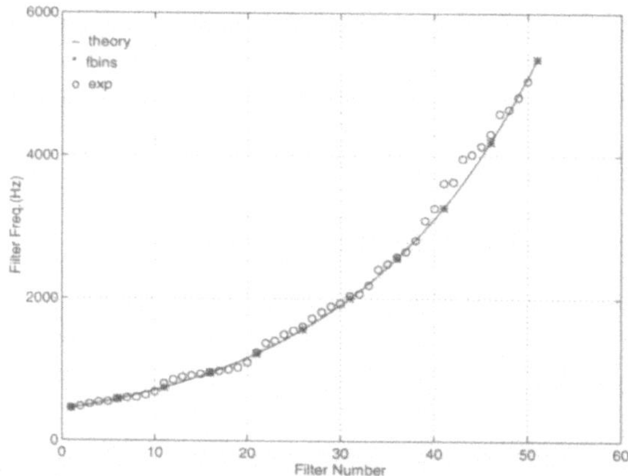

Figure 6. Center frequency versus filter number. The solid line is theoretical result; the mark '*' is $fbin$ used in histogram Figure 5; the mark 'o' is the experimental data used to build the model.

The impulse response of the resulting gerbil cochlea model is shown in Figure 7 in a three-dimensional spectrograph with axes of time, filter center frequency, and amplitude of response. The lack of coherence between responses of neighboring filters is conspicuous. This very likely reflects, in part, the effect of sampling over a heterogenous population of filter types in individual subjects and, in part, the effect of sampling over many individuals.

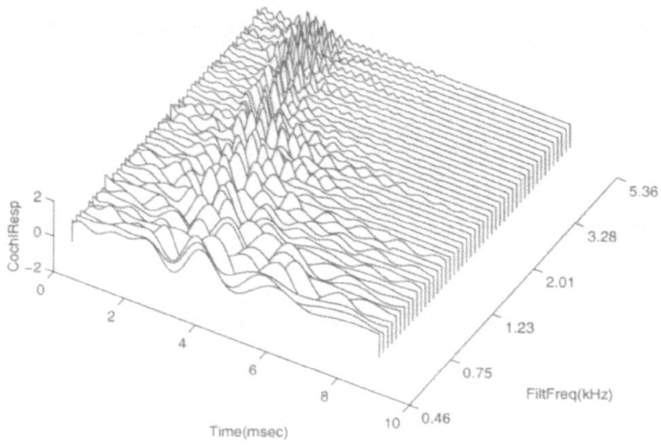

Figure 7. Impulse response of the gerbil cochlea model

RESPONSE OF THE GERBIL COCHLEA MODEL

The first example is the model response to a dual-tone signal with frequency components at 1kHz and 2kHz and is shown in a two-dimensional plot; the horizontal axis is time and the vertical is filter frequency (Figure 8, bottom). Gray level indicates the model's amplitude response with darker indicating higher amplitude. The top plot in the figure is the dual-tone stimulus. The maximal responses appear around each of these frequencies.

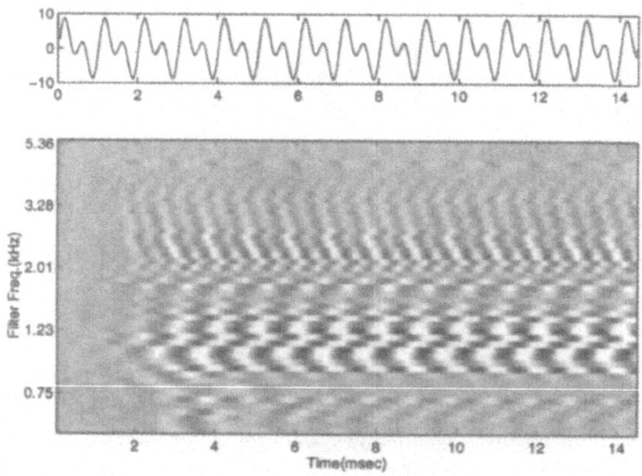

Figure 8. The gerbil cochlea model response to a dual-tone (1kHz, 2kHz) signal (top), plotted in two dimensions spectrograph (bottom).

In the next example, a natural sound (frog call) stimulus was used to investigate the model's response. Figure 9 clearly shows that the model responds to the input signal's temporal variations and also to its corresponding frequency features. This can be seen in the strong responses around 0-120msec and 220-360msec and 1.2-2kHz and 2-4.5kHz. This provides us with a first, crude estimate of the image of the sound that would be sent to the gerbil auditory brainstem nuclei.

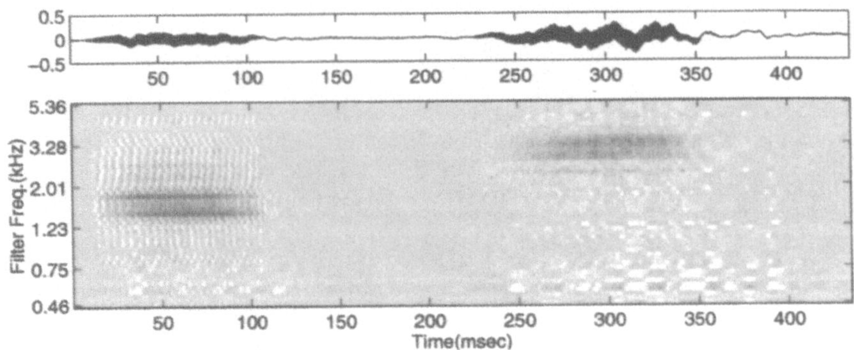

Figure 9. The gerbil cochlea model response to a frog call (top), plotted in two dimensions spectrograph (bottom).

CONCLUSION

In this paper we introduced a method for generating a cochlea model from experimental data. The result produces what might be called a cochlear spectrograph of sounds.

In most cases, SSM proved to be a good method for smoothing REVCOR functions. The lack of coherence between responses of neighboring channels in our model, however, suggests that one must consider the effects and implications of heterogeneity in one's cochlear nerve fiber samples. We are presently doing this. We have considerable faith in the REVCOR process, however, because in fiber after fiber it has been shown that the estimated impulse response consistently is an excellent predictor of the time courses of the instantaneous spike rate in response to novel stimuli of arbitrary complexity (at the same power level as that used to obtain the REVCOR function) [5].

ACKNOWLEDGEMENT

The authors would like to thank Eva Poinar for assistance in data analysis and Dr. Michael Stiber for his valuable advice.

This work is supported by NIH grant number DC-00112 to E.R.Lewis

REFERENCES

1. B. R. Parnas, Studies with a neuronal modeling system for the mammalian auditory pathway, *PhD thesis*, Electrical Engineering and Computer Science, Univ. of Calif., Berkeley (1992).
2. E. R. Lewis and K. R. Henry, Cochlear nerve responses to waveform singularities and envelope corners, *Hear. Res.*, vol. 39, pp. 209–24 (1989).
3. E. de Boer and H. R. de Jongh, On cochlear modeling: Potentialities and limitations of the reverse-correlation technique, *J. Acoust. Soc. Am.*, vol. 63, pp. 115–35 (1978).
4. G. Wolodkin, System identification methods for a class structured nonlinear systems, *PhD thesis*, Electrical Engineering and Computer Science, Univ. of Calif., Berkeley (1996).
5. W. M. Yamada, Second order Wiener kernel analysis of auditory afferent axons of the North American bullfrog and Mongolian gerbil responding to noise, *PhD thesis*, Neurobiology Graduate Group, Univ. of Calif., Berkeley (1997).

RANK ORDER CODING

Simon Thorpe and Jacques Gautrais

Centre de Recherche Cerveau et Cognition (UMR 5549)
Faculté de Médecine de Rangueil
133 route de Narbonne
31062 Toulouse, France

INTRODUCTION

The idea that neurones transmit information using a rate code is extremely entrenched in the neuroscience community. The vast majority of neurophysiological studies simply describe neural responses in terms of firing rate, and while studies using Peri-Stimulus Time Histograms (PSTHs) are fairly common, only rarely does one get to see the underlying spikes in the form of a raster display. Even rarer are studies that provide information about how spikes are generated across a population of neurones.

One consequence of this strong bias is that many alternative coding schemes, and particularly those involving patterns of activity distributed across populations of neurones, have simply not been considered seriously. It is now virtually 30 years since the publication of Perkel and Bullocks' review of Neural Coding in which a whole range of candidate coding schemes were discussed[1]. Few of these various candidate codes have been disproved experimentally. Even today, when increasing numbers of researchers are interested in the potential of temporal coding schemes and in particular the role played by synchrony[2,3], few question the underlying assumption that this synchrony is imposed on an underlying rate code.

PROBLEMS WITH RATE CODING

Although using rate coding to encode analog values seems reasonable, problems arise when one tries to implement such coding in networks of biologically realistic neurons, because one is forced to deal with the fact that real neurons generate spikes, not floating point numbers! As we have argued in more detail elsewhere[4,5], the assumption that rate coding involves a pseudo-Poisson spike generation mechanism (as is often the case), is difficult to reconcile with the extreme rapidity of processing in sensory pathways, where it appears that sophisticated processing can be performed on the basis of only 10 ms of activity at each stage [6-8]. To take a simple example, suppose that the retina contains two pools of neurones, A and B, firing at rates of 100 spikes.s^{-1} and 75 spikes.s^{-1}, and that we need to know which is which. Suppose that spike generation is Poisson and that we need to make a decision on the basis of 10 ms of activity with an error rate of no more than 5%. How many cells would we need in each pool? A simple calculation[5] reveals that we would need no less than 76 neurones in each pool. Note that this is a situation where we are not even able to recover 1 bit of information about the stimulus.

TEMPORAL CODING

Recent data indicates that the initial transient response of a neuron may in fact be temporally quite precise and reliable [9]. This opens up possibilities for alternative coding schemes that depend on the timing of spikes [10-12]. One such alternative considers a neuron not so much as an *analog-to-frequency* converter, but rather as an *analog-to-delay* converter. The idea is very simple and is based on the fact that the time taken for an integrate-and-fire neuron to reach threshold depends on the strength of the input (see figure 1A). This would lead to the sort of latency-intensity function illustrated in figure 1B. Consider now what would happen with several receptor cells and an intensity profile as illustrated in figure 1C. Because of the intensity-delay transformation, the neurons will tend to generate spikes in an order which reflects the input distribution.

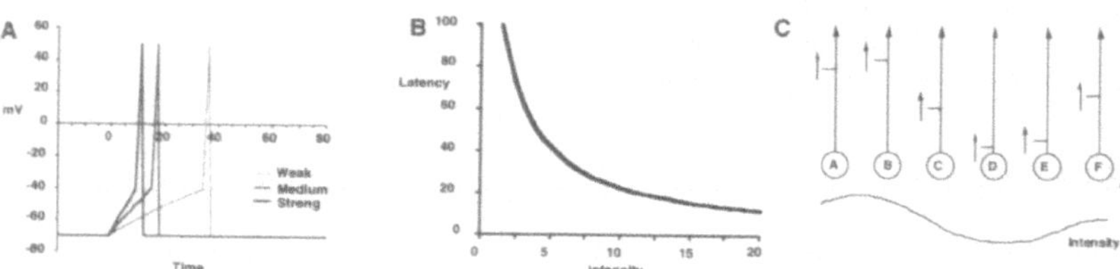

Figure 1. A. The response of a leaky integrate-and-fire neuron to inputs of different strengths. B. A typical intensity-latency response function. C. Generation of an asynchronous spike wave in a population of receptors.

Given this asynchrony in spike generation, there are actually various ways that one could develop a coding scheme. One would be to use the relative timings of spikes across the different cells [13]. This certainly allows a large amount of information to be transmitted within a very short time. Suppose that we have N neurons, and that precise arrival times of spikes can be measured to the nearest millisecond. Within a time window of T milliseconds one can transmit a maximum of $\log_2(T^N)$ bits of information. Thus with 6 neurons, one can transmit up to $\log_2(10^6) \approx 20$ bits of information. Of course, in order to be useable, this information has to be decoded, and although it is possible to conceive of biological circuits capable of responding selectively to particular temporal patterns across a set of inputs, it is likely that this would require a great deal of specialized hardware. Although sensory systems often rely on timing differences between spikes coming from different sources[14], in general, these tend to involve only two sources as in the case of calculating interaural delay times or the timing differences between electrical signals arriving on the left and right hand sides of the body in electric fish [15]. Even these require very large amounts of specialized neural hardware. As far as we are aware, there is little or no direct evidence for neural circuits sensitive to delays arising from several sources at once.

RANK ORDER CODING

However, it appears that a simpler coding scheme can be used which also uses the relative timing of spikes across a population of cells, but which is considerably easier to compute. The idea is to throw away the precise timing information, and use a code that depends only on the order in which the spikes arrive. Returning to the six neurons in figure 2, we can now code the stimulus by the ordering B > A > F > C > E > D. Since there are factorial N possible orderings, a rank order code can transmit up to $\log_2(N!)$ bits of information under conditions where each input neuron can only emit one spike.

Information capacity

Figure 2A compares the maximum amount of information that can be transmitted as a function of the number of input neurons for a range of candidate codes. We have included a simple binary code in which the state of each input is either 0 or 1. Here the maximum information transfer is $\log_2(2^N) = N$ bits. We have also included values for a rate code in which the only information that is available comes from the number of neurons that have fired. In that case, the maximum transfer is $\log_2(N+1)$ bits ($N+1$ because we include the case where none of the units fires). From this graph it is clear that the two codes that rely on spike timing have much greater potential, and that although the timing code is clearly the most powerful, rank order coding is nevertheless a good second best.

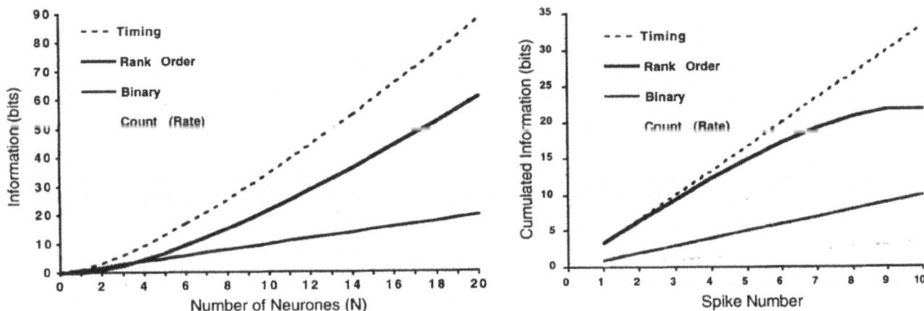

Figure 2. Left. Maximum information transfer for four types of coding schemes as a function of the number of inputs (N) and on the assumption that each input can generate a maximum of one spike. Right. Information accumulation over time for four different codes assuming 10 inputs, a temporal window of 10 ms, and a temporal precision of 1 ms.

Figure 2B compares the way in which information accumulates over time for each of the four codes, under the assumption that we have 10 input neurons, a 10 ms time window, and a temporal resolution of 1 ms. After one spike has arrived, the rate (count) code can provide only a tiny amount of information, because only 1 of the N+1 possibilities has been eliminated, namely, the situation where the final count would have been zero. In contrast, the rank order code can already eliminate 90% of the 10! (i.e. 3628800) possible orderings, which means that the very first spike can transmit up to $\log_2(N)$ bits of information. It can be seen that with rank order coding the amount of information provided by each spike drops progressively, until at the end, the final spike provides no information at all. Thus although the absolute timing code can ultimately provide more information ($\log_2(10^{10}) \approx 33$ bits) than a rank order code ($\log_2(10!) \approx 22$ bits), this difference only becomes apparent towards the end of the propagation sequence. Thus, when decisions need to be made rapidly, a rank order code can be almost as good as a code in which the precise spike timings are preserved.

Invariance

A further advantage of rank order coding is its invariance to changes in both input intensity and contrast. Figure 3 illustrates the fact that although changing either the intensity or the contrast of the input pattern will change the latencies of the neurons, there will be no change in the rank ordering of the units. This automatic normalization of inputs is a further advantage of rank order coding that is difficult if not impossible to obtain using other coding schemes. In this respect it is worth noting that using rank orders rather than, for example, firing rates or precise latencies, confers exactly the same advantages that non-parametric statistics have when compared with parametric statistical tests, namely, the fact that no hypotheses need to be made concerning the nature of the input distributions.

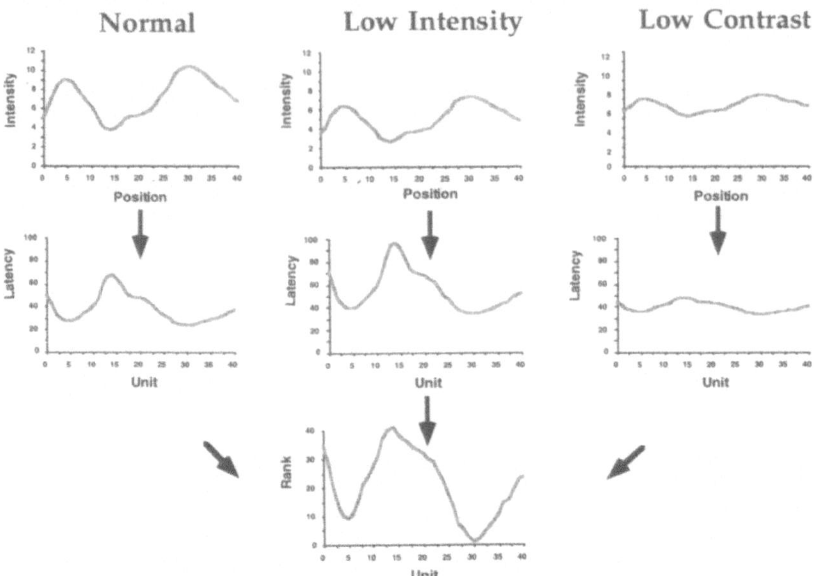

Figure 3. Invariance of Rank Order Coding. The three graphs on the left illustrate how an input intensity profile can be converted first into a latency profile across 40 units, and then into a rank profile. The center and right-hand panels show that changes in either intensity (center) or contrast (right) will have no effect on the rank ordering of the units, although latencies will change.

RANK ORDER DECODING

Rank order coding would be of no use if there was no practical way of decoding the information. There is, however, a very simple way of making neurons sensitive to the order in which their inputs fire and which does not depend on complicated delay lines. The idea takes a simple integrate and fire neuron with inputs that have differing synaptic weights and adds a mechanism which makes the neuron increasingly less sensitive over time as a function of the number of spikes that have already arrived. This could be done by a variety of mechanisms, including some form of feed-forward shunting inhibition. But whatever the precise mechanism, rank order decoding requires that the effect of activating a synapse depends not just on its synaptic weight, but also on a general sensitivity term which gets progressively smaller over time. More precisely, let $A = \{ a_1 , a_2 , a_3 \ldots a_{m-1} , a_m \}$ be the ensemble of afferent neurons of neuron i, with $W = \{ w_{1,i} , w_{2,i} , w_{3,i} \ldots w_{m-1,i} , w_{m,i} \}$ the weights of the m corresponding connections; let mod be an arbitrary modulation factor between 0 and 1. The activation level of neuron i at time t is given by

$$\text{Activation}(i,t) = \sum_{j \in [1,m]} \text{mod}^{\text{order}(aj)} \, w_{j,i}$$

where order(a_j) is the firing rank of neuron a_j in the ensemble A. By convention, order(a_j) $= +\infty$ if neuron a_j has not fired at time t, setting the corresponding term in the above sum to zero.

Figure 4 shows how a the activation level of such a neuron would change as a function of the order in which its 16 inputs (A to P) are activated. The synaptic weights are distributed linearly with input A having a weight of 16 arbitrary units, input B a weight of 15 units and so on down to input P which has a weight of 1. The value for mod was set to 0.85 so that the first input to fire produces an activation equal to the synaptic weight, the second to the weight*0.85, the third equal to the weight*0.85^2 and so on. The upper curve illustrates how the neuron would react to the inputs firing in alphabetical order. In that case, the final activation value is maximal (71.7 units). The bottom curve gives the worst possible result, i.e. when the order of activation is PONMLKJIHGFEDCBA. The other lines correspond to a variety of random sequences taken from the 16! (i.e. over 2 x 10^{13}) possible orderings of 16 inputs, none of which produce activation values close to the optimal. In fact, both mathematical analysis and

Monte Carlo simulations can be used to demonstrate that the distribution of final activation values is very close to a normal distribution, with the mean value situated exactly halfway between the minimum and maximum values. This is illustrated by the dotted lines which indicate the percentage of random sequences that will be expected to exceed a certain final activation value. It can be seen that only 10% of all possible orderings will produce values above 59.0 units, and only 1% will go beyond 64.0 units. The low variance of this distribution means that such a neuron is actually extremely sensitive to the order of its inputs, so much so that if one was to set a threshold for firing at 69.0, it would only respond to a random input about 0.001% of the time!!

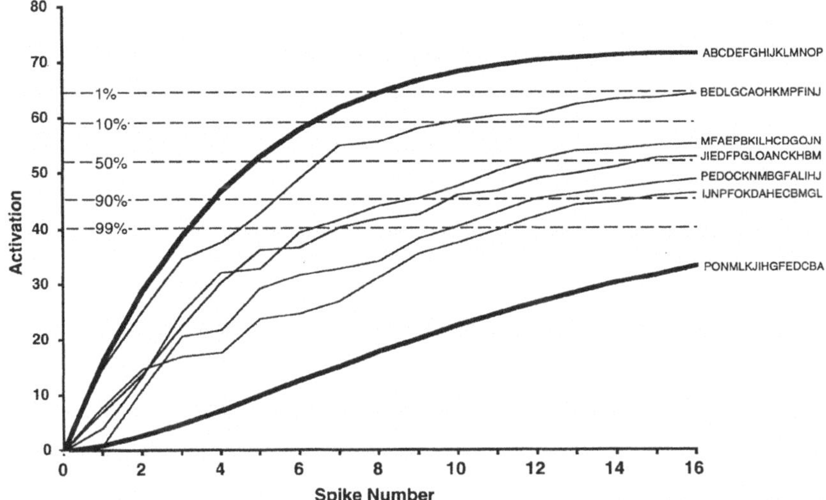

Figure 4. Response of a rank order-sensitive neuron with 16 inputs to variations in the order in which its inputs fire. The upper solid line shows how its activation increases as spikes arrive when the order of arrival exactly matches the weights of the inputs. The lower solid line shows the response to the exact opposite order. Intermediate lines give examples of the response to 5 randomly selected input patterns. The five dotted lines specify the proportion of random patterns that will exceed a given final activation value.

Another interesting feature of the Rank Order decoding mechanism proposed here is that by adjusting the threshold level, the output cell will fire at different times depending on how well the input pattern matches the weights. If we set the threshold at around the 50% level in figure 4, the neuron would fire after only five inputs have been activated if the order was perfect (i.e. ABCDE). With random inputs, firing when only 5 inputs have been activated would only occur roughly 0.01% of the time, and the few sequences that do work would be very close indeed to optimal. As more and more spikes arrive, the proportion of patterns capable of resulting in a spike increase, from 4.4% after 6 inputs have fired, 14% after 8 have fired and 50% when all 16 inputs have fired.

Note that varying the threshold provides a simple way of adjusting the speed-accuracy trade-off of such a mechanism. With a high threshold (e.g. corresponding to the 1% line), the mechanism is very selective, but will take a relatively long time to respond. In contrast, with a low threshold (e.g. the 99% line), the neuron will respond to virtually all possible inputs but it will do so much more quickly in the case of a "good" pattern (e.g. "ABCD..."). Note that this ordering information could be reused again by the next stage of processing. Imagine what would happen with a whole array of neurons each receiving inputs from the same 16 inputs, but where each cell has a different pattern of weights. Even with the threshold set so low that essentially all the cells fire to virtually all the input patterns, a cell in the next layer could still recover information from the order in which the cells fire. To put it another way, the rank order coding scheme that we propose is cascadable.

DEVELOPMENTS

Rank Order Coding appears to have considerable potential. We have developed a neural network simulator called SPIKENET, which has been designed to simulate large networks of relatively simple integrate-and-fire neurons[16] [4]. The simulator incorporates the rank order decoding mechanism just described which makes the neurons sensitive to the order in which their inputs fire. With Rufin van Rullen and Arnaud Delorme, we have used it to construct simple feed-forward networks capable of performing challenging tasks such as the detection and localization of faces in natural images[17]. Strikingly, this was possible in a network where the cells in the input layer each fire just one spike, effectively ruling out any possibility of rate encoding at the single cell level, and demonstrating that rank order coding is a viable alternative coding scheme which needs to be taken seriously.

REFERENCES

1.   D. H. Perkel and T. H. Bullock, Neural Coding,*Neuroscience Research Program Bulletin* 63:221-348 (1968).
2.   W. Singer and C. M. Gray, Visual feature integration and the temporal correlation hypothesis,*Annu Rev Neurosci* 18:555-586 (1995).
3.   M. Abeles, *Corticonics. Neural Circuits of the Cerebral Cortex.* Cambridge University Press, Cambridge, (1991).
4.   J. Gautrais, Théorie et simulations d'un nouveau type de codage impulsionnel pour le traitement visuel rapide: Le codage par l'ordre d'activation,*University Thesis, Ecole des Hautes Etudes en Sciences Sociales, December* (1997).
5.   J. Gautrais and S. J. Thorpe, Rate coding vs temporal order coding: A theoretical approach,*Biosystems* submitted (1997).
6.   S. Thorpe, D. Fize and C. Marlot, Speed of processing in the human visual system,*Nature* 381:520-522 (1996).
7.   S. J. Thorpe and M. Imbert, Biological constraints on connectionist models., in :*Connectionism in Perspective.* R. Pfeifer, Z. Schreter, F. Fogelman-Soulié, L. Steels, eds. Elsevier, Amsterdam, (1989) pp. 63-92.
8.   M. Fabre-Thorpe, G. Richard and S. Thorpe, Rapid categorization of natural images by rhesus monkeys,*NeuroReport* 9:in press (1998).
9.   Z. F. Mainen and T. J. Sejnowski, Reliability of spike timing in neocortical neurons,*Science* 268:1503-6 (1995).
10.   F. Rieke, D. Warland, R. Ruyter van Steveninck and W. Bialek, *Spikes: Exploring the neural code.* MIT Press, Cambridge, MA, (1997).
11.   J. J. Hopfield, Pattern recognition computation using action potential timing for stimulus representation,*Nature* 376:33-36 (1995).
12.   W. Maass, Fast sigmoidal networks via spiking neurons,*Neural Comput* 9:279-304 (1997).
13.   S. J. Thorpe, Spike arrival times: A highly efficient coding scheme for neural networks., in :*Parallel processing in neural systems* R. Eckmiller, G. Hartman, G. Hauske, eds. Elsevier, North-Holland, (1990) pp. 91-94.
14.   C. E. Carr, Processing of temporal information in the brain,*Annu Rev Neurosci* 16:223-243 (1993).
15.   M. Kawasaki, G. Rose and W. Heiligenberg, Temporal hyperacuity in single neurons of electric fish. ,*Nature* 336:173-176 (1988).
16.   S. J. Thorpe and J. Gautrais, Rapid visual processing using spike asynchrony, in :*Advances in Neural Information Processing Systems 9* M. C. Mozer, M. I. Jordan, T. Petsche, eds. MIT Press, Cambridge, (1997) pp. 901-907.
17.   R. van Rullen, J. Gautrais, A. Delorme and S. Thorpe, Face processing using one spike per neurone,*Biosystems, submitted* (1997).

NEUROMODULATION OF HIPPOCAMPAL POPULATION CODING: PLACE FIELD DEVELOPMENT AND PHASE PRECESSION

Gene V. Wallenstein and Michael E. Hasselmo

Department of Psychology, Program in Neuroscience
Harvard University, Cambridge, MA 02138

INTRODUCTION

It has been shown repeatedly that a broad class of theoretical models known as "associative memory networks" have the ability to learn and recall static patterns of information stored across their spatial extent.[1,2] Hebbian-type learning in conjunction with a pronounced anatomical background of recurrent excitatory synapses has served as a foundation for many models of associative memory in hippocampal region CA3.[3-7] A problem becomes apparent, however, when one attempts to store and recall *temporal* patterns of information in associative networks using realistic approximations of long-term potentiation (LTP). Experimental observations have shown that LTP of a synapse occurs preferentially when a pre-synaptic cell fires prior to post-synaptic cell activation within a time window of approximately 50 milliseconds.[8,9] How, then, are temporal associations learned in the hippocampus if the time interval spanning two distinct events is greater than this value? As has been remarked on elsewhere, most hippocampal models of sequence learning have failed to address this issue without relying on mechanisms unsupported by existing physiological observations.[10] The question of how temporal sequences are learned and recalled in the hippocampus provides a general framework for asking how relational memories are biologically supported. For example, if one makes the analogy between a particular pattern in a temporal sequence given to a model network and a rat's spatial location in a maze during a memory task, then the temporal sequence of such patterns is analogous to the rat's path through space. Thus, understanding how temporal sequences are learned and recalled in biophysically realistic models of the hippocampus, may shed light on questions concerning how organisms remember navigational routes. A physiological basis for solving this problem may reside in several lines of converging evidence involving the modulation of cellular behavior in the hippocampus by gamma-aminobutyric acid (GABA) and how this modulation may alter firing activity on a time scale comparable to the endogenous 4 - 9 Hz theta rhythm. [6,11]

Recent electrophysiological experiments have shown that the GABA-B receptor agonist baclofen selectively decreases the amplitude of excitatory post-synaptic potentials (EPSPs) in hippocampal CA1 pyramidal cells induced by Schaffer collateral stimulation but does not markedly alter perforant path transmission.[12,13] Baclofen has also been found to reduce inhibitory post-synaptic potentials (IPSPs) in the hippocampus.[14] The suppression of excitatory and inhibitory synaptic transmission by GABA-B receptor activation is an interesting effect in the hippocampus because it is possible that such modulation occurs in a rhythmical fashion during theta activity, and that it may influence the way single-units fire relative to population oscillations.

METHODS

Single-cell and Network Biophysics

The model of region CA3 consisted of 1000 pyramidal cells and 200 inhibitory interneurons. Each pyramidal cell was a reduced six-compartment version of the Traub model,[15] which included a fast sodium current, a delayed rectifier potassium current, a high-threshold calcium current, two calcium-dependent potassium currents, a transient potassium current (A-current) and a potassium leak current. Calcium buffering was performed in each compartment using a first-order diffusion process.[15] Each interneuron consisted of a fast sodium current and a delayed rectifier potassium current located at the soma, with the four remaining compartments (2 basal and 2 apical) were passive.[5,6]

Recurrent excitatory synapses located at the apical dendrites of pyramidal cells included Hebbian modification of NMDA conductances in conjunction with a voltage-dependent Mg^{2+} block of these channels. The network also included recurrent inhibitory synapses (GABA-A and GABA-B) at the soma and proximal dendrites of interneurons. Feedforward inhibition of pyramidal cells occurred via GABA-A receptors situated at the soma and proximal dendrites, while slower, GABA-B receptor-mediated inhibition was located at distal dendrites. Feedforward excitation occurred through NMDA and AMPA receptors located at the soma of interneurons. Both cholinergic and GABAergic projections from the medial septum were also included in the model. The GABAergic suppression of EPSPs and IPSPs (was modeled as a down-regulation of synaptic release due to the activation of GABA-B autoreceptors on the terminal endings of pyramidal cells and interneurons.[16]

Learning and recall

During the learning period, the network was presented with a sequence of nonorthogonal spatial patterns. Each pattern was delivered to the model as a fast AMPA receptor-mediated excitatory input to the apical dendrites of pyramidal cells for a period of 20 milliseconds. A different spatial pattern was presented to the network every 100 milliseconds. Thus, the entire sequence was delivered to the model as a sequence of spikes riding on the crest of each theta cycle. The entire sequence was repeated five times, constituting the learning period of the task.

RESULTS AND DISCUSSION

Learning Sequence Information and the Development of Place Fields

A completion task was employed to determine if the model was capable of learning and recalling temporal sequence information. During the learning period, it became apparent that many of the cells which fired independently of the input pattern for a brief period

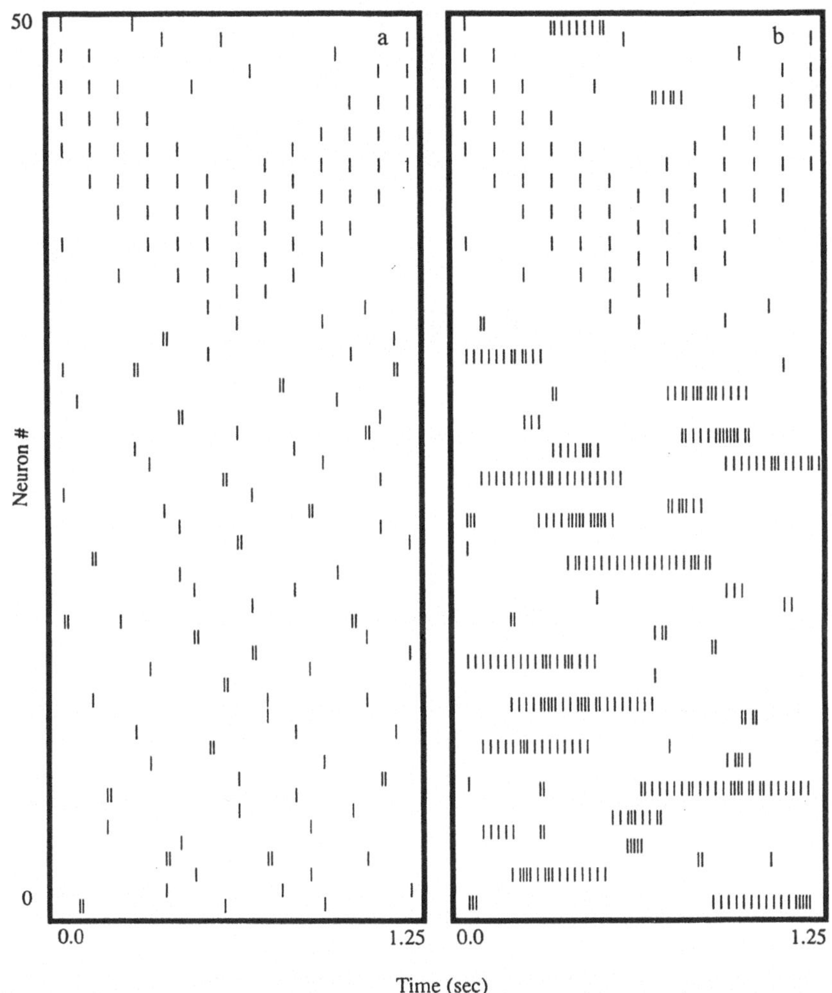

Figure 1. The activity of a subset of 50 pyramidal cells (from a total of 1000) in the model in response to a sample input sequence (a) before learning and (b) on the fourth learning trial. Initially, background activity (15 % of the total pyramidal cell population) appears randomly -different cells fire at different times. After several learning trials, however, this activity becomes more persistent. Some cells begin to fire in a sustained fashion over contiguous segments of the input sequence. This sustained firing resembles the formation of place fields observed in hippocampal *in vivo* recordings.[17]

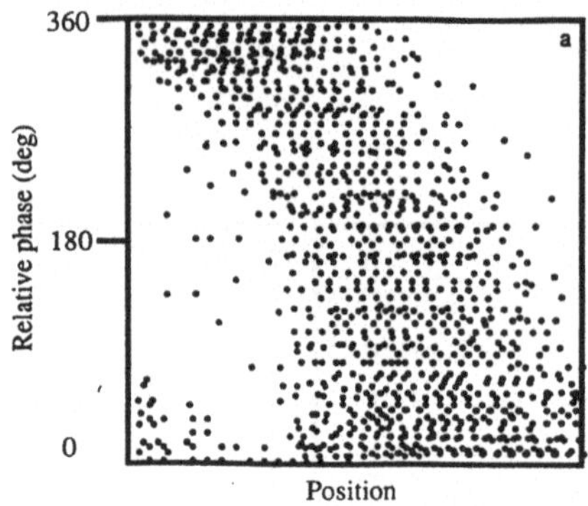

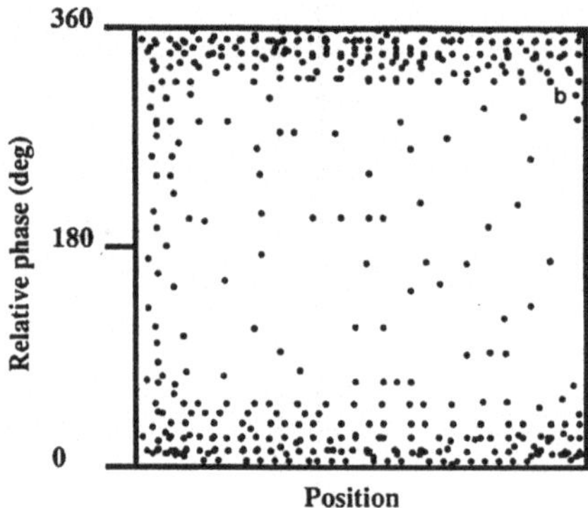

Figure 2. Phase precession -- (a) A representative firing pattern from a single place cell as the simulated rat passed through the place field of the cell. Note the systematic advance from firing late in a theta cycle when the rat first entered the field, to firing earlier in the cycle as the field is exited. This effect has recently been observed in hippocampal *in vivo* recordings.[18,19] (b) Phase precession failed to occur when GABAergic modulation was omitted from the model.

of time (ie. during a single pattern in the sequence) in the initial stages of learning began to fire continuously over a portion of the full sequence pattern after several exposures. Thus, with repeated exposure to the full sequence, these cells learned to respond to particular subsequences as shown in Figure 1. A visual inspection of Figure 1 indicates that the entire sequence is reconstructible from the appropriate interdigitation of these cells representing particular subsequences. This result is somewhat analogous to observations of place-specific firing of single-units.[17] Place field development was markedly curtailed when GABAergic modulation was not included in the model.

Recalling Sequence Information and the Phase Precession Effect

Sequence recall was tested by placing the simulated rat at a random location on the track and determining if the remaining sequence (navigational path) was completed. Because GABA-B receptor-mediated effects modulated pyramidal cell firing with the theta rhythm, cells directly related to the sequence pattern as well as those exhibiting place cell behavior tended to preferentially fire toward the latter portion of each theta cycle. The initial cell firing as the rat was situated at a random location on the track was always dominated by afferent activity related to the present location because intrinsic fibers at recurrent collaterals were suppressed early in each theta cycle. Information about future locations became available later in each cycle once this suppression was attenuated. At the start of the next theta cycle, the present location of the rat was again updated by a dominant afferent activity pattern and the scenario repeated. This typically resulted in a place cell firing late in a theta cycle as the rat initially entered that cell's field. Several locations were then rapidly recalled late in a theta cycle with sustained place cell firing across a contiguous segment of locations. As the rat exited a place field, activity of the associated place cell was then usually shunted during the early portion of the next theta cycle due to GABA-B receptor-mediated suppression of EPSPs at recurrent collaterals. Thus the phase of the theta cycle at which a place cell fired, systematically advanced as the rat passed through the cell's place field. This is illustrated in Figure 2a, which shows the activity of a single place cell plotted as a function of the phase in the theta cycle and spatial location of the rat when it fired. The observations that as a rat enters a place field, activity in the associated cell typically begins at the same point in a theta cycle, and advances toward earlier phases as the rat exits the field, are both supported by recent *in vivo* hippocampal recordings.[18,19] The phase precession of unit firing relative to the theta rhythm was not observed when GABAergic modulation was omitted from the model (Figure 2b).

Taken together, these effects point to the possibility of a common physiological mechanism for the successful development of place fields and the phase precession of single-unit firing relative to the theta rhythm during navigational tasks.

REFERENCES

1. Marr, D. (1971) Simple memory: A theory for archicortex. *Phil. Trans. Roy. Soc. B*, 262, 23-81.
2. Hasselmo, M.E. (1995) Neuromodulation and cortical function: modeling the physiological basis of behavior. *Behav. Brain Res.*, 67, 1-27.
3. Hasselmo, M.E., Schnell, E. and Barkai, E. (1995) Dynamics of learning and recall at excitatory recurrent synapses and cholinergic modulation in rat hippocampal region CA3. *J. Neurosci.*, 15, 5249-5262.
4. Hasselmo, M.E., Wyble, B.P. and Wallenstein, G.V. (1997) Encoding and retrieval of epsodic memories: Role of cholinergic and GABAergic modulation in the hippocampus. *Hippocampus*, 6, 693-708.
5. Wallenstein, G.V. and Hasselmo, M.E. (1997a) Functional transitions between epilptiform-like activity and associative memory in hippocampal region CA3. *Brain Research Bulletin*, 43, 485-493.
6. Wallenstein, G.V. and Hasselmo, M.E. (1997b) GABAergic modulation of hippocampal population activity: Sequence learning, place field development, and the phase precession effect. *J. Neurophysiol.*, 78, 393-408.
7. Levy, W.B. (1996) A sequence predicting CA3 is a flexible associator that learns and uses context to solve hippocampal-like tasks. *Hippocampus*, In press.
8. Levy, W.B. and Steward, O. (1983) Temporal contiguity requirements for long-term associative potentiation/depression in the hippocampus. *Neuroscience*, 8, 791-797.
9. Larson, J. and Lynch, G. (1989) Theta pattern stimulation and the induction of LTP: the sequence in which synapses are stimulated determines the degree to which they potentiate. *Brain Res.*, 489, 49-58.
10. Skaggs, W.E. and McNaughton, B.L. (1995) The hippocampal theta rhythm and memory for temporal sequences of event. *Int. J. Neural Sys., Supplementary Issue:* Proceedings of the Third Workshop on Neural Networks: From Biology to High Energy Physics, 101-105.

11. Wallenstein, G.V., Eichenbaum, H. and Hasselmo, M.E. (1998) The hippocampus as an associator of discontiguous events. *Trends in Neurosciences*, In press.

12. Ault, B. and Nadler, J.V. (1982) Baclofen selectively inhibits transmission at synapse made by axons of CA3 pyramidal cells in the hippocampal slice. *J. Pharmacol. Exp. Ther.*, 223, 291-297.

13. Colbert, C.M. and Levy, W.B. (1992) Electrophysiological and pharmacological characterization of perforant path synapses in CA1: mediation by glutamate receptors. *J. Neurophysiol.*, 63, 1-8.

14. Kamiya, H. (1991) Some pharmacological differences between hippocampal excitatory and inhibitory synapses in transmitter release: an in vitro study. *Synapse*, 8, 229-235.

15. Traub, R.D., Wong, R.K.S., Miles, R. and Michelson, H. (1991) A model of a CA3 hippocampal pyramidal neuron incorporating voltage-clamp data on intrinsic conductances. *J. Neurophysiol.*, 66, 635-650.

16. Howe, J.R., Sutor, B. and Zieglgansberger, W. (1987) Baclofen reduces post-synaptic potentials of rat cortical neurons by an action other than its hyperpolarizing action. *J. Physiol.*, 384, 539-569.

17. O'Keefe, J. and Dostrovsky, J. (1971) The hippocampus as a spatial map: preliminary evidence from unit activity in freely moving rats. *Brain Res.*, 34, 171-175.

18. O'Keefe, J. and Recce, M.L. (1993) Phase relationship between hippocampal place units and the EEG theta rhythm. *Hippocampus*, 3, 317-330.

19. Skaggs, W.E., McNaughton, B.L., Wilson, M.A. and Barnes, C.A. (1996) Theta phase precession in hippocampal neuronal populations and the compression of temporal sequences. *Hippocampus*, 6, 149-172.

CORTICAL SYNCHRONIZATION AND PERCEPTUAL SALIENCE

Shih-Cheng Yen, Elliot D. Menschik, Leif H. Finkel

Department of Bioengineering
University of Pennsylvania
3320 Smith Walk
Philadelphia, PA 19104

ABSTRACT

We present a striate-cortical model which proposes a direct relationship between cellular synchronization and perceptual salience. The model focuses on the role of the long-range horizontal connections between oriented simple cells in striate cortex and is able to account for current physiological and psychophysical results on contour salience. We demonstrate that horizontal connections between realistically-modeled multi-compartment pyramidal cells and interneurons can generate robust context-dependent synchronization. Closed contours induce better synchronization in the network than open contours, and closure thus increases perceptual salience, as observed psychophysically by Kovács and Julesz. This result is a general topological property of synchronization. The model supports a temporal synchronization solution to the binding problem, in that changes in synchronization are directly linked to changes in visual perception.

INTRODUCTION

Recent theoretical and experimental studies suggest that temporal synchronization mechanisms may be involved in solving the binding problem[1]. A missing link in these studies, is the demonstration that synchronization/de-synchronization affects perception, and that switching which sets of cells are synchronized results in a switch in percepts, and vice versa. We believe that a critical experiment in this regard has been performed by Kovács and Julesz[2] who showed that, in complex visual displays, closed contours have greater perceptual salience than open contours. The stimuli used by Kovács and Julesz consist of an array of Gabor patches placed at random positions and orientations, as shown in Figure 1. A subset of the Gabor patches are arranged so as to form either an open or closed contour. They find that the salience of the contour is governed by, Δ, the average separation between the contour elements. At the threshold of detection, elements on a closed contour can be separated farther apart than those on an open contour. Most interestingly, at the threshold separation for open contours, Δ_{Open}, adding additional elements increases the salience of the contour monotonically. However, for elements on a closed contour separated by Δ_{Closed}, additional elements provide no increase in the salience until the last element is added and the contour becomes closed, at which point there is a large increase in salience.

We propose that the results of their experiment can be explained in terms of cortical synchronization in striate cortex, and that their results reflect a change in *perception* that accompanies changes in cortical synchronization.

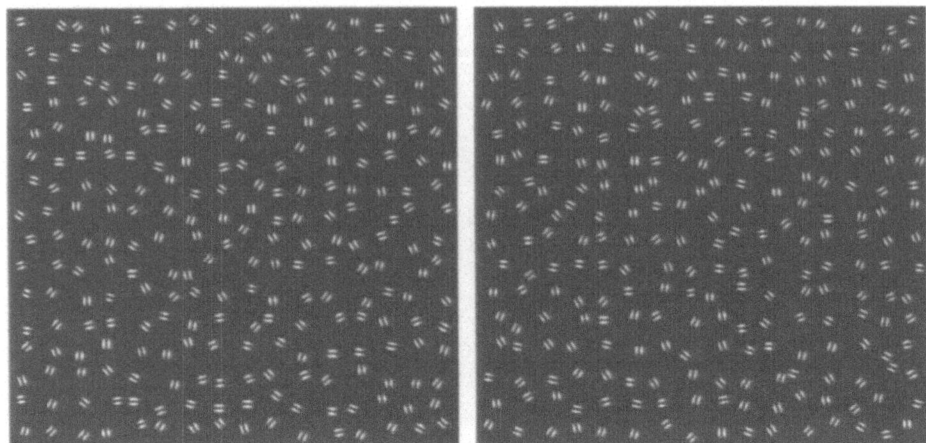

Figure 1. Stimuli used by Kovács and Julesz[2]. The stimulus on the left contains an open contour made up of a subset of Gabor elements, while the stimulus on the right contains a closed contour.

MODEL

The model[3, 4] consists of oriented simple cells, interconnected with long distance horizontal connections in a pattern similar to that suggested by Parent and Zucker[5] and Field *et al.*[6]. The long-distance horizontal inputs mediate both facilitation and inhibition between cortical cells (depending upon their relative positions and orientation preferences), and cause cells lying along a smooth contour to become facilitated relative to the background elements. The facilitation mediates the transition of the cortical cells to a high-frequency bursting mode (as seen in the "chattering" cells described by Gray and McCormick[7]). In this bursting state, cortical cells can synchronize with other similarly bursting cells via the horizontal connections. The salience of the contour is then represented by the sum of synchronized activity.

Synchronization was studied using realistically-modeled, multi-compartment models of pyramidal cells and interneurons. The synchronization mechanism was adapted from Traub *et al.*[8], in which coupled interneurons mediate the synchronization of pyramidal cells[*]. We used Traub's 66-compartment pyramidal cell[11] and the 51-compartment interneuron of Traub and Miles[12]. Both cells contained the following currents: fast sodium (I_{Na}), delayed rectifier (I_{K_DR}), calcium-dependent potassium (I_{K_Ca}), transient potassium (I_{K_A}), calcium-dependent afterhyperpolarizing potassium (I_{K_AHP}), and high-threshold calcium (I_{Ca}) currents.

Each orientation column stimulated by a Gabor element was modeled as a local group of 8 pyramidal cells and 8 interneurons (see Figure 2). Input to the pyramidal cells was modeled as a current injection of 1.6 ± 0.2 nA. Interneurons received a small (0.05 nA)

---

[*]Similar mechanisms have also been proposed by Wang and Buzsaki[9] and Bush and Sejnowski[10].

hyperpolarizing current to suppress spontaneous firing. Pyramidal cells make both AMPA and NMDA mediated synapses onto apical and basal arbors of interneurons and other pyramidal cells. Interneurons make "fast" $GABA_A$-ergic synapses (decay constant of 10 ms) onto the perisomatic region of pyramidal cells and interneurons and slower "dendritic" $GABA_A$-ergic synapses (decay constant of 50 ms) on the apical arbor of pyramidal cells. The cells within a group had all-to-all connectivity with negligible synaptic delays and peak synaptic conductances that remained unchanged for all simulations. Each cell was also connected to half of the cells in each of its two neighboring groups, except the cells at the ends of the open contour, which had connections to only one neighboring group. These inter-group connections had axonal delays of 5 ms and peak AMPA, NMDA, and $GABA_A$ conductances inversely proportional to the separation between the Gabor elements (this serves to simulate the decrease in the density of synaptic connections with increasing separation of the orientation columns). Simulations were carried out using the Parallel GENESIS[13] simulation environment.

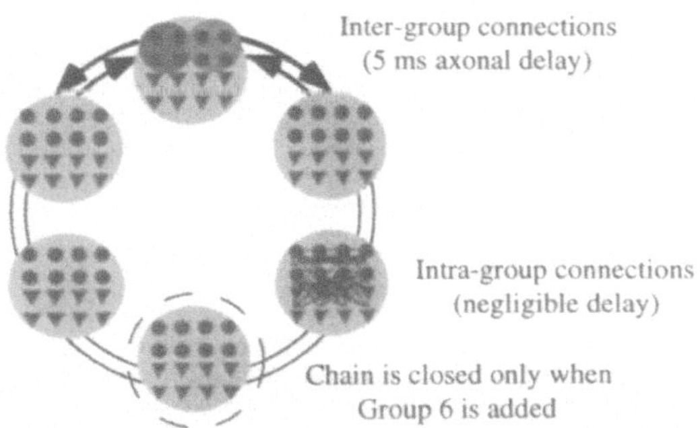

Figure 2. Connections within and between orientation columns. Each group in the figure represents cells in one orientation column. In this set of simulations, a closed chain is made up of six groups. Only nearest neighbor groups are connected; thus there are no connections between Group 1 and Group 5 in an open chain.

RESULTS

We were able to reproduce the synchronization reported in Traub *et al.*[8] which depends upon the production of "doublet" spikes in the interneurons. The interneurons control network synchronization when the time delay between groups matches the local, inter-doublet time delays within a group. The voltage traces of all the pyramidal cells in the simulation are shown in Figure 3. The cells within each group are tightly synchronized across all stimulus configurations, with an inter-burst interval of approximate 25 ms. The cells in each group are thus oscillating at 40 Hz and it is the inter-group connections that mediates the global synchrony across groups. When the Gabor elements form a closed contour, with the elements separated at Δ_{Open}, the pyramidal cells are tightly synchronized across all groups. This is due to the fact that the high peak conductances of the inter-group connections (AMPA: 8 nS, $GABA_A$: 1.5 nS, NMDA: 2 nS) contributes to the production of spike doublets in the interneurons, which are responsible for creating global synchrony. As the separation between the elements of the closed contour approaches Δ_{Closed}, the pyramidal cells are still synchronized across all groups, but not as tightly as before. The reduction in the peak conductances of the inter-group connections (AMPA: 0.8 nS, GABAA: 0.15 nS, NMDA: 0.2 nS) reduces the frequency of the spike doublets, which reduces the global synchrony. When one of the contour elements is removed such that the contour is no longer closed, the inter-group connections to the groups on the ends are effectively reduced by half (since they now have connections to only one neighboring group). Spike doublets are rarely seen in the interneurons across the groups and the synchrony between the groups falls apart. The cells on the open contour are able to synchronize once again when the separation between the elements is reduced to Δ_{Open} such that the increase in the conductances of the inter-group connections is able to compensate for the loss in half the inter-group connections for the groups on the ends of the open contour.

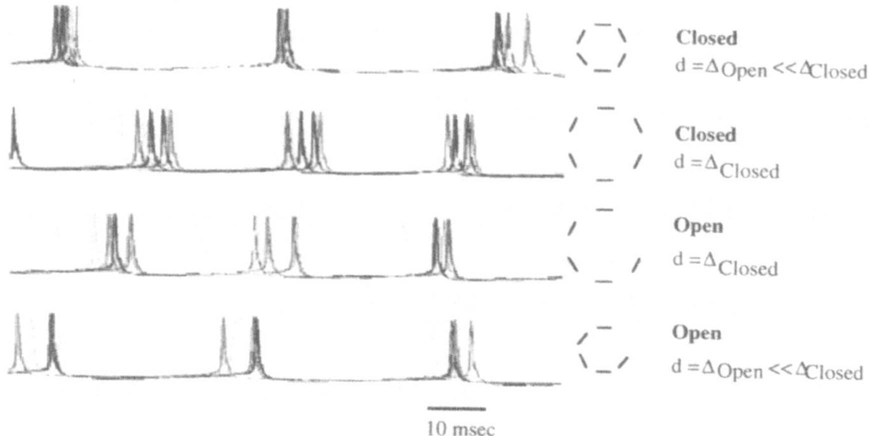

Figure 3. Synchronization using realistic, multi-compartment cells. The voltage traces of all the pyramidal cells (48 cells on closed chains, and 40 on open chains) in each simulation are plotted in the figure. Spikes from each group are shaded differently. The peak conductances of the intra-group and inter-group connections in the first and fourth traces were AMPA: 8 nS, GABAA: 1.5 nS, NMDA: 2 nS. This corresponds to the strong connection strengths between groups when the contour elements are closely spaced (Δ_{Open}). In the second and third traces, the inter-group conductances were: AMPA: 0.8 nS, GABAA: 0.15 nS, NMDA: 0.2 nS, while the intra-group conductances remained unchanged. This corresponds to widely-spaced contour elements (Δ_{Closed}).

We can therefore account for Kovács and Julesz's experiment as follows. Elements on closed contours may be separated further than elements on open contours ($\Delta_{Closed} > \Delta_{Open}$) because of the boundary conditions affecting the synchrony (and thus the salience) of cells on open and closed contours. At Δ_{Open}, the cells on the contour are synchronized; this means that each additional contour element adds to the sum of synchronized activity. The monotonic increase in salience of the open contour may thus be the perceptual correlate of this increase in synchronized activity. At Δ_{Closed}, the cells on the contour are not synchronized unless the contour is closed. This means that each additional contour element is <u>not</u> going to add to the sum of synchronized activity (and thus the salience), until the last element (which causes the contour to become closed) is added. At that point, all the cells on the contour are able to synchronize and the salience of the contour increases dramatically.

DISCUSSION

Temporal synchronization serves several roles in this model. As in previous models[1], synchronization serves as a means of distinguishing different contours in the scene. Synchronization also serves as the representation of the contour--a "contour" is perceived only when the orientation columns responding to individual contour elements are synchronized. Most importantly, we propose that the amount of synchronized activity correlates with the salience of the contour.

Changes in synchronization parallel changes in salience, and thus provide an explanation for the difference in salience between open and closed contours. This finding appears to be a general property of synchronization, independent of the exact mechanism. For example, we have previously demonstrated that the same result holds for synchronization of chains of phase-coupled oscillators[3]. Kopell and Ermentrout[14] have proven analytically that synchronization in chains of coupled neural oscillators is enhanced when the chain is closed. Similar results also hold with relaxation oscillators[15]. In all cases, as a consequence of the boundary conditions, closure yields better synchronization.

The synchronization mechanism in the model depends upon the existence of long-range inhibitory connections in cortex[16-18], which extend up to 10 mm. Conduction delays along these connections are less than 5 ms[19, 20]. Even connections between cortical areas, or hemispheres have conduction delays within the required time window[21]. We predict that even longer delays can be consistent with synchronization if, instead of doublets, the local networks produce triplets or longer bursts.

An alternative explanation of the Kovács and Julesz result is to assume that salience is defined solely on the basis of activity. Cells on smooth contours may have higher activity due to recurrent facilitatory activity. Cells on closed contours may have higher activity due to the absence of weakly supported end elements. An increase in activity across a hypercolumn should be perceived as an increase in stimulus contrast. Thus, an activity-based mechanism might lead to confusions between the encoding of contrast and salience. In addition, several studies have shown that there is no change in apparent contrast for Gabor elements on a contour [22, 23]. It remains possible that salience is encoded by a selective increase in the activity of a subpopulation of cells in the hypercolumn. Such a profile of enhanced activity might not generate a change in perceived contrast, particularly if contrast perception depends upon a population code.

In any activity-based model, the detection of the contour is presumably mediated by cells in higher cortical areas that are able to differentiate the enhanced activity of the facilitated cells in striate cortex. Although higher visual areas are undoubtedly involved in the perception of contours, a synchronization mechanism allows an implicit representation of contours to be maintained at the level of striate cortex, the area with the highest spatial resolution. Higher visual areas may modulate synchronization, perhaps via cholinergic effects on interneurons[24] and/or pyramidal cells[7], thereby altering the salience of particular stimuli.

These two models predict opposite physiological results for cells whose receptive fields lie along a contour. An activity model predicts increased firing rates, at least in cells whose preferred orientation matches the local contour orientation. The synchronization model predicts little or no increase in activity, but a significant change in correlated firing.

Acknowledgments

We thank Ilona Kovács and Nancy Kopell for helpful discussions of this work. This work was supported by ONR, NSF (CRI) and the Whitaker Foundation.

REFERENCES

1. W. Singer and C.M. Gray, Visual feature integration and the temporal correlation hypothesis, *Annu Rev Neurosci*, 18:555-86 (1995).
2. I. Kovács and B. Julesz, A closed curve is much more than an incomplete one: effect of closure in figure-ground segmentation, *Proc Natl Acad Sci U S A*, 90:7495-7 (1993).
3. S.-C. Yen and L.H. Finkel, Cortical synchronization mechanism for "pop-out" of salient image contours. In J. Bower (Ed.), *Computational Neuroscience: Trends in Research 1997*, Plenum, New York, 1997, pp. 553-560.

4. S.-C. Yen and L.H. Finkel, Extraction of perceptually salient contours by striate cortical networks, *Vision Research* (in press).
5. P. Parent and S.W. Zucker, Trace inference, curvature consistency, and curve detection, *IEEE Transactions on Pattern Analysis and Machine Intelligence*, 11:823-839 (1989).
6. D.J. Field, A. Hayes and R.F. Hess, Contour integration by the human visual system: evidence for a local "association field", *Vision Res*, 33:173-93 (1993).
7. C.M. Gray and D.A. McCormick, Chattering cells: superficial pyramidal neurons contributing to the generation of synchronous oscillations in the visual cortex, *Science*, 274:109-13 (1996).
8. R.D. Traub, M.A. Whittington, I.M. Stanford and J.G. Jefferys, A mechanism for generation of long-range synchronous fast oscillations in the cortex, *Nature*, 383:621-4 (1996).
9. X.J. Wang and G. Buzsaki, Gamma oscillation by synaptic inhibition in a hippocampal interneuronal network model, *Journal of Neuroscience*, 16:6402-13 (1996).
10. P. Bush and T. Sejnowski, Inhibition synchronizes sparsely connected cortical neurons within and between columns in realistic network models, *Journal of Computational Neuroscience*, 3:91-110 (1996).
11. R.D. Traub, J.G. Jefferys, R. Miles, M.A. Whittington and K. Toth, A branching dendritic model of a rodent CA3 pyramidal neurone, *J Physiol (Lond)*, 481:79-95 (1994).
12. R.D. Traub and R. Miles, Pyramidal cell-to-inhibitory cell spike transduction explicable by active dendritic conductances in inhibitory cell, *J Comput Neurosci*, 2:291-8 (1995).
13. N.H. Goddard and G. Hood, Large Scale Simulation with PGENESIS. In J. M. Bower and D. Beeman (Eds.), *The Book of GENESIS: Exploring Realistic Neural Models with the GEneral NEural SImulation System*, Springer-Verlag, in press.
14. N. Kopell and G.B. Ermentrout, Symmetry and phaselocking in chains of weakly coupled oscillators, *Communications on Pure and Applied Mathematics*, 39:623-660 (1986).
15. D. Somers and N. Kopell, Rapid synchronization through fast threshold modulation, *Biological Cybernetics*, 68:393-407 (1993).
16. Z.F. Kisvarday and U.T. Eysel, Functional and structural topography of horizontal inhibitory connections in cat visual cortex, *European Journal of Neuroscience*, 5:1558-72 (1993).
17. C.T. McDonald and A. Burkhalter, Organization of long-range inhibitory connections with rat visual cortex, *Journal of Neuroscience*, 13:768-81 (1993).
18. C. Morin and S. Molotchnikoff, Influences of horizontal connections on visual responses in rabbit striate cortex, *European Journal of Neuroscience*, 6:1063-71 (1994).
19. A. Mason, A. Nicoll and K. Stratford, Synaptic transmission between individual pyramidal neurons of the rat visual cortex in vitro, *Journal of Neuroscience*, 11:72-84 (1991).
20. A.M. Thomson, D. Girdlestone and D.C. West, Voltage-dependent currents prolong single-axon postsynaptic potentials in layer III pyramidal neurons in rat neocortical slices, *Journal of Neurophysiology*, 60:1896-907 (1988).
21. M.E. McCourt, J. Thalluri and G.H. Henry, Properties of area 17/18 border neurons contributing to the visual transcallosal pathway in the cat, *Visual Neuroscience*, 5:83-98 (1990).
22. M.W. Pettet and P. Verghese, Enhanced contour detection is not mediated by apparent contrast. , *27th Annual Meeting of the Society for Neuroscience, Vol. 23*, New Orleans, LA, USA, 1997, pp. 176.
23. W.H. McIlhagga and K.T. Mullen, Contour integration with colour and luminance contrast, *Vision Research*, 36:1265-1279 (1996).
24. T.F. Freund and G. Buzsaki, Interneurons of the hippocampus, *Hippocampus*, 6:347-470 (1996).

RESOLVING THE PARADOXICAL EFFECT OF ACTIVITY ON SYNAPSE ELIMINATION

Michael J. Barber[1] and Jeff W. Lichtman[2]

[1]Department of Physics
Washington University
Saint Louis, MO 63130
[2]Department of Anatomy and Neurobiology
Washington University School of Medicine
Saint Louis, MO 63110

INTRODUCTION

The circuitry of the nervous system emerges by a process of selection during development (Colman and Lichtman, 1993). Synaptic connections are both established and removed based on patterns of neural activity. Such structural changes have been directly observed at the neuromuscular junction, a simple and accessible synapse (Balice-Gordon and Lichtman, 1993; Balice-Gordon et al., 1993). Studies of synaptic plasticity at the neuromuscular junction avoid the technical difficulties of locating specific synapses in the central nervous system, but are aimed at understanding synaptic changes throughout the nervous system.

One change in the connections at the neuromuscular junction occurs in the first few weeks of neonatal life of mammals and most other vertebrates. Each neuromuscular junction undergoes a transition from innervation by several motor axons to innervation by a single motor axon. The removal of presynaptic terminals is mirrored by a loss of postsynaptic acetylcholine receptors (AChRs) from the muscle fiber membrane at the same sites. Focal blockade experiments in mice by Balice-Gordon and Lichtman (1994) suggest that differential activity of the AChRs is necessary for synaptic changes. They found that focal application *in vivo* of α-bungarotoxin to block neurotransmission in a small region of a neuromuscular junction in an adult animal causes long-lasting synapse elimination at that site, but blockade of neurotransmission throughout a junction does not cause synapse elimination. This indicates that a motor axon activating the AChRs underlying it has the ability to eliminate synaptic connections of inactive synapses on the same postsynaptic cell. The area occupied by the competing terminal was also found to be important; an active site with a large synaptic area eliminated synapses at inactive sites, but when the active area was small and the inactive area large, no loss occurred.

The foregoing makes it tempting to conclude that motor neurons with greater activities have an advantage at eliminating connections from their less active competi-

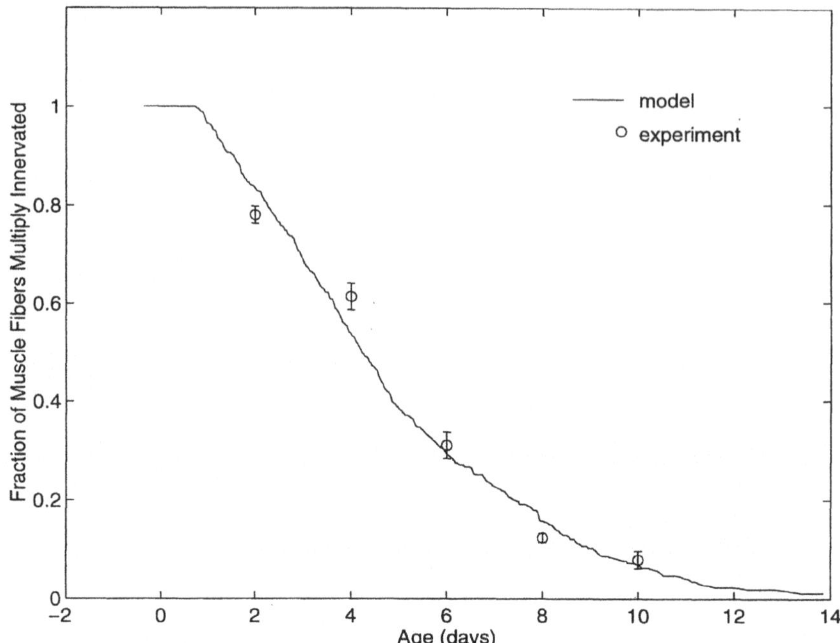

Figure 1. The process of synapse elimination steadily decreases the fraction of muscle fibers that are multiply innervated. The results of our model are here compared to data from physiological experiments (Colman *et al.*, 1997).

tors via a mechanism in which active synaptic sites destabilize inactive sites (Jennings, 1994). However, in at least some adult muscles, the number of muscle fibers with which an axon maintains contact seems to suggest the opposite conclusion: inactive axons out-compete active axons. In particular, motor units are recruited in an orderly manner according to Henneman's size principle (Henneman, 1985). Those motor units which have the lowest recruitment threshold, and hence are presumably the most active, have the smallest motor unit sizes, suggesting that during synapse elimination they lost more muscle fiber contacts than did less active axons. In addition, some experiments that tested the role of activity by nerve blockade have reached a similar conclusion (Callaway *et al.*, 1989). The question we wish to address is whether the implied contradiction in these experiments can be reconciled.

MODEL FORMULATION

We have modeled the changes in synaptic areas and connectivity of several motor neurons competing for a population of muscle fibers. The model we have developed focuses on the competition occurring at each of many neuromuscular junctions. The outcome at each junction is influenced by the neural resources available for maintenance and growth of synaptic area and by the susceptibility of each axon to elimination by its competitors. Conceptually, for each input to a doubly innervated neuromuscular

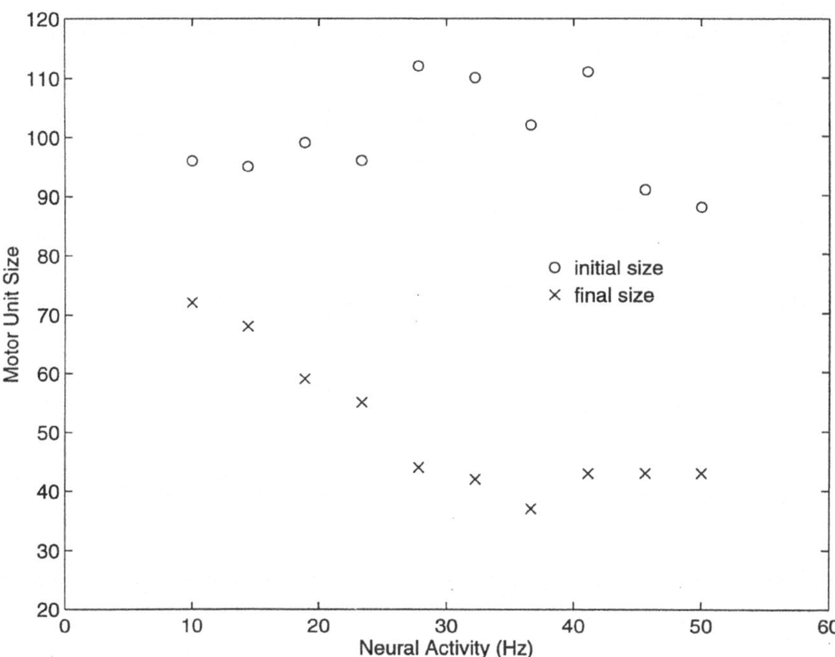

Figure 2. The desired relation between activity and motor unit size is obtained from this model. Although the initial motor unit sizes vary independently of neural activity, the final motor unit sizes are related to the activities. Similar results are obtained for Gaussian distributed activities as for the uniformly distributed activities shown here.

junction

$$\left\{ \begin{array}{c} \text{synaptic} \\ \text{area change} \end{array} \right\} = - \left\{ \begin{array}{c} \text{effect of} \\ \text{competition} \end{array} \right\} + \left\{ \begin{array}{c} \text{use of neural resources} \\ \text{for maintenance} \end{array} \right\}$$

Consider two neurons n_1 and n_2 with activities f_{n_1} and f_{n_2} competing at a neuromuscular junction on muscle fiber m. Each neuron's metabolism is constrained such that it makes a resource available at a limited rate R. This limiting resource, which is distributed amongst the neurons' synapses, is used to maintain synaptic area and activity. Neurotransmitter or synaptic vesicle proteins are possible candidate molecules for this resource. Assuming that the periods of synchronous activity of the two neurons have a small effect and may be neglected, the synaptic areas develop by

$$\frac{dA_{n_1 m}}{dt} = -\alpha f_{n_2} A_{n_2 m} + \frac{\beta}{M_{n_1}} \left(R - f_{n_1} \sum_{m'} (A_{n_1 m'})^\gamma \right)$$

where M_{n_1} is the motor unit size of neuron n_1 and α, β, and γ are constant parameters. The parameter γ allows for an economy of scale in the production and usage of the neural resource. Axons whose synaptic areas diminish below a small value withdraw and the corresponding equation is eliminated from the model, consistent with the observation that an axon's synaptic area diminishes to a very small amount prior to its withdrawal (Balice-Gordon *et al.*, 1993; Colman *et al.*, 1997).

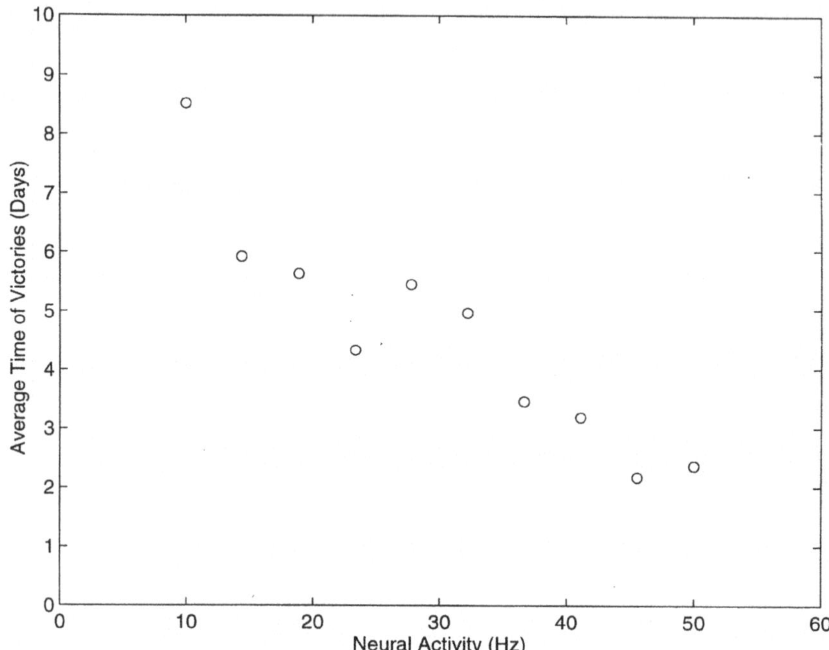

Figure 3. Different levels of neural activity are favored at different times in the development period. The most active neurons eliminate their competitors early, if at all, while the less active neurons obtain their larger number of victories later.

These equations impose a competitive elimination which directly affects the axons at only a single neuromuscular junction. However, the resource constraint is determined at the level of the neuron, and thus the outcome of the competition on one fiber can have profound indirect effects on the outcomes of the competitions on other fibers contacted by the same neurons.

SIMULATIONS

To investigate the properties of the model equations, we employ computer simulations. We consider here a model muscle with 500 muscle fibers innervated by 10 motor neurons. Each muscle fiber is initially innervated by two motor neurons, selected at random with equal likelihood for each neuron. The initial areas of the synaptic connections are all identical, with a value equal to one-fourth of the synaptic area of an average adult neuromuscular junction to accommodate the growth of junctions that have been measured in early postnatal life (Balice-Gordon $et\ al.$, 1993). Average activity levels (firing rates) are assigned to the motor neurons. The activity values are held fixed during the simulations; the detailed structure of neural activity patterns is not considered.

We set $R = 30$ so that, with the activities we specify, the average total synaptic area maintained by a neuron is unity. With the synaptic areas measured this way, activity measured in Hz, and time measured in days, we must set $\alpha = 0.419$, $\beta = 0.013$, and $\gamma = 0.75$ to match the experimental time course of the elimination of multiple innervation. We took the minimum synaptic area before elimination to be three percent of the average adult synaptic area at a neuromuscular junction.

Figure 1 shows how the number of multiply innervated fibers changes over time. Our simulation results are presented together with experimental findings (Colman *et al.*, 1997). Multiple innervation is steadily eliminated until each muscle fiber is innervated by only a single motor unit; innervation is not completely eliminated at any muscle fiber .

Figure 2 shows the development of the number of synaptic connections that each motor neuron makes with muscle fibers. From an initial set of connections in which the motor unit size of each neurons is independent of neural activity, a set of connections develops in which there is a strong relation between motor unit size and neural activity. The most active motor neurons maintain fewer connections than less active motor neurons.

CONCLUSIONS

The contradictory role of activity observed experimentally (a motor axons must activate AChRs to eliminate synaptic connections of competing motor axons on the same postsynaptic cell, but the most active neurons maintain the fewest connections) is duplicated in this model. Importantly, the model explains how this apparent paradox can be resolved. As synapses of a neuron are eliminated, the resources of the neuron are distributed to fewer neuromuscular junctions. This allows the remaining synapses to increase in area and efficacy. Elimination of synaptic connections alters the competitive environment allowing less active, but more resource-rich, synapses to dominate (figure 3). Neural activity provides an advantage early in the competition, but synaptic efficacy provides an advantage late in the competition.

REFERENCES

Balice-Gordon, R.J., Chua, C.K., Nelson, C.C., and Lichtman, J.W., 1993, Gradual loss of synaptic cartels precedes axon withdrawal at developing neuromuscular junctions, *Neuron*, 11:801–815.

Balice-Gordon, R.J., and Lichtman, J.W., 1993, *In vivo* observations of pre- and postsynaptic changes during the transition from multiple to single innervation at developing neuromuscular junctions, *J. Neurosci.* 13:834–855.

Balice-Gordon, R.J., and Lichtman, J.W., 1994, Long-term synapse loss induced by focal blockade of postsynaptic receptors, *Nature* 372:519–524.

Callaway, E.M., Soha, J.M., and Van Essen, D.C., 1989, Differential loss of neuromuscular connections according to activity level and spinal position of neonatal rabbit soleus motor neurons, *J. Neurosci.* 9:1806-1824.

Colman, H., and Lichtman, J.W., 1993, Interactions between nerve and muscle: synapse elimination at the developing neuromuscular junction, *Dev. Biol.* 156:1–10.

Colman, H., Nabekura, J., and Lichtman, J.W., 1997, Alterations in synaptic strength preceding axon withdrawal, *Science* 275:356–361.

Henneman, E.,1985, The Size-principle: a deterministic output emerges from a set of probabilistic connections, *J. exp. Biol.* 115:108–112.

Jennings, C., 1994, Death of a synapse, *Nature* 372:498-499.

CELLULAR MECHANISMS OF CALCIUM ELEVATION INVOLVED IN LONG TERM MEMORY

K.T. Blackwell[1], T.P. Vogl[1,2], D.L. Alkon[3]

[1]Institute for Computational Sciences and Informatics
George Mason University, Fairfax, VA 22030
[2]Environmental Research Institute of Michigan,
1101 Wilson Blvd. Arlington, VA 22209
[3]Laboratory of Adaptive Systems, NINDS/NIH, Bethesda, MD 20892

ABSTRACT

Hermissenda crassicornis is a shell-less marine snail that can be classically conditioned to associate light (the conditioned stimulus) with turbulence (the unconditioned stimulus). Acquisition of the association is critically dependent on elevation of intracellular calcium in the B cell soma, the location of memory storage. The source of calcium elevation has implications for cellular mechanisms underlying memory storage. To investigate sources of calcium elevation, a model of calcium dynamics in the photoreceptor was developed. Mechanisms of calcium diffusion, buffering and light activated release from intracellular stores were included in the model. Simulations show that release of calcium from intracellular stores is a significant source of calcium in the soma and may be the activating signal for the light-induced potassium current.

INTRODUCTION

Hermissenda crassicornis is a small, shell-less marine snail that can be classically conditioned to associate light (the conditioned stimulus) with turbulence (the unconditioned stimulus). The memory of the association is stored in the medial type B photoreceptor, which both transduces the light stimulus, and receives direct inhibitory synaptic input from the statocyst, which transduces the turbulence. Input resistance is increased and voltage- and calcium-dependent potassium currents are reduced in B cells from animals trained with paired light and turbulence as compared to unpaired controls. These cellular changes are critically dependent on elevation of intracellular calcium, and subsequent activation of protein kinase C (Alkon 1984).

Connor and Alkon (1984) showed that light stimulation causes an increase in intracellular calcium concentration, but that calcium elevation was greatly reduced in the presence of cadmium or hyperpolarizing current injection, suggesting that the source of calcium elevation is influx through voltage-dependent channels. However, Sakakibara et al. (1986) demonstrated that a single injection of IP_3 causes a reduction in the potassium currents that are reduced in classical conditioning, suggesting that IP_3 induced calcium release (IICR) is a significant source of calcium elevation in the B photoreceptor. Additional evidence for IICR comes from existence of a light-induced calcium-dependent potassium current. This current may be activated with a light flash under voltage clamp conditions (Chen and Alkon 1990) in which case influx through voltage-dependent channels cannot be the activating signal.

Given that light stimulation causes an increase in intracellular calcium concentration, why is there no long term reduction in potassium currents consequent to light alone? Either memory storage requires a non-linear combination of light-induced and turbulence induced calcium, or unique second messengers, contributed by light and turbulence are required for memory storage. A model of the B photoreceptor which includes ionic currents (Sakakibara et al. 1993, Blackwell et al. 1993, Fost et al. 1996) and intracellular calcium dynamics can distinguish these two alternatives.

For the present study, a model of calcium dynamics in the B photoreceptor was developed. This model, without ionic currents included, was used to investigate two questions: (1) what are the sources of calcium in the soma, and (2) is the light-induced potassium current equivalent to the light induced calcium release followed by calcium activation of the calcium-dependent potassium current?

METHODS

Calcium dynamics are modeled using the GENESIS simulation software. Additional objects were written in C for simulating general biochemical reactions (e.g. Michaelis-Menten enzyme kinetics, and calcium buffering), calcium release from intracellular stores, diffusion, and the effect of these processes and ionic current flow on the change in concentration of a molecule (e.g. calcium).

The Hermissenda B cell is approximately 40 μm in length and consists of a rhabdomere connected to the larger soma by a thin neck (Stensaas et al. 1969, Eakin et al. 1967, Crow et al. 1979). In the model, the rhabdomere cylinder is 12 μm in diameter and 12 μm long (4 compartments of 3 μm length), the soma cylinder is 20 μm in diameter and 24 μm long (8 compartments of 3 μm length). The thin neck connecting the rhabdomere and soma is modeled as a cylinder 1 μm long. The diameter of the connection between the rhabdomere and the soma is a parameter.

Calcium diffusion between compartments is modeled as a one dimensional processes in the axial dimension. This is implemented using the following discrete time equation:

$$\frac{\Delta[Ca^{++}]}{\Delta t} = \frac{-D_{Ca}\sum_i ([Ca^{++}] - [Ca^{++}]_i)\ Area_i}{Vol\ Len_i} \tag{1}$$

where [Ca^{++}], and *Vol* are calcium concentration and volume of the compartment; [Ca^{++}]$_i$ is calcium concentration of an adjacent compartment i; *Area$_i$*, *Len$_i$* are cross-sectional area and distance between the compartment and adjacent compartments; D$_{Ca}$ is the calcium diffusion coefficient, 6 x 10^{-6} cm^2/sec. The effect of calcium buffering was modeled as a second order reaction between free calcium and an immobile buffer (Eqn 2). The concentration of buffer, as well as forward (k$_f$)and backward (k$_b$) rate constants were taken from Blumenfeld et al. (1992).

$$Ca^{++} + B \underset{k_b}{\overset{k_f}{\rightleftharpoons}} Ca\text{-}B \qquad (2)$$

The discrete time equation describing this reaction is:

$$\frac{\Delta[Ca^{++}]}{\Delta t} = k_f\,[Ca^{++}]\,[B] - k_b[Ca\text{-}B] \qquad (3)$$

Calcium release from the ER occurs through the IP$_3$ receptor channel, which is gated by both calcium and IP$_3$. Kinetic equations and rate constants governing channel gating are taken from Li and Rinzel (1994). The permeability of the IP$_3$R channel is its maximum permeability times the open probability of the channel. Calcium stores in the Endoplasmic Reticulum (ER) are replenished by a SERCA-ATPase pump modeled using Michaelis-Menton kinetics. The value of passive calcium leakage out of the ER was adjusted to balance the pump rate when cytosolic calcium concentration is equal to the resting value of 50 nanoMolar (nM).

Production of IP$_3$ by Phospholipase C (PLC) occurs in the rhabdomeric compartment furthest away from the soma and is governed by Michaelis-Menton kinetics. Because the concentration of substrate is much greater than the K$_d$ of PLC, the kinetics may be approximated as d[IP$_3$]/dt = kf [PLC]. The value of kf is 0.167 x 10^{-3} mmol sec^{-1} mg^{-1} of PLC (Mitchell et al. 1995); PLC concentration is non-zero only when the light is on. IP$_3$ is allowed to diffuse into all other compartments, and degradation of IP$_3$ occurs in all compartments.

To investigate the source of calcium elevation in the soma, simulations were run with and without an ER in the soma. Parameters explored include the ratio of maximum permeability of the IP$_3$R channel (MAXCR) to maximum rate of the SERCA pump (MAXPUMP); the neck diameter, the rate of IP$_3$ degradation, and the value of PLC.

RESULTS

The first set of simulations explored the effect of MAXCR and MAXPUMP on calcium concentration, and tried to qualitatively reproduce the data of Connor and Alkon (1984). They measured calcium concentration in the B photoreceptor in response to a light stimulus and found that calcium concentration in the rhabdomere oscillated in response to bright lights. Figure 1 reproduces their finding, and shows that calcium concentration oscillates for large values of Phospholipase C when the ratio of MAXPUMP to MAXCR was between 200 and 800 (MAXPUMP = 0.731 sec^{-1}.) When the ratio was too low, calcium concentration peaked, and then settled to a steady state value of approximately 1 micromolar (μM); when the ratio was too high, calcium concentration remained low for the duration of the light flash. The ratio was fixed at 400 for the remainder of the simulations.

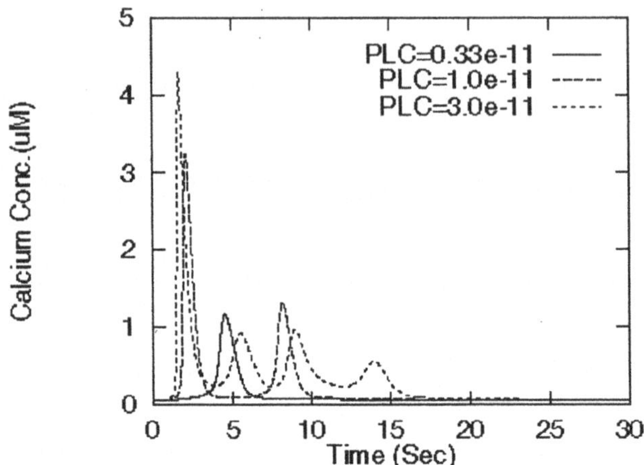

Figure 1. Calcium concentration in rhabdomere oscillates for large values of PLC.

The second set of simulations investigated two sources of calcium in the soma. One source is diffusion from the rhabdomere, the second source is release of calcium from intracellular stores through the IP$_3$ receptor channel. Figure 2 shows that diffusion of calcium from the rhabdomere causes only a small increase in soma calcium concentration. When no calcium release occurs in the soma compartment, calcium concentration in the soma remains less than 400 nM in the compartment closest to the rhabdomere, even with a rhabdomeric calcium concentration which exceeds 4 μM. Calcium concentration in the soma compartment furthest from the rhabdomere never exceeds 200 nM. Figure 2 shows the results for a neck diameter of 6 μm, half that of the rhabdomere diameter. When the neck diameter is smaller, somatic calcium concentration is still smaller.

Figure 3 shows that calcium release from the ER into the soma produces a large increase in somatic calcium concentration, with calcium oscillations occurring for higher values of PLC. The delay between light flash and calcium peak in the soma decreases as the level of PLC increases. Calcium release in the soma requires diffusion of IP$_3$ from the rhabdomere. IP$_3$ diffuses into the soma faster than calcium, and reaches a higher concentration than calcium because calcium is heavily buffered. In figures 1 through 3, the IP$_3$ degradation was 0.17 sec^{-1}. If the value is increased by a factor of 10, the same pattern of results is obtained: The major source of calcium in the soma is release from the ER.

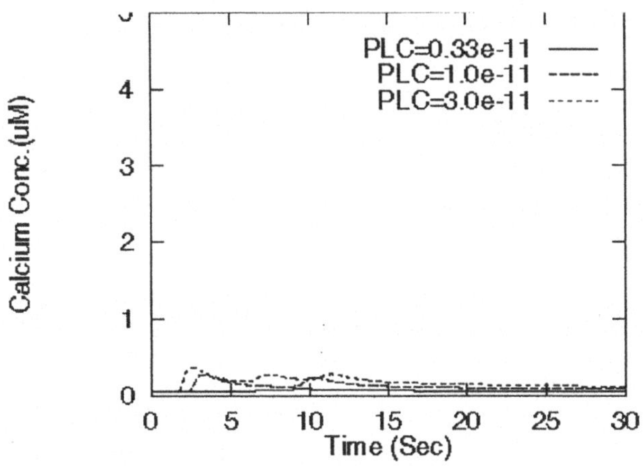

Figure 2. Calcium concentration in soma. Calcium source is diffusion from rhabdomere.

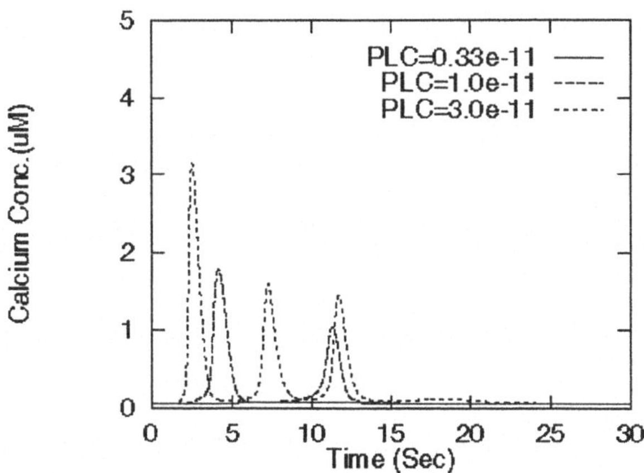

Figure 3. Calcium concentration in soma. Calcium sources are diffusion and release from the ER through IP$_3$R

The results in figure 3 begin to answer the question of whether the light-induced potassium current is equivalent to light-induced calcium release followed by calcium-activation of the calcium-dependent potassium current Figure 3 shows that a calcium elevation in the soma has a delay of 1000 msec (value – 200 nM) and peaks at 1540 msec after light onset. This delay is longer than, but comparable to that of the light-induced potassium current recorded under voltage clamp conditions, which has a delay of approximately 500 msec and peaks at 2200 msec (Chen and Alkon 1990, Sakakibara et al. 1986).

These simulations show that the release of calcium from stores is a significant source of calcium in the soma, and may be the activating signal for the light-induced potassium current. Although the delay from light onset to simulated somatic calcium elevation is larger than the delay to light-activated potassium currents, several assumptions made in the simulations lengthens the delay. The first assumption is that baseline [IP$_3$] is zero during dark adaptation. Biochemical assays show that PLC has some activity in the dark (Suzuki et al. 1995, Bhatia et al. 1996). Thus, baseline [IP$_3$] should be greater than zero. The second assumption is that there is no influx through the voltage-dependent calcium channel (I_{Ca}). Measurements show that I_{Ca} in Hermissenda activates at voltages greater than -45 mV and does not inactivate. Thus, when the cell is clamped to potentials greater than -45 mV, baseline calcium concentration may be greater than 50 nM. Both an increase in baseline [IP$_3$] and an increase in baseline calcium concentration would increase the fraction of IP$_3$R in the conducting state and reduce the time between light flash and calcium elevation.

ACKNOWLEDGMENT

KTB is supported by a Scientist Development Award from NIMH. TPV is supported by NIH/NINDS contract N01NS32304 and ONR contract N00014-92-C-0018.

REFERENCES

Alkon D.L. (1984) *Memory Traces in the Brain* Cambridge University Press, New York

Bhatia J., Davies A., Gaudoin J.G., Saibil H.R. (1996) Rhodopsin, Gq and phospholipase C activation in cephalopod photoreceptors. *J. Photochem Photobiology* B: Biology 35:19-23

Blumenfeld H., Zablow L., Sabatini B. (1992) Evaluation of cellular mechanisms for modulation of calcium transients using a mathematical model of fura-2 Ca^{2+} imaging in Aplysia sensory neurons. *Biophys J.* 64:1146-1164

Chen C.C., Alkon D.L. (1990) Voltage dependence of light-induced ionic currents of a molluscan photoreceptor. Unpublished manuscript.

Connor J., Alkon D.L. (1984) Light- and voltage-dependent increases of calcium ion concentration in molluscan photoreceptors. *J. Neurosphysiology* 51:745-752

Crow T., Heldman E., Hacopian V., Enos R., Alkon D.L. (1979) Ultrastructure of photoreceptors in the eye of *Hermissenda* labelled with intracellular injections of horseradish peroxidase. *J. Neurocytology* 8:181-195

Eakin R.M., Westfall J.A., Dennis M.J. (1967) Fine structure of the eye of a nudibranch mollusc, *Hermissenda crassicornis. J. Cell Sci.* 2:349-358

Fost J.W., Clark G.A. (1996) Modeling *Hermissenda* Differential contributions of I_A and I_C to type-B cell plasticity *J. Computational Neuroscience* 3:127-154

Li Y.-X., Rinzel J. (1994) Equations for InsP$_3$ receptor-mediated [Ca^{2+}]$_i$ oscillations derived from detailed kinetic model: a Hodgkin-Huxley like formalism. *J. Theor. Biol.* 166:461-473

Mitchell J., Gutierrez J., Northup J.K. (1995) Purification, characterization, and partial amino acid sequence of G protein-activated phospholipase C from squid photoreceptors. *J. Biol. Chem.* 270:854-859

Sakakibara M., Alkon D.L., Neary J.T., Heldman E., Gould R. (1986) Inositol trisphosphate regulation of photoreceptor membrane currents. *Biophys. J.* 50:797-803

Sakakibara M., Ikeno H., Usui S., Collin C., Alkon D.L. (1993) Reconstruction of ionic currents in a molluscan photoreceptor. *Biophysical J.* 65:519-527

Stensaas L.J., Stensaas S.S., Trujillo-Cenoz O. (1969) Some morphological aspects of the visual system of *Hermissenda crassicornis* (Mollusca: Nudibranchia) *J. Ultrastructure Research* 27:510-532

Suzuki T., Terakit A., Narita K., Nagai K., Tsukahara Y., Kito Y. (1995) Squid photoreceptor phospholipase C is stimulated by membrane Gqα but not by soluble Gqα. *FEBS Letters* 377:333-337

TEMPORAL CHARACTERISTICS OF V1 CELLS ARISING FROM SYNAPTIC DEPRESSION

Frances S. Chance, Sacha B. Nelson, and L. F. Abbott

Volen Center and Department of Biology
Brandeis University
Waltham, Massachusetts 02254-9110

INTRODUCTION

Neurons in the primary visual cortex (V1) exhibit response characteristics qualitatively different from their LGN afferents. Some of these characteristics, such as the oriented structure of simple cell receptive fields, can be explained as arising from linear combinations of LGN receptive fields.[1] However, many aspects of V1 responses, especially in the temporal domain, reflect underlying nonlinear mechanisms. Here we study the idea that synaptic depression may give rise to many of these nonlinearities. Synaptic depression is observed in intracortical[3–5] as well as thalamocortical[6,7] synapses and has been proposed as a possible mechanism for contrast adaptation and direction selectivity.[2,9] We construct a model V1 simple cell and explore the effects of synaptic depression on the temporal dynamics of its responses. Our results indicate that synaptic depression may play an important role in the enhancement of responses to transient stimuli, direction selectivity, and contrast adaptation.

METHODS

The model V1 simple cell we consider is a single-compartment integrate-and-fire neuron with feedforward excitatory and inhibitory LGN-like inputs. Because the purpose of this study is to explore the effects of synaptic depression, we model the inputs as simply as possible to avoid the additional temporal nonlinearities introduced by a recurrent network. The arrangement of input receptive fields determines the spatial features of the model simple-cell receptive field. A typical three-lobed receptive field for the simple cell arises from a central region of excitatory ON-center and inhibitory OFF-center afferent receptive fields flanked by regions of excitatory OFF-center and inhibitory ON-center afferent receptive fields.

Synaptic inputs are modeled as transient synaptic conductance changes. With each presynaptic spike, the total excitatory synaptic conductance, g_E, and the total inhibitory synaptic conductance, g_I, increase: $g_E \to g_E + G_i D_i S_i$ and $g_I \to g_I + G_i D_i S_i$, where G_i represents the strength of synapse i and D_i and S_i represent the degree of fast and slow depression. Between presynaptic action potentials, g_E and g_I

exponentially decay to zero with time constants τ_E (2ms) and τ_I (10ms):

$$\tau_E \frac{dg_E}{dt} = -g_E \ \text{ and } \ \tau_I \frac{dg_I}{dt} = -g_I.$$

We base our model of synaptic depression on previous fits of experimental data[3,5]. With each presynaptic action potential, the factors D_i and S_i are reduced by multiplicative factors d_i and s_i ($0 \le d_i, s_i \le 1$) as follows: $D_i \to d_i D_i$ and $S_i \to s_i S_i$. Between presynaptic action potentials, D_i and S_i recover exponentially,

$$\tau_D \frac{dD_i}{dt} = 1 - D_i \ \text{ and } \ \tau_S \frac{dS_i}{dt} = 1 - S_i.$$

Thus d_i and s_i determine the onset rates of depression and τ_D and τ_S determine the rates of recovery. In this report $\tau_D = 300$ms, $\tau_S = 20$s, $0.4 \le d \le 1$, and $s = 1$ in all figures but figure 4 where $s = 0.99$.

Afferent input is modeled as Poisson spiking at a rate computed from a previously proposed LGN model[10],

$$R = R_b \pm A(C) \int dx\,dy\,dt'[W_c(x,y)K_c(t-t') - 0.6 * W_s(x,y)K_s(t-t')]I(x,y,t').$$

The center-surround receptive field structure is a difference of Gaussians; the temporal response, $K_{c,s}$, of each sub-region is a difference of alpha functions. The stimulus, $I(x,y,t)$, is defined as the difference between the mean luminance and the stimulus luminance at point (x,y) and time t. The contrast amplitude factor, $A(C)$, is a fit to data recorded by Ohzawa et al.[11],

$$A(C) = (172 Hz)\ln(67C).$$

For zero contrast stimuli, we set $A(C) = 0$; otherwise we use contrasts for which $A(C)$ is positive. The G_i are scaled from figure to figure to match firing rates seen in experimental data. All other parameter values are the same for all figures (except where specified for figure 1).

RESULTS

We have previously explored the effects of synaptic depression in this model.[9] Here we review some of our results. We first discuss the basic temporal response characteristics arising from synaptic depression. In figure 1, we directly manipulate the afferent firing rates (as depicted in the legend) rather than simulating afferent cell responses to visual stimuli to further isolate the effects of depression on the postsynaptic cell response. This demonstrates the effects of synaptic depression on the responses to periodic (circles) and transient (triangles) stimuli. At low frequencies the rise of the afferent firing rate is slow. When depression is present, the afferent synapses become fairly depressed before the afferents reach their peak firing rates, reducing the response (compare open symbols with filled symbols at frequencies below 1 Hz). At higher frequencies the afferent firing rates rise faster so the synapses are less depressed during peak input rates and the response is stronger. Above a few Hz, however, the synapses do not have enough time to recover from depression between pulses of the periodic stimulus so the response is decreased. This effect is not seen for single-pulse stimuli. Thus the frequency response for single-pulse stimuli peaks at a higher frequency than for periodic stimuli. In all cases, the decline in responses at higher frequecies is also due to the RC filtering properties of the postsynaptic cell. The peak in response at a few Hz for periodic stimuli and the increase in both

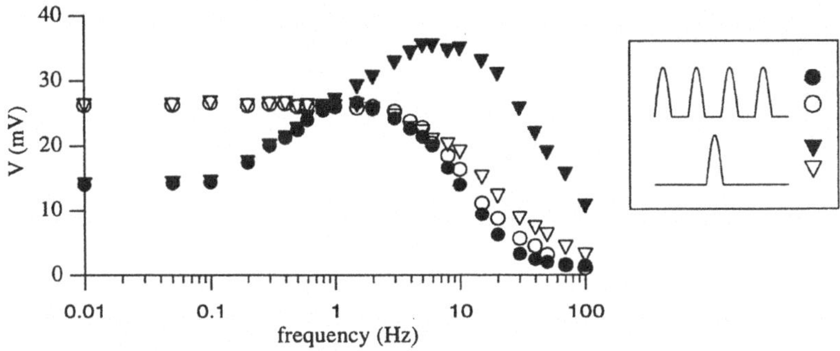

Figure 1: Peak-to-peak membrane potential fluctuations (action potentials blocked) in response to the input patterns shown in the legend. Afferent firing rates either followed a rectified sine wave (circles) or one half cycle of a sine wave (triangles) with a peak of 100Hz. Filled symbols correspond to results with depression ($d = 0.75$) and open symbols correspond to results without depression ($d = 1$). Synaptic strengths were scaled to make the maximum periodic amplitudes match with and without depression.

amplitude and frequency of the peak for transient stimuli are characteristic of cortical responses.[12,13]

Existing models of direction selectivity rely upon a combination of spatial and temporal phase shifts. The temporal phase shift may arise from a variety of sources, including delays due to intracortical connections[10,14] and lagged thalamocorical inputs.[15] Our model uses instead a phase advance introduced by synaptic depression. The direction-selective simple cell receives inputs from two populations of afferents, arranged so that the receptive fields of one population are shifted from those of the other by half the size of a center region of a receptive field (see figure 2A). The two populations connect to the model simple cell through two types of synapses: synapses that depress ($d = 0.4$) and synapses that do not ($d = 1.0$). Figure 2D shows the membrane potential oscillation evoked by driving each afferent population separately with a stationary oscillating grating. The response evoked by nondepressing inputs is approximately sinusoidal. In contrast, the response evoked by depressing inputs has a sawtooth shape and is advanced in phase by approximately 90°.

The top panels of figure 2B show the membrane potential fluctuations evoked in the postsynaptic cell by each afferent population in response to a drifting 2Hz grating. When the grating moves in the preferred direction (left to right in the figure), it aligns first with the nondepressing inputs and then the depressing inputs. Because of the phase advance due to synaptic depression, the contributions from the two populations sum in the postsynaptic cell, eliciting a strong response (lower left of figure 2B). When the grating moves in the nonpreferred direction it aligns first with the depressing inputs and then the nondepressing inputs. In this case, synaptic depression enhances the delay between the two contributions and the response is weak (lower right).

Jagadeesh et al.[16] recorded intracellular responses of direction selective cells to stationary gratings at different positions in the receptive field. The responses of the model direction selective cell are qualitatively similar to the intracellular data as seen in figure 2C. In both cases, the response at 0° is approximately sinusoidal and at 90° has a sawtooth shape. Examination of the contribution from each afferent population explains the response of the model; at 0° the grating aligns with the receptive field structure of the nondepressing afferents and at 90° the grating aligns with the receptive fields of the depressing afferents. Kontsevich analyzed the

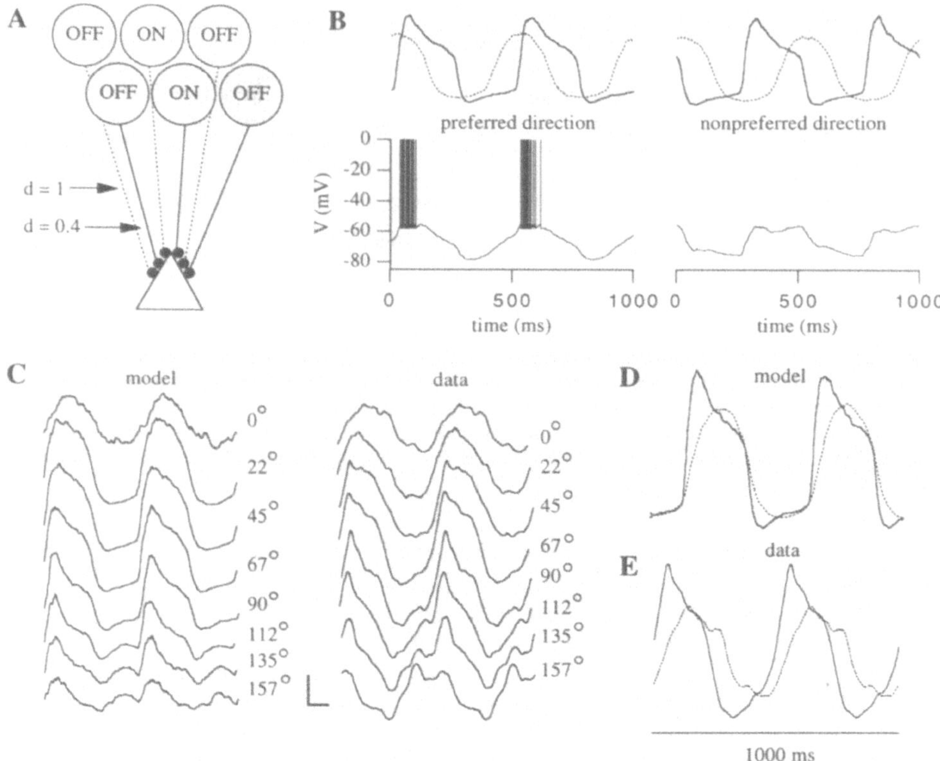

Figure 2: A) Each row of circles corresponds to one afferent population. Depressing afferents are shaded and indicated by solid lines; nondepressing afferents are clear and indicated by dashed lines. The two rows are horizontally displaced from each other by one half the width of the center subregion of a receptive field. The model V1 receptive field is one-dimensional; the vertical displacement is drawn only for clarity. B) Upper panels represent the membrane potential fluctuations (spikes blocked) induced by either the depressing (solid curves) or the nondepressing (dashed curves) inputs. Lower panels represent the response of the neuron when the two components are combined. C) Model V1 membrane potential fluctuations (left) compared with intracellular recordings (right) in response to stationary 2Hz gratings positioned at various spatial phases. The data is redrawn from Jagadeesh et al.[16] using DataThief. Horizontal scale bar = 100ms. Vertical scale bar = 10mV for model and 2.5 mV for data. D) Model V1 cell membrane potential fluctuations evoked by driving the depressing (solid curve) and the nondepressing (dashed curve) afferent populations separately with a stationary oscillating grating, positioned to evoke the maximal response from each population. E) The two principal components redrawn from Kontsevich's analysis[17] of the data in C.

intracellular data and developed a phenomenological model based upon two principal components[17](figure 2E). The traces of membrane potential fluctuations evoked by the two afferent populations (figure 2D) bear a striking resemblance to the two principal components extracted by Kontsevich.

Direction selective neurons remain selective over a wide range of stimulus contrasts[18]. Although the responses of direction selective neurons to gratings moving in the preferred direction (R_+) and the nonpreferred direction (R_-) both increase with stimulus contrast, the direction index ($1 - R_-/R_+$) remains more or less constant. Direction selective neurons also maintain selectivity over a great range of stimulus temporal frequencies.[19] Figure 3 demonstrates that our model maintains direction

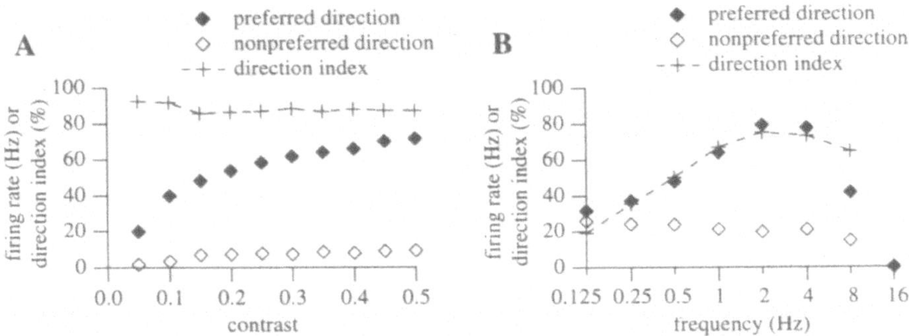

Figure 3: Responses to gratings moving in the preferred direction are denoted by solid circles and responses to gratings moving in the nonpreferred direction are denoted by open circles. Direction index (expressed as a percentage) is denoted by the + symbols and dashed lines. A) The firing rate of the model direction selective cell in response to drifting 2Hz gratings of various contrasts. B) The firing rate in response to a 100% contrast drifting gratings at various temporal frequencies.

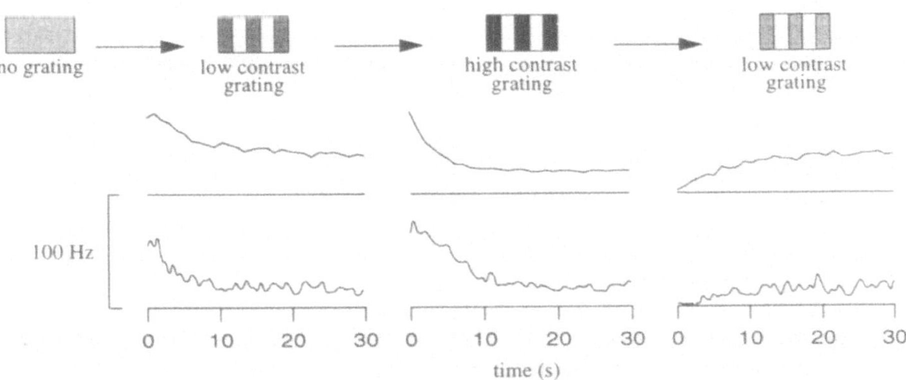

Figure 4: The sequence of presented stimuli is drawn at the top of the figure. Each stimulus was a 30s presentation of a 2Hz grating moving in the preferred direction. Before each stimulus sequence, both the model and experimental cells were adapted to a 0% grating; the responses to the 0% grating are not shown. The top set of traces depict the model's responses to each stimulus. The bottom set of traces are recordings from cat area 17, replotted from Ohzawa et al.[11] using DataThief.

selectivity over a range of stimulus contrasts (figure 3A) and over a range of stimulus temporal frequencies (figure 3B).

We examine the effects of the slow component of depression on contrast adaptation in figure 4. The model and the data[11] show essentially no response to the 0% contrast grating (not shown). Each time the contrast of the stimulus increases, both the model and experimental cells show a strong but transient response which settles down to a weaker steady-state response. In the model, the transient response occurs because the initial response to the new stimulus is reduced as the slow component depresses. When the contrast of the grating is lowered, both model and experimental responses are initially weak but slowly recover.

SUMMARY

Synaptic depression provides a possible mechanism for nonlinear dynamics in V1 simple cell responses. A model V1 simple cell with synaptic depression on its afferent synapses has a realistic temporal frequency response curve with enhanced responses to transient high-frequency stimuli. Correlation between the amount of depression and the location of afferent receptive fields can give rise to direction selectivity. The direction selective model constructed in this way demonstrates direction selectivity over a range of stimulus contrasts and temporal frequencies similar to those in experimental data. Model membrane potential fluctuations in response to stationary gratings closely resemble those of intracellular recordings. Finally, a slower component of depression can reproduce effects seen in contrast adaptation experiments.

REFERENCES

1. Hubel, D. H. and Wiesel, T. N., Receptive fields, binocular interaction and functional architecture in the cat's visual cortex, *J. Neurophysiol.* 160: 106-154 (1962).

2. Markram, H. and Tsodyks, M. V., Redistribution of synaptic efficacy between neocortical pyramidal neurones, *Nature* 382: 807-809.

3. Abbott, L. F., Sen, K., Varela, J. A., and Nelson, S. B., Synaptic depression and cortical gain control, *Science* 275: 220-222 (1997).

4. Tsodyks, M. and V., Markram, H., The neural code between neocortical pyramidal neurons depends on neurotransmitter release probability, *Proc. Natl. Acad. Sci. USA* 94: 719-723, (1997).

5. Varela, J., Sen, K., Gibson, J., Fost, J., Abbott, L. F., and Nelson, S. B., A quantitative description of short-term plasticity at excitatory synapses in layer 2/3 of rat primary visual cortex, *J. Neurosci.* 17: 7926-7940 (1997).

6. Stratford, K. J., Tarczy-Hornuch, K., Martin, K. A. C., Bannister, N. J., Jack, J. J. B., Excitatory synaptic inputs to spiny stellate cells in cat visual cortex, *Nature* 382: 258-261 (1996).

7. Gil, Z., Douglas, R., Marlin, S., Cynader, M., Differential regulation of neocortical synapses by neuromodulators and activity, *Neuron* 19: 679-686 (1997).

8. Nelson, S. B., Varela, J. A., Sen, K., and Abbott, L. F., Functional significance of synaptic depression between cortical neurons, in: *Computational Neuroscience, Trends in Research 1997*, J. Bower, ed, Plenum, NY (1997).

9. Chance, F. S., Nelson, S. B., Abbott, L. F., Synaptic depression and the temporal response characteristics of V1 cells, submitted (1998).

10. Maex, R. and Orban, G. A., Model circuit of spiking neurons generating directional selectivity in simple cells, *J. Neurophysiol.* 75: 1515-1545 (1996).

11. Ohzawa, I., Sclar, G., and Freeman, R. D., Contrast gain control in the cat visual cortex, *Nature* 298: 266-268 (1982).

12. Dean, A. F., Tolhurst, D. J., and Walker, N. S., Non-linear temporal summation by simple cells in cat striate cortex demonstrated by failure of superposition, *Exp. Brain Res.* 62: 143-151.

13. Reid, R. C., Victor, J. D., and Shapley, R. M., Broadband temporal stimuli decrease the integration time of neurons in cat striate cortex, *Vis. Neurosci.* 9: 39-45.

14. Suarez, H., Koch, C., and Douglas, R., Modeling direction selectivity of simple cells in striate visual cortex within the framework of the canonical microcircuit, *J. Neurosci.* 15: 6700-6719 (1995).

15. Saul, A. B. and Humphreys, A. L., Spatial and temporal properties of lagged and nonlagged cells in the cat lateral geniculate nucleus, *J. Neurophysiol.* 68: 1190-1208 (1990).

16. Jagadeesh, B., Wheat, H. S., and Ferster, D., Linearity of summation of synaptic potentials underlying direction selectivity in simple cells of the cat visual cortex, *Science* 262: 1901-1904 (1993).

17. Kontsevich, L. L., The nature of the inputs to cortical motion detectors, *Vision Res.* 35: 2785-2793 (1995).

18. Tolhurst, D. J. and Dean, A. F., Evaluation of a linear model of directional selectivity in simple cells of the cats striate cortex, *Vis. Neurosci.* 6: 421-428 (1991).

19. Saul, A. B. and Humphreys, A. L., Temporal-frequency tuning of direction selectivity in cat visual cortex. *Vis. Neurosci.* 8: 365-372 (1992).

SYNAPTIC PRUNING IN DEVELOPMENT:
A NOVEL ACCOUNT IN NEURAL TERMS

Gal Chechik [1], Isaac Meilijson , [1] and Eytan Ruppin [2]

[1]School of Mathematical Sciences
[2]Departments of Computer Science & Physiology
Tel Aviv University, Tel Aviv 69978, Israel

INTRODUCTION

A fundamental phenomenon in brain development is the reduction in the amount of synapses that occurs between early childhood and puberty. Recent studies in humans, substantiated by studies in monkeys [1, 2], show that synaptic density grows steadily until the age of 2 years, remains constant for a few years and then decreases continuously until puberty [3, 4]. The peak level of synaptic density in early childhood is up to 50% to 100% higher than the adult level depending on the brain region.

What advantage could such a seemingly wasteful developmental strategy offer? Some researchers have treated the phenomenon as an inevitable result of synaptic maturation, lacking any computational significance. Others have hypothesized that synapses which are strengthened at an early stage might be later revealed as harmful to overall memory function, when additional memories are stored. Thus, they claimed, synaptic elimination may reduce the interference between memories, and yield better overall performance [5].

This paper shows that in associative memory networks models these previous explanations do not hold, and puts forward a different explanation. Our proposal is based on the assumption that synapses are a costly resource whose efficient utilization is a major optimization goal guiding brain development. This assumption is motivated by the observation that the changes in synaptic density along brain development are highly correlated to the temporal course of changes in energy consumption [6], and by the fact that the brain consumes a large fraction of total energy consumption of the resting adult [6]. By analyzing the network's performance under various synaptic constraints such as limited number of synapses or limited total synaptic strength, we show that if synapses are properly pruned, the performance decrease due to synaptic deletion is small compared to the energy saving. Deriving optimal synaptic pruning strategies, we show that efficient memory storage in the brain requires a specific learning process characterized by initial synaptic over-growth followed by judicious synaptic pruning.

The next section describes the models studied and our analytical results, which are verified numerically and discussed in the last section.

Computational Neuroscience
edited by Bower, Plenum Press, New York, 1998

ANALYTICAL RESULTS

In order to investigate synaptic elimination, we address the more general question of optimal modification of a Hebbian memory matrix. Given previously learned Hebbian synapses we apply a function which changes the synaptic values, and investigate the effect of such a modification function. First, we analyze the way the memory performance depends on a general synaptic modification function. Then, we proceed to derive optimal modification functions under different constraints.

The Models

We investigate synaptic modification in two Hebbian models. The first model is a variant of the canonical Hopfield model. M memories are stored in a N-neuron network forming approximate fixed points of the network dynamics. The synaptic efficacy J_{ij} between the jth (pre-synaptic) neuron and the ith (post-synaptic) neuron is

$$J_{ij} = g(W_{ij}) = g\left(\frac{1}{\sqrt{M}} \sum_{\mu=1}^{M} \xi_i^\mu \xi_j^\mu\right), \quad 1 \le i \ne j \le N; \quad W_{ii} = 0 \quad , \tag{1}$$

where $\{\xi^\mu\}_{\mu=1}^{M}$ are ± 1 binary patterns representing the stored memories, and g is a general modification function over the Hebbian weights, such that $g(z)$ has finite moment if z is normally distributed. The updating rule for the state X_i^t of the ith neuron at time t is

$$X_i^{t+1} = \theta(f_i), \quad f_i = \sum_{j=1}^{N} J_{ij} X_j^t \quad , \tag{2}$$

where f_i is the neuron's input field, and θ is the function $\theta(f) = sign(f)$. The overlap m^μ (or similarity) between the network's activity pattern X and the memory ξ^μ is $m^\mu = \frac{1}{N} \sum_{j=1}^{N} \xi_j^\mu X_j$.

The second model is a variant of a low activity biologically-motivated model [7], in which synaptic efficacies are described by

$$J_{ij} = g(W_{ij}) = g\left(\frac{1}{p(1-p)\sqrt{M}} \sum_{\mu=1}^{M} (\xi_i^\mu - p)(\xi_j^\mu - p)\right) . \tag{3}$$

where ξ^μ are $\{0,1\}$ memory patterns with coding level p (fraction of firing neurons), and g is a synaptic modification function. The updating rule for the state of the network is similar to Eq.(2), with $\theta(f) = \frac{1+sign(f)}{2}$ and $f_i = \sum_j J_{ij} X_j^t - T$ where T is the neuronal threshold set to its optimal value. The overlap m^μ in this model is defined by $m^\mu = \frac{1}{Np(1-p)} \sum_{j=1}^{N} (\xi_j^\mu - p) X_j$.

Optimal Synaptic Modification Functions

To evaluate the impact of synaptic modification on the network's performance, we study its effect on the signal to noise ratio (S/N) of the neuron's input field in the modified models. The S/N is known to be the primary determinant of retrieval capacity [8] (ignoring higher order correlations in the neurons input fields), and is calculated by analyzing the moments of the neuron's field. In the modified Hopfield model this analysis yields

$$\frac{S}{N} = \frac{1}{\sqrt{\alpha}} m_0^\mu \rho(g(z), z) , \tag{4}$$

where $\alpha = M/N$ is the memory load, m_0^μ is the initial overlap between the presented pattern and the stored memory pattern, z is a random variable with standard normal distribution and ρ denotes the correlation coefficient. Similar analysis in the low activity model yields

$$\frac{S}{N} = \frac{1}{\sqrt{\alpha p}} \, m_0^\mu \, \rho(g(z), z). \tag{5}$$

In both models S/N is a product of *independent* terms of the load, the initial overlap and a correlation term which depends solely on the modification function. Thus, the only effect of the modification function g is through the correlation coefficient and the behavior of the two different models under synaptic modification could be investigated by analyzing $\rho(g(z), z)$ only *.

The immediate consequence of Eqs. (4) and (5) is that there is no local synaptic modification function that can improve the performance of the Hebbian network, since ρ has values in the range $[-1, 1]$, and the identity function $g(z) = z$ already gives the maximal possible value of $\rho = 1$. In particular, no deletion strategy can yield better memory performance than the intact network. A similar result was previously shown by [9] in the Hopfield model. The use here of signal-to-noise analysis enables us to proceed and derive optimal functions under different constraints on modification functions, and evaluate the performance of various modification functions.

When no constraints are involved, pruning has no beneficial effect. However, since synaptic activity is strongly correlated with energy consumption in the brain, its resources may be inherently limited in the adult, and synaptic modification functions should satisfy various synaptic constraints. Motivated by theses considerations we proceed and study deletion under two different constraints originating from the assumed energy restriction: limited number of synapses, and limited total synaptic efficacy. We find that the optimal synaptic modification strategy when the amount of synapses is restricted is *"minimal value"* pruning: delete all synapses whose magnitude is smaller than some threshold, and leave all others unmodified

$$g_t(z) = \begin{cases} z & when & z > t \\ 0 & when & |z| < t \\ z & when & z < -t \end{cases} \tag{6}$$

where t is the deletion threshold. This strategy is illustrated in figure 1(a).

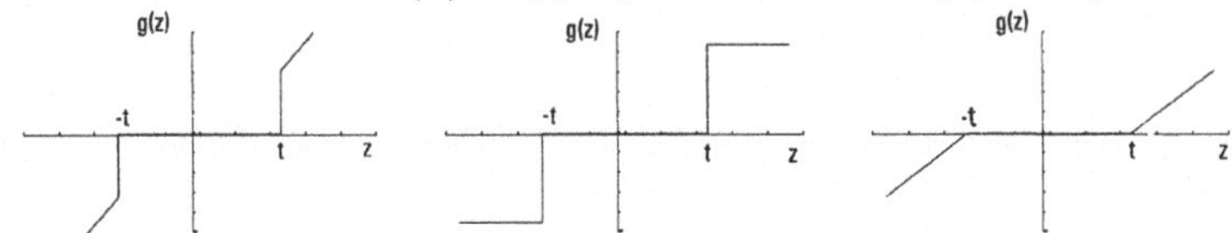

Figure 1. Different synaptic modification strategies. (a) Minimal value deletion: $g(z) = z$ for all $|z| > t$ and zero otherwise (see Eq. 6). (b) Clipping: $g(z) = sign(z)$ for all $|z| > t$ and zero otherwise. (c) Compressed synapses: $g(z) = z - sign(z)t$ for all $|z| > t$ and zero otherwise (see Eq. 7).

As synapses differ by their strength, a different optimization goal may be implied by the energy consumption constraints: minimizing the overall synaptic strength in the network. The optimal modification function under this criterion is

$$g_t(z) = \begin{cases} z - t & when & z > t \\ 0 & when & |z| < t \\ z + t & when & z < -t \end{cases} \tag{7}$$

---

*These results remain valid even if memories are embedded in a noisy synaptic matrix.

We denote this function *"compressed deletion"*, and illustrate it in figure 1(c).

NUMERICAL RESULTS

To quantitatively evaluate the performance gain achieved by the strategies described in the previous section, we measure the network's performance by calculating the networks' capacity as a function of synaptic deletion levels. The capacity is measured as the maximal number of memories which can be stored in the network and retrieved almost correctly ($m^\mu \geq 0.95$), starting from patterns with an initial overlap of $m_0^\mu = 0.8$, after one or ten iterations.

Figure 2 compares three modification strategies: minimal value deletion (Eq. 6), random deletion (independent of the weights strengths) and a clipping deletion strategy previously investigated by [9], and illustrated in figure 1(b). Simulations of the compressed deletion strategy yield similar results and are brought in [10]. Minimal-value deletion is indeed significantly better than the other deletion strategies, but in high deletion levels it is almost equaled by the clipping strategy.

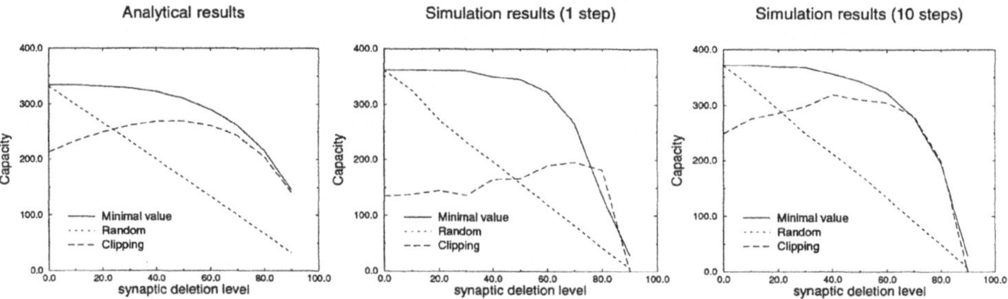

Figure 2. Capacity of a network with different synaptic modification strategies as a function of the synaptic deletion level. Both analytic and simulation results of single step and multiple step dynamics are presented, showing a fairly close correspondence. Simulations were performed in a low activity network with $N = 800$ neurons and coding level $p = 0.1$.

The above results show that if a network must be subjected to synaptic deletion, minimal value deletion will minimize the damage. Yet deletion reduces performance and is hence unfavorable. The question remains then - what is the role of synaptic elimination ? Several computational answers were suggested: Some had hypothesized that synaptic elimination can improve network performance, but this paper proves this argument is incorrect in several associative memory models. Others have claimed that the brain can be viewed as a cascade of filters which can be modeled by feed forward networks models. In these models it is known that a reduction in the amount of free parameters may improve the ability of the network to generalize if the size of the network is too large [11].

Our proposal is that synaptic over-growth and deletion emerge because synaptic resources must be scrupulously utilized, due to metabolic energy consumption constraints. If we have to use a restricted amount of synapses in the adult, better performance is achieved if the synapses are first over-grown, and then cleverly pruned after more memories are stored. Figure 3 compares the performance of networks with the same synaptic resources, but with varying number of neurons. The smallest network is fully connected while larger networks are pruned according to the minimal value deletion strategy to end up with the same amount of synapses. The optimally pruned network is not advantageous over the undeleted network (which has many more synapses), but over other pruned networks, with the same total number of synapses. This optimal network, can

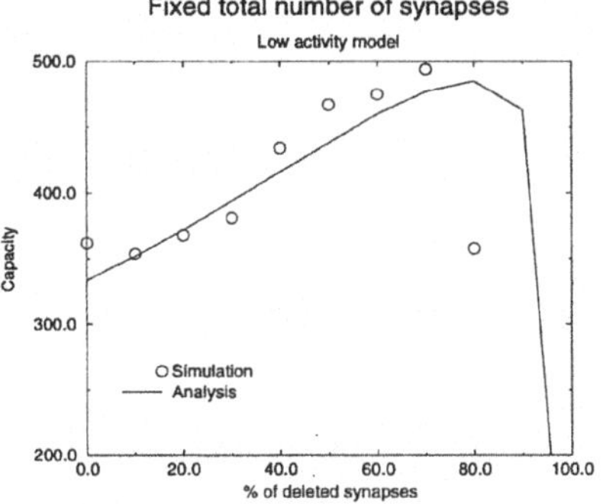

Figure 3. Performance of networks with different number of neurons but the same total number of synapses as a function of network connectivity. As the connectivity decreases the network's size (number of neurons) is inversely increased to keep the total number of synapses (800^2) constant. Capacity at the optimal (80%)deletion is 45% higher than in the fully connected network.

store three times more information than the fully connected network [†].

Implementing minimal-value deletion suggests interesting cognitive predictions: Figure 4. traces the network's performance during continuous memory storage of new patterns accompanied by changes in synaptic density that roughly mimic those taking place during human development. Two observations should be noted: the first is a deterioration in the performance of the "teenage" network as more recent memories are recalled (see the decline in the dashed line). This result is in accordance with psychological data [12].

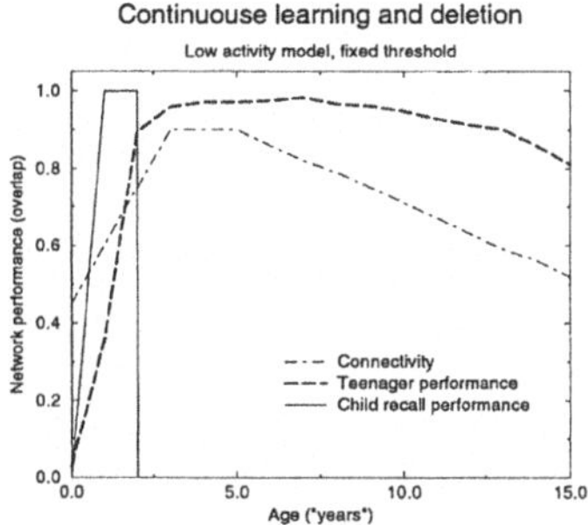

Figure 4. Memory retrieval as a function of storage period. The figure displays both synaptic density and memory performance data. At each time step ("year") 10 memories are stored in the network, and network's connectivity is changed following human data (dot-dashed line). Performance is tested both at "age" 3 when network connectivity has reached its peak (solid line), and at "age" 15 after more memories have been stored in the network and connectivity was reduced (dashed line).

Second, a comparison of retrieval quality of early memories by the teenager network versus the infant network (see figure 4 at the age of 2 years) shows a sharp change in retrieval quality of early memories. This is reminiscent of the infantile amnesia phenomenon, which is the inability of the adult to recall events from infancy that he could previously recall. In our model, this pattern arises from the fact that earlier memories are stored in sparsely-connected networks (that is, embedded in less synapses) and

---

[†]Assuming sub-optimal thresholds, noisy dynamics, or non-zero energy cost for neurons, the optimum drifts to lower deletion levels of $50 - 60\%$ which fits the experimental data.

hence are more sensitive to the noise known to accumulate in the network as additional memories are stored in it. This scenario may provide a network-level perspective to infantile-amnesia, complementing theories suggesting that maturation of memory related structures such as the Hippocampus are responsible for the amnesia [13].

SUMMARY

We have analyzed the effect of modifying Hebbian synapses in an optimal way that maximizes performance, while keeping constant the overall number or total strength of the synapses. Our results predict that during the elimination phase in the brain synapses undergo weight-dependent pruning in a way that deletes the weak synapses. Interestingly, although rising from different basic principles, recent studies in the neuromuscular junction have found synapses that are indeed pruned according to their initial synaptic strength [14].

Synaptic elimination is a broad phenomenon found throughout different brain structures and not restricted to associative memory areas. We believe however that our explanation may be generalized to other network models. For example, feed forward Hebbian projections between consecutive networks share similar properties with a single step of synchronous dynamics of associative memory networks analyzed here. There is also some evidence that synaptic growth followed by deletion is not limited to the developmental stage, but may have a more general scope, and can be found in adults [15]. These interesting open issues await further studies in the future.

REFERENCES

1. J.P. Bourgeois and P. Rakic", Changing of synaptic density in the primary visual cortex of the Rhesus monkey from fetal to adult age, J. Neurosci., 13, 2801-2820, (1993).
2. P. Rakic, J.P. Bourgeois and P.S. Goldman-Rakic, Synaptic development of the cerebral cortex: implications for learning, memory and mental illness", Progress in Brain Research, 102, 227-243, (1994).
3. P.R. Huttenlocher, Synaptic density in human frontal cortex. Development changes and effects of age, Brain Res., 163, 195-205, (1979).
4. P.R. Huttenlocher and C. De Courten, The development of synapses in striate cortex of man, J. Neuroscience, (1987).
5. J.R. Wolff, R. Laskawi, W.B. Spatz and M. Missler, Structural dynamics of synapses and synaptic components, Behavioural Brain Research, 66, 13-20, (1995).
6. Per E. Roland, "Brain Activation", Willey-Liss, NY, 474-476, (1993).
7. M.V. Tsodyks and M. Feigel'man, Enhanced storage capacity in neural networks with low activity level., Europhys. Lett., 6, 101-105, (1988).
8. I. Meilijson and E. Ruppin, Optimal firing in sparsely-connected low-activity attractor networks., Biological cybernetics, 74, 479-485, (1996).
9. H. Sompolinsky, Neural networks with non linear synapses and static noise, Phys Rev A., 34, 2571-2574, (1988).
10. G. Chechik, I. Meilijson and E. Ruppin, Synaptic pruning in development: a computational account, Neural Computation, in press, (1998).
11. R. Reed, Pruning algorithms - a survey, IEEE transactions on neural networks, 4(5), 740-747, (1993).
12. K. Sheingold and J. Tenney, Memory for a salient childhood event, in "Memory Observed", U. Neisser ed., W.H. Freeman and co., San Francisco, (1982).
13. L. Nadel, Infantile amnesia: a neurobiological Perspective, in "Infant Memory; Its Relation To Normal And Pathological Memory In Humans And Other Animals", M. Moscovitch ed., Plenum Press, New-York, (1986).
14. E. Frank, Synapse elimination: for nerves it's all or nothing, Science, 275, 324-325, (1997).
15. W.T. Greenough, J.E. Black and C.S. Wallace, Experience and brain development, Child Development, 58, 539-559, (1987).

A NONLINEAR SYSTEMS APPROACH OF CHARACTERIZING AMPA AND NMDA RECEPTOR DYNAMICS

Sunil S. Dalal,[1,3,4] Vasilis Z. Marmarelis,[1,3] and Theodore W. Berger [1,2,3]

[1]Department of Biomedical Engineering
[2]Program in Neuroscience
[3]Center for Neural Engineering
University of Southern California
Los Angeles, CA 90089-1451

[4]Matsushita Electric Works Research and Development Laboratory, Inc.
Mountain View, CA 94040

ABSTRACT

The nonlinear properties of evoked glutamatergic intracellular dentate granule cell EPSPs were characterized via the Laguerre expansions of the kernels of a Volterra power series. Using pharmacological manipulations in the context of a nonlinear systems approach, we decomposed the glutamatergic EPSP dynamics into its AMPA and NMDA-receptor mediated components. A model that involved simple summation of the AMPA and NMDA kernels power spectra could not recreate the global nonlinear behavior of the intact system. However, a model that included "proportional-plus-first-derivative-plus-second-derivative" positive feedback of the intracellular membrane potential onto the NMDA subsystem was able to do so precisely. The decomposed feedback transfer function, which was found to resemble the mathematical "triplet," represents the nonlinear interaction between the two receptor subtypes.

INTRODUCTION

The perforant pathway provides glutamatergic input to the granule cells of the dentate gyrus. Electrical stimulation of this pathway yields a granule cell EPSP (excitatory postsynaptic potential) comprised of both AMPA and NMDA receptor-mediated components, each of which is facilitated or depressed depending on the pattern of prior stimulation. Hence, both components of the system are nonlinear with respect to interstimulus interval. Furthermore, a functional interaction exists between the AMPA and NMDA receptor channels due to the voltage-dependent blockade inherent to the NMDA channel. The nonlinearity of each subsystem coupled with their interaction creates the problem of quantifying the nature of the interaction.

Computational Neuroscience
edited by Bower, Plenum Press, New York, 1998

PHYSIOLOGICAL BACKGROUND

Excitatory amino acid mediated synaptic transmission in the dentate gyrus is initiated by release of glutamate from presynaptic terminals and its subsequent binding to glutamate receptors on the postsynaptic membrane, causing the opening of channels coupled to those receptors. Perforant path fibers arise from the entorhinal cortex and synapse onto granule cell dendrites in the dentate molecular layer, providing the primary source of excitatory input to dentate granule cells. The dentate molecular layer contains high densities of both AMPA and NMDA glutamate receptor subtypes. These receptor channels are colocalized (Bekkers and Stevens, 1989), cross-modulating the global EPSP in response to stimulation. NMDA receptors mediate a slow component of EPSPs. The NMDA-receptor conductance exhibits a nonlinear current-voltage (I-V) relation, a consequence of a voltage-dependent magnesium blockade of the channel (Mayer et al., 1984). Calcium entering through the NMDA channel initiates signal transduction cascades that influence a wide variety of calcium-dependent mechanisms, including those related to short-term and long-term plasticity (Bliss and Collingridge, 1993). AMPA receptors mediate the fast component of glutamatergic synaptic transmission and have a linear I-V relation (Keller et al., 1991). AMPA receptors provide feedback to NMDA receptors, because the postsynaptic depolarization which accompanies their activation relieves the voltage-dependent blockade of the NMDA receptor/channels.

ANALYTICAL METHODS

The dynamics of pharmacologically isolated AMPA and NMDA receptor-mediated EPSPs were represented in the form of input/output functions. The linear and higher-order nonlinear components of the input/output relations were determined experimentally by stimulating the perforant path afferents to the granule cells with random signals in order to generate a wide range of interactions among the intracellular elements, while simultaneously recording the output of the neurons (Berger et al., 1994). Nonlinearities of the synaptic responses were represented as the kernels of a functional power series, expanded on a set of Laguerre functions. The asymptotic exponential form of the Laguerre functions facilitated this parsimonious representation due to the similar exponential relaxation properties of the subthreshold EPSPs evoked in our study (Marmarelis and Orme, 1993).

Volterra showed that for a nonlinear system, the following series can describe the relation between the input $x(t)$ and the output $y(t)$

$$y(t) = k_0 + \int\limits_0^\infty k_1(\tau)x(t-\tau)d\tau + \int\limits_0^\infty \int\limits_0^\infty k_2(\tau_1, \tau_2)x(t-\tau_1)x(t-\tau_2)d\tau_1 d\tau_2$$

where k_0, k_1, and k_2 are the zero, first, and second order kernels. k_0 represents the output of the system when there is no external input, which in our case is the resting membrane potential of the granule cell. The first order kernel is interpreted as the average of the evoked responses during a given observation period. It represents the best linear model of the system. The second order kernel represents the modulatory effect of a preceding stimulus.

We have chosen the Laguerre expansion technique (Marmarelis and Orme, 1993) to model our system. This technique utilizes the orthonormal basis of discrete-time Laguerre functions to expand the kernels and reduce the number of unknown parameters that need to be estimated. It employs least-squares fitting of the unknown expansion coefficients. After expanding the Volterra kernels on the Laguerre basis $\{L_j\}$, the Volterra series (in discrete time) transforms into the following multinomial power series expression:

$$y(n) = c_0 + \sum_j c_1(j)v_j(n) + \sum_{j_1}\sum_{j_2} c_2(j_1, j_2)v_{j_1}(n)v_{j_2}(n) + \dots$$

where c_0, c_1, and c_2 are the Laguerre expansion coefficients of the zero, first, and second order kernels respectively. The Laguerre expansion coefficients are estimated by linear regression of the output data on the terms of the multinomial power series expression. The term $v_j(n)$ represents the convolution of the input data with the jth order discrete-time orthonormal Laguerre function, $L_j(m)$. It is defined as:

$$v_j(n) = \sum_M L_j(m)x(n - m)$$

where M is the kernel memory extent in the number of lags. The discrete-time Laguerre functions themselves are expressed as:

$$L_j(m) = \alpha^{\frac{m-j}{2}}(1 - \alpha)^{\frac{1}{2}} \sum_{k=0}^{j}(-1)^k \binom{m}{k}\binom{j}{k} \alpha^{j-k}(1 - \alpha)^k \quad (m \geq 0)$$

where α is the discrete-time Laguerre parameter ($0 < \alpha < 1$) which determines the rate of exponential decline of these functions (Marmarelis and Orme, 1993).

EXPERIMENTAL METHODS

Intracellular sharp microelectrode recordings of granule cell EPSPs were performed in the rabbit hippocampal slice preparation. The slices consisted of 500 μm tissue sections and were kept at 33°C in physiological medium that consisted of (mM): $MgSO_4$ (1), NaCl (126), KCl (5), NaH_2PO_4 (1.25), $NaHCO_3$ (26), glucose (10), ascorbic acid (2), and $CaCl_2$ (2). The intracellular microelectrodes were filled with 1 M potassium acetate (KAc). To prevent IPSPs (inhibitory postsynaptic potentials) from contaminating EPSPs, bicuculine (20 μM) and saclofen (250 μM) were included in the perfusate to block GABAergic inhibition.

Bipolar stimulating electrodes were positioned to activate medial perforant path fibers with a train of 1024 pulses having interpulse intervals determined by a Poisson process with a mean frequency of 2 Hz. The random impulse train (RIT) had a range of interstimulus intervals that varied from 1 ms to 5000 ms, the mean interval being 500 ms. Throughout delivery of the train, the synaptic responses from a single granule cell were recorded. In this manner, the intact system comprised of both the AMPA and NMDA components was characterized.

The AMPA and NMDA components then were pharmacologically isolated by adding either D-APV (50 μM) or CNQX (10 μM), respectively, to the medium (Keller et al., 1991). First, the AMPA receptor dynamics were assesed by applying D-APV and then recording the responses to RIT stimulation. D-APV was then fully washed out before applying CNQX to isolate the NMDA mediated component. Finally, the NMDA responses to RIT stimulation were recorded.

MODELING METHODS

The block diagram illustrating the control system representing excitatory granule cell dynamics is shown in Figure 1. We derive time domain numerical methodical

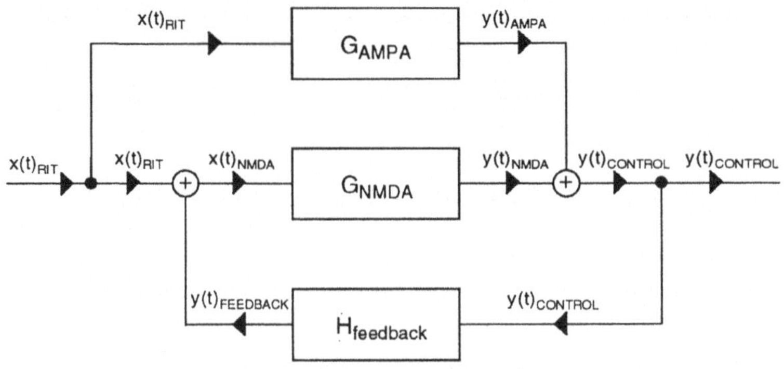

Figure 1. Positive feedback systems model of glutamatergic synaptic transmission. G_{AMPA} and G_{NMDA} are the experimentally based kernels representing the AMPA and NMDA conductances, and $H_{FEEDBACK}$ is the unknown transfer function of the feedback element that has to be computationally determined. $x(t)_{RIT}$ is the input sequence that was experimentally applied to obtain the kernels, and $y(t)_{CONTROL}$ represents the output of a granule cell to perforant path stimulation. $x(t)_{NMDA}$ is the "closed-loop" NMDA input driving the "open-loop" NMDA kernels. Hence, $y(t)_{NMDA}$ is the "closed-loop" NMDA output to RIT stimulation. $y(t)_{CONTROL}$ drives the feedback transfer function, and $y(t)_{FEEDBACK}$ represents the output of the feedback element.

procedures in this section to solve for the feedback transfer function representing the nonlinear interaction between AMPA and NMDA receptors. We made the assumption that under "closed-loop" conditions, the input to the NMDA subsystem is modulated, not the properties of the NMDA subsystem itself. We defined the closed-loop output of the NMDA subsystem as the differential between the control and AMPA signals:

$$y(n) = y_c(n) - y_a(n)$$

We obtained the closed-loop NMDA input signal, $x(n)$, using a combination of a globally convergent version of Newton's method and singular value decomposition on individual segments of length equal to memory size, M (Press et al., 1992). We formulated the problem as the set of equations

$$F(x, n) = 0$$

where the nth equation is defined by

$$F(x, n) = -y(n) + k_0 + \sum_{m=1}^{M} k_1(m)x(n + 1 - m)$$

$$+ \sum_{m_1=1}^{M} \sum_{m_2=1}^{M} k_2(m_1, m_2)x(n + 1 - m_1)x(n + 1 - m_2) + \ldots$$

Hence, we determined the closed-loop stimulus time-history for the NMDA subsystem, $x(t)_{NMDA}$, by deconvolving in the time-domain. Next, we determined the predicted response time-history of the feedback branch, $y(t)_{FEEDBACK}$, by subtracting the input stimulus, $x(t)_{RIT}$, from $x(t)_{NMDA}$. Finally, we determined the unknown feedback kernels, $H_{FEEDBACK}$, via the Laguerre expansion technique. The predicted control response time-history, $y(t)_{CONTROL}$, was taken as the input stimulus to the feedback subsystem, and $y(t)_{FEEBACK}$ was used as the output (Dalal, 1997).

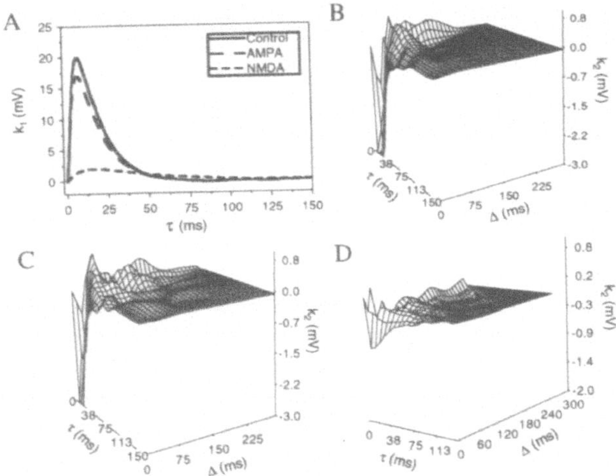

Figure 2. A: First order kernels of the control, AMPA, and NMDA subsystems. B: Control second order kernel. C: AMPA second order kernel. D: NMDA second order kernel. The second order kernels are presented in τ-Δ format, where Δ is the time interval between input impulses ($\Delta = \tau_2 - \tau_1$), and τ is the observation period. (n=5 cells)

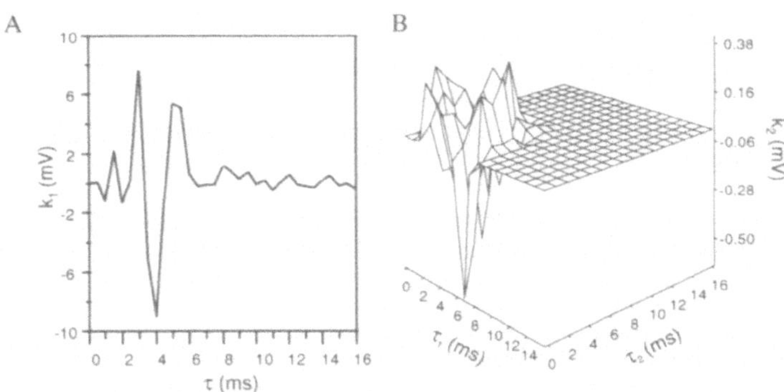

Figure 3. Decomposed feedback element transfer function. A. First order kernel. B. Second order kernel. Note that both the first order kernel and diagonal of the second order kernel have the characteristic shape of the mathematical "triplet."

ANALYTICAL AND MODELING RESULTS

The kernels for the control, AMPA, and NMDA subsystems were calculated (Figure 2). A second order model was sufficient to represent each component of the system, as the cumulative mean square error (the difference between the measured output and that predicted by the estimated kernels) was only approximately 1.2% for the 1024 inputs in each case. Next, the kernels of the feedback element, $H_{FEEDBACK}$, were determined through time-domain decomposition techniques (Figure 3). A delay of 2 ms had to be incorporated into the feedback in order to achieve mathematical convergence. The first order feedback kernel was found to have the mathematical resemblance of a "triplet," and the diagonal of the second order kernel seemed to mimic that characteristic shape also. The "triplet" finding indicates that the feedback is "proportional-plus-first-derivative-plus-second-derivative" in nature (i.e. a PDD controler).

DISCUSSION

We reconstructed the global nonlinear behavior of evoked granule cell EPSPs by first characterizing the nonlinearities of the AMPA and NMDA subcomponents individually and then devising a model that accounted for their nonlinear interaction. There are several mechanisms that are responsible for the biological processes underlying the feedback element regulating granule cell dynamics. Because the feedback kernels are a nonparametric assessment, it includes all mechanisms, both known and unknown, underlying its function. We plan to extend the model to explain how certain patterns of activity induce AMPA-dependent long-term potentiation (LTP) while other patterns induce NMDA-dependent LTP or long-term depression (LTD). By augmenting the gain of either the AMPA or NMDA transfer functions, we can perform simulation studies to test how such alterations affect the global input/output relationship of the granule cell. Such studies would provide valuable insight into how the hippocampus transforms sensory information into both short-term and long-term memories.

ACKNOWLEDGEMENT

Supported by ONR, NCRR, and NIMH.

REFERENCES

Bekkers, J. M., and Stevens, C. F., 1989, NMDA and non-NMDA receptors are colocalized at individual excitatory synapses in cultured rat hippocampus, *Nature.* 341:230-233.

Berger, T. W., Chauvet, G., and Sclabassi, R. J., 1994, A biologically based model of functional properties of the hippocampus, *Neural Networks.* 7:1031-1064.

Bliss, T. V. P., and Collingridge, G. L., 1993, A synaptic model of memory: long-term potentiation in the hippocampus, *Nature.* 361:31-39.

Dalal, S. S., 1997, "A Nonlinear Systems Identification Approach to Characterizing AMPA and NMDA Receptor Dynamics in Dentate Gyrus: Delineation of a Positive Feedback Loop Regulating Granule Cell Dynamics," Doctoral dissertation, University of Southern California.

Keller, B. U., Konnerth, A., and Yaari, Y., 1991, Patch clamp analysis of excitatory synaptic currents in granule cells of rat hippocampus, *J. Physiol. (Lond.)* 435:275-293.

Marmarelis, V. Z., and Orme, M. E., 1993, Modeling of neural systems by use of neuronal modes, *IEEE Trans Biomed Eng.* 40:1149-1158.

Mayer, M. L., Westbrook, G. L., and Guthrie, P. B., 1984, Voltage-dependent block by Mg^{2+} of NMDA responses in spinal cord neurons, *Nature.* 309:261-263.

Press, W. H., Teukolsky, S. A., Vetterling, W. T., and Flannery, B. P., 1992, "Numerical Recipes in FORTRAN," Cambridge University Press, New York.

DETAILED MODEL OF RYANODINE RECEPTOR-MEDIATED CALCIUM RELEASE IN PURKINJE CELLS

Erik De Schutter

Born-Bunge Foundation
University of Antwerp - UIA
B2610 Antwerp
Belgium
erik@bbf.uia.ac.be

ABSTRACT

Ryanodine receptor-mediated calcium release was modeled in a compartmental model of the Purkinje cell with a detailed representation of buffered calcium diffusion. For several important parameters no constraining experimental data were available. The importance of the threshold of activation of release and of the unbinding rate of the main calcium buffer in the stores (calsequestrin) is described. Using reasonable assumptions for these parameters, calcium influx during voltage-gated calcium spikes can activate ryanodine receptors in the submembrane region without noticeably changing the firing pattern of the model.

INTRODUCTION

It has been known for a long time that dendrites of cerebellar Purkinje cells contain high densities of intracellular Ca^{2+} stores,[1] making these cells a popular preparation for the study of Ca^{2+} release mechanisms. Calcium release from stores can be activated by IP_3 receptors, subsequent to synaptic stimulation of metabotropic glutamate receptors,[2] or by Ca^{2+} itself activating ryanodine receptors.[3,4] But the physiological role of these processes remains unclear and positive evidence for Ca^{2+} release under normal *in vivo* conditions is still lacking. As a first step towards a better understanding, we have modeled the effect of ryanodine receptor-mediated Ca^{2+} release, also called Ca^{2+}-induced Ca^{2+} release (CICR), in Purkinje cells.

METHODS

These modeling studies used a new version of a highly detailed compartmental model of the Purkinje cell with voltage-gated channels in the soma and dendrites.[5] In particular, the dendrite contained a P-type and a T-type Ca^{2+} channel and two types of Ca^{2+}-activated K^+ channels and a persistent K^+ channel, while in addition the soma also contained a fast and a persistent Na^+ channel, a delayed rectifier and A-current K^+ channel, and an anomalous rectifier channel.[5] The new version of the model incorporated a detailed representation of calcium dynamics based on buffered, radial diffusion of calcium.[6] It also computed the electrogenic effect of the plasma membrane Ca^{2+}-ATPase pump and of the Na^+-Ca^{2+} exchanger.[6]

Radial diffusion of Ca^{2+} was modeled using concentric shells with a Δr of 0.2 μm, resulting in a total of 7471 shells for the 1604 compartments in the model. We used a D_{Ca} of 2.10^{-6} cm^2sec^{-1}, based on measurements of Ca^{2+} diffusion rates in cytoplasm.[7] A very high buffer capacity has been estimated for Purkinje cells.[3,8] The non-mobile buffer capacity was put at 2080 by including 4 μM of a slow buffer with a K_d of 1.9 μM. Additionally fura-2 was simulated, represented as 75 mM of a mobile buffer (D_{fura} 2.10^{-6} cm^2sec^{-1} with a K_d of 0.2 μM). The fura-2 estimated $[Ca^{2+}]$ was computed using standard methods.[6,9]

Calcium stores were modeled as pools occupying 20 % of the volume of each diffusion shell, based on an electron microscopic reconstruction of the ER in Purkinje cells.[1] This store volume was converted into store surface using the assumption that the ER could be represented as a long tube with a diameter of 0.050 μm.[10] Calcium uptake by the Ca^{2+}-ATPase pump and Ca^{2+} release by ryanodine receptor activation were modeled using equations based on those of Goldbeter et al.[11] These Hill-function based equations were modified to include a time constant of activation[12] and release was proportional to the concentration gradient between store and cytoplasm.[6] Stores contained a moderately high Ca^{2+} concentration at rest (200 μM) and 100 mM of the low affinity buffer calsequestrin (K_d 1 mM).

RESULTS

Tuning of Free Parameters

CICR is in first analysis a regenerative system as the released Ca^{2+} can activate more release. Most modeling studies of CICR have assumed that CICR continues on timescales of seconds to minutes till the stores are empty.[11,13] It is unlikely that the complete ER in a region of the Purkinje cell dendrite empties, as in the model the stores contain 99.985% of the Ca^{2+} present in the cell. In fact, instantaneous release would cause a cytoplasmic $[Ca^{2+}]$ of more than 100 μM! Slower complete emptying of stores might of course be possible, but the experimental evidence argues against this in Purkinje cells[3,4] and in other preparations.[14] We have assumed that CICR operates only in a limited range of cytoplasmic $[Ca^{2+}]$; below an activation threshold there was no CICR[4,12] and at higher concentrations the Ca^{2+}-ATPase uptake (SERCA3[15]) dominated. The parameters for uptake into the stores were: V_{max} of 4.10^{-9} $μM^{-2}msec^{-1}$ and K_d of 1 μM; and for CICR: V_{max} of 10^{-8} $μM^{-2}msec^{-1}$ or 3.10^{-8} $μM^{-2}msec^{-1}$, time constant of 1.2 msec and K_d of 0.3 μM.[6] Another potential mechanisms limiting the ryanodine-receptor mediated release would be inactivation of the receptor,[12] but this inactivation is not complete and was not included in the model.

Even when constrained by the above assumptions, the model was very sensitive to the values of its parameters, most of which were not better than educated guesses. In particular, CICR produced reasonable results only if calsequestrin was assumed to have extremely slow backward binding rates (1 sec^{-1} or 10 sec^{-1}). Unfortunately these binding rate are not known as only the steady state kinetics of calsequestrin have been measured, though indirect evidence exists for slow unbinding.[16] With faster unbinding rates Ca^{2+} release dumped the complete contents of the stores fast enough to raise the cytoplasmic $[Ca^{2+}]$ beyond 20 μM.

We used also higher threshold values for induction of Ca^{2+} release (120 nM or 200 nM) than reported for Purkinje cells,[4] because the experimental values are distorted by the spatial averaging of the fura-2 signal.[6] In fact, the apparent $[Ca^{2+}]$ threshold for CICR (80 nM) that could estimated from the computed fura-2 signal in a spiny dendrite of both the 120 nM and 200 nM threshold models was very close to the experimentally observed values.[4]

Finally, we tuned the parameters of the CICR model so that CICR was activated during somatic current injection without causing noticeable changes to the firing pattern of the model. This assumption was based on the experimental observation that Purkinje cell firing patterns look normal in cells where subsequent addition of low concentrations of caffeine evokes large Ca^{2+} signals due to CICR[4,17] and that Ca^{2+} influx through voltage-gated channels can activate CICR.[3] The model could reproduce the effects of caffeine application on spontaneous firing patterns,[4,17] simulated by increasing the V_{max} of CICR (results not shown).

In the rest of this paper we will compare two versions of the release model: the low-threshold model (120 nM of cytoplasmic [Ca^{2+}] needed to cause CICR in the same shell) had a V_{max} of 10^{-8} μM^{-2}msec^{-1} and a 1 sec^{-1} backward binding rate for calsequestrin, while the high-threshold model (200 nM) had a V_{max} of 3.10^{-8} μM^{-2}msec^{-1} and a 10 sec^{-1} unbinding rate.

Calcium Release During Spontaneous Dendritic Spiking

Under these conditions the model showed several interesting results. Figure 1 demonstrates that the firing pattern of the two release models with CICR (1B and C) looked very similar to that of a Purkinje cell model without Ca^{2+} stores (1A). In fact, the only effect was a decreased delay to the first dendritic spike after onset of the somatic current injection and a slightly sharper dendritic spike as recorded in the soma. While these differences may seem significant when these models are compared to each other, it should be noted that they fall within the natural variability as observed during recordings of Purkinje cell activity in slice[18]. In other words, the traces produced by the two models with CICR look like any normal Purkinje cell, demonstrating that significant Ca^{2+} release is possible (Figure 2) without causing a specific change in the cell's firing pattern.

The ryanodine receptor-mediated calcium release caused complex changes to the cytoplasmic Ca^{2+} levels during the dendritic spikes. Each dendritic spikes evoked CICR in the smooth dendritic compartment shown, causing prolonged increases of the submembrane [Ca^{2+}] (Figure 2, compare to Figure 6.6 of De Schutter and Smolen[6] for the [Ca^{2+}] changes in a model without CICR). A constant observation in our simulations was that this CICR, which was triggered by the voltage-gated Ca^{2+} entry, always remained localised to the submembrane

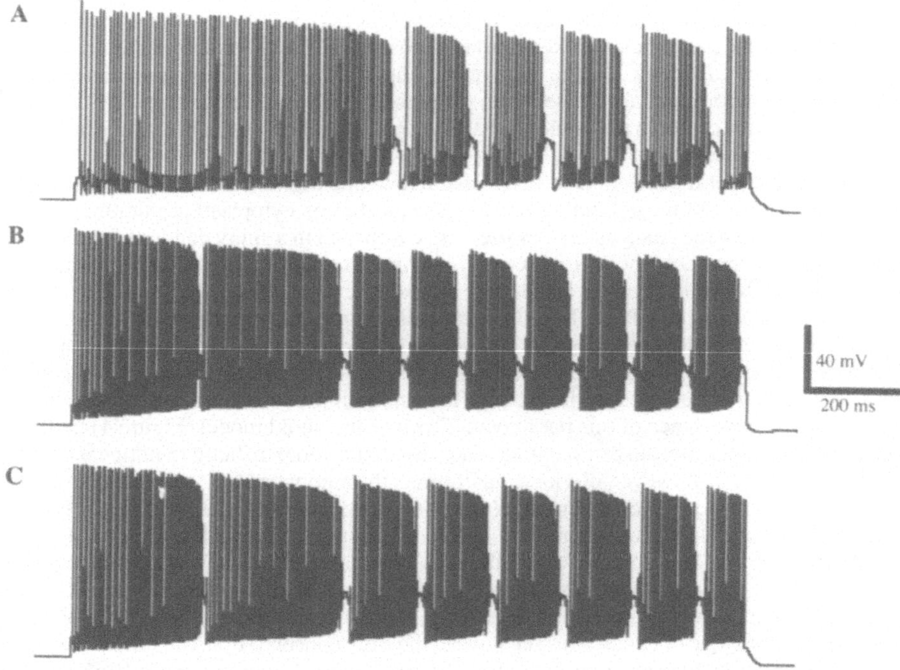

Figure 1. The activation of ryanodine receptor mediated calcium release may have little effect on membrane potential recordings. Somatic recordings from models without (**A**), with high-threshold (**B**) or low-threshold (**C**) calcium stores during a 1.5 nA current injection in the soma.

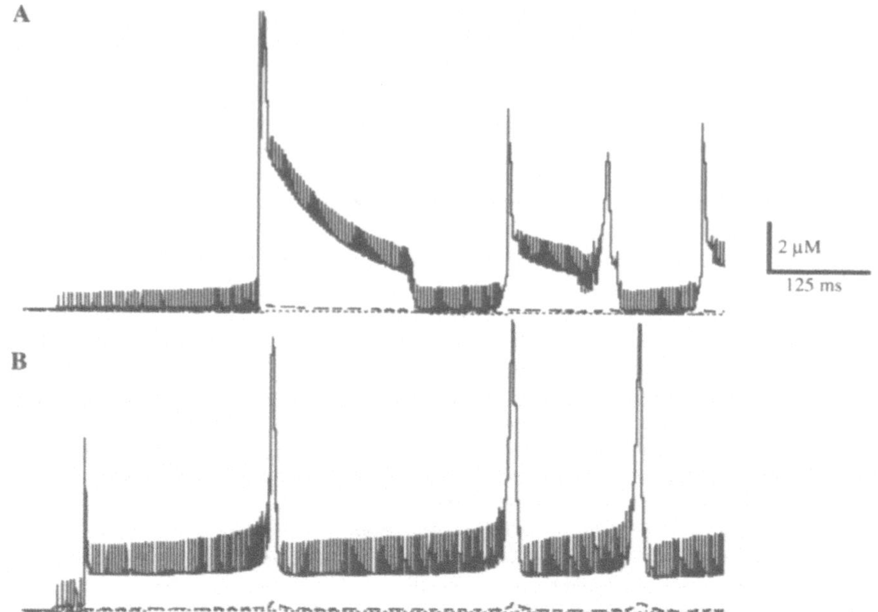

Figure 2. Changes in cytoplasmic calcium concentration in the smooth dendrite (diameter 8.6 μm) during activation of ryanodine receptor mediated calcium release. Recordings from models with high-threshold (**A**) or low-threshold (**B**) calcium stores during a 1.5 nA current injection in the soma. The concentrations in the submembrane shell (full line), in a shell 2 μm lower (broken line) and in the core (stippled line) are shown.

region of large compartments. This can be observed in Figure 2, where the trace from the shell 2 μm below the membrane (broken line) is almost flat. This lack of propagation towards deeper regions can be explained by the relative high threshold for CICR initiation and suggests that the conditions for quantal release[14] may be present. We could not evoke Ca^{2+} waves[13] traveling towards the center in the model, as has been observed in the somata of some neurons[19] but not in Purkinje cells.[4]

Next we consider the differences between the low- and high-threshold models and more in particular the effect of the unbinding rate of calsequestrin on cytoplasmic calcium profiles. Figures 3 and 4 show the changes in the stores and cytoplasm in a spiny dendrite for the high-threshold and low-threshold model respectively. An important difference with the much larger smooth dendrite shown in Figure 2 is that here almost no cytoplasmic $[Ca^{2+}]$ gradient existed both before and during the CICR. Looking first at the high-threshold model in Figure 3 it is obvious that, after the initial peak in $[Ca^{2+}]$ caused by each dendritic spike, the $[Ca^{2+}]$ transients followed the inverse of the free calsequestrin curve. In other words, the evolution of Ca^{2+} release to a steady state was rate-limited by the unbinding of Ca^{2+} from the store buffer, emphasising the importance of this parameter. The low-threshold model (Figure 4) had even slower buffer kinetics. Because of the time scales used, it is more difficult to notice the buffer unbinding effect. It is, however, obvious that it limits the amount of Ca^{2+} that can be released from the stores in the low-threshold model (which also had a lower V_{max} for release). The low threshold also caused low levels of release before the first dendritic spike (Figure 2**B**).

Two observations are common to both models. First, the Ca^{2+} uptake was not homogenous over the shells. The submembrane stores removed a large part of the voltage-gated Ca^{2+} influx into the compartment, causing a constant rise in their concentration before the phase of release. Because the rate of release was sensitive to the store concentration,[6] this rise contributed to the inititation of the first release phase. Second, the steady state cytoplasmic $[Ca^{2+}]$ rose to rather high levels in spiny dendrites that are probably unrealistic, though data from such tiny dendrites are lacking.[4] This high $[Ca^{2+}]$ can be explained by either too high a V_{max} for release in the model or, more likely, by the absence of a mechanism which can end CICR.

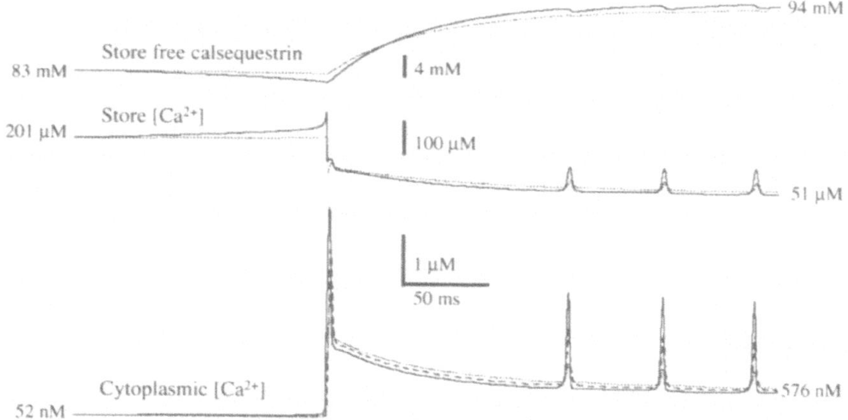

Figure 3. Dynamics in a spiny dendrite (diameter 1.3 μm) of the high-threshold release model with slow store buffer kinetics. Changes in (from top to bottom) the free buffer and calcium concentrations in the stores and in the cytoplasmic calcium concentration during the current injection shown in Figure 1B. For each location the concentrations in the submembrane shell (full line) and in the core (stippled line) are shown. The fura-2 estimated cytoplasmic calcium concentration is shown by the broken line. For each set of traces the initial concentration in the submembrane shell is shown at the left and the final concentration at the right.

DISCUSSION

Our earlier model of the Purkinje cell,[5,20] with simple Ca²⁺ dynamics, was much more constrained by experimental data than the current model. As such, the earlier model has been quite successful in predicting new experimental results, like the importance of inhibition[21,22] and the dendritic Ca²⁺ inflow caused by focal parallel fiber activation.[23,24] The current model should be considered to be much more explorative; it inspires us to think about how CICR could function under normal conditions.

Nevertheless, the current model reproduced many experimental results, like the effect of high CICR rates on firing patterns[17] and the threshold for activation and the activation by complex spikes.[4,6] Moreover, it showed that CICR may occur during bursting without visibly affecting the membrane potential record (Figure 1). This is a very important result, because intuitively this was not expected. Similarly, the presence of a threshold in cytoplasmic [Ca²⁺] for CICR activation hinders propagation to low [Ca²⁺] regions (Figure 2).

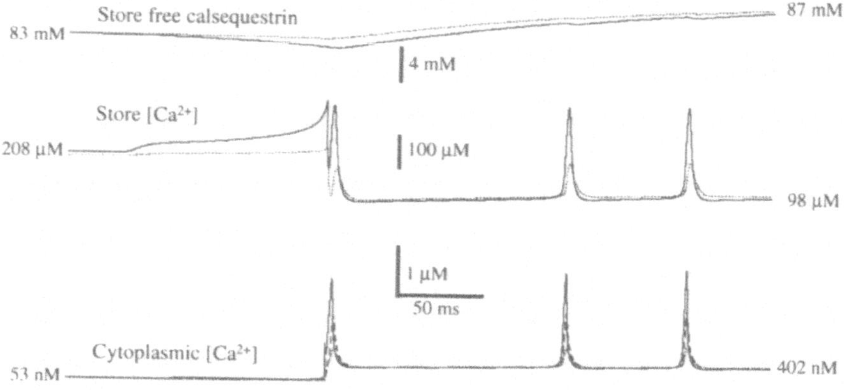

Figure 4. Dynamics in a spiny dendrite of the low-threshold release model with very slow store buffer kinetics during the current injection shown in Figure 1C. Same conventions as in Figure 3.

Another important conclusion is that if CICR persists beyond the first few milliseconds of release, it will be dominated by the unbinding kinetics of the store buffers (Figures 3, 4). Unfortunately this important parameter has never been measured directly! Indirect evidence for very slow unbinding can be found, however, in Ikemoto et al.[16] who show that caffeine-induced release from sarcoplasmic reticulum vesicles begins with a fast rate followed by a slow release rate of about 2 sec^{-1}. The authors show that this second phase is related to binding to the calsequestrin buffer and we propose that the slow rate probably reflects the buffer unbinding rate, while the faster initial rate follows free Ca^{2+} being released at V_{max}.

Supported by the Fund for Scientific Research (Flanders) and by NIMH 1-R01-MH52903.

REFERENCES

1. M.E. Martone, Y. Zhang, V.M. Simpliciano, B.O. Carragher, and M.H. Ellisman, Three-dimensional visualization of the smooth endoplasmatic reticulum in Purkinje cell dendrites, *J. Neurosci.* 13:4636 (1993).
2. I. Llano, J. Dreessen, M. Kano, and A. Konnerth, Intradendritic release of calcium induced by glutamate in cerebellar Purkinje cells, *Neuron* 7:577 (1991).
3. I. Llano, R. DiPolo, and A. Marty, Calcium-induced calcium release in cerebellar Purkinje cells, *Neuron* 12:663 (1994).
4. M. Kano, O. Garaschuk, A. Verkhratsky, and A. Konnerth, Ryanodine receptor-mediated intracellular calcium release in rat cerebellar Purkinje neurones, *J. Physiol.* 487:1 (1995).
5. E. De Schutter, and J.M. Bower, An active membrane model of the cerebellar Purkinje cell. I. Simulation of current clamps in slice., *J. Neurophysiol.* 71:375 (1994).
6. E. De Schutter, and P. Smolen, Calcium dynamics in large neuronal models in *Methods in Neuronal Modeling: from Synapses to Networks*, C. Koch, and I. Segev, eds., MIT Press, Cambridge, MA (1998).
7. M.L. Albritton, T. Meyer, and L. Stryer, Range of messenger action of calcium ion and inositol 1,4,5-triphosphate, *Science* 258:1812 (1992).
8. L. Fierro, and I. Llano, High endogenous calcium buffering in Purkinje cells from rat cerebellar slices, *J. Physiol.* 496:617 (1996).
9. H. Blumenfeld, L. Zablow, and B. Sabatini, Evaluation of cellular mechanisms for modulation of calcium transients using a mathematical model of fura-2 Ca^{2+} imaging in *Aplysia* sensory neurons, *Biophys. J.* 63:1146 (1992).
10. S.L. Palay, and V. Chan-Palay. *Cerebellar Cortex*, Springer-Verlag, New York (1974).
11. A. Goldbeter, G. Dupont and M.J. Berridge, Minimal model for signal-induced Ca^{2+} oscillations and for their frequency encoding through protein phosphorylation, *Proc. Natl. Acad. Sci. USA* 87:1461 (1990).
12. S. Györke, and M. Fill, Ryanodine receptor adaptation: control mechanism of Ca^{2+}-induced Ca^{2+} release in heart, *Science* 260:807 (1993).
13. J. Sneyd, A.C. Charles, and M.J. Sanderson, A model for the propagation of intracellular calcium waves, *Amer. J. Physiol.* 266:C293-C302 (1994).
14. T.R. Cheek, M.J. Berridge, R.B. Moreton, K.A. Stauderman, M.M. Murawsky, and M.D. Bootman, Quantal Ca^{2+} mobilization by ryanodine receptors is due to all-or-none release from functionally discrete intracellular stores, *Biochem. J.* 301:879 (1994).
15. F. Baba-Aissa, L. Raeymaekers, F. Wuytack, G. Callewaert, L. Dode, L. Missiaen, and R. Casteels, Purkinje neurons express the SERCA3 isoform of the organellar type Ca^{2+}-transport ATPase, *Molec. Brain Res.* 41:169 (1996).
16. N. Ikemoto, M. Ronjat, L.G. Mészáros, and M. Koshita, Postulated role of calsequestrin in the regulation of calcium release from sarcoplasmic reticulum, *Biochemistry* 28:6764 (1989).
17. J.R. Brorson, D. Bleakman, S.J. Gibbons, and R.J. Miller, The properties of intracellular calcium stores in cultured rat cerebellar neurons, *J. Neurosci.* 11:4024 (1991).
18. R.R. Llinás, and M. Sugimori, Electrophysiological properties of *in vitro* Purkinje cell somata in mammalian cerebellar slices, *J. Physiol.* 305:171 (1980).
19. Y.M. Usachev, and S.A. Thayer, All-or-none Ca^{2+} release from intracellular stores triggered by Ca^{2+} influx through voltage-gated Ca^{2+} channels in rat sensory neurons, *J. Neurosci.* 17:7404 (1997).
20. E. De Schutter, Modelling the cerebellar Purkinje cell: Experiments in computo, *Prog. Brain Res.* 102:427 (1994).
21. D. Jaeger, E. De Schutter, and J.M. Bower, The role of synaptic and voltage-gated currents in the control of Purkinje cell spiking: a modeling study, *J. Neurosci.* 17:91 (1997).
22. D. Jaeger, and J.M. Bower, The function of background synaptic input in cerebellar Purkinje cells explored with dynamic current clamping, *Abstr. Soc. Neurosci.* 22:494 (1996).
23. E. De Schutter, and J.M. Bower, Simulated responses of cerebellar Purkinje cells are independent of the dendritic location of granule cell synaptic inputs, *Proc. Natl. Acad. Sci. USA* 91:4736 (1994).
24. J. Eilers, G.J. Augustine, and A. Konnerth, Subthreshold synaptic Ca^{2+} signaling in fine dendrites and spines of cerebellar Purkinje neurons, *Nature* 373:155 (1995).

SOMATO-DENDRITIC INTERACTIONS UNDERLYING ACTION POTENTIAL GENERATION IN NEOCORTICAL PYRAMIDAL CELLS IN VIVO

Alain Destexhe,[1] Eric J. Lang[2] and Denis Paré[1]

[1] Laboratoire de Neurophysiologie, Université Laval,
Québec G1K 7P4, Canada

[2] Department of Physiology and Neuroscience, New York University,
New York, NY 10016, USA

Action potential (AP) generation and propagation through the dendritic tree of pyramidal neurons has received much attention lately[1]. Experiments indicate that inhibitory postsynaptic potentials (IPSPs) have a decisive effect in controlling action potential invasion in dendrites[2, 3]. However, these phenomena were observed in slices, and remain to be investigated under *in vivo* conditions where neurons are subject to intense synaptic activity. Here, we have combined computational models with intracellular recordings of neocortical pyramidal cells in order to investigate how spikes can be controlled by IPSPs *in vivo*.

EXPERIMENTAL OBSERVATIONS

We obtained intracellular recordings of morphologically-identified pyramidal neurons (n=42) from the parietal cortex (areas 5-7) in barbiturate-anesthetized cats and compared APs evoked by synaptic inputs, antidromic invasion, or direct current injection (methods were described in detail elsewhere[4]). The neurons were recorded within 1-2 mm of a ten-electrode array (Fig. 1A).

Evidence was obtained that cortical stimuli delivered at different depths activate partially segregated sets of afferents that preferentially end at corresponding cortical depths, thus exerting maximal effects on different compartments of pyramidal neurons. As shown in Fig. 1B, IPSPs elicited by deep cortical shocks were larger in amplitude (12.4 ± 2.25 mV compared to 4.8 ± 1.89 mV), had a shorter peak-latency (23.4 ± 2.93 ms compared to 41.3 ± 5.33 ms), a more positive reversal potential (-73.8 ± 1.64 mV compared to -80.3 ± 1.75 mV) and were associated with larger decreases in input resistance ($63 \pm 8.4\%$ compared to $27 \pm 5.1\%$) than those elicited by superficial cortical stimuli. These differences were statistically significant (paired t-test, $p < 0.05$).

Using this paradigm, we investigated the effect of synaptic activity occurring in different somatodendritic compartments on the shape of the somatic action potential.

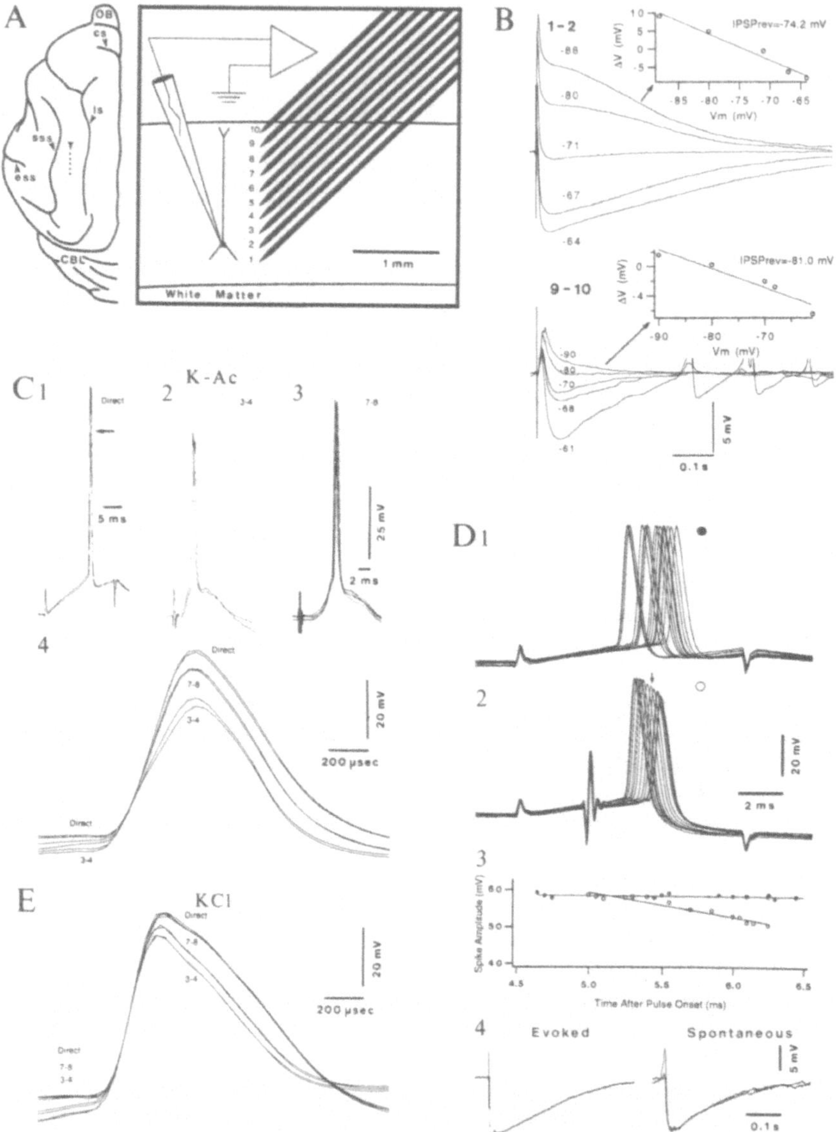

Figure 1. Spike attenuation in neocortical pyramidal cells *in vivo*.
A. Experimental paradigm. B. EPSP/IPSP sequences for proximal (1-2) and superficial (9-10) microstimulation. C. Spike attenuation with K-Acetate-filled electrodes (Direct = spike evoked by current
injection). D. Spikes evoked by current injection. 1: control, 2: in the presence of evoked IPSP, 3:
progressive attenuation of amplitude with latency; 4: comparison between evoked and spontaneous
IPSPs. E. Spike attenuation with Chloride-filled electrodes. Modified from a previous study[4].

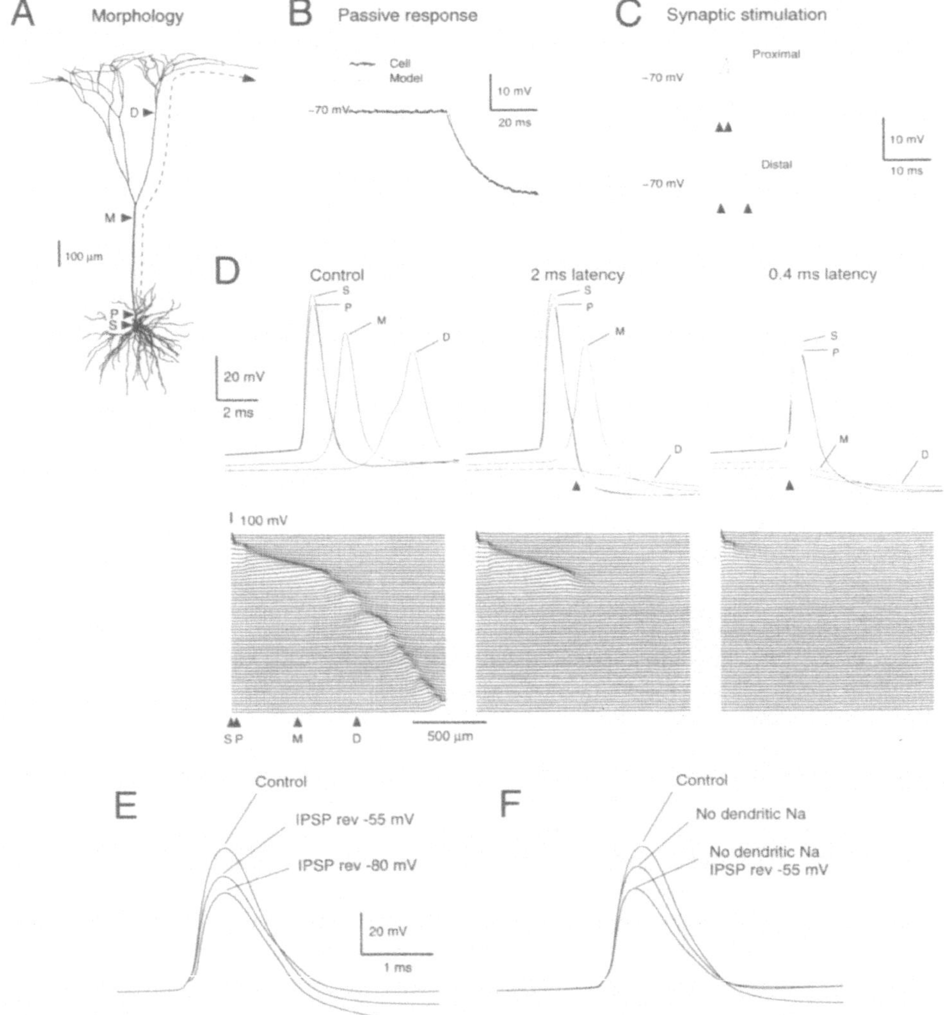

Figure 2. Model of spike attenuation in cortical pyramidal cells.
A. Morphology of a reconstructed layer V pyramidal cell. B. Passive response of the model. C. EPSP/IPSP sequences evoked by proximal and distal stimulation. D. Attenuation of spikes by IPSPs. The spike evoked by current injection is shown in control conditions and with IPSP at 2 different latencies. Bottom graphs in D show the spatial profile of membrane potential (path indicated by dotted line in A). These plots were made over a period of 12 ms in steps of 0.2 ms (from top to bottom). E. Effect of IPSP reversal on spike amplitude. F. Effect of removing dendritic Na^+ channels on spike amplitude. Modified from a previous study[4].

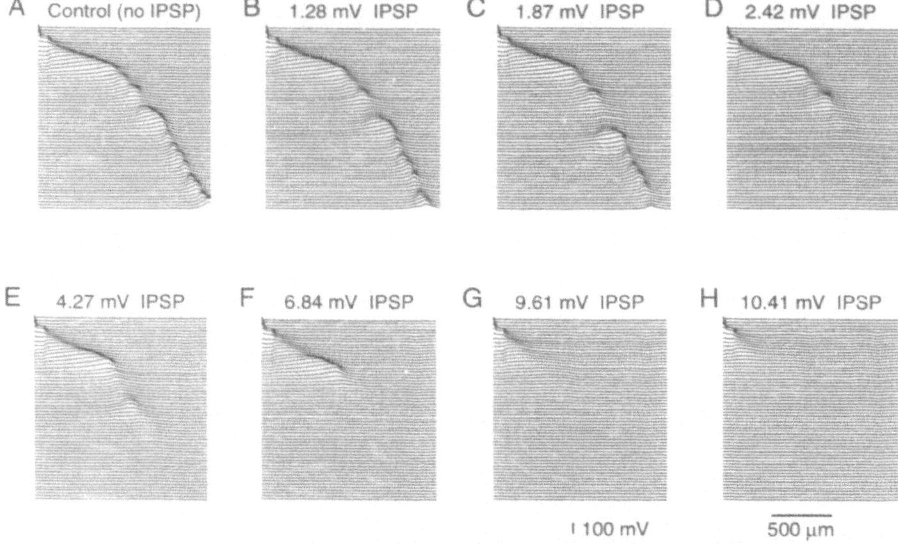

Figure 3. Simulated IPSPs are very effective in suppressing backpropagating action potentials. Each plot represents dendritic membrane potential profiles similar to Fig. 2D (0.4 ms latency). A-H: IPSPs of increasing somatic amplitude (indicated in each graph) influenced the velocity of backpropagating spikes (B-C), then suppressed backpropagation (D-H). Modified from a previous study[4].

Control). The amplitude and duration of action potentials was markedly affected by IPSPs. IPSPs occurring with small latencies with respect to the spike led to significant reductions up to 25 mV (Fig 2D), with maximal attenuations for IPSPs simulated at the perisomatic level.

To simulate the effect of Chloride injection (Fig. 1E), the IPSP reversal potential was changed to -55 mV. The significant attenuation observed in these conditions (Fig 2E) shows that the increased conductance due to IPSPs (shunt) only accounts for about 60% of spike attenuation. Removing dendritic Na^+ channels (Fig 2F) accounted for the remaining 40 % of spike attenuation, showing that dendritic Na^+ currents significantly contribute to action potential amplitude. The full attenuation of spike amplitude could be accounted by combining this effect with the shunt described above (Fig 2F, lowermost curve).

The model therefore suggests that spike amplitudes are reduced proportionally to the dendritic area that is "shut-off" by IPSPs, resulting in a spike that contains less or no participation from dendritic Na^+ channels. In addition, there is an important current leak through the open channels underlying the IPSPs, resulting in further attenuation. These two factors diminish in importance with more superficial stimuli.

A corollary observation is that, by preventing the participation of dendritic Na^+ channels, IPSPs suppress spike backpropagation in distal dendrites (Fig 3). Backpropagating action potentials are fragile[2, 3, 14, 15] and the present simulations show that they can be suppressed by small-amplitude IPSPs. As these IPSPs are well within the range of spontaneous IPSPs occurring *in vivo* (see Fig 1D4), our results corroborate recent optical imaging evidence that spike-related calcium transients remains confined to proximal dendrites *in vivo*[16].

The intensity of the cortical stimuli was adjusted to just above the spike threshold from rest. In pyramidal cells recorded with KAc pipettes (n=29), orthodromic spikes were reduced in amplitude and duration (Fig. 1C2-3) compared to APs elicited by short depolarizing current pulses (Fig. 1C1), consistent with previous findings in the amygdala[5]. The magnitude of the amplitude decrement was a function of the cortical stimulation depth relative to the position of the recorded cells (Fig. 1C). In both deep and superficial cells, maximal spike reductions were obtained with stimuli applied at the soma level.

To ensure that these differences between current-evoked and orthodromic spikes did not reflect differences in the slopes of the depolarization triggering the APs, we investigated the influence of sub-threshold cortical shocks on APs evoked by intracellular current pulses adjusted to elicit spikes in 50% of the trials. As shown in Fig. 1D, the amplitude of the direct spikes remained constant (Fig. 1D1), whereas the amplitude of spikes affected by IPSPs was progressively decreased (by up to 8 mV) as their latency increased (Fig. 1D2).

To assess the relative importance of local changes in input resistance and membrane potential in these phenomena, pyramidal neurons were recorded with KCl pipettes (n=30). Whereas in cells recorded with KAc pipettes the IPSP reversals averaged -78.1 $\pm$ 0.77 mV (n=4), in neurons recorded with KCl pipettes they averaged -52 $\pm$ 2.89 mV (n=13). Chloride diffusion inside the cells thus caused a 20-25 mV shift in the IPSP reversal potential, bringing it close to the spike threshold. In these conditions, high intensity cortical shocks applied at the soma level produced spike amplitude reductions of 7.45 $\pm$ 1.22 mV (n=10; Fig. 1E) compared to 15.58 $\pm$ 1.24 (n=9) in cells recorded with KAc pipettes (Fig. 1C).

COMPUTATIONAL MODELS

Computational models were based on a cellular reconstruction provided by R. Douglas and K. Martin (Fig 2A) and were simulated using NEURON (methods were described in detail elsewhere[4]). The cell was corrected for spines and an axon hillock and initial segment were included. Passive parameters were estimated by fitting the model to a voltage trace from a layer V cortical pyramidal cell recorded in the present series of experiments (Fig 2B). Active currents were inserted into the soma, dendrites and axon with different densities in accordance with available experimental evidence[6]. Na^+ channel density was low in soma and dendrites (70 pS/μm^2 as in adult pyramidal cells[6] – range tested 20-100 pS/μm^2), but was high in the axon, as indicated by experimental[7] and modeling studies[8, 9]. The delayed-rectifier was also distributed uniformly in the dendrites and no calcium currents were included. All currents were described by Hodgkin-Huxley equations, and two different kinetic models were tested for Na^+ and K^+ channels[8, 10] with negligible influence on the present results.

The relative density of glutamatergic and GABAergic synapses in different regions of the cell was distributed based on morphological data reviewed previously[11, 12]. Glutamatergic synapses were located exclusively in dendrites at more than 40 μm away from the soma. GABAergic synapses were located in all compartments of the neuron, with the highest density in the somatic region[11, 12]. Synaptic currents were simulated using kinetic models of postsynaptic receptors [13]. The depth-dependent features of evoked EPSP/IPSP sequences (Fig 2C) could be reproduced assuming that stimuli activated synapses preferentially in a relatively localized dendritic region[4].

In the absence of other stimuli, spikes evoked by current injection initiated in the initial segment and backpropagated successively to soma and dendrites (Fig 2D,

CONCLUDING REMARKS

In conclusion, by combining computational models with intracellular recordings of neocortical pyramidal cells *in vivo*, we provided a plausible explanation for our experimental observations on how action potential are controlled by IPSPs in these cells. Models and experiments suggest that IPSPs affect action potentials by two mechanisms: a shunt effect due to the opening of ion channels underlying the IPSPs, and a voltage-dependent effect by preventing dendritic Na^+ channels to participate to the somatic spike. Further, we suggest that under conditions of synaptic activity that occurs during active states *in vivo*, the conductance shunt and the voltage-dependent effect of synaptic inputs do not provide favorable conditions for backpropagating action potentials[17].

REFERENCES

1. Stuart, G., Spruston, N., Sakmann, B. and Hausser, M. Action potential initiation and back-propagation in neurons of the mammalian CNS. *Trends Neurosci.* **20**: 125-131 (1997) .

2. Kim H.G., Beierlein M. and Connors B.W. Inhibitory control of excitable dendrites in neocortex. *J. Neurophysiol.* **74**: 1810-1814 (1995).

3. Tsubokawa H. and Ross W.N. IPSPs modulate spike backpropagation and associated $[Ca^{2+}]$ changes in the dendrites of hippocampal CA1 pyramidal neurons. *J. Neurophysiol.* **76**: 2896-2906 (1996).

4. Paré, D., Lang, E.J. and Destexhe, A. Inhibitory control of somatic and dendritic sodium spikes in neocortical pyramidal neurons *in vivo*: an intracellular and computational study. *Neuroscience*, in press (1998).

5. Lang, E.J. and Paré, D. Synaptic and synaptically activated intrinsic conductances underlie inhibitory potentials in cat lateral amygdaloid projection neurons in vivo. *J. Neurophysiol.* **77**: 353-363 (1997)

6. Magee J.C. and Johnston D. Characterization of single voltage-gated Na^+ and Ca^{2+} channels in apical dendrites of rat CA1 pyramidal neurons. *J. Physiol.* **487**: 67-90 (1995).

7. Black J.A., Kocsis J.D. and Waxman S.G. Ion channel organization of the myelinated fiber. *Trends Neurosci.* **13**: 48-54 (1990).

8. Mainen Z.F., Joerges J., Huguenard J.R. and Sejnowski T.J. A model of spike initiation in neocortical pyramidal neurons. *Neuron* **15**: 1427-1439 (1995).

9. Rapp, M., Yarom, Y., and Segev, I. Modeling back propagating action potential in weakly excitable dendrites of neocortical pyramidal cells. *Proc. Natl. Acad. Sci. USA* **93**: 11985-11990 (1996).

10. Traub R.D. and Miles R. *Neuronal Networks of the Hippocampus*. Cambridge University Press, Cambridge (1991).

11. DeFelipe J. and Fariñas I. The pyramidal neuron of the cerebral cortex: morphological and chemical characteristics of the synaptic inputs. *Progress Neurobiol.* **39**: 563-607 (1992).

12. White E. L. *Cortical circuits*. Birkhauser, Boston, MA (1989).

13. Destexhe, A., Mainen, Z.F. and Sejnowski, T.J. Kinetic models of synaptic transmission. In: *Methods in Neuronal Modeling* (2nd ed), edited by C. Koch and I. Segev. MIT Press, Cambridge, MA, pp. 1-26 (1998).

14. Callaway J.C. and Ross W.N. Frequency-dependent propagation of sodium action potentials in dendrites of hippocampal CA1 pyramidal neurons. *J. Neurophysiol.* **74**: 1395-1403 (1995).

15. Spruston N., Schiller Y., Stuart G. and Sakmann B. Activity-dependent action potential invasion and calcium influx into hippocampal CA1 dendrites. *Science* **268**: 297-300 (1995).

16. Svoboda, K., Denk, W., Kleinfeld, D. and Tank, D.W. In vivo calcium dynamics in neocortical pyramidal neurons. *Nature* **385**: 161-165 (1997).

17. Research supported by grants from MRC, NSERC and FRSQ. E.J.L. was supported by a NINDS fellowship.

MODELLING THE EVOKED RELEASE OF QUANTA AT ACTIVE ZONES: THEORETICAL INVESTIGATION OF THE SECRETOSOME HYPOTHESIS

William G. Gibson,[1] Max R. Bennett,[2] and John Robinson[1]

[1]School of Mathematics and Statistics
[2]Department of Physiology
University of Sydney
New South Wales, 2006
Australia

INTRODUCTION

The release of transmitter at synapses is mediated by calcium. The opening of calcium channels in the presynaptic membrane allows the influx of calcium ions and the formation of discrete peaks of calcium ('calcium domains'[15,9]) immediately adjacent to each channel. This calcium can then bind to the proteins (synaptotagmin, syntaxin, etc) that are associated with vesicles[14,10]; the binding of a sufficient number of calcium ions, usually four, then leads to exocytosis of this vesicle and hence to the release of a quantum of transmitter.

We have previously modeled the spontaneous release of quanta, under the assumptions that a single calcium channel opens for a random (exponentially distributed) time and that the subsequent influx of calcium ions can effect one or more vesicles[1,5]. We now turn our attention to evoked release, where the calcium channels open under the depolarization resulting from the arrival of an action potential at the presynaptic terminal. A number of calcium channels may open more or less simultaneously under this impulse; however, in this investigation we assume that each such channel is associated with precisely one vesicle and that the calcium domains formed by the simultaneous opening of several channels do not overlap. Thus we have a system of functionally independent channel-vesicle units[13] which have been called synaptosecretosomes[10] and will here be called secretosomes (Figure 1). The evoked release of quanta at an active zone can be modelled by investigating the behaviour of one of these.

The modeling process is more complicated that in the spontaneous case, since the potential change due to the arrival of the action potential at the presynaptic terminal effects both the channel open time and the magnitude of the single-channel current.

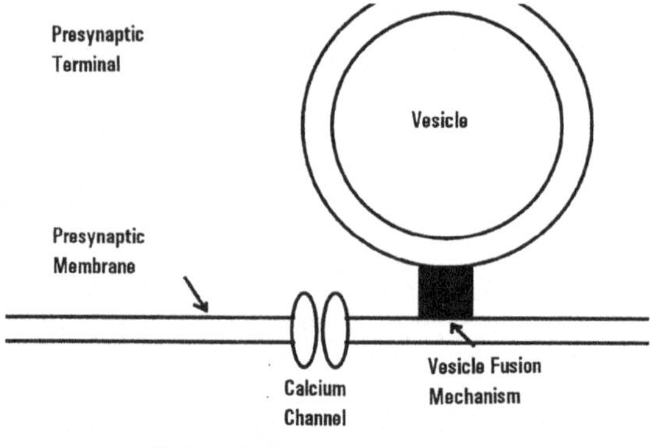

Figure 1. Schematic diagram of a secretosome. It consists of a vesicle attached to the presynaptic membrane by a complex which includes protein calcium sensors for triggering exocytosis and a single calcium channel in close proximity. (The channel-vesicle distance is taken to be 20 nm in the calculations reported here.) The calcium channel opens stochastically under the potential change due to an action potential allowing calcium ions to diffuse into the presynaptic region; these bind reversible to the calcium sensors and when four sites are occupied there can be a further configurational change leading to exocytosis of the vesicle.

METHOD

The action potential is obtained by solving the standard Hodgkin-Huxley equations and is shown in Figure 2 (solid curve). The calcium channel in the secretosome is assumed to open and close stochastically with forward and backward rate parameters having the standard Boltzmann forms $k_1 = a_1 \exp(b_1 V)$, $k_2 = a_2 \exp(b_2 V)$, where V is the membrane potential [8]; the values of the constants a_i, b_i are obtained by fitting to experimental data for N-type calcium channels operating in the low-p_0 mode [4]. The opening and closing of the channel is thus a (non-homogeneous) Poisson process and the times can be generated by standard means [3]; histograms of opening times (shaded) and opening duration (unshaded), generated by 10,000 simulations under the action potential of Figure 2, are shown in Figure 3. While the channel is open, calcium ions enter at a rate which is well-described by the single channel current

$$i(t) = \kappa(V - \overline{V})/(1 - \exp(0.075(V - \overline{V}))), \tag{1}$$

where again the constants are fitted using experimental data [4]. The form of $i(t)$, as given by equation (1), is approximately the mirror image of the action potential of Figure 2, and so the calcium current is smallest during the period of depolarization.

The calcium concentration $c(r, t)$ at time t at a point distant r from a calcium channel is found by solving the differential equation describing diffusion into a half-space from a source with a Gaussian spatial profile [15,11,1]. The functional form of $c(r, t)$, for the case where a channel opens for 1 ms and passes a constant current of 1 pA, is shown in Figure 4.

The calcium ions now bind to the vesicle-associated protein in the secretosome. The simplest case involves a single-affinity scheme, in which there is reversible, non-cooperative binding to four independent sites, followed by a configurational change giving exocytosis. Also considered is a dual-affinity scheme, in which the detachment rates are different, two sites having high affinity and the other two having low affin-

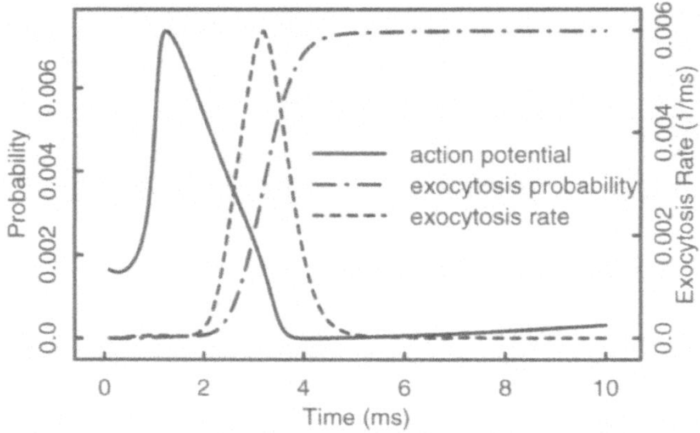

Figure 2. Time-dependence of exocytotic events. The solid curve shows the time course of the Hodgkin-Huxley action potential used to initiate the calcium channel openings (no vertical scale is given, but the potential peaks at about 40 mV and its resting level is about -60 mV). Also shown is the probability of the vesicle undergoing exocytosis (dot-dash line) and the exocytosis rate (broken line); both these curves are the average of 1000 simulations using the action potential.

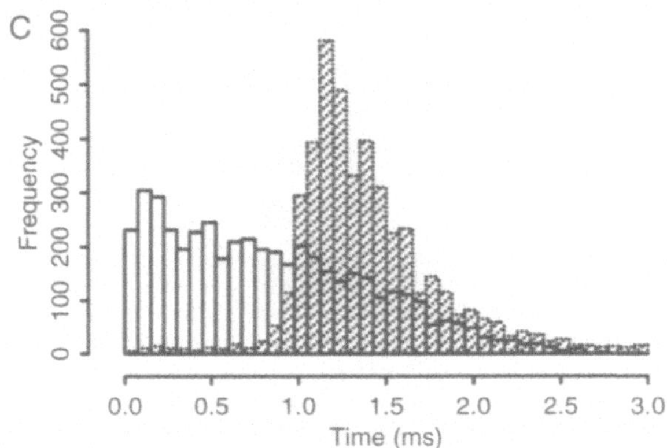

Figure 3. Opening times and the opening durations of a calcium channel under an action potential. The shaded histogram shows the opening times of the channel for a run of 10,000 impulses; there were 4638 actual openings. The unshaded histogram gives the corresponding open times. Note that the distribution of opening times approximately follows the timecourse of the action potential of Figure 2.

ity for calcium. The attachment rates are taken to be proportional to the calcium concentration and the detachment rates to be constants. (For experimental values of coefficients, see [7,6].) These schemes lead to sets of differential equations which can be solved for the exocytotic probability $P_{ex}(t)$ and the exocytosis rate dP_{ox}/dt.

RESULTS

The result of 1000 simulations, using the action potential of Figure 2 and the single-affinity binding scheme are shown in Figure 2, where the probability of exocytosis.

Table 1. Total release probabilities for multiple secretsomes at various active zones

Number of secretosomes	Probabilities p_k of k releases at active zone			
n	p_0	p_1	p_2	p_3
120	0.414	0.366	0.161	0.046
80	0.556	0.327	0.095	0.018
40	0.745	0.220	0.032	0.003

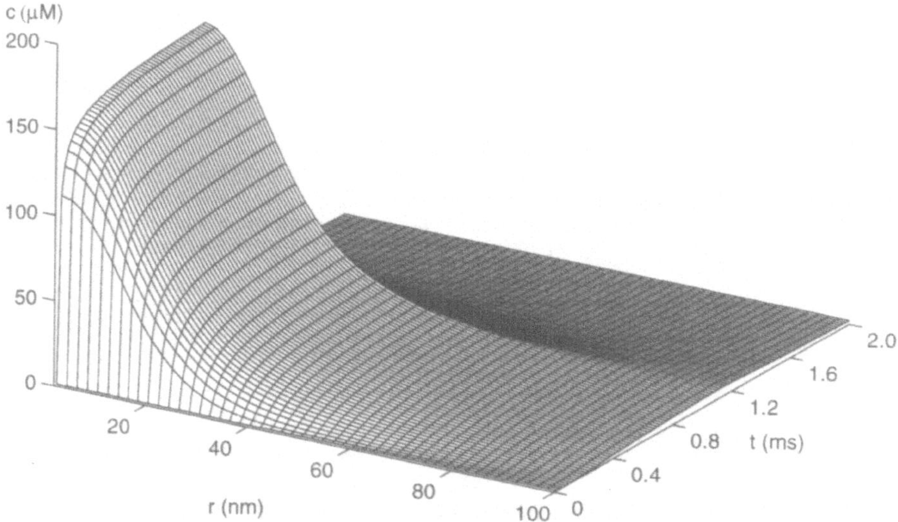

Figure 4. Calcium concentration due to the opening of a single calcium channel. Shown is the calcium concentration (in μM) at a distance r nm from the center of the channel at time t ms after the opening of the channel. The channel opens deterministically for a period of 1 ms and admits a constant calcium current of magnitude 1 pA. The concentration peaks at 1 ms when the channel closes and then decreases rapidly to zero. If the channel were to open under an action potential two main differences would be observed: first, the channel opening time and open duration would be stochastic, as shown in Figure 3; second, the calcium current through the open channel would not be constant but would vary according to equation (1), being much larger during the repolarizing phase of the action potential. These effects have been incorporated into the calculations reported here, by averaging over a large number of runs.

P_{ex} (dot-dash curve) and the exocytosis rate, dP_{ex}/dt (broken curve) are plotted as functions of time. The probability P_{ex} attains its final value shortly after the end of the action potential; the exocytosis rate peaks somewhat after the peak depolarization of the action potential. These timecourses can be understood by observing that although a calcium channel is most likely to open near the peak of the action potential (Figure 3) the calcium current is then very small; it increases during the repolarizing phase, so a channel that remains open during this late phase contributes most to the calcium influx.

As explained in the Introduction, an active zone is assumed to contain a number of independent secretosomes. Thus the probability that a single action potential will result in the release of k quanta will be given by the binomial distribution bin(k, n, p) where n is the total number of releases at a single secretosome. For the single-affinity case, $p = 0.0073$ and estimates of the number of secretosomes are $n = 120$ for the case of a varicosity at an autonomic neuromuscular junction, $n = 80$ for a bouton in the central nervous system, and $n = 40$ for a somatic motor-nerve terminal. Release probabilities calculated on this basis are given in Table 1; note that this predicts very little multiquantal release at the active zone of a motor-nerve terminal, as compared to the other cases.

The effect of varying the external calcium concentration on release probability has also been calculated. Results are that the probability increases steadily with increasing calcium concentration; at low concentrations this increase approximates a fourth-power law but at higher concentrations it falls considerably below this.

The phenomenon of facilitation occurs when two or more impulses arrive at the terminal within a short time interval. In the simplest case, an action potential initiated at time $t = 0$ is followed at time $t = t_1$ by a second impulse and the facilitation is defined as P_{ex}[second impulse]$/P_{ex}$[first impulse]; this will be greater than unity because of residual binding of calcium ions due to the first impulse. Calculations using the single-affinity binding scheme give only a small effect in normal external calcium concentration: use of the dual-affinity scheme considerably increases this, since the two high-affinity sites bind more residual calcium [2]. The calculations also show that facilitation decreases with increasing external calcium concentration, in accord with experiment [12].

CONCLUSION

A model incorporating the stochastic elements of calcium channel kinetics and calcium binding to vesicle-associated proteins and using physiologically realistic values for parameters can account for observed rates of evoked quantal release at active zones. This model incorporates the secretosome hypothesis, which is that the relevant vesicles and channels occur in associated pairs, with each pair functioning as an independent unit. The assumption that some binding sites have a high calcium affinity gives significant facilitation of release when a second impulse follows a few milliseconds after a first impulse.

REFERENCES

1. M. R. Bennett, W. G. Gibson and J. Robinson, Probabilistic secretion of quanta: spontaneous release at active zones of varicosities, boutons and endplates, *Biophysical J.* 69: 42-53 (1995).
2. R. Bertram, A. Sherman and E. F. Stanley, Single-domain/bound calcium hypothesis of transmitter release and facilitation, *J. Neurosci.* 75:1919-1931 (1996).
3. J. R. Clay and L. J. DeFelice, Relationship between membrane excitability and single channel open-close kinetics, *Biophys. J.* 42:151-157 (1983).
4. A. H. Delcour, D. Lipscombe and R. W. Tsien, Multiple modes of N-type calcium channel activity distinguished by differences in gating kinetics, *J. Neuroscience* 13:181-194 (1993).
5. W. G. Gibson, M. R. Bennett and J. Robinson, Modeling the spontaneous release of quanta at active zones of varicosities, boutons and endplates, *in* " Computational Neuroscience. Trends in Research 1995", J. M. Bower, ed., Academic Press, San Diego (1996).
6. R. Heidelberger, C. Heinemann, E. Neher and G. Matthews, Calcium dependence of the rate of exocytosis in a synaptic terminal, *Nature* 371:513-515(1994).
7. C. Heinemann, R. H. Chow, E. Neher and R. S. Zucker, Kinetics of the secretory response in bovine chromaffin cells following flash photolysis of caged calcium, *Biophys. J.* 67:2546-2557 (1994).
8. B. Hille, "Ionic channels of excitable membranes", Sinauer Associates, Massachusetts (1992).
9. R. Llinás, M. Sugimori and R. B. Silver, Time resolved calcium microdomains and synaptic transmission, *J. Physiol. Paris* 89:77-81(1995).

10. V. M. O'Connor, O. Shamotienko, E. Grishin and H. Betz, On the structure of the 'synaptosecretosome', Evidence for a neurexin/synaptotagmin/syntaxin/Ca^{2+} channel complex. *FEBS Letters* 326:255-60 (1993).

11. H. Parnas, G. Hovav and I. Parnas, Effect of Ca^{2+} diffusion on the time course of neurotransmitter release., *Biophys. J.* 55:859-874 (1989).

12. R. Rahamimoff, A dual effect of calcium ions on neuromuscular facilitation, *J. Physiol.* 195:471-480 (1968).

13. E. F. Stanley, Single calcium channels and acetylcholine release at a presynaptic nerve terminal, *Neuron* 11:1007-1011 (1993).

14. T. C. Südhof, The synaptic vesicle cycle: a cascade of protein-protein interactions, *Nature* 375:645-653 (1995).

15. R.S. Zucker, and A. L. Fogelson, Relationship between transmitter release and presynaptic calcium influx when calcium enters through discrete channels, *Proc. Natl. Acad. Sci. USA* 83: 3032-3036 (1986).

ANALYSIS OF SENSORY CODING IN THE LATERAL SUPERIOR OLIVE

Charlotte M. Gruner and Don H. Johnson

Computer and Information Technology Institute
Department of Electrical & Computer Engineering
Rice University, MS 366
Houston, TX 77005–1892

INTRODUCTION

The fundamental questions of sensory neuroscience are how neurons encode information and how neurons extract (process) information. For example, in the auditory system, sound location is determined by a combination of amplitude and time cues. The Lateral Superior Olive (LSO), one of the first nuclei to receive inputs from both ears, is thought to extract high-frequency localization information based on interaural level difference (ILD) and interaural time difference (ITD) cues. However, while the optimal localization processing system produces the same response for the same ILD regardless of source level,[1] LSO responses are affected by source level.[2] This sensitivity questions the presumption that the LSO is strictly an angle processor, and suggests that the LSO may be coding other stimulus features such as sound level. We have used a new analysis technique to analyze the significance of the LSO sensitivity to sound level compared to that of an optimal angle processor. We also examined the processing of two features: azimuthal angle and source level by the LSO network by comparing the response sensitivity of network inputs with that of its outputs.

METHODS

A general neural sensory system is shown in figure 1. The vector inputs, $\mathbf{X}$, are functionally related to the stimulus parameters $\mathbf{S}$, and the output of the neural system, $\mathbf{Y}$, is functionally related to the inputs, $\mathbf{X}$. This system can be considered a Markov chain $(\mathbf{S} \to \mathbf{X} \to \mathbf{Y})$ if $\Pr\{\mathbf{Y}$ given $\mathbf{X}$ and $\mathbf{S}\} = \Pr\{\mathbf{Y}$ given $\mathbf{X}\}$. If this condition holds, then according to the Data Processing Inequality of information theory[3] the mutual information between the stimulus and the input is greater than or equal the mutual information between the output and the stimulus,

$$I(\mathbf{X}, \mathbf{S}) \geq I(\mathbf{Y}, \mathbf{S}).$$

The system's output cannot have more information about the stimulus than the inputs. Evidence of this inequality condition is readily observed in neural systems.[4] Because of this

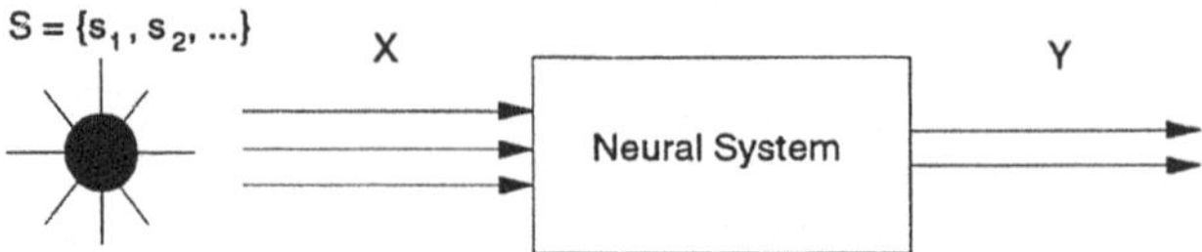

Figure 1. A typical sensory system. A stimulus with parameters $\{s_1, s_2, \ldots\}$ is coded by the neural inputs $\mathbf{X}$ to the neural system. The neural system, consisting of one or and ensemble of neurons processes the inputs to produce the outputs $\mathbf{Y}$.

fundamental result, many similar processing bounds can be derived. In particular,

$$D(p_{\mathbf{S}_1}(\mathbf{X})\|p_{\mathbf{S}_2}(\mathbf{X})) \geq D(p_{\mathbf{S}_1}(\mathbf{Y})\|p_{\mathbf{S}_2}(\mathbf{Y}))$$

where

$$D(p_{\mathbf{S}_1}(\mathbf{X})\|p_{\mathbf{S}_2}(\mathbf{X})) = \sum_{\mathbf{X}} p_{\mathbf{S}_1}(\mathbf{X}) \log \frac{p_{\mathbf{S}_1}(\mathbf{X})}{p_{\mathbf{S}_2}(\mathbf{X})}$$

is the Kullback Leibler (KL) distance[3] between the response of the ensemble of inputs to stimulus conditions $\mathbf{S}_1$ and $\mathbf{S}_2$ and $D(p_{\mathbf{S}_1}(\mathbf{Y})\|p_{\mathbf{S}_2}(\mathbf{Y}))$ is the KL distance between the output ensemble responses to the same two stimuli. Thus the "distance" between the outputs under two stimulus conditions must be less than or equal the "distance" between the inputs under the same stimulus conditions. According to Stein's lemma, the KL distance, $D(p_0\|p_1)$, is asymptotically related to the miss probability: The probability that in a hypothesis test between H_0 and H_1, H_0 is chosen given that H_1 is the correct choice.[3] The KL distance is a measure of the intrinsic difficulty of the classification problem. The process for empirical KL distance calculation between the responses of a single neuron to two stimuli and extension of the technique to calculation of the distance between the responses of neural ensembles is discussed in a companion paper in this volume.[5]

The KL distance can be used to quantify the stimulus parameter sensitivity of a system's inputs and outputs to examine the processing done by the system. Consider a parameter space consisting of two parameters, $\mathbf{S} = \{\theta, \zeta\}$. The neural system with inputs $\mathbf{X}$ and outputs $\mathbf{Y}$ may

- extract information pertaining to one of the parameters but not the other

- extract information pertaining to both parameters

- suppress information pertaining to both parameters

To determine which option is true, we examine the system's inputs and outputs as the stimulus parameters are changed by small increments. Assuming that the input changes in response to perturbations of both parameters and the system extracts θ but not ζ, the output should change in response to a change in stimulus θ but not ζ. Similarly, if the system extracts information pertaining to both parameters, perturbation in either parameter should change the output.

RESULTS

We evaluated the Kullback-Leibler distance for $\mathbf{S} = \{\theta, \zeta\}$ where θ is the azimuthal angle of the stimulus and ζ is source level. We used a previously developed Hodgkin Huxley-based model implemented with NEURON[6] to simulate the sustained output response of a single LSO neuron to tone burst stimuli having various parameters. PST histograms from these simulations are shown in figure 2. These PST histograms show evidence of LSO sus-

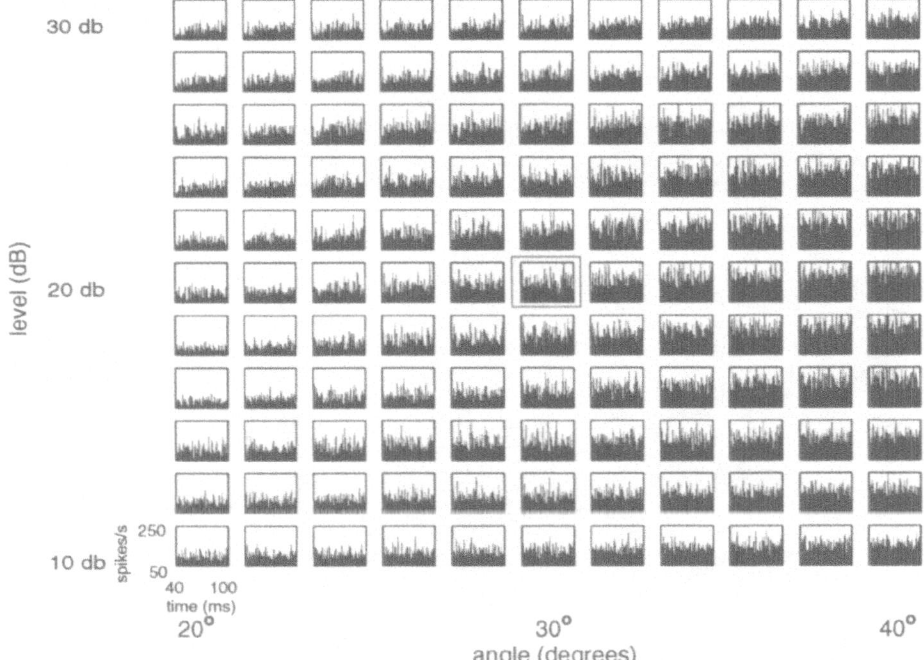

Figure 2. Sustained response of model output to changes in stimulus angle and level. At least 400 stimulus repetitions were simulated at each combination of stimulus parameters. The sustained response is shown as beginning approximately 40 ms after the stimulus onset and lasting 60 ms until the stimulus is turned off. This LSO simulation models a "fast chopper" response. The fast chopper transient PST histogram response during the first 40 ms is characterized by a series of closely spaced peaks.

tained response sensitivity to stimulus level as well as angle. The inputs to the model are stochastically generated EPSP and IPSP occurrence times of the numerous inputs to each LSO cell. The inputs statistically match the primary-like characteristics of the afferents to the LSO. Figure 3 shows example input PST histograms for two different stimuli as well as the resulting simulated output and indicates the KL distance calculations made from the data.

Figure 4(a) shows the KL surface for the inputs to the model. The surface is noticeably parabolic the level plane. (It is also parabolic in the angle plane, but on a smaller scale.) The surface therefore resembles a bowl – at the center of the bowl the KL distance is 0, $(D(p_{\mathbf{S}_0}||p_{\mathbf{S}_0}) = 0)$ and the distance increases as the stimulus parameters are perturbed farther from $\mathbf{S}_0$. Figure 4(b) shows the KL surface that results from comparing the model neuron output at $\mathbf{S}_0 = \{\theta_0 = 30°, \zeta_0 = 20\text{dB}\}$ (the PST histogram response for this stimulus is boxed in figure 2) with the model output at each stimulus condition shown in figure 2. In figure 4(c) the model neuron has been replaced by an optimal angle system that optimally estimates the stimulus angle from observations of the input,[1] then optimally codes the angle with a firing rate code.[7] The output KL surfaces shown in figure 4(b,c) completely change the orientation of the KL surface, emphasizing the angle sensitivity, suppressing the level sensitivity. The optimal system's KL surface is parabolic in the angle plane, indicating that the response changes little in response to changes in stimulus level. The KL surface of the model neuron is also largely parabolic in the angle plane, however, this surface has evidence of non-trivial level dependent features. For example, the saddle-like nature of the surface,

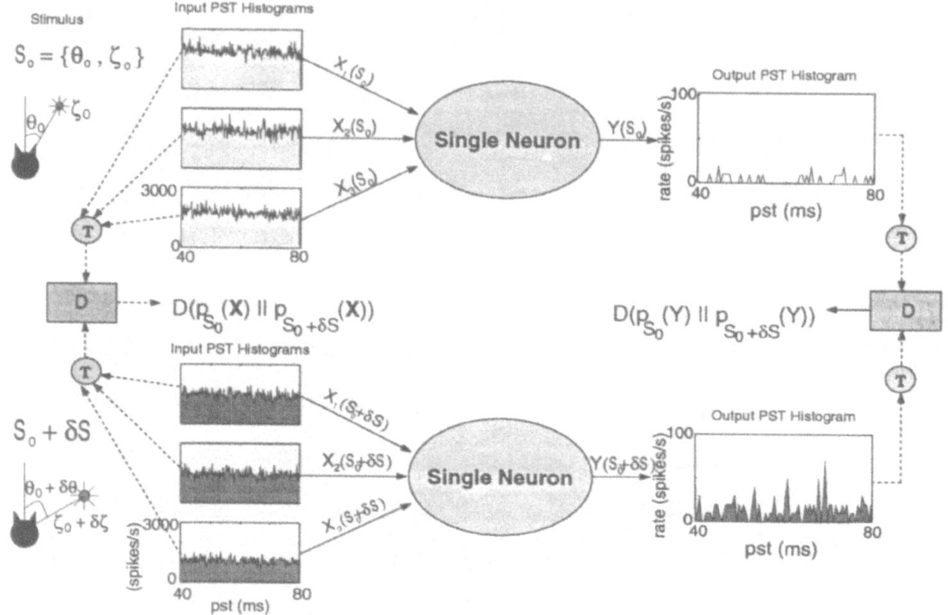

Figure 3. Computing input and output distances with the biophysical model. The input to the biophysical LSO model is statistically generated and depends on the stimulus conditions. The multiple inputs represent the convergence of numerous afferents to each LSO cell. The generated inputs produce a single output. A change in stimulus parameters changes the model input and the model output is also changed. The KL distance between inputs is calculated by comparing the ensemble responses as indicated. The KL distance between outputs is calculated by comparing the single outputs. The procedure is similar to forming a PST histogram but is generalized to allow for representation of ensembles and time correlation. Empirical histograms called "types" are computed (the circles marked "T") and these histograms are compared by the KL distance.[5]

especially for angles $\theta \geq 30°$ seems to indicate a direction of greatest sensitivity for this neuron along the $\zeta = \zeta_0$ axis, in contrast to the optimal system, for which the sensitivity is independent of ζ. Also, the bowl shape of the surface indicates the possibility of some non-trivial level information available at the output.

Figure 5 shows the KL surfaces for an LSO model cell that only receives input from one ear. Clearly the output is no longer sensitive to changes in angle. The output, like the input is mainly sensitive to changes in level. In contrast to the binaural system that performs processing, the monaural system is merely relaying the same information from input to output.

DISCUSSION

The analysis shows that even though the inputs to the LSO are much more sensitive to changes in sound level than changes in azimuthal angle, LSO binaural processing reverses the output sensitivity, almost eliminating dependence on level. The observed remnant level sensitivity may be an insignificant by-product of imperfect simulations or of processing that is as optimal as possible under some unknown neural processing constraints. The comparison with the "optimal" angle processor may be unfair: The optimal processor is permitted to observe the inputs for 60 ms before producing any output. LSO integration times are an order of magnitude shorter. It is also possible that the observed sensitivity is significant and indicates a specific receptive field property of this neuron at this specific operating point.

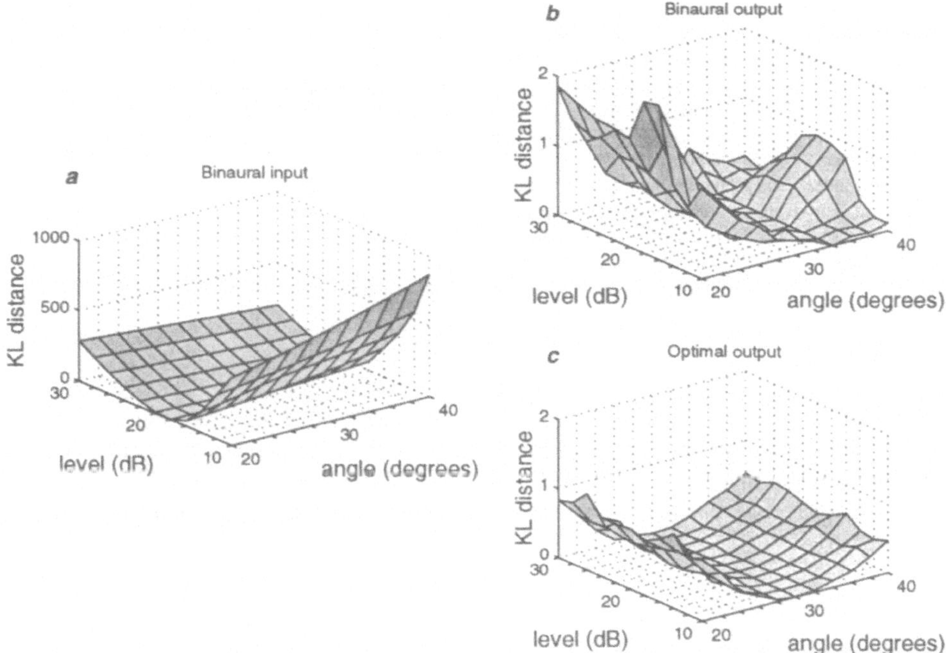

Figure 4. (a) The KL surface for the binaural inputs to the LSO model. (b) The KL surface plot for the outputs of the LSO model. (c) The KL surface plot for the optimal angle system described in the text. The difference in scale between the inputs and the outputs because the large number of convergent inputs to the LSO model has the capability for both larger overall rates and larger swings in rates than the single output. Although it is not obvious from these figures, the Data Processing Inequality for KL holds for this data; the output distance is always less than the input distance. Although the input does not appear to change much with angle, it is in fact parabolic on a much smaller scale than in the level direction. For example, $D(p_{\mathbf{S}=\{20°,20\mathrm{dB}\}}||p_{\mathbf{S}_o}) = 8.0$.

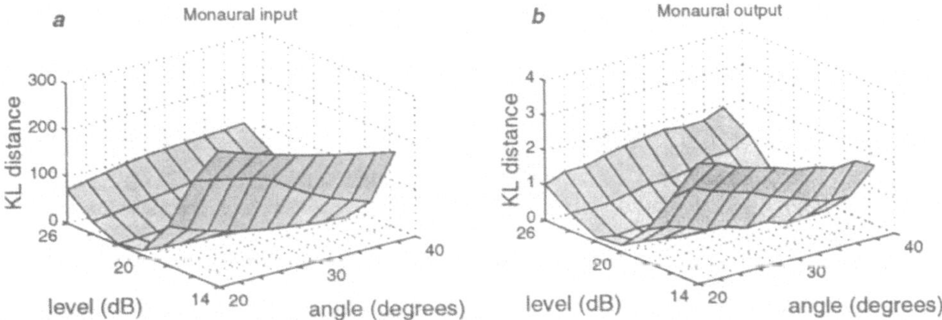

Figure 5. (a) The KL surface for the monaural inputs to the LSO model. (b) The KL surface plot for the output of the monaural model.

REFERENCES

1. D. H. Johnson, A. Dabak, and C. Tsuchitani. Function-based modeling of binaural interactions: Interaural level, *Hearing Research*, 49:301-320, 1990.
2. C. Tsuchitani. The inhibition of cat lateral superior olivary unit excitatory response to binaural tone bursts: I. The transient chopper discharges. *Journal of Neurophysiology*, 59:164-183, 1988.
3. T. M. Cover and J. A. Thomas. "Elements of Information Theory," John Wiley & Sons, New York (1991).
4. F. Gabbiani, W. Metzner, and C. Koch. From stimulus encoding to feature extraction in weakly electric fish, *Nature* 384:564–567 (1996).
5. D. H. Johnson and C. M. Gruner. Neural Ensemble Processing with Types, *in*: "Computational Neuroscience, Trends in research 1998" (this volume).
6. M. Zacksenhouse, D. H. Johnson, J. Williams, and C. Tsuchitani. Single-neuron modeling of LSO unit responses, *Journal of Neurophysiology* to appear, (1998).
7. R. B. Stein. The information capacity of nerve cells using a frequency code, *Biophysical Journal* 7:797–826 (1967).

Acknowledgments

This publication was supported in part by the Keck Center and grant number 1T15 LM07093 from the National Library of Medicine to the W. M. Keck Center for Computational Biology, Houston, Texas. Its contents are solely the responsibility of the authors and do not necessarily represent the official views of the National Library of Medicine.

DYNAMICS OF SPIKE GENERATION MAY UNDERLY *IN VIVO* SPIKE TRAIN STATISTICS

Boris Gutkin,[1] and G.Bard Ermentrout[2]

[1]Program in Neurobiology and Center for Neural Basis of Cognition
[2]Mathematics Department
University of Pittsburgh
Pittsburgh, Pa, 15205

INTRODUCTION

In this work we propose a novel biophysical explanation for the highly variable statistics of *in vivo* cortical spike trains reported in Softky [1], Dean[2], McCormick[3] . Previously, numerous explanations have been proposed: random spike generation[4,6,5], active dendrites [1], balanced inhibition/excitation [7,8], network dynamics[9,10] and high gain integrate-and-fire neurons[11].

We focus on the non-linear dynamics of the spike generating mechanism in cortical neurons and propose a biophysical explanation for the spike train statistics. We argue that *in vitro* spike generating properties in cortical pyramidal neurons are consistent with type I membrane dynamics [12]. We show that dynamical properties of type I neural membrane coupled with irregular inputs yield firing statistics indistinguishable from those observed in *in vivo* experiments. The key property of type I membrane is the delay to spike whose length is dependent on the size of the input.

The path to sustained repetitive firing (oscillations) in type I membranes is through a saddle-node bifurcation, with highly non-uniform motion around a given limit cycle and oscillations with arbitrarily long periods. Type I membranes also produce all-or-none spike of roughly constant shape. This is quite different from Type II neurons, where the oscillations arise through a sub-critical Hopf bifurcation, consequently appearing with a non-zero frequency and exhibiting no variable spike delay. We note that most popular models of cortical cells are type I and that the delay-to-spike characteristic of these is absent in type II (e.g. Hodgkin-Huxley and Fitzhugh-Nagumo) as well as integrate and fire neurons.

We studied the Morris-Lecar neural model and a canonical phase model with Type I dynamics. The Morris-Lecar model can be put into both type I and type II regimes depending on the parameters. The phase model results from a mathematical reduction of any conductance based model that exhibits type I dynamics. The methods for reduction and full discussion of the canonical nature of the model are presented in Ermentrout[14] and Hoppensteadt[15] while the discussion of how to include noise in

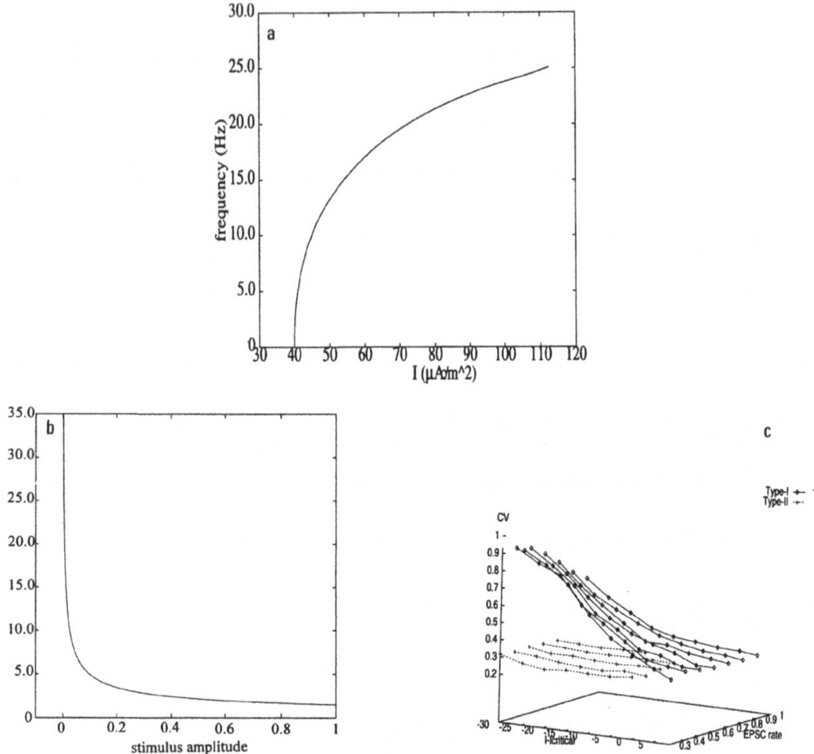

Figure 1. a. Firing frequency to input current (f-i) plot for type I membrane shows oscillations appearing with arbitrarily low frequencies. b. The delay to spike in the type I model depends on the amplitude of the supra-threshold pulse stimulus c. CV results for Morris-Lecar models, the upper trace shows type I behavior, lower trace is type II. Note that IPSC rate was 0.3 for all simulations, EPSC rate varied.

the phase model is in a forthcoming report Gutkin and Ermentrout[16]. All models discussed here can be moved from excitable to oscillating regimes by adding a constant depolarizing bias current.

MORRIS-LECAR RESULTS

The random inputs to the Morris-Lecar models consist of Poisson timed epsc's and ipsc's of small amplitude. The equations and parameters for the type I and type II models are in Appendix A. Simulations for the type I regime Morris-Lecar show highly variable behavior for a wide range of inhibition-to-excitation ratios; results are summarised in figure 1. The type I model yields a wide range of firing rates for different excitability and input parameters, with a clear dependence of the cv on the bias current (upper trace in figure 1). When firing is driven purely by the random inputs, cvs near 1.0 are observed. The cv decreases to 0.3 in the intrinsically oscillating regime where the model dynamics dominate the inputs. The response of type II Morris-Lecar to noisy inputs exhibits a narrow range of firing rates with the cv never rising above 0.5, regardless of input and bias characteristics (see lower trace in figure 1).

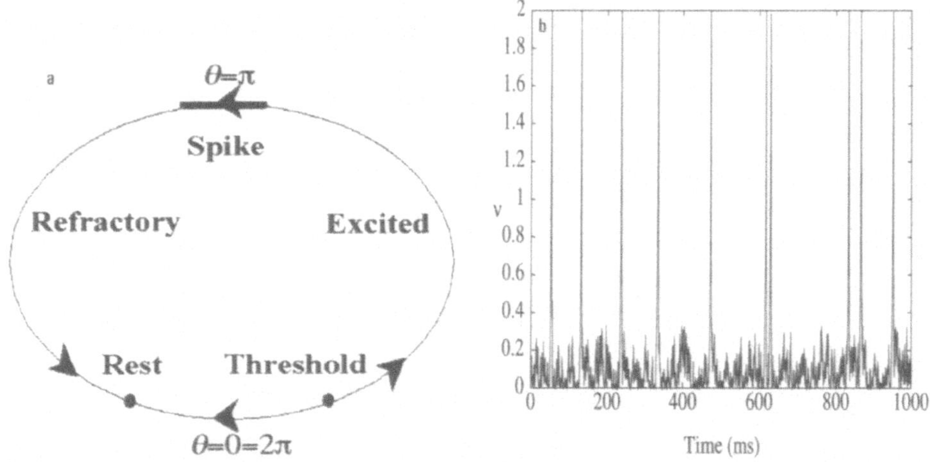

Figure 2. a. Phase evolution on a circle and its analogue in state of the membrane. Note that the spike occupies a small region near π and the model spends most of its time near the threshold. A supra-threshold stimulus pushes θ past the threshold and into the excited region. Here the regenerative dynamics that summarize active conductances carry the phase through the spike. b. A representative spike train for the phase model excited by a random stimulus, here we plot not θ but $\nu = 1 - \cos(\theta)$ which makes the spikes apparent.

THE CANONICAL θ-NEURON

We show that the highly variable spike train is a generic property of noise-driven type I membrane by exhibiting such in a reduced neural phase model (θ-neuron). The θ-neuron is based on a formal mathematical reduction of multi-dimensional neural models. The reduction, based on standard bifurcation methods for saddle-node dynamics[13], shows precisely how to include noisy inputs into the phase dynamics and produces a one-dimensional spiking model. We describe the neurons activity by a phase variable θ:

$$d\theta = (\beta(1 + \cos\theta) + (1 - \cos\theta))dt + \sigma(1 + \cos\theta)dW_t, \theta \in [0, 2\pi] \qquad (1)$$

For white noise dW_t with intensity σ The β is the bias parameter reflecting the strength of the constant biasing current. Figure 2 is an illustration of the model structure (rest state, threshold and the intrinsic spike).

RESULTS FOR THE θ-NEURON

The deterministic dynamics of the model have been described in Ermentrout [13]. For negative values of bias the model has a stable rest state and an unstable node that acts as a threshold for firing. A supra-threshold impulse induces a procession around the circle back to the stable node, thus producing a spike, i.e. θ moves through π (see figure 2a). The amplitude of the supra-threshold stimulus determines the spike latency. Oscillations arise in the model as the bias current is increased past zero. The rest state and the threshold come together and anihilate each other leaving a periodic. The amplitude of the bias sets the period of the oscillation.

When $\beta > 0$, the stochastic θ-neuron exhibits intrinsic oscillations with low cvs (near 0.3) . When $\beta < 0$, the firing is induced purely by the influence of the random inputs and is extremely irregular, with cvs near 1.0 (see figure 3). With β fixed below

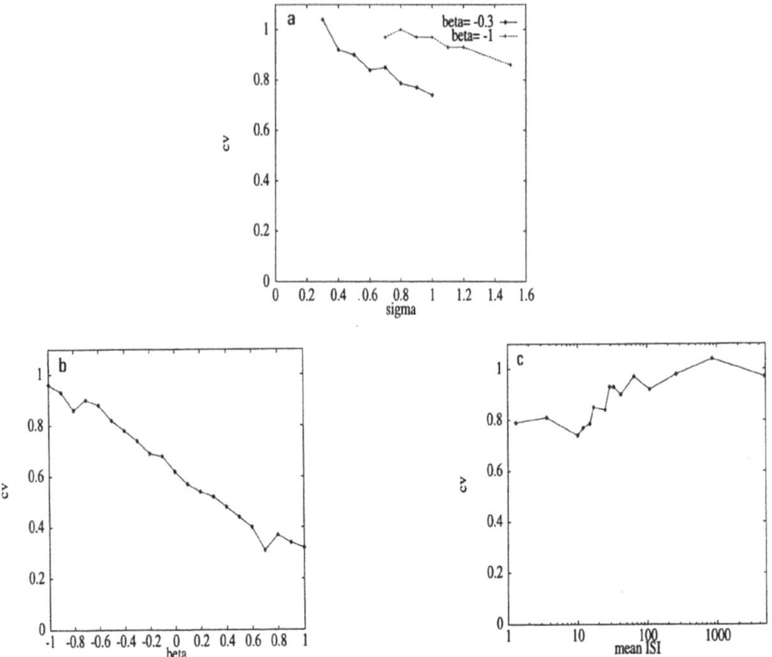

Figure 3. CV results for θ-model. a.cv remains high across wide range of noise parameters. b. cv decreases linearly with β. c. cv remains high across a wide range of firing rate, when such is controlled by noise, here $\beta = -1$, σ varied

0, increasing σ decreases the mean ISI but the cv remains invariant. When we fix σ and increase β (simulating an intracellular current injection) the cv decreases as the model moves into the oscillating regime. Thus θ-neuron behavior parallels the Morris-Lecar results.

CONCLUSION

Our results suggest that the high cv is an intrinsic property of type I membrane in the excitable regime. That is, the highly irregular firing is not dependent on the details of a particular type I model. The saddle node bifurcation underlies the observed results. We propose that the high cvs appear because the spike generating mechanism exhibits a delay-to-spike that is strongly dependent on the size of the input. Thus in cells with type I membrane dynamics, a temporally structured input that provides suprathreshold shocks of random amplitude will evoke strongly variable spike timing. This is in contrast to type II dynamics, where there is no delay-to-spike, and generically low cv values are observed.

The major point here is that when the firing in the type I model is induced by a constant injected current it is extremely regular, with frequency response characteristics similar to those observed in *in vitro* pyramidal cells and in simulations[11]. On the other hand when the type I neuron responds to noisy inputs, its firing is extremely irregular, regardless of balance in the inputs.) Our theory accounts for both the highly variable *in vivo* as well as the regular *in vitro* behavior. This work also implies that spike generation in cortical neurons is characterised by type I saddle-node bifurcation dynamics.

APPENDIX A. MORRIS-LECAR EQUATIONS

The Morris-Lecar equations that we have used are based on the model that appeared in [Rinzel and Ermentrout, 1990]. They have the form:

$$C\frac{dV}{dt} = -g_{Ca}m_\infty(V)(V - V_{Ca}) - g_K w(V - V_K) - g_L(V - V_L) + I$$

$$\frac{dw}{dt} = \phi(w_\infty(V) - w)/\tau_w(V)$$

$$m_\infty(V) = .5(1 + \tanh((V - V_1)/V_2))$$

$$w_\infty(V) = .5(1 + \tanh((V - V_3)/V_4))$$

$$\tau_w(V) = 1/\cosh((V - V_3)/V_4)$$

Values for type I membrane are $V_K = -80$mV, V_L=-60mV,V_{Ca}=120mV,C=20$\mu F/cm^2$, g_L=2$\mu S/cm^2$,$g_K = 8\mu S/cm^2$, $V_1 = -1.2$mV, $V_2 = 18$mV, $V_3 = 12$mV,$V_4 = 17.4$mV,$\phi =$ .067, and $g_{Ca} = 4.0\mu S/cm^2$ For type II membrane parameters are the same except, $V_3 = 2$mV, $V_4 = 30$mV,$\phi = 0.04$, and $g_{Ca} = 4.4\mu S/cm^2$.

ACKNOWLEDGMENTS The authors thank Drs. Satish Iyengar and John Horn for fruitful discussions and Dr. Harold Kyriazy for valuable comments on earlier versions of this manuscript. Work on this project was supported by NSF

REFERENCES

1. Softky, W., Koch, K. The highly irregular firing of cortical cells is inconsistent with temporal integration of random EPSPs. *J. Neurosci.* 13:334 (1993)
2. Dean, A. The variability of discharge of simple cells in the cat striate cortex. Exp. Brain Res.44:437 (1972)
3. McCormick, D.A., Connors, B.W., Lighthall, J.W., Prince, D.A. Comparative electrophysiology of pyramidal and sparsely spiny stellate neurones of the neocortex. *J. Neurophysiol.* 54 (1985)
4. Wilbur, W.J., Rinzel, J. A theoretical basis for large coefficient of variation and bimodality in neuronal interspike interval distribution *Journal of Theoretical Biology* 105:345 (1983)
5. Ricciardi, L.M. Diffusion models of single neurons *in* "Neural Modelling and Neural Networks", Ventriglia, F. ed., Pergamon Press, Oxford (1994)
6. Smith, C. E. A Heuristic Approach to Stochastic Models of Single Neurons. *in* "Neural Nets: Foundations to Applications" T. McKenna et al. (eds.) Academic Press. (1992)
7. Shadlen, M.N. and Newsome W.T. Noise, neural codes and cortical organization. *Current Opinions in Neurobiology* 4:569 (1994)
8. Bell, A.,Mainen, Z.,Tsodyks, M., Sejnowski, T. "Balancing" of conductances may explain irregular cortical spiking Institute for Neural Computation, UCSD, San Diego Technical report INC-9502 (1995)
9. Usher, M., Stemmler, M.,Koch, C., Olami, Z. Network amplification of local fluctuations causes high spike rate variability, fractal firing patterns and oscillatory local field potentials *Neural Computation* 6:795 (1994)
10. van Vreeswejk, C. and Sompolinsky, H. Chaos in neuronal networks with balanced excitatory and inhibitory activity. *Science.* 274(5293):1724 (1997)
11. Troyer T.W., Miller K.D. Physiological gain leads to high ISI variability in a simple model of a cortical regular spiking cell. *Neural Computation* 9:971 (1997)
12. Rinzel J. & Ermentrout, G.B. Analysis of neural excitability and oscillations *in* Koch K. & Segev I. (eds) "Methods in Neuronal Modeling" MIT Press, Cambridge, Mass. (1989)
13. Ermentrout, G.B. "XPPAUT1.8 -the differential equations tool", available at www.pitt.edu/ bardware (1996a)
14. Ermentrout G.B. Type I Membranes, Phase resetting Curves,and Synchrony. *Neural Computation* 8:979 (1996b)
15. Hoppensteadt, F.C. & Izhkevich, E.M. "Weakly Connected Neural Networks". Springer-Verlag, New York (1997)
16. Gutkin, B.S., Ermentrout, G.B. *Neural Computation* (in press)

MODELING THE CONTRIBUTIONS OF CALCIUM CHANNELS AND NMDA RECEPTOR CHANNELS TO CALCIUM CURRENT IN DENDRITIC SPINES

William R. Holmes and Ildikó Aradi

Neurobiology Program
Department of Biological Sciences
Ohio University
Athens, OH 45701

ABSTRACT

A model of a dentate granule cell that explicitly included all dendritic spines was used to explore the relative contributions of voltage-gated calcium channels and NMDA receptor channels to calcium influx in dendritic spines following tetanic stimulation. Up to 10 T, N, or L calcium channels were placed in each dendritic spine. It was found that calcium currents through calcium channels in spines had a much shorter duration than the NMDA calcium current, except for T channels, and had an amplitude that was in many cases comparable in magnitude to the NMDA calcium current.

INTRODUCTION

Experimental evidence suggests that voltage-gated calcium channels are present on dendritic spines[1]. However, it is not known what types of calcium channels or how many calcium channels exist on spines. Nor is it known how much of the calcium influx into dendritic spines comes from calcium channels compared to that through NMDA receptor channels. Since calcium influx through NMDA receptor channels has been shown to be essential for LTP and LTD in many cells[2], it is important to understand the contributions of the different sources that affect calcium concentration in dendritic spines.

To explore these issues appropriately, we developed a detailed model of a dentate granule cell with complete morphology including approximately 1,000 spines. Voltage- and calcium-dependent conductances were distributed in the cell with densities chosen to be consistent with experimental data. It was necessary to model all spines explicitly to incorporate correctly the effects of voltage dependent activation of calcium channels in the cell and to allow calcium channels on spines to be isolated from calcium-dependent conductances in dendrites.

This model was used to explore the relative contributions of voltage-gated calcium channels and NMDA receptor channels to calcium influx following tetanic stimulation. Up to 10 T, N, or L calcium channels were placed in each dendritic spine. It was found that calcium currents through calcium channels in spines had a much shorter duration than the NMDA calcium current, except for T channels, and had an amplitude that was in some cases comparable in magnitude or larger than the NMDA calcium current.

METHODS

Neuron Model

A fully reconstructed dentate granule cell was used in the simulations. Simulations were done with the GENESIS simulator[3]. Nine different ionic conductances were included: fast sodium, fast potassium, slow potassium, T-, N-, and L-type calcium, K_A, fast-AHP (BK), and slow-AHP (SK). Kinetics were taken or modified from a number of different published reports[4,5,6,7]. Densities were chosen to match action potential shape and qualitative responses to current injection given in the literature[8,9,10] while also providing a stable resting potential. Because widely varying density values and distributions were consistent with these data, several sets of density values were tested.

Inputs

A strong tetanic input was modeled as 8 pulse 400 Hz activation of 273 synapses in the middle third of the dendritic tree (representing medial perforant path input) or 248 synapses in the distal third of the dendritic tree (lateral perforant path input). Both NMDA and AMPA receptor mediated conductances were modeled at all activated synapses. There were approximately 19 NMDA and 63 AMPA receptor channels at each spine synapse. NMDA receptor channels were subject to voltage dependent magnesium block. Inhibition produced as a result of tetanic input was modeled by reducing R_m in the soma and in the middle and distal thirds of the dendritic tree[11].

Calcium Currents In Dendritic Spines

The model included approximately 4000 dendritic spines, the approximate number of spines thought to exist on these cells. Between 0 and 10 calcium channels of type T, N, or L were added to each dendritic spine with single channel conductances of 8, 14, and 25 pS respectively. Kinetics were taken from Jaffe et al.[5] The calcium current through NMDA receptor channels was computed as in Holmes and Levy[12] and was compared in magnitude and duration to the total calcium current through the voltage-gated calcium channels in the dendritic spine.

RESULTS

Medial perforant path tetanus

An 8 pulse 400 Hz tetanus was applied to 273 medial perforant path synapses in the middle third of the dendritic tree when there were 1, 5, or 10 T, N, or L calcium channels in each spine head. The calcium current through NMDA receptor channels was compared to the current through calcium channels at a representative medial perforant path synapse.

With one calcium channel in every spine head, the calcium current through NMDA receptor channels was larger and wider than the calcium current through the calcium

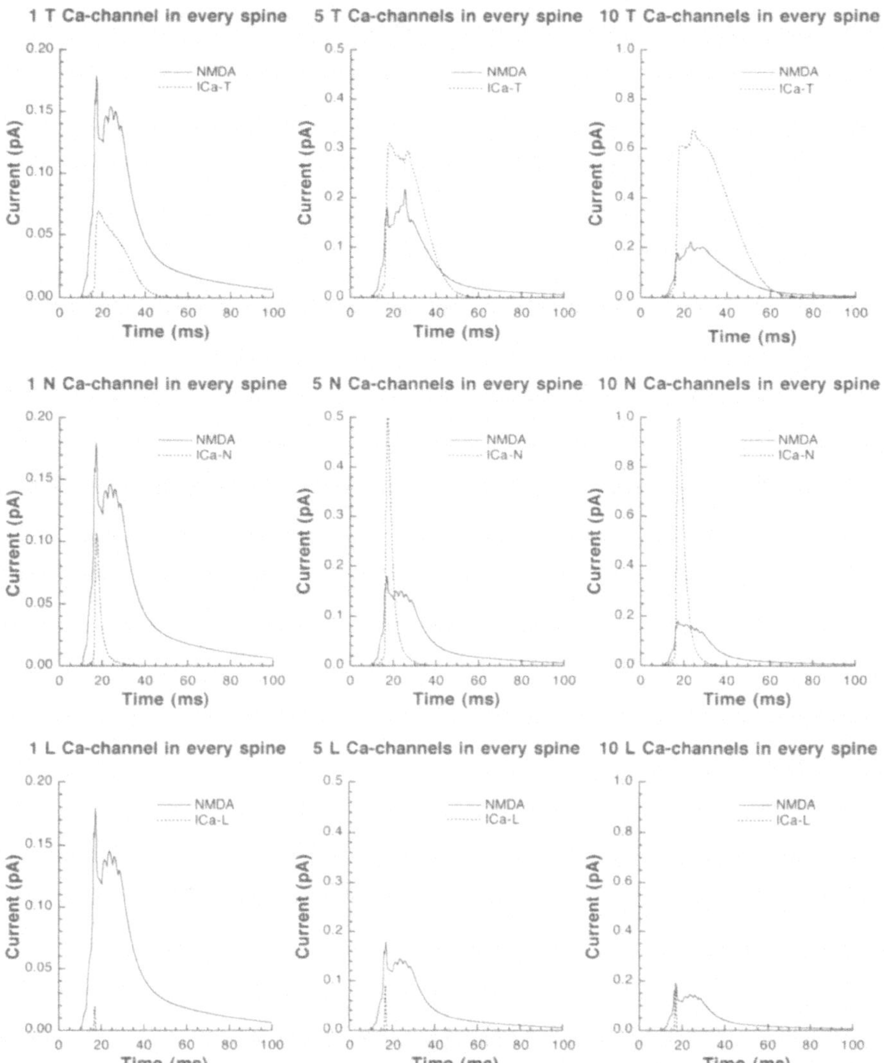

Figure 1. The calcium component of the NMDA current and the calcium channel calcium current are compared at a medial perforant path synapse following an 8 pulse 400 Hz tetanus of 273 medial perforant path synapses. The tetanus produced one action potential in the cell in all cases illustrated except for 5 or 10 T channels in every spine. In these two cases there were two action potentials. The sharp spikes in some of the traces correspond to times when the backpropagating action potential reached the synapse.

channel (Fig. 1, left column). The NMDA receptor mediated calcium current was virtually identical regardless of which calcium channel type was present in all of the spines; one channel in each spine was not sufficient to affect significantly the voltage, and hence the voltage-dependent magnesium block of NMDA receptor channels, at the observed synapse. The NMDA calcium current also had a very long tail not seen with any of the calcium channel calcium currents. With a T channel in every spine, the T channel current was slightly narrower and had half the amplitude of the NMDA calcium current. With an

N channel in every spine, the N channel current amplitude was 60% of the peak amplitude of the NMDA calcium current, but the N channel current had a much briefer duration. With an L channel in every spine, the L channel current was extremely brief and had a very small amplitude.

With five calcium channels in every spine head, the peak amplitude of the calcium channel calcium current exceeded that of the NMDA calcium current when the spines contained T or N channels (Fig. 1, middle column). The NMDA calcium current was slightly larger when five T or five N channels were in every spine than when spines had only one calcium channel because of the voltage boost provided by these extra channels. With five T channels in every spine, the T channel current was 30% larger than the NMDA calcium current and durations were similar (although the NMDA calcium current again had a long tail). With five N channels in every spine, the N channel current was more than twice as large as the NMDA calcium current, but its duration was only one-third as long. With five L channels in every spine, the L channel current, although five times larger than with one channel, was again small and very brief and had a negligible effect on voltage at the observed spine.

With ten calcium channels in every spine head, calcium channel currents dwarfed the NMDA calcium current when the spines contained T or N channels (Fig. 1, right column). The voltage boost provided by the extra T or N channels caused an additional small increase in the NMDA calcium current amplitude. With ten T channels in every spine, the T channel current was three times larger than the NMDA calcium current, and again, durations were similar. With ten N channels in every spine, the N channel current amplitude was five times larger than the peak NMDA calcium current, but its duration was only half as long. With ten L channels in every spine, the peak L channel amplitude matched that of the NMDA calcium current, but again, the duration was extremely brief.

Lateral perforant path tetanus

An 8 pulse 400 Hz tetanus was applied to 248 lateral perforant path synapses in the distal third of the dendritic tree when there were 1, 5, or 10 T, N, or L calcium channels in each spine head. The calcium current through NMDA receptor channels was compared to the current through calcium channels at a representative lateral perforant path synapse. Results with L channels are not shown because the L channel current was negligible at this distal synapse, even with ten L channels in every spine.

Results with 1, 5, and 10 T channels in every spine followed the same pattern found above with the medial perforant path tetanus except that the T channel currents were larger (Fig. 2, top row). T channel calcium current was much smaller than the NMDA calcium current when there was one T channel in every spine, but was 50% larger and nearly four-fold larger than the NMDA calcium current when there were five and ten T channels in every spine, respectively. The T channel current duration was shorter than the NMDA calcium current duration when there was one T channel in every spine, but was comparable to the NMDA calcium current duration when five or ten T channels were in every spine.

N channel calcium currents following the lateral perforant path tetanus were much smaller, but wider, than those found with the medial perforant path tetanus (Fig. 2, bottom row). This was consistent with the smaller amplitude, longer duration voltage change found at the observed synapse in simulations with the lateral perforant path tetanus. With one N channel in every spine, the N channel current amplitude was one-fourth as large as the NMDA calcium current amplitude, and its duration was brief. With five N channels in every spine, the N channel current had the same amplitude as the NMDA calcium current, but its duration was only one-half as long. With ten N channels in every spine, the N channel current was twice as large as the NMDA calcium current, and its duration was only slightly shorter.

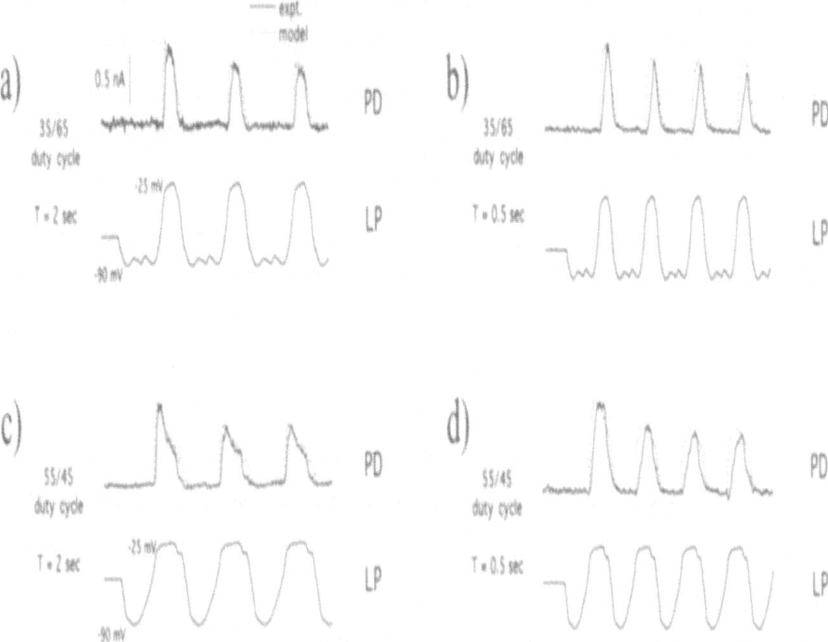

Figure 2. The calcium component of the NMDA current and the calcium channel calcium current are compared at a lateral perforant path synapse following an 8 pulse 400 Hz tetanus of 248 lateral perforant path synapses. The tetanus failed to produce an action potential except when there were 10 T channels in every spine. The rounded peaks in the NMDA current traces occurred with different pulses in the tetanus.

DISCUSSION

The simulations done here are necessarily exploratory and speculative in nature because densities of voltage-dependent channels in different parts of the dendritic tree and the types of calcium channels present in dendritic spines are not known. Nevertheless, we tested several different sets of conductance densities consistent with experimental data and found qualitatively similar results. Perhaps a larger issue is that calcium channels in spines are not likely to be all of one type and distal spines and proximal spines may have different numbers and different types of calcium channels. Thus, the results presented here can be viewed as limiting cases. Despite the highly exploratory nature of the study, the results suggest that calcium channel calcium current may be comparable in magnitude to the NMDA calcium current in dendritic spines of dentate granule cells.

The results make suggestions regarding likely numbers and types of calcium channels in dendritic spines at different locations in dentate granule cells. The number of calcium channels in a spine is likely to be small, perhaps no more than five. For comparison, the number of NMDA channels open and not blocked by magnesium at any given moment in time during the tetanus was usually one or less. Of the channels studied, the N channel seems to be the best overall candidate for spines. T channels may be located more distally, but in small numbers because T channels cause bursting that is difficult to control without nearby calcium dependent potassium channels. Only proximal spines seem likely to have L channels because L channels are not significantly activated beyond proximal regions of this cell. The L channel calcium current might have been larger compared to the NMDA current if we had looked at a spine in the inner third of the molecular layer.

The contribution of calcium channel calcium current to calcium concentration changes in dendritic spines compared to that of the NMDA calcium current needs to be studied with models of calcium dynamics in spines. In the present simulations the NMDA calcium current had a tail that was much prolonged, while calcium channel calcium currents ended more abruptly and in most cases much earlier. This difference in current duration may reflect a difference in the roles calcium channels and NMDA receptor channels play in long-term potentiation and long-term depression.

ACKNOWLEDGEMENTS. This work was supported by NIMH grant NS51081. We thank W.B. Levy and N.L. Desmond for the morphology of the cell used in the simulations.

REFERENCES

1. D.B. Jaffe, S.A. Fisher and T.H. Brown, Confocal laser scanning microscopy reveals voltage-gated calcium signals within hippocampal dendritic spines, *J. Neurobiol.* 25:220-233 (1993).

2. R.C. Malenka and R.A. Nicoll, NMDA-receptor-dependent synaptic plasticity: Multiple forms and mechanisms, *Trends Neurosci.* 16:521-527 (1993).

3. J.M. Bower and D. Beeman. *The Book of GENESIS: Exploring Neural Models with the GEneral NEural SImulation System*, Telos Springer, Santa Clara, CA (1995).

4. G.L.F. Yuen and D. Durand, Reconstruction of hippocampal granule cell electrophysiology by computer simulation, *Neuroscience* 41:411-423 (1991).

5. D.B. Jaffe, W.N. Ross, J.E. Lisman, N. Lasser-Ross, H. Miakawa and D. Johnston, Model for dendritic Ca^{2+} accumulation in hippocampal pyramidal neurons based on fluorescence imaging measurements, *J. Neurophysiol.* 71:1065-1077 (1994).

6. E.N. Warman, D.M. Durand and G.L.F. Yuen, Reconstruction of hippocampal CA1 pyramidal cell electrophysiology by computer simulation, *J. Neurophysiol.* 71:2033-2045 (1994).

7. E. DeSchutter and J.M. Bower, An active membrane model of the cerebellar Purkinje cell. I. Simulation of current clamps in slice, *J. Neurophysiol.* 71:375-400 (1994).

8. K.J. Staley, T.S. Otis, and I. Mody, Membrane properties of dentate gyrus granule cells: Comparison of sharp microelectrode and whole-cell recordings, *J. Neurophysiol.* 67:1346-1358 (1992).

9. N. Spruston and D. Johnston, Perforated patch-clamp analysis of the passive membrane properties of three classes of hippocampal neurons, *J. Neurophysiol.* 67:508-529 (1992).

10. L. Zhang, T.A. Valiante and P.L. Carlen, Contribution of the low-threshold T-type calcium current in generating the post-spike depolarizing afterpotential in dentate granule neurons of immature rats, *J. Neurophysiol.* 70:223-231 (1993).

11. W.R. Holmes and W.B. Levy, Quantifying the role of inhibition in associative long-term potentiation in dentate granule cells with computational models, *J. Neurophysiol.* 78:103-116 (1997).

12. W.R. Holmes and W.B. Levy, Insights into associative long-term potentiation from computational models of NMDA receptor-mediated calcium influx and intracellular calcium concentration changes, *J. Neurophysiol.* 63:1148-1168 (1990).

COMPUTATIONAL PROPERTIES OF A NEURONAL MODEL FOR NOISY SUBTHRESHOLD OSCILLATIONS

Martin T. Huber[1], Hans A. Braun[2], Mathias Dewald[2], Karlheinz Voigt[2] and Jürgen C. Krieg[1]

[1]Zentrum für Nervenheilkunde, Klinik für Psychiatrie, Universität Marburg, Rudolf Bultmannstr. 8, D-35033 Marburg;
[2]Institut für Normale und Pathologische Physiologie, Universität Marburg, Deutschhausstr. 2, D-35033 Marburg;

INTRODUCTION

Intrinsic subthreshold oscillations are a rather common feature of a variety of neurons in the peripheral and central nervous system. Examples reach from neurons in the amygdala, entorhinal and frontal cortex to sensory receptors such as mammalian cold receptors or multimodal ampullary sensory receptors of teleosts[1-6]. The phenomenon is characterized by oscillatory changes in the membrane potential which are below or close to the spike threshold. In this situation, naturally occuring stochastic influences due to membrane or synaptic noise seem to be an essential component for signal encoding. The reason is, that now the noise actually determines whether a spike is triggered during an oscillation cycle or not. Typical mixed patterns result consisting of spike-triggering and subthreshold oscillations (figure 1, see also ref. 3).

Functionally, subthreshold oscillations in the CNS have attracted interest because they might serve as cellular mechanisms for synchronisation and resonance in neuronal networks[1,4-8]. A different view emerges from studies on peripheral receptors where it has been shown that such oscillations in combination with noise can also have useful signal encoding properties. For example, stimulus-dependent modulation of intrinsic noisy oscillations allows shark electroreceptors a very sensitive and differential encoding of electrical and thermal stimuli[3]. This is because the spiking probability per oscillation, on the one hand, is essentially determined by the noise but, on the other hand, can be effectively and selectively be modulated by small stimuli which slightly change the oscillation amplitude or frequency. We assume that similar effects can contribute to encoding in the CNS. In the CNS, subthreshold oscillating neurons are found in input regions such as the basolateral nucleus of the amygdala or layer II of the entorhinal cortex where a variety of signals converges and where neuromodulatory control is important[1,6]. Therefore, sensitive and fine-tuned signal encoding and neuromodulatory properties as described for peripheral sensory receptors might also contribute to signal encoding and neuromodulation in these CNS areas.

In this paper, we outline principle encoding characteristics of noisy subthreshold oscillations by using a computational approach. Specifically, we demonstrate that differential stimulus encoding properties, as observed in sensory receptors, can be accounted for by a simplified but noisy Hodgkin/Huxley-model for subthreshold oscillations.

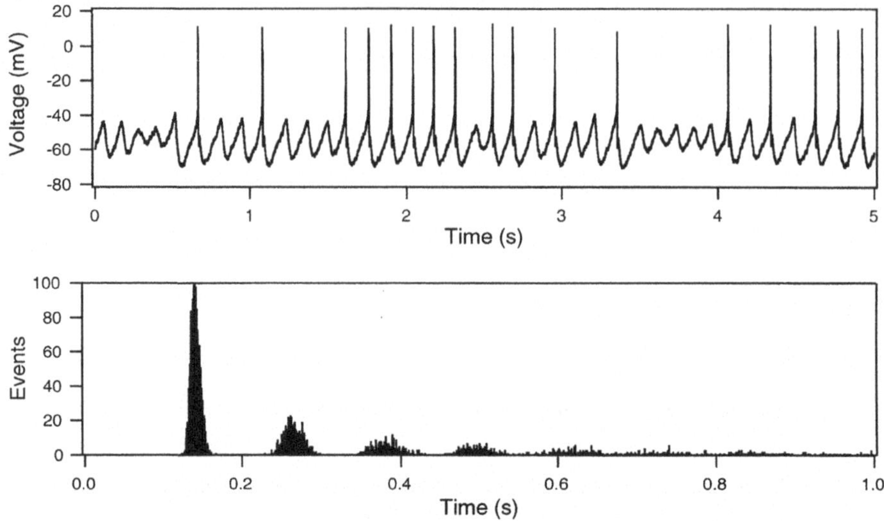

Figure 1: A) Simulated voltage trace showing a mixed pattern of subthreshold oscillations and spikes (I_{app} = 2.3 μA/cm^2). B) The corresponding interspike interval histogram exhibit different distribution peaks located at approximate integer multiples of the oscillation period (first peak).

THE MODEL

The model is based on experimental and theoretical data from cortical neurons[1,4,12]. We use two sets of simplified sodium and potassium conductances which operate at two different membrane potentials and on two different time scales: action potentials are represented by rapid, high-voltage activating Hodgkin/Huxley-like g_{Na} and g_K, whereas the subthreshold oscillations essentially are generated by the interaction of a persistent sodium conductance, g_{Nap}, and a subthreshold potassium conductance, g_{Ks}. The model neuron is represented by the membrane equation

$$C\, dV/dt = -g_l\,(V-V_l) -I_{Nap} - I_{Ks} - I_{Na} - I_K + I_{app} + gw \tag{1}$$

with C = 1μF/cm^2 the membrane capacity, g_l = 0.1 mS/cm^2 the leak conductance and V_l = -60 mV the leak potential. I_{app} is injected current and gw is white noise with zero mean and intensity D. Noise intensity D is set to 0.1 and kept constant for all simulations.

The voltage dependent currents I_{Nap}, I_{Ks}, I_{Na} and I_K are given by $I_i = g_i\, a_i^n\, (V - V_i)$, with g_i the respective maximum conductances (i denotes Na, Nap, K, Ks), a_i^n the voltage-dependent activation variables (n=2 for I_K, n=1 otherwise), V the membrane potential and V_i the respective Nernst potentials. The activation variables are given as

$$dai\, /\, dt = \phi\, (a_{i\infty} - a_i)/\tau_i \tag{2}$$

where the τ_i are voltage-independent time constants and ϕ is a temperature-like scaling factor. The steady-state activations $a_{i\infty} = 1/\{1 + \exp[-s_i\,(V - V_{0,i})]\}$ with s_i the steepness and $V_{0,i}$ the half-activation potentials. We set activation of I_{Na} as instantaneous, thus $a_{Na} = a_{Na\infty}$. For simplicity, we have ignored any inactivation of ionic conductances.

In contrast to previous models for subthreshold oscillations[4,12] we, in addition to the deterministic equations, have accounted for stochastic influences by including gaussian white noise into the membrane equation (see e.g. reference 13 for a detailed discussion of neuronal noise). The system of equations has been solved numerically by use of the forward Euler integration method with stepsize adjusted to 0.1 ms according to the implementation given in reference 14. The numerical parameter values are: V_{Na} = 50, V_K = -90, g_{Na} = 1.5, g_K = 2.0,

$g_{Nap} = 0.3$, $g_{Ks} = 1.8$, $s_{Na} = s_K = s_{Nap} = s_{Ks} = 0.25$, $\tau_K = 2.0$, $\tau_{Nap} = 10$, $\tau_{Ks} = 50$, $V_{0Na} = V_{0K} = -25$, $V_{0Nap} = V_{0Ks} = -40$. Systems of units is ms, mV, mS/cm², μA/cm².

ROLE OF NOISE

With the basic set of parameters the deterministic model exhibits a stable resting membrane potential at about -60 mV. Applying a constant depolarizing current I_{app} leads to subthreshold oscillations in the membrane potential. An increase in I_{app} results in tonic firing. On transition to tonic firing, the mean discharge frequency suddenly increases from zero to a frequency value which equals the oscillation frequency. With further increases of depolarizing current, the mean discharge frequency and oscillation frequency rise monotonically.

Adding noise significantly alters the response properties of the model especially with respect to the nonlinearity, appearing at the transition from subthreshold oscillations to tonic firing. The noise introduces a probabilistic component in spike initiation, which, together with the oscillatory state, determines the response behaviour. The effect of noise is that, on the one hand, stochastic variations in the membrane potential can lead to initiation of spiking even when oscillations are subthreshold in the deterministic mode and that, on the other hand, noise can also lead to "skipping" - suppression of spiking - in deterministically tonic firing states. Noise-induced spiking and skipping is demonstrated in figure 2 (upper and lower traces, respectively). In the following we will concentrate on this response range. The resulting effect is, that the former step-like response behaviour with respect to the firing rate is replaced by a continous, approximately sigmoidal relation between firing rate and applied depolarizing current. Accordingly, the mean firing rate F is not only related to the oscillation frequency OF but also to the spiking probability per oscillation cycle SOgiven by the relation F = OF * SO (see ref. 2).

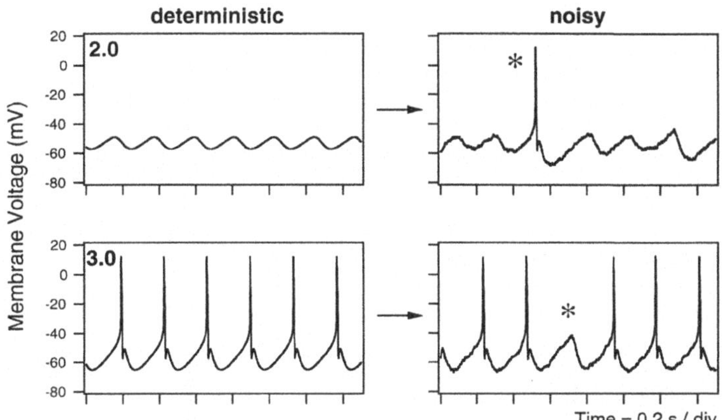

Figure 2: Effects of noise on spike generation. Upper traces: noise can induce spiking in former subthreshold oscillations. Lower traces: noise can suppress spiking in previously tonically firing states. Numeric values indicate the value of applied current.

VARIATION OF SPIKING PROBABILITY

In the range where the spiking probability, in dependence on the depolarizing current, varies between zero and one, the oscillation frequency remains almost unaffected. Thus, the firing rate is directly proportional to the spiking probability (F = SO * OF, with OF = const.). Variation of the depolarizing current I_{app} in this range almost exclusively affect the amplitude of the subthreshold oscillations and therefore result in changes of the spiking probability which is demonstrated with the interval interspike histograms shown in figure 3. Here, the locations of the distribution peaks are not altered by different levels of the applied current. However, an increasing I_{app} decreases the number of subthreshold oscillations. Correspondingly, the number of interspike intervals representing higher multiples of the oscillation period are reduced.

This behaviour is not only of theoretical interest but resembles the behaviour of certain cortical neurons and sensory receptors. For example, fronto-cortical layer IV interneurons exhibit fast subthreshold oscillations which, on depolarizing input, remain in a narrow frequency-range ("narrow-frequency oscillators")[5]. In addition, the selective modulation of the spiking probability as response to applied current represents the encoding mechanism used by ampullary electroreceptors to encode electrical stimuli[3].

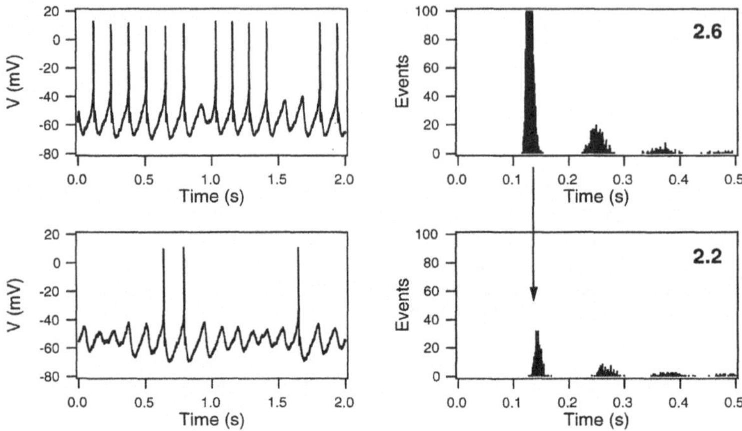

Figure 3: Voltage traces and corresponding ISIH´s determined at different values of applied current (I_{app} = 2.2 and 2.6 μA/cm^2 in lower and upper traces respectively). The applied current varies the spiking probability but does not affect the oscillation frequency as represented by the location of the distribution peaks in the ISIH´s.

VARIATION OF OSCILLATION FREQUENCY

As a response range exists where applied current selectively changes the spiking probability, a second stimulus can be encoded by modulation of the oscillation frequency. In analogy to the response properties of thermosensitive receptors (see e.g refs. 2,3,13) we use temperature-scaling of the ionic currents to acchieve frequency modulation of the oscillations of the model neuron. In doing so, we applied a temperature-like scaling factor to the time constants of the ionic conductances ($\phi = Q_{10}^{(T - 37)/10}$). A physiological Q_{10} value of 3.0 was chosen (see also refs. 13, 15,16).

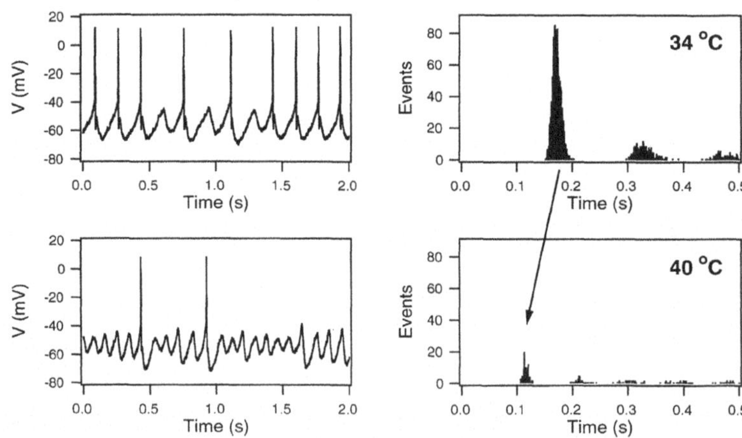

Figure 4: Voltage traces and corresponding ISIH´s at different "temperatures" (T = 34 and 40 °C in upper and lower traces, respectively; I_{app} = 2.3 μA/cm^2). Temperature modulates the oscillation frequency which also changes the spiking probability. Not only the amplitude but also the location of the distribution peaks is changed.

The effect of temperature-dependent frequency modulation on the simulated voltage traces and corresponding ISIH´s is shown in figure 4. It can be seen that increasing temperature accelerates the oscillation frequency due to more rapidly activating ionic kinetics. In addition, as an indirect effect of the enhanced oscillation frequency, the spiking probability per oscillation cycle can also vary and, as shown in the example provided in figure 4, can even decrease when the oscillation frequency increases above a critical value. Accordingly, the distribution peaks in the ISIH´s are shifted and the number of intervals per distribution peaks are varied. The effect on the mean discharge frequency thereby, given by $F = SO * OF$, is more complex because now both, the OF and the SO change. In our model, similar as shown in several studies on temperature transduction in sensory receptors (see e.g. references 2,3,13), this relation, with respect to a broad temperature range follows a maximum curve and thus is non-monotonous.

CONCLUSION

We have shown that a neuron model which contains an only minimal set of ionic conductances can generate subthreshold oscillations. When naturally occuring stochastic variations are accounted for by addition of noise, the firing range, on average, is extended to lower values of the applied current. In the range where the spiking probability, in dependence on the applied current, varies between zero and one, alterations of depolarizing current almost exclusively alter the spiking probability and not the oscillation frequency. In contrast, temperature-like scaling of ionic conductances modulates the oscillation frequency and, as an indirect effect, also the spiking probability.

The two different conditions, I_{app}-dependent modulation of SO and temperature-dependent modulation of OF and SO, have their physiological counterparts in the encoding properties of shark electroreceptors which in that way use noisy intrinsic oscillations to differentially encode electrical and thermal stimuli[3]. Our simulation data is in good agreement with the experimental data from shark electroreceptors. This indicates that a simplified ionic conductance model based on findings from cortical neurons[1,4-6,12] can represent the principle encoding properties, provided that stochastic influences are included.

It seems justified from these findings to assume, that similar encoding principles might also be an important functional principle in central nervous system neurons, allowing for sensitive and specific signal encoding or neuromodulation. In this context it is of particular interest that subthreshold oscillations have frequently been recorded in regions such as for example input regions of the entorhinal cortex or amygdala where a variety of signals converges for further information processing[1,6]. In addition to serving as cellular substrates for synchronisation in neuronal networks, subthreshold oscillations could in addition be cellular substrates for sensitive and differential neuromodulatory control by just using their intrinsic oscillatory dynamics but optimized by naturally occuring noise sources (for a discussion of noise-*tuning* effects see e.g.refs. 9 to 11). We conclude that further studies on subthreshold oscillating neurons should keep in mind the interesting neuromodulatory and encoding properties which arise from cooperative effects of oscillations with noise.

ACKNOWLEDGEMENT

This work was supported by the Stiftung Volkswagenwerk and the Friedrich-Naumann Stiftung.

REFERENCES

1. Alonso, A., and Klink, K., 1993, Differential electroresponsiveness of stellate- and pyramidal-like cells of medial entorhinal cortex layer II. J. Neurophysiol. 70,128-143.
2. Braun, H.A., Bade, H. and Hensel, H., 1980, Static and dynamic discharge patterns of bursting cold fibers related to hypothetical receptor mechanisms. Pflügers Arch. 386,1-9.

3. Braun, H.A., Wissing, H., Schäfer, K. and Hirsch, M.C., 1994, Oscillation and noise determine signal transduction in shark multimodal sensory cells. Nature 367,270-273.

4. Gutfreund, Y., Yarom,Y. and Segev, I., 1995, Subthreshold oscillations and resonant frequencies in guinea-pig cortical neurons: physiology and modelling. J. Physiol. 483.3,621-640.

5. Llinas, R.R., Grace, A.A. and Yarom, Y., 1991, In vitro neurons in mammalian cortical layer 4 exhibit intrinsic oscillatory activity in the 10- to 50-Hz frequency range. Proc. Natl. Acad. Sci. USA 88,897-901.

6. Pare, D., Pape, H.C. and Dong, J., 1995, Bursting and oscillating neurons of the cat basolateral amygdaloid complex in vivo: electrophysiological properties and morphological features. J. Neurophysiol. 74,1179-1191.

7. Eckhorn, R., Bauer, R., Jordan, W., Brosch, M., Kruse, W., Munk, M. and Reitboeck, H.J., 1988), Coherent oscillations: a mechanism of feature linking in the visual cortex? Biol. Cybern. 60,121-130.

8. Engel, A.K., Kreiter, A.K., König, P. and Singer, W., 1991, Synchronization of oscillatory neurona l responses between striate and extrastriate cortical areas of the cat. Proc. Natl. Acad. Sci. USA 88,6048-6052.

9. Bulsara, A. and Gammaitoni, L., 1996, Tuning in to noise. Physics Today March, 39-45.

10. Wiesenfeld, K. and Moss, F., 1995, Stochastic resonance and the benefits of noise: from ice ages to crayfish and SQUIDs. Nature 373,33-36.

11. Longtin, A., 1993, Stochastic resonance in neuron models. J. Stat. Phys. 70,309-327.

12. Wang, X.J., 1993, Ionic basis for intrinsic 40Hz neuronal oscillations. NeuroReport 5,221-224.

13. Longtin, A. and Hinzer, K., 1996, Encoding with bursting, subthreshold oscillations, and noise in mammalian cold receptors. Neural. Comp. 8,215-255.

14. Fox, R.F., Gatland, I.R., Roy, R. and Vemuri, G., 1988, Fast, accurate algorithm for numerical simulation of exponentially correlated colored noise. Phys. Rev. A 38,5938-5940.

15. Braun, H.A., Huber, M.T., Dewald, M., Schäfer, K. and Voigt, K., 1998, Computer simulations of neuronal signal transduction: the role of nonlinear dynamics and noise. Int. J. Bifurcation and Chaos. In press.

16. Braun, H.A., Huber, M.T., Dewald, M. and Voigt, K., 1997: The neuromodulatory properties of noisy neuronal oscillators. In: Kadtge JB, Bulsara A, editors. Applied Nonlinear Dynamics and Stochastic Systems near the Millenium, The American Institute of Physics, pp 281-286.

ANALYSIS OF LIGHT RESPONSES OF THE RETINAL BIPOLAR CELLS BASED ON IONIC CURRENT MODEL

Akito Ishihara,[1] Yoshimi Kamiyama,[2] and Shiro Usui[1]

[1]Department of Information and Computer Sciences
[2]Computer Center
Toyohashi University of Technology, Toyohashi, 441-8580, JAPAN

Introduction

The outer retina which consists of photoreceptor, horizontal cell and bipolar cell is considered as a fundamental neural circuit for spatial and color information processings (Kaneko, 1987). We have developed mathematical models of those cells based on their ionic current mechanisms to analyze the information processings in outer plexiform layer (OPL) of retina (Kamiyama et al., 1996; Usui et al., 1996a,b). In order to further understand information processing in bipolar cells, mathematical models of synaptic connection in OPL have to be constructed. In general, the following sequence of events at the OPL synapse during neurotransmission is involved: 1) At rest in the dark, the synaptic terminal of photoreceptor is depolarized. The depolarization of the terminal activates voltage-gated calcium channels, allowing the entry of calcium ions. 2) The entry of calcium ions into the terminal near the release sites triggers some unknown sequence of events leading to the fusion to the plasma membrane of vesicles containing neurotransmitter glutamates. 3) The glutamates diffuse across the synaptic cleft and make contact with the postsynaptic membrane of bipolar cells. 4) The binding of glutamate to the receptors of the dendrite of bipolar cell causes changes in glutamate induced currents. The release of glutamate is suppressed when photoreceptors are hyperpolarized by light. In the present study, we modeled ionotropic- and metabotropic-type of glutamate induced current of bipolar cell based on available physiological data. We also reconstructed a neural circuit of photoreceptor and bipolar cell by integrating the synaptic models into the model of the bipolar cell body, and analyzed light response properties of a bipolar cell.

Glutamate induced currents of bipolar cells

There are two types of glutamate receptors in the bipolar cells, i.e. ionotropic and metabotropic receptors (Attwell et al., 1987; Sheills & Falk, 1994). The current mediated by the ionotropic glutamate receptor (I_{Glu}) flows through non-selective cation channels. I_{Glu} is the main input from photoreceptors to OFF-center bipolar cells (Murakami et al., 1975). On the other hand, the current mediated by the metabotropic glutamate receptor (I_{APB}) flows through the cGMP sensitive non-selective cation channels. The metabotropic receptor is selective for the glutamate agonist, APB (Nawy & Jahr, 1990). I_{APB} is the main input from photoreceptors to ON-center bipolar cells (Slaughter & Miller, 1985; Nawy & Jahr, 1991). In the following sections, we model I_{Glu} and I_{APB} based on the available experimental results.

The current mediated by the ionotropic receptors (I_{Glu})

I_{Glu} responds to the glutamate with inward current at negative voltage-clamp potentials which reverses around 0 mV to outward current more positive potentials (Attwell et al., 1987; Shiells & Falk, 1994). The dose response characteristics of I_{Glu} shows saturation at high glutamate concentrations, with a limiting slope of two at low concentration (Shiells & Falk, 1994). Since this current is similar in characteristics to the glutamate sensitive current of the horizontal cells, we modeled the current by the same formulation as the glutamate sensitive current of the horizontal cell model (Usui et al., 1996a). We modeled the voltage and glutamate concentration dependencies of I_{Glu} by the following equations:

$$I_{Glu} = g_{Glu} \left\{ \exp\left(\frac{V - 5.82}{115}\right) - 1 \right\} \tag{1}$$

$$g_{Glu} = \overline{g_{Glu}} \frac{[\mathrm{Glu}]^2}{[\mathrm{Glu}]^2 + K_{Glu}^2} \tag{2}$$

The parameters are estimated from the measurement by Shiells & Falk (1994). Figure 1 shows I-V relationship and dose response characteristic of simulated I_{Glu}. The experimental observations are well reproduced by the proposed model.

The current mediated by the metabotropic receptors (I_{APB})

I_{APB} is an outward membrane current in the physiological potential range. I-V relationship for I_{APB} is roughly ohmic, with reversal potential about 0 mV (Attwell et al., 1987; Shiells & Falk, 1994). The peak amplitude of the current in response to glutamate increases linearly with glutamate concentration, and shows saturation above 200 μM glutamate (Shiells & Falk, 1994). The receptors which modulate I_{APB} are coupled to G-proteins. Activation of G-protein stimulates cGMP-phosphodiesterase (PDE) activity resulting in the closing of cGMP sensitive cation channels, similar to phototransduction of rod outer segment (Nawy & Jahr, 1990, 1991). In order to describe the kinetics of I_{APB}, we modeled the cascade of biochemical reactions by the following equations, which are almost identical to the model of phototransduction (Torre et al., 1990).

$$\frac{d[\mathrm{PDE}]}{dt} = \alpha_p \cdot ([\mathrm{PDE}]_{max} - [\mathrm{PDE}]) \cdot [\mathrm{Glu}] - \beta_p \cdot [\mathrm{PDE}] \tag{3}$$

$$\frac{d[\mathrm{cGMP}]}{dt} = \alpha_{cg} \cdot ([\mathrm{cGMP}]_{max} - [\mathrm{cGMP}]) - \beta_{cg} \cdot [\mathrm{PDE}] \cdot [\mathrm{cGMP}] \tag{4}$$

$$g_{APB} = \overline{g_{APB}} \frac{[\mathrm{cGMP}]^{n_{cg}}}{[\mathrm{cGMP}]^{n_{cg}} + K_g^{n_{cg}}} \tag{5}$$

$$I_{APB} = g_{APB} \cdot (V - E_{APB}) \tag{6}$$

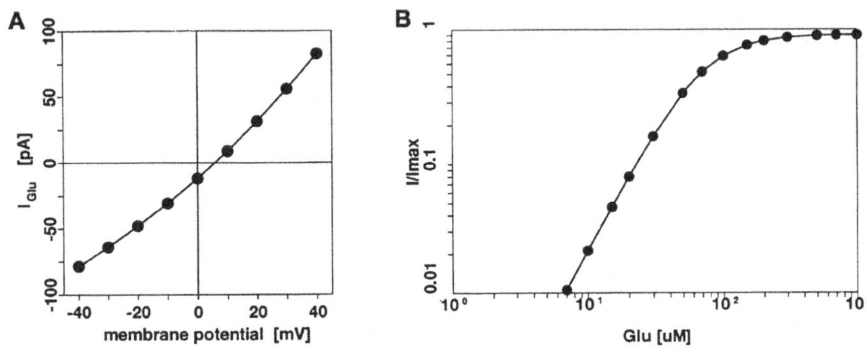

Figure 1. Simulated responses of the I_{Glu} model. A: I-V relationship of current response of glutamate. B: Dose response characteristics.

Table 1. Parameter of the glutamate induced current models

Model	Parameter	Value	Reason for choice
I_{Glu}	$\overline{g_{Glu}}$	240 [nS]	from Shiells & Falk (1994)
	K_{Glu}	68 [μM]	
I_{APB}	α_p	0.1 [sec$^{-1}\mu$M^{-1}]	from Shiells & Falk (1994) and Nawy & Jahr (1990, 1991)
	β_p	542.1 [sec^{-1}]	
	$[PDE]_{max}$	43.2 [μM]	
	α_{cg}	4.5 [sec$^{-1}\mu$M^{-1}]	
	β_{cg}	23.9 [sec^{-1}]	
	$[cGMP]_{max}$	1500 [μM]	
	n_{cg}	3.0	from the phototransduction model (Torre *et al.*, 1990)
	K_g	1500 [μM]	
	$\overline{g_{APB}}$	6.7 [nS]	from Shiells & Falk (1994)
	E_{APB}	0 [mV]	

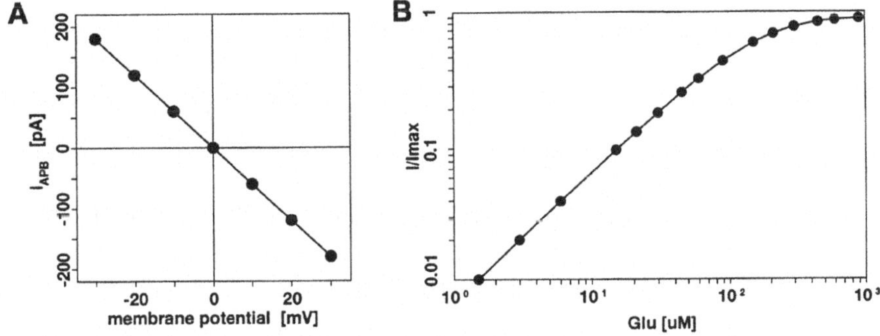

Figure 2. Simulated response of the I_{APB} model. A: I-V relationship of current response of glutamate. B: Dose response characteristics.

We estimated the model parameters from the measurements by Shiells & Falk (1994) and Nawy & Jahr (1990, 1991). As shown in Figure 2, the model well reproduces the results of Shiells & Falk (1994).

The light response model of bipolar cell

The mathematical models of I_{Glu} and I_{APB} in dendrite of bipolar cells were proposed. Using the models and the ionic current models of photoreceptor and bipolar cell (Kamiyama *et al.*, 1996, Usui *et al.*, 1996b), it is possible to analyze light responses of bipolar cells. The metabotropic and ionotropic glutamate receptors may mediate input from rods and cones, respectively, both of which are located on dendrite of rod-dominant ON-center bipolar cells in the carp retina (Saito *et al.*, 1979). Thus, we reconstruct a network model of rod-dominant ON-center bipolar cells to evaluate the glutamate induced current models.

The network model

Figure 3 shows the equivalent circuit of rod-dominant ON-center bipolar cell model which consists of rod, cone and bipolar cell. Each cell was modeled as follows.

Cone and rod photoreceptor Responses to light of cone (V_{Cone}) were modeled by the following equation:

$$C_m \frac{dV_{Cone}}{dt} = -G_m(V_{Cone} - E_m) - I_{photo} \tag{7}$$

where I_{photo} is the photocurrent, C_m is the membrane capacitance, G_m is the membrane conductance and E_m is the reversal potential. I_{photo} was modeled as a simple linear third-order system by the following equation:

$$\left(\frac{d}{dt} + \tau_\gamma\right)^3 \cdot I_{photo} = \tau_\gamma^3 \cdot L(t) \tag{8}$$

where $L(t)$ is the light stimulus and τ_γ is the time constant. To fit the time course of cone response, we used values of $C_m = 0.0234$, $G_m = 0.75$, $E_m = -33.3$, $\tau_\gamma = 34.04$.

The responses of rod were described by the ionic current model (Kamiyama *et al.*, 1996).

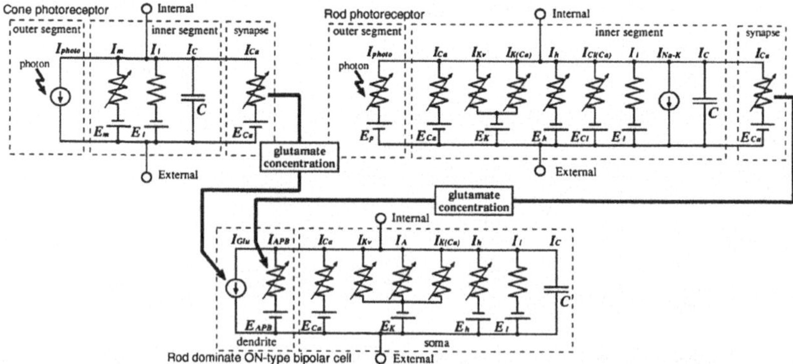

Figure 3. The equivalent circuit of rod-dominant ON-center bipolar cell model

Presynaptic terminal of photoreceptor A change in concentration of glutamate in the synaptic cleft between photoreceptor and bipolar cell is a function of calcium current (I_{Ca}) in presynaptic terminal of photoreceptor:

$$[\text{Glu}] = \frac{(-I_{Ca})^{n_{Glu}}}{(-I_{Ca})^{n_{Glu}} + K_{Glu}^{n_{Glu}}}, \tag{9}$$

where I_{Ca} was given by:

$$I_{Ca} = C_{m_{Ca}} \cdot m_{Ca}^4 \cdot j_{Ca} \tag{10}$$

$$\frac{dm_{Ca}}{dt} = \alpha_{m_{Ca}} \cdot (1 - m_{Ca}) - \beta_{m_{Ca}} \cdot m_{Ca} \tag{11}$$

$$\alpha_{m_{Ca}} = \frac{100 \cdot (270 - V)}{\exp\left(\frac{270-V}{50}\right) - 1} \tag{12}$$

$$\beta_{m_{Ca}} = \frac{150}{1 + \exp\left(\frac{34+V}{6}\right)} \tag{13}$$

$$j_{Ca} = \frac{\beta_j K_j ([\text{Ca}^{2+}]_o \exp(-80V) - [\text{Ca}^{2+}]_i)}{K_j \cdot [\text{Ca}^{2+}]_o \exp(-80V) + 1.0} \tag{14}$$

We estimated parameters from voltage-clamp data by Korenbrot & Maricq (1989). The values are as follows: $n_{Glu} = 2$, $K_{Glu} = 4.5$, $C_{m_{Ca}} = 90$, $K_j = 10$, $\beta_j = -1$, $[\text{Ca}^{2+}]_o = 2500$, $[\text{Ca}^{2+}]_i = 0.1$.

Bipolar cell The ionic current model of the soma of bipolar cell has been proposed by Usui *et al.* (1996b). We introduced the flow of the glutamate induced currents into the soma of bipolar cell. The membrane potential is determined as:

$$C\frac{dV_{BC}}{dt} = -(I_{APB} + I_{Glu} + I_{Kv} + I_h + I_{Ca} + I_{K(Ca)} + I_l) \tag{15}$$

I_{APB} is activated by synaptic transmission from the rod to the bipolar cell; I_{Glu} is activated by inputs from the cone.

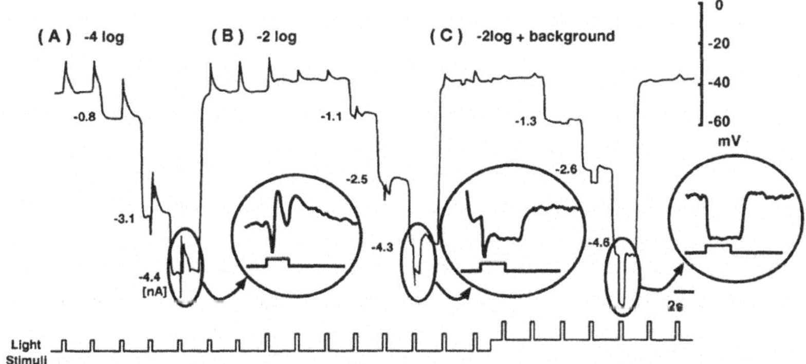

Figure 4. Light responses of rod-dominant ON-center bipolar cell in isolated carp retina (Saito *et al.*, 1979). Responses on −4.0 log units (A), −2.0 log units (B) and −2.0 log units with 500 nm background illumination (C). The membrane is hyperpolarized by injecting current. Numbers at the beginning of each polarization indicate the strength of injected current in nA.

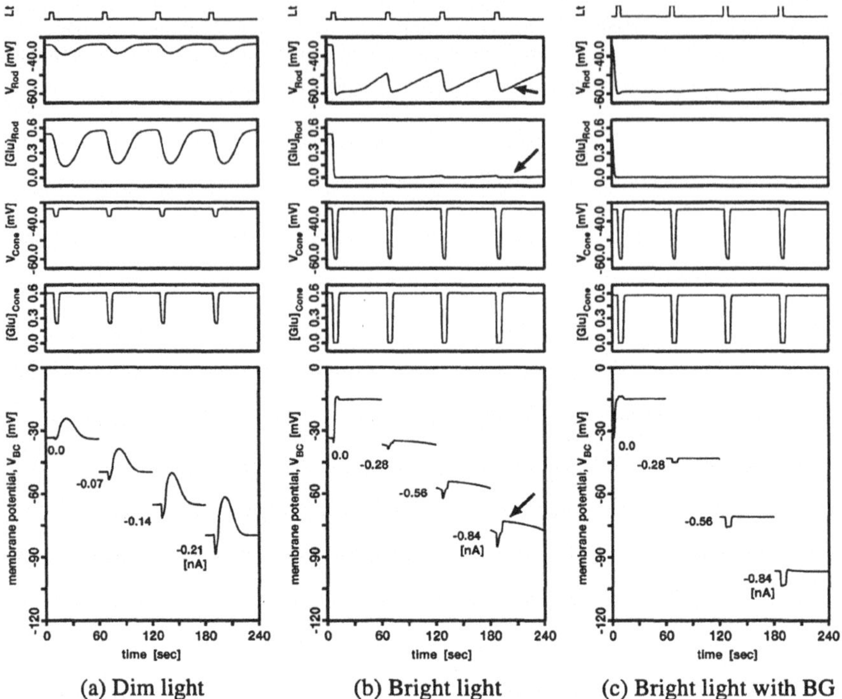

(a) Dim light (b) Bright light (c) Bright light with BG

Figure 5. Simulated responses of the rod-dominant ON-center bipolar cell model. Light stimulus (top) and light responses of rod, glutamate concentration in synaptic cleft between rod and bipolar cell, cone responses, glutamate concentration in synaptic cleft between cone and bipolar cell, bipolar cell responses (from upper side). In response of bipolar cell, numbers at the beginning of each polarization indicate the strength of current in nA.

Experimental observations

Figure 4 shows experimental results which were made to isolate the rod and cone inputs by changing stimulus conditions (Saito *et al.*, 1979). When the membrane is hyperpolarized by injecting current, the response to −4.0 log units shows the initial transient hyperpolarization followed by depolarization. The stronger the membrane hyperpolarization, the more enhanced the hyperpolarizing component. When the light intensity is increased to −2.0 units, there is a marked change in the ratio of the depolarizing and the hyperpolarizing components. The depolarizing component is completely suppressed under background illumination of 500 nm.

Figure 5 is the simulated response of the model, which shows the responses of rod, cone and bipolar cell under three different light conditions. The model clearly reproduced the experimental results. From these conditions, we can explain the complicated waveform of the light responses as follows. The bipolar cell responses to dim light stimulus show both the initial transient hyperpolarizing component and the following depolarization, which are enhanced by hyperpolarizing current injection (Figure 5a). When the light intensity is increased, the depolarizing component is getting smaller than that of dim light (Figure 5b), because the rod is too hyperpolarized which blocks the glutamate release even though the rod is responding enough (Figure 5b, arrow).

Under the background illumination (Figure 5c), the depolarizing component is completely suppressed because the rod response is saturated, and the response has only hyperpolarizing component generated by the input from cone. There is a delayed hyperpolarizing component that is not reproduced. This component is supposed to be generated by the input from other cell(s), e.g. horizontal cell.

Conclusion

We modeled two types of glutamate induced currents at dendrite of bipolar cells. Since glutamate has been identified as the major neurotransmitter in central nervous system, the models can be also applied to other neurons in the central nervous system by estimating appropriate parameter set. Using these models, we also simulated the rod-dominant ON-center bipolar cell which consists of rod, cone and bipolar cells so as to analyze the light responses. The model well reproduced the effect of hyperpolarizing current on the responses of the rod-dominant ON-center bipolar cell under different light intensities.

Acknowledgment

This work was supported in part by the Grant-in-Aid for Scientific Research (No.08279106, No.08650486 and No.09750493) from the Ministry of Education, Science, Sports and Culture of Japan.

REFERENCES

Attwell, D., Mobbs, P., and Tessire-Lavigne, M., 1987, *J. Physiol.* 387:125–161.
Kaneko, A., 1987, *Jpn. J. Physiol.*, 37:341–358.
Kamiyama, Y., Ogura, T., and Usui, S., 1996, *Vision Res.*, 36:4059–4068.
Korenbrot, J. I., and Maricq, A. V., 1989, *Neuron*, 1:503–515.
Murakami, M., Ohtsuka, T., and Shimazaki, H., 1975, *Vision Res.*, 15:456–458.
Nawy, S., and Jahr, C. E., 1990, *Nature*, 346:269–271.
Nawy, S., and Jahr, C. E., 1991, *Neuron*, 7:677–683.
Saito, T., Kondo, H., and Toyoda, J., 1979, *J. Gen. Physiol.*, 73:73–90.
Shiells, R. A., and Falk, G., 1994, *Visual Neurosci.*, 11:1175–1183.
Slaughter, M. M., and Miller, R. F., 1985, *J. Neurosci.*, 5:224–233.
Torre, V., Straforini, M., and Campani, M., 1990, *Cold Spring Harbor Symposia on Quantitative Biology*, LV:563–573.
Usui, S., Kamiyama, Y., Ishii, H. and Ikeno, H., 1996a, *Vision Res.*, 36:1711–1719.
Usui, S., Ishihara, A., Kamiyama, Y., and Ishii, H., 1996b, *Vision Res.*, 36:4069–4076.

CABLE PROPERTIES OF MOTONEURONS IN RAT SPINAL CORD SLICE CULTURES

J. Kleinle[12], M. Larkum[1], N. Buchs[2], W. Senn[12] and H.-R. Lüscher[1]

[1] Physiologisches Institut, Bühlplatz 5,
[2] Institut für Informatik und angewandte Mathematik, Universität Bern, Neubrückstr. 10, CH-3012 Bern, Switzerland

INTRODUCTION

With the rediscovery of active conductances (cf. e. g. [1]) in the dendritic membrane the description of dendrites as simple passive electrical cables has expired. For small voltage fluctuations, however, the electrotonic architecture of the cell remains the decisive determinant for the spatio-temporal evolution of the voltage profile in the somato-dendritic apparatus. By combining several experimental and theoretical approaches we addressed the question of how propagation properties of synaptic events can be described by cable parameters and whether or not the passive membrane properties are homogeneous. 3-site patch clamp recording was done in whole cell configuration at the soma and dendrites of 3 motoneurons in rat spinal cord slice cultures. This technique enabled us to inject current at one electrode and record simultaneously at the other two electrodes without affecting the accuracy of measurement by the application of current. We chose a cylinder as morphological approximation of a dendritic section and compared the experimentally recorded signals with solutions of the cable equation. Cable properties were investigated separately for the two dendritic sections between the 3 recording electrodes which opened the possibility to address the question of membrane inhomogeneities.

EXPERIMENTAL METHODS

Spinal cord slices were made at embryonic day 12 and left in culture for 2-3 weeks as previously described [2] and perfused with a Ringer solution containing (in mM): 150 NaCl, 4 KCl, 1 MgCl$_2$, 2 CaCl$_2$, 5 HEPES, 2 Pyruvic acid, 5 Glucose (pH 7.4). Ventrally located neurons in the organotypic culture of rat were chosen on the basis of their morphological similarity to motoneurons. Very fine patch electrodes (15-30 MΩ resistance) were filled with the following solution (in mM): 120 K-gluconate, 20 KCl, 10 HEPES, 10 EGTA, 2 MgCl$_2$, 2 Na$_2$ ATP (pH 7.3). Data was digitized at 5 kHz and sampled for up to 5 minutes. Simultaneous whole-cell recordings were made from the soma and 2 dendritic sites. Current was injected at either electrode. Both de- and hyper-polarizing α-current (a function of the form $f(t) = t \exp\{-\alpha t\}$) was applied to the cells. Spontaneous synaptic activity was recorded as well and transients with small amplitudes were taken for analysis. Phase contrast video images were taken from the living cells and used for reconstruction. Only incomplete reconstruction could be made as shown in Fig.1(a). The 3 recording sites (s, m, e) are indicated.

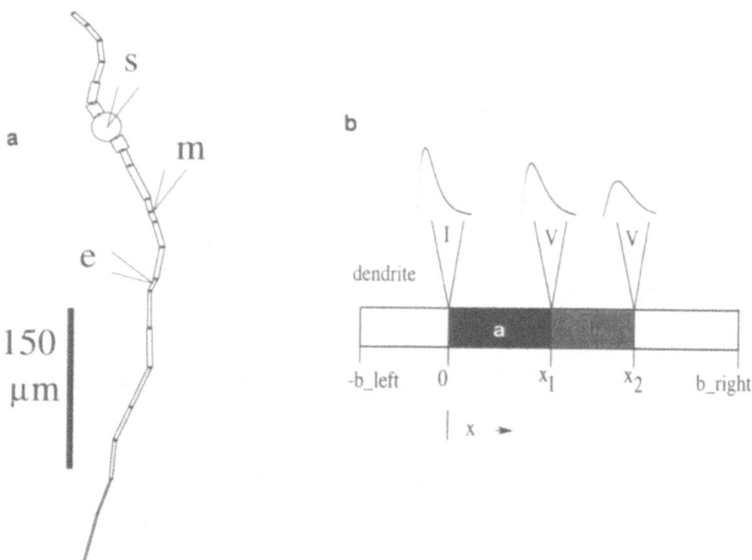

Figure 1: Experiment and Model. (a) Experiment: The incomplete reconstruction of the cell contains the soma and two dendritic electrodes. The electrodes are indicated by s (soma), m (mid-dendritic) and e (end-dendritic). The distance between s and e was $80.5\mu m$ and between m and e $57.5\mu m$. (b) Model: Bound dendrite: The total cable length is $|b_{left}| + |b_{right}|$. A current $I(t)$ is applied at the origin. x_1 and x_2 are the recording sites of the voltage transient caused by $I(t)$. The dendrite is assumed to have sealed end boundary conditions at $x = b_{left}$ and $x = b_{right}$, respectively.

CABLE ANALYSIS

Recorded EPSPs The recorded signal trains at the 3 recording electrodes were decomposed into the presumed single excitatory postsynaptic potentials (EPSPs): Voltage transients following the application of an α-current had a defined onset and could be easily fitted. The amplitude and time course of transients following spontaneous synaptic activity were not as easy to determine. Spontaneous transients with different amplitudes at all recording sites were taken for propagation analysis. Presumed single EPSPs were fitted by a function of the form $f(t) = t^a \exp\{-bt\}$ (generalized α-function with appropriate parameters a and b) and subtracted from the recorded data. The fitted single EPSPs were transformed into the Laplace domain.

Calculated EPSPs A geometrical approximation of the cell morphology was made: The cell geometry (e. g. in Fig.1(**a**) was reduced to a single cylinder in Fig.1(**b**): The unknown dendritic arborization distal to electrode e and s in Fig.1(**a**) was collapsed into a finite cylinder. The diameter of section (a) in Fig.1(**b**) was extended to the bounds b_{left} and b_{right} with sealed end boundary conditions. A set of values was taken for b_{left} and b_{right}, ranging from 0 to infinity. This was also done for section (b) with a correspondingly different diameter. For this simplified geometry the cable equation was solved analytically closely following [3]. The solutions did depend on the choice of b_{left} and b_{right}. Applying the convolution theorem [4] for pairs of solutions of the cable equation in the Laplace domain, the convolution of the Green's function with the synaptic current source could be factorized. For all calculations, passive propagation of the voltage transients along the dendritic section has been assumed. **Comparison between Experiment and Model** The ratio of the transformed EPSPs at the two V-electrodes was compared to the ratio of transformed Green's functions (calculated transients). Assuming a fixed membrane time constant, the graphical comparison between model and experiment revealed an estimate of the steady state electrotonic length constant (λ_{DC}) for each dendritic section and for each direction of signal propagation. By dividing calculated transients which correspond to the same current source in the Laplace domain the current cancels out. We used this strategy to predict the cellular λ_{DC} independently of the site of current application.

STEADY STATE LENGTH CONSTANT (λ_{DC})

Table: λ_{DC}in [μm] for cell A for spontaneous events. The subscripts λ_{sm} indicate that the postsynaptic event did propagate from site s (soma) to site m (mid). b_{right} and b_{left} denote the upper and lower bound of the cell morphology in the model approximation (cf. Fig.1(b)), infinity denotes $|b_{left}| = |b_{right}| = \infty$. The origin is chosen at the somatic recording site, x_1 is the site of the mid-dendritic electrode and x_2 that of the end-dendritic electrode. from mid denotes spontaneous voltage transients generated between the two recording electrodes.

length unit: [μm], $x_1 = 80.5 \mu m$, $x_2 = 138 \mu m$				
b_{right}/b_{left}	λ_{sm}	λ_{ms}	λ_{me}	λ_{em}
infinity	(951 ± 199)	(809 ± 305)	(410 ± 128)	(544 ± 268)
from mid	-	(900 ± 151)	-	-
$b_{right} = 368$	(774 ± 167)	-	(396 ± 108)	-
$b_{left} = -106$	-	(573 ± 149)	-	(481 ± 179)
from mid	-	(590 ± 99)	-	-

The estimates of λ_{DC} are independent of the direction in which the signal did propagate $\lambda_{sm} \sim \lambda_{ms}$ and $\lambda_{me} \sim \lambda_{em}$. The boundary conditions of the cable equation are responsible for the direction dependence of attenuation. The independence of λ_{DC} of the bounds reflects that the boundary conditions (b_{right} and b_{left} in Fig.1a) were chosen appropriately to describe the attenuation along section (a) in Fig.1(b), although the cell morphology is tapering from the soma towards the end-dendritic electrode[1] (cf. Fig.1(a)). EPSPs elicited close to a recording electrode showed less attenuation than signals generated further away (cf λ_{ms} (from mid) in Table 1). Simulated postsynaptic potentials were evoked by an alpha current injected at a third electrode. No difference in λ_{DC} between de- and hyperpolarizing signals (tested on 2 cells, ca. 10^3 transients) was observed, confirming our assumption that these small-amplitude transients propagate passively. The values of λ_{DC} did not change for different amplitudes and time constants of the injected currents (decay times from 0.5 to $2ms$ in steps of $0.5ms$, data not shown).

CRITICAL RISE TIME (CRT)

Neurons obeying the passive cable equation are low pass frequency filters. The critical time constant of a neuron for conduction and filtering of electrical signals turned out to be smaller than its membrane time constant (cf W. Rall & I. Segev in [5], [6]). Our results revealed a critical rise time (CRT) which is up to ten times smaller than the membrane time constant. Voltage transients with rise times shorter than the CRT were attenuated by a factor of 1.2 to 2 depending strongly on the rise time (for all electrode distances studied, $58\mu m \leq$ electrode distance $\leq 152\mu m$). Transients with longer rise times were attenuated by a factor of 1.1 to 1.6 (nearly) independent of their rise time. For fixed membrane time constant the CRT depends on the distance between the electrodes and the electrotonic length constant. The theoretically predicted CRT for an infinite cable with constant diameter turned out to be independent of the distance between current source and recording electrodes as well as time course of the simulated synaptic event. Transients propagating along one dendritic

---

[1]And may be branched in a complicated manner.

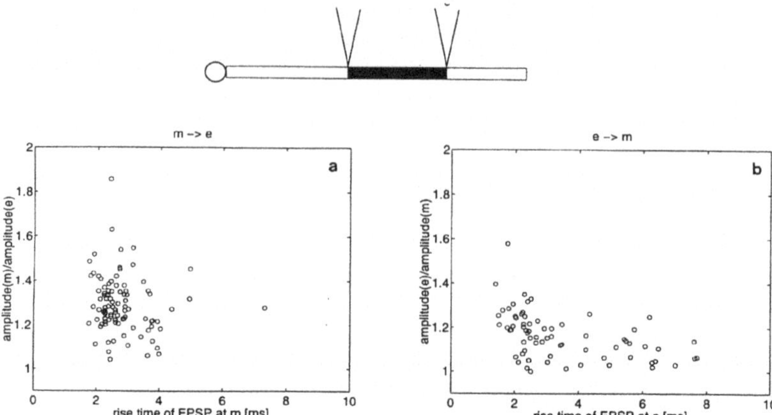

Figure 2: Attenuation vs. rise time: Experimental Results. (a) Plot of the amplitude ratio vs. rise time of the signal at m of recorded spontaneous synaptic potentials. EPSPs propagating from recording electrode m(mid-dendritic) to e(end-dendritic) electrode (somatofugal). (b) Amplitude ratio vs. rise time of the signal at e. EPSPs propagating from e to m (somatopetal).

section in one direction show a characteristic attenuation profile. This could be used to estimate the λ_{DC} characteristic for this section of dendritic membrane. As an example we present recorded (Fig.2) and estimated (Fig.3) attenuation profiles in parallel. In Fig.2(a) the attenuation of somatofugally propagating transients is shown. Each circle corresponds to a pair of EPSPs recorded simultaneously from the two electrodes m and e.

Transients were passing form the mid-dendritic electrode (m, cf. Fig.1(a)) towards the end-dendritic (e) electrode. The amplitude ratio is plotted of the transient at m divided by the amplitude of the corresponding transient at e versus the rise time of the transient at m. In Fig.2(b) the attenuation of somatopetally propagating transients is shown (from e to m). In Fig.3 attenuation profiles of calculated transients are shown. Each plus is a plot of the amplitude ratio of 2 solutions of the cable equation (distance in x: $58\mu m$ corresponding to the electrode distance in the experiment) versus the rise time of the first transient. The calculated amplitude ratios (plus-signs) were connected by linear interpolation (dotted lines). Fig.3(a) shows the attenuation profile for infinite cylinder approximation (see section cable analysis, ($|b_{left}| = |b_{right}| = \infty$ in Fig.1(b)). Each line corresponds to an attenuation profile with one choice of λ_{DC}. The upper line corresponds to $\lambda_{m \to e}$ of $200\mu m$ (strong attenuation). The lower lines are for $\lambda_{m \to e}$ from $300\mu m$ to $900\mu m$ in steps of $100\mu m$. The membrane time constant was fixed ($\tau_m = 13.9ms$). Fig.3(b) shows the attenuation profile for the bound cylinder approximation. The curves are decaying faster towards the value 1 than for the infinite cylinder approximation (no attenuation at all). The smaller attenuation is due to the close sealed end boundary conditions ($-b_{left} = -106\mu m$ and $b_{right} = 368\mu m$. Cf. Fig.1(a)) acting as accumulators of charge. A comparison of the recorded attenuation profiles Fig.2(a) with the theoretical expectation Fig.3(a) allows an estimate of $\lambda_{m \to e}{}^2$. This estimate relies on the assumption that the propagation along section (b) in Fig.1(b) can be approximated as a section of an infinite cylinder with the same length constant. More than 63 % of the data points lie within a range of ($300\mu m \leq \lambda_{me} \leq 500\mu m$, (Fig.2(a)). This corresponds well with the estimate of λ_{me} with a different method (cf. Table). The attenuation profile in somatopetal direction in Fig.2(b)[3] was better approximated by a

---

[2] A different method of analysis was used than in section: steady state length constant. The cable equation was solved with the method of reflection. This is an independent check of our method of analysis („temporal" versus „spatial" analysis)

[3] Propagation of transients: from $e \to m$ showed less attenuation as expected from Fig.3(a), lower values of amplitude ratio corresponding to less attenuation.

214

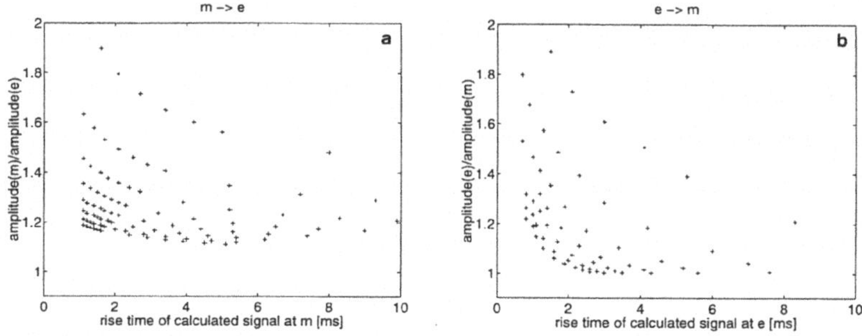

Figure 3: Attenuation vs. rise time: Theoretical estimates. (a) Maximal amplitude ratio of calculated transients from m to e versus rise time of transient at m. Each λ_{DC} has a characteristic attenuation profile. Plotted lines are from λ_{DC} of $200\mu m$ (top) to $\lambda_{DC} = 900\mu m$ (bottom). $\tau_m = 13.9ms$, infinite cylinder approximation. (b) Maximal EPSP amplitude ratio vs. rise time of the signal at e. Bound cylinder approximation: $-b_{left} = -106\mu m$ and $b_{right} = 368\mu m$.

bound rather than an unbound cylinder geometry. Also for somatopetal transients the range of $\lambda_{e \to m}$ ($480\mu m \leq \lambda_{me} \leq 550\mu m$ (cf. Table) was confirmed by the attenuation profile.

CONCLUSION

In agreement with cable theory [7] we showed that the dendrite acts as a low pass filter for typical synaptic input. Each section of dendritic membrane shows a critical rise time for attenuation from which its λ_{DC} can be estimated. Eliminating the dependence of λ_{DC} on the direction of propagation, one could ask for homogeneity of membrane parameters. For the ratio $\frac{\lambda_{me}}{\lambda_{em}}$ we obtained (0.56 ± 0.28). This number should equal the diameter ratio $\sqrt{\frac{d_{me}}{d_{em}}} = 0.69$ (ratio of „average diameter" of cell A). The difference in these numbers is not significant to account for a possible inhomogeneity of passive membrane properties of cell A. For the other two cells this comparison could not be performed due to a lack of data.

Acknowledgments We thank Joel Tabac for helpful discussion. This work was supported by Swiss National Science Foundation grant No. 3100-042055.94.

References

[1] G. J. Stuart and B. Sakmann. Active propagation of somatic action potentials in cerebellar Purkinje cells. *Nature*, 367:69–72, 1994.

[2] M. E. Larkum, M. G. Rioult, and H.-R. Lüscher. Propagation of action potentials in the dendrites of neurons from rat spinal cord slice cultures. *J. Neurophysiol.*, 75(1):154–170, 1996.

[3] J. J. B. Jack and S. J. Redman. The propagation of transient potentials in some linear cable structures. *J. Physiol. (London)*, 215:283–320, November 1971.

[4] C. Koch, T. Poggio, and V. Torre. Retinal ganglion cells: a functional interpretation of dendritic morphology. *Philosophical Transactions of the Royal Society*, 298:227–264, July 1982.

[5] T. G. Jr Smith, H. Lecar, and S. J. Redman, editors. *Voltage and patch-clamping with microelectrodes.* Am. Physiol. Soc., 1985.

[6] C. Koch, M. Rapp, and I. Segev. A brief history of time constants. *Cerebral Cortex*, 6:93–101, March 1996.

[7] W. Rall. Branching dendritic trees and motoneuron membran resisitivity. *Exp. Neurol.*, 1:491–527, 1959.

ACTIVE DENDRITIC CONDUCTANCES INFLUENCE THE RELATIONS BETWEEN SYNAPTIC INPUT AND THE CURRENT-VOLTAGE RELATION OF ADULT SPINAL MOTONEURONS.

Robert H. Lee and C.J. Heckman

Department of Physiology, M211,
Northwestern University Medical School,
303 E. Chicago Ave, Chicago, IL 60611.

INTRODUCTION

Techniques for recording directly from dendrites of CNS neurons require in vitro preparations (e.g. Spruston et al. 1995). However, we have developed a combined computational/experimental approach that allows us to obtain quantitative information about the role of voltage-sensitive dendritic conductances in an in vivo preparation.   The basis of this technique is to take advantage of the well known lack of space clamp of dendritic regions when a voltage clamp is applied at the soma of the cell (Rall and Segev 1985).   When synaptic input is applied to the cell, the voltage clamp applied via an electrode in the soma holds the behaviors of voltage-sensitive conductances in that region constant.   However, the lack of space clamp means that synaptic input to the dendrites can produce depolarizations that alter the state of dendritic voltage-sensitive conductances (Lee and Heckman 1996; Schwindt and Crill 1995).   Thus, the voltage clamp has the advantage of holding the behavior of somatic voltage-sensitive conductances constant while allowing the synaptic input to depolarize dendritic voltage sensitive conductances.

EXPERIMENTAL DATA FROM PREVIOUS STUDIES

In the presence of the monoamines serotonin or noradrenalin, spinal motoneurons in the decerebrate cat preparation can exhibit bistable firing patterns, which is a tendency for the cell to continue stable firing after a brief period of input ceases (Hounsgaard et al. 1988). We have applied the single electrode voltage clamp technique to such motoneurons to show that the plateau potential that produces bistable firing is, to a large degree, produced by voltage-sensitive dendritic conductances (Lee and Heckman 1996; Schwindt and Crill 1995).   Just as recently found in an in vitro study of turtle motoneurons (Svirskis and Hounsgaard 1997), bistable firing in motoneurons in the cat is associated with a high degree of hysteresis in the current-voltage (I-V) relationship (Heckman and Lee 1996). Application of a triangular shaped voltage command reveals a large "N" in the I-V relation during the upward trajectory, but this N is shifted to a much lower voltage on the descending trajectory.

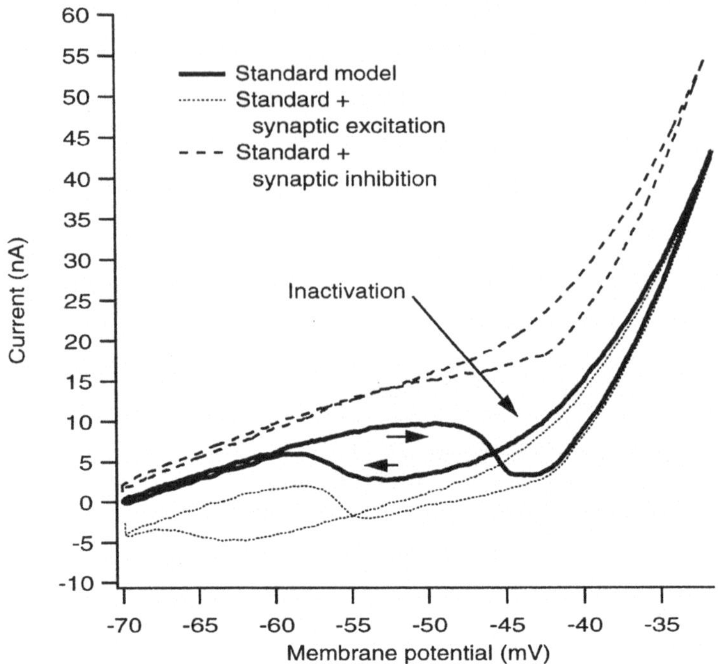

Figure 1: Computer simulation of hysteresis in I-V functions in cat spinal motoneurons.

COMPUTER SIMULATION RESULTS

A computer simulation of this I-V hysteresis is illustrated in Fig. 1 (the pair of arrows on the thick trace marks the region of hysteresis). The model consisted of two compartments, one for the soma and one for the dendrites. A coupling conductance linked the two compartments. The somatic compartment contained the basic channels known to exist in spinal motoneurons (i.e. Na^+ and K^+ spike channels, high threshold Ca^{++} channels, a Ca-mediated channel to produce the AHP, an H channel, and a persistent Na^+ channel) along with a pump and buffer system to control intracellular Ca^{++} concentration. The experimental data could be accurately matched when the dendrite compartment contained conductances the fast Na^+ and K^+ channels (Larkum et al. 1996), an LVA Ca^{++} channel, and a Ca^{++}-activated non-selective cation channel (Rekling and Feldman 1996; Zhang et al. 1995). As in the experiments, the clamp was only applied to the somatic compartment. Analysis of the model indicated that the hysteresis along the voltage axis (i.e. the change in position of the onset of the N) is due to the differential polarization of the dendritic vs. somatic compartments. A similar result was previously attained by Gutman (Gutman 1991). The reduction in current amplitude between the ascending N and descending N was due to inactivation of the dendritic Na^+ and LVA Ca^{++} channel (single arrow in Fig. 1).

Because the hysteresis in the N shape was primarily due to the presence of the voltage-sensitive channels in the dendritic compartment, it was expected that applying steady synaptic input during the course of an applied triangular voltage command at the soma would strongly impact the shape of the I-V relationship. Simulations of an excitatory input (modeled as a conductance change in the dendritic compartment with a reversal potential of

0 mV) combined with the triangular voltage command showed that the N in the I-V relation was shifted to a much lower voltage as seen from the electrode in the soma. However, the overall pattern of hysteresis remained qualitatively the same (lower thin trace in Fig. 1). In contrast, addition of steady inhibition (modeled by a conductance change with a reversal potential of -80 mV) virtually eliminated the N (upper thin trace in Fig. 1).

DISCUSSION AND PREDICTIONS FOR FURTHER STUDIES

The reason for the dramatic effect of the inhibition was primarily that the N was shifted positively into the region were there was strong activation of the delayed rectifier K channel, which overwhelms the N. Because the large N appears to be necessary for motoneurons to maintain bistable firing patterns, these data predict that the presence of a large background inhibition will prevent motoneurons from exhibiting bistable firing even with large amounts of compensating current injected into the soma. We are presently testing these model predictions in experimental studies. However, there is an important caveat to this prediction. If the synaptic inputs themselves have neuromodulatory actions, then the changes in the shape of the I-V function would result in part from actions of second messenger systems on the properties of the voltage sensitive channels. Our current experimental plan is to use synaptic inputs which are thought to be solely ionotropic, i.e. the monosynaptic excitation produced by muscle spindle Ia afferents and the reciprocal inhibition this same input generates in antagonist motoneurons. Our expectation is that the experimental results will conform closely to the computer simulation predictions. Preliminary data in two experiments are consistent with this expectation.

REFERENCES:

Gutman, A.M. Bistability of dendrites. *Int. J. Neural Sys.* 1: 291-304, 1991.

Heckman, C.J. and Lee, R.H. Dendritic plateau potentials and bistable firing in spinal motoneurons: in vivo voltage clamp studies. *Soc. Neurosci. Abstr.* 22: 1845, 1996.

Hounsgaard, J., Hultborn, H., Jespersen, B., and Kiehn, O. Bistability of alpha-motoneurones in the decerebrate cat and in the acute spinal cat after intravenous 5-hydroxytryptophan. *J. Physiol.* 405: 345-67, 1988.

Larkum, M.E., Rioult, M.G., and Luscher, H.R. Propagation of action potentials in the dendrites of neurons from rat spinal cord slice cultures. *J. Neurophysiol.* 75: 154-170, 1996.

Lee, R.H. and Heckman, C.J. Influence of voltage-sensitive dendritic conductances on bistable firing and effective synaptic current in cat spinal motoneurons in vivo. *J. Neurophysiol.* 76: 2107-2110, 1996.

Rall, W. and Segev, I. Space clamp problems when voltage clamping branched neurons with intracellular microelectrodes. In: *Voltage and Patch Clamping with Microelectrodes.* Bethesda MD: Am. Physiol. Soc., 1985,

Rekling, J.C. and Feldman, J.L. Plateau potentials in ambiguus neurons in newborn mouse brainstem *in vitro. Soc. Neurosci. Abstr.* 22: 792, 1996.

Schwindt, P.C. and Crill, W.E. Amplification of synaptic current by persistent sodium conductance in apical dendrite of neocortical neurons. *J. Neurophysiol.* 74: 2220-2224, 1995.

Spruston, N., Jonas, P., and Sakmann, B. Dendritic glutamate receptor channels in rat hippocampal CA3 and CA1 pyramidal neurons. *J. Physiol.* 482: 325-352, 1995.

Svirskis, G. and Hounsgaard, J. Depolarization-induced facilitation of a plateau-generating current in ventral horn neurons in the turtle spinal cord. *J. Neurophysiol.* 78: 1740-2, 1997.

Zhang, B., Wooten, J.F., and Harris-Warrick, M. Calcium plateau potentials in crab stomatogastric ganglion motor neuron. II. Calcium-activated slow inward current. *J. Neurophysiol.* 74: 1938-1946, 1995.

EMULATION OF HOPFIELD NETWORKS WITH SPIKING NEURONS IN TEMPORAL CODING

Wolfgang Maass and Thomas Natschläger

Institute for Theoretical Computer Science, Technische Universität Graz,
Klosterwiesgasse 32/2, A-8010 Graz, Austria
E-mail: {maass, tnatschl}@igi.tu-graz.ac.at

ABSTRACT

A theoretical model for analog computation with temporal coding is introduced
and tested through simulations in GENESIS. It turns out that the use of multiple
synapses yields very noise robust mechanisms for analog computations with temporal
coding in networks of detailed compartmental neuron models. One arrives in this way
at a method for emulating arbitrary Hopfield nets with spiking neurons in temporal
coding, yielding new models for associative recall of spatio-temporal firing patterns. A
corresponding *layered* architecture yields a refinement of the synfire-chain model that
can assume a fairly large set of different firing patterns for different inputs.

INTRODUCTION

We present a new model for associative memory in networks of spiking neurons
that is based on analog computation with *temporal coding*. We consider network inputs
which encode an analog pattern $\mathbf{x} \in [-1, 1]^n$ (the "stimulus") through relative *time
delays* of the initial firing of all (or almost all) neurons in the network. In such a code
any $\langle x_1, \ldots, x_n \rangle \in [-1, 1]^n$ can be represented by a firing pattern where the i-th neuron
fires at time $T - c \cdot x_i$ (for suitable constants c and T). We refer to this scheme as
temporal coding in the following. This type of neural code is discussed for example by
Hopfield[1], Thorpe and Imbert[2] and Kjaer *et al*[3]. It has been argued that such neural
code is especially useful for *very fast* neural computations with low firing rates.

We exhibit both through a rigorous mathematical result and through simulations
of compartmental neuron models in GENESIS a computational mechanism that allows
us to emulate any given Hopfield net with a network of spiking neurons with temporal
coding. This mechanism can be implemented in a surprisingly noise robust way.

This article also addresses an important general question regarding the modeling
of neural computation: To what extent are mechanisms and results that have been
demonstrated analytically for networks of relatively simple mathematical models for

leaky integrate-and-fire neurons also valid for networks of substantially more complex compartmental neuron models?

THEORETICAL RESULT

In this section we outline the construction of a network $\mathcal{S}_{\mathcal{H}}$ of spiking neurons which approximates the computation of an arbitrary given Hopfield network $\mathcal{H}$. As a model for a spiking neuron we take the common model of a *leaky integrate-and-fire neuron with noise*, respectively the somewhat more general *spike response model*[4, 5]. The only specific assumption that is needed for the construction of $\mathcal{S}_{\mathcal{H}}$ in theorem 1 is that both the beginning of the rising part of an EPSP and the beginning of the descending part of an IPSP can be described by a linear function.

Theorem 1 *Let $\mathcal{H}$ be an arbitrary given Hopfield net with graded response[6] and synchronous update. We assume that $\mathcal{H}$ consists of n sigmoidal neurons u_i for $i \in \{1, \ldots n\}$ with arbitrary weights $w_{ij} \in \mathbb{R}$ for $i, j \in \{1, \ldots n\}$ and a piecewise linear activation function σ ($\sigma(x) = -1$ if $x < -1$, $\sigma(x) = x$ if $-1 \leq x \leq +1$ and $\sigma(x) = +1$ if $x > +1$).*

Then one can approximate any computation of $\mathcal{H}$ by a recurrent network $\mathcal{S}_{\mathcal{H}}$ of n spiking neurons (with $O(1)$ auxiliary spiking neurons) in temporal coding. An input, internal state, or output of $\mathcal{H}$ of the form $\langle x_1, \ldots, x_n \rangle \in [-1, 1]^n$ is represented in $\mathcal{S}_{\mathcal{H}}$ by temporal coding, i.e. by a firing pattern of $\mathcal{S}_{\mathcal{H}}$ in which its i-th neuron fires at time $kT - c\tilde{x}_i$, where $|x_i - \tilde{x}_i|$ can be made arbitrarily small. The reference time points kT for $k = 0, 1, \ldots$ are defined by $O(1)$ periodically firing auxiliary neurons in $\mathcal{S}_{\mathcal{H}}$. Any fixpoint of $\mathcal{H}$ corresponds to a stable periodic firing pattern of $\mathcal{S}_{\mathcal{H}}$.

The *proof* of theorem 1 is based on a construction of Maass[5] for leaky integrate-and-fire neurons an is given in more detail elsewhere[7]. It relies on the observation that for a certain range of parameters the firing delay of a neuron v can be expressed as a weighted sum $\sum_{i=1}^{n} w_i x_i$ of the firing delays $c \cdot x_i \in [-c, c]$ of n presynaptic neurons (where the parameter w_i describes the efficacy of the synapse from the i-th presynaptic neuron, and c is some constant). The weights w_{ij} in $\mathcal{S}_{\mathcal{H}}$ are the same (properly scaled) as for the corresponding edges in $\mathcal{H}$.

SIMULATIONS WITH GENESIS

In the remainder of this article we take this theoretical construction as basis for a case study. We want to find out to what extent mechanisms for computations with temporal coding that have been verified theoretically for integrate-and-fire neurons correspond to stable computational mechanisms for substantially more detailed compartmental models of biological neurons simulated in GENESIS[8].

Surprisingly, it turns out that the most essential computational mechanism that underlies the proof of theorem 1 (and the construction of Maass[5]) works in the more detailed neuron model of GENESIS *even better*. The previously mentioned computational mechanism for computing a weighted sum $\sum_{i=1}^{n} w_i x_i$ is theoretically sound as long as the range $[0, 2c]$ of the differences in firing times of the n presynaptic neurons $a_1, \ldots, a_n$ is so small that there is a time point at which — in spite of their temporal differences — the resulting postsynaptic potentials are at the soma of v *all* in their initial *linear* phase. However, since non-NMDA EPSP's rise very fast, the theory would suggest that the length $2c$ of this interval would have to be chosen around $1\,\mathrm{ms}$. This is so small

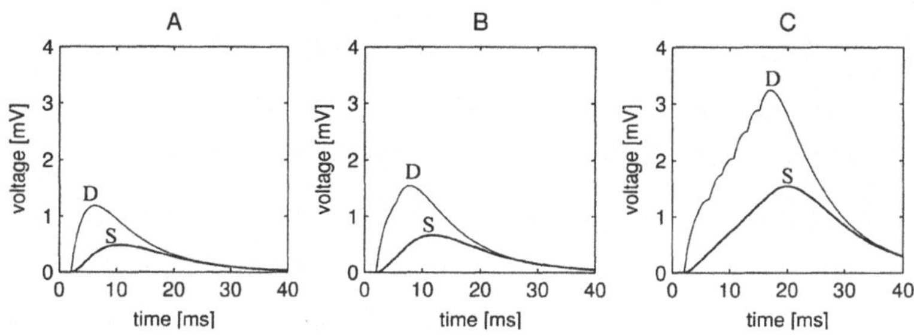

Figure 1. Superposition of several non-NMDA EPSP's caused by a single spike from a presynaptic neuron v_j. The membrane voltage of neuron v_i is measured at the dendrite (D) and at the soma (S). Three cases are shown: a single synapse (A), three synapses (B) and six synapses (C) from the presynaptic neuron. In all subsequent simulations we use three synapses, which results in a time interval of 7 ms during which the superposition of these EPSP's increases linearly.

that one has to be concerned about the effect of various sources of temporal jitter on the precision of this temporal coding. Figure 1 shows that for GENESIS neurons the value $2c$ can be chosen much larger.

One can extend this length by replacing each synapse between a presynaptic neuron v_j and v_i by $l > 1$ synapses. In this way a single spike from v_j causes a superposition of l EPSP's at the soma of v_i. This superposition may have a substantially longer increasing phase if the signal pathways along the l synapses between v_j and v_i have a reasonable difference in their total delays[9, 10] (time from generation of a spike at neuron v_j to the onset of the PSP at the soma at neuron v_i).

For our GENESIS-simulations we started out from a Hopfield net $\mathcal{H}$ with $n = 60$ neurons, whose weights were computed with the projection method[11] so that 9 arbitrarily chosen vectors from $\{-1, 1\}^{60}$ (see Figure 2) are fixpoints of $\mathcal{H}$. We have simulated the network $\mathcal{S}_{\mathcal{H}}$ of spiking neurons (that simulates $\mathcal{H}$ according to theorem 1) in GENESIS with a network $\mathcal{G}_{\mathcal{H}}$ of 61 neurons. In $\mathcal{G}_{\mathcal{H}}$ each neuron is modeled with 122 compartments, and there are 3 synapses between each pair of neurons (whose delays differ by up to 3.5 msec). $\mathcal{G}_{\mathcal{H}}$ has the same architecture and the same weights (properly scaled) as $\mathcal{S}_{\mathcal{H}}$ (30% of the weights were rounded to 0). Weights with negative values are modeled by inhibitory synapses.

In the protocols of our simulations (Figures 3 and 4) of $\mathcal{G}_{\mathcal{H}}$ the bars to the left ("i-error") indicate the difference between the firing times of the *first* wave of firing and the closest memory pattern (both interpreted as vectors from $[-1, +1]^{60}$ via temporal coding). Correspondingly the bars to the right ("o-error") indicate the difference between the firing times of the *last* shown firing wave and the same memory pattern. In both cases a non-firing of a neuron is treated like a firing at the very end of the firing

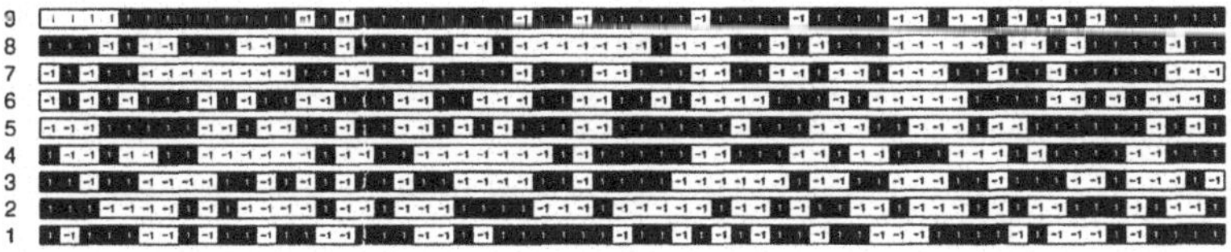

Figure 2. 9 randomly chosen "memory patterns" from $\{-1, +1\}^{60}$.

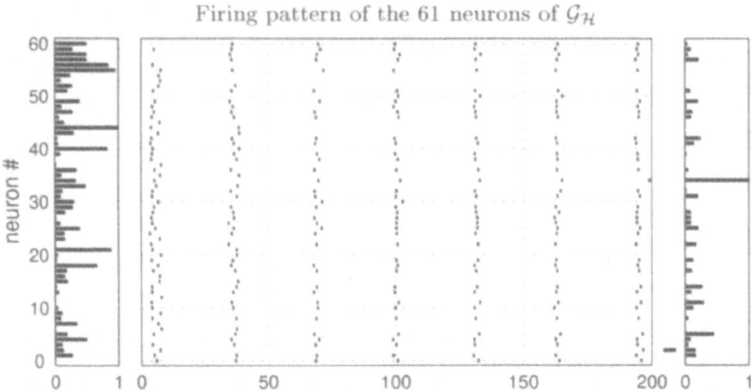

Figure 3. GENESIS-simulation of $\mathcal{G}_\mathcal{H}$ with a corrupted version $\mathbf{x}$ of memory pattern 1 as input (see text for details). The resulting effective difference between the *input* vector $\mathbf{x} \in [-1, +1]^{60}$ that is given to $\mathcal{G}_\mathcal{H}$ in temporal coding and memory pattern 1 is indicated to the left ("i-error"). For each neuron i an error bar of length in $[0, 1]$ indicates the size of the deviation between the i-th component of $\mathbf{x}$ and the i-th component of memory pattern 1. The difference between the *output* vector $\mathbf{y} \in [-1, +1]^{60}$ in temporal coding and memory pattern 1 is indicated to the right ("o-error").

wave. This is justified by the observation that both scenarios reflect equivalent ways of encoding the smallest possible analog value "-1"[5]. A neuron v_j that fires *very late* during a firing wave has an equally negligible impact an the firing times during the next firing wave as if v_j would not have fired at all. It is interesting to note that in all our simulation both types of encoding "-1" are present.

In the space right before "o-error" in Figures 3 and 4 we have marked with a horizontal bar each neuron i of $\mathcal{G}_\mathcal{H}$ which does *not* satisfy following property: Neuron i fires during the last firing wave *if and only if* the i-th component of the memory pattern is a "+1".

To investigate the noise robustness of our model two internal sources of noise are considered (in addition to the noise by which the network inputs are perturbated): a noisy membrane potential and failing synapses. These two types of noise are underlying all our simulations, which shows that our construction is robust against a substantial level of noise. We assume the existence of a random current at the soma of each neuron (Gaussian distribution with a variance of $10^{-18}\,\mathrm{A}^2$), and that each synapse fails with a probability of 15%.

The results of the simulation of $\mathcal{G}_\mathcal{H}$ with a *very* noisy version of memory pattern 1 as input is shown in Figure 3. The input $\mathbf{x}$ was constructed as follows from memory pattern 1: randomly chosen 15% of the components of memory pattern 1 were multiplied with -1. On top of this, the value of each component of $\mathbf{x}$ was moved by a random value from $[-0.4, 0.4]$. This input $\mathbf{x}$ was presented to $\mathcal{G}_\mathcal{H}$ in *temporal coding*, although not precisely: randomly chosen 10% of the neurons in $\mathcal{G}_\mathcal{H}$ were prevented from firing (corresponding to an input value -1). Further simulations[7] verify that in fact *all arbitrarily chosen fixpoints of $\mathcal{H}$* can be associatively recalled by $\mathcal{G}_\mathcal{H}$. Figure 3 shows that after 6 firing waves almost no "digital" errors in associative recall in terms of firing/non-firing occur.

It turns out that in contrast to the theoretical model $\mathcal{S}_\mathcal{H}$ no separate "oscillator" is needed in $\mathcal{G}_\mathcal{H}$ to define the reference times kT for $k = 0, 1, 2, \ldots$. Instead, neuron 0

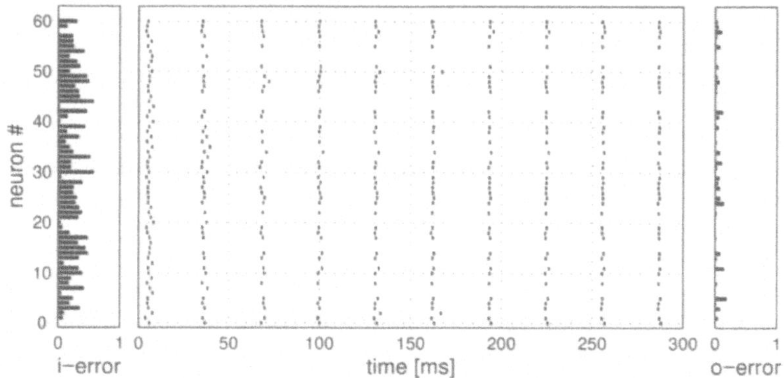

Figure 4. Here the input to $\mathcal{G}_\mathcal{H}$ is a linear combination of memory pattern 1 with factor 0.5 and of memory pattern 4 with factor 0.3, presented in temporal coding, with noise added as in Figure 3. The error bars to the left indicate the difference between the actual input and memory pattern 1 – which is its stronger component. The error bars to the right indicate the difference between the output and memory pattern 1 in temporal coding.

of $\mathcal{G}_\mathcal{H}$, whose firing times provide the reference times kT, is a neuron like all the others, and it receives inputs from all other neurons in $\mathcal{G}_\mathcal{H}$ through synapses with equal weight. It also has outgoing synapses to all other neurons in $\mathcal{G}_\mathcal{H}$. Thus its firing contributes to the triggering of the next firing wave.

The network $\mathcal{G}_\mathcal{H}$ is not only capable of an associative recall of memory patterns if the input consists of a noisy version of one of the stored patterns. It also has the ability to find the stronger component if the stimulus is a combination of two of the stored memory patterns. This fact is demonstrated through simulations (see Figure 4). This is in contrast to a pattern recognition system comprised of spiking neurons proposed by Hopfield[1].

In $\mathcal{G}_\mathcal{H}$ we employ large transmission delays of 25 ms in order to avoid refractory effects and we modeled each negative weight w_{ij} form the network $\mathcal{H}$ through separate inhibitory synapses. One may argue that these two assumptions are biologically rather implausible. To overcome this we have replaced the completely connected network $\mathcal{G}_\mathcal{H}$ by a layered network $\mathcal{G}_\mathcal{H}^{synfire}$, with transmission delays of 5 ms between neurons in successive layers[7]. In $\mathcal{G}_\mathcal{H}^{synfire}$ negative weights w_{ij} are not modeled explicitly. Instead an inhibitory bias (equal for all neurons) is applied. This "global" inhibition is produced by synapses formed between the additional neuron 0 and each neuron i. Thus all information about the weights w_{ij} of the simulated Hopfield net $\mathcal{H}$ is stored exclusively in the efficacies of *excitatory* synapses in $\mathcal{G}_\mathcal{H}^{synfire}$. Simulations have shown that the layered network, has the same ability as $\mathcal{G}_\mathcal{H}$ for associative recall of any of the 9 given memory patterns[7].

CONCLUSIONS

We have exhibited a new way of emulating a Hopfield net in temporal coding with spiking neurons, that differs in essential aspects from previous models for associative memory in networks of spiking neurons[12, 13, 14, 4, 15, 16]. The resulting networks of spiking neurons carry out their computation very fast, even with biologically realistic low firing rates. This emulation is based on a rigorous theoretical result, and our GENESIS-simulations show that it is quite noise-robust.

The layered version of the resulting network of spiking neurons provides a refinement of the synfire-chain model[17]. This refined version of a synfire chain overcomes one of the essential bottlenecks of the original synfire chain model: its low storage capacity. Whereas the original version of a synfire chain has only two states (active and inactive), the here considered layered networks (with a similar architecture, but a more complex weight-assignment) exhibit a fairly *large reservoir of different stable firing patterns*.

We show that this can even be achieved by storing all information about these "memory patterns" in the efficacies of *excitatory* synapses. To the best of our knowledge all arguments that support the existence of synfire chains in higher cortical areas (on the basis of the consistency of their firing pattern with recorded data[18]) also support the existence of our refined version in the same biological neural system.

REFERENCES

1.   J. Hopfield. Pattern recognition computing using action potential timing for stimulus representation. *Nature*, 376:33–36, 1995.
2.   S. T. Thorpe and M. Imbert. Biological constraints on connectionist modelling. In R. Pfeifer, Z. Schreter, F. Fogelman-Soulie, and L. Steels, editors, *Connectionism in Perspective*. Elsevier, 1989.
3.   T. W. Kjaer, T. J. Gawne, and B. J. Richmond. Latency: Another potential code for feature binding in striate cortex. *Journal of Neurophysiology*, 76(2):1356–1360, 1996.
4.   W. Gerstner and J. L. van Hemmen. Associative memory in a network of "spiking" neurons. *Network*, 3:139–164, 1992.
5.   W. Maass. Fast sigmoidal networks via spiking neurons. *Neural Computation*, 9:279–304, 1997.
6.   J. J. Hopfield. Neurons with graded response have collective computational properties like those of two-state neurons. *Proceedings of the National Academy of Sciences, USA*, 81:3088–3092, 1984.
7.   W. Maass and T. Natschläger. Networks of spiking neurons can emulate arbitrary Hopfield networks in temporal coding. *Network: Computation in Neural Systems*, 8(4):355–372, 1997.
8.   J. M. Bower and D. Beeman. *The Book of GENESIS: Exploring Realistic Neural Models with the GEneral NEural SImulation System*. Springer-Verlag, Inc. Published by TELOS, New York, 1995.
9.   A. Zador, H. Agmon-Snir, and I. Segev. The morphoelectronic transform: A graphical approach to dendritic function. *The Journal of Neuroscience*, 15(3):1669–1682, 1995.
10.   Y. Manor, C. Koch, and I. Segev. Effect of geometrical irregularities on propagation delay in axonal trees. *Biophysical Journal*, 60:1424–1437, 1991.
11.   J. Hertz, A. Krogh, and R. G. Palmer. *Introduction to the Theory of Neural Computation*. Addison-Wesley, 1991.
12.   E. Fransén. *Biophysical Simulation of Cortical Associative Memory*. PhD thesis, Stockholm University, October 1996.
13.   M. W. Simmen, E. T. Rolls, and A. Treves. Rapid retrieval in an autoassociative network of spiking neurons. In J. M. Bower, editor, *Computational Neuroscience*, pages 273–278. Academic Press, London, 1995.
14.   R. Ritz, W. Gerstner, U. Fuentes, and J. L. van Hemmen. A biologically motivated and analytically soluble model of collective oscillations in the cortex. *Biological Cybernetics*, 71:349–358, 1994.
15.   A. Lansner and E. Fransén. Modelling hebbian cell assemblies comprised of cortical neurons. *Network*, 3:105–119, 1992.
16.   A. V. M. Herz, Z. Li, and J. L. van Hemmen. Statistical mechanics of temporal association in neural networks with transmission delays. *Physical Review Letters*, 66(10):1370–1373, 1991.
17.   M. Abeles. *Corticonics: Neural Circuits of the Cerebral Cortex*. Cambridge University Press, 1991.
18.   M. Abeles, H. Bergman, E. Margalit, and E. Vaadia. Spatiotemporal firing patterns in the frontal cortex of behaving monkeys. *Journal of Neurophysiologie*, 70(4):1629–1638, October 1993.

BINOCULAR DISPARITY TUNING IN CORTICAL 'COMPLEX' CELLS: YET ANOTHER ROLE FOR INTRADENDRITIC COMPUTATION?

Bartlett W. Mel,[1] Kevin A. Archie,[2] and Daniel L. Ruderman[3]

[1] Biomedical Engineering Department
University of Southern California
Mail Code 1451
Los Angeles, CA 90089
mel@quake.usc.edu

[2] Neuroscience Program
University of Southern California
Mail Code 2520
Los Angeles, CA 90089
karchie@lnc.usc.edu

[3] The Salk Institute
P.O. Box 85800
San Diego, CA 92186
ruderman@salk.edu

INTRODUCTION

Studies of information flow in the individual neuron have generally been based on the idea that the depolarization due to an activated excitatory synapse is passively propagated through the dendrite to the soma (Rall 1964). Recent evidence, however, demonstrates that the dendrites of cortical pyramidal cells contain significant concentrations of voltage-dependent channels (Stuart and Sakmann 1994).

Earlier biophysical modeling studies (Mel 1992a, 1992b, 1993) have shown that the presence of such active conductances in the dendrites leads to nonlinear interactions between nearby synapses. If these interactions are approximated as pairwise multiplications between nearby synapses, then the total response can be abstracted as a sum of first-order terms (the individual synaptic inputs) and second-order terms (the intersynaptic interactions).

This form is similar to energy models that have been used to describe several computations of interest, including direction selectivity (Adelson and Bergen 1985) and orientation tuning (Pollen and Ronner 1983). The similarity leads us to hypothesize

Table 1. Simulation Parameters.

Parameter	Value
R_m	10k Ωcm^2
R_a	200 Ωcm
C_m	1.0 μF/cm^2
V_{rest}	-70 mV
Number of Compartments	615
Somatic $\bar{g}_{\text{Na}}$, $\bar{g}_{\text{DR}}$	0.20, 0.12 S/cm^2
Dendritic $\bar{g}_{\text{Na}}$, $\bar{g}_{\text{DR}}$	0.05, 0.03 S/cm^2
Input Frequency	0 – 100 Hz
$\bar{g}_{\text{AMPA}}$	0.027 nS – 0.295 nS
τ_{AMPA} (on, off)	0.5 ms, 3 ms
$\bar{g}_{\text{NMDA}}$	0.27 nS – 2.95 nS
τ_{NMDA} (on, off)	0.5 ms, 50 ms
E_{syn}	0 mV

that dendrites with active conductances could be the locus for performing computations that fit the form of an energy model.

Recently, we have shown (Mel et al. 1997) that one such computation, the phase-invariant orientation tuning characteristic of a visual "complex" cell, could be performed within the dendrites of a single neuron.

Binocular Disparity Tuning

Hubel and Wiesel (1962) first demonstrated that many neurons in primary visual cortex are driven by input from both eyes. It has since been shown that some cells are finely tuned for binocular disparity and are largely invariant to the absolute position of the stimuli (Pettigrew et al. 1968; Ohzawa et al. 1990; Ohzawa et al. 1997).

Notably, the responses of such disparity-tuned complex cells have been described as energy models (Ohzawa et al. 1990, 1997). This suggests that disparity tuning could be another computation to which active dendrites contribute.

METHODS

Compartmental simulations of a pyramidal cell from cat visual cortex (morphological data provided by Rodney Douglas and Kevan Martin) were carried out in NEURON (Hines 1989). The parameters used for the simulations are summarized in Table 1.

Both soma and dendrites contained Hodgkin-Huxley-type, voltage-dependent Na$^+$ and K$^+$ channels. The channel model, adapted from a channel implementation provided by Öjvind Bernander and Rodney Douglas, is described in detail in Mel (1993). All input to the cell was provided by excitatory synapses (both NMDA and AMPA types) from the modeled LGN. The peak conductance of each synapse was scaled by the local input resistance so that the EPSP size was approximately the same at every synapse. The binocular LGN consisted of two copies of the monocular LGN model used previously (Mel et al. 1997), each consisting of a superimposed pair of 64x64 arrays of ON and OFF cells. Each LGN cell was modeled as a linear, half-rectified center-surround filter, with a center 7 pixels wide. A random subsample of each LGN array was taken to provide a total of 1,024 inputs to the cortical pyramidal cell.

The arrangement of inputs onto the dendrites was determined through a correlational learning rule: if the activity of two LGN cells is correlated when a preferred stimulus is shown, the synapses from those two cells will tend to be close together on the

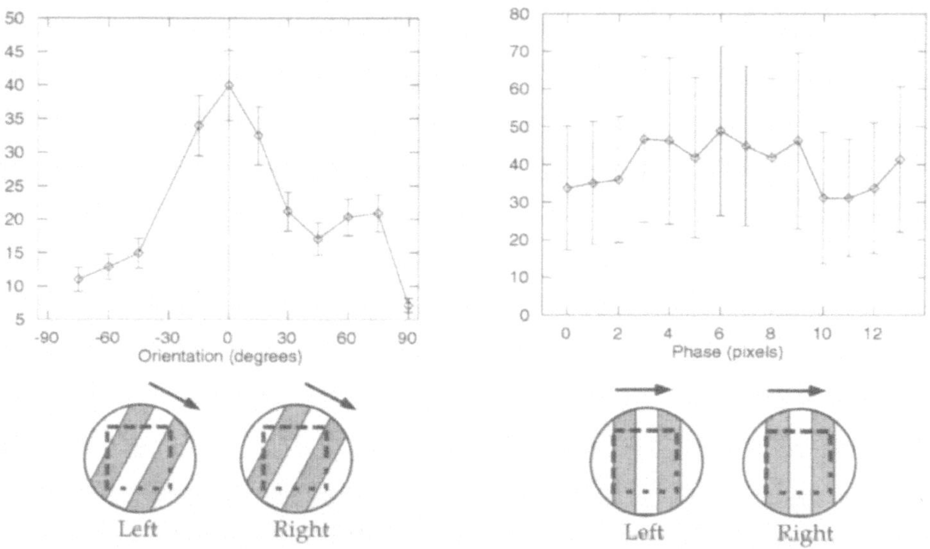

Figure 1. Orientation tuning and phase invariance in the response of model binocular cell.

dendrites. Thus, if we present an optimal stimulus — a zero-disparity pair of optimally oriented bars — the resulting active synapses are grouped in a few discrete "clusters" on the dendritic tree, so the voltage-dependent conductances near these groups are strongly activated. Other (nonpreferred) stimuli yield more diffuse patterns of active synapses.

RESULTS

The modeled cell exhibits the phase-invariant orientation tuning characteristic of complex cells, shown in Figure 1. The left graph shows the response of the cell to a grating as a function of orientation, averaged over all phases. The cell responds preferentially to the optimal orientation (0 degrees). The right graph shows the response of the cell to a grating shown at optimal orientation, as a function of phase; the response is largely independent of phase.

The modeled cell also exhibits disparity tuning. Figure 2A shows the response to a dark bar shown to each eye (left figure), and to a dark bar shown to one eye and a light bar to the other (right figure). The response pattern is similar to that of the typical disparity-tuned complex cell from cat visual cortex, shown in Figure 2B (Ohzawa et al. 1997).

DISCUSSION

Interactions between nearby synapses in dendrites with active conductances could provide a substrate for computations fitting the form of an energy model. Our simulations demonstrate that intradendritic interactions can compute at least two such responses: the phase-invariant orientation tuning of the complex cell, and the disparity tuning of binocular cells.

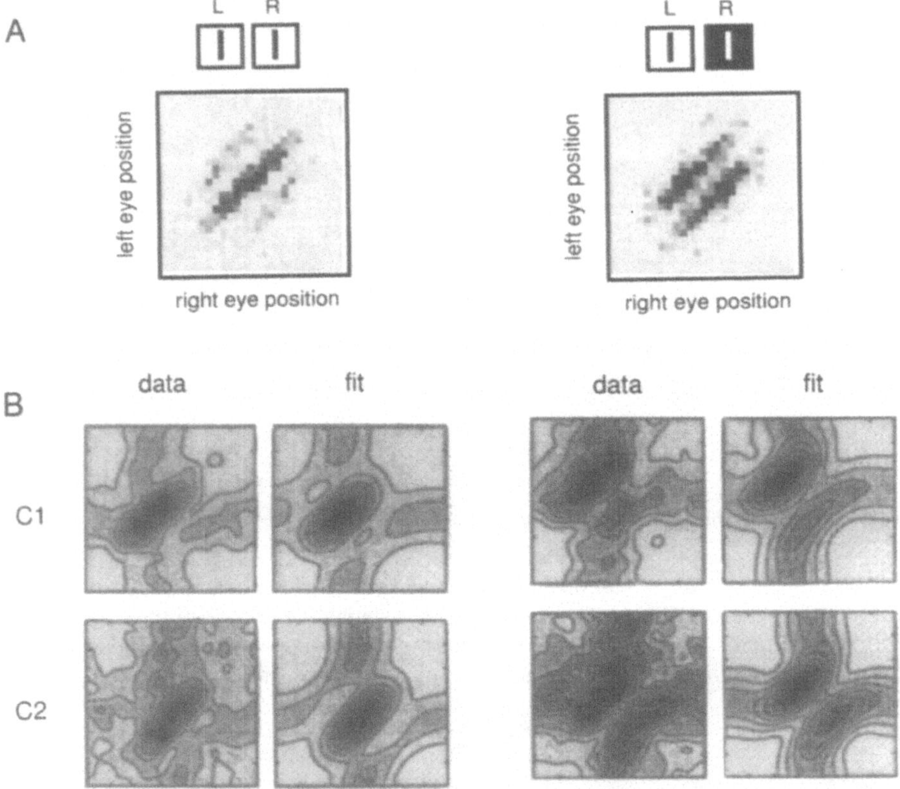

Figure 2. Disparity tuning in the response of binocular complex cell. **A.** Response of model cell to constrast-matched stimuli (left figure) and opposing-contrast stimuli (right figure). **B.** Response of two binocular disparity-tuned cells in cat visual cortex, reprinted from Ohzawa et al. (1997). Each row shows the response for a single cell. Within each row, the leftmost graph is the response of the cell to contrast-matched stimuli, while the second graph is the best fit of an energy model to the cell response. The third graph in each row is the response of the cell to opposing-contrast stimuli, and the rightmost is again the best fit of an energy model to the cell response.

These results emphasize the possible importance of dendritic nonlinearities in cortical computation. While intracortical network effects are certainly important in shaping the responses of disparity-tuned complex cells, the basic form of the response could be computed within the dendrites of the individual cell.

Acknowledgements

Thanks to Ken Miller, Allan Dobbins, and Christof Koch for many helpful comments on this work. This work was funded by the National Science Foundation and the Office of Naval Research, and by a Sloan Foundation Fellowship (D.R.).

REFERENCES

Adelson, E. and Bergen, J., 1985, Spatiotemporal energy models for the perception of motion, *J. Opt. Soc. Amer., A 2*, 284–299.

Hines M., 1989, A program for simulation of nerve equations with branching geometries, *Int. J. Biomed. Comput, 24*, 55–68.

Hubel, D. and Wiesel, T., 1962, Receptive fields, binocular interaction and functional architecture in the cat's visual cortex, *J. Physiol., 160*, 106–154.

Mel, B., 1992a, The clusteron: Toward a simple abstraction for a complex neuron, *in:* "Advances in Neural Information Processing Systems, vol. 4," J. Moody, S. Hanson, and R. Lippmann, eds., Morgan Kaufmann, San Mateo, CA.

Mel, B., 1992b, NMDA-based pattern discrimination in a modeled cortical neuron, *Neural Computation, 4*, 502–516.

Mel, B., 1993, Synaptic integration in an excitable dendritic tree, *J. Neurophysiol., 70:3*, 1086–1101.

Mel, B., Ruderman, D., and Archie, K., 1997, Complex-cell responses derived from center-surround inputs: the surprising power of intradendritic computation, *in:* "Advances in Neural Information Processing Systems, vol. 9," MIT Press, Cambridge, MA.

Movshon, J., Thompson, I., and Tolhurst, D., 1978, Receptive field organization of complex cells in the cat's striate cortex, *J. Physiol., 283*, 79–99.

Nishihara, H, and Poggio, T., 1984, Stereo vision for robotics, *in:* "Proceedings of the First International Symposium of Robotics Research", Brady and Paul, eds., MIT Press, Cambridge, MA.

Ohzawa, I., DeAngelis, G., and Freeman, R., 1990, Stereoscopic depth discrimination in the visual cortex: Neurons ideally suited as disparity detectors, *Science, 249*, 1037–1041.

Ohzawa, I., DeAngelis, G., and Freeman, R., 1997, Encoding of binocular disparity by complex cells in the cat's visual cortex, *J. Neurophysiol, 77:6*, 2879–2909.

Orban, G., 1984, "Neuronal Operations in the Visual Cortex," Springer Verlag, New York.

Pettigrew, J., Nikara, T., and Bishop, P., 1968, Responses to moving slits by single units in cat striate cortex, *Exp. Brain Res., 6*, 373–390.

Pollen, D., and Ronner, S., 1983, Visual cortical neurons as localized spatial frequency filters, *IEEE Trans. Sys. Man Cybern., 13*, 907–916.

Rall, W., 1964, Theoretical significance of dendritic trees for neuronal input-output relations, *in:* "Neural Theory and Modeling," R. F. Reiss, ed., Stanford University Press, Stanford, CA.

Stuart, D., and Sakmann, B., 1994, Active propagation of somatic action potentials into neocortical pyramidal cell dendrites, *Nature, 367*, 69–72.

DENDRITIC CALCIUM CURRENTS IN THALAMIC RELAY CELLS

Mike Neubig,[1] Daniel Ulrich,[2] John Huguenard[2] and Alain Destexhe[1]

[1] Laboratoire de Neurophysiologie, Université Laval,
Québec G1K 7P4, Canada

[2] Department of Neurology and Neurological Sciences,
Stanford University Medical Center, Stanford CA 94305, USA

Techniques of *in vitro* whole-cell recording and compartmental modeling were combined to investigate dendritic structure and calcium currents in individual thalamocortical neurons. Voltage-clamp recordings of I_T obtained in intact and dissociated ventrobasal neurons were used to constrain I_T conductances and passive parameters incorporated in morphorealistic models. High dendritic I_T densities were found necessary to establish congruent model behavior. Several methodologies based on axial resistance conservation were developed to algorithmically reduce thalamocortical morphology into a behaviorally congruent low-order compartmental model. This type of simplified model is suitable for investigating the functional role played by distal T-current localization at the network level in sleep oscillations and epilepsy.

INTRODUCTION

Rhythmic bursting is one of two characteristic firing patterns observed in individual thalamocortical (TC) neurons. This firing pattern is an essential player in the induction and perpetuation of certain oscillatory modes of cortical circuits, such as sleep oscillations and epileptic hypersynchrony[1]. In the present study, we combined experimental and computational approaches in an investigation of: i) the dendritic localization of the low-threshold Ca^{2+} current (I_T) that underlies this rhythmic bursting, and ii) the role played by the dendritic morphology in producing this firing pattern.

EXPERIMENTAL OBSERVATIONS

Voltage-clamp recordings were first obtained in dissociated ventrobasal TC neurons. The dissociation procedure removes most of the dendritic tree and leaves behind a soma with several trunks of proximal dendrite[2]. In these dissociated neurons, I_T showed

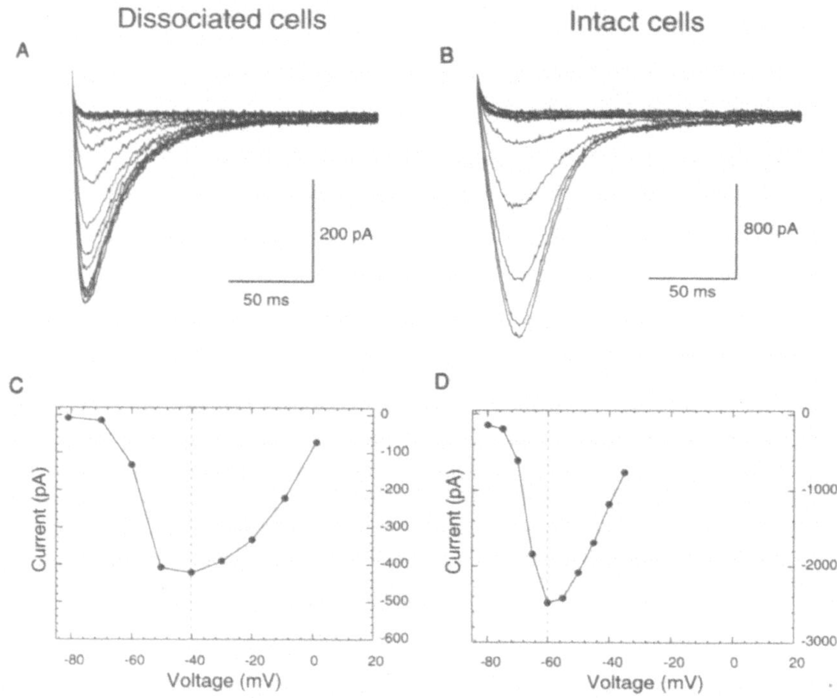

Figure 1. Voltage-clamp recordings of I_T in thalamocortical cells.
A. Inactivation protocol in a dissociated TC cell (holding voltages of V_h=-125 to -60 mV; command voltage V_c=-30 mV). B. I-V curve of I_T activation in the same cell (V_h=-100; V_c=-80 to +12 mV). C. Inactivation protocol in an intact TC cell (V_h=-105 to -40 mV; V_c=-65 mV). D. I-V curve of activation in the same cell (V_h=-100; V_c=-80 to -35 mV). Dotted lines in B-D indicate V_c used in A,C. All recordings correspond to 24 °C. Methods described in detail in[2]. B and D were modified from a previous study[3].

a bell-shaped I-V curve with peak current amplitude of about 400 pA at around -40 mV (Fig. 1A-B).

Voltage-clamp recordings of I_T were then obtained in intact TC cells in ventrobasal thalamic slices. In this case, by contrast to dissociated cells, the maximal T-current amplitude was found to be much larger, from 2 to 8.3 nA (of which one example is shown in Fig. 1C), and there was a marked shift in the I-V curve, with the peak now occurring at -70 to -60 mV (Fig. 1D).

COMPUTATIONAL MODELS

Of the group of intact TC neurons that were recorded and filled with biocytin, one was reconstructed using a tracing system and incorporated into the NEURON simulator (Fig. 2, intact-cell model). Passive responses recorded in this neuron were used to constrain the fitting of the model's passive parameters by running successive iterations of a simplex algorithm. The iterations were repeated until the model converged to this experimental recording (Fig. 2B). This procedure yielded optimized parameters that were found to be consistent with typical values for TC neurons. I_T was simulated using Goldman-Hodgkin-Katz equations[4], with activation and inactivation kinetic parameters based on voltage-clamp experiments on dissociated TC cells[2], which were described in detail in a previous paper[3].

Next, a dissociated-cell model was constructed by removing most of the dendritic tree of the intact-cell model. The soma and two proximal segments were retained so as to match the morphology and the input capacitance of dissociated TC neurons. Using this geometry and a uniform perisomatic distribution of T-channels, simulated voltage-clamp protocols established that I_T must have a permeability of 0.17 μm/s to yield an I-V curve with a maximum current amplitude around 400 pA and peak occurring around -40 mV (Fig. 2D). This T-current density was used in the proximal region of the cell ("perisomatic density") for all subsequent simulations.

In voltage-clamp simulations run on the intact-cell model, this perisomatic T-current density could not reproduce the peak I_T amplitude of intact neurons. More importantly, the perisomatic density, when extended to the whole neuron, was insufficient to produce bursts of action potentials in current-clamp simulations[3]. On the other hand, the range of peak T-current in intact neurons (2 to 8.3 nA) and the range of I-V curve shifts could be resolved by inserting T-channels in the outlying dendrites, with a permeability of 0.25 μm/s (Fig. 2E). The range of peak T-current amplitudes observed in different intact TC cells could be resolved with this model using dendritic T-current densities between 0.17 and 0.65 μm/s, which are about 1.5 to 4 times greater than the perisomatic density. Even a high but non-physiologic T-channel density, distributed strictly perisomatically, could not account for both the peak amplitude and the I-V curve shift simultaneously.

Next, the Bush-Sejnowski theme of axial resistance conservation[5] was used in developing several algorithms for reduction of the dendritic arbors of thalamocortical neurons. These algorithms were applied to the geometry of the intact-cell model in order to obtain length, diameter and axial resistance constraints for several low-order compartmental models (Fig. 3A).

The remaining model parameters were then optimized against experimental voltage-clamp recordings. Optimization against input resistance data and against recordings of passive responses to voltage-clamp protocols yielded virtually identical ratios of leak conductance density for the intact-cell and reduced models. Further, these ratios differed by less than one percent from the corresponding dendritic surface area ratios.

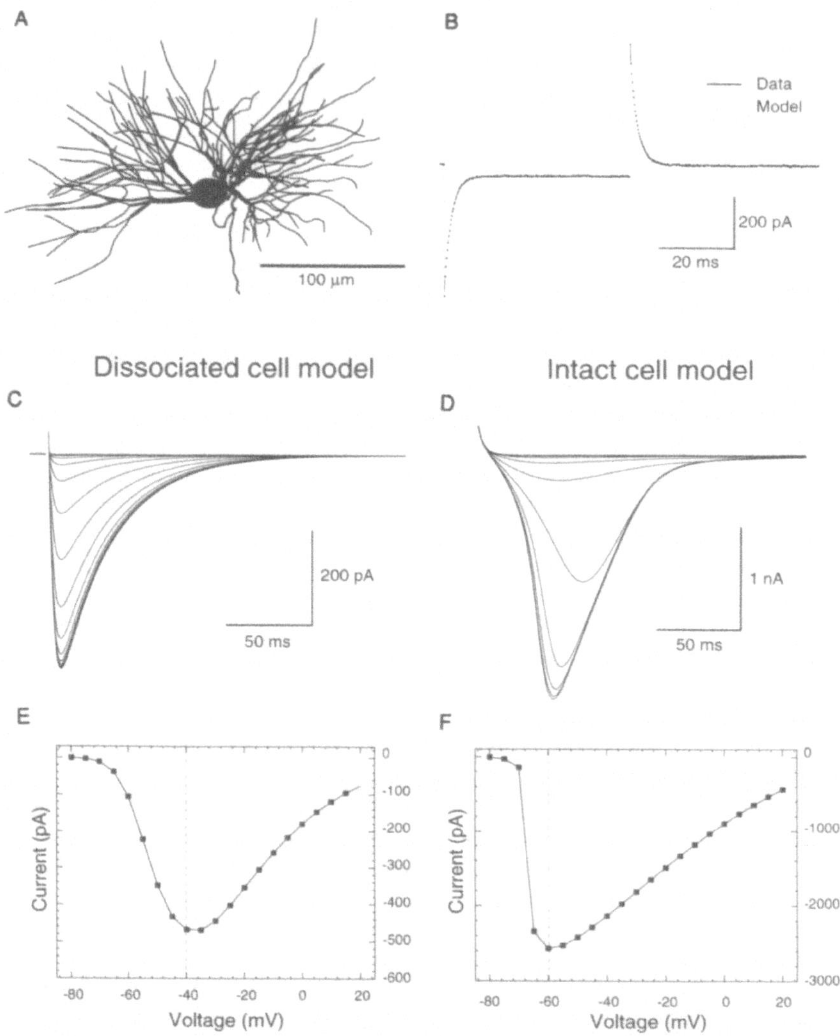

Figure 2. Multicompartmental modeling of the T-current in thalamocortical cells.
A. Reconstructed TC cell from rat ventrobasal nucleus. The cell was stained with biocytin during recording in a slice preparation[2], reconstructed using a tracing system (EUTECTIC) and incorporated into the NEURON simulator. Models with 200 compartments were used. B. Simplex fitting of the multicompartment model to passive responses in the same cell. Dots: 5 mV voltage-clamp steps during 50 ms from a holding potential of -80 mV (average of 20 traces). Continuous line: best simulation obtained after 260 iterations of the simplex procedure. C,D: Simulated voltage-clamp protocol of inactivation (C) and activation I-V curve of I_T (D) in dissociated cells (same protocols as Fig. 1A-B). E,F show the same protocols and I-V curves as in Fig 1C-D, simulated using the intact TC cell model with dendritic T-current. All simulations correspond to 24 $^\circ C$. A-C were modified from a previous study[3].

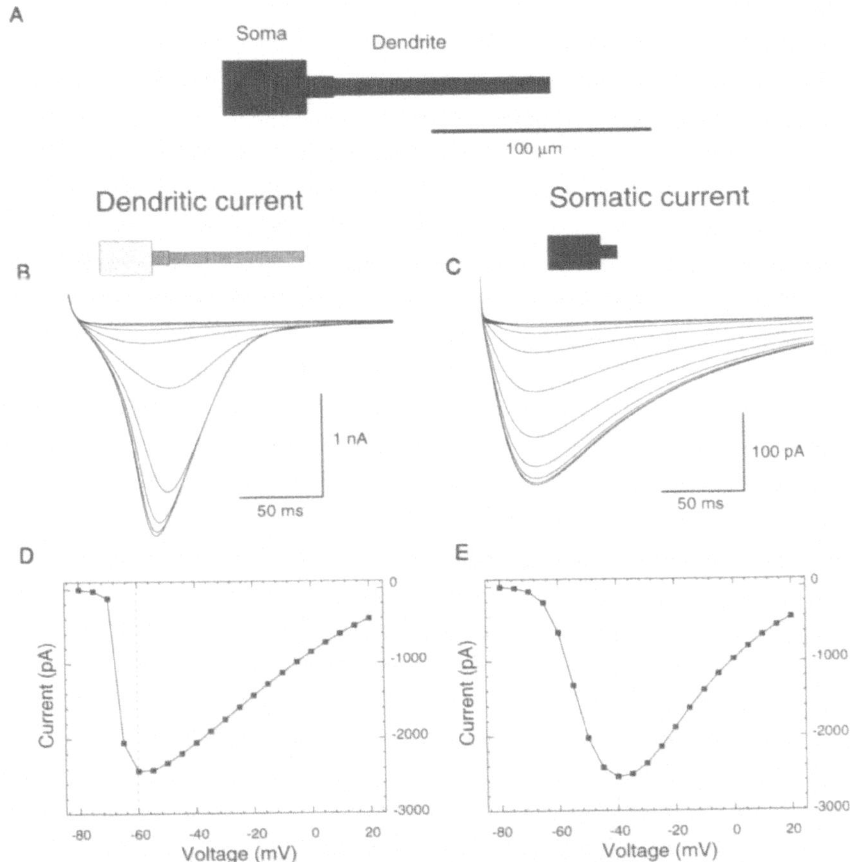

Figure 3. Low-order compartmental modeling of thalamocortical cells.
A. Simplified dendritic morphology obtained from a reduction algorithm based on the conservation of axial resistance. B,C: simulation runs of the reduced model with dendritic T-current, using the same voltage-clamp protocols as in Fig. 1C,D. D,E: same protocols run with T-current limited exclusively to the perisomatic region.

To determine the distribution of T-channels appropriate for the reduced models, permeability values from the intact-cell model were adopted after being scaled with the leak conductance ratio. With this somatodendritic distribution, the reduced models performed in accord with the intact-cell model (Fig. 3B-C). On the other hand, a strictly perisomatic distribution of T-current (1.05 μm/s) was not able to reproduce both the peak current and the I-V curve shift at the same time (Fig. 3D-E).

CONCLUDING REMARKS

In conclusion, morphorealistic modeling and precise matching of model parameters to voltage-clamp recordings provides strong evidence that the major portion of the T-current is localized in the dendrites of TC cells, as is similar to thalamic reticular (RE) cells[6]. Our evidence is also consistent with recent optical imaging data gathered from TC cell's dendrites 50 μm from the soma, showing significant T-current in these dendritic regions[7]. Investigations using the reduced models in TC-RE and thalamocortical networks should yield insight into the network-level role played by these extra-perisomatic portions of I_T in the induction and perpetuation of sleep oscillations and epilepsy[8].

REFERENCES

1.. Steriade, M., McCormick, D.A. and Sejnowski., T.J. Thalamocortical oscillations in the sleeping and aroused brain. *Science* **262**: 679-685 (1993).
2. Huguenard, J.R. and Prince, D.A. A novel T-type current underlies prolonged calcium-dependent burst firing in GABAergic neurons of rat thalamic reticular nucleus. *J. Neurosci.* **12**: 3804-3817 (1992).
3. Destexhe, A., Neubig, M., Ulrich, D. and Huguenard, J.R. Dendritic low-threshold calcium currents in thalamic relay cells. *J. Neurosci.*, submitted (1997)
4. Hille, B. *Ionic Channels of Excitable Membranes.* Sinauer Associates INC, Sunderland, MA (1992).
5. Bush, P.C. and Sejnowski T.J. Reduced compartmental models of neocortical pyramidal cells. *J. Neurosci. Methods* **46**: 159-166 (1993).
6. Destexhe A., Contreras D., Steriade M., Sejnowski T.J. and Huguenard J.R. In vivo, in vitro and computational analysis of dendritic calcium currents in thalamic reticular neurons. *J. Neurosci.* **16**: 169-185 (1996).
7. Zhou, Q., Godwin, D.W., O'Malley, D., Adams, P.R. Visualisation of calcium influx through channels that shape the burst and tonic firing modes of thalamic relay cells. *J. Neurophysiol.* **77**: 2816-2825 (1997).
8. We thank Dr. D. Amaral for kindly allowing us to use his tracing system.
Research supported by the Medical Research Council of Canada (MT-13724) and the National Institutes of Neurological Disorders and Stroke (NS-06477 and NS-34774).

A MODEL OF HOW RAPID CHANGES IN LOCAL INPUT RESISTANCE OF SHARK ELECTROSENSORY NEURONS MAY ENABLE DETECTION OF SMALL SIGNALS

Michael Paulin[1], Walter Senn[2], Yosef Yarom[3], Hanoch Meiri[3] and Dana Cohen[3]

[1] Department of Zoology, University of Otago, Dunedin, New Zealand.
mpaulin@otago.ac.nz

[2] Physiological Institute, University of Berne, Switzerland.
wsenn@iam.unibe.ch

[3] Department of Neurobiology, Hebrew University of Jerusalem, Israel.
yarom@vms.huji.ac.il, hanoch@lobster.huji.ac.il, dana@lobster.huji.ac.il

INTRODUCTION

Sharks and other elasmobranchs hunt by detecting weak electric fields produced by their prey (Kalmijn, 1982). The behavioural threshold for initiating an electrically-guided attack is a few nanovolts (Kalmijn, 1982), but the sensitivity of primary afferent neurons is in the order of a few spikes per second per microvolt (Montgomery, 1984a; Conley and Bodznick, 1994). Thus the change in afferent firing rate caused by prey at the threshold is in the order of 1 spike per minute, or about 0.1% of the spontaneous rate (~15/sec, Montgomery, 1984a).

We conducted electrophysiological experiments on an *in vitro* brain preparation of *Iago omanensis*, a dogfish found in the Red Sea. We found that when a pair of stimulus pulses is delivered to the afferent nerve, the response of secondary neurons to the second pulse is attenuated. As seen in field potentials recorded among secondary neurons (Figure 1), the attenuation is about 90% for 40msec, then recovers with a time constant in the order of 100 msec. Intracellular recordings in secondary neurons show that the first stimulus pulse produces an excitatory current closely followed by a prolonged inhibitory current. The time course of the inhibition matches the time course of the attenuation in field potentials. This indicates that the attenuation is caused by shunting inhibition reducing net current injected by the second volley of excitatory synaptic activations.

The time course of inhibition is such that the responsiveness of secondary neurons would be reduced by about 50% at spontaneous afferent firing rates. If the afferent impulses arrive more quickly each will cause a smaller depolarisation. If they arrive more slowly each will cause a larger depolarisation. Thus the shunting inhibition provides automatic gain control that stabilises the firing rate of the secondary neuron. After a step change in input rate it will take several inputs for the secondary neuron to approach a new equilibrium level. Therefore there will be a high-pass filter effect, with slower variations in input rates attenuated more than faster variations. Secondary electrosensory neurons do behave like high-pass filters.

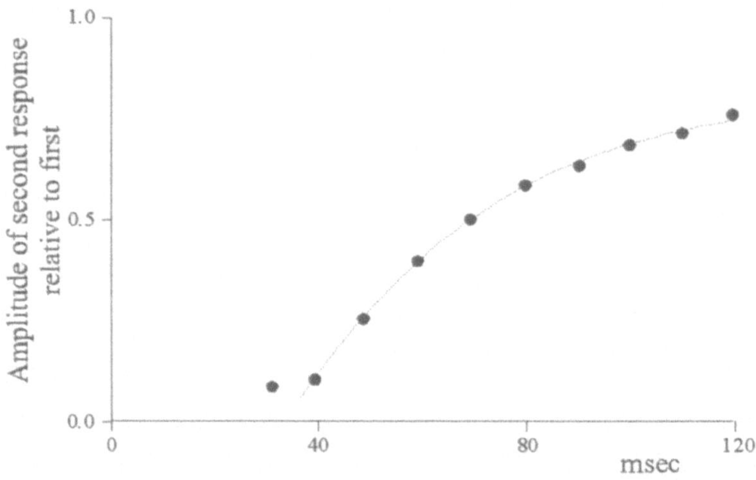

Figure 1: Depression of field potential produced by the second shock to the afferent nerve.

However, this is accompanied by a substantial increase in sensitivity to localised fast stimulus changes (Montgomery, 1984b).

The remarkable selectivity of secondary neurons for spatially and temporally localised stimuli has previously been explained in terms of additive cancellation mechanisms in the secondary nucleus (Montgomery, 1984b; Montgomery and Bodznick, 1994; Nelson and Paulin, 1995). Our experimental results suggest an alternative mechanism. Rapid decline and recovery of synaptic input gain following activation means that variations in time differences between successive synaptic activations are expressed in terms of variations in EPSP amplitude. In this paper we show, using a computer model, how this could make secondary neurons hypersensitive to localised rapid changes in stimulus levels.

SIMULATION METHOD

The computer model is written using MATLAB 5.0 (The Mathworks, Inc., MA). There is an array of synapses each with resting weight ω_i. When a particular synapse is activated the shunting inhibition is modelled by reducing the strength of that synapse by a factor $r_i < 1$, which tends to 1 with a time constant of 100 msec. This implies localised shunting, i.e. only currents entering via the previously excited synapse are affected. Each synapse produces a conductance change $g(t) = \alpha^2 t e^{-\alpha t}$ when activated, and current $\omega_i r_i g(t)$ enters the cell (cf Nelson and Paulin, 1995). Currents are summed over all synapses and translated to initial segment potential using a single membrane time constant of 5 msec. This is converted to a spike train using an integrate-and-fire mechanism (Paulin, 1993). A similar integrate-and-fire mechanism is used to convert stimulus waveforms to afferent spike trains. Spike train variability is introduced by varying the threshold for spike generation with a gamma distribution. We set γ=25 for all afferents, giving each a coefficient of variation of 0.2.

The resting weight ω_i of each synapse is continuously adjusted according to an anti-Hebbian learning rule as follows: If a postsynaptic action potential is generated while current is flowing through synapse i, then decrease ω_i by a small amount $\Delta\omega_i$. If no action potential is generated while the synapse is activated, then increase ω_i by the same amount. This learning rule tunes the model so that it demonstrates the computational implications of rapid activity-dependent changes in synaptic input gains. We make no claims about whether a rule of this kind may be present at these synapses in the real system, we have simply used it to adjust the parameters of the model.

RESULTS

One synapse

Figure 2(a) shows activity of a single secondary neuron receiving random excitation through one synapse. The learning rule has tuned resting synaptic weights ω_i so that the largest 50% of EPSPs cause the neuron to fire while the remainder do not. Because EPSP size increases as a function of time since the previous spike, median EPSP size corresponds to median interspike interval length. Therefore the neuron acts as a "longer than expected interspike interval" detector.

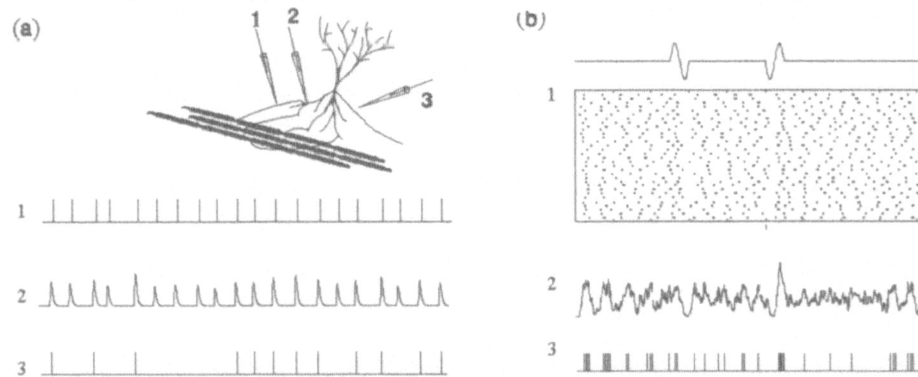

Figure 2: (a) "long interval detector" using single synapse with variable input gain. Traces show (1) presynaptic spike train, (2) postsynaptic potential and (3) response spike train. (b) Hypersensitivity with many variable-gain synapses. (1) Raster plot of input trains, (2) summed postsynaptic potentials (3) response spike train.

Many Synapses

A long-interval detector could be used to detect stimuli causing small delays in afferent spike arrival times. Because such stimuli must be rare compared to spontaneous afferent firing rates this would produce many false positives. However, if many of these detectors operate in parallel then false positives on one detector can be recognised by the failure of other detectors to respond at a higher than chance level. Stimuli can then be detected reliably by coincidence detection with an appropriate threshold.

To test this idea we ran the model with an array of independent input spike trains generated from a single stimulus waveform, converging onto a single secondary neuron. We compared responses of a model in which excitation was followed by shunting inhibition, against a control model with only excitation.

Across a realistic range of parameters the simulation results indicated that in general the inhibition provides no signal enhancement, although as expected the secondary neuron responded less to slower variations in stimulus levels. However, we then tested the model with single-cycle sinusoidal stimuli, resembling what would be seen at a receptor as a dipole source moved past it. The secondary neuron responded strongly when the stimulus period matched spontaneous inter-spike intervals (figure 2b).

The voltage recorded along a trajectory parallel to a dipole at distance r cm has a biphasic waveform with peaks separated by $r/\sqrt{2}$ cm. At velocity v cm.s^{-1} this produces a biphasic ripple with peaks separated by $r/\sqrt{2}\,v$ seconds. A dogfish such as *Iago* swimming at speeds in the order of tens of centimeters per second, a few centimeters from small prey items, will experience receptor voltage ripples with peaks separated by about the same time as intervals between spontaneous afferent spikes (50-100 msec). Thus the secondary neurons may be specifically tuned to detect receptor voltage fluctuations of the kind generated by prey as the animal swims past them.

DISCUSSION

The computer model illustrates a computational mechanism that could be responsible for target signal selectivity and hypersensitivity in the elasmobranch electrosensory system. Target detection is accomplished by a parallel computation on sensory spike trains, in which the target produces essentially no change in the firing rate or the number of spikes. The presence of prey is indicated by small, but spatially correlated delays in afferent spike arrival times. Hypersensitivity is a consequence of the rapid increase in input gains occurring across the dendritic tree during the delay. Thus convergence is supplemented by a transient increase in effective synaptic weight.

DON is generally regarded as a device for detecting unexpected sensory inputs. According to existing models, central expectations about sensory inputs are passed to DON via parallel fibres in terms of a predicted input waveform that is subtracted from the actual input waveform (Montgomery and Bodznick, 1994; Nelson and Paulin, 1995). Our model hints at an alternative possibility, that central expectations about incoming spikes may be passed to secondary electrosensory neurons by parallel fibber modulation of membrane conductance. There is a variety of direct and circumstantial evidence that the DON molecular layer may provide descending gain control of this kind (Bastian, 1986; Conley, 1995).

REFERENCES

Bastian, J (1986) Gain control in the electrosensory system mediated by descending inputs to the electrosensory lateral line lobe. *J. Neurosci.* 6(2):553-562, 1986.

Conley, RA (1995) Descending input from the vestibulolateral cerebellum supresses electrosensory responses in the dorsal octavolateralis nucleus of the elasmobranch, *Raja erinacea. J. Comp. Physiol.* A 176:325-335, 1995.

Conley, RA and Bodznick, D (1994) The cerebellar dorsal granular ridge in an elasmobranch has proprioceptive and electroreceptive representations and projects homotopically to the medullary electrosensory nucleus. J. Comp. Physiol. A 174:707-721, 1994.

Kalmijn, AJ (1982) Electric and magnetic field orientation in elasmobranch fishes. *Science* 218:916-918, 1982.

Montgomery, JC (1984a) Frequency response characteristics of primary and secondary neurons in the electrosensory system of the thornback ray. *Comp. Biochem. Physiol.* 79A; 189-195, 1984.

Montgomery, JC (1984b) Noise cancellation in the electrosensory system of the thornback ray; common mode rejection of input produced by the animal's own ventilatory movements. *J. Comp. Physiol.* A 155:103-111.

Montgomery JC and Bodznick, D (1994) An adaptive filter that cancels self-induced noise in the electrosensory and lateral-line mechanosensory systems of fish. *Neurosci. Lett.* 174:145-148, 1994.

Nelson ME and Paulin, MG (1995) Neural simulations of adaptive reafference supression in the elasmobranch electrosensory system. *J. Comp. Physiol.* A 177:723-736, 1995.

DYNAMICS OF THE ELECTRORECEPTORS IN THE PADDLEFISH, *POLYODON SPATHULA*

Xing Pei, David F. Russell, Lon A. Wilkens and Frank Moss

Center for Neurodynamics,
University of Missouri-St. Louis.
St. Louis, MO 63121

Introduction

The paddlefish, *Polyodon spathula*, inhabits the major rivers of the Midwest and feeds on zooplankton, such as the water flea *Daphnia* sp., in murky water where vision is of limited effectiveness. Small paddlefish capture the zooplankton individually while adult paddlefish filter feed in large quantity. The paddlefish has evolved an elongated (about 1/3 the animal's length) paddle-shaped rostrum which extends anterior to the head from just above the animal's mouth. On both upper and lower rostrum surfaces and extending onto the head and opercular flaps, the skin contains numerous electrosensitive ampullae (Jørgensen *et al.*, 1972), indicating the existence of an electrosensory system (New and Bodznick, 1984). It has been shown recently in this laboratory (Wilkens *et. al.*, 1997) that the paddlefish feeding relies on electrosensory detection of the oscillating electric fields, characteristic of its planktonic prey, rather than relying on vision. The ampullary electroreceptors are innervated by primary afferents of the anterior lateral line nerves that project to the medulla through dorsal and ventral roots (New and Bodznick, 1984).

Ampullary electroreceptor cells have been studied extensively in sharks (dogfish), skates, rays, and catfish (for reviews, see Bretschneider and Peters 1992; Bullock 1974). The receptor cells fire endogenously in high frequency and the information of electrical stimuli is coded as the change of firing rates and patterns (Braun *et al.*, 1994; Schäfer *et al.*, 1995; Wessel *et al.*, 1996; Gabbiani *et al* 1996). We show here that the response characteristics of paddlefish electrosensory primary afferents are low-frequency band pass, and overlap the frequencies of the oscillatory electric field produced by the aforementioned plankton prey

Method and Preparation

Juvenile paddlefishes 35-40cm in length were anesthetized initially in Tricane Methanesulfonate (0.15g/l, Sigma) and Trizma Base (0.075g/l, Sigma), and artificially respired by irrigation of the gills. The fish was then mounted in a recording tank and 'decerebrated' under anesthesia by crushing the juncture between the diencephalon and the optic tectum. Anesthetic was then discontinued and D- tubocurarine (0.05-0.1ml, 3mg/ml, Lilly) was injected intramuscularly to abolish the body movements. Sensory roots and ganglia of the anterior lateral line nerve (ALLn) were exposed near their juncture with the medulla and single unit activities were recorded extracellularly with metal microelectrodes from primary afferents. Action potentials were amplified, filtered and fed into a data acquisition system (Cambridge Electronic Design, 1401plus) with a Pentium computer and spikes were sorted for further offline analysis. Electrical stimuli were applied through two Ag/AgCl electrodes that were driven by a stimulus isolation unit (Linear Stimulus Isolator A395, WPI). To measure the frequency sensitivity characteristics of the primary afferents, stimuli of sinusoidal waveform were delivered with a function generator (Global specialties). The primary afferent responses to the sinusoidal stimulus were calculated as the amplitude of the first harmonic of the power spectra of the spike trains

Results

The primary afferents, which receive inputs directly from the receptor cells, produce action potentials at fairly regular intervals without external stimulation. Most of the primary afferents fire endogenously at frequencies around 70 spikes/sec (range from 25-85 spikes/sec, depending on the temperature and individual cells). The spontaneous firing pattern is approximately periodic as shown in the figure 1 (upper traces). The fundamental peaks of the interval histograms during spontaneous activity are often narrow (Fig 2) with mean width 5.6 ± 3ms (n=22). There is a tendency that as the discharge rates decrease the peaks of the interval histograms become broader (compare Fig 2A and 2B). However, the correlation is weak. The regular discharge patterns of the cells are likely the result of an endogenous periodic driving force that is at- or supra-threshold. In some cells, the noise abruptly interrupts the endogenous oscillation, pushing the membrane potential below the firing threshold, resulting in skipping of the discharges in the spike trains. The skipping

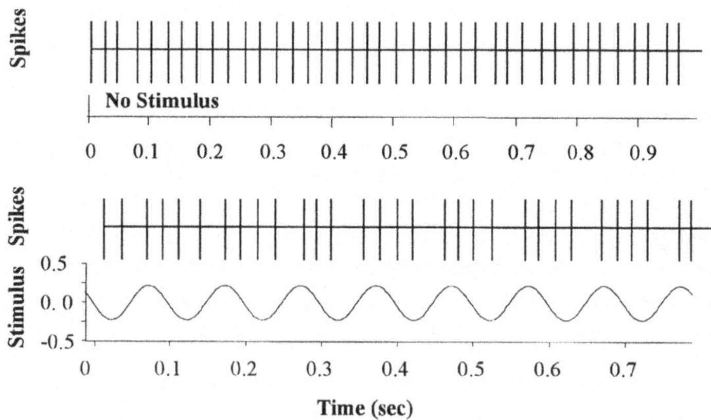

Figure 1. Spike trains of a primary afferent during resting state (upper trace) and with a sinusoidal electrical stimulation (lower traces).

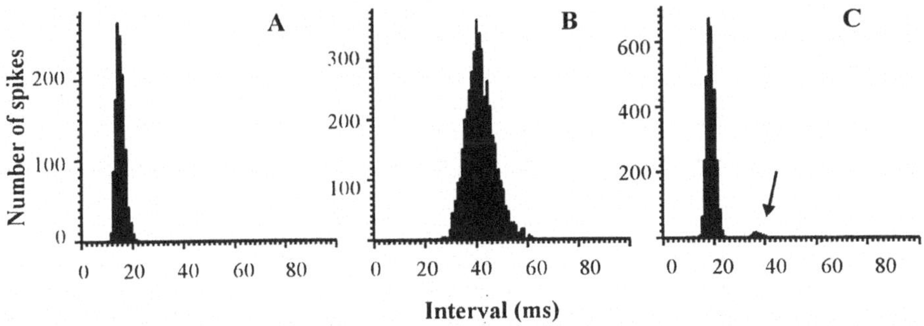

Figure 2. Interval histograms from three primary afferents of paddlefish. The cells fired periodically, in (A) at 62Hz with a narrow peak (width of half height 5ms) and in (B) at 25Hz with a peak of 15ms width. In (C) the cell's firing was at 53Hz with a narrow first peak (width of half height 5ms) but noise-induced skipping which resulted in the second peak of the interval histogram (arrow in C)

results in a bimodal distribution as shown by the second peak in Fig. 2C (arrow), where the interval of the second peak is twice of the interval of the fundamental peak.

Our data are indicative of an endogenous, noisy, sub- or just supra-threshold oscillator, which fires the afferent neuron by means of a threshold crossing process. A similar dynamical system has been previously observed in the dogfish, *Scyliorhinus canicula*, a salt water shark with similar electroreceptors, by Braun *et al*. (1994).

In response to DC step stimulation, the primary afferents are excited by cathodal stimulus currents and inhibited by anodal currents, with adaptation. To measure the frequency response we applied sinusoidal stimulus waveforms from a nearby dipole electrode, and recorded the responses. In the presence of periodic stimulation, the lower trace in Fig. 1 shows the primary afferent response to the electrical stimulus as a

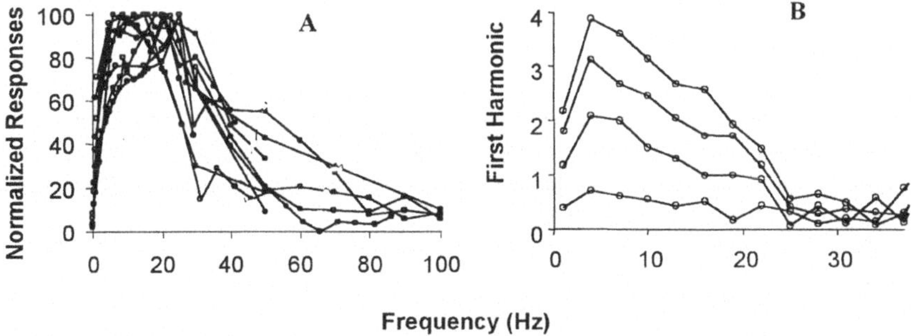

Figure 3. Frequency characteristics of paddlefish primary afferents. (A) Frequency tuning of 10 afferents. The responses were measured as the first harmonic of the spike trains. For each curve, the amplitude is normalized to its maximum. (B) Frequency tuning of a single primary afferent with four different stimulus intensities, from top to the bottom, 0.6uA, 0.4uA, 0.2uA and 0.08uA, with a dipole electrode.

modulation of the firing pattern. The power spectra of the spike trains were calculated, and the first harmonics were taken as the amplitudes of the responses and used to draw the tuning curves (Fig. 3). The response characteristics of the primary afferents are low-frequency bandpass (Fig. 3A). There is little sensitivity below 0.1Hz or above 100Hz and the peak sensitivity is in the 5-25 Hz range, for the 10 different afferents. The frequency tunings are scattered and broad (Fig. 3A), and overlap the oscillation frequencies of the electric fields produced by the zooplanktonic pray, such as *Daphnia*, (Wilkens *et al.*, 1997).

The stimulus intensity does not affect the shape of the frequency-tuning curve. Instead, it scaled the whole response up or down, as shown in the Fig. 3B, indicating that frequency tuning is an internal property of the receptor cell. Also, when driven with a sinusoidal electrical stimulus, complicated response patterns were observed in the afferent cells as a result of the non-linear interaction between the two oscillators. At some frequencies and amplitudes of the stimuli, we also observed different states of synchronization of the afferent discharges to the stimuli.

Discussion

The experimental data are comparable qualitatively to a Hodgkin-Huxley model applied to the electroreceptor and with an electronic circuit model consisting of a sub- or just supra-threshold, forced, nonlinear oscillator with additive noise (Longtin *et al.*, 1991). The implications for the range of preferred frequencies and noise widths (examples, Fig. 2) remain to be explored.

The frequency characteristics of the primary afferents observed electrophysiologically match closely the frequency range of artificial electrical stimuli which induce feeding behavior in the paddlefish (Wilkens *et al.*, 1997). The frequency sensitivity also matches the frequency characteristics of the weak electric potentials (around 0.1mV when recorded at a short distance) associated with the swimming movements of natural zooplankton prey (unpublished observation from this laboratory). These weak electric potentials have been shown to be sufficient to evoke responses from the electrosensitive primary afferents of the paddlefish (Wilkens *et al.*, 1997). Thus, the frequency tuning sensitivity in the paddlefish electroreceptor matches the signals in the environment from its natural prey, which provides evidence that the paddlefish is well evolved to feed on them.

Acknowledgement

Paddlefish were graciously supplied by Mr. J Hamilton, Blind Pony Fish Hatchery, Missouri Department of Conservation. The project was supported in part by grants from Department of Defense, the University of Missouri Research Board and Department of Energy, and by the Whitehall Foundation.

References

Braun HA, Wissing H, Schafer K and Hirsch MC, 1994, Oscillation and noise determine signal transduction in shark multimodal sensory cells. Nature 367:270-273.

Bullock TH, 1974. In Electroreceptors and other specialized receptors in lower vertebrates. Ed. Fessard, A (springer-Verlag, New York).

Gabbiani F, Metzner W, Wessel W and Koch C (1996) From stimulus encoding to feature extraction in weakly electric fish. Nature 384: 564-576.

Jørgensen JM, Flock A and Wersäll J., 1972, The Lorenzinian ampullae of *Polyodon spathula*, Z. Zellforsch., 130:362-377.

Longtin A, A. Bulsara and F. Moss, 1991. Time-interval sequences in bistable systems and the noise-induced transmission of information by sensory neurons. *Phys. Rev. Lett.* 67:656-659.

New J and Bodznick D., 1984, Segregation of electroreceptive and mechanoreceptive lateral line afferents in the hindbrain of chondrostean Fishes. Brain Res. 336: 89-98.

Schäfer K, Braun HA, Peters RC and Bretschneider F (1995). Periodic firing pattern in afferent discharges from electroreceptor organs of catfish. Pflüger Arch - Eur J Physiol 429: 378385.

Schneider F and Peters RC , 1992, Transduction and transmission in ampullary electroreceptors of catfish. Comp. Biochem. Physiol. 103:245-252.

Wessel W, Koch C and Gabbiani F (1996). Coding of time-varying electric field amplitude modulations in a wave-type electric fish. J Neurophysiol. 75:2280-93.

Wilkens L.A., Russell D.F., Pei X and Gurgens C., 1997, The paddlefish rostrum functions as an electrosensory antenna in plankton feeding. *Proc Royal Society Lond- B: Biol.* 264:1723.

COMPUTATIONAL MECHANISMS UNDERLYING
THE SECOND-ORDER STRUCTURE OF CORTICAL COMPLEX CELLS

Ko Sakai and Shigeru Tanaka

Laboratory for Neural Modeling,
Brain Science Institute,
The Institute of Physical and Chemical Research (RIKEN)
2-1 Hirosawa, Wako, Saitama 351-01, Japan
ko@yugiri.brain.riken.go.jp

ABSTRACT

We investigate what computational mechanisms give rise to the nonlinearity of the complex cell receptive field in the primary visual cortex. Complex cells are characterized by their nonlinear spatial properties such as spatial phase invariance and nonlinear response to multiple bar presentations. We carry out network simulations to estimate the second-order Wiener-like kernels for several different models. Models with nonlinear spatial pooling of simple-cell-like linear subunits reproduce the second-order kernels in good agreement with physiologically estimated kernels, while models without the pooling mechanism fail to reproduce the kernel. The structure of the kernels is independent of specific nonlinear transfer functions such as half-wave rectification and squaring. The results support the cascade mechanism consisting of simple cells' local feature extraction followed by nonlinear spatial pooling.

INTRODUCTION

Complex cells in the primary visual cortex, in contrast to simple cells, are characterized by their nonlinear spatial properties. For instance, the response of complex cells to the simultaneous presentation of two bars fails to follow the spatial superposition, i.e. the response will be different from the simple sum of the responses to the two bars presented individually. This nonlinearity might be important for texture processing since multiple elements of texture tend to fall onto the receptive field of a single cell. This spatial interaction is determined by the Gaussian white noise technique and is described by higher-order Wiener kernels [1].

Szulborski and Palmer [2] experimentally revealed the second-order kernels which consist of a central augmented region and flanking suppressed regions along the preferred orientation of the cell. However, the significance and the underlying mechanisms of the second-order spatial structure are not clearly known. A number of complex cell models have been proposed which reproduce several important aspects of complex cells' spatial properties. Cascade models, such as an energy model, consist of simple-cell-like receptive fields followed by a nonlinear transfer function and a spatial pooling mechanism[3,4]. Non-cascade models consist of LGN-cell-like receptive fields directly followed by a complex-cell-unit which contains local nonlinear processing in a dendric tree[5]. These models reproduce several major aspects of complex-cells' spatial properties including invariance to spatial phase and contrast direction, and specificity to orientation and spatial frequency. However, it has not been revealed whether these models capture the higher-order properties of receptive field structure.

We investigated which computational mechanisms play a key role in the higher-order spatial structure of the complex cell receptive field. We carried out network simulations to estimate the second-order Wiener-like kernels for several different models including cascade and non-cascade models, and compared the results with physiologically estimated kernels. Models with nonlinear spatial pooling of simple-cell-like linear receptive fields reproduce the second-order kernels regardless of their specific nonlinear transfer functions such as half-wave rectification and squaring, while models without the pooling mechanism fail to reproduce the kernel.

MODELS AND METHODS

Simulating the various models including cascade and non-cascade models, we investigated which computational mechanisms reproduce the second-order Wiener-like kernels revealed by physiological experiments. The basic architecture of the cascade models tested here consisted of three sequential stages, linear sub-units, a transfer function, and a pooling mechanism. This structure, illustrated in figure 1, is able to simulate various complex cell models by the appropriate selections of linear filters for the sub-units, the transfer function, and the pooling mechanism. For example, an energy model is realized by choosing Gabor functions with various phases for the sub-units, squaring for the transfer function, and linear summation for the pooling.

For the cascade architecture, we tested four types of the model each of which had a unique combination of a transfer function and a pooling mechanism. All cascade models have eight Gabor filters with increments of a 45 degree spatial phase for the first stage sub-units. The transfer function and the pooling mechanism of the models are: (a) half-wave rectification followed by linear summation, (b) squaring followed by linear summation, (c) a single process of selecting the maximum over eight sub-units for the sake of both the transfer function and pooling, and (d) half-wave rectification followed by no pooling. Half-wave rectification is a well-known property of cortical neurons. Squaring has been proposed by Bergen and Adelson in the energy model [3] and by Emerson *et. al.* in the cascade model which reproduces spatiotemporal nonlinearity[6]. Selecting the maximum involves a simple and biologically plausible mechanism which is easily implemented by winner-takes-all circuits[4]. The model (d) consists of a single linear sub-unit followed by half-wave rectification, thus this model, unlike the others, exhibits simple cell properties.

A non-cascade model takes direct input from LGN cells which has unoriented ON- and OFF-center receptive fields. The architecture of the non-cascade model consists of three stages as illustrated in figure 2. The first stage is a set of circular Difference Of Gaussian (DOG) filters modeling LGN cells. The second stage takes inputs from the LGN cells aligned along the preferred orientation of the complex cell. This stage includes the local multiplication among the outputs of aligned LGN cells together with the local summation over the cells. The ratio of the multiplication and summation are so chosen that the model reproduces similar first-order responses to the cascade models. The local multiplication corresponds to the assumptions that the axon of the aligned LGN cells terminates nearby locations on the dendrite of a complex cell, and that voltage dependent channels such as NMDA- and AMPA-type channels are dominant around these axon terminals. The last stage is the global summation of the local computation over the complex cell.

The receptive field of all models consist of 65x65 units with the vertical axis aligned along the preferred orientation of the cell. We designed the models so that one unit length corresponded to 0.1 degree visual angle. The stimuli consist of simultaneously presented pairs of a square with 0.3x0.3 degrees. For each pair, a single square is always fixed at the reference point, while the location of the other varies over the entire receptive field. We computed the second-order Wiener-like kernel from the series of the cell's output. Computation of the Wiener-like kernel is described elsewhere [1].

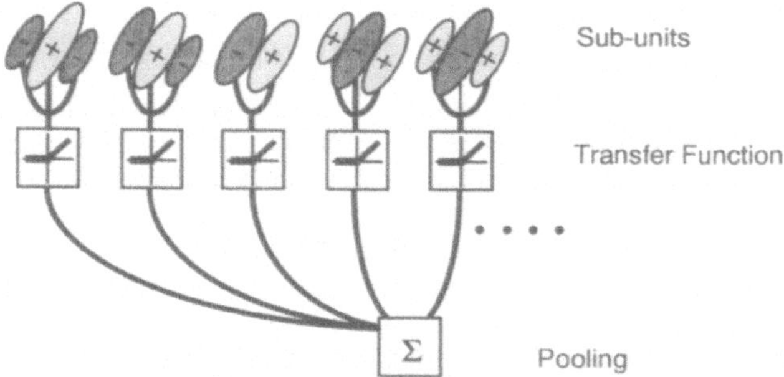

Figure 1. An example of cascade models consisting of three sequential stages. The first stage consists of Gabor filters. The output of the first stage passes through the transfer function of half-wave rectification. The third stage is the summation over the channels. We tested several models with various transfer functions and pooling mechanisms in the same sequential architecture.

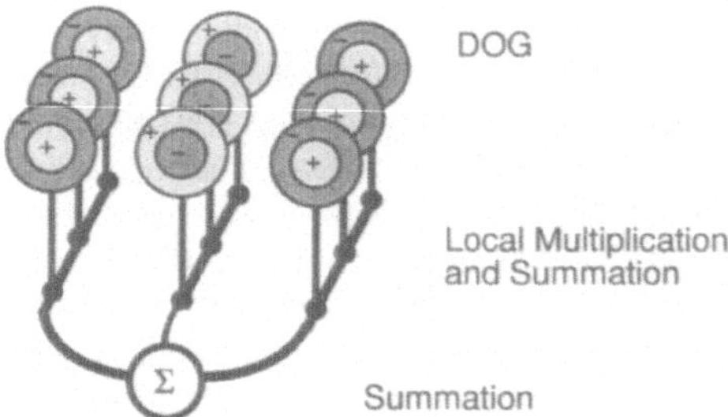

Figure 2. An illustration of the non-cascade model's architecture. The first stage models LGN cells whose receptive fields are circular DOG functions. The second stage is the local multiplication and summation over the outputs of the DOG filters aligned along the preferred orientation of the complex cell. The last stage is the global summation of the local computation over the cell.

SIMULATION RESULTS

Figure 3 shows the network estimated second-order kernels of the five models. The first four panels show the kernels for the cascade models: (a) half-wave rectification followed by linear summation, (b) squaring followed by linear summation, (c) taking the maximum over eight sub-units, and (d) half-wave rectification without pooling. The second-order kernels of a complex cell estimated from a similar physiological experiment [2] are shown in panel (f) for comparison. The first three models with a pooling mechanism, summation or taking the maximum, reproduced the physiologically estimated kernels regardless of their specific nonlinear transfer function, half-wave rectification, squaring, or taking the maximum. On the other hand, the model without pooling did not reproduce the physiologically estimated kernels. Computed kernels for the non-cascade model are shown in panel (e) in which physiologically estimated kernels are not reproduced. These results suggest that spatial pooling of simple-cell-like-units is crucial for the second-order structure while the structure is independent of specific nonlinear transfer functions.

DISCUSSION

We carried out network simulations of various complex cell models in order to estimate the higher-order structure of the receptive fields. The models with spatial pooling of simple-cell-like linear subunits reproduce the second-order kernels in good agreement with physiologically estimated kernels. However, the models without the pooling mechanism fail to reproduce the kernels. These results support that the cascade mechanism consisting of simple cells' local feature extraction followed by nonlinear spatial pooling might be crucial for the spatial properties of complex cells. This conclusion agrees with a recent physiological study using cross-correlogram which suggests excitatory afferent connections from simple cells to complex cells selective to the same retinotopy and orientation [7].

The models with spatial pooling mechanisms reproduce the physiologically estimated kernels independent of specific nonlinear transfer functions such as half-wave rectification, squaring, and taking maximum. It is noteworthy that simple mechanisms, such as half-wave rectification and taking maximum, reproduce the nonlinear spatial property of complex cells. This result suggests that rectification is a crucial role of the transfer function. Detailed properties of transfer functions might be required for other characteristics of complex cells, such as contrast sensitivity and orientation selectivity.

The evidence presented suggests that the nonlinearity of spatial summation is originated from the rectification of oriented filters followed by spatial pooling. There are two cases showing the nonlinearity when two dots are presented simultaneously. The first case is that two dots have the same polarity in contrast, and each falls onto the receptive field subregion with different polarity. The second case is that two dots have different polarities in contrast, and both fall onto the subregion of the same polarity. In these cases, the rectification affects the result of filtering and introduces the nonlinearity. This process produces an elongated region of augmentation in the second-order kernel if the receptive field is elongated, but not if the receptive field is circular. Flanking suppressed regions in the kernel are produced if the positive and negative regions of the receptive field are spatially alternated such as a Gabor function. Pooling the results of all rectified filters with different spatial phases, the cell determines polarity and magnitude of the nonlinearity. This pooling is crucial for the invariance of the kernel shape to the spatial position of the reference point in a series of stimuli.

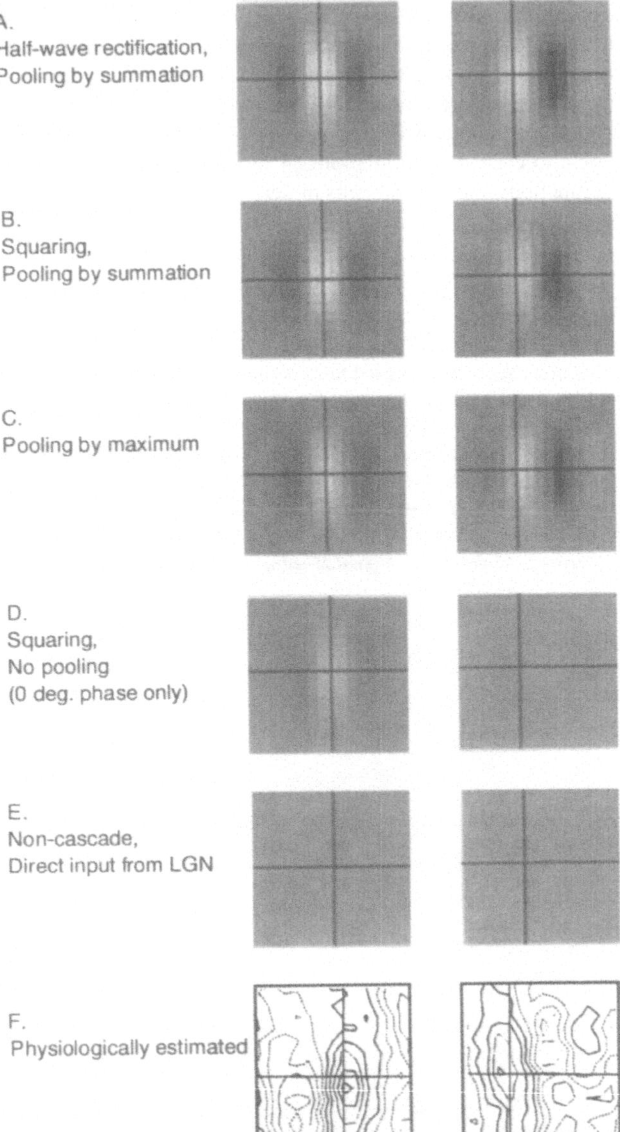

Figure 3. The second-order Wiener-like kernels estimated for models (A) to (E). The panel indicated correspond to those used in the text. Gray darkness indicates the magnitude of normalized response; darker grays show a suppressive response and lighter grays show an augmented response. For each model, the response is normalized by the largest response among the stimuli presented. The left column shows kernels with a fixed reference stimulus at the center of the receptive field indicated by the intersection of the axes. The right column shows those 0.4 degrees left of the center, as indicated by the intersection. The bottom row (F) shows the kernels estimated from a similar physiological experiment (reproduced from [2]). Kernels are drawn in contour plots with solid lines denoting augmented regions and dashed lines denoting suppressive regions. Models with a pooling mechanism reproduced the physiologically estimated kernels regardless of their specific nonlinear transfer function, while a model without pooling did not reproduce the physiologically estimated kernels.

It is natural to find that the process of oriented filters with rectification followed by spatial pooling agrees with the idea of cascade architecture. However, biological correlate of the process is not clearly known. A non-cascade model with a particular mechanism may be able to realize the process and reproduce the second-order kernel. Complex cells in superficial layers and deeper layers have different anatomical connections and show somewhat different properties. Investigation on such differences may also lead to further understanding of complex cells. We plan to carry out simulations of biophysically detailed models in order to further investigate the functions and mechanisms underlying the nonlinear properties of cortical cells.

REFERENCES

1. R. C. Emerson, M. C. Citron, W. J. Vaughn, and S. A. Klein, *J. Neurophysiology*, 58, 33-65, (1987)
2. R. G. Szulborski and L. A. Palmer, *Vision Research*, 30, 249-254, (1990)
3. J. R. Bergen and E. H. Adelson, *Nature*, 333, 363-364, (1988)
4. K. Sakai and L. H. Finkel, *J. Optical Society of America*, A, 12(6), 1208-1224, (1995)
5. B. W. Mell, D. L. Rudermen, and K. A. Archie, *Advances in Neural Information Processing Systems 9*, (Ed.) M. C. Mozer, et. al., 83-89, MIT Press, Cambridge, MA (1997)
6. R. C. Emerson, M. J. Korenberg, and M. C. Citron, *Biological Cybernetics*, 66, 291-300, (1992)
7. J.-M. Alonso and L. Martinez, *Abstracts for Annual Meeting, Society for Neuroscience*, 1668, (1997)

A CALCIUM DIFFUSION-REACTION MODEL FOR FACILITATION

Thomas Schlumpberger
Department of Molecular and Cell Biology
University of California at Berkeley
Berkeley, CA 94270

INTRODUCTION

At many synapses, repeated presynaptic action potentials often evoke increased neurotransmitter release to successive spikes. This enhancement of released neurotransmitter accumulates in a train of action potentials, rises with time constants of 30 and 300 ms, decays with similar time constants after the last spike in the train and is called (homosynaptic) facilitation. An early hypothesis for facilitation (Katz & Miledi, 1968), the "single-site/nonlinear-summation/residual-calcium" hypothesis, states that facilitation is due to a residuum of calcium remaining in the terminal from the first action potential that summates with calcium that enters during the second action potential, which facilitates the release of neurotransmitter due to the nonlinear dependence of transmitter release on calcium. However, this simple model (facilitation occurs at the secretory trigger) cannot account for the large magnitude of facilitation (several fold increase for a few action potentials). Other models (Yamada & Zucker, 1992) propose that calcium acts at distinct facilitation sites, with slow unbinding kinetics determining the time constants of facilitation. However these models contradict experimental results of Kamiya & Zucker (1994) implying that a small residual calcium acts with fast kinetics and high affinity to strongly facilitate release.

A facilitation site with kinetics and affinities consistent with experimental results at the same location as the secretory trigger is saturated by each action potential, and cannot cause facilitation. We propose that facilitation is caused by free residual calcium acting with fast kinetics and high affinity at a facilitation site that is located much further from the calcium channels than the secretory trigger. A facilitation site with high affinity (3 μm) and fast kinetics ($\tau = 0.8$ ms) located 60 nm away from the nearest calcium channel produces facilitation similar to what is observed experimentally.

MATERIALS AND METHODS

In fast synapses, transmission begins in a fraction of a millisecond (< 200 μs) after the presynaptic calcium current starts to flow (Llinas et al.,1981), suggesting very fast kinetics of the secretory trigger and a short distance between secretory trigger and the nearest calcium channel (20 - 50 nm). Additional evidence for fast calcium-binding kinetics at the secretory trigger comes from the effect of injecting exogenous buffer into the squid giant presynaptic terminal. Fast buffers based on BAPTA effectively block transmitter release (Neher, 1986) whereas the slow buffer EGTA almost did not affect transmission (Adler et al., 1991). The calcium affinity of the secretory trigger must be low because transmission is not saturated by local $[Ca^{2+}]_i$ reaching 100 μM or more.

Evidence for a Second Site

Simulations of calcium diffusion (Yamada & Zucker, 1992) using the "single-site/nonlinear-summation/residual-calcium" model failed to generate as much facilitation as is observed experimentally. These results led to the suggestion that, in addition to summating with peak calcium transients at release sites, $[Ca^{2+}]_i$ acts to augment transmission at one or more targets that are distinct from the release site.

Van der Kloot (1994) found that facilitation grows while secretion decays after one action potential at temperatures at 0°C at frog neuromuscular junction and Zengel & Magleby (1980) showed that Sr^{2+} and Ba^{2+} selectively enhance facilitation and augmentation of evoked transmitter release at frog neuromuscular junctions, giving additional evidence for separate sites of calcium action in synaptic plasticity.

Kamiya & Zucker (1994) found that rapid reduction of residual calcium by photolytic release of a presynaptic calcium buffer rapidly eliminated facilitation, implying that a small residual calcium (less than 1 μM) acts with fast kinetics and high affinity to strongly facilitate release elicited by local brief rises in $[Ca^{2+}]$ to about 100 μM.

Compartmental Representation of the Cell

We are using a two-stage model, the first stage simulates the calcium dynamics in the active zone, the second stage calculates the reaction schemes of the secretory and facilitation sites and delivers the transmitter release over time. To simulate the temporal and spatial calcium concentration we prefer to use a numerical approach rather than an analytical solution, because the numerical approach is more flexible in incorporating nonlinear processes such as calcium buffering and extrusion.

The numerical solution in rectilinear coordinates requires a discretization in space and time, so that we have to divide the nerve terminal into a large number of compartments (spatial resolution is 10 nm). Each compartment represents the $[Ca^{2+}]_i$ of a specific spatial location at a certain time in the cell. The dimensions of active zones (0.8 μm $\times$ 0.8 μm) and the spacing between them was based on ultrastructural measurements (Fogelson, A.L. & R.S. Zucker, 1985). Calcium ions that enter the cell through calcium channels and diffuse away from the channel mouth can occupy binding sites that trigger release and/or generate facilitation, can be up taken by internal organelles or bind to intracellular buffer, can be extruded by pumps or increase $[Ca^{2+}]_i$.

Calcium influx occurs through channels that are embodied in a regular array (Pumplin & Reese, 1978) located on the surface layer of the compartments, an open channel is represented as a constant flux (depending on the voltage) into the corresponding compartment. Calcium extrusion is represented as a surface pump located at the front and rear terminal surfaces of the nerve. The effect of intracellular buffer and internal organelles is enacted by the calcium-buffer-reaction scheme. For each compartment the increase/decrease in $[Ca^{2+}]_i$ per time step due to diffusion from neighboring compartments is calculated by the numerical approximation of the diffusion equation (Fick's Law).

The increase/decrease in $[Ca^{2+}]_i$ and in Buffer ([B]) in each compartment due to binding/unbinding of calcium to the intracellular buffer and the uptake by intracellular organs is calculated by the discretization of the differential equation of the calcium-buffer reaction scheme.

The uptake by the pump in the compartments of the front and rear surface can be calculated by the numerical approximation of the corresponding differential equation.

RESULTS

A 20 Hz tetanus, consisting of 5 spikes, is used to establish facilitation, test pulses 5 - 1000 ms after the conditioning stimulus are used to display the decay of facilitation. This stimulus pattern was similar to the one used in experiments (Zucker, 1974). The spatio-temporal calcium concentration drives the binding and reaction schemes of the secretory trigger and the facilitation site to calculate the time course of transmitter release. Being located 20 nm from the nearest calcium channel, the secretory trigger has four rapidly equilibrating low-affinity calcium binding sites with an on-rate $k_{on} = 0.5$ ms^{-1} µM^{-1} and an off-rate $k_{off} = 100$ ms^{-1} (200 µM dissociation constant).

The facilitation site has one high-affinity calcium binding site (the dissociation constant was varied between 1 µM and 15 µM), the distance to the nearest calcium channel was varied between 20 nm and 100 nm. Facilitation increases with increasing distance between the facilitation site and the nearest calcium channel.

The best match with facilitation observed experimentally (20 fold increase in the postsynaptic response, two phase exponential decay of facilitation with time constants of 10 and 200 ms) was achieved with a facilitation site located 60 nm from the nearest calcium channel with high affinity (dissociation constant 3 µM) and fast kinetics ($\tau = 0.8$ ms).

This distance may reflect the mean free path of calcium ions from the nearest open calcium channel to a physically obscured binding site on the backside of the proteins that dock versicles to release sites.

REFERENCES

Adler, E.M., G.J. Augustine, S.N. Duffy, & M.P. Charlton, 1991. Alien intracellular calcium chelators attenuate neurotransmitter release at the squid giant synapse. J. Neurosci. 11:1496-1507.

Fogelson, A. L. & R. S. Zucker, 1985. Presynaptic calcium diffusion from various arrays of single channels: implications for transmitter release and synaptic facilitation. Biophys. J. 48:1003-1017

Kamyia H. & R. S. Zucker, 1994. Residual Ca^{2+} and short-term synaptic plasticity. Nature 371:603-606

Katz B. & R. Miledi, 1968. The role of calcium in neuromuscular facilitation. J. Physiol. (Lond.) 195:481-492

Llinas R., Steinberg Iz., Walton K., 1981. Relationship between presynaptic calcium current and postsynaptic potential in squid giant synapse. Biophys J 33:323-352.

Neher, E., 1986. Concentration profiles of intracellular calcium in the presence of a diffusible chelator. Exp. Brain Res. 14:80-96

Pumplin, D. W. & T. S. Reese, 1978. Membrane ultrastructure of the giant synapse of the squid Loligo pealei. Neurosci. 3:685-696

Van der Kloot, W. 1994. Facilitation at the frog neuromuscular junction at 0° C is not maximal at time zero. J. Neurosci. 14:5722-5724.

Yamada, W.M. & R. S. Zucker, 1992. Time course of transmitter release calculated from simulations of a calcium diffusion model. Biophys. J. 61:671-682

Zengel, J.E. & K.L. Magleby. 1980. Differential effects of Ba^{2+}, and Sr^{2+}, and Ca^{2+} on stimulation-induced changes in transmitter release at the frog neuromuscular junction. J. Gen. Physiol. 76:175-211.

Zucker, R.S. 1974. Characteristics of crayfish neuromuscular facilitation and their calcium dependence. J.Physiol. (Lond.). 241:91-110.

SPIKE TIMING RELIABILITY IN A STOCHASTIC
HODGKIN-HUXLEY MODEL

Elad Schneidman,[1,2] Barry Freedman,[1] and Idan Segev[1]

[1]Department of Neurobiology, Institute of Life Sciences and Center for
Neural Computation
[2]Institute of Computer Science
Hebrew University
Jerusalem 91904, Israel

INTRODUCTION

The Hodgkin-Huxley (HH) formulation [1] describes the excitable nature of neurons, through a set of deterministic differential equations for the neuron's ion conductances. These conductances, which range continuously from zero to some maximum, are made out of individual ion channels which are inherently discrete and stochastic. Correspondingly, the electrical activity of nerve cells would be more accurately described by a set of biophysically-inspired stochastic equations, rather than by deterministic equations.

Following Fitzhugh [2], a few groups have used a Markovian-based stochastic model for ion channel kinetics to demonstrate when the collective behavior of the stochastic ion channels can be well approximated by the continuous deterministic HH system [3, 4]. These studies have also shown that, for a small or even intermediate number of ion channels, the stochastic model could differ from the deterministic one in the firing rate, the regularity of the spike train, and the tendency to fire spontaneously when no input is presented [3, 4, 5]. Simulation of propagation of action potentials in axons [6, 7] have shown that the channel stochasticity may also have an effect on the spike propagation delay.

Recently, Mainen and Sejnowski [8] demonstrated that the reliability and precision of action potential (AP) generation in cortical neurons is input-dependent. The reliability of spike firing time in response to 'synaptic-like' fluctuating input current, injected repeatedly to the soma, was much higher compared to DC depolarizing current steps (see also [9]).

We have investigated the reliability of spike firing in a stochastic HH model to both DC and fluctuating stimuli, in order to examine whether the reliability of spike firing time observed experimentally could be the consequence of the stochastic nature of voltage-dependent ion channels. The immediate intuition would be that the number of channels in the axon hillock and axon's initial segment (where the AP is believed to be initiated) is sufficiently large (probably several tens of thousands) so that, the individual channel noise would 'average-out' to produce an essentially deterministic (reliable) behavior. Our simulations show that membrane patches of areas in the order of a few hundred micrometers squared, with a few tens of thousands of ion channels replicate, at least qualitatively, the reliability properties of cortical neurons. The channel 'noise' also produces additional experimentally observed phenomena, namely subthreshold membrane voltage oscillations, 'spontaneous spikes' and 'missing' spikes.

THE STOCHASTIC HH MODEL

The stochastic HH model is based on the membrane dynamics of the HH equation,

$$C_m \frac{dV}{dt} = -g_L(V - V_L) - g_K(V,t)(V - V_K) - g_{Na}(V,t)(V - V_{Na}) + I \qquad (1)$$

where V is the membrane potential, V_L, V_K and V_{Na} are the reversal potentials of the leakage, potassium and sodium currents, respectively, g_L, $g_K(V,t)$ and $g_{Na}(V,t)$ are the corresponding ion conductances, C_m is the membrane capacitance and I is the injected current. In the stochastic HH model, the equations used to describe the ion channel conductances are replaced with an explicit voltage-dependent Markovian kinetic model for single ion channels [2, 10]. Based on the activation and inactivation variables of the deterministic HH model, each Na channel can be in one of eight different states, and the rates for transition between these states are given in the following diagram,

$$
\begin{array}{ccccccc}
[\mathbf{m_0 h_1}] & \overset{3\alpha_m}{\underset{\beta_m}{\rightleftharpoons}} & [\mathbf{m_1 h_1}] & \overset{2\alpha_m}{\underset{2\beta_m}{\rightleftharpoons}} & [\mathbf{m_2 h_1}] & \overset{\alpha_m}{\underset{3\beta_m}{\rightleftharpoons}} & [\mathbf{m_3 h_1}] \\
\alpha_h \updownarrow \beta_h & & \alpha_h \updownarrow \beta_h & & \alpha_h \updownarrow \beta_h & & \alpha_h \updownarrow \beta_h \\
[\mathbf{m_0 h_0}] & \overset{3\alpha_m}{\underset{\beta_m}{\rightleftharpoons}} & [\mathbf{m_1 h_0}] & \overset{2\alpha_m}{\underset{2\beta_m}{\rightleftharpoons}} & [\mathbf{m_2 h_0}] & \overset{\alpha_m}{\underset{3\beta_m}{\rightleftharpoons}} & [\mathbf{m_3 h_0}]
\end{array}
$$

$$(2)$$

where $[\mathbf{m_i h_j}]$ refers to the number of channels which are currently in the stable state $m_i h_j$. Here $m_3 h_1$ labels the single open state, and α_h, β_h, α_m and β_m are the voltage-dependent transfer rate-functions from the Hodgkin-Huxley formalism. Similarly, the kinetic scheme of the potassium channel is given by,

$$[\mathbf{n_0}] \overset{4\alpha_n}{\underset{\beta_n}{\rightleftharpoons}} [\mathbf{n_1}] \overset{3\alpha_n}{\underset{2\beta_n}{\rightleftharpoons}} [\mathbf{n_2}] \overset{2\alpha_n}{\underset{3\beta_n}{\rightleftharpoons}} [\mathbf{n_3}] \overset{\alpha_n}{\underset{4\beta_n}{\rightleftharpoons}} [\mathbf{n_4}]$$

$$(3)$$

where $[\mathbf{n_4}]$ labels the single open state of the potassium channel. The potassium and sodium membrane conductances are given by,

$$g_K(V,t) = \gamma_K [\mathbf{n_4}] \qquad g_{Na}(V,t) = \gamma_{Na} [\mathbf{m_3 h_1}] \qquad (4)$$

where γ_K and γ_{Na} are the potassium and sodium ion channel open state conductances, respectively (both were taken to be $20pS$ in the present study).

By switching from the 'usual' HH model to the Markovian kinetics model it is possible to incorporate the single channel stochastic dynamics into the voltage dynamics. Unless otherwise mentioned, in the present study we assumed that the density of Na^+ and K^+ channels per μm^2 of membrane is 60 and 18, respectively. Temperature used was $6.3°C$.

RESULTS

The response of a $200\mu m^2$ stochastic HH membrane patch to repeated presentation of suprathreshold currents is shown in Figure 1. When the same DC input is repeatedly presented (A), the resulting spike trains are very different from one another, i.e., spike firing time is unreliable. On the other hand, when the input is fluctuating (B), imitating the current that the soma supposedly gets from the dendritic synapses, the reliability of the spike timing is relatively high.

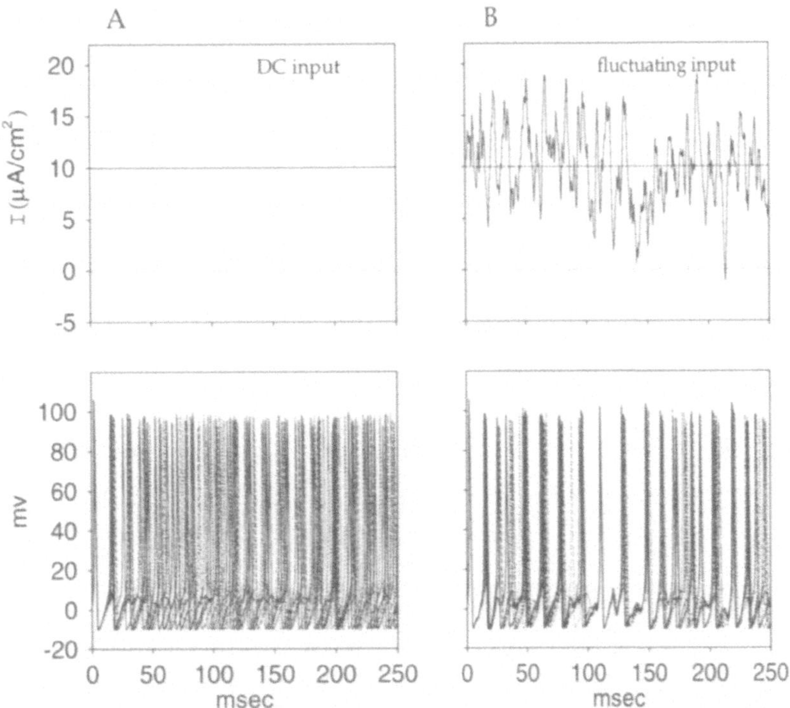

Figure 1. Reliability of spike firing in a simulated isopotential stochastic Hodgkin-Huxley membrane patch. (A) A suprathreshold DC current pulse (10 $\mu A/cm^2$, 250 $msec$; top frame), evoked trains of action potentials (responses of 10 repeated trials are superimposed, lower frame). (B) The same patch was again 'stimulated' repeatedly, this time with a fluctuating stimulus (as in [8] we use low-pass Gaussian white noise with a mean of $10\,\mu A/cm^2$, and std of $7\,\mu A/cm^2$ which was convolved with an alpha-function with $\tau_\alpha = 1\,msec$, top frame). As can be clearly seen, the 'jitter' in spike timing in B (lower frame) is significantly smaller than in A (i.e., increased reliability for the fluctuating current input). Patch area used was $200\,\mu m^2$ bearing $4000K^+$ channels and $12000Na^+$ channels (For qualitative comparison to experimental results, see Figure 1 in [8]).

The reliability and precision of the spike train grow as a function of the amplitude of the fluctuations in the current input (up to a sub-millisecond precision) [11], in close agreement with the results of Mainen and Sejnowski [8]. The reason for the surprisingly large effect of the channel stochasticity is that near the threshold for spike firing, only a small fraction of the total ion channel population is open. Hence, in this voltage regime, the fluctuations in the number of open channels are sufficiently large to have a marked effect on the exact timing of the spike firing in the case of DC input, whereas, in the case of highly fluctuating input, this channel noise is largely overridden.

Together with the effect of channel fluctuations on the exact spike timing, the inherent channel 'noise' in the stochastic HH model is the origin of the considerable subthreshold oscillations in the membrane voltage, and of 'missing' spikes, as shown in Figure 2. The membrane voltage oscillations around resting voltage (A) may give rise to 'spontaneous' spikes in the case of DC depolarizing current inputs (B), which would not yield spikes in the corresponding deterministic HH model (threshold of the deterministic model is $7\mu A/cm^2$).

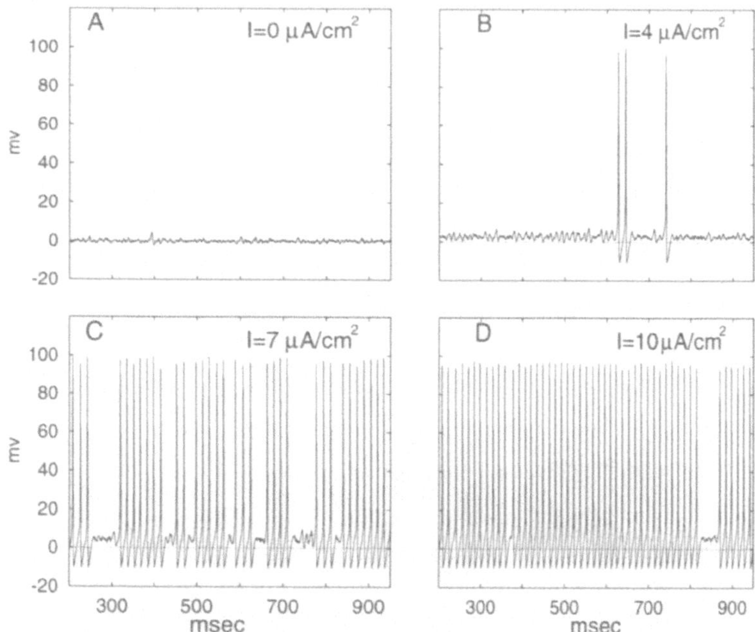

Figure 2. The response of the stochastic model to injected DC current inputs. DC currents with different amplitudes were injected to an isopotential Hodgkin-Huxley membrane patch model of area $600\,\mu m^2$ (channel populations used are approximately 12000 K^+ channels and 36000 Na^+ channels). The stochastic ion channel fluctuations in the model result in oscillation of the membrane voltage which are voltage-dependent. In (A,B) subthreshold oscillations are the dominant effect of the stochastic nature of the ion channels, with occasional spontaneous spiking. In (C,D), suprathreshold DC input currents are injected to the cell, resulting in non-regular spiking, 'missing' spikes, and subthreshold oscillations in membrane voltage between spikes. This is not expected in the corresponding deterministic HH model where the threshold is $7\,\mu A/cm^2$. Below this value, steady subthreshold voltage is reached (not shown); above this value, regular firing output is obtained (not shown).

The amplitude and frequency of these subthreshold oscillations (Figure 2 C,D) are voltage-dependent, just as in the case for cortical and other types of neurons [12, 13]. A careful investigation of the model has revealed that the channel fluctuations are responsible for 'missing' of spikes in the case of suprathreshold inputs, as well as for spontaneous spiking, by flipping the system between the firing and non-firing stable states of the HH model (see also [11] and experimental results [14]).

The effect of inherent channel 'noise' on the degree the spike train regularity is quantitatively demonstrated using the correlation of variation (CV) of the inter-spike intervals (ISI) of the stochastic model's response to DC current inputs (Figure 3). While the deterministic model responds with regular firing to DC inputs (not shown), the stochastic model yields highly irregular spike trains. For a wide range of DC input values around the deterministic (vertical dashed line) threshold, the CV is quite high, as was observed experimentally [15], implying that the inherent channel stochasticity may be one of the sources for that high CV seen in cortical neurons [16, 17].

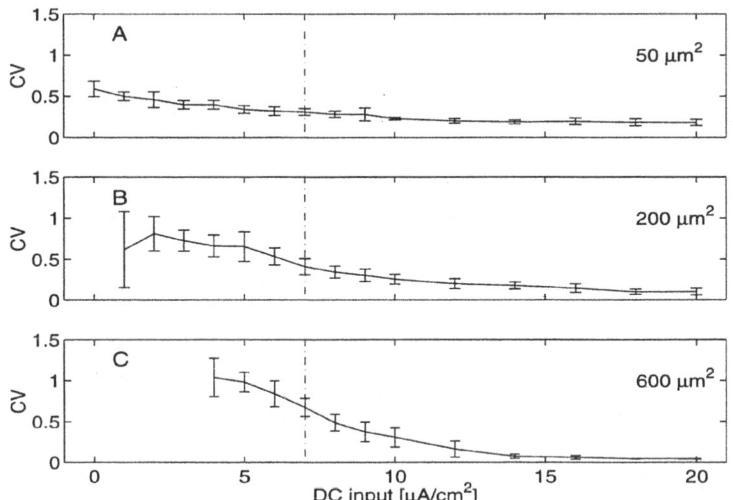

Figure 3. Stochastic HH model shows high CV in response to DC current inputs. (A) The CV of the ISI in a $50\mu m^2$ membrane patch in response to $1sec$ DC inputs. Ten epochs of $1sec$ for each DC input were used to calculate the error bar. The vertical dashed line marks the deterministic threshold for DC inputs. (B,C) Same as in A, for $200\mu m^2$ and $600\mu m^2$ patches, respectively.

Increasing the temperature, thereby accelerating the rate-functions of the model, results in a dramatic change in the nature of the spike firing (Figure 4). The regular spiking in the deterministic model (A) turns, in the stochastic case (B) into a highly irregular and 'bursty' spike trains [18]. Again, this emphasizes the significant difference between the deterministic model and the more realistic stochastic one.

CONCLUSION

We have shown that incorporating the inevitable stochasticity into an HH model with a realistically large number of ion channels yields an input-dependent reliability behavior of the system, and gives rise to subthreshold oscillations, spontaneous spikes, 'missing' spikes and high irregularity of spike patterns, all of which were observed experimentally. The effect of channel stochasticity on the reliability of the spike trains is very likely to repeat itself for more realistic HH-like models for cortical cells (as our initial simulation results show). However, it remains to be seen to what extent would the subthreshold oscillations, spontaneous and missing spikes appear in such models, since the nature of the bistability of the HH model is a key player in their existence. We conclude that channel stochasticity plays an important role in determining the exact temporal nature of spike firing in neurons.

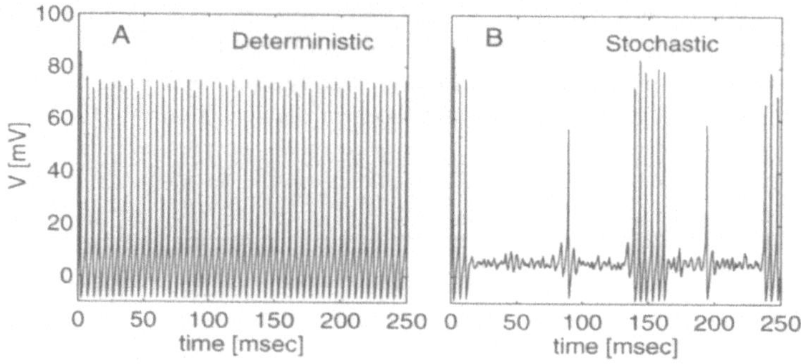

Figure 4. Temperature effect on the SHH model. (A) The response of a $200\mu m^2$ deterministic HH membrane patch to a $10\mu A/cm^2$ DC input with the temperature raised to $20°C$. A high frequency regular spiking is seen. (B) The corresponding stochastic model responds with a highly irregular, 'bursty' spike train and subthreshold oscillations between the spikes.

REFERENCES

1. A.L. Hodgkin and A.F. Huxley. A quantitative description of membrane current and its application to conduction and excitation in nerve. *J. Physiol. (London)*, 117:500–544, 1952.
2. R. Fitzhugh. A kinetic model of the conductance changes in nerve membrane. *J. Cell. Comp. Physiol.*, 66:111–118, 1965.
3. E. Skaugen and L. Walløe. Firing behavior in a stochastic nerve membrane model based upon the Hodgkin-Huxley equations. *Acta Physiol. Scand.*, 107:343–363, 1979.
4. A.F. Strassberg and L.J. DeFelice. Limitations of the Hodgkin-Huxley formalism: Effects of single channel kinetics on transmembrane voltage dynamics. *Neural Computation*, 5:843–856, 1993.
5. C.C. Chow and J.A. White. Spontaneous action potentials due to channel fluctuations. *Biophys. J.*, 71:3013–3021, 1996.
6. H. Horikawa. Noise effects on spike propagation in the stochastic Hodgkin Huxley models. *Biol. Cybern.*, 66:19–30, 1991.
7. J.T. Rubinstein. Threshold fluctuations in an N sodium channel model of the node of Ranvier. *Biophys. J.*, 68:779–785, 1995.
8. Z. F. Mainen and T.J. Sejnowski. Reliability of spike timing in neocortical neurons. *Science*, 268:1503–1508, 1995.
9. L.G. Nowak, M.V. Sanches-Vives, and D.A. McCormick. Influence of low and high frequency inputs on spike timing in visual cortical neurons. *Cerebral Cortex*, 7:487–501, 1997.
10. J.R. Clay and L.J. DeFelice. Relationship between membrane excitability and single channel open-close kinetics. *Biophys. J.*, 42:151–157, 1983.
11. E. Schneidman, B. Freedman, and I. Segev. Ion channel stochasticity may be critical in determining the reliability and precision of spike timing. *submitted*, 1997.
12. Y. Gutfreund, Y. Yarom, and I. Segev. Subthreshold oscillations and resonant frequency in guinea-pig cortical neurons: physiology and modeling. *J. Physiol.*, 483:621–640, 1995.
13. R. Klink and A. Alonso. Ionic mechanisms for the subthreshold oscillations and differential electroresponsiveness of medial enthorinal cortex layer II Neurons. *J. Neurophysiol.*, 70:144–157, 1993.
14. R. Guttman, S. Lewis, and J. Rinzel. Control of repetitive firing in squid axon membrane as a model for a neuroneoscillator. *J. Physiol.*, 305:377–395, 1980.
15. W.R. Softky and C. Koch. The highly irregular firing of cortical cells is inconsistent with temporal integration of random EPSPs. *J. Neuroscience*, 13:334–350, 1993.
16. A.J. Bell, Z.F. Mainen, M. Tsodyks, and T.J. Sejnowski. What are the causes of cortical variability. *Preprint*, 1996.
17. T.W. Troyer and K.D. Miller. Physiological gain leads to high ISI variability in a simple model of a cortical regular spiking cell. *Neural Computation*, 9:733–745, 1997.
18. A. Longtin and K. Hinzer. Encoding with bursting, subthreshold oscillations, and noise in mammalian cold receptors. *Neural Comp.*, 8:215–255, 1996.

CAN STOCHASTIC NEURONS SUPPORT SPATIO-TEMPORAL CODES

Harel Shouval and Ömer B. Artun
Department of Physics and
The Institute for Brain and Neural Systems
Box 1843, Brown University
Providence, R. I., 02912
Email: hzs@cns.brown.edu, artun@cns.brown.edu

Introduction

The single cell firing rate interpretation of the neural code has long dominated the field of electro-physiology. The major reason for this is that it has proved to be an extremely successful interpretation. There have also been several different theoretical arguments as to why it is the most reasonable interpretation (Barlow, 1972; Shadlen and Newsome, 1994). Recent advances, such as multi electrode recording have made it possible to look at the spike trains of many neurons simultaneously and see how they relate to behavior or stimulus. Furthermore, recent work has tried to examine whether spike trains in single cells carry more information than carried by the firing rate of the cells (Richmond et al., 1987; Kajaer et al., 1994; Bialek et al., 1991; Victor and Purpura, 1996). A temporal code is much richer than a firing rate code, many suggestions have been made about the usefulness of such codes (Abeles, 1982; Bienenstock, 1995; Bialek et al., 1991). Many researchers are skeptical about the ability of the neural machinery to to support a temporal code. Neurons are considered too slow and too noisy to carry out such computations (Shadlen and Newsome, 1994). Recently direct evidence that neurons indeed have a quick reliable response to brief stimuli at the cell body has been presented by Mainen and Sejnowski (1995) . Even if cell bodies are fast and reliable noise in synaptic transmission (Allen and Stevens, 1994) can degrade their effective reliability. Noise in synaptic transmission could be averaged out by slow membrane constants, but will be significant with fast membrane constants. Thus even if membrane properties allow quick computations they may do so with large levels of noise. The question posed in this paper is how detrimental is the stochastic nature of the neurons to spatio-temporal codes. We will do so using two examples: a random spatio-temporal code which is easy to analyze but does not have the familiar features of physiological neural codes and a Hierarchical Neural Multiplexing (HNM) code that has non-stationary Poisson statistics. We show that for the random code a number of patterns exponential in the dimensionality of the code can be stored for large levels of noise. Although this is not a biologically plausible code the result shows that in principle noise is not detrimental to spatio-temporal coding. The HNM code conveys a similar results, that despite noisy time varying Poisson statistics, information can be stored in the temporal structure of the neural code, given a large number of neurons, time bins, and a population coding scheme. These, somewhat counter intuitive, results stem from the inherent sparsity of high dimensional codes could therefore be called **the bless of dimensionality** and we therefore

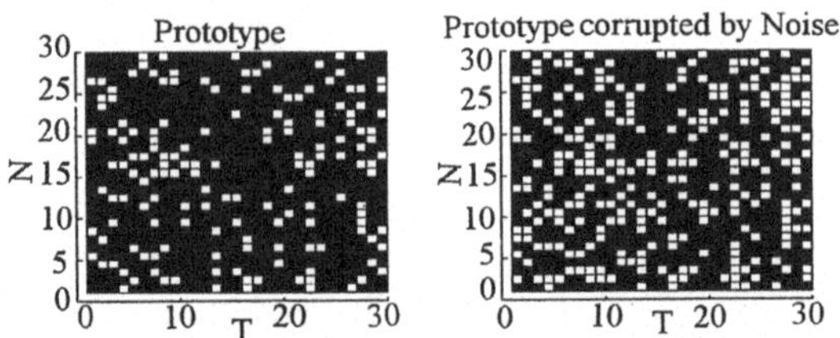

Figure 1. On the left an example of a random template (**S**) with an average activity $a = 0.2$. This template consists of 30 neurons over 30 time steps. A light square represents an action potential. On the right a randomly corrupted version of the template (ξ) with a noise level of $q = 0.2$.

Random Spatio-Temporal codes.

A spatio-temporal code is the firing pattern of N neurons over T time bins. We use 1 ms time bins to encode action potentials. This results in a very simple binary code with dimensionality of NT. We denote by S^l the $l'th$ template symbol represented in this code. These templates can be corrupted by noise which are denoted by ξ^l. The templates S_i^l and corrupted versions ξ^l are chosen at random, with mean activity level a and q respectively.

$$S_i^l = \begin{cases} 1 & \text{with probability } a \\ 0 & \text{with probability } (1-a) \end{cases} \tag{1}$$

The corrupted template ξ_i^l is found by flipping the bin content with probability q. The overlap between the template and the corrupted version is defined as $C^{ln} = \sum_j S_j^l \xi_j^n$. There are two distinct cases the overlap between a template and it's own corrupted version (C^{ll}), and the overlap between a template and a corrupted version of another template (C^{ln}). Correct retrieval occurs when $C^{ll} > C^{ln}$ for every $n \neq l$. The probability of correct retrieval can be exactly calculated in this case.

In Appendix A we show that when the size of the code (NT) is large the number of stored templates is

$$M < \epsilon \rho \sqrt{\frac{NT}{\pi}} e^{NT\rho^2} \tag{2}$$

where ϵ is the average retrieval error and ρ is a constant that depends on the activity level a and the noise parameter q. It has the form $\rho = \left((r-x)/\sqrt{|r(1-r)| + |x(1-x)|} \right)$, $r = a(1-q)$ and $x = a(a(1-q) + q(1-a))$. Thus we see that for large NT and for $\rho > 0$ the number of robust templates can increase exponentially with NT. The level of the noise q determines the size of ρ. If $1 \gg \rho > 0$ it will reduce the growth of M with NT. If $\rho < 0$ which happens when $q > 0.5$ then $M \to 0$, that indicates the noise level too high to sustain robust templates. We will call the region where $\rho > 0$ the *retrieval phase*. This result is a special case of Shannon's noisy coding theorem (Shannon and Weaver, 1949; Ramon, 1992). E.g. If noise level, $q = .3$, activity level, $a = .1$, the number of neurons, $N = 50$ sampled over $T = 50$ time bins, and $\epsilon = 0.01$, we obtain that $M \leq 7 \cdot 10^{12}$. This indicates a high capacity with an error level of 1% with a relatively high noise level and moderate number of neurons and time bins.

Hierarchical Neural Multiplexing

The random coding has extremely high capacity yet it does not seem to resemble known properties of cortical neurons. Furthermore, even in the coding community these codes are not perceived as practical since the decoding process requires lengthy computations [for example see Ramon, 1992]. We now propose a family of codes in which the neurons exhibit time varying *Poisson statistics*. We attempt to code D different variables $\theta_{i=1...D}$ such as orientation, contrast, etc. Variables θ_i are coded by modulating the rate of the Poisson process in time. This could be done by expanding the firing rate $\lambda(t)$ in an orthogonal basis*. where the basis vectors $h_i(t)$ are Haar type wavelets and the expansion coefficients α_i are function(s) of θ_i.

$$\lambda^j(t) = \sum_i \alpha_i(C_1^j \ldots C_i^j) h_i(t). \tag{3}$$

The C_k^i variables are a generalization of the firing rate of the neurons they depend through a tuning curve on the coding parameters θ_k. The upper index of the firing rate variables (C) and the rate (λ) specifies the neuron.

In order to be more concrete we will first describe a $D = 2$ case with 256 discrete time bins in which the basis functions have the form, $h_1(t) = 1/16, \quad h_2(t) = sign(128 - t)/16 \quad t \in [0\ 256]$, which are the first two basis functions in orthonormal Haar wavelet basis. The variable α_1 sets the overall firing rate $(\alpha_1 = C_1)$, α_2 modulates this firing rate on the first and the second half. In order to prevent negative firing rates, the modulation (α_2) is dependent on the overall firing rate set by α_1 thus we have chosen $\alpha_2 = C_2 C_1 / C_{max}$. In figure 2 we show that the reconstruction error of θ_1 for a firing rate code and a HNM code are identical, since the process is a sum of two Poisson processes with a different rate and it has the same mean and standard deviation as a Poisson process with a rate fixed to the mean of these two rates. As expected, the errors for the second variable θ_2 are larger than for the first variable, thus revealing the hierarchical nature of this code. The standard deviation of C_2, seems independent of the value of

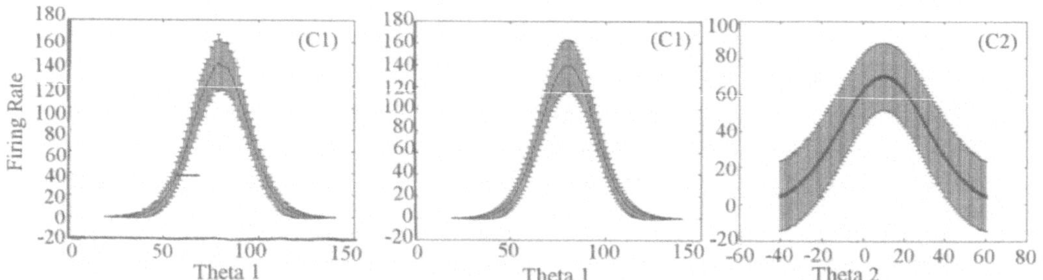

Figure 2. Reconstruction of single neuron tuning curves from a Poisson process. On the left, the firing rate tuning curve for a constant rate Poisson process is shown. In the center, a reconstructed tuning curve of variable C_1 coded with HNM scheme. It is identical to the curve on the left. On the right, a reconstructed tuning curve for variable C_2 coded with HNM scheme.

θ_2, unlike the case of the C_1. We can approximately derive this form, as well as the

---

*Using an orthogonal basis is merely a convenience not a fundamental feature of these types of codes

more general case of $D > 2$. In Appendix B we carry out this calculation. This scheme of coding can be used in a population of HNM neurons with different tuning curve centers, which can be used to estimate the variables θ_1 and θ_2 similar to population coding of (Georgopoulos et al., 1986). On the left of figure 3 we show the reconstruction of θ_1 as a function of the number of neurons, as expected it improves as the number of neurons increases. The reconstruction of θ_1 is identical for the HNM case and for a simple Poisson firing rate case, this could be seen already from the reconstruction of the tuning curves in figure 2, above. Thus coding extra, higher order, variables in the HNM scheme does not reduce the coding fidelity of the previous, lower order, variables. The reconstruction error of θ_2 is larger than for θ_1 but it too decreases with the size of the network and achieves very reasonable values for network sizes that on neurobiological scale are very small. It is evident from the single neuron results described above, that the reconstruction of θ_1 is identical in both cases This coding scheme is similar to the

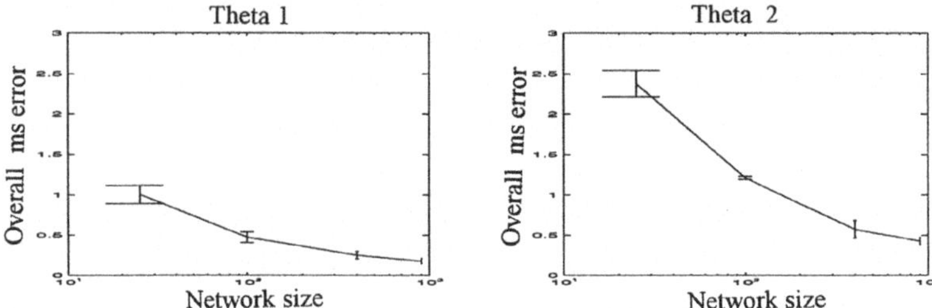

Figure 3. RMS reconstruction errors using a population of HNM neurons. Variables θ_1 and θ_2 were sampled from range [0 180]. We can see that the errors for θ_1 (left) is smaller than θ_2 (right) and that both estimates improve with increasing network size.

latency, orientation coding scheme described by Gawne et. al. (1996) in visual cortex neurons, where orientation is coded by the analog of C_1 and contrast by the analog of C_2. As we have shown adding more variables to the code with higher order temporal modulation does not reduce the fidelity of previous variables.

Discussion

The stability of spatio-temporal codes has been investigated in the simplified context of random binary codes. We have found that they are stable to large amounts of noise given that the dimensionality (NT) is large, the noise is smaller than a critical value and the number of stored templates M is small enough. However the possible number of templates is still very large and increases exponentially with NT. The idea of multiplexing in neural codes was advocated in much of the experimental work of Barry Richmond and his collaborators, and the properties of the code presented in section are qualitatively similar to those found in visual cortical neurons (Gawne et al., 1996). We have shown that an HNM temporal code can represent information about several dimensions, without reducing at all the information carried by the firing rate code. In contrast, the HNM coding scheme does not become less exact when more dimensions are added. As the number of neurons increases more dimensions can be represented by the networks with high fidelity. These examples show that noise and the stochastic nature of the neural machinery are not necessarily detrimental to spatio-temporal coding, provided the number of neurons and the sampling time are large enough.

Appendix A: Noisy Coding

Eq. 2 can be derived by calculating $P(C^{ll} > C^{ln})$, Due to the independence of templates the probability that it is larger than all the other templates (all n such that $n \neq l$) is P^{M-1} where M is the number of templates. If we allow a fraction of errors ϵ then $P^M > 1 - \epsilon$. Thus, taking the log from both sides of the equation we obtain that

$$M < \frac{log(1 - \epsilon)}{log(P)} \approx \frac{-\epsilon}{log(P)} \tag{4}$$

where the approximation holds for small values of ϵ. Both of the variables C^{ll} and C^{ln} have a binomial distribution with a different mean and variance obtained from the probabilities $r = 1 - q$ and $x = a(a(1 - q) + (1 - a)q)$ respectively.

$$P(C^{ll} = C) = \binom{NT}{C} r^C (1 - r)^{(NT - C)} \tag{5}$$

$$P(C^{ln} = C) = \binom{NT}{C} x^C (1 - x)^{(NT - C)} \tag{6}$$

The probability for correct retrieval is therefore $P = P(C^{ll} > C^{ln}) = \sum_{k=1}^{NT} P(C^{ll} = k) \sum_{j=1}^{k} P(C^{ln} = j)$.

For large values of NT and for finite values of r and x these distributions can be approximated by Gaussian distributions with means $\mu^{ll} = \mu_1 = NTr$, $\mu^{ln} = \mu^2 = NTx$ and variances $(\sigma^{ll})^2 = \sigma_1^2 = NTr(1 - r)$, $\sigma^{ln} = \sigma_2^2 = NTx(1 - x)$ respectively.

Thus the Gaussian approximation for $P = \Phi(\mu_1, \mu_2; \sigma_1, \sigma_2)$ with a simple change of coordinates to a rotated coordinate system reduces to

$$P = 1 - \Phi\left(\frac{\mu_1 - \mu_2}{\sqrt{\sigma_1^2 + \sigma_2^2}}\right) = 1 - \Phi\left(\sqrt{NT} \frac{r - x}{\sqrt{(|r(1 - r)| + |x(1 - x)|}}\right) = 1 - \Phi\left(\sqrt{NT}\rho\right) \tag{7}$$

where Φ is the accumulative Gaussian distribution $\Phi(x) = \frac{1}{\sqrt{2\pi}} \int_x^\infty e^{-(1/2)y^2} dy$. and $\rho = \left((r - x)/\sqrt{|r(1 - r)| + |x(1 - x)|}\right)$. Using equation 4 and the approximation that $log(1 - \Phi) \approx -\Phi$, which is valid for small Φ, we obtain that $M < \epsilon\rho\sqrt{\frac{NT}{\pi}} e^{NT\rho^2}$.

Appendix B: Approximating the standard deviation of the HNM model

We use the following, wavelet type notation, In variable X_{ij}, i denotes the decimation level and $j = 1..2^{i-1}$ denotes the bin number. Thus $T_{01} = T$ is the total time of level zero and T_{ij} is the time of level i index j and is equal to $T_{ij} = T_i = T/2^{i-1}$ index j, F_{ij} are firing rates in time slot (ij), C_{ij} firing rate modulation coefficients, α_{ij} and h_{ij} expansion coefficients and Haar wavelets respectively. The father wavelet which sets the total firing rate (the DC component) is set as h_{00} with the corresponding coefficient α_{00}. Thus the firing rate λ could be written as $\lambda(t) = \alpha_{00}h_{00} + \Sigma_{i=1}^n \Sigma_{j=1}^{2^{i-1}} \alpha_{ij}h_{ij}(t)$. The firing rate modulation C_{ij} is defined as

$$C_{ij} = \frac{\frac{1}{2}[F_{i+1,2j-1} - F_{i+1,2j}]}{F_{ij}} = \frac{\frac{1}{2}[F_{i+1,2j-1} - F_{i+1,2j}]}{\frac{1}{2}[F_{i+1,2j-1} + F_{i+1,2j}]} = \frac{\alpha_{ij}\sqrt{T_i}}{F_{ij}} = \frac{\alpha_{ij}}{F_{ij}}\sqrt{\frac{T}{2^{i-1}}} \tag{8}$$

$\sqrt{T_i}$ term is the normalization constant. This method for choosing λ insures that $\lambda(t) \geq 0$. The time modulated firing rate $\lambda(t)$ is then used for producing a time varying

Poisson process. From individual trials we get $\tilde{C}_{ij}$ and $\tilde{\theta}_{ij}$ which are the estimates of C_{ij} and θ_{ij} respectively. If we denote $d = F_{i+1,2j-1} - F_{i+1,2j}$ and $s = F_{i+1,2j-1} + F_{i+1,2j}$, then $C_{ij} = d/s$. Since the variables d and s are independent, the first order approximation to the variance $\sigma^2(\tilde{C}_{ij}) \approx (1/<s>)^2 \sigma_d^2 + (<d>/<s>^2)^2 \sigma_s^2$. We have denoted the mean of the numerator as $<d>$ and the denominator as $<s>$ for simplicity omitting the (ij) indexes. The variables d and s have the general form $d = a - b$ and $s = a + b$, if a and b are uncorrelated then $\sigma(d) = \sigma(s)$. For Poisson spike trains, $<s> = T_i \sigma(d) = T_i \sigma(s)$. Using the relation that $<s> = F_{ij}$ and that $<d> = \alpha_{ij}/\sqrt{T_i}$, we obtain that

$$\sigma^2(\tilde{C}_{ij}) \approx 1/F_{ij}T_i \left(1 + C_{ij}^2\right) = 2^{i-1}/F_{ij}T \left(1 + C_{ij}^2\right) \tag{9}$$

References

Abeles, M. (1982). *Local Cortical Circuits: An Electrophysiological Study.* Springer-Verlag, Berlin.

Allen, C. and Stevens, C. F. (1994). An evaluation of causes of unreliability of synaptic transmission. *Proc. Natl. Acad. Sci. U.S.A*, 91:10380.

Barlow, H. B. (1972). Single units and sensation: A neuron doctrine for perceptual psychology. *Perception*, 1:371–.

Bialek, W., Rieke, H., de Ruyter van Steveninck, R. R., and Warland, D. (1991). Reading the neural code. *Science*, 252:1854–1857.

Bienenstock, E. (1995). A model of neocortex. *Network: Computation in Neural Systems*, 6:1–45.

Gawne, T. J., Kjaer, T. J., and Richmond, B. J. (1996). Latency: Another potential code for feature binding in striate cortex. *Journal of Neurophysiology*, 76:1356–1360.

Georgopoulos, A. P., Schwatz, A., and Kettner, R. (1986). Neuronal population coding of movement direction. *Science*, 233:1416–1419.

Kajaer, T. W., Hertz, J. A., and Richmond, B. J. (1994). Decoding cortical neuronal signals: Network models, information estimation and spatial tuning. *Journal of Computational Neuroscience*, 1(1/2):109–140.

Mainen, Z. F. and Sejnowski, T. J. (1995). Reliability of spike traines in neocortical neurons. *Science*, 268:1503–1506.

Ramon, A. (1992). *Coding and information theory.* Springer-Verlag, Berlin.

Richmond, B. J., Optican, L. M., Podell, M., and Spitzer, H. (1987). Temporal encoding of two-dimensional patterns by single units in primate inferior temporal cortex. i. response character istics. *Journal of Neurophysiology*, 57(1):132–146.

Shadlen, M. N. and Newsome, W. T. (1994). Noise, neural codes and cortical organization. *Current opinion in Neurobiology*, 4:569–579.

Shannon, C. E. and Weaver, W. (1949). *The Mathematical Theory of Communications.* The University of Illinois press, Urbana, Ill.

Victor, J. D. and Purpura, K. P. (1996). Nature and precision of temporal codes in visual cortex: A metric space analysis. *Journal of Neurophysiology*, 76:1310–1326.

MODELLING THE CONTROL OF CALCIUM OSCILLATIONS BY PHOSPHORYLATION OF METABOTROPIC GLUTAMATE RECEPTORS

Volker Steuber and David J. Willshaw

Centre for Cognitive Science,
University of Edinburgh,
Edinburgh EH8 9LW,
Scotland, U.K.
{V.Steuber, D.Willshaw}@cns.ed.ac.uk

INTRODUCTION

Stimulation of metabotropic glutamate receptors (mGluRs) triggers release of Ca^{2+} from intracellular stores through activation of the phospholipase C (PLC) $\rightarrow$ inositol (1,4,5)-trisphosphate (IP3) cascade (figure 1). Recent experiments have shown that the exact pattern of the Ca^{2+} response depends on the mGluR subtype.[1] Glutamate stimulation of the *mGluR5a* subtype results in damped Ca^{2+} oscillations, whereas stimulation of the *mGluR1α* subtype causes a single-peaked response (figure 2). Given that the *mGluR5a* (but not the *mGluR1α*) subtype is a substrate for phosphorylation by protein kinase C (PKC), this indicates that the oscillations are generated by inactivation of the receptors by Ca^{2+} dependent PKC phosphorylation.

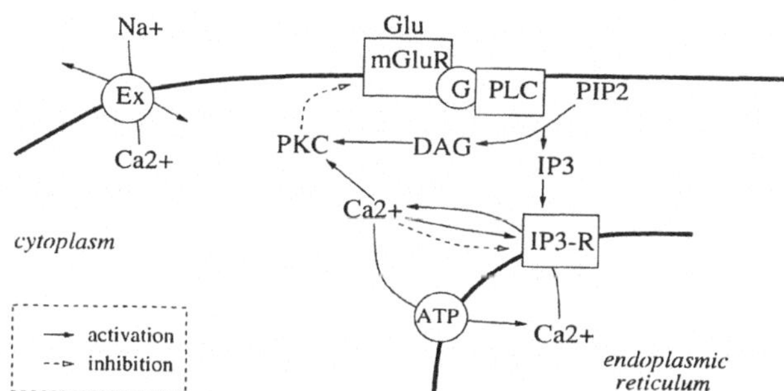

Figure 1. The mGluR second messenger network. *Glu* glutamate, *G* G-protein, *PLC* phospholipase C, *PIP2* phosphatidylinositol (4,5)-bisphosphate, *DAG* diacylglycerol, *IP3* inositol (1,4,5)-trisphosphate, *PKC* protein kinase C, *ATP* Ca ATPase, *Ex* Na/Ca Exchanger.

Here we describe two models of the mGluR second messenger network which support this hypothesis. We have modified elements of existing intracellular signalling models[2,3] to construct a detailed model which uses 8 ordinary differential equations (ODEs) to represent the mGluR network. The detailed model has then been simplified to a 2 ODE model which can be analysed with phase plane methods. Understanding the mGluR signalling network is a prerequisite for understanding phenomena like Long-Term Potentiation (LTP) and Long-Term Depression (LTD) which are thought to be involved in learning.

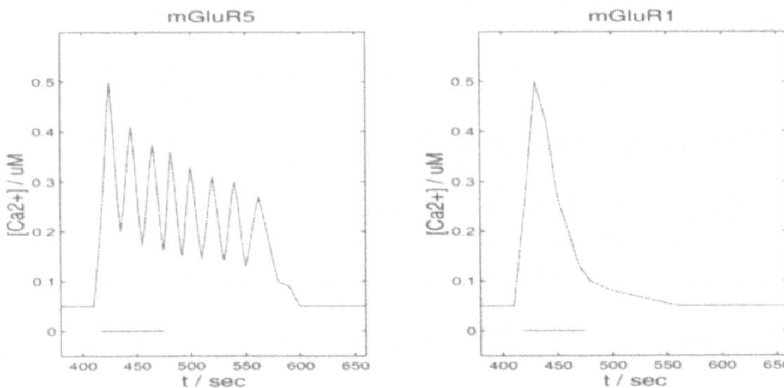

Figure 2. Experimentally observed[1] calcium oscillations (mGluR5, left) and single-peaked calcium response (mGluR1, right) after stimulation of the receptors with a glutamate pulse (bar).

A DETAILED MODEL OF METABOTROPIC RECEPTOR SIGNALLING

The mGluR second messenger network is represented by 8 ODEs. Glutamate (Glu) binds to inactive mGluRs to produce an active receptor R. The active receptor R can be inactivated by dissociation of glutamate or phosphorylation by PKC (C) which produces a phosphorylated receptor P. With a maximum receptor concentration R_{max}, the change of R and P is given by:

$$\frac{dR}{dt} = k_1 \left(R_{max} - R - P \right) Glu - k_2\, R - k_3\, RC \tag{1}$$

$$\frac{dP}{dt} = k_3\, RC - k_4\, P \tag{2}$$

Activated mGluRs trigger the activation of G-proteins which in turn stimulate the production of IP3 and diacylglycerol (DAG) by PLC. All steps involving G-proteins are very rapid.[4] Thus, we represent the production of IP3 (I) and DAG (D) by:

$$\frac{dI}{dt} = k_5\, R - k_6\, I \tag{3}$$

$$\frac{dD}{dt} = k_5\, R - k_7\, D \tag{4}$$

The IP3 dependent release of Ca^{2+} from intracellular stores and the activation of PKC by Ca^{2+} and DAG are modelled in a similar way.[5,6]
We implemented the model in C++, using a 5th order Runge-Kutta algorithm with adaptive stepsize control.

Parameters were taken from the literature, where possible.[3,6,7,8,9,10] However, we were still left with 17 unknown parameter values. Fitting these parameters by hand gave us Ca^{2+} oscillations which agreed in frequency and damping, but not in their amplitude with the experimental data (figure 3).

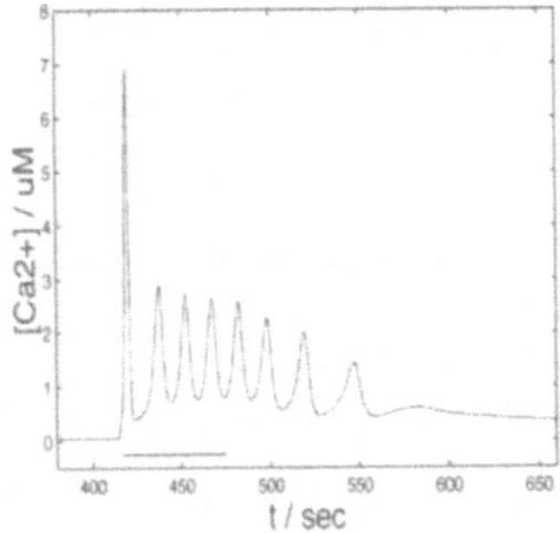

Figure 3. Damped calcium oscillations after glutamate stimulation (bar) in the detailed mGluR5 model with PKC feedback.

We are currently using an evolutionary algorithm to automate the parameter search. Preliminary results indicate the best performance for uniform crossover, fitness ranking and a constant mutation probability of 0.1.

THE SIMPLIFIED MODEL

In order to understand the relative importance of the different parts of the model for the generation of the oscillations, a simplified model was developed. We assume constant activation and calcium (c) dependent inactivation of the mGluRs (r):

$$\frac{dr}{dt} = k_1 - k_2\, r\, c \tag{5}$$

Calcium increases its own release in an autocatalytic way. Thus, the time course of calcium is modelled by:

$$\frac{dc}{dt} = k_3\, r\, c - k_4\, c \tag{6}$$

This gives us a steady state

$$(\bar{r}, \bar{c}) = (\frac{k_4}{k_3}, \frac{k_1\, k_3}{k_2\, k_4}) \tag{7}$$

with a Jacobian

$$\mathbf{J} = \begin{pmatrix} -k_2\, \bar{c} & -k_2\, \bar{r} \\ k_3\, \bar{c} & k_3\, \bar{r} - k_4 \end{pmatrix} \tag{8}$$

The eigenvalues are

$$\lambda_{1,2} = 0.5\, (Tr\, \mathbf{J} \pm \sqrt{disc\, \mathbf{J}}) = 0.5\left(-\frac{k_1\, k_3}{k_4} \pm \sqrt{\frac{k_1^2\, k_3^2}{k_4^2} - 4\, k_1\, k_3} \right) \tag{9}$$

The steady state will be a stable spiral for complex eigenvalues with a negative real part, i.e. for

275

$$disc\,\mathbf{J} = \frac{k_1^2\,k_3^2}{k_4^2} - 4\,k_1\,k_3 < 0 \tag{10}$$

In this case the system will show damped oscillations. The time dependence of the amplitude is

$$exp\left(\frac{Tr\mathbf{J}}{2}\,t\right) = exp\left(-\frac{k_1\,k_3}{2\,k_4}\,t\right) \tag{11}$$

Thus, the damping of the oscillations decreases for large k_4, i.e. if Ca^{2+} is removed rapidly from the cytoplasm.

For the example of $k_1 = k_2 = k_3 = 1$, we find a negative discriminant and damped oscillations if $k_4 > 0.5$. The dependence of disc $\mathbf{J}$ (oscillations) and Tr $\mathbf{J}$ (damping) on k_4 are shown in figure 4 (a). A phase-plane diagram and the temporal evolution of the system for $k_4 = 3$ are shown in figures 4 (b) and (c).

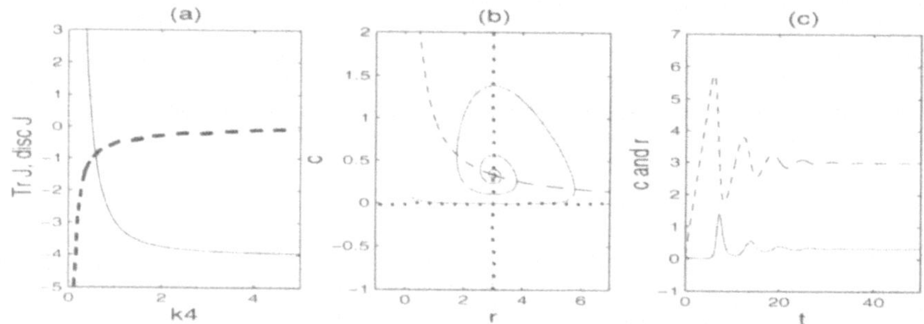

Figure 4. (a) Tr $\mathbf{J}$ (dashed) and disc $\mathbf{J}$ (solid) for varying k_4. (b) phase-plane diagram and (c) time plot of $(r(t), c(t))$ starting from $(r(0), c(0)) = (0.1, 0.1)$ (c solid, r dashed).

Similar analyses were performed for models without glutamate pulse, without negative Ca^{2+} feedback and without Ca^{2+} autocatalysis. It was found that the presence of both a glutamate pulse and negative Ca^{2+} feedback is necessary to get a negative dicriminant leading to complex eigenvalues and a stable spiral. In models without Ca^{2+} autocatalysis, a large negative Tr $\mathbf{J}$ leads to high damping and causes the system to settle very rapidly into the steady state.

CONCLUSIONS

We have described two different models of the mGluR second messenger network. Both models support the hypothesis that Ca^{2+} dependent PKC phosphorylation of mGluRs can lead to damped Ca^{2+} oscillations after mGluR stimulation. Both models have different advantages. The detailed high-granularity model employs biophysically meaningful parameters,[11] and if these parameters are correctly tuned, it can be used to make quantitative predictions to be tested in further experiments. The simplified low-granularity model is useful in understanding the relative contribution of the different reactions and in isolating the importance of Ca^{2+} dependent negative feedback and autocatalysis. This is another clear demonstration of the benefits of a multi-level modelling approach.

ACKNOWLEDGEMENTS

Thanks to Arjen van Ooyen for helpful discussions.

REFERENCES

1. S. Kawabata, R. Tsutsumi, A. Kohara, T. Yamaguchi, S. Nakanishi, and M. Okada, Control of calcium oscillations by phosphorylation of metabotropic glutamate receptors, *Nature* 383:89-92 (1996).

2. J. C. Fiala, S. Grossberg, and D. Bullock, Metabotropic glutamate receptor activation in cerebellar Purkinje cells as substrate for adaptive timing of the classically conditioned eye-blink response, *J. Neurosci.* 16(11):3760-3774 (1996).

3. S. H. Wang, A. A. Alousi, and S. H. Thompson, The lifetime of inositol 1,4,5-trisphosphate in single cells, *J. Gen. Physiol.* 105:149-171 (1995).

4. T. M. Vuong, M. Chabre, and I. Stryer, Millisecond activation transducin in the cyclic nucleotide cascade of vision, *Nature* 311:659-661 (1984).

5. E. Carafoli, Intracellular calcium homeostasis, *Ann. Rev. Biochem.* 56:395-433 (1987).

6. W. M. Yamada, C. Koch, and P. R. Adams, Multiple channels and calcium dynamics, *in Methods in Neuronal Modeling: from Synapses to Networks*, C. Koch and I. Segev, eds., MIT Press, Boston (1989).

7. J. Eilers, G. J. Augustine, and A. Konnerth, Subthreshold synaptic calcium signalling in fine dendrites and spines of cerebellar Purkinje neurons, *Nature* 373:155-158 (1995).

8. C. Thomsen, E. R. Mulvihill, B. Haldeman, D. S. Pickering, D. R. Hampson, and P. D. Suzdak, A pharmacological characterization of the mGluR1α subtype of the metabotropic glutamate receptor expressed in a cloned baby hamster kidney cell line, *Brain Res.* 619:22-28 (1993).

9. O. Baumann, B. Walz, A. V. Somlyo, and A. P. Somlyo, Electron probe microanalysis of calcium release and magnesium uptake by endoplasmic reticulum in bee photoreceptors, *Proc. Natl. Acad. Sci. USA* 88:741-744 (1991).

10. Y. Nishizuka, The molecular heterogeneity of protein kinase C and its implications for cellular regulation, *Nature* 334:661-665 (1988).

11. E. De Schutter, and P. Smolen, Calcium dynamics in large neuronal models, *in Methods in Neuronal Modeling: from Synapses to Networks (2nd edition)*, C. Koch and I. Segev, eds., MIT Press, Boston (in press).

MONTE CARLO SIMULATION OF NEURO-TRANSMITTER RELEASE USING MCell, A GENERAL SIMULATOR OF CELLULAR PHYSIOLOGICAL PROCESSES

Joel R. Stiles[1], Thomas M. Bartol, Jr.[2],
Edwin E. Salpeter[3], and Miriam M. Salpeter[1]

[1]Section of Neurobiology & Behavior
Cornell University, Ithaca, NY 14853
[2]Computational Neurobiology Laboratory
The Salk Institute, La Jolla, CA 92037
[3]Departments of Physics and Astronomy
Cornell University, Ithaca NY 14853

INTRODUCTION

Issues surrounding synaptic current efficacy, variability, plasticity, and possible cross-talk are presently of great interest and lend themselves well to computational investigations. One important factor that impacts on all of these issues is the time course of neurotransmitter exocytosis from a synaptic vesicle.[1] We have recently reported excellent quantitative agreement between highly accurate recordings of the fast rising phase of miniature endplate currents (mEPCs), and Monte Carlo simulations of vesicular acetylcholine release and postsynaptic mEPC generation.[2] The simulations were performed using an early version of our program MCell,* a generalized and highly optimized outgrowth of our original Monte Carlo programs specifically tailored to simulation of mEPC generation.[3,4] In this paper we first briefly introduce the present version of MCell and its capabilities, and then next discuss simulation of neurotransmitter exocytosis. We focus on: (1) the theoretical and computational factors underlying simulation accuracy: (2) how inadvertent use of seemingly appropriate input parameters can lead to orders-of-magnitude errors; and (3) how some of MCell's features can reduce the computation time required for simulations by orders of magnitude.

---

* MCell's simulation design features and program code were created and written by T.M. Bartol Jr. and J.R. Stiles. Monte Carlo random walk and chemical reaction algorithms were designed by J.R. Stiles, T.M. Bartol Jr., and E.E. Salpeter. Continuing development and support includes J.R. Stiles (NIH K08NS01776 and NSF IBN-9603611), M.M. Salpeter (NIH GM10422), E.E. Salpeter, and T.M. Bartol Jr. and T.J. Sejnowski (Salk Institute, NSF IBN-9603611).

MCell OVERVIEW

MCell is best described as a Monte Carlo simulator of microcellular physiology, and was developed for use with the UNIX operating system. In the near future, a version for Windows NT will be available as well. As indicated schematically in Fig. 1, simulations

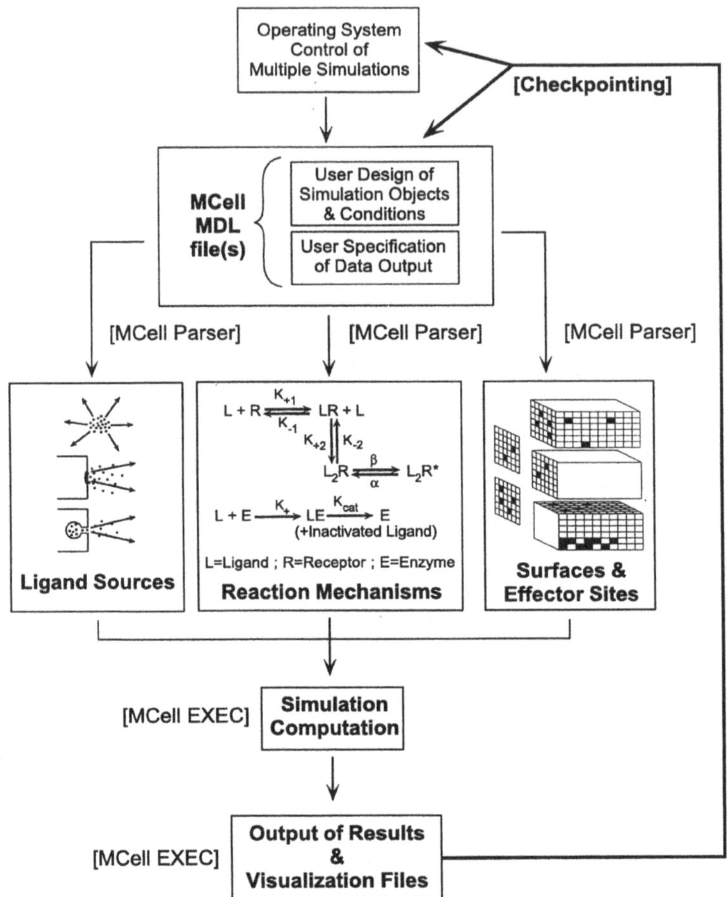

Figure 1. Schematic overview of MCell simulations. Model Description Language (MDL) input files are written by the user and then interpreted by the MCell parser at run-time. Parsed simulation objects are passed to the MCell EXEC, which runs the simulation and outputs results to data files that can be used for 2-D and 3-D plots. Simulations can also be stopped and restarted (checkpointing), incorporating one or more input parameter changes during each cycle. In addition, lengthy checkpoint sequences can be controlled by using operating system command files.

are designed and controlled by means of a Model Description Language (MDL), in essence a simple programming language the user employs to write files that tell MCell how to run the simulation. Because of the MDL's generality and features, it is possible to: (1) Use different types of ligand molecules, such as multiple neurotransmitters or neurotransmitters together with neuromodulators; (2) Use different initial distributions of diffusing ligand molecules; (3) Define patterns of ligand release, e.g. to simulate a train of nerve impulses; (4) Define arbitrary locations of surfaces (diffusion boundaries) representing cell or organelle membranes; (5) Create moving surfaces such as an expanding pore between a vesicle and a cell membrane; (6) Define multiple types of ligand-binding sites (effector sites) on permeable and impermeable surfaces, and use them to simulate molecules such as receptors, enzymes, and transporters; (7) Define arbitrary chemical reaction mechanisms for different effector sites and ligands; (8) Select specific simulation data and events to be output as results, e.g. ligand binding and unbinding transitions, fluxes through 3-D regions, or the number of effector sites in a particular bound state; (9) Specify the format and content of data files to be output and used for visualizations.

When MCell is started it first parses (interprets) one or more MDL input files (Fig. 1). It then initializes the simulation by creating the necessary ligands, surfaces, effector sites, and chemical reactions. Initialized objects subsequently are used in the actual simulation, in which the calculated probabilities of individual molecular events are compared with random numbers in order to decide which of the events take place during each time-step.[3,4] Aside from the MDL, one of the most important MCell features is 'checkpointing', indicated by the feedback loop in Fig. 1. Checkpointing involves stopping and restarting a simulation as many times as desired, and each trip around the loop can incorporate one or more modifications to MDL file input parameters. Using checkpointing: (1) Surfaces can be moved or removed, and new surfaces can be added; (2) Surface permeability can be changed; (3) New ligand molecules and effector sites can be added; (4) New reaction mechanisms can be added; (5) Existing reaction mechanisms can be modified; (6) The simulation time-step can be changed; (7) The type and amount of information to be output as results can be changed.

MCell was released to an initial group of registered beta-test neuroscience laboratories in June 1997. Those interested in becoming beta-testers should send email to J.R. Stiles and T.M. Bartol at mcell@salk.edu. At present, beta-test labs include: Philippe Ascher (Ecole Normale Superieure), Anthony Auerbach (SUNY Buffalo), David Colquhoun (Univ. Coll. London), Don Faber (Med. Coll. of Pennsylvania), Kristen Harris (Children's Hosp., Boston), Craig Jahr (Vollum Inst., Oregon Health Science Univ.), Peter Jonas (Univ. Freiburg), Read Montague (Baylor Coll. of Med.), Bert Sakmann (Max Planck Institut fur Medizinische Forschung), Eric Schwartz (Univ. Chicago), Peter Sterling (Univ. Pennsylvania), Ken Stratford (Oxford Univ.), Larry Trussell (Univ. of Wisconsin Med. School), and Gary Westbrook (Vollum Inst., Oregon Health Science Univ.).

SIMULATION OF NEUROTRANSMITTER EXOCYTOSIS

In MCell simulations of ligand diffusion, the Brownian movements of single molecules are approximated using a 3-D random walk. One model system we have studied is neurotransmitter exocytosis, i.e. the diffusion of transmitter molecules through an expanding fusion pore that connects a synaptic vesicle to a synaptic cleft space.[2] The critical relationship for accurate simulation of exocytosis is the ratio L_d/r, where L_d is the mean random walk distance traveled by each molecule during each time-step dt (specifically the mean ($\pm$) x, y, and z components of a net radial displacement[4]), and r is the radius of the fusion pore. As detailed below, numerical accuracy requires L_d/r to be significantly less than unity. The bounding shape of the vesicle is unimportant so long as the vesicle radius R is much greater than r, since the intravesicular concentration gradient is significant only over a distance of several r from the inner pore opening. We used a cube-shaped vesicle (equivalent volume $V = 4\pi R^3/3$, $R = 18.6$ nm) rather than a faceted sphere.[2] This greatly reduces computation time because checking for collisions between diffusing ligand molecules and surfaces is one of the most time-consuming parts of a simulation.

The cross-sectional shape of the fusion pore itself is important, because it influences the 3-D concentration gradients that form inside the vesicle and synaptic cleft spaces. To nearly approximate a circular cross-section, we used a faceted cylinder composed of 16 rectangular polygons of height h (in cross-section a regular 16-gon of area πr^2).

To obtain a uniform initial distribution of neurotransmitter inside the vesicle, and to simulate radial expansion of the fusion pore, we used MCell's checkpointing feature to run a series of simulations in succession. At the beginning of each series, the ligand molecules originated from a point source inside the vesicle. With the vesicle walls impermeable, several hundred time-step (dt) iterations were executed using a value of dt on the μs scale. This first checkpoint run in the series ended when the ligand molecules were uniformly distributed inside the vesicle. From the beginning of the second run in the series, the fusion pore of height h (generally from 1 to 2 thicknesses of plasma membrane, or 6 to 12 nm) connected the vesicle to the synaptic cleft space (50 nm high as at the vertebrate neuromuscular junction[2]). The pore radius r began at some minimum value corresponding to a conductance of ~300 pS,[2,5] and increased incrementally at rate b (nm·ms^{-1}) over the remaining ~100 runs in the checkpoint series. With the introduction of the pore, the value of dt was reduced to the sub-ns scale in order to maintain L_d small relative to r, and over the course of the checkpoint series dt was increased along with r to maintain an approximately constant ratio L_d/r (since $L_d = \sqrt{(4Ddt/\pi)}$,[4] where D is the ligand's diffusion constant, dt increased as the square of the relative change in L_d).

The impact of L_d/r on numerical accuracy is illustrated in Fig. 2, which shows 2 pairs of vesicle emptying curves obtained with 2 different pore opening rates. For each pair, the 2 curves differ in the value of L_d/r used during the simulations. For $b = 56$ nm·ms^{-1}, the

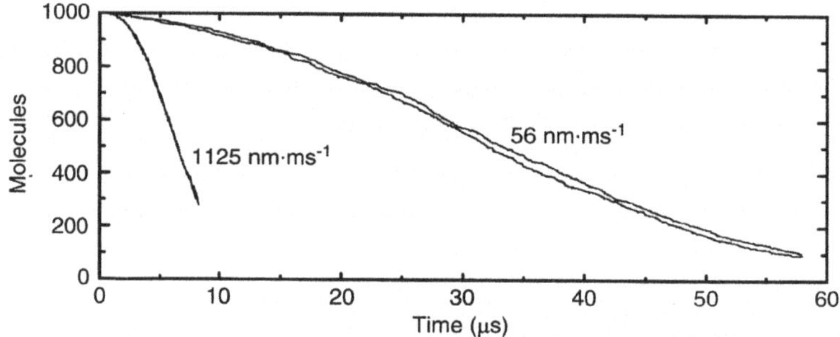

Figure 2. Illustrative results from simulation of neurotransmitter exocytosis through a fusion pore expanding radially at indicated rate b. Pore height h was 6 nm, the initial radius r was 0.75 nm, and the ligand diffusion constant D was 6 x 10^{-6} cm^2·s^{-1}.

maximal L_d/r value was either 2/3 (upper curve) or 1/3 (lower curve). For $b = 1125$ nm·ms^{-1}, the maximal L_d/r value was either 1/3 or 1/10, which gave 2 indistinguishable curves, i.e. the results are fully converged (such convergence was independent of pore opening rate). Thus, L_d/r values $\leq 1/3$ were required to maintain optimal numerical accuracy. It is important to note that with $L_d/r > 1/3$, the error is in the direction of *slowed* vesicle emptying. This results because the random walk approximation of Brownian motion uses a straight line trajectory from originating point P_1 to destination point P_2, in place of the molecule's actual tortuous movement path from P_1 to P_2 over time dt. As dt increases, the average apparent molecular velocity *decreases* in the simulation because of the nonlinear relationship between L_d and dt given above. Thus, with large dt the apparently "slowly" moving ligand molecules cannot "find" the pore rapidly enough, cannot move rapidly enough through it, and also cannot set up the appropriate concentration gradients over small scales comparable to r. If neurotransmitter exocytosis is simulated with dt on the μs scale,[6] then L_d would be of order 25 nm. This is appropriate once the ligand molecules have entered the synaptic cleft space.[2,3,4] However, for diffusion through the fusion pore, an order-of-magnitude error can be introduced because L_d/r would greatly exceed unity.

Lengthy checkpoint sequences run with thousands of ligand molecules and hundreds of thousands of time-step iterations can involve hours to days of computation time. On the other hand, MCell's MDL can be used to release ligand molecules periodically during the course of a single simulation, and this feature can be set up to mimic the time course of exocytotic release and thereby reduce run lengths to minutes. To take advantage of this feature, we needed an empirical function reproducing efflux curves such as those shown in Fig. 2. To derive such a function, we first simulated efflux through unchanging pore dimensions over a wide range (<< 1 to 10) of the ratio r/h. The ratio L_d/r was maintained $\leq 1/5$, which maintained numerical accuracy. As outlined below and illustrated in Fig. 3, we then used our results to determine one unknown (f) in a flux equation that led to the empirical release function.

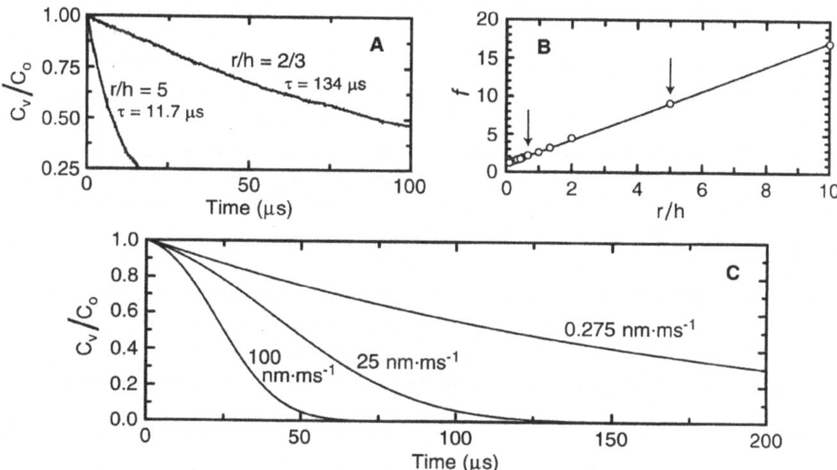

Figure 3. A. Examples of exponential efflux curves obtained with two different but constant sets of pore dimensions (r/h). Average intravesicular concentration (C_v) normalized to initial concentration (C_o). B. The empirical parameter f as a function of r/h, fitted by simple linear regression. At the limit of $r << h$ (tall thin pore), f approaches unity because the concentration gradients inside the vesicle and cleft become insignificant compared to the gradient across the pore. Arrows indicate points calculated from curves shown in panel A. C. Efflux curves plotted from Eq. 2 for indicated pore expansion rates b. Parameter values: c, 1.61; h, 9 nm; r_o, 0.93 nm; V, 2.7 x 10^4 nm^3; D, 6 x 10^{-6} cm^2·s^{-1}.

The empirical equation for net flux (J_t) out of the vesicle at time t is given by $J_t = -dN/dt = D(C_v/hf)(\pi r^2)$, where N, the number of molecules in the vesicle, is equal to the product of vesicle volume V and the average vesicle concentration (C_v) at time t. The unknown parameter f ($\geq$ unity) depends on the ratio r/h and accounts for the concentration gradients inside the vesicle and cleft spaces. After substitution and rearrangement, the flux equation yields:

$$\int \frac{dC_v}{C_v} = -\frac{D\pi r^2}{V}\int \frac{1}{hf}\,dt \tag{1}$$

For unchanging pore dimensions (constant h and r, and also therefore f), integration of Eq. 1 yields an exponential decay of C_v with time constant $\tau = Vhf/D\pi r^2$. We determined τ by fitting our simulation results (Fig. 3A), solved for f, and then plotted f as a function of r/h (Fig. 3B). We found empirically that $f = 1 + c \cdot r/h$, where $c = 1.61$ (accurate to within $\pm 3\%$ as long as R is 2-3 times larger than r). Using this linear dependence of f on r/h, Eq. 1 can be integrated numerically for any dr/dt and dh/dt. For the simpler case where h is fixed and r increases linearly with time from an initial value r_o ($r_t = r_o + bt$), Eq. 1 can be integrated analytically, giving:

$$\frac{C_v}{C_o} = \exp\left\{-\frac{Dr_o\pi}{cV}\left[\left(1-\frac{h}{cr_o}\right)t + \left(\frac{b}{2r_o}\right)t^2 + \left(\frac{h^2}{c^2 r_o b}\right)\ln\left(1+\frac{cbt}{h+cr_0}\right)\right]\right\} \tag{2}$$

where C_o is the initial concentration of neurotransmitter in the vesicle. Example plots of vesicle emptying obtained from Eq. 2 are shown in Fig. 3C. As the curves illustrate, rapid pore expansion (≥ 25 nm·ms^{-1}, as expected for neurotransmitter exocytosis[2]) yields a sigmoidal time course of vesicle emptying, while the result for extremely slow expansion (< 0.5 nm·ms^{-1} as measured during endocrine exocytosis from mast cells[5]) simplifies to an apparent exponential efflux. Straightforward numerical integration of curves such as those shown in Fig. 3C can be used to obtain the amount of ligand released during successive time-steps, and such results then can be used together with MCell's MDL to introduce diffusing ligand molecules into a running simulation. In the setting of neurotransmitter exocytosis, this method eliminates the actual vesicle and pore from the simulation, and therefore also eliminates the need for a sub-ns time-step. For simulations of miniature endplate current generation,[2] the savings in computation time amounts to orders of magnitude.

Monte Carlo simulations of ligand diffusion and receptor interactions are becoming an increasingly utilized research tool. In this study we have shown how simulations can be used to accurately predict the time course of neurotransmitter release from a synaptic vesicle, and how the inappropriate design of such simulations can lead to markedly inaccurate results. We have also used this study to introduce the capabilities of our own simulation program, MCell, which will continue to be developed and improved as it is applied in similar and other areas of physiological research.

REFERENCES

1. J.D. Clements, Transmitter timecourse in the synaptic cleft: its role in central synaptic function. *TINS* 19:163 (1996).
2. J.R. Stiles, D. Van Helden, T.M. Bartol Jr, E.E. Salpeter, and M.M. Salpeter, Miniature endplate current rise times <100 µs from improved dual recordings can be modeled with passive acetylcholine diffusion from a synaptic vesicle. *Proc. Natl. Acad. Sci. USA* 93:5747 (1996).
3. J.R. Stiles, Ph.D. thesis, Univ. of Kansas, Lawrence, KS (1990).
4. T.M. Bartol Jr., B.R. Land, E.E. Salpeter, and M.M. Salpeter, Monte Carlo simulation of MEPC generation in the vertebrate neuromuscular junction. *Biophys. J.* 59:1290 (1991).
5. A.E. Spruce, L.J. Breckenridge, A.K. Lee, and W. Almers, Properties of the fusion pore that forms during exocytosis of a mast cell secretory vesicle. *Neuron* 4:643 (1990).
6. L.M. Wahl, C. Pouzat, and K.J. Stratford, Monte Carlo simulation of fast excitatory synaptic transmission at a hippocampal synapse. *J. Neurophysiol.* 75:597 (1996).

NON-LINEAR PARAMETER ESTIMATION OF MEMBRANE PROPERTIES
IN XENOPUS EMBRYONIC NEURONS

Laurence Prime,[2] Joel Tabak,[1,2] François Tiaho,[2]
Benoit Saint-Mleux,[1] Yves Pichon,[2] C. R. Murphey,[3] and L. E. Moore[1,3]

[1]Laboratoire de Physiologie de la Perception et de l'Action
CNRS-Collège de France, UMR C-9950
15 Rue de l' Ecole de Médecine
75270 Paris Cedex 06, France

[2]Equipe Canaux et Récepteurs Membranaires
UPRESA-A 6026
Université de Rennes 1
35042 Rennes Cedex, France

[3]Department of Physiology and Biophysics
University of Texas Medical Branch
Galveston, Texas 77555, U.S.A.

INTRODUCTION

In this paper, we have combined a linear kinetic analysis with parameter estimation methods applied to non-linear differential equations. The linearized form of the equations is a critical part of the analysis and is especially important in the estimation of distributed parameters, both active and passive. Thus, our strategy is to constrain the parameter values by simultaneously estimating parameters using both non-linear real time and linear frequency domain data. Constraints of this nature significantly improve the possibilities of obtaining a more unique description of real neurons than either method alone.

In order to explicitly take into account the cable properties of Xenopus neurons, a frequency domain method was used to estimate the electrotonic properties. The properties of the potassium conductance present in intact Xenopus embryonic neurons over a limited potential range were determined. Combined fits of constant current measurements using non-linear compartmental models and analytical models for frequency domain data of the same cell, showed good agreement with the marked rectification that occurs in these neurons. Parameter estimates of compartmental models with constant current data were used for a range of polarizations relative to the resting potential.

METHODS

The developmental stage, 42, of Xenopus larvae obtained after hormonally induced fertilization was investigated.[1,2] Larvae were reared in the following manner: At stage 30 the embryos were placed in a cold room at 4 C for a maximum of one week and then placed at 20 C until stage 42 was reached. These larvae showed the expected fictive locomotion behavior, namely, bursting during each cycle. However, beginning with stage 37, the embryos appeared to be more sensitive to APV than those reared at room temperature. The larvae were anesthetized in MS222 (0.5 mg/Ml, tricaine, Sigma) and immobilized with d-tubocurarine by slitting the dorsal fin to allow access to the neuromuscular junctions. The preparations were then pinned to a Sylgard coated chamber and bathed in a normal Ringer's solution containing in mM: 115 NaCl, 2.5 KCl, 1.0 $MgCl_2$, 2.0 $CaCl_2$, 2.5 $NaHCO_3$, 10 Hepes buffered at pH 7.2. The experiments were done at room temperature (17- 21 degrees centigrade).

A combination frequency domain and voltage or current clamp was used[3,4,5]. The amplitude of the superimposed white noise stimulus was below 3 mV rms to obtain a linear response at the steady state potential that was reached during the evoked non-linear response to a step. All measurements were done under physiological conditions as normal as possible with no compensation whatsoever in the whole cell clamp situation.

Cells were voltage or current clamped with patch electrodes filled in mM with 90 K-Gluconate, 20 KCl, 2.0 $MgCl_2$, 10 EGTA, 3 ATP, 0.5 GTP, and 10 Hepes buffer at pH 7.2. The d.c. resistance of the electrodes was 5-10 Megohms. Recordings were made with either an Axoclamp-2A in or a Biologic Clamp Amplifier for continuous single electrode voltage or current patch clamp measurements. Ventral root recordings of locomotor activity were done by placing glass suction electrodes onto ventral roots between myotomes at their emergence from the spinal cord.

A whole cell patch clamp technique[6,7] on neurons from partially dissociated spinal cords was used. The duration of the enzymatic treatment was controlled so that the integrity of the network connections was maintained. Two findings verified that synaptic junctions were maintained in our preparation, (1) synaptic currents were observed during the whole cell clamp, and (2) bath perfusion of 10-50 micromolar NMDA produced oscillatory patterns indicative of fictive swimming.

RESULTS

In order to take full advantage of both the frequency and time domain methods we have used parameter estimation procedures[8] that combine frequency and time domain data. The procedures consist of restricting the frequency domain records to potentials near rest for structural and resting parameters, while including a wider range of potentials for the constant current or data for voltage dependent kinetic behavior. Spatial resolution is assured in the frequency domain by using analytical models,[9] however, the time domain solutions have been done with restricted compartmental models that preserve the electrotonic structure found with the analytical formulation.

Figure 1 illustrates the constant current responses of such a neuron having a resting potential of -60 mV and an action potential overshoot of more than 20 mV. These neurons showed repetitive firing to rectangular current pulses typical of embryos at this stage.[10] The constant current responses show a dramatic rectification illustrating that the potassium conductance is significantly activated at the resting potential. Therefore, the electrotonic properties at the resting potential must include a consideration of the voltage dependent potassium conductance.

Current clamp data at large depolarizations were not fitted because of possible errors introduced by the design of patch clamp amplifiers.[11] Such errors do not apply to the frequency domain measurements since they were done under voltage clamp. In addition, magnitude and phase functions measured under either voltage or current clamp were virtually identical. In the analysis presented here, the same set of model parameters describe all three sets of data, namely constant current, voltage clamp and frequency domain.

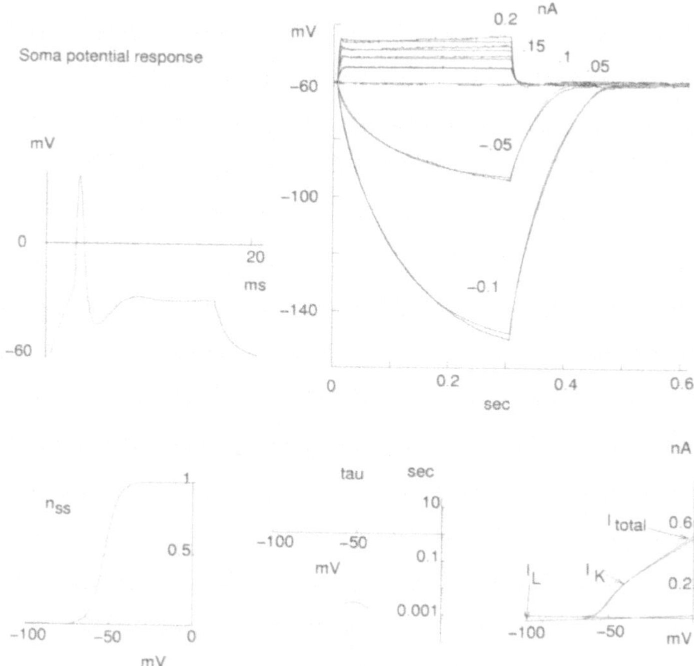

Figure 1. Whole cell patch current clamp recordings. Superimposed model fits are shown for the subthreshold responses obtained for currents from 0.1 to .1 nA. The action potential was measured at the threshold current of 0.7 nA. The steady state properties of the gating variable (nss), the potassium time constant (tau), and membrane currents are given in the lower row. I_K, I_L, and I_{total} are the potassium, leakage, and total current, respectively. The parameters for the model were cs=55 pF, gs=0.0005 µS, vs=-37 mV, gk= 0.008 µS, vk= -74 mV, vn=-51 mV, sn=0.055/mV, tn=0.0018 sec, rn= -0.08/mV, area-ratio = .75, and electrotonic length = 0.52. The estimated electrode resistance, re, was 4.2 Mohms. Mean values with standard deviations for three neurons were cs = 37±16 pF, gs = 0.0012±.0008 µS, L= 0.57± 0.08, area-ratio= 0.60±0.15, gk = .007 ±.00008 µS and tn = .0008 ± .0008 sec. Cs is the soma capacitance, gs is the soma conductance with a reversal potential of vs, gk is the potassium conductance, area-ratio is the ratio of total dendritic to soma area, vn is the half-activation voltage, sn is the slope of nss at half-activation, tn is the time constant at half-activation, and rn is the normalized slope of tn at half-activation.[12]

Whole cell recordings in Figures 1 and 2 illustrate that a simple cable model with a single gating variable, namely a potassium conductance, describes the frequency domain data as well as the voltage clamp and constant current behavior over a restricted range of membrane potentials. The impedance data of Figure 2 (right panels) measured with patch electrodes is markedly distorted by the electrode resistance despite the relative large difference between it and the cell d. c. impedance. This is reflected in both the magnitude reaching a high frequency asymptotic value of about 4 Mohms and the phase returning towards 0. Also there is a bend to the right of the complex admittance plot. Visual observation of complex admittance from whole cell clamp data does not reveal the inflection characteristic of a cable because of the electrode distortion. It is therefore necessary to curve fit the data to demonstrate the cable structure. Switching current clamp results (not shown) with patch electrodes do show the inflection indicating a cable structure independent of the electrode.

In order to analyze more rigorously the impedance functions at depolarized levels, we have done parameter optimization in which the potential profile is taken into account for a three compartmental model. However, this procedure is limited by the validity of a small number of compartments when the membrane is depolarized. Under such conditions, the overall membrane impedance is low and the space constant becomes shorter such that spatial resolution is compromised. In the lower right panel of Figure 2, at a membrane potential of -47 mV, the

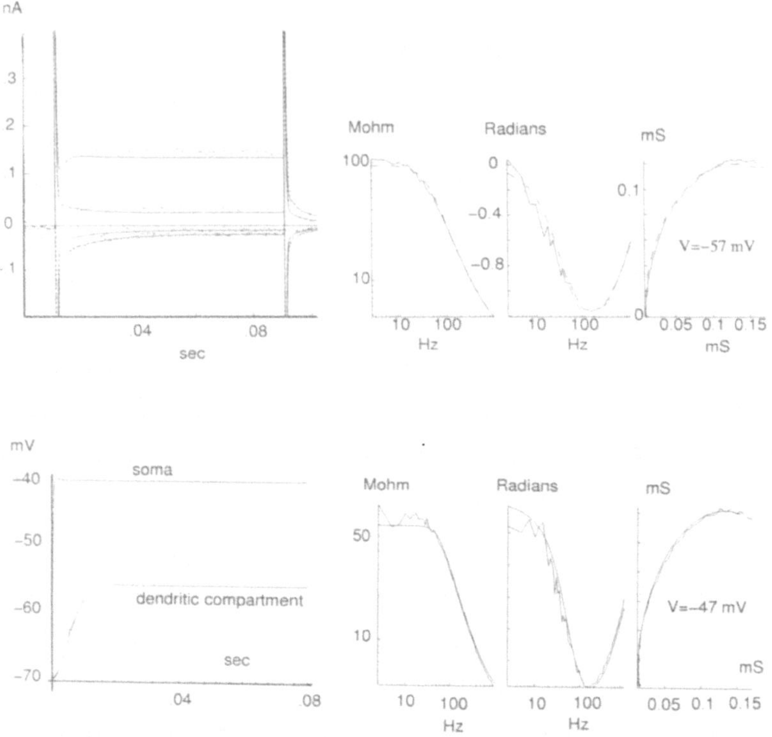

Figure 2. Whole cell voltage clamp frequency domain measurements. Fits made with the model parameters obtained from constant current data of Figure 1 are superimposed on the whole cell voltage clamp currents recorded from the soma for potentials from −90 to −50 mV in 10 mV intervals. Magnitude, phase and complex impedance plots are shown for two membrane potentials. A three dendritic compartmental model having electrotonic parameters determined by the analytical model was used to estimate the membrane potential profiles of the soma and the most peripheral dendritic compartment. The soma was depolarized to -40 mV.

analytical model has impedance amplitude that is slightly below the data at low frequencies. This is a relatively small effect, in part due to an excess activation of the dendritic potassium caused by the uniform model. The error is small most likely because the cable properties of the dendrites attenuate the effect.

The steady state value of the gating variable, n_∞ and τ_n, the potassium conductance time constant versus the membrane potential are shown in Figure 1. The best fit of the data required the presence of a dendritic cable structure. The electrotonic structure of the measured neurons appears consistent with previous anatomical studies. The area ratio obtained in this analysis provides an indication of relative dendritic area compared to the soma if the equivalent cable structure approximation is valid. These findings would suggest that neuronal models used in simulations of the Xenopus locomotion system should contain some dendritic structure. Mean values for these and other parameters are given in the legend of Figure 1.

Figure 2 shows a soma voltage clamp recording of the same neuron over a subthreshold range of membrane potentials indicating the presence of a delayed rectification at the depolarized levels. The current transients observed after a hyperpolarizing voltage clamp pulse show a multi-exponential time course demonstrating that the neuron is a multi-compartmental structure.[13] The model predictions shown in Figure 2, using parameter values from the constant current and impedance fits of this cell, describe the soma voltage clamp data. Thus, the parameters obtained in this analysis quantitatively describe three different data types, namely, piece-wise linear results in the frequency domain, non-linear constant current responses in the time domain and voltage clamp current transients measured from the soma of an intact neuron.

An accurate voltage clamp analysis of these neurons is complicated by the dendritic cable. The parameters obtained for the model fits were used to generate the potential profile of the dendritic compartments. The lower right panel of Figure 2 shows the membrane potential profile in the soma and the most peripheral compartment for a soma voltage clamp step of -40 mV. As one would expect from a length constant of 0.5 there is a significant potential drop along the dendritic tree that will clearly change the admittance functions associated with each compartment. It is evident that the actual measured currents, even after a correction for the leakage transient and steady state currents, originates from the soma and all dendritic compartments. Thus, since the fitting of the measured current responses requires the solution of the full set of non-linear differential equations, there is little advantage of the voltage over current clamp of distributed neuronal structures. In addition, the voltage clamp circuitry can introduce additional distortions.

DISCUSSION

The complex admittance measurements dramatically illustrate the effect of the electrode resistance. At low values of L the complex admittance shows a bend to the right that is similar in shape to the effect of a 4 Megohm electrode. Thus, the electrode properties must be distinguished in whole cell patch recordings from those due to dendritic cables. In conclusion, this analysis indicates that low resistance patch electrodes contribute to the whole cell clamp responses over a wide frequency range emphasizing the importance of verifying that series resistance compensation procedures need to be extremely accurate to avoid major errors in analyzing kinetic data during voltage clamp experiments. This problem could be especially acute in perforated patch measurements where series resistance compensations are of the order of 30-100 Megohms[14] or with patch clamp recordings from small dendritic fragments requiring compensations of 30-40 Megohms.[15] Since the value of the electrode series resistance often changes during a whole cell patch clamp experiment, it seems not advisable to rely on compensation methods.

Although the use of the whole cell patch clamp significantly reduces the electrode resistance, it is still the limiting factor in the interpretation of electrophysiological data and must be modeled accurately. The strategy used to analyze the electrode properties in this paper was to estimate all of the passive properties over the highest frequency range measured. As illustrated by the semi-circle in the complex admittance plot of Figure 2, the series electrode dominates the admittance function at high frequencies (2.5-1 kHz) and can be accurately determined. Furthermore, if the solution level is relative high, the effect of an electrode capacitance will be readily observed. As mentioned above, the shape of the plot for short electrotonic lengths without the series resistance can be similar to that observed with a series resistance. Thus, it is apparent that the determination of the electrotonic structure can be easily confounded with the properties of the electrode. Estimating these properties over different frequency ranges reduces this error.

REFERENCES

1. P.D. Nieuwkoop and J. Faber, ed., Normal Tables of *Xenopus laevis* (Daudin), Second edition, first reprint, North-Holland Publishing Company, Amsterdam, (1995).
2. P. Van Mier, J. Armstrong, and A. Roberts, Development of early swimming in Xenopus laevis embryos:myotomal musculature, its innervation and activation. Neuroscience 32:113 (1989).
3. L.E. Moore and R.N. Christensen, White noise analysis of cable properties of neuroblastoma cells and lamprey central neurons. J. Neurophysiology, 53: 636 (1985).
4. L.E. Moore, R.H. Hill, and S. Grillner, S., Voltage clamp analysis of lamprey neurons - Role of N- methyl-D-aspartate receptors in fictive locomotion. Brain Res. 419: 397 (1987).
5. L.E. Moore, R.H. Hill, and S. Grillner, S. Voltage clamp frequency domain analysis of NMDA activated neurons. J. Exp. Biol. 175:59 (1993).
6. D. Desarmenien, E, B, Clendening, and N, C, Spitzer., 1993, In vivo development of voltage dependent ionic currents in embryonic Xenopus spinal neurons. The Journal of Neurosciences, 13: 2575 (1993).
7. N. Dale, The isolation and identification of spinal neurons that control movement in the Xenopus embryo. European J. Neuroscience 3:1025 (1992).

8. W. H. Press, B. P. Flannery, S. A. Teukolsky, and W. T. Vetterling. Numerical Recipes in C, Cambridge University Press, London (1988).

9. W. Rall, R.E. Burke, W.R. Holmes, J.J.B. Jack, S.J. Redman, and I. Segev, Matching dendritic neuron models to experimental data. Physiol. Rev. 72:159 (1992).

10. K.T. Sillar, J.F. Wedderburn, and A.J. Simmers, 1991, The development of swimming rhythmicity in post-embryonic Xenopus laevis. Proc. R. Soc. Lond. Biol. 246:147 (1991).

11. J. Magistretti, M. Mantegazza, E. Guatto, and E. Wanke, Action potentials recorded with patch-clamp amplifiers: are they genuine? TINS 19(No. 12) 12:530 (1996).

12. C.R. Murphey, L.E. Moore, and J.T. Buchanan, Quantitative analysis of electrotonic structure and membrane properties of NMDA activated lamprey spinal neurons. Neural Computation 7:486 (1995).

13. S.R. Soffe, Active and passive membrane properties of spinal cord neurons that are rhythmically active during swimming in Xenopus embryos. Eur. J. Neurosci. 2:1 (1990).

14. N. Spruston and D. Johnston, Perforated patch-clamp analysis of the passive membrane properties of three classes of hippocampal neurons. J. Neurophysiology 67:508 (1992).

15. E.T. Kavalali, M. Zhuo, H. Bito,. and R.W. Tsien, Dendritic Ca 2+ channels characterized by recordings from isolated hippocampal dendritic segments. Neuron, 18:651 (1997).

CHOLINERGIC MODULATION OF SPIKE TIMING AND SPIKE FREQUENCY ADAPTATION IN NEOCORTICAL NEURONS

A.C. Tang,[1,2] A.M. Bartels,[1,3] and T.J. Sejnowski[1]

[1]Computational Neurobiology Laboratory, Howard Hughes
Medical Institute, Salk Institute, San Diego, CA 92037
[2]Department of Psychology, University of New Mexico
Albuquerque, NM 87131
[3]Wellcome Department of Cognitive Neurology
University College London. London, UK

INTRODUCTION

Neocortical neurons *in vivo* are spontaneously active and intracellular recordings have revealed strongly fluctuating membrane potentials arising from the irregular arrival of excitatory and inhibitory synaptic potentials. In addition to these rapid fluctuations, more slowly changing influences from diffuse activation of neuromodulatory systems alter the excitability of cortical neurons by modulating a variety of potassium conductances. In particular, acetylcholine, which affects learning and memory, reduces the slow afterhyperpolarization, which contributes to spike frequency adaptation. We used whole cell patch clamp recordings of pyramidal neurons in neocortical slices and computational simulations to show, first, that when fluctuating inputs were added to a constant current pulse, spike frequency adaptation was reduced as the amplitude of the fluctuations was increased. High-frequency, high-amplitude fluctuating inputs that resembled *in vivo* conditions [1] elicited only weak spike frequency adaptation. Second, bath application of carbachol, a cholinergic agonist, significantly increased the firing rate in response to a fluctuating input but minimally displaced the spike times by less than 3 milliseconds, comparable to the spike jitter observed when a visual stimulus is repeated under *in vivo* conditions (see [2] for more details on experiments and simulations). These results suggest that cholinergic modulation may preserve information encoded in precise spike timing, but not in interspike intervals, and that cholinergic mechanisms other than those involving adaptation may contribute significantly to cholinergic modulation of learning and memory.

EFFECTS OF CHOLINERGIC MODULATION ON SPIKE FREQUENCY ADAPTATION.

Neocortical neurons in a slice preparation of the rat visual cortex were stimulated with both conventional constant current pulses and fluctuating inputs that resembled the impact of synaptic inputs recorded *in vivo*. In previous studies using constant current pulses, acetylcholine significantly enhanced the firing rate by blocking the currents underlying spike frequency adaptation [3, 4, 5]. In contrast to the strong spike frequency adaptation evoked by a constant current pulse (Fig. 1a), injection of a fluctuating current into neocortical neurons reduced or completely eliminated pronounced spike frequency adaptation at room temperature (Fig. 1b). To obtain a simple index to quantify adaptation under both constant current pulse and fluctuating inputs, we counted spikes during the first and second halves of the current injection interval, marked by the solid and white bars below the voltage traces in Fig. 1. The adaptation index, A, defined as the normalized difference between the spike count during the early and late portions of the stimulation, was large for a constant current pulse, indicating strong adaptation, but was much smaller for fluctuating inputs (compare Figs. 1e and 1f control). For blocks of 20 trials in 10 neurons the difference was highly significant (paired t test, t=4.923, p $\leq$ .001, N=10). With a constant current stimulus, cholinergic modulation enhanced excitability mostly in the second half of the spiketrain, due to the blockade of adaptation (Fig. 1a, 1c and 1e). In contrast, excitability was increased more uniformly throughout the spiketrain evoked by the fluctuating input (Fig. 1b, 1d and 1f).

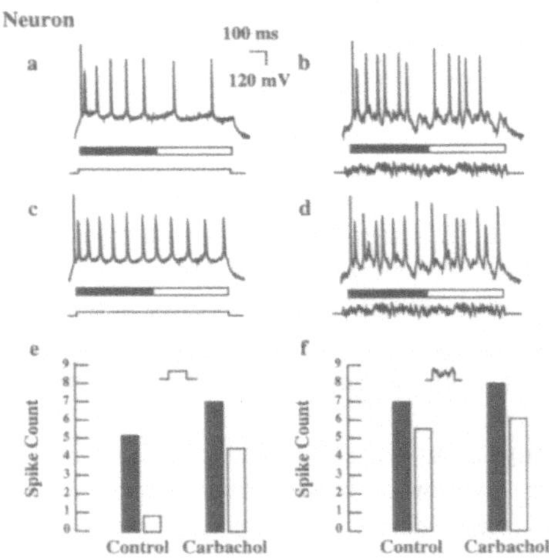

Figure 1. Cholinergic modulation of spike trains elicited by constant current pulses and fluctuating inputs in a neocortical neuron. a and b: control. c and d: 5 μM carbachol. e and f: Effects of stimulus fluctuation on adaptation and effects of cholinergic modulation.

EFFECTS OF CHOLINERGIC MODULATION ON SPIKE TIMING.

To examine the influence of cholinergic modulation on the accuracy of spike timing we directly compared the responses of neurons to both constant current pulses and fluctuating inputs before and after the bath application of carbachol. The results in Fig. 2 show that large displacements of spikes occurred for constant current pulses, as expected from the large decrease in the spike frequency adaptation (Fig. 2a); however, for fluctuating inputs the spikes were preserved with relatively small displacements when carbachol was added and additional spikes were inserted between existing ones (Fig. 2b). One way to assess the precision of the spike initiation mechanisms is to compute the average post stimulus time histogram, as shown in Fig. 2c. The displacement in spike timing for each event in the histogram, d_i, is defined as the absolute value of the time difference between the nearest peak of the event under carbachol and control conditions (see [1]). The weight for each event, w_i, is determined by the height of the event (the greater the height, the less the spike jitter). Define mean displacement, $D = \sum d_i w_i / \sum w_i$, where $i = 1, 2, \ldots$ is the event index in the control condition. For each cell, D was measured using identical fluctuating input. For 5 and 7.5 μM carbachol (mean±sem): 2.76±.38 ms, N=15; for 15 μM carbachol: D=3.33 ms, N=1; for 30 μM carbachol: D=9.3 ms, N=2. For all 18 cells examined, the mean current injection ranged from 50 to 120 pA; the standard deviation of current fluctuation ranged from 50 to 100 pA. The spike jitter, measured as the mean half width of the events, was not changed significantly by carbachol under repeated identical stimulation (standard deviation for control (mean±sem) was 0.9 ± 0.3 ms and for the carbachol condition was 0.9 ± 0.4 ms).

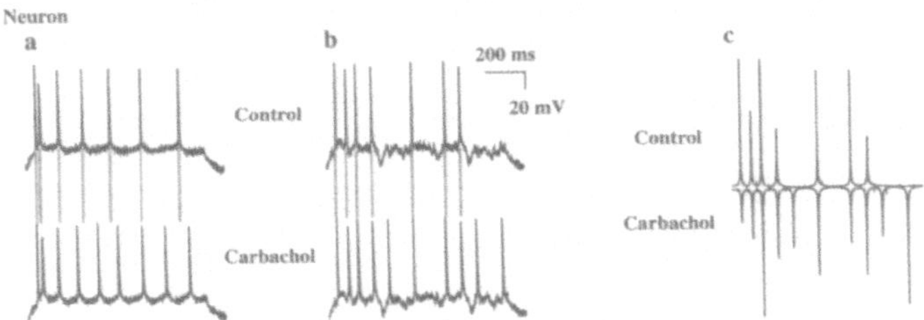

Figure 2. Cholinergic modulation preserved spike timing in response to the same fluctuating input. a: constant pulse current injection. b: fluctuating current injection. c: smoothed histogram (based on 20 spike trains per condition).

SIMULATIONS OF THE CHOLINERGIC MODULATION OF SPIKE FREQUENCY ADAPTATION.

To show that the reduced adaptation due to increased stimulus fluctuation observed in the slice experiments is a general property of cortical neurons over a wide range of conditions, we constructed a model neuron that exhibited the basic response characteristics of neocortical neurons and varied the parameters in the model to determine which were essential for reproducing the observed responses from the cortical neurons *in vitro*. Noise was added to the membrane potential to make the variability of the responses of the computer model quantitatively match the variability observed in recordings from slice preparations. Details of the compartmental model are given elsewhere [2]. The model neocortical neuron was stimulated with the same conventional constant current pulses and fluctuating inputs used in the *in vitro* studies reported above. Compare the results of the experiments in Figs. 1 with those of the model in Figs. 3. As in the experiments, pronounced spike frequency adaptation was evoked in model neurons by a constant current pulse (Fig. 3a), but injection of a fluctuating current reduced spike frequency adaptation (Fig. 3b). Cholinergic modulation was simulated by reducing three potassium conductances in the model: $g_{K(Ca)}$, g_{K_M}, and $g_{K,leak}$. Cholinergic modulation increased excitability for the later period of stimulation due to a blockade of adaptation (Fig. 3c and 3e), but cholinergic modulation increased the excitability more uniformly for the fluctuating input (Fig. 3d and 3f).

In the model, spike frequency adaptation decreased as a function of increasing amplitude of stimulus fluctuation (Fig. 4a) . Similar results were also obtained for different levels of mean synaptic inputs (mean current injections from 100 pA to 300

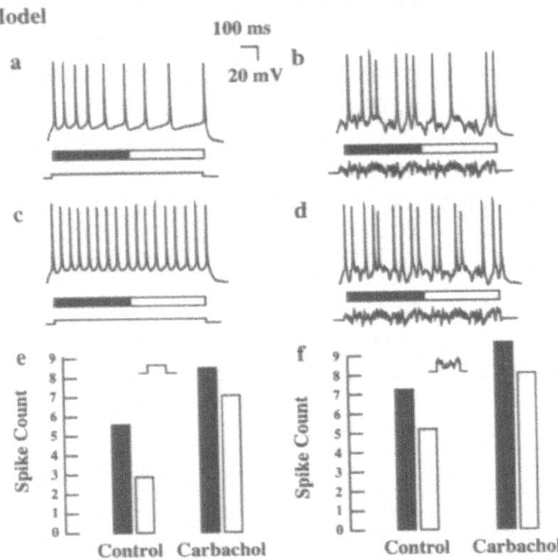

Figure 3. Simulations of cholinergic modulation of spike trains elicited by constant current pulse and fluctuating inputs. a-d as in Fig. 1 caption.

pA; 150 pA shown in Fig 4a), a range of initial degrees of adaptation ($g_{K(Ca)}$ = 0.2–1.6 pS/μm^2, shown for 0.8 pS/μm^2 in Fig. 4a), and a variety of model geometries (1 and 10 compartment layer 2/3 pyramidal cell models, shown in Fig 4a for the 10-compartment model). I_{AHP}, which is generated by $g_{K(Ca)}$, is the primary current underlying adaptation in cortical neurons *in vitro* [6].

If reduction of adaptation contributes less to cholinergic control of excitability *in vivo*, then $g_{K(Ca)}$ should have correspondingly less influence on increasing excitability. Consistent with this hypothesis, simulations at multiple levels of cholinergic reduction of $g_{K(Ca)}$ produced less increase in neuronal excitability when the stimulus fluctuation was increased (Fig. 4b), indicating that reducing I_{sAHP} became less effective in enhancing excitability as input fluctuation increased. The relatively uniform increase of excitability throughout the stimulation interval during fluctuating inputs in Fig. 3 suggests an important role for potassium currents other than $g_{K(Ca)}$. This observation is supported by simulations showing that modulation of I_{leak} and I_M significantly potentiated the effects of cholinergic modulation of I_{sAHP} on excitability. Reducing I_M (Fig. 4c) or I_{leak} (Fig. 4d) has only a weak effect on the modulation of the firing rate when $g_{K(Ca)}$ is at its maximum value. When $g_{K(Ca)}$ has been reduced, reducing these other potassium currents can significantly boost the firing rate.

SIMULATIONS OF THE EFFECTS OF CHOLINERGIC MODULATION ON SPIKE TIMING.

In recordings from neocortical neurons, the cholinergic modulation had a much greater effect on the accuracy of spike timing for constant current pulses compared with fluctuating inputs. The experimental results in Fig. 2 showing that carbachol induced large displacements of spikes for constant current pulses but relatively small displacements for fluctuating inputs was replicated in the model pyramidal neuron without changing any of the parameters that were determined from the previous simulations (Fig. 5). The average spike timing displacement due to cholinergic modulation was 2.34 ± 0.73 ms (mean ± standard deviation) (N=10 seeds). The average spike jitter for

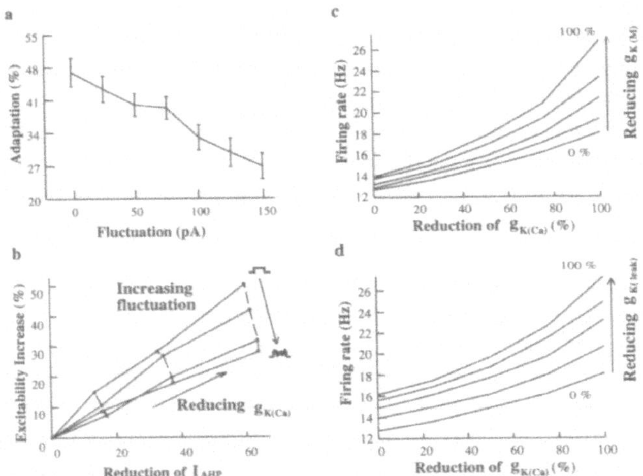

Figure 4. Simulations of adaptation and cholinergic control of excitability in a model cortical neuron. a: Adaptation decreased as a function of the standard deviation of the stimulus fluctuation. b. Increased excitability as a function of reduction in I_{sAHP}. The solid lines represent different levels of stimulus fluctuation (from 0, 50, 100 to 150 pA). The reduction in $g_{K(Ca)}$ was set to 0.8, 0.6, 0.4, and 0.2 pS/μm^2 respectively for control (origin) and increasing carbachol concentrations (successive points on the solid line). c and d: other potassium currents potentiate the effect of I_{sAHP} modulation on cholinergic control of excitability. c: Accompanied by a complete I_M blockade, the effect of I_{sAHP} modulation almost doubled (14 to 27 Hz). d: Accompanied by a modulation of I_{leak}, associated with an increase in resting membrane potential by 10 mV, the effects of I_{sAHP} modulation increased by 69% (16 to 27 Hz).

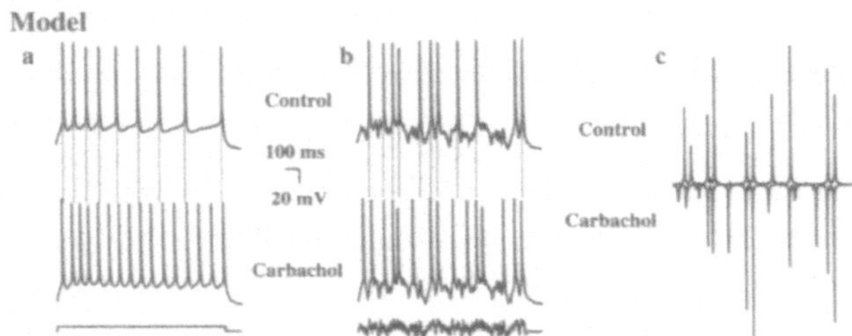

Figure 5. Simulations of the effects of cholinergic modulation on spike timing in response to the same fluctuating input. a-c as in Fig. 2.

the control condition was not significantly changed between the control (0.914 ± 0.27 ms) and carbachol (0.963 ± 0.14 ms) conditions. The smoothed histogram (20 trials) in Fig. 5c showing control (above) and the carbachol condition (below) can be compared with a similar histogram in Fig 2c.

DISCUSSION

In this paper we explored the responses of neurons in neocortical slices to inputs that resembled the ongoing spontaneous activity and the effects of cholinergic modulation on these responses. In particular, we examined the influence of a range of concentrations of bath-applied carbachol on the spike timing of neocortical pyramidal neurons in response to fluctuating current injection designed to resemble the fluctuations observed *in vivo* [7, 8].

The cholinergic modulation of spike frequency adaption to two different types of current injection were compared in this study: constant current pulses and fluctuating current inputs that were repeated under stationary conditions in a slice preparation. Spike frequency adaptation induced by a constant current pulse has been considered a mechanism for mediating cholinergic enhancement of neuronal excitability which also results in apparent changes in spike timing; however, the reduction of adaptation observed with fluctuating inputs suggests that the contribution of spike frequency adaptation to cholinergic control of excitability *in vivo* needs to be reconsidered. The decreased impact of I_{AHP} and the strong interaction among multiple potassium currents with fluctuating inputs suggests that other cholinergic mechanisms, such as a reduction in leak potassium or M currents [3, 4, 6, 9], or selective suppression of recurrent synaptic transmission [10] should be included in models of learning and memory and targeted for drug intervention to alleviate cognitive deficits.

The effects of cholinergic modulation on spike timing depended strongly on the temporal structure of the current input. The greater the amplitude of the stimulus fluctuation, the more resistant was the precision of spike timing to neuromodulation. For stimulus fluctuations with properties similar to *in vivo* conditions [7], the spike time displacement over a wide range of cholinergic excitation was on the order of a few ms, comparable to spike jitter observed *in vivo* [11, 12]. The preservation of spike timing suggests that although acetylcholine enhances neuronal excitability, a stable code based on spike-timing could be maintained under varying concentrations. Thus, neuromodulators may actually enhance spike timing information by recruiting additional spikes to code excitatory fluctuations that previously were subthreshold. Spike timing could complement information conveyed by the firing rate and take advantage of the high precision with which synaptic plasticity is apparently regulated by the relative timing of presynaptic and postsynaptic events.

REFERENCES

[1] Z. F. Mainen and T. J. Sejnowski. Reliability of spike timing in neocortical neurons. *Science*, 268:1503–6, 1995.

[2] A.C. Tang, A.M. Bartels, and T.J. Sejnowksi. Effects of cholinergic modulation on responses of neocortical neurons to fluctuating inputs. *Cerebral Cortex*, pages 502–509, 1997.

[3] D. A. McCormick. Actions of acetylecholine in the cerebral cortex and thalamus and implications for function. *Prog. Brain Res.*, 98:303–308, 1993.

[4] R.A. Nicoll. The coupling of neurotransmitter receptors to ion channels in the brain. *Science*, 241:545–550, 1988.

[5] D. A. McCormick and A. Williamson. Convergence and divergence of neurotransmitter action in human cerebral cortex. *Proc. Natl. Acad. Sci*, 86:8098–8102, 1989.

[6] D. V. Madison, B. Lancaster, and R. A. Nicoll. Voltage clamp analysis of cholinergic action in the hippocampus. *J. Neurosci.*, 7(3):733–741, 1987.

[7] D. Ferster and B. Jagadeesh. EPSP-IPSP interactions in cat visual cortex studied with *in vivo* whole-cell patch recording. *J. Neurosci.*, 12(4):1262–1274, 1992.

[8] L.G. Nowak, M.V. Sanchez-Vives, and D.A. McCormick. Influence of low and high frequency fluctuations on spike timing in visual cortical neurons, 1997. this volume.

[9] K. Krnjevic, R. Purain, and L. Renaud. The mechanism of excitation of acetylcholine in the cerebral cortex. *J. Physiol.*, 215:447–465, 1971.

[10] M. E. Hasselmo and J. M. Bower. Cholinergic suppression specific to intrinsic not afferent fiber synapses in rat piriform (olfactory) cortex. *J. Neurophysiol.*, 67(5):1222–1229, 1992.

[11] W. Bair and C. Koch. Temporal precision of spike trains in extrastriate cortex of the behaving macaque monkey. *Neural Computation*, 8(6):1184–1202, 1996.

[12] M.J. Berry, W.K. Warland, and M. Meister. The structure and precision of retinal spike trains. *Proc. Natl. Acad. Sci. USA*, 94:5411–5416, 1997.

NOISE REMOVAL BY NONLINEAR SYNAPSES

M. C. W. van Rossum[1] and R. G. Smith

Department of Neuroscience
University of Pennsylvania
Philadelphia, PA 19104-6058
[1]e-mail: vrossum@retina.anatomy.upenn.edu

INTRODUCTION

When many synaptic inputs convergence onto a neuron the pooled noise can overwhelm sparse input signals. However, thresholding the inputs can remove noise and signal quality can be maintained. We present here how this mechanism applies to photon detection in the retina.

Mammalian rods respond to a single photon with a hyperpolarization of $1mV$ which is superposed on continuous noise with a standard deviation of $0.2mV$[1,2]. The retinal circuitry preserves the quantal identity of the single photon signal as the signal has been observed in ganglion cell spike trains[3,4] and psychophysically. At low scotopic intensities (starlight) only one rod in 1000 absorbs a photon per second, so to concentrate the signal the second order neuron, the rod bipolar cell, collects input from 20-100 rods.

However, if the rod signals were linearly summed by the bipolar cell, noise from surrounding rods would overwhelm the single photon signal from one rod[1]. The reason is that the variance of the summed noise is proportional to the number of independent noise sources, hence the noise from the convergence of 25 rods is comparable to the signal. An additional noise source is the synaptic noise caused by the random character of vesicle release. A nonlinear threshold somewhere at the rod $\rightarrow$ rod bipolar synapse was therefore proposed to resolve this problem[1].

A likely candidate for the thresholding mechanism is the mGluR6 receptor and its second-messenger cascade of the rod bipolar cell. The response of an mGluR6-driven bipolar cell to glutamate is nonlinear: at high concentrations of glutamate all the postsynaptic ion channels close (in the dark the rod release glutamate, photon absorption suppresses release). Below a threshold glutamate concentration the synaptic conductance strongly increases[6].

We measure the performance of the rod $\rightarrow$ rod bipolar circuit using a computer simulation. The objectives are 1) to estimate the necessary vesicle release rate, 2) to explore the contribution of the different noise sources on single photon detection, and 3) to discover whether the mGluR6 threshold nonlinearity would allow single photon detection.

Computational Neuroscience
edited by Bower, Plenum Press, New York, 1998

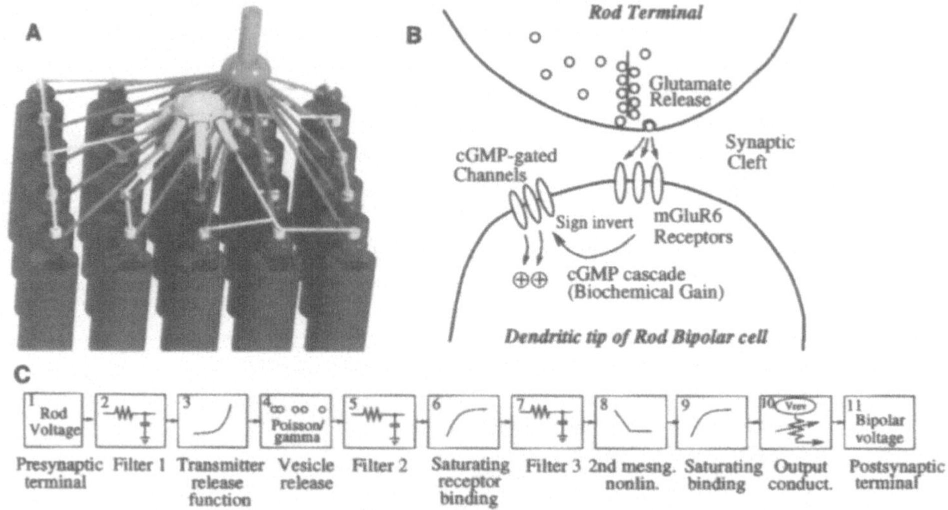

Figure 1. A) Model of the rod bipolar circuit, containing 25 rods, one horizontal cell (light gray center cell) and on top a rod bipolar cell. B) Schematic of the rod → rod bipolar synapse. C) Simulation of the synapse. Box 2: voltage of the rod is temporally filtered. 3: Exponential release function. 4: Vesicles are randomly released. 5: The pulse of transmitter from a vesicle is shaped in the second filter which represents diffusion and the dynamics of the binding to the receptor. 6: The transmitter binds, saturating according the Michaelis-Menten law. 7: Delay in the action of the bound transmitter. 8: The cGMP concentration is a nonlinear function of the amount of receptor bound glutamate. 9 and 10: saturating binding of cGMP to ion channels 11, modulating the postsynaptic voltage.

MODEL

Using the simulation language "NeuronC"[7] we simmulate a model consisting of an array of rods, one bipolar cell, and one horizontal cell axon terminal (Fig.1A). The horizontal cell is included to provide a feedback conductance onto the rods to appropriately set the rod resting voltage. The horizontal cell voltage is manually set at a constant value. See [7] for the photon transduction model. The model for the synapse from rod to rod bipolar is shown in Fig.1B,C.

The filters in the synapse consists of a number of low-pass filters with equal time constants in series. The third filter one is the most important as it has the longest time constant. It filters out the high frequency components of the vesicle noise before they are passed through the threshold. The filter is third order[8], the time constant is 50 ms, compatible with the bipolar response.

The rod voltage V_{rod} modulates the rate of vesicle release, $\rho(t) = \rho_0\, e^{[V_{\mathrm{rod}}(t) - V_{\mathrm{dark}}]/V_e}$, where V_{dark} is the average dark voltage, ρ_0 is the release rate in the dark, V_e gives the ratio of the release rates in the dark and with one photon absorbed. We assume that the single photon signal almost completely stopped vesicle release. Given the rate, vesicle release is modeled as a modulated Poisson process.

The rod bipolar of mammals utilizes the mGluR6 receptor. Glutamate binds to the postsynaptic receptors and leads to a reduction in cyclic GMP (cGMP) which closes

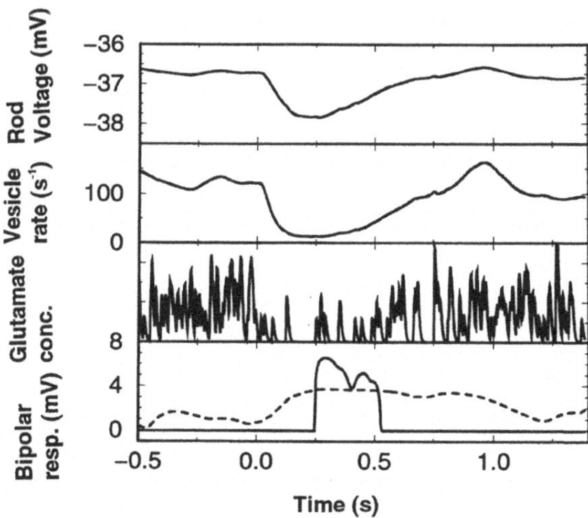

Figure 2. Trace of a simulation with one rod connected to the bipolar cell. At $t = 0$ a photon is absorbed by the rod. Note the continuous dark noise fluctuations in the rod voltage, and the vesicle noise in the glutamate concentration (about 100 ves/s). The rod bipolar response is shown for the linear (dashed line) and the thresholding synapse (solid line).

the depolarizing channels in the rod bipolar dendritic membrane[6, 9] (temporal filters are omitted here for clarity):

$$[cGMP] = 1 - c_{\text{gain}} \cdot \frac{[Glu]_{\text{cleft}}}{[Glu]_{\text{cleft}} + k_1}, \qquad \text{but } [cGMP] \geq 0 \tag{1}$$

$$g = g_{\text{syn}} \frac{[cGMP]}{[cGMP] + k_2} \tag{2}$$

where $[Glu]_{\text{cleft}}$ is the glutamate concentration in the synaptic cleft resulting from the random release, k_1 sets saturation of the glutamate binding to the receptor, c_{gain} is the biochemical gain of the cascade. k_2 sets the saturation for cGMP, finally, g and g_{syn} represent the synaptic conductance of the bipolar cell and the maximal synaptic conductance. If the glutamate concentration is high, the cGMP concentration is zero as it can not become negative. At lower concentrations the cGMP concentration follows inversely the variations of the glutamate concentration. It thus shows a thresholding behavior at high glutamate concentrations and fluctuations in the rod voltage in the dark do not cause a response in the bipolar cell. To examine the effect of the nonlinearity we set the synapse parameters in two ways so the transfer function has 1) a central linear zone, and 2) a strong thresholding nonlinearity (Fig. 2).

The voltage from the rod bipolar soma is integrated with a matched filter over 100 ms, representing the integration time of the bipolar cell. The filtered voltage is compared to a preset discrimination voltage, if it is above (below) discrimination voltage it is decided that a (no) photon was detected. After an optimized latency comparison with the stimulus yields the error rate. Because of the signal's binary character, we quantify the performance of the system in false positive and false negative rates. A false positive is the detection of a photon when there is none (measured in false positives per second), a false negative is missing a photon (measured in fraction of photons missed).

The obvious goals of the retina are to maximize efficiency (reduce false negative rate) and minimize errors (reduce false positive rate). The efficiency of a ganglion

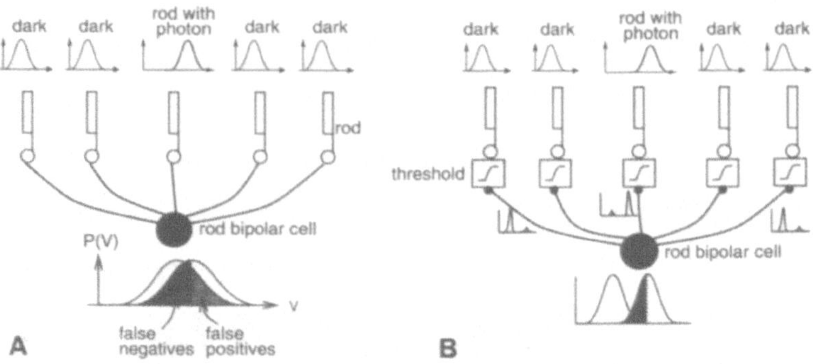

Figure 3. How thresholding reduces errors. A) The rod voltage distribution is broadened by the continuous dark noise. The voltage distribution in the bipolar cell (lower curve) is shown for two cases: none of the rods absorbs a photon and one of the rods absorbs a photon. There is considerable overlap resulting in a high false positive rate (shaded area). B) If prior to summation the signal is thresholded, the distributions remain narrow and the summation does not strongly increase the false positive rate.

cell is estimated at 70% which corresponds to a false negative rate of 30%[3]. To set a permissive value we pose the condition (the limits are likely more stringent since noise from higher order neurons was not included): *1) the false negative rate caused by intrinsic noise should be smaller than 50 % missed photons.*

The limit on the false positive rate is provided by the dark event rate, which originates from spontaneous, thermal rhodopsin isomerizations causing rod responses indistinguishable from photon responses. From direct rod recordings the number of dark light events is 0.0063 Rh*/rod/s[1], psychophysical experiments yield 0.01 events/rod/s[10]. Apparently, the retinal circuitry itself introduces few errors. As a permissive criterion we pose the condition that the errors are less than the spontaneous dark events: *2) false positive rate caused by dark continuous noise and vesicle noise should be smaller than 0.01 events/rod/s.*

In the following the false negative rate is adjusted to 50% and the false positive rate of the circuit is determined.

RESULTS

Simulation of single rod First, just one rod was connected to the rod bipolar cell. A stimulus consisting of a series of dim flashes was presented. For a single rod connected to the bipolar cell there was no significant difference in performance between the linear and the thresholding synapse. With only one input source, the classification of responses into "photon" or "no photon" is unchanged whether or not the signal passes through a prior nonlinear threshold.

For a standard deviation of the continuous rod dark noise equal to 19% of the peak response, a Poisson rate of at least 80 vesicles/s was necessary (Fig. 4A). The value is comparable to literature[11, 12]. Higher rates of vesicle release would improve the performance of the system, but much higher continuous rates are thought to be biologically implausible.

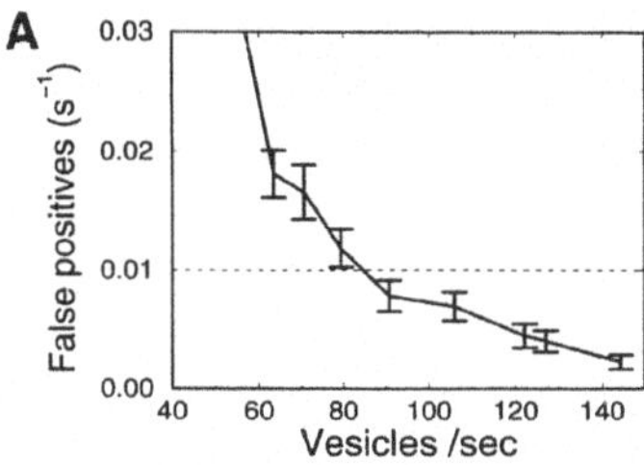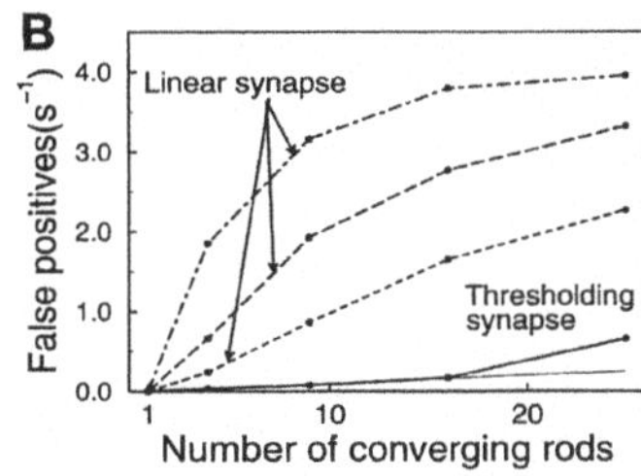

Figure 4. A) The false positive rate for a single rod connected to the rod bipolar cell, Poisson release. About 80 vesicles/sec minimum are necessary to yield a low enough false positive rate (horizontal line). B) False positive rate in the rod bipolar for multiple rods converging. The solid line: thresholding synapse, both noise sources present. Dotted-dashed: a linear synapse, both noise sources present. Dotted line: linear synapse, no vesicle release noise. Dashed line: linear synapse, no continuous dark noise. The dark light in the rod bipolar cell equals the single rod dark light times the number of rods converging (thin line). The values for one rod are comparable to the dark light (0.01/s).

Simulation of multiple converging rods Next, the number of rods connected to the bipolar cell was increased. The dark vesicle rate was set to 100 vesicles/s and the dark noise was set to 19%. Without thresholding, the standard deviation of the voltage distribution in the bipolar was much larger than the standard deviation for a single rod. When the threshold was present, it transformed each rod bipolar voltage distribution into two narrow peaks and allowed the single photon signals to remain detectable (Fig.3).

As a result the false positive rate increased strongly with the number of rods converging for the linear synapse (Fig. 4B). When the simulations included only vesicle release noise or continuous dark noise the false positive rate for the linear synapse was still larger than for the thresholding synapse.

For the thresholding synapse the false positive rate was much lower and proportionate to the number of rods converging. Only for 25 rods there is a deviation and the false positive rate is somewhat larger than the dark light. Similar results can be obtained with a simple analytical model[13].

CONCLUSION

We found that if rod signals were summed linearly by the rod bipolar, it could not reliably transmit the single photon signal as the false positive rate increases rapidly with the number of converging rods to an unacceptably high level. A thresholding nonlinearity is effective for reducing the false positives to a rate consistent with physiological measurements.

In the model the threshold voltage needs to be set accurately, which we did by hand. In the retina the type B horizontal cell axon terminal is a likely candidate to set the threshold. It is thought to provide feedback to rods and it receives input from 2000-3000 rods, so even in the dark it receives a reasonably steady input from spontaneous isomerizations.

Note that the threshold does not imply a nonlinearity in the scotopic light response of the rod bipolar because the thresholding precedes the summation of the single photon signals in the rod bipolar. At mesopic intensities, however, where 2 or more photon events pass simultaneously through a single rod synapse, the first photon response is thresholded but the superimposed second photon response is not. Therefore the two photon event would have an amplitude greater than twice the single photon event. This would be most evident shortly after the stimulus where the response has just exceeded threshold. Evidence for just such a "supra-linearity" in the rod bipolar response has recently been reported (Fig. 8A of [8]). At later times the bipolar cell showed a saturating response.

The nonlinear summation described above is a general method to improve the performance of a neural circuit that sums spatially-localized signals. For example, the complex cell sums its many synaptic inputs in a nonlinear fashion. The nonlinearity can be described with a threshold prior to the summation very similar to ours[14] suggesting that a possible function of such nonlinear processing in the visual cortex could be noise removal.

Acknowledgments

The authors thank Drs. Michael Freed, Loren Haarsma, Rukki Rao-Mirotznik, Peter Sterling, and Noga Vardi for helpful discussion. This research was supported by grant MH48168.

REFERENCES

1. D. A. Baylor, B. J. Nunn, and J. L. Schnapf. The photocurrent, noise and spectral sensitivity of rods of the monkey *macaca fascicularis*. *J. Physiol.*, 357:575–607, 1984.

2. D. M. Schneeweis and J. L. Schnapf. Photovoltage in rods and cones in the macaque retina. *Science*, 268:1053–1056, 1995.

3. D. N. Mastronarde. Correlated firing of cat retinal ganglions cells: II. responses of X- and Y-cell to single quantal events. *J. Neurophysiol.*, 49:325–349, 1983.

4. H. B. Barlow, W. R. Levick, and M. Yoon. Responses to single quanta of light in retinal ganglion cells of the cat. *Vision Res. Suppl.*, 3:87–101, 1971.

5. A. Kaneko. Physiology of the retina. *Annu. Rev.Neurosci.*, 2:169–191, 1979.

6. R. A. Shiells and G. Falk. Responses of rod bipolar cells isolated from dogfish retinal slices to concentration jumps of glutamate. *Vis. Neurosci.*, 11:1175–1183, 1994.

7. R. G. Smith. NeuronC: a computational language for investigating functional architecture of neural circuits. *J. Neurosc. Methods*, 43:83–108, 1992 (http://retina.anatomy.upenn.edu/~rob).

8. J. G. Robson and L. J. Frishman. Response linearity and kinetics of the cat retina: The bipolar component of the dark-adapted electroretinogram. *Vis. Neurosci.*, 12:837–850, 1995.

9. P. de la Villa, T. Kurahashi, and A. Kaneko. L-glutamate-induced responses and cGMP-activated channels in three subtypes of retinal bipolar cells dissociated from the cat. *J. Neurosci.*, 15:3571–3582, 1995.

10. L. J. Frishman, M. G. Reddy, and J. G. Robson. Effects of background light on the human dark-adapted electroretinogram and psychophysical threshold. *J. Opt. Soc. Am. A*, 13:601–612, 1996.

11. R. Rao, G. Buchsbaum, and P. Sterling. Rate of quantal transmitter release at the mammalian rod synapse. *Biophys. J.*, 67:57–64, 1994.

12. J. F. Ashmore and D. R. Copenhagen. An analysis of transmission from cones to hyperpolarizing bipolar cells in the retina of the turtle. *J. Physiol.*, 340:569–597, 1983.

13. M. C. W. van Rossum and R. G. Smith. Noise removal at the rod synapse of mammalian retina. submitted.

14. K. Sakia and S. Tanaka. Computational mechanisms underlying the second-order structure of cortical complex cells. *In these proceedings.*

TEMPORAL CODING WITH OSCILLATORY SEQUENCES OF FIRING.

Michael Wehr and Gilles Laurent

Computation and Neural Systems Program
California Institute of Technology
139-74 Caltech, Pasadena, CA 91125
email: mike@cns.caltech.edu

Abstract

Stimulus-evoked oscillatory synchronization of neuronal activity has been observed in many systems[1-5], yet the possible functions of such rhythmic synchronization in neural coding remain largely speculative[6-8]. In the locust, odors appear to be represented by dynamic ensembles of transiently synchronized neurons[9-11]. We now report that the components of these ensembles change in a stimulus-specific manner and with a high degree of reliability on a cycle-by-cycle basis during an odor response. Thus, information about an odor is contained in the precise temporal sequence in which these ensembles are updated during an odor response.

Although stimulus-evoked oscillations of brain potentials (caused by synchronized neural activity) have now been known for about fifty years, the roles, if any, that they might play in information coding remain largely speculative. We tested two possible functional hypotheses of temporal codes that might use oscillations. The first proposes that oscillations, by virtue of their periodic nature, allow the phase of neural signals (i.e., the timing of action potentials relative to a specific or common reference during each cycle) to be a coding parameter. Different (and possibly co-existing) stimuli could thus be represented by different neural assemblies respectively defined by a common phase of activity[3,7,8]. If true, this hypothesis predicts that the neural assemblies activated by different stimuli should synchronize at different phases. The second hypothesis rather proposes that oscillations allow the rank order (e.g., cycles 1, 2 and 4 of the oscillation) of action potentials produced by participating neurons to be coding parameters. In this scheme, the order of recruitment of neurons in an oscillating assembly should be stimulus-specific.

To test these hypotheses, we used the olfactory antennal lobe of the locust *Schistocerca americana* (anatomy shown in Fig. 1), in which individual odors are represented combinatorially by oscillating and dynamic neural assemblies formed by about 10% of a total of about 1,000 neurons[9].

We recorded simultaneously the responses of small ensembles (2-5) of these projection neurons (PNs)— the analogs of the mitral-tufted cells of the vertebrate olfactory bulb— *in vivo*, to airborne odorants. Fig. 2 (top) illustrates the temporally modulated reponses of two PNs to a cherry odor. While both neurons responded partly with an increased firing rate to this odor, their peaks of activity did not coincide and the durations of increased firing differed between them. The fine temporal structure of these responses, shown at the bottom of Fig. 2, reveals that the firing probability of each PN varied rhythmically and that the timing of either neuron's activity was locked to the other's and to the field potential oscillation.

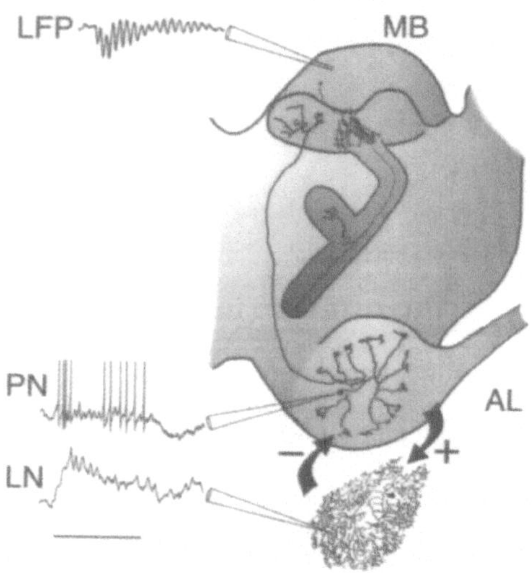

Figure 1: The locust olfactory system. Receptor afferents project from the antenna (not shown) to the antennal lobe (AL). The AL consists of projection neurons (PNs) which are spiking, excitatory cells with discrete glomerular arborization patterns, and local neurons (LNs) which are nonspiking, inhibitory cells with global arborization patterns. Both cell types experience odor-evoked membrane potential oscillations as shown in representative intracellular recordings at left (odor delivery indicated by solid bar, 1 sec.). Projection neurons project to the mushroom body, where local field potential (LFP) oscillations were recorded.

The coherence (normalized cross power) of firing of these two PNs over the 1 second response period was significantly ($p<0.001$) high at 20Hz, the frequency of the field potential oscillations (Fig 3), indicating very tight phase-locking.

We presented 13 odors and their mixtures and measured the phase of activity of each neuron relative to the field potential (not shown) or to the other neuron (Fig. 4), for all odors with which these neurons phase-locked to the field potential, using cross-correlation techniques. Fig. 4, for example, indicates that, while these two neurons fired and phase-locked in response to cherry, apple, or a mixture of both, the slight phase difference between their action potentials was constant and thus, not odor specific. This experiment was repeated with 5 pairs of PNs and 17 'neuron pair-odor' combinations for which coherence values were significant at $p<0.01$ or better. In no case did the phase of action potentials relative to the field potential or to action potentials of other participating neurons appear to vary with, and thus participate in encoding, the stimulus.

We next considered the second hypothesis. Fig. 2, for example, indicates that, while both neurons were synchronized in response to cherry, coincident firing of both PNs occurred only during certain cycles of the oscillatory ensemble response (e.g., cycles 6-7, 10-13), and never during other cycles (e.g., 1-4). This information is generally not extracted by a cross-correlation (Fig. 4), for this analysis averages events that occur at successive times, thus obscuring temporal details of correlated activity. Are such temporal patterns stimulus specific? Figure 5 illustrates the responses of a different pair of neurons to the odor apple. The fine temporal structure during the initial part of the response (shown underneath) consists of simultaneous action potentials fired during each of the first three cycles of the LFP oscillation. Figure 6 shows the response of the same pair to the mixture <cherry+apple>. The fine temporal structure is different for PN2: Whereas both PNs 1 and 2 fired during cycles 1-3 in response to apple (Fig. 5), PN2 failed to fire during cycle 1 when the mixture was presented, resulting in an apparent relative shift of the two PNs' firing by one cycle (Fig. 6).

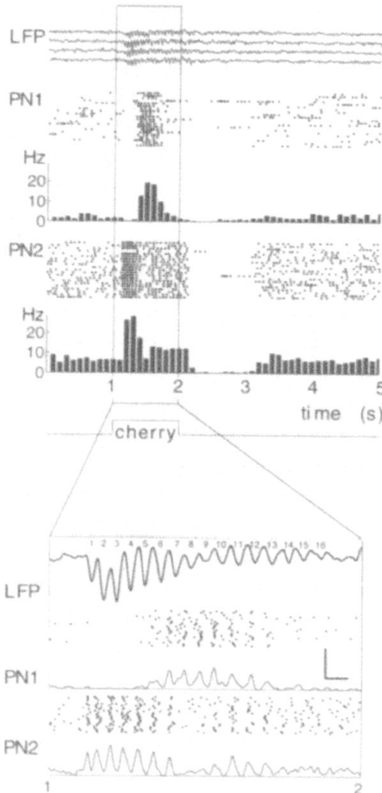

Figure 2: Top: Rasters and peri-stimulus histograms (PSTHs) of the responses of two simultaneously recorded antennal lobe PNs to the odor cherry (21 trials at 0.1Hz, aligned on the odor delivery pulse). The local field potential (LFP) recorded in the ipsilateral mushroom body during trials 1-4 are shown. Each PN and the LFP were recorded with separate extracellular glass microelectrodes. Calibration (LFP): 500 mV. Bottom: Fine temporal structure of the PNs' responses during the 1-2 s epoch in figure 1. Spikes produced by each PN over the 21 trials were convolved one at a time with a 5 ms-wide Gaussian function (continuous lines), giving a smooth firing probability distribution free of binning artifacts. The LFP trace is the average of the 21 LFP traces, indicating a very tight stimulus-locking with this odor. Calibration: horizontal, 50 ms; vertical, 30 Hz or p=0.03 of firing within a 1 ms bin.

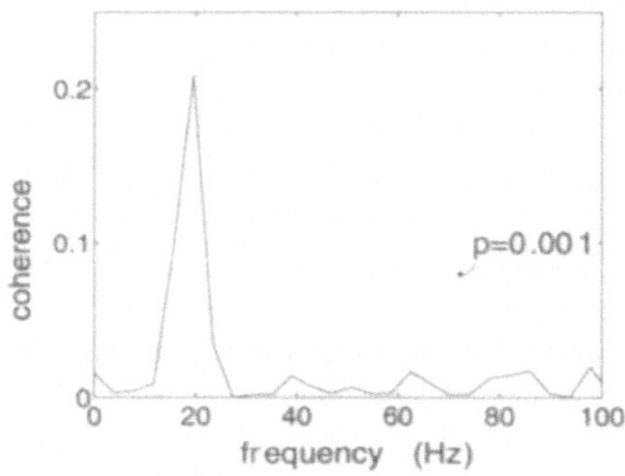

Figure 3: Coherence function (varies between 0 and 1) calculated for the spikes produced by the two PNs over the 1 s epoch in figure 2.

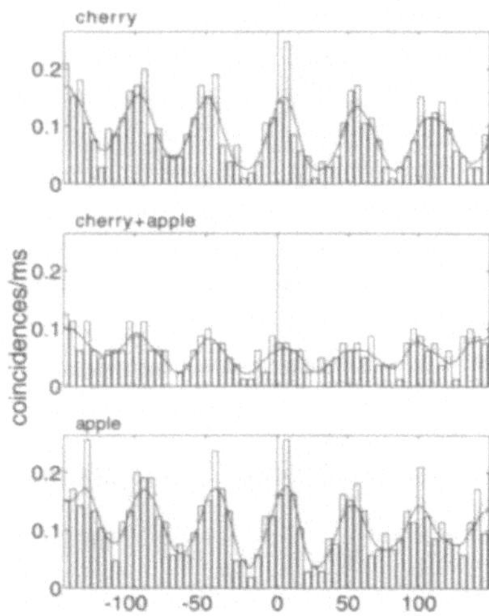

Figure 4: Cross-correlation functions calculated between the spike trains produced by both PNs in figure 1-2 in response to two odors and their 1:1 mixture. Spike coincidences were binned into 5 ms bins or convolved with 5 ms wide gaussian (solid lines). Dotted lines indicated one standard deviation from the mean.

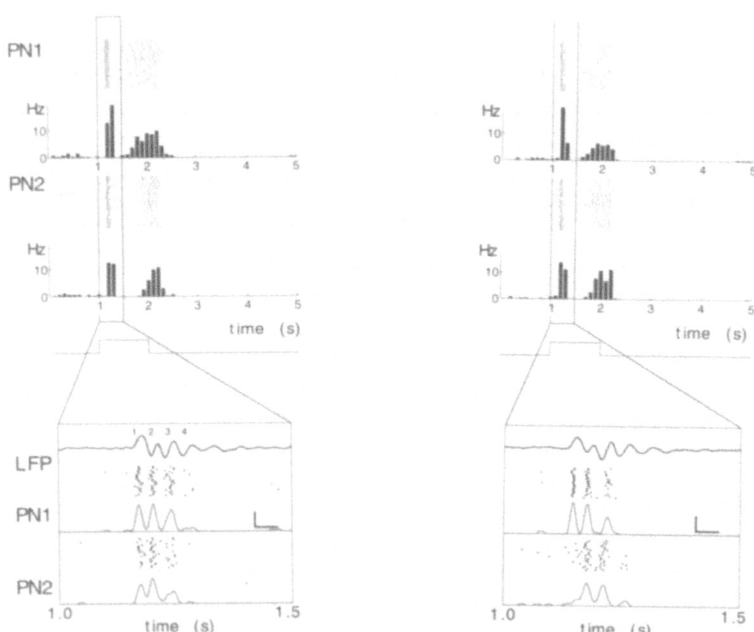

Figure 5 (left): Responses of two PNs (different animal than in Fig. 2) to apple odor. Top: Rasters and PSTHs constructed as in Fig. 2. Bottom: Fine temporal structure of the PNs' responses during the 1-1.5 s epoch (representation as in Fig. 3; bin size: 10 ms; vertical calibration: 20 Hz or p=0.02 of firing within a 1 ms bin).

Figure 6 (right): Responses of the same two PNs in Fig. 6 to a 1:1 mixture of apple and cherry odor. (representation as in Fig. 5).

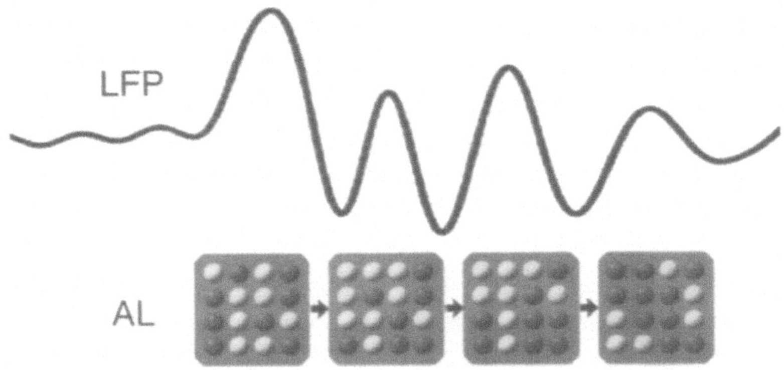

Figure 7. Depiction of odor-specific sequence of synchronized ensembles of projection neurons in the antennal lobe (AL). During each cycle of the LFP oscillation (solid curve at top), a "snapshot" depicts the state of the antennal lobe. Dark circles represent quiescent PNs, while light circles represent the ensemble of PNs firing a synchronous action potential during that cycle.

These fine temporal patterns were highly reliable. The firing probability of a PN during a given cycle could exceed 0.9 (for example, cycle 1 for PN1 in response to cherry+apple, Fig 6), indicating participation of this PN in the odor-encoding assembly during this cycle of almost every trial. In summary, we have shown that odors can be represented by ensembles of synchronized neurons that are updated non-randomly at each cycle of an oscillatory activation. Information about the odor smelled is present not in the phase of the active neurons, which is invariant, but rather in their identity and in the temporal sequence in which they are recruited. We propose that oscillations are a means to encode stimuli in a temporal combinatorial fashion. Stimuli can thus be represented both by spatial (instantaneous) ensembles— *i.e.*, the neurons activated together during any one cycle— and by temporal sequences (which are expressed by individual or groups of neurons). We conclude that this sequence of synchronized ensembles (depicted in Fig. 7) is stimulus-specific, and suggest that it may, in fact, be part of the neural code for odors in this system.

References

1. Adrian, ED. *J. Physiol. (Lond.)* **100**, 459-473 (1942)
2. Freeman, WJ. *J. Neurophysiol.* **23**, 111-131 (1960)
3. Gray, CM. & Singer, W. *Proc. Natl. Acad. Sci.* **86**, 1698-1702 (1989)
4. Gelperin, A. & Tank, DW. *Nature* **345**, 437-440 (1990)
5. Laurent, G. & Naraghi, M. *J. Neurosci* **14**, 2993-3004 (1994)
6. Milner, P. *Psychol. Rev.* **81**, 521-535 (1974)
7. von der malsburg, C. & Schneider, W. *Biol. Cybern.* **54**, 29-40 (1986)
8. Hopfield JJ. *Nature* **376**, 33-36 (1995)
9. Laurent, G. & Davidowitz, H. *Science* **265**, 1872-1875 (1994)
10. Laurent, G., Wehr, M., & Davidowitz, H. *J. Neurosci.* **16**, 3837-47(1996)
11. Wehr, M. & Laurent, G. *Nature* **384**, 162-166 (1996)

An Oscillating Cortical Network Model of Sensory-Motor Timing and Cordination

Bill Baird
Dept Mathematics,
U.C.Berkeley, Berkeley, Ca. 94720,
baird@math.berkeley.edu

Abstract

We report on the preliminary development of a model of sensory-motor control
and rhythmic motor coordination that proposes a solution to the "timing" problem in
the theory of motor control and a unification the "motor program" and the "systems
dynamics" points of view. Subsets of oscillating associative memories modeling cortical
columns are coupled to form a rhythmic time base of counters that change 40 Hz
attractors on the peaks of thalamic cycles to divide down the thalamic clock rate from
10 to .5 Hz. We train a Jordan motor control network to tap the end of a planar three
joint arm at target points in time specified by a "plan" vector representing a target
state of this time base. The biomechanics of the arm are modeled by an invertible
equilibrium point model, and the arm and controler after learning form an integrated
analog feedback system modelling primary cortex and below. The arm dynamics are
continuous and behave functionally as a "system dynamics" type nonlinear oscillator
during rhythmic motor activity. A "motor program" is specified by a 40 Hz plan vector
in premotor cortex unaffected by immediate proprioceptive feedback. These plans may
change on a thalamic cycle when activated and entrained by learned cortico-cortico
connections to an ongoing 40 Hz "attentional stream" of synchronized sensory-motor
loops. Sequences of plan vectors are activated by the output of a nested higher order
level "Elman" network. It's plan (input) vector is given in supplementary motor area
by a further nested association cortex Elman net with dorsolateral prefrontal working
memory holding the highest "master" plan. A simple rhythmic tapping sequence is
implemented where tap and return programs alternate in premotor cortex according
to a plan vector in supplementary motor area held in place by the plan in working
memory.

Oscillatory Computational Mechanisms

We have developed a neural network cortical architecture that implements a theory
of attention, learning, and communication between cortical areas by adaptive synchronization of 5-20 Hz and 30-80 Hz oscillations2, 1. Using dynamical systems theory, the
architecture is constructed from recurrently interconnected oscillatory associative mem-

ory modules that model higher order sensory and motor areas of cortex. The modules learn connection weights between themselves which cause the system to evolve under a 5-20 Hz clocked sensory-motor processing cycle by a sequence of transitions of synchronized 30-80 Hz oscillatory attractors within the modules2. The architecture employs selective"attentional" control of the synchronization of the 30-80 Hz "gamma band" oscillations between modules to direct the flow of computation. It has been used to recognize and generate sequences to solve a grammatical inference problem2.

The 30-80 Hz attractor amplitude patterns code the information content of a cortical area, whereas phase and frequency are used to "softwire" the network, since only the synchronized areas communicate by exchanging amplitude information. The system works like a broadcast network where the unavoidable crosstalk to all areas from previous learned connections is overcome by frequency coding to allow the moment to moment operation of attentional communication only between selected task-relevant areas2. The behavior of the time traces in different modules of the architecture models the temporary appearance and switching of the synchronization of 5-20 and 30-80 Hz oscillations between cortical areas that is observed during sensorimotor tasks in monkeys and humans.

There is considerable evidence for the claim of the model that the 30-80 Hz gamma band activity in the brain accomplishes attentional processing, since it appears in cortex when and where attention is required. For example, it is found in somatosensory, motor and premotor cortex of monkeys when they must pick a rasin out of a small box, but not when a habitual lever press delivers the reward. The architecture illustrates the notion that synchronization of gamma band activity not only"binds" the features of inputs in primary sensory cortex into "objects", but further binds the activity of an attended object to oscillatory activity in associational and higher-order sensory and motor cortical areas to create an evolving attentional network of intercommunicating cortical areas that directs behavior.

The architecture models the 5-20 Hz evoked potentials seen in the EEG as the control signals which determine the sensory-motor processing cycle. The 5-20 Hz clocks which drive these control signals in the architecture model thalamic pacemakers which are thought to control the excitability of neocortical tissue through similar nonspecific biasing currents that cause the cognitive and sensory evoked potentials of the EEG. The 5-20 Hz cycles "quantize time" and form the basis of derived somato-motor rhythms with periods up to seconds that entrain to each other in motor coordination and to external rhythms in speech perception1.

Nested Elman Net Architecture

The basic structure of each Elman net section involves the mapping of two input systems through a hidden layer to an output. One input set called the "context" contains the values of the hidden units from the previous time step (clock cycle) and is fast changing. The other, called the "plan", specifies the target of the motor act and changes only after its completion.

The solid arrows in figure 1 show pattern transmission from clamped and unclamped modules on the negative peaks of the theta-alpha processing cycle to unclamped modules on the postitve antiphase whose states can be affected by this input – ie. which are open to recieve it. The dotted arrows show connective mappings that cannot presently affect the state of clamped modules, but which will become effective in the antiphase of the next cycle of the thalamic reticular clocks when the roles of broadcasting and recieving areas are reversed.

In the system shown in figure 1, we provisionally map the architecture onto cortical areas, with primary, higher-order, and associative levels having distinct nested roles. The output and the hidden layers of the outer Elman motor control network are viewed as parts of primary motor cortex, while the context module is seen as part of primary

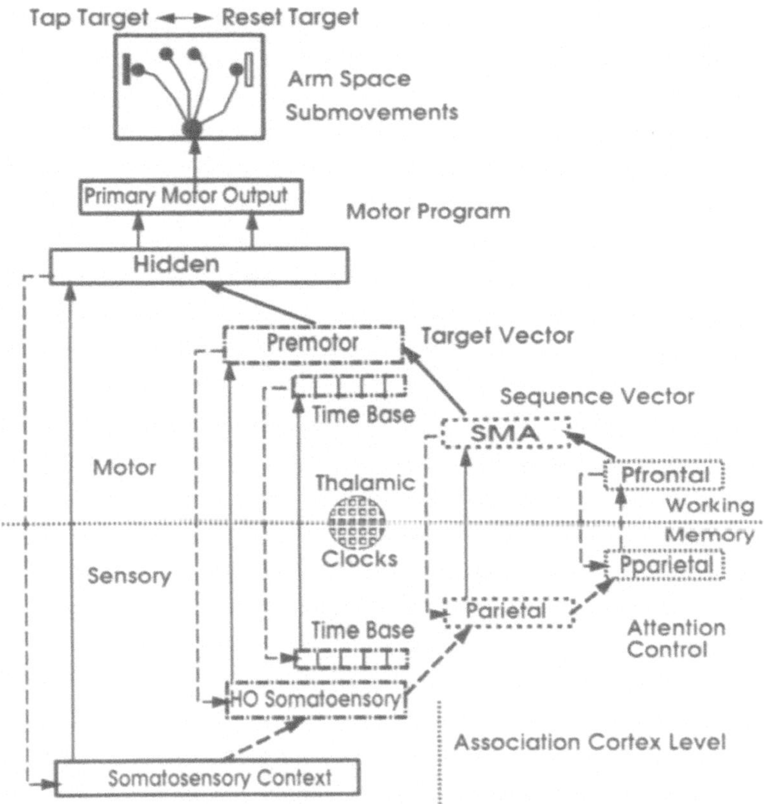

Figure 1. Nested Elman Net Architecture

somatosensory cortex. These areas operate in analog mode without attractors and can operate without gamma oscillation during movement. Modules of the architecture can perform a continuous mapping from analog input to analog output states at certain parameter values. This can be with or without gamma oscillation, depending on the levels of local feedback inhibition.

The plan vector is an attractor in premotor cortex which can be held constant while the output and feedback context evolve through the movement. At this higher order cortical level and above, the system operates by sequential attractor transitions in a discrete time fashion. Modules still take on analog oscillatory attractor states, but can be held clamped while inputs change.

The premotor target attractor is itself the hidden state of a higher order Elman network whose context area is higher order somatosensory cortex, and whose plan vector input is from an attractor in the suplementary motor area. This network generates the sequence (alternation) of tap and reset premotor targets to be reached by the outer motor control network. Finally the attractor in SMA is itself selected as the hidden state of the innermost Elman net with parietal association cortex as the context area. The highest level master plan input comes from prefrontal working memory areas which can change states in alternation with posterior parietal areas as working memory tasks are satisfied.

There is a cascade of hidden states on the motor side of the architecture with the higher level hidden module driving the next lower level hidden module as its output.

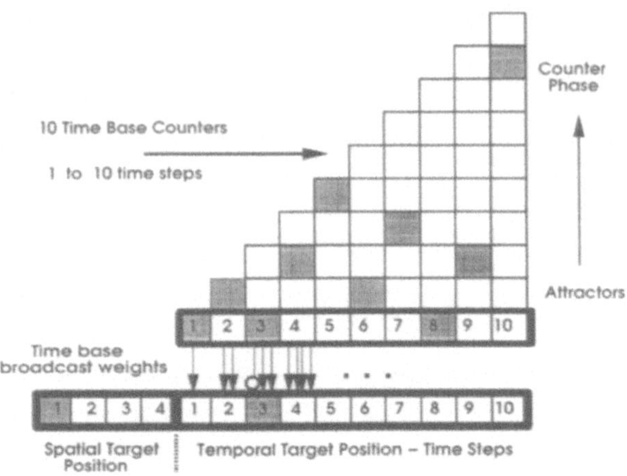

Figure 2. Time Base and Spatio-temporal Target Vector

A temporal hierarchy is also established where analog primary areas are continuously evolving in submovement program steps under control of a sequence of discrete higher order premotor programs. The sequence of these in turn is set by an association level (supplemtary motor area) sequence program selected by a working memory goal state that changes only when the tapping task is finished. Arrows in figure 1 show connections from lower to higher sensory context module levels which are required for sensory processing and to convey information about arrival at goal states of motor actions.

Attentional control of gamma synchrony between selected regions within functional areas is controled as well by the prefrontal/parietal cortex modules in the model. In the brain, this would be through direct frontal connections to thalamic reticular areas as well as indirect connections via the basal ganglia. Only the synchronized areas transmit and recieve spatial oscillation amplitude patterns which can effect the learned sequence transitions. Unsynchronized activity cancels at input summations because of its random phase2. Only the spatially coherent activity from a region can survive this summation filter. This is an important special property of communication within oscillating computational systems.

MEG tomographic observations show large scale rostral to caudal motor-sensory sweeps of coherent thalamo-cortical 40Hz activity accross the entire brain, the phase of which is reset by sensory input in waking, but not in dream states4. This suggests that an inner higher order "attentional stream" is constantly cycling between motor and sensory areas as in this architecture.

Time Base

An internal time base is hypothesized to be formed in the brain from diverse regions of premotor and higher-order somatosensory cortex which become stereotyped to cycle consistently through repetitive sequences of attractors of different lengths under the driving of the thalamic sensory-motor cycle. These form a set of counters dividing down the theta-alpha base clock rate to span 100-2000 msec periods. Kinesthetic cortical images of subcycles of body postures during locomotion might evolve to such a hierarchical system of clocks.

Each such derived clock in the architecture is created by a premotor and somatosensory associative memory module pair that has been specialized to act stereotypically as a counter or shift register by repeatedly cycling through all its attractors at the rate of one for each time step of its clock. Its overall cycle time is therefore determined by the number of attractors, and each cycle is guaranteed to be identical, as required for clock function, because it has strong attractors that correct the perturbing effect of noise. This time base is used for rhythmic expectancy in the auditory stream formation model1. In this motor application adaptive 5-20 Hz base clock frequencies allow global time warping so that relative timing is learned. Behavior may speed up or slow down as locomotion does in the rat with changing septo-hippocampal theta.

Figure 2 illustrates the timebase and shows its interaction with the spatio-temporal target vector in premotor and somatosensory areas. The vertical columns represent the number of distinct attractor amplitude patterns or "phases" in each cycle through all of a counter's states. The dark blocks indicate the active attractor of the moment. This is a snapshot of the state of the whole timebase.

These counters broadcast their attractor patterns to other cortical areas to act as a temporal frame of reference locked to the internal processing cycles for timing control. The phase state signature of each paeticular counter is broadcast to a particular module in the temporal target vector. When the target vector is selected in premotor cortex with one temporal position activated, it also activates a fast weight from the presently active phase of the corresponding counter by a Hebbian learning rule. This is shown in figure 2 as the unfilled circle from the three step cycle which happens to be at phase 1 when the three step interval is activated in the target vector. After the full cycle of steps of this particular counter, the first phase returns and a match occurs between the programmed target interval and the internally generated interval. This is wired to enable reset of the target vector to the next position and time interval to commence the next motor program.

The spatial target is minimally represented here by four possible spatial positions (In figure 2, 1 and 4 are the tap target and the reset target postures respectively). Both spatial and temporal target positions are specified and must be matched in error free performance by feedback from arm proprioception and the internal timebase reference. During learning in the brain, when the spatial target is reached the error in the temporal position can be used to alter the velocity profile magnitude (speed) by scaling the output muscle activations to achieve match in both degrees of freedom.

Biomechanics of the Arm

Inclusion of a model of the physical muscle/joint plant to be controled is essential to replicating the results of systems dynamics experiments6 where the physical characteristics of the motor system interact intimately with the network control systems to give rise to the observed phenomena. We simulated a planar three joint arm following the biologically based approach of Massione and Bizzi5. They modeled the arm as a mass and spring system which moves to an equilibrium position given by the oposing forces applied by command signals sent to agonist and antagonist muscle groups. This model is invertible at these equilibrium points, and the complexities of the general inverse problem for redundant motor degrees of freedom are avoided. We also avoided coordinate transformation issues as they did by assuming arm position was represented in body centered coordinates.

A desired trajectory and its bell shaped velocity profile can be specified, and the required muscle activations obtained by inverting the model. Training the control network becomes simply the problem of training the output units to generate this required sequence of activations. For this purpose the Elman network that is integral to our clocked models is sufficient, as opposed to the Jordan network used by Massone and Bizzi. Our feedback is from the hidden instead of the output units, and the "state units" are replaced by the "context units" of the Elman net.

Since our focus is on temporal targeting and hierarchical neural programming, we simplified the postition control problem further. Rather than assuming the duration of movements to be constant (six equal time steps), and training on different spatial trajectories, as Massone and Bizzi did, we took the same spatial trajectory and trained on different total durations with different numbers of equal intermediate intervals or "via points" (3-10 steps). In this case, the bell shaped velocity profiles are of different amplitude, but of equal areas to move the same distance at different speeds.

There is evidence that voluntary motor output (when drawing a curved line for example) is quantized into a series of submovements (motor program steps) at roughly theta cycle intervals of time just like visual saccades. Training was therefore done by specifying the via points at submovement theta clock cycle intervals (200ms) along a one dimensional trajectory between the reset and the tap target postures. Two motor programs were learned – one from reset to tap position, and one from tap back to reset posture. Errors were calculated at the output muscle activation units and the Elman net weights updated by backpropagation until performance was satisfactory.

We have not yet tested for generalization nor employed the coarse coding in either the spatial or time domain that Massone and Bizzi found necessary. Our interest was first in demonstrating the simplest operation of the system. Considerable further development is required before we may quantitatively address issues in motor control. The arm dynamics are continuous and can behave functionally as a "system dynamics" type nonlinear oscillator during rhythmic behaviors. The architecture thus illustrates a solution to the "timing problem" and a unification of "the motor program" 3 and the "systems dynamics" points of view6 in the theory of movement control. The model will be developed and applied to explain experimental work on internal coordination of motor and sensory systems and their resonant entrainment to external rhythms in synchronized tapping to a metromome, cortical phase transition phenomena in the switch from antiphase to inphase synchrony, and coordinated rhythmic motor tasks6.

REFERENCES

[1] B. Baird. A cortical network model of cognitive attentional streams, rhythmic expectation, and auditory stream segregation. In J. Bower, editor, *Computational Neuroscience '96*, pages 67-74, New York, 1997. Plenum Press.

[2] B. Baird, T. Troyer, and F. H. Eeckman. Attention as selective synchronization of oscillating cortical sensory and motor associative memories. In F. H. Eeckman, editor, *Neural Systems Analysis and Modeling*, pages 167-175, Norwell, Ma, 1994. Kluwer.

[3] S. W. Keele. Learning and control of coordinated motor patterns. In S. J. A. Kelso, editor, *Human Motor Behavior*, volume 423, pages 143–161. Laurence Erlbaum Associates, 1982.

[4] R. Llinas and U. Ribary. Coherent 40-hz oscillation characterizes dream state in humans. *Proc. Natl. Acad. Sci. USA*, 90:2078–2081, 1993.

[5] L. Massone and E. Bizzi. A neural network model for limb trajectory formation. *Biological Cybernetics*, 61:417–425, 1989.

[6] P. J. Treffner and M. T. Turvey. Resonance constraints on rhythmic movement. *Journal of Experimental Psychology: Human Perception and Performance*, 19(6):1221–1237, 1993.

PATTERN-GENERATOR-DRIVEN DEVELOPMENT IN SELF-ORGANIZING MODELS

James A. Bednar and Risto Miikkulainen

Department of Computer Sciences
The University of Texas at Austin
Austin, TX 78712
jbednar, risto@cs.utexas.edu

ABSTRACT

Self-organizing models develop realistic cortical structures when given approximations of the visual environment as input. Recently it has been proposed that internally generated input patterns, such as those found in the developing retina and in PGO waves during REM sleep, may have the same effect. Internal pattern generators would constitute an efficient way to specify, develop, and maintain functionally appropriate perceptual organization. They may help express complex structures from minimal genetic information, and retain this genetic structure within a highly plastic system. Simulations with the RF-LISSOM orientation map model indicate that such preorganization is possible, providing a computational framework for examining how genetic influences interact with visual experience.

INTRODUCTION

Many self-organizing computational models of cortical development have been proposed in recent years[1,2]. The most common type of such models shows that simple activity-dependent learning processes can result in the development of realistic cortical structures. For instance, the RF-LISSOM model[3-6] is trained using Hebbian learning with simulated visual inputs, and the initially undifferentiated neurons and connections in the model self-organize into orientation, ocular dominance, and size-selective columns with patchy lateral connections between them. Models of this type give computational support to the idea that the cortex organizes to represent and process regularities in the visual input[7,8].

Despite a common visual environment, different species develop cortical architectures with a variety of functional properties[9]. Such idiosyncrasies are generally beneficial to each species. For instance, a predator such as a cat develops a preponderance of cortical motion detectors, while fruit-eating animals such as monkeys devote more neurons to the representation of form. Cortical areas within a single animal also develop similar specializations, even though some receive their primary input from a common source.

This article considers how such a variety of architectures could develop in a self-organizing system. In the RF-LISSOM approach, the large-scale structure of the network is geneti-

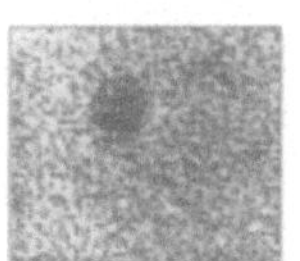 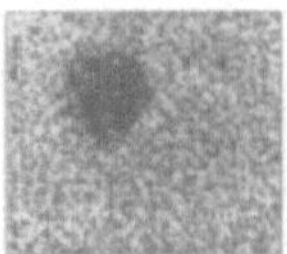 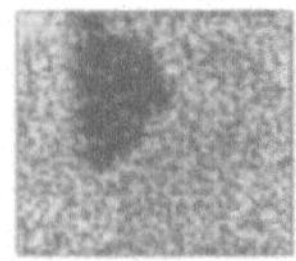  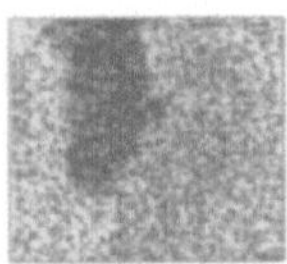

Figure 1. Spontaneous wave in the ferret retina. Each of the frames shows approximately 1 mm^2 of ferret retina using calcium concentration imaging, a measure of the activity of the retinal cells. Dark areas indicate increased activity. From left to right, the frames form a 2-second sequence showing the start and expansion of a spontaneous retinal wave. (Reprinted with permission from Feller et al. "Requirement for cholinergic synaptic transmission in the propagation of spontaneous retinal waves", *Science*, 272:1182–1187, 1996. Copyright 1996 American Association for the Advancement of Science.)

cally specified. Self-organization operates within this structure and develops local functional properties based on input activation. A variety of results can be obtained if one of the three basic components of the self-organizing system is varied: the initial configuration, the learning mechanism, or the training inputs.

The initial configuration may be an important source of variation. Highly-specific connection patterns have been found at birth in many cortical areas[10]. However, there is clearly not enough space available in the genome of a mammal (on the order of 10^5 genes) to specify every connection in its nervous system (as many as 10^{15} connections)[11]. Thus specific genetic hardwiring cannot fully explain systematic inter-species differences[12]. Neither can differences in learning mechanisms, since these mechanisms appear to be highly similar among mammalian species and brain areas[13].

Cortical training inputs, on the other hand, are quite likely to differ substantially between species and brain areas. For one thing, retinal cells of different species have different distributions of spatial and temporal response properties, which would cause different patterns of activity to be seen by higher areas. In addition, different species might have genetic predispositions to attend to certain features of the visual input, thus automatically selecting different training distributions.

Even more systematic differences in training inputs are possible, however, if training inputs are generated internally under genetic control[12,14–18]. Instead of precisely specifying cortical organization, the genome may simply encode a developmental process that is based on patterns presented to self-organizing mechanisms. Candidate patterns have been found in two regions that send input to the visual cortex: the developing retina and the brain stem. These two sources may serve different purposes in a self-organizing system: preorganization into genetically specified structures before birth, and maintenance of these structures in later life while simultaneously allowing a large degree of cortical plasticity. Simulations using the RF-LISSOM model demonstrate how the cortex could self-organize using retinal waves, and represent a step towards understanding genetic expression within a highly adaptive system.

SPONTANEOUS RETINAL WAVES

The best-documented source of generated patterns is the developing retina, where the patterns take the form of intermittent spatially-coherent activity waves across groups of ganglion cells[19,20]. Similar waves have also been documented in other early sensory areas, such as the auditory systems of birds[21]. A number of recent experiments strongly suggest that the pattern of retinal activity waves is responsible for the segregation of the LGN into eye-specific layers before birth[12,18].

The spontaneous retinal waves may also represent the earliest activity seen by the de-

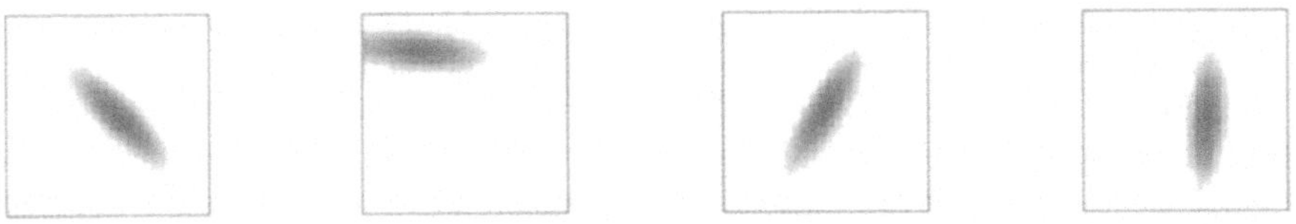

Figure 2. Sample RF-LISSOM training inputs plotted on the retina. Simple Gaussian patterns similar to retinal waves (figure 1) are used to train the model. Each presentation represents a single retinal wave, with position and orientation chosen at random. The temporal behavior of the waves is not currently modeled, only the approximate shape.

veloping visual cortex[22]. Visual input is known to have dramatic effects on the organization of the cortex[23], a process which has also been demonstrated in self-organizing models[3,24,25]. One would expect that similar reorganization also occurs in response to the early, non-visual inputs. One purpose for the retinal waves could be to substitute for visual experience before eye-opening, thus minimizing the amount of time required to develop visual competence[18]. The precise shape and temporal behavior of the waves might also serve to guide development in a species-specific way by exploiting the same learning mechanisms which later incorporate environmental influences[18].

PGO WAVES IN REM SLEEP

Once appropriate structures have been developed, internal pattern generation may help ensure that genetically-specified architectures remain, even though the system is continually adapting to the visual environment[26]. Presumably, such patterns would need to be presented when the system is not processing visual input, such as during sleep. Interestingly, the amount of rapid eye movement (REM) sleep is strongly correlated with the degree of neural plasticity across phylogeny and ontogeny[17,26]. Thus one function of REM sleep may be to present such training patterns for genetic expression and maintenance during neural adaptation[17,26].

During and just before REM sleep, internally generated phasic waves called pontogeniculo-occipital (PGO) waves can be measured in the LGN, V1, and many other cortical areas[26,27]. The waves originate in the pons of the brain stem and travel via direct pathways to the LGN and visual cortex[27], and appear to be relayed to many other areas of the cortex[26]. PGO waves are strongly correlated with eye movements as well as with vivid visual imagery in dreams, suggesting that they activate the visual cortex as if they were visual inputs[16].

Furthermore, blocking these waves has been shown to *heighten* the effect of abnormal visual experience during development[16]. When the visual input to one eye of a kitten is blocked for a short time during a critical period, the cortical area devoted to signals from the other eye increases[10]. This effect is even stronger if REM sleep is also interrupted[16], suggesting that PGO waves or other aspects of REM sleep ordinarily limit or counteract the effects of visual experience. As discussed below, one interpretation of these results is that PGO waves are part of a system for establishing and reinforcing genetically-specified architectures[26].

DEVELOPMENTAL EXPERIMENTS WITH RF-LISSOM

Several preliminary experiments with internally-generated patterns have been conducted using the RF-LISSOM self-organizing model. When trained on randomly oriented Gaussian inputs similar to the spontaneous retinal waves (see figure 2), the model develops a realistic columnar orientation map (figure 3)[3,4]. Thus the retinal patterns may explain how kittens have a crude version of such a map even when raised entirely in the dark[10].

In animals, subsequent visual experience sharpens the map and greatly increases the number of orientation-selective cells[10]. This suggests that the visual environment has a larger percentage of strongly oriented inputs than are present in the retinal waves. Similar sharpening is observed in the RF-LISSOM model if it is first trained on less oriented inputs, then on more strongly oriented inputs[6]. Simulations with multiple simultaneous Gaussian inputs also suggest that early self-organization is improved if the Gaussians do not overlap[6]. This feature has subsequently been found in the retinal waves, which (unlike visual patterns) tend to occur singly in well-defined spatial domains[20]. Thus the internally-generated patterns and visual inputs together can provide a better basis for self-organization than either alone.

DISCUSSION AND FUTURE WORK

The simulations show that internally-generated patterns can be used successfully for self-organization. The pattern-generation approach represents a middle ground between full genetic specification of neural connections, at one extreme, and purely sensory-driven self-organization, at the other. Full genetic specification of structures as complex as the cortex would require an enormous number of genes, while specifying simple input patterns and learning mechanisms would take relatively few. For instance, the complicated structures seen in the self-organized model cortex (figure 3*b*) were generated using very simple patterns (figure 2). Furthermore, specification via pattern generation would be robust to random differences in the details of the external environment and the internal arrangement of cells and connections, unlike a literal encoding of specific connections.

From an evolutionary perspective, a literal encoding would require an implausibly large number of coordinated genetic changes before a viable yet functionally distinct architecture could evolve. Pattern generation, on the other hand, would allow species to differentiate via relatively small changes in the genome (i.e., only those portions encoding the input patterns). Jouvet[26] has proposed that PGO waves are part of such a system for expression of individual and species-specific characteristics.

At the opposite extreme, a pure learning system would eventually overwrite any genetically encoded starting point. This would be undesirable since the information that is available to the organism directly from the environment is quite sparse compared to that from the millions of years contributing to the formation of the genome. Thus it would be better to retain some genetic guidance, regardless of how much visual experience is obtained — there will always be situations that an organism has not yet encountered. The genetic influences may act as a constraint upon the amount of learning that can occur, preventing the organism from becoming too specifically adapted to its particular circumstances at the expense of generality.

Thus the pattern-generation approach represents an efficient way to combine learning with genetic expression to develop a complex system. Taking into account the fact that a vertebrate genome is only somewhat larger than those of much simpler organisms[9], one can even speculate that the combination of learning and pattern-generator-driven development was the key step that enabled the evolution of the complex nervous systems of higher vertebrates. At some point, evolution may have discovered how to trick learning into being a general mechanism for genetic and environmentally-controlled development.

Future RF-LISSOM simulations using patterns approximating PGO waves and other internally-generated waves should help clarify their effects upon cortical organization. In particular, different wave shapes and temporal characteristics could explain differences in functional properties among species. Through RF-LISSOM simulations, it is possible to determine what features in the input are crucial for self-organization. By comparing those

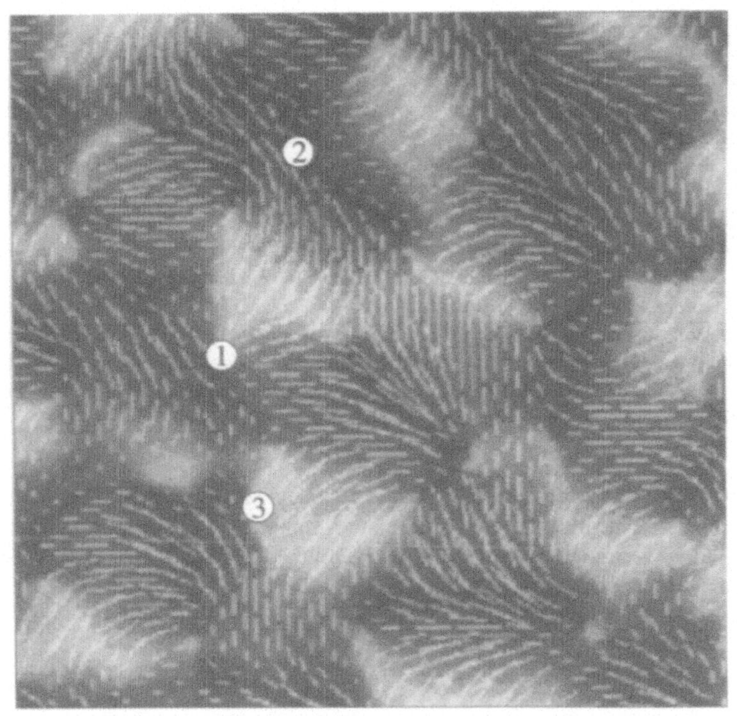

(*a*) Macaque orientation map

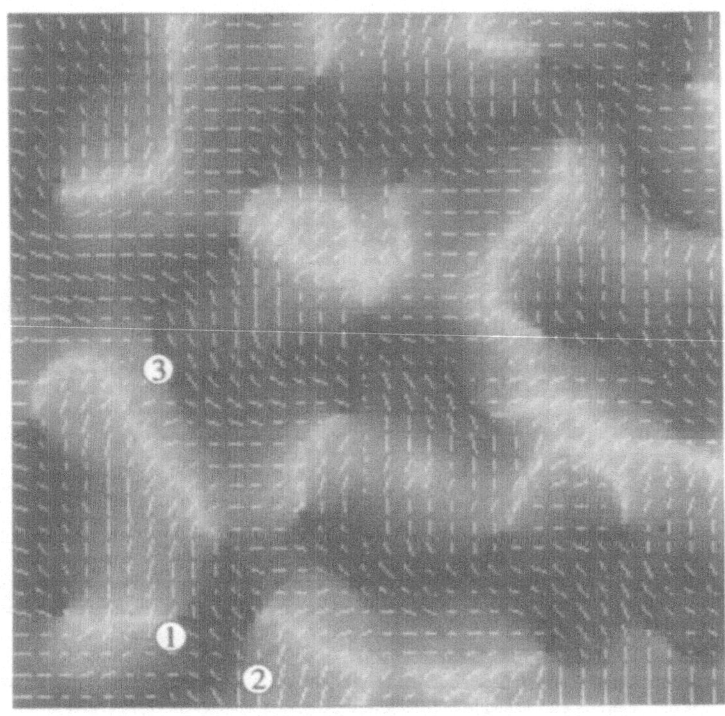

(*b*) RF-LISSOM orientation map

Figure 3. Orientation Maps. (*a*) Orientation preferences of neurons in a 2.7 × 2.7 mm area of the primary visual cortex of a macaque monkey. (Adapted with permission from figure 10A, Blasdel, "Orientation Selectivity, Preference, and...", *The Journal of Neuroscience* 12 (8): 3150, 1992.) (*b*) Orientation preferences of neurons in a 96 × 96 unit area of an RF-LISSOM map self-organized using inputs like those in figure 2[4]. In both figures, the white lines indicate the preferred orientation of the neuron, and the length of each line indicates the degree of orientation specificity of that neuron. The shading varies from black (45° to the left of a vertical line pointing up) to light gray (45° to the right). Both maps share important features (marked with numbered circles) such as (1) *pinwheel centers*, around which orientation preference changes through 180°, (2) *linear zones*, where orientation preference changes almost linearly, and (3) *fractures*, where there is a discontinuous change of orientation preference. The similarity of the model and experimental maps was also demonstrated using Fourier transforms, autocorrelation functions, and correlation angle histograms. A general discussion of these methods[1], the relevant measurements for RF-LISSOM[6], and an animation of the self-organizing process[3] are available.

features with ones known to be present in the visual environment or in internally generated patterns, more detailed hypotheses about the genetic and environmental basis of cortical structures can be formulated. Eventually, such simulations should help explain how and why the brain develops as it does in different animals.

CONCLUSION

We propose that internally-generated patterns represent a general mechanism that allows an organism to specify, develop, and maintain functional structures. RF-LISSOM simulations have shown how a crude initial orientation map could develop from non-visual inputs. Future work will examine how species differentiate, and how these genetic factors are maintained in the adult. Such experiments with self-organizing models should greatly improve our understanding of the balance between environmental and genetic determinants of individuality.

ACKNOWLEDGMENTS

This research was supported in part by the National Science Foundation under grant #IRI-9309273. Computer time for the simulations was provided by the Pittsburgh Supercomputing Center under grant #IRI940004P.

REFERENCES

1. E. Erwin, K. Obermayer, and K. Schulten, Models of orientation and ocular dominance columns in the visual cortex: A critical comparison, *Neural Computation*, 7(3):425–468 (1995).
2. N. V. Swindale, The development of topography in the visual cortex: A review of models, *Network*, 7:161–247 (1996).
3. J. Sirosh, R. Miikkulainen, and J. A. Bednar, Self-organization of orientation maps, lateral connections, and dynamic receptive fields in the primary visual cortex, in: *Lateral Interactions in the Cortex: Structure and Function*, J. Sirosh et al., eds., The UTCS Neural Networks Research Group, Austin, TX (1996). Electronic book, ISBN 0-9647060-0-8, http://www.cs.utexas.edu/users/nn/web-pubs/htmlbook96.
4. R. Miikkulainen, J. A. Bednar, Y. Choe, and J. Sirosh, Self-organization, plasticity, and low-level visual phenomena in a laterally connected map model of the primary visual cortex, in: *Perceptual Learning*, R. L. Goldstone, P. G. Schyns, and D. L. Medin, eds., volume 36 of *Psychology of Learning and Motivation*, 257–308, Academic Press, San Diego, CA (1997).

5. J. Sirosh and R. Miikkulainen, Cooperative self-organization of afferent and lateral connections in cortical maps, *Biological Cybernetics*, 71:66–78 (1994).

6. J. Sirosh, *A Self-Organizing Neural Network Model of the Primary Visual Cortex*, Ph.D. thesis, Department of Computer Sciences, The University of Texas at Austin, Austin, TX (1995). Technical Report AI95-237.

7. H. B. Barlow, The twelfth Bartlett memorial lecture: The role of single neurons in the psychology of perception, *Quarterly Journal of Experimental Psychology*, 37A:121–145 (1985).

8. D. J. Field, What is the goal of sensory coding?, *Neural Computation*, 6:559–601 (1994).

9. D. Purves. *Body and brain: A trophic theory of neural connections*, Harvard University Press, Cambridge, MA (1988).

10. C. Blakemore and R. C. van Sluyters, Innate and environmental factors in the development of the kitten's visual cortex, *Journal of Physiology (London)*, 248:663–716 (1975).

11. E. R. Kandel, J. H. Schwartz, and T. M. Jessell. *Principles of Neural Science*, third edition, Elsevier, New York (1991).

12. C. J. Shatz, Emergence of order in visual system development, *Proceedings of the National Academy of Sciences, USA*, 93:602–608 (1996).

13. A. Kirkwood and M. F. Bear, Hebbian synapses in visual cortex, *Journal of Neuroscience*, 14(4):1634–1645 (1994).

14. M. Constantine-Paton, H. T. Cline, and E. Debski, Patterned activity, synaptic convergence, and the NMDA receptor in developing visual pathways, *Annual Review of Neuroscience*, 13:129–154 (1990).

15. L. Maffei and L. Galli-Resta, Correlation in the discharges of neighboring rat retinal ganglion cells during prenatal life, *Proceedings of the National Academy of Sciences, USA*, 87:2861–2864 (1990).

16. G. A. Marks, J. P. Shaffery, A. Oksenberg, S. G. Speciale, and H. P. Roffwarg, A functional role for REM sleep in brain maturation, *Behavioural Brain Research*, 69:1–11 (1995).

17. H. P. Roffwarg, J. N. Muzio, and W. C. Dement, Ontogenetic development of the human sleep-dream cycle, *Science*, 152:604–619 (1966).

18. C. J. Shatz, Impulse activity and the patterning of connections during CNS development, *Neuron*, 5:745–756 (1990).

19. M. Meister, R. O. L. Wong, D. A. Baylor, and C. J. Shatz, Synchronous bursts of action-potentials in the ganglion cells of the developing mammalian retina, *Science*, 252:939–943 (1991).

20. M. B. Feller, D. P. Wellis, D. Stellwagen, F. S. Werblin, and C. J. Shatz, Requirement for cholinergic synaptic transmission in the propagation of spontaneous retinal waves, *Science*, 272:1182–1187 (1996).

21. W. R. Lippe, Rhythmic spontaneous activity in the developing avian auditory system, *Journal of Neuroscience*, 14(3):1486–1495 (1994).

22. R. Mooney, A. A. Penn, R. Gallego, and C. J. Shatz, Thalamic relay of spontaneous retinal activity prior to vision, *Neuron*, 17:863–874 (1996).

23. J. A. Movshon and R. C. van Sluyters, Visual neural development, *Ann. Rev. Psych.*, 32:477–522 (1981).

24. G. Goodhill, Topography and ocular dominance: A model exploring positive correlations, *Biological Cybernetics*, 69:109–118 (1993).

25. C. von der Malsburg, Self-organization of orientation-sensitive cells in the striate cortex, *Kybernetik*, 15:85–100 (1973).

26. M. Jouvet, Paradoxical sleep and the nature-nurture controversy, in: *Adaptive Capabilities of the Nervous System*, P. S. McConnell, G. J. Boer, H. J. Romijn, N. E. van de Poll, and M. A. Corner, eds., volume 53 of *Progress in Brain Research*, 331–346, Elsevier, New York (1980).

27. M. Steriade, D. Paré, D. Bouhassira, M. Deschênes, and G. Oakson, Phasic activation of lateral geniculate and perigeniculate thalamic neurons during sleep with ponto-geniculo-occipital waves, *Journal of Neuroscience*, 9(7):2215–2229 (1989).

AN EMPIRICAL MODEL DESCRIBING THE DYNAMICS OF GRADED TRANSMISSION IN THE LOBSTER PYLORIC NETWORK

J. T. Birmingham, Y. Manor, F. Nadim, L. F. Abbott, E. Marder

Volen Center for Complex Systems
415 South Street
Brandeis University
Waltham, MA 02254

INTRODUCTION

Graded synaptic transmission plays an important role in both invertebrate and vertebrate systems. Models that describe graded synaptic transmission can be valuable tools to understand network dynamics. Few studies have previously attempted to develop such models (De Schutter et al., 1993). In this paper, we present a model for graded synaptic transmission that relies on simplifying (and experimentally justified) assumptions and requires few experimental measurements.

We focus on the synaptic connection from the Lateral Pyloric (LP) neuron to the Pyloric Dilator (PD) neurons in the stomatogastric ganglion of the spiny lobster *Panulirus interruptus*. This synapse, which has an important role in the frequency regulation of the pyloric rhythm, has a large graded component that shows depression. Previous work has shown that transmitter is released by the presynaptic cell at voltages above -50 mV (Manor et al., 1997). Because the membrane potential of LP oscillates well below and above this voltage during normal pyloric activity, the amount of transmitter release depends on the shape and frequency of the presynaptic waveform. Our model captures the essential dynamics of the synapse in response to realistic waveforms.

CONSTRUCTION OF THE MODEL

The model consists of two parts. The first part is the response to a single cycle of the presynaptic waveform. The second part addresses the response to multiple cycles and accounts for the depressing properties of the synapse observed in the second and later cycles.

Constructing the Response to a Single Cycle: the General Strategy

To predict the PD response to a single cycle of a realistic LP waveform, we decompose the waveform into a series of voltage steps, as shown in Fig. 1. Responses to these individual steps are used to construct the postsynaptic response to the waveform. For the rising phase, we assume that the postsynaptic response is equal to the sum of responses to individual voltage steps, where the response to each individual step is the sum of transient and sustained currents with an exponential rise. This assumption is justified for the LP to PD synapse by the observation that the amplitude of the postsynaptic response is a nearly linear function of the presynaptic potential (Fig. 2). We measure the PD responses to voltage steps in LP, for different baselines and amplitudes, and we derive an empirical formula that describes the response to an arbitrary voltage step.

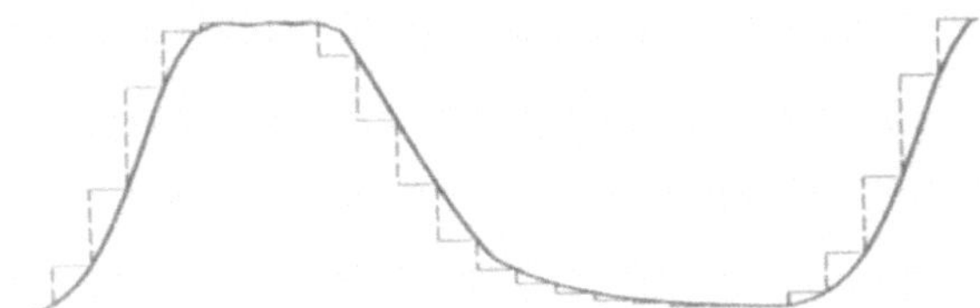

Figure 1. Schematic discretization of an LP waveform. Each cycle of the presynaptic waveform is sampled at t_i, $0 \leq i \leq n$, with Δt of 10 ms. The waveform is broken into a series of voltage steps from $V(t_i)$ to $V(t_{i+1})$.

During each hyperpolarizing step of the falling phase, the accumulated postsynaptic response decays with three different time constants in a manner that depends on the voltage step size. The two slower time scales describe the decays of the transient components that result from the depression of the synapse. The faster time scale describes the decays of both the persistent and transient components that result from the fall in the presynaptic LP potential.

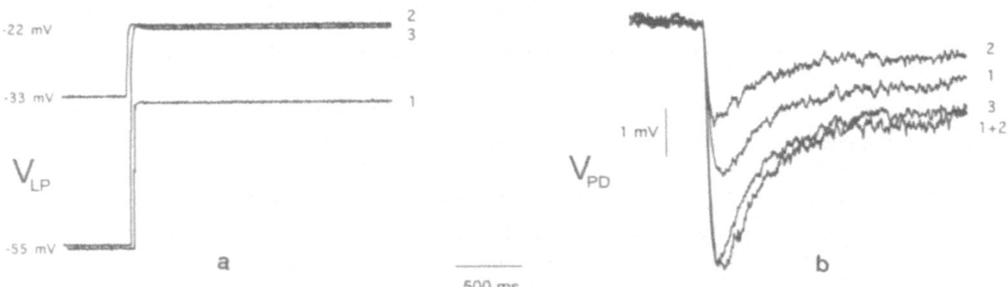

Figure 2. LP step depolarizations (**a**) and corresponding PD responses (**b**). The responses to voltage steps from -55 to -33 mV (trace 1) and -33 to -22 mV (trace 2) sum approximately (trace 1+2) to the response to a single voltage step from -55 to -22 mV (trace 3).

Constructing the Response to a Single Cycle: Determining Parameters

The PD response to a depolarizing step in LP has a persistent and a transient part. The persistent component is given by a single exponential rise (with time constant τ_{act} and amplitude A_{per}). The transient part consists of two distinct components, governed by a single exponential rise (with time constant τ_{act} and amplitude A_{trans}) and two exponential decays (with time constants $\tau_{decay,1}$ and $\tau_{decay,2}$). These time constants are indicated in Fig. 3.

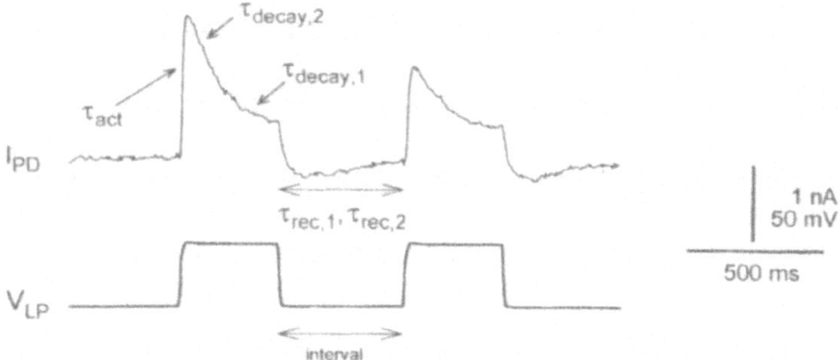

Figure 3. The PD synaptic response to a train of LP square pulses. See text for further description.

All model time constants are approximately independent of the LP membrane potential. We assume that the rising phase of the persistent and transient currents can be described by a single time constant τ_{act} that is obtained by fitting the rise to peak of the PD response to a voltage step in LP. The decay of the PD response during the prolonged depolarization could not be fit with a single exponential function. Two time constants were used to fit that decay: a relatively fast $\tau_{decay,2}$ (estimated to be 100 ms), and a slower $\tau_{decay,1}$. This latter time constant was measured experimentally by fitting the decay rates of the PD responses to LP square pulses of different amplitudes (Fig. 4a). These fits yield a range of values for $\tau_{decay,1}$ without any systematic dependence on the LP amplitude. We therefore approximate $\tau_{decay,1}$ by the mean of the measurements. We assumed that the decay of the PD response at the end of the pulse resulted from deactivation of the persistent and transient components, and we therefore modeled this decay with τ_{act}.

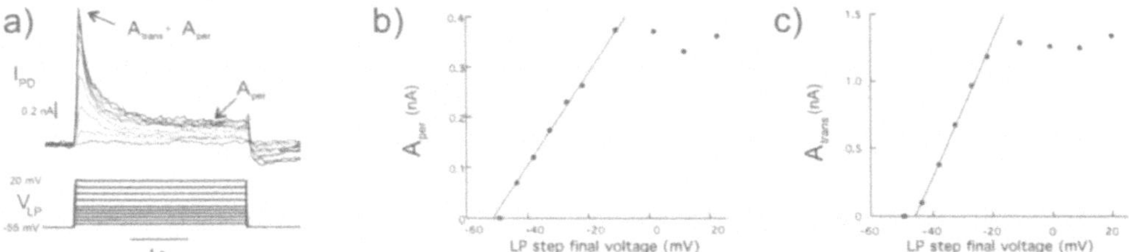

Figure 4. a) PD response to a series of LP square pulses. Exponential fits to the decaying portions of the PD responses yielded $\tau_{decay,1}$ values ranging between 200 and 300 ms. **b,c)** Amplitude of the PD persistent ($A_{per}(V)$) and transient ($A_{trans}(V)$) currents as function of the LP membrane potential. These amplitudes correspond to the traces shown in a). Solid lines are linear fits to values below -25 mV.

Constructing the Response to a Single Cycle: the Algorithm

The rising phase. During the rising phase, the LP waveform is approximated by a series of discrete positive voltage steps at times t_i, where $0 \le i \le n$ and for all i, $t_{i+1} - t_i = \Delta t = 10$ ms (Fig. 1). For each step from V_i to V_{i+1}, the PD response is:

$$I_{PD,step} = (I_{per,step} + I_{trans,step}(t))(1 - e^{-t/\tau_{act}})$$

where

$$I_{per,step} = (A_{per}(V_{i+1}) - A_{per}(V_i))$$

$$I_{trans,step}(t) = C_0(A_{trans}(V_{i+1}) - A_{trans}(V_i))(xe^{-t/\tau_{decay,1}} + (1-x)e^{-t/\tau_{decay,2}}) \qquad (1)$$

where x is the fraction of the transient response that decays with time constant $\tau_{decay,1}$. This parameter cannot be computed from the response to a single step, because the two transient components decay with time constants $(\tau_{decay,1}$ and $\tau_{decay,2})$ of comparable size. However, the two transient components recover with very different time constants, as explained below. We therefore postpone the calculation of x to the following sections. C_0 is a correction factor (typically between 1.2 and 1.6) that is used to optimize the fit. The need for the correction results from errors in the measurement of τ_{act} and estimation of $\tau_{decay,2}$.

The response to the total depolarization during the first cycle is the sum of the contributions from each depolarizing step:

$$I_{PD}(t) = I_{per}(t) + I_{trans}(t) = \sum_{t_i} I_{PD,step}(t - t_i)$$

The falling phase. During the falling phase, the LP waveform is approximated by a series of discrete negative voltage steps at times t_i. For each step, we assume that the PD response decays exponentially:

$$I_{per}(t_{i+1}) = I_{per}(t_i)e^{-\Delta t / \tau_{act}} \tag{2}$$

$$I_{trans}(t_{i+1}) = I_{trans}(t_i)(a_0 e^{-\Delta t / \tau_{act}} + a_1 e^{-\Delta t / \tau_{decay,1}} + a_2 e^{-\Delta t / \tau_{decay,2}}) \tag{3}$$

where a_0, a_1 and a_2 are weighting factors that sum to 1. The first decay term in Eq. (3) corresponds to the deactivation of the two transient components; the latter two terms correspond to the depression of these components. When ΔV_i is close to 0, it is reasonable to assume that a_0 is nearly 0. When the magnitude of ΔV_i is large, a_0 approaches 1. For an intermediate ΔV_i, a_0 is calculated from a simple linear interpolation:

$$a_0 = (V_{LP}(t_i) - V_{LP}(t_{i+1})) / (V_{LP}(t_i) - V_0)$$

where V_0 is the threshold for the transient response. a_1 and a_2 are determined by

$$a_1 = x(1 - a_0)$$
$$a_2 = (1 - x)(1 - a_0)$$

Constructing the Response to Multiple Cycles: the General Strategy

The PD response to a train of square pulses in LP shows depression in amplitude (Fig. 3). We assume that the response to the first pulse is not subject to depression. For subsequent pulses, the depression of each transient component is represented by a depression factor that describes how much of this component recovered during the preceding interval. We therefore construct the response to a train of pulses from the response to the first pulse (using the procedures described in the previous sections) and these depression factors. The depression factors are calculated from the decay time constants $(\tau_{decay,1}$ and $\tau_{decay,2})$ and corresponding recovery time constants $(\tau_{rec,1}$ and $\tau_{rec,2})$, as explained in the next section.

Constructing the Response to Multiple Cycles: Determining Parameters

We assume that the recovery from depression of both transient components begins when the LP membrane potential falls below some threshold V_{rec}. Because the transitions between the trough and peak of the LP waveforms are relatively sharp, the model is insensitive to the particular value of V_{rec}. We set V_{rec} to be equal to the threshold for the transient response (around -50 mV). To determine $\tau_{rec,1}$, $\tau_{rec,2}$ and x, we use the responses to trains of square pulses in LP, with various intervals between pulses (Fig. 5a,b). We measure the amplitude of the PD peak as a function of the interval, for the first (no depression), second (partial depression) and sixth (maximal depression) pulses in the train. Using the values of $\tau_{decay,1}$, $\tau_{decay,2}$, $A_{per}(V)$, and $A_{trans}(V)$, we fit the amplitudes of the peak responses (Fig. 5c) to obtain $\tau_{rec,1}$, $\tau_{rec,2}$ and x.

A single recovery time constant cannot describe the behaviors seen in Fig. 5. Because $\tau_{rec,2} \gg 1$ sec $\gg \tau_{rec,1}$, we use trains of pulses with inter-pulse intervals of several seconds (Fig. 5b) to measure $\tau_{rec,2}$ and x, independent of $\tau_{rec,1}$. During the long inter-pulse intervals the first component completely recovers. We obtain $\tau_{rec,1}$ using pulse trains with short inter-pulse intervals (Fig. 5a), during which the second component recovers very little between pulses. In our measurements the component that decays with the fast time constant ($\tau_{decay,2}$) recovers more slowly than the component that decays with the slow time constant ($\tau_{decay,1}$).

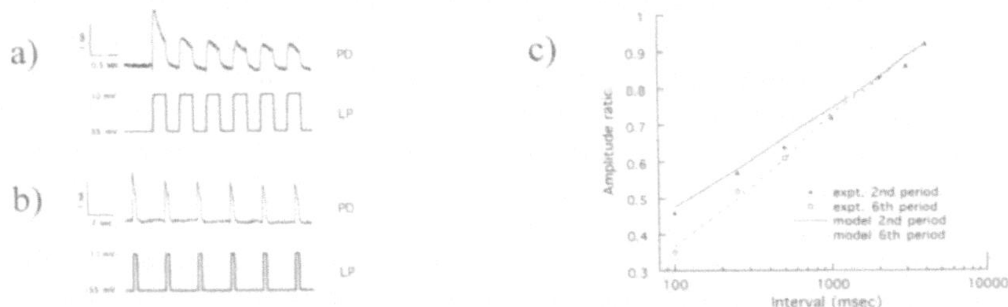

Figure 5. **a)** PD response to train of 250 ms LP pulses with 250 ms inter-pulse intervals. **b)** PD response to train of 250 ms LP pulses with 2000 ms inter-pulse intervals. **c)** Measured ratios (n^{th} cycle/1st cycle, n=2,6) of PD peak response and fit to a model of two transient currents with parameters $\tau_{decay,1}$= 250 ms, $\tau_{rec,1}$ = 300 ms, $\tau_{decay,2}$ = 100 ms, $\tau_{rec,2}$ = 2600 ms, and $x = 0.5$. See text for description of parameters.

Constructing the Response to Multiple Cycles: the Algorithm

Depression results from incomplete recovery of the two transient currents during the time that the LP waveform is below V_{rec}. At the beginning of the second cycle, depression factors for each of the transient currents are calculated by considering the time spent above and below threshold during the previous cycle:

$$d_{1,2} = e^{-T_{up}/\tau_{decay,1}} + (1 - e^{-T_{up}/\tau_{decay,1}})(1 - e^{-T_{down}/\tau_{rec,1}})$$

$$d_{2,2} = e^{-T_{up}/\tau_{decay,2}} + (1 - e^{-T_{up}/\tau_{decay,2}})(1 - e^{-T_{down}/\tau_{rec,2}})$$

where T_{up} and T_{down} are the time intervals during which V_{LP} is above and below V_{rec}, respectively. The depression factors $d_{1,n}$ and $d_{2,n}$ for the nth period are calculated using the following recursive procedure:

$$d_{1,n} = d_{1,n-1}e^{-T_{up}/\tau_{decay,1}} + (1 - d_{1,n-1}e^{-T_{up}/\tau_{decay,1}})(1 - e^{-T_{down}/\tau_{rec,1}})$$

$$d_{2,n} = d_{2,n-1}e^{-T_{up}/\tau_{decay,2}} + (1 - d_{2,n-1}e^{-T_{up}/\tau_{decay,2}})(1 - e^{-T_{down}/\tau_{rec,2}})$$

To compute the PD response for the nth cycle, we incorporate the depression factors into the calculation for the single cycle response (Eq. (1)), as follows:

$$I_{trans,step,n}(t) = C_0(A_{trans}(V_{i+1}) - A_{trans}(V_i))(d_{1,n}xe^{-t/\tau_{decay,1}} + d_{2,n}(1-x)e^{-t/\tau_{decay,2}})$$

COMPARISON BETWEEN THE MODEL AND THE BIOLOGICAL SYNAPSE

Fig. 6 compares PD responses predicted by the model with actual PD responses, when realistic LP waveforms with duty cycles 0.35 and 0.55 and periods of 0.5 and 2 sec are used. These results are from the same preparation as in Figs. 4-5. The model faithfully reproduces most of the postsynaptic (PD) response, including the synaptic depression. It predicts well the shape of the responses for two different waveforms when the cycle period is 2 sec (Fig. 6a,c). When the cycle period is 0.5 sec, the predicted response rises and falls more rapidly than the measured response (Fig. 6b,d). These discrepancies are due to errors in the measurement of τ_{act} and estimation of $\tau_{decay,1}$.

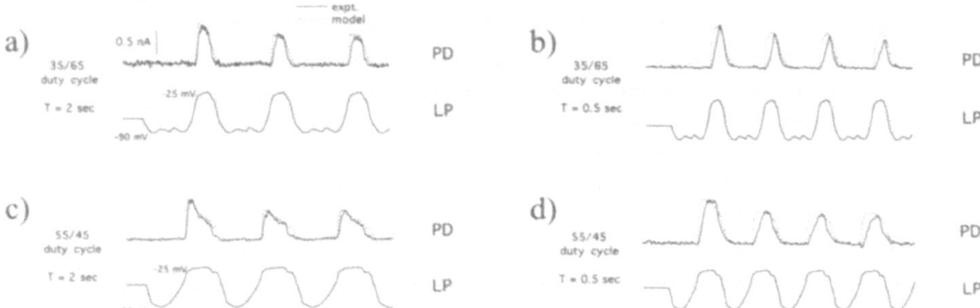

Figure 6. Actual (solid line) and predicted (dashed line) PD responses to realistic LP waveforms with **a)** Period = 2 sec, Duty cycle = 0.35, **b)** Period = 0.5 sec, Duty cycle = 0.35, **c)** Period = 2 sec, Duty cycle = 0.55, **d)** Period = 0.5 sec, Duty cycle = 0.55. Predicted responses computed using voltage steps with 10 ms duration.

Previously, we built a "conventional" biophysical model of the synapse with one activation and two decay variables (Nadim et al., 1997). This model was built by fitting the IPSPs in response to square pulses, but it underestimated the response to realistic waveforms, particularly when the cycle period was long. The model described in this paper drastically improves the predicted response to realistic waveforms by directly measuring all components for trains of square pulses, decomposing the realistic waveforms into steps, and constructing the postsynaptic response by using the linearity of the step response.

ACKNOWLEDGMENTS

This research was supported by NS17813, MH46742, and the Sloan Center for Theoretical Neurobiology at Brandeis University. We thank K. Sen for useful discussions.

REFERENCES

De Schutter, E., J. D. Angstadt and R. L. Calabrese, 1993, A model of graded synaptic transmission for use in dynamic network simulations. *J Neurophysiol.* **69**: 1225-1235.

Manor, Y., F. Nadim, L. F. Abbott and E. Marder, 1997, Temporal dynamics of graded synaptic transmission in the lobster stomatogastric ganglion. *J Neurosci.* **17**: 5610-5621.

Nadim, F., Y. Manor, L. F. Abbott and E. Marder, 1997, Strength and timing of graded synaptic transmission depend on frequency and shape of the presynaptic waveform., pp. 391-394 in *Computational Neuroscience: Trends in Research, 1997*, edited by J. Bower. Plenum Press, New York.

Novel Frequency Control in a Population of Bursting Neurons with Excitatory Synaptic Coupling

Robert J. Butera, Jr.,[1]* John Rinzel,[2,1] and Jeffrey C. Smith[3]

[1]Mathematical Research Branch, NIDDK, NIH, Bethesda, MD 20892
[2]Center for Neural Science and Courant Inst. for Mathematical Sciences,
New York University, New York, NY 10003
[3]Laboratory of Neural Control, NINDS, NIH, Bethesda, MD 20892

INTRODUCTION

Excitatory synaptic coupling is commonly described as a form of input that depolarizes a neuron, often leading to an increase in neuronal spike frequency. In this study we show that the effect of excitatory coupling on the oscillation frequency of coupled bursting neurons is context-dependent. In a model system consisting of a population of bursting neurons, we show that increasing extrinsic excitatory input increases the frequency of the bursting oscillation, while increasing the strength of the excitatory coupling among the bursting neurons decreases the frequency of the bursting oscillation.

This study is motivated by ongoing experimental investigations into the origin of respiratory rhythmogenesis in the mammalian central nervous system. Breathing movements in mammals arise from networks of neurons in the brain stem that produce a complex rhythmic pattern. The primary neuronal kernel for rhythm generation is located in the pre-Bötzinger complex (pre-BötC), a subregion of the ventrolateral medulla.[1] This rhythm is currently hypothesized to be generated by a dynamic interaction of cellular pacemaker and synaptic properties (the hybrid pacemaker-network model[2]): the pacemaker kernel lies within a population of neurons with intrinsic oscillatory bursting properties, where the cells are synaptically coupled by excitatory connections to synchronize the oscillation. This population is embedded in a network that provides synaptic inputs for the control of the oscillation frequency.

Pre-BötC neurons with bursting pacemaker-like properties have been identified *in vitro*.[1,3] Experimental evidence suggests that this population of neurons is coupled by excitatory amino acid (EAA) mediated synapses and receives EAA-mediated tonic drive[4] as one mechanism for frequency control. Quantitative data regarding the specific ionic currents of cells in the pre-BötC are only recently becoming available. We are developing models to complement experimental work in investigating possible mechanisms of rhythm generation.[2] In this present study we summarize the results of a

---

*To whom correspondence should be addressed. E-mail: butera@helix.nih.gov

Computational Neuroscience
edited by Bower, Plenum Press, New York, 1998

"

modeling investigation of the effects of excitatory synaptic drive and excitatory synaptic coupling on the burst dynamics of individual neurons and populations of bursting cells. The respiratory oscillator provides an interesting model system to investigate general principles underlying population behavior of coupled bursting neurons.

The Model

We begin by describing a minimal model of the bursting neurons within the pre-BötC that are believed to be responsible for rhythmogenesis. Experimental evidence suggests that a persistent Na^+ current (I_{NaP}) plays a significant role in bursting, e.g. bursting persists in the absence of extracellular Ca^{2+}.[3] We postulate that an inward cationic current such as I_{NaP} is responsible for burst initiation and consider two plausible models for burst termination: slow inactivation of I_{NaP} (Model I) or a slowly activating K^+ current (Model II). In both models the slow process is postulated to be voltage-dependent.

Model I consists of the following ionic currents: I_{NaP} with instantaneous voltage-dependent activation and slow voltage-dependent inactivation, based on the data of French et al.[5] and Fleidervish et al.[6, 7]; a K^+-dominated leakage current with a reversal potential E_L(nominally -65 mV); and fast Na^+ and delayed-rectifier K^+ currents based on a reduced two-variable Hodgkin-Huxley mechanism to produce action potentials. Model II consists of the same currents and *identical* parameters as Model I, with two modifications: the persistent Na^+ current does not inactivate and there is an additional K^+ current (I_{KS}) with slow voltage-dependent activation. In both cases the resulting model consists of only three state variables. Complete equations describing the dynamics of both models are presented elsewhere.[8] The dynamics of both models may be qualitatively described as "square-wave" (type 1) bursting,[9, 10] similar to many models of electrical bursting oscillations in pancreatic β-cells. To our knowledge this is one of the first examples of square-wave bursting in a neuronal preparation.

Excitatory input is mediated by non-NMDA glutamatergic synapses with a reversal potential of 0 mV.[11] I_{tonic} represents synaptic excitation from tonically spiking (beating) neurons and is modeled as

$$I_{tonic} = g_{tonic}(V - E_{syn}), \tag{1}$$

where g_{tonic} represents a mean level of synaptic input from a population of beating neurons. In our models of populations of neurons, each cell also has a synaptic current (I_{syn}) corresponding to input from other bursting neurons, modeled as

$$I_{syn} = \bar{g}_{syn}s(V - E_{syn}), \tag{2}$$

where the gating variable s is driven by presynaptic depolarization according to

$$\dot{s} = ((1 - s)s_\infty(V_{pre}) - k_r s)/\tau_s. \tag{3}$$

This formulation is based on the assumption that the binding of transmitter is rapid compared to the rate of the conformational change of the channel[12] and is similar to the formulation described in Wang and Rinzel.[13] The postsynaptic responses ($\tau_s = 5$ msec, k_r=1) are fast due to non-NMDA glutamatergic receptor activation. The function $s_\infty(V_{pre})$ is a steep sigmoidal function whose half-activation value is set such that it is only activated by the firing of a presynaptic action potential.

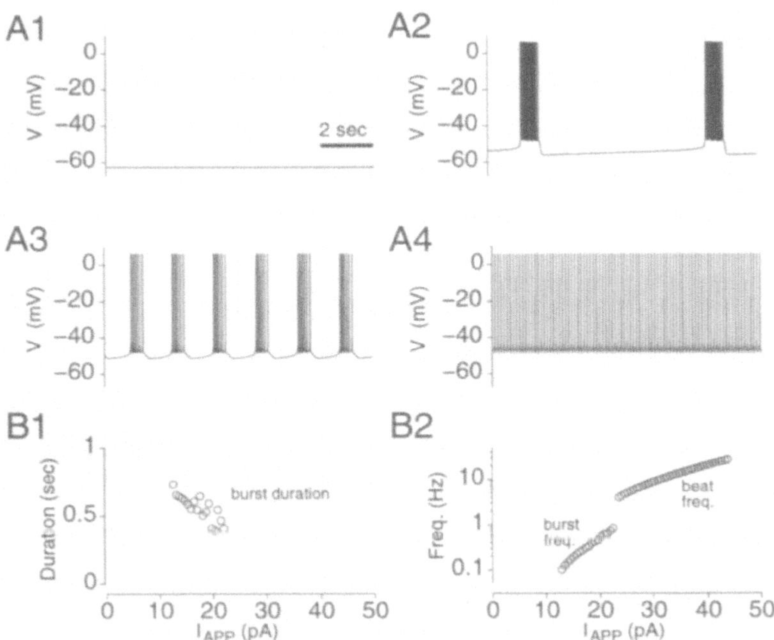

Figure 1. Voltage-dependent bursting properties of Model I. **A1-A4**: Modes of activity of Model I for $I_{app} = 0$, 14, 21, and 30.8 pA, respectively (A1: quiescence; A2,A3: oscillatory bursting; A4: beating). **B1-B2**: Burst duration and frequencies of both bursting and beating as I_{app} is varied. Similar results are obtained if g_{tonic} is varied instead of I_{app}.

RESULTS

Single Cell Simulations

Oscillatory bursting pre-BötC neurons possess several distinct features: voltage-dependent behavior with quiescent, bursting, and beating regimes as the baseline membrane potential of the cell is depolarized by an applied current (I_{app}); a decaying spike frequency throughout the burst; burst durations of approximately 0.5-1 second; burst frequencies that vary over an order of magnitude ($\sim$0.1 to 1 Hz) as I_{app} is increased; and a relatively flat membrane potential trajectory during the interburst interval with a depolarizing shift of the baseline potential as I_{app} is increased. As illustrated in Fig. 1, all of the properties of oscillatory bursting pre-BötC neurons just described are present in Model I. These voltage-dependent model dynamics persist across a wide range of choices of model parameters. Similar voltage-dependent control of both the modes of activity and frequency of bursting and beating is achieved if E_L or g_{tonic} are varied, representing intrinsic and synaptic mechanisms for voltage-dependent control of burst frequency.

The dynamics of Model II (not shown) satisfy most of the above criteria, with two exceptions: the membrane potential trajectory during the interburst interval is not as flat, and the burst duration does not decrease with depolarization. These dynamics persist across a wide range of parameter choices, varying not only maximal conductances but parameters associated with the activation kinetics of both I_{NaP} and I_{KS}. We have identified parameter sets where Model II has a relatively flat membrane

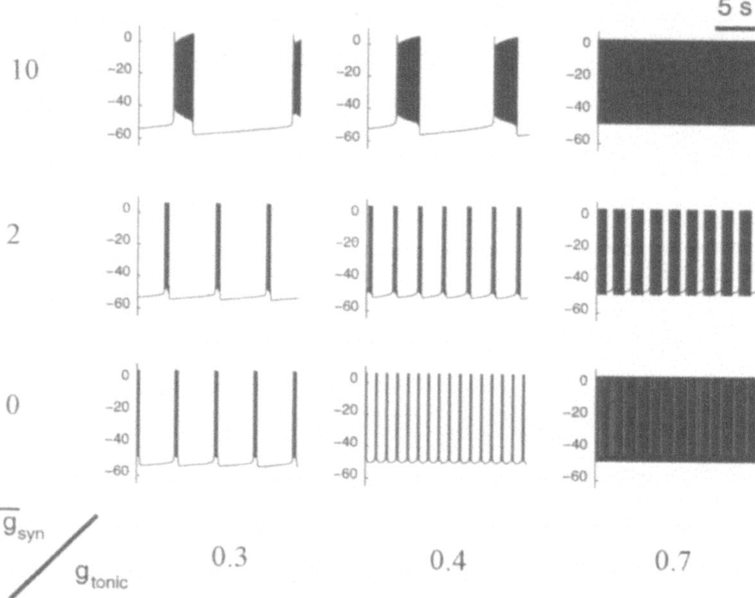

Figure 2. Effects of excitatory coupling on the burst dynamics of pairs of modeled bursting cells. Panels illustrate dynamics of a single member of the pair as the level of tonic drive (g_{tonic}) and synaptic coupling ($\bar{g}_{syn}$) are varied. Values of $\bar{g}_{syn}$ and g_{tonic} are indicated in bold. During bursting activity the other cell is bursting in synchrony. Individual action potentials cannot be visually resolved at this time-scale. During bursting, action potentials of the two cells are nearly in-phase at the beginning of a burst and drift to anti-phase firing by the end of the burst. During beating the action potentials are firing anti-phase.

potential trajectory in the interburst interval and the burst duration decreases as the baseline potential is depolarized. However, for these effects to occur, we must make the assumption that E_K is approximately -55 mV.

Although neither model has been refuted experimentally, we will consider only Model I for the remainder of this paper, which is concerned with the dynamics of populations of coupled bursting cells.

Two Cell Simulations

To model the effects of excitatory coupling on the dynamics of bursting, we consider the simple case of a pair of neurons. Each neuron of the pair uses the ionic current equations and parameters of Model I. These neurons are reciprocally coupled by an excitatory synaptic current (I_{syn}). Each neuron in the pair also receives a mean level of tonic synaptic drive (I_{tonic}). As described previously, this current has a similar depolarizing effect on the burst dynamics of single neurons as I_{app}. We studied the dynamics of this pair of bursters as $\bar{g}_{syn}$ and g_{tonic} were varied in order to assess the control of bursting frequency by tonic drive (g_{tonic}) and synaptic coupling ($\bar{g}_{syn}$). In all cases g_{tonic} was identical for each cell and $\bar{g}_{syn}$ was identical for each cell. These results are summarized in Fig. 2.

Both g_{tonic} and $\bar{g}_{syn}$ are excitatory mechanisms. However, increasing g_{tonic} at a given value of $\bar{g}_{syn}$ *increases* the frequency of bursting, while increasing $\bar{g}_{syn}$ at a given value of g_{tonic} *decreases* the frequency of bursting. Increasing $\bar{g}_{syn}$ also increases the frequency of firing of action potentials during a burst.

Increasing $\bar{g}_{syn}$ increases burst duration and decreases burst frequency. These effects are not independent: increasing excitatory coupling, which is triggered by the firing of action potentials, causes action potentials in each cell to further excite the other cell, with a net increase in intensity of firing in both cells. The higher spike frequency increases excitatory synaptic input during the active phase of a burst. In order for the burst to terminate, I_{NaP} must be inactivated further than it would in the absence of synaptic input. Thus the burst duration is extended. The additional inactivation of I_{NaP} prolongs recovery from inactivation as well, increasing the duration of the silent phase of the burst cycle. This effect of excitatory coupling on burst duration and burst frequency has been previously described more generally using relaxation oscillators.[14]

Another effect of moderate levels of synaptic coupling (e.g. $\bar{g}_{syn}=2$ nS in Fig. 2) is that it prolongs the range of values of g_{tonic} over which bursting is supported. Thus it is possible that excitatory coupling makes bursting more robust: bursting may occur over a wider range of single-cell parameters than occurs with individual cells in isolation. Similar coupling effects on the parameter range where oscillatory bursting is supported have been reported in models of gap-junction coupled β-cells.[15]

We are presently extending this analysis to examine the dynamics of large populations of bursting neurons with excitatory synaptic coupling. Specifically, we are interested in determining how heterogeneity in both intrinsic parameters (ionic current properties) and synaptic parameters (coupling strength and connectivity schemes) affects the generation of a synchronous bursting oscillation within the population. We have found that heterogeneity in a population of bursting neurons is sufficient to account for the variety of spike-frequency histograms of bursting cells found in the pre-BötC when recording from brain stem slice preparations.[3] Preliminary results have revealed that even with significant heterogeneity among intrinsic parameters such that less than 20% of the neurons exhibit oscillatory bursting in isolation, weak excitatory coupling will transform a majority of neurons in the network to a mode of synchronous bursting. Furthermore, increasing g_{tonic} increases the frequency of the synchronous bursting oscillation, whereas increasing $\bar{g}_{syn}$, while enhancing synchronization of bursting, reduces the population oscillation frequency. Thus the effect of tonic drive and synaptic coupling on the burst frequency of a large population of neurons is similar to that reported here for the simple case of two cells.

SUMMARY

The mammalian respiratory oscillator provides a model system to analyze how interactions of intrinsic burst-generating mechanisms and synaptic properties govern the population dynamics of coupled bursting neurons. We have shown here that two forms of synaptic excitation have opposing effects on the oscillation frequency in a synaptically coupled population of square-wave type bursting neurons. A mean level of excitation via g_{tonic} that is present throughout the burst cycle depolarizes the cell and *increases* burst frequency. Phasic excitation during the active phase of a burst via $\bar{g}_{syn}$ increases the firing frequency of action potentials, but *decreases* the burst frequency. The effect of $\bar{g}_{syn}$ on burst frequency is due to the interaction of the phasic synaptic current via an increase in spike frequency with the intrinsic current responsible for burst termination. Current experiments are directed toward testing these proposed roles of ionic current and synaptic mechanisms in the control of cellular and network oscillations.

REFERENCES

1. J. C. Smith, H. H. Ellenberger, K. Ballanyi, D. W. Richter, and J. L. Feldman. Pre-Bötzinger Complex: A brainstem region that may generate respiratory rhythm in mammals. *Science*, 254:726–729, 1991.

2. J. C. Smith. Integration of cellular and network mechanisms in mammalian oscillatory motor circuits: Insights from the respiratory oscillator. In P. Stein, S. Grillner, A. I. Selverston, and D. G. Stuart, editors, *Neurons, Networks, and Motor Behavior*, pages 97–104. MIT Press, Cambridge, MA, 1997.

3. S. M. Johnson, J. C. Smith, G. D. Funk, and J. L. Feldman. Pacemaker behavior of respiratory neurons in medullary slices from neonatal rat. *J. Neurophysiol.*, 72:2598–2608, 1994.

4. G.D. Funk, J. C. Smith, and J. L. Feldman. Generation and transmission of respiratory oscillations in medullary slices: Role of excitatory amino acids. *J. Neurophysiol.*, 70:1497–1515, 1993.

5. C. R. French, P. Sah, K. Buckett, and P. W. Gage. A voltage-dependent persistent sodium current in mammalian hippocampal neurons. *J. Gen. Physiol.*, 95:1139–1157, 1990.

6. I. A. Fleidervish, A. Friedman, and M. J. Gutnick. Slow inactivation of Na^+ current and slow cumulative spike adaption in mouse and guinea-pig neocortical neurones in slices. *J. Physiol. (London)*, 493:83–97, 1996.

7. I. A. Fleidervish and M. J. Gutnick. Kinetics of slow inactivation of persistent sodium current in layer V neurons of mouse neocortical slices. *J. Neurophysiol.*, 76:2125–2129, 1996.

8. R. J. Butera, J. Rinzel, and J. C. Smith. Models of the neuronal kernel for respiratory rhythm generation in the pre-Bötzinger complex of mammals. I. Bursting pacemaker neurons. In Preparation.

9. R. Bertram, M. J. Butte, T. Kiemel, and A. Sherman. Topological and phenomenological classification of bursting oscillations. *Bull. Math. Biol.*, 57:413–439, 1995.

10. J. Rinzel. A formal classification of bursting mechanisms in excitable systems. In E. Teramoto and M. Yamaguti, editors, *Mathematical Topics in Population Biology, Morphogenesis, and Neurosciences*, volume 71 of *Lecture Notes in Biomathematics*, pages 267–281. Springer-Verlag, Berlin, 1987.

11. J. C. Smith, K. Ballanyi, and D. W. Richter. Whole-cell patch-clamp recordings from respiratory neurons in neonatal rat brainstem in vitro. *Neurosci. Let.*, 134:153–156, 1992.

12. K. L. Magleby and C. F. Stevens. A quantitative description of end-plate currents. *J. Physiol. (London)*, 223:173–197, 1972.

13. X.-J. Wang and J. Rinzel. Alternating and synchronous rhythms in reciprocally inhibitory model neurons. *Neural Comp.*, 4:84–97, 1992.

14. D. Somers and N. Kopell. Rapid synchronization through fast threshold modulation. *Biol. Cybern.*, 68:393–407, 1993.

15. P. Smolen, J. Rinzel, and A. Sherman. Why pancreatic islets burst but single β cells do not: The heterogeneity hypothesis. *Biophys. J.*, 64:1668–1680, 1993.

A TWO-LAYER MODEL DESCRIBES THE SPATIOTEMPORAL PROPERTIES OF SPONTANEOUS RETINAL WAVES

Daniel A. Butts,[1] Marla B. Feller,[2] Holly L. Aaron,[2] Carla J. Shatz,[2] and Daniel S. Rokhsar[1]

[1]Physical Biosciences Division, Lawrence Berkeley National Laboratory and Department of Physics, University of California, Berkeley, CA 94720
[2]Howard Hughes Medical Institute and Department of Molecular and Cell Biology. University of California, Berkeley, CA 94720

SUMMARY

In the developing mammalian retina, spontaneous waves of action potentials are present in the ganglion cell layer weeks before vision. Fluorescence imaging reveals that this highly correlated activity spans spatially restricted domains, which cover the entire retina with shifting mosaic patterns. Here we present a novel dynamical model consisting of two coupled populations of cells that quantitatively reproduces the experimentally observed domain sizes, interwave intervals, and wavefront velocity profiles. Model and experiment together show that the highly correlated activity generated by retinal waves can be explained by a combination of random spontaneous activation of cells and the past history of local retinal activity. These results were originally published in *Neuron* 19: 293-306 (1997).[1]

INTRODUCTION

Spontaneous activity generated in immature circuits is present in various regions of the central nervous system, and participates in the development and differentiation of synaptic circuitry.[2] Long before the onset of photoreceptor function, mammalian retinal ganglion cells exhibit spontaneous, periodic bursts of action potentials that propagate across the retina in a wavelike manner, such that highly correlated firing is induced among neighboring ganglion cells.[3,4,5] This activity can be monitored over large areas of the retina through the use of multi-electrode arrays[3,6] or fluorescence imaging of calcium indicators.[4,5] These studies reveal that this patterned, correlated firing does not spread freely across the retina, but is instead confined to spatially restricted domains, which tile the entire retina over time.[5]

Since the boundaries of these domains are different for each wave, discrete functional units of cells evidently cannot account for the spatial properties of the waves.[5] The generation of this spontaneous activity requires the activation of a horizontal synaptic circuit, thought to involve both amacrine cells and ganglion cells.[4,5] At the earliest times when waves are present, there are no evident synapses from bipolar to amacrine or to ganglion cells, though there are likely to be functional synapses between amacrine cells or from amacrine cells to ganglion cells.[7] Further evidence implicates cholinergic synaptic transmission in the initiation and propagation of waves, implying that starburst amacrine cells are directly involved.[5]

Although there is strong evidence that a network of amacrine and retinal ganglion cells is involved in producing retinal waves, exactly how the two cells cooperate to generate the complex spatiotemporal dynamics is unknown. In Feller *et al.*,[1] we use the observed spatiotemporal features of these domains to infer the nature of the underlying circuitry in the early retina. We present a model of the developing retina that reproduces the spatiotemporal dynamics observed in experiment, and offers a simple understanding of the apparent complexity of retinal waves.

RESULTS

Retinal waves initiate from local areas of tissue and propagate over spatially restricted regions, which we have called domains.[5] To examine the spatial configuration of domains,

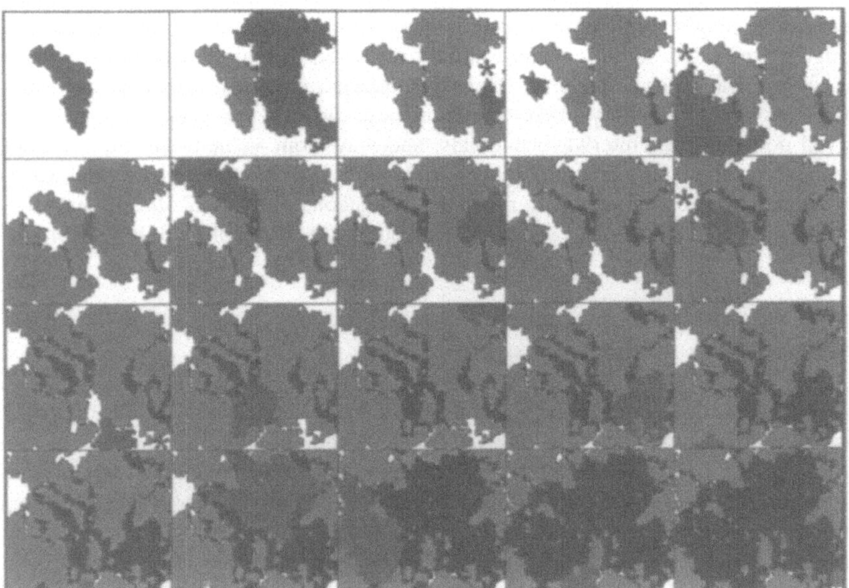

Figure 1: Spontaneous waves measured with fluorescence imaging in a P2 ferret retina loaded with fura-2AM. A mosaic of domains is created by sequential spontaneous waves. The dark gray domain in each frame corresponds to a new wave arising in the same region of retina. Overlapping regions in which more than one wave occurs are shown in black. The entire sequence corresponds to 2 minutes of recording (from left to right, top row first). Asterisks indicate regions where waves terminate despite lack of recent activity. The total field of view is 1.4 mm × 1.2 mm. Reprinted from Feller *et al.*[1] with the permission of Cell Press.

consider a sequence of waves that was imaged over a period of two minutes from a P2 ferret retina (Figure 1). As previously mentioned, the boundaries of each domain (outlined areas) shift over time, such that the entire retinal ganglion cell layer is eventually tiled by these domains (bottom right frame). Over periods of time less than one minute (first two rows), there is very little overlap (shown in black) between new domains (shown in darker gray) and the domains of earlier waves (shown in lighter gray). Over longer times, new waves encroach significantly into domains defined by earlier waves (bottom two rows). This implies that the retina has a refractory period which restricts wave propagation.[5] We define the refractory period as the minimum time interval following the activation of an area of retina during which it cannot participate in a wave. Similar analysis of 11 retinas shows that this refractory period ranges from 40 to 60 seconds (data not shown, see also Feller *et al.*[5]). Waves in the ganglion cell layer can also occasionally terminate in regions *without* detectable recent activity (see asterisks); these boundaries do not appear to be set by the refractory period. The presence of shifting domain boundaries suggests that there are no fixed structures within the retina that limit wave propagation. This point has been more rigorously demonstrated through statistical analyses of the locations of initiation sites and wave boundaries.[1]

A careful characterization of the spatiotemporal properties of the waves demonstrates that the retina is homogeneous with respect to wave-initiation sites, domain boundaries, and observed variations in wavefront velocity. The behavior of the retinal waves can be further quantified through the use of two statistical measures characterizing their spatiotemporal properties (Figure 2). First, we used the time interval between successive waves traversing a given retinal location to study the temporal regularity of the activity. The distribution of the experimental interwave intervals is broad (Figure 2A, top), showing that while there is a characteristic time interval between waves, they are not strictly periodic. The lower bound of the interwave interval distribution is determined by the minimum refractory period, yet the interwave interval distribution peaks near 2 min, which

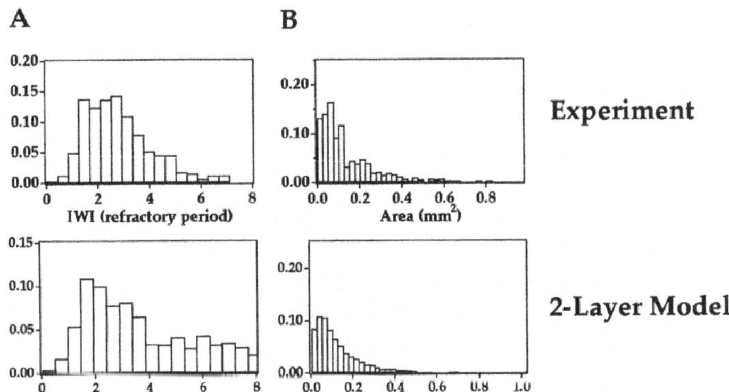

Figure 2: Interwave interval and domain area distributions of experiment and model. (A) shows the interwave interval distributions, and (B) shows the domain size distributions. The experimental results presented are cumulative from 5 retinas (results from each retina are shown separately in Feller *et al.*[1]) Interwave interval distributions for the simulated waves were measured at six equally-spaced points within the model retina. Spatial dimensions for the model domain size distributions were determined by allowing 250 μm^2 for each cell. All distributions are expressed as a fraction of the total number of waves. Results from 4876 simulated waves were included. Reprinted from Feller *et al.*[1] with the permission of Cell Press.

is two to three times the measured refractory period. The ratio of the interwave interval peak to the refractory period is a critical parameter and severely restricts the possible circuitry responsible for wave generation.

The second statistical property we characterized is the domain size distribution. We determined the domain size distributions by measuring the area of the retina covered by each retinal wave over the course of an hour for six retinas (Figure 2B, top). Like the interwave interval distributions, the area distributions are extremely broad, showing that a given retina supports a wide range of domain sizes. A distinctive property of the waves that must be taken into account by any model of this phenomenon is the compact (*i.e.* not sparse) nature of the domains.

Can the observed complexity of retinal wave patterns – namely, the periodicity, domain size, compactness, and wavefront velocity – be reproduced by a simple neuronal network model that is consistent with the known anatomy and physiology of the retina? Here we introduce a two-layer readout model (Figure 3), comprised of two distinct populations of cells having properties that correspond to the ganglion and amacrine cells known to be involved in the neuronal circuit that generates waves.[4,5]

Since our goal is to describe the system at time scales related to the waves (seconds) rather than the underlying cellular events (milliseconds), the model is correspondingly coarse, and treats each cell within the network as a simple unit whose excitation level (analogous to depolarization away from resting membrane potential) is increased by excitatory inputs and decays exponentially with time. (For a model using a different approach – based on biophysical properties of ganglion cells – see Burgi & Grzywacz[8]). Horizontal coupling in the network is determined by connections between amacrine cells, and between amacrine and ganglion cells, as illustrated schematically in Figure 3. As much as possible, the characteristic lengths and times of the model have been chosen to correspond to available anatomical and physiological data (see Feller *et al.*[1], for details).

Both model cell types have a characteristic threshold. A model amacrine cell can reach its threshold in two ways: either through an intrinsic spontaneous depolarization or through the cumulative effect of excitatory inputs from other nearby amacrine cells.

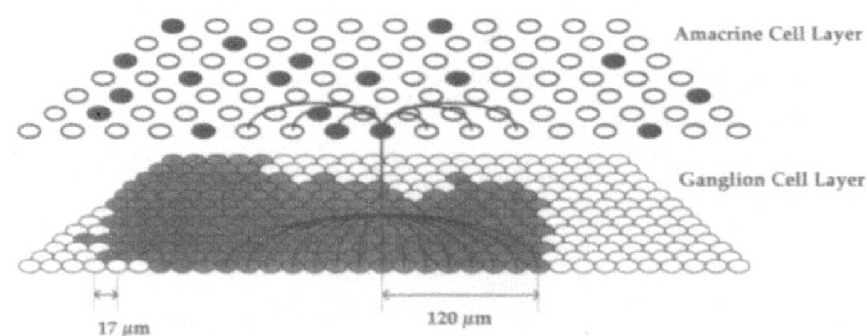

Figure 3: Model Schematic. The model consists of two cell types: amacrine cells and ganglion cells. The top layer of cells corresponds to amacrine cells and the bottom layer to ganglion cells. Connections from one amacrine are shown (black lines). Note that each retinal ganglion cell and each amacrine cell receives inputs from many other amacrine cells. Activity in the amacrine cell layer (gray ovals, top layer) can be both spontaneous or stimulated. When a sufficiently high density of amacrine cells becomes active, ganglion cells in that region will be stimulated above threshold to create a compact domain (gray ovals, bottom layer). Reprinted from Feller *et al.*[1] with the permission of Cell Press.

Afterwards, it enters a refractory period during which it cannot be depolarized. Only when an amacrine cell exceeds its threshold can it cause other nearby amacrine cells and ganglion cells to depolarize via synaptic coupling. A model ganglion cell receives inputs only from amacrine cells, and has neither a refractory period nor the ability to depolarize spontaneously. The ganglion cell layer serves as a passive filter of the amacrine cell layer activity (that is, there is no coupling between ganglion cells), and we therefore refer to the ganglion cell layer as a "readout" layer. We also assume that a ganglion cell participates in a wave only when its excitation level exceeds a threshold. In our model, subthreshold amacrine and ganglion cell activity is not detectable.

The model and the experiment are qualitatively similar in that the properties of the simulated waves – their shape, frequency, rate of movement across the retina, *etc.* – are essentially indistinguishable from those observed experimentally. In addition, the statistical distributions of simulated waves (*i.e.*, interwave interval and domain area distributions) have the same features as those of the observed retinal waves (Figure 2, bottom row).

DISCUSSION

We have presented a coarse-grained model of the developing retina that shows that a simple network involving only two cell types can robustly reproduce the spatiotemporal patterns of retinal waves. The predictions of the model, in combination with our experimental observations, allow us to describe how the physiological properties of individual neurons can cooperatively generate intrinsic large-scale patterns of correlated activity.

The model predicts that wave initiation and propagation are strongly influenced by the past history of amacrine cell activity, both associated with and independent of waves. In the context of our model, this past history is represented by the density of *non-refractory* amacrine cells in a region. This constantly-changing density directly influences the wave dynamics in the retina. First, waves are more likely to initiate in regions with a high density of non-refractory amacrine cells, since they contain a larger pool of available amacrine cells that could spontaneously depolarize. Second, wave boundaries tend to occur in regions with a low density of non-refractory amacrine cells, since a wave passing through such a region would not be able recruit enough cells to stimulate ganglion cells or other amacrine cells above threshold. Finally, this density of non-refractory amacrine cells is dynamically modified by the waves themselves: waves tend to occur in regions with a high density of non-refractory amacrine cells, but a passing wave causes these cells to become refractory, leaving behind a low density region.

This analysis suggests a possible origin of the refractory period measured in the ganglion cell layer. Recall that in our model, the ganglion cells do not have an intrinsic refractory period. Rather, the intrinsic refractory period of the amacrine cells, combined with a ganglion cell threshold for firing, confers an effective refractory period to the visualized tissue. In the model, the refractory period measured in the ganglion cell layer is therefore a collective property of the amacrine-ganglion network.

The two-layer model makes few assumptions about the biophysical properties of the cells involved, and yet is capable of generating the spatiotemporal properties of real waves. While this tends to rule out simpler classes of models, we cannot rule out models based on more complicated computational units. Nevertheless, any successful model must be able to account for (a) the source of spontaneous activity, (b) the propagation of this activity in a compact, finite wave, and (c) the discrepancy between the peak of the interwave interval distribution and the refractory period. Our model accounts for these properties by making the following assumptions, which are consistent with the known properties of the immature retina. First, we include spontaneous activity in the amacrine cell layer, since some source of initiation must be present in the network, and ganglion cells are known not to be spontaneously active. Second, activity in the amacrine cell layer is subject to a refractory period, but can also propagate from cell to cell over the known dendritic spread of these cells. This accounts for the minimal size of the waves and the minimal interval between waves at a given location. Third, the ganglion cell layer acts as a passive filter of amacrine cell activity. Filtering the amacrine cell activity is necessary to simultaneously account for the compactness of the waves and the discrepancy between measured interwave intervals and the refractory period.

Though the exact nature of the cellular mechanisms underlying the model is not known, the model is able to reproduce the observed wave properties over a range of parameter choices (see Feller *et al.*[1]). This point is significant on two levels. First, it shows that the reproduction of experimental properties is a general feature of the architecture of the two-layer model and not a unique property of the parameters that we have used. There is also a second and more profound implication: if a simple model neural circuit can generate wave behavior without requiring fine adjustment of parameters, then we may expect that the wave behavior in the real circuit is not sensitive to the precise biophysical properties of neuronal networks. The approach of using crude, early-forming circuits to generate precise patterns of activity may be a general strategy used by a variety of circuits throughout the nervous system to refine imprecise wiring within the circuits themselves and their targets.

ACKNOWLEDGEMENTS

This work was supported by National Science Foundation grant IBN9319539 and National Institutes of Health grant MH 48108 to C.J.S., a Miller Postdoctoral Fellowship to M.B.F., and an ARCS Fellowship award to D.A.B.   D.S.R. and D.A.B. were supported by Lawrence Berkeley National Laboratory grant LDRD-3669-57 and National Science Foundation grant DMR-91-57414 to D.S.R. and acknowledge the use of the Cray T3E at the National Energy Research Scientific Computing Center. C.J.S. is an Investigator of the Howard Hughes Medical Institute.

REFERENCES

1. M.B. Feller, D.A. Butts, H.L. Aaron, D.S. Rokhsar, and C.J. Shatz, Dynamic processes shape spatiotemporal properties of retinal waves, *Neuron* 19: 293-306 (1997).
2. R. Yuste, Spontaneous activity in the developing central nervous system, *Sem. in Cell & Devel. Bio.* 8: 1-4 (1997).
3. M. Meister, R.O. Wong, D.A. Baylor, and C.J. Shatz, Synchronous bursts of action potentials in ganglion cells of the developing mammalian retina, *Science* 252: 939-43 (1991).
4. R.O. Wong, A. Chernjavsky, S.J. Smith, and C.J. Shatz, Early functional neural networks in the developing retina, *Nature* 374: 716-8 (1995).
5. M.B. Feller, D.P. Wellis, D. Stellwagen, F.S. Werblin, and C.J. Shatz, Requirement for cholinergic synaptic transmission in the propagation of spontaneous retinal waves, *Science* 272: 1182-7 (1996).
6. R.O. Wong, M. Meister, and C.J. Shatz, Transient period of correlated bursting activity during development of the mammalian retina, *Neuron* 11: 923-38 (1993).
7. J.B. Hutchins, J.M. Bernanke, and V.E. Jefferson, Acetylcholinesterase in the developing ferret retina, *Exp Eye Res* 60: 113-25 (1995).
8. P.Y. Burgi and N.M. Grzywacz, Model for the pharmacological basis of spontaneous synchronous activity in developing retinas, *J. Neurosci.* 14: 7426-39 (1994).

THE INHIBITORY CONTROL OF PYRAMIDAL CELL DISCHARGE IN A NEURAL NETWORK SIMULATION OF A LOCAL CIRCUIT IN HIPPOCAMPUS AREA CA1.

Allan Coop[1] and Stephen Redman[2]

[1]The Neurobiology Laboratory, Department of Physiology,
The University of Sydney, Sydney, New South Wales 2006, Australia.
Email: allanc@physiol.usyd.edu.au.
[2]The Neurophysiology Group, Division of Neuroscience,
The John Curtin School of Medical Research, Canberra 0200, Australia.
Email: Steve.Redman@anu.edu.au.

INTRODUCTION

To reduce either the effect of the lumping of parameters on the behaviour of a canonical[1] circuit model, or the implementational problems associated with large scale and detail[2, 3], an alternative approach to the simulation of neural circuits at the level of the local, or "unit" circuit is proposed. This circuit is an abstraction assumed to represent the smallest possible functional network of cells within an area of interest. It is defined as the circuit which contains monosynaptically connected neurones such that the least frequent cell type is represented at least once, and the number of cells of each other type are represented in integer proportions.

We define a unit circuit that represents a putative local circuit in hippocampus area CA1 and employ it to explore the control of pyramidal cell discharge by both fast ($GABA_A$ mediated) and slow ($GABA_B$ mediated) inhibitory postsynaptic potentials (fIPSPs and sIPSPs respectively).

THE UNIT CIRCUIT MODEL

The circuit was composed of one excitatory and three inhibitory cell types[4]. The excitatory cells were pyramidal (PY) neurones. Two types of inhibitory interneurone generated $GABA_A$ mediated inhibition – the *stratum oriens-alveus* (OA) and basket (BA) cells, whereas $GABA_B$ mediated inhibition was generated by the *stratum lacunosum-moleculare* (LM) cell. The ratio of excitatory to inhibitory cells was set to $1:9^{(5)}$, and in the absence of emperical data, the simplest ratio of $1:1:1$ was assumed for the inhibitory cell types. The unit circuit contained 30 cells; 27 PY cells and one OA, BA, and LM cell. It was assumed that each afferent fibre made a single synaptic contact with each cell in the circuit. When fewer than the total number of afferent fibres converged at an interneurone, synaptic contacts were choosen at random. Excitatory synapses between PY cells were not included because of their relatively low probability of occurrence[6]. Except for the absence of these synapses, and excitatory synapses between PY and the LM cell, the circuit was fully connected.

Cells were represented as single somatic compartments by an appropriate equivalent circuit with synaptic conductances modelled as alpha functions. Cells incorporated a passive leak conductance, a voltage- and time-dependent lumped potassium conductance ($g_{K_{Lump}}$), and two time-dependent fast synaptic conductances, one excitatory (g_{Glu}), and the other inhibitory (g_{GABA_A}). All cells except the LM cell also included a slow inhibitory conductance (g_{GABA_B}).

PY cell discharge in area CA3 was assumed to provide inputs to the circuit. The temporal distribution of impulses following a single stimulus of all afferent fibres was assumed to be gaussian with a coefficient of variation (CV) of 0.3. The weights associated with the afferent synapses contacting each PY cell were drawn at random from an exponential distribution such that the excitatory postsynaptic potentials (EPSPs) generated had a mean amplitude of $200\,\mu$V, and a range of 80–$1000\,\mu$V[7]. In the absence of emperical data it was assumed that EPSP amplitudes in the interneurones were similar to those described for area CA3, and the amplitudes of IPSPs were similar for those described for PY cells in area CA1. Three EPSP amplitudes were tested in the interneurones, 200, 800, and $1500\,\mu$V[8]. When generated by the discharge of afferent fibres, these EPSPs are referred to explicitly by amplitude as the interneurone EPSPs. but when generated by the discharge of PY cells within the circuit, they are referred to as *low*, *medium*, or *high* respectively. Similarly, three amplitudes of fIPSP and sIPSP were tested[9], approximately 1.5 (*low*), 5.0 (*medium*), or 8.5 mV (*high*), and 1 (*low*), 2 (*medium*), or 3 mV (*high*), respectively. In any one simulation, all intrinsic synaptic weights were scaled to give either *low*, *medium*, or *high* postsynaptic potentials.

The small size of the circuit allowed temporal delays due to the cable properties of the dendritic tree and impulse propagation in connecting axons to be ignored. A simplified scheme was employed to generate the action potential. At the discharge threshold (15, 8.5, 8.5, and 12.5 mV for PY, OA, BA, and LM cells respectively). the membrane potential was increased by an amount representing the peak depolarization of the action potential and $g_{K_{Lump}}$ was activated to assist with membrane repolarization. In the each of the interneurones, the kinetics of $g_{K_{Lump}}$ was scaled to generate a fast afterhyperpolarization (fAHP) of approximately 7, 5 and 11 mV for the OA[11], BA[12] and LM[13] cell respectively. A fAHP appeared in a PY cell following an action potential when either afferent excitation was increased, or the cell was depolarized. All cells exhibited an absolute refractory period of 4–5 ms and a relative refractory period determined by the fAHP. In PY cells the fAHP was extended by an slow afterhyperpolarization (sAHP). This reduced the amplitude of an EPSP occuring immediately after an impulse by a factor of 0.3, with subsequent EPSPs returned to their normal amplitude over a period of 1500 ms[10].

PY cell discharge was explored by varying the frequency of afferent stimulation. the extent of convergence, and synaptic weights. These simulations were used to define a representative unit circuit (control circuit) that was maximally sensitive to realistic changes in parameter values. This circuit was used to assess the effect on PY cell discharge of varying the temporal CV of afferent impulses, the extent of afferent convergence at PY cells, and the PY cell mean EPSP amplitude. Further simulations examined the effect of cell ablation on PY cell discharge.

RESULTS

The temporal distribution of afferent EPSPs associated with low frequency stimulation required 2,000 afferent fibres converged at a PY cell to reliably generate discharge at a stimulation frequency of 2.5 Hz. All simulations employed this number of afferent fibres unless otherwise noted. Because of their larger EPSPs and lower thresholds, interneurones required less convergence than PY cells to discharge. In the control circuit, 180, 160, and 320 afferent fibres converged at the OA, BA, and LM cells respectively, the interneurone EPSP was set to 800 μV, the intrinsic IPSPs were set to *medium*, and the PY cell sAHP was scaled such that at each frequency of afferent stimulation tested (2.5–20 Hz) PY cells generated a single impulse in response to a single gaussian stimulus of all afferent fibres. In this circuit the LM cell did not discharge at an interneurone EPSP of 200 μV, or until contacted by more than the number of fibres converged at the OA and BA cells.

At an interneurone EPSP of 200 μV, feed-back inhibition generated by OA and BA cell discharge scaled the frequency of PY cell discharge with the frequency of stimulation. An increase in afferent convergence at the interneurones from 160.160.320 to 320.320.320 (for the OA.BA.LM cell respectively) and an increase in the frequency of stimulation to greater than 10 Hz converted the effect of OA and BA cell discharge on the PY cells from feed-back to feed-foward, and further increases in the frequency of stimulation increasingly decoupled PY cell discharge from the frequency of stimulation (Fig. 1, Panel A). At an interneurone EPSP of 800 μV, the frequency of PY cell discharge was determined by the extent of convergence at the OA and BA cells. In summary, when convergence at the OA and BA cells was 80, increased convergence at the LM cell inhibited PY cell discharge. Doubling the convergence at the OA and BA cells (to 160) decoupled PY cell discharge from the frequency of stimulation until convergence at the LM cell was raised above that of the other interneurones. Under these conditions, inhibition of OA and BA cell discharge by the LM cell converted their inhibitory effect on PY cells from feed-foward to feed-back. This recoupled PY cell discharge to the frequency of stimulation (Fig. 1, Panel B). Further increases in the interneurone EPSP typically lowered the frequency of PY cell discharge (Fig. 1, Panel C).

PY cells also exhibited two modes of discharge distinguished by the level of intrinsic inhibition. The greatest changes in PY cell discharge occured when synaptic weights were *low*. Under these conditions PY cell discharge scaled with the frequency of stimulation. However, a sufficient increase in inhibition converted the control of PY cell discharge by the OA and BA cells from feed-back to feed-foward inhibition and PY cell discharge was decoupled from the frequency of stimulation.

A clear picture of the influence of changes in intrinsic synaptic weights on PY cell discharge was obtained when PY cell discharge was normalized by the mean frequency of PY cell discharge generated by the equivalent control simulations. At an interneurone EPSP of 200 μV, PY cell discharge showed a large increase when intrinsic weights were *low* and the frequency of stimulation was raised from 2.5 to 7.5 Hz (Fig. 2, Panel A). In contrast, at an interneurone EPSP of 800 μV, the frequency of PY cell discharge showed a large increase when the frequency of stimulation was raised from 7.5 to 12.5 Hz (Fig. 2, Panel B). A further increase in the interneurone EPSP reduced the discharge behaviour of the PY cells in proportion (Fig. 2, Panel C).

Variation of stimulation and circuit parameters such as, the temporal CV of afferent impulses (0.15, 0.6), the extent of convergence at PY cells (1000, 4000), and the mean amplitude of the PY cell EPSP (200, 400μV), showed the extent of afferent convergence at the PY cells to have the greatest influence on determining the effect

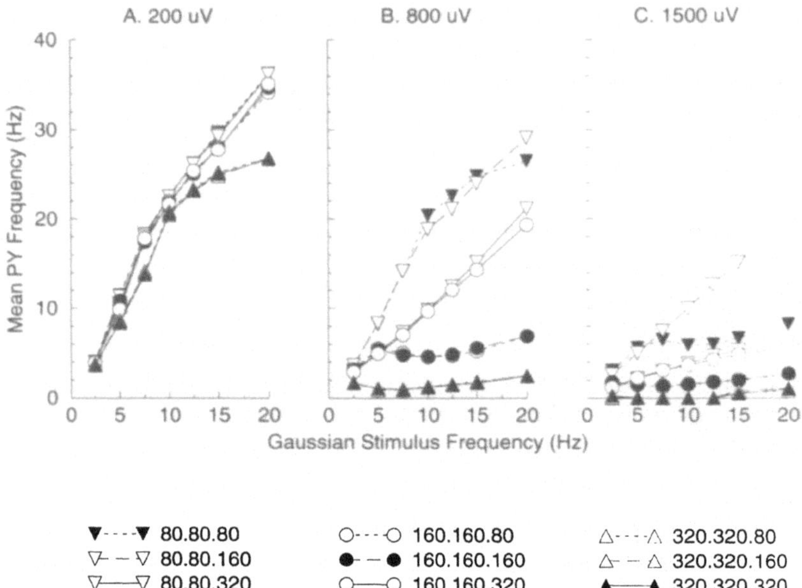

Figure 1. Changes in afferent convergence at the interneurones alters pyramidal cell discharge. The influence of an increase in the amplitude of afferent EPSP in each interneurone on pyramidal cell discharge. 2000 afferent fibres converged at the pyramidal cells, and the intrinsic PSPs were medium. The legend gives the number of afferent fibres converged at each interneurone for the OA.BA.LM cells respectively. Filled symbols indicate pyramidal cell discharge when convergence at each interneurone was the same.

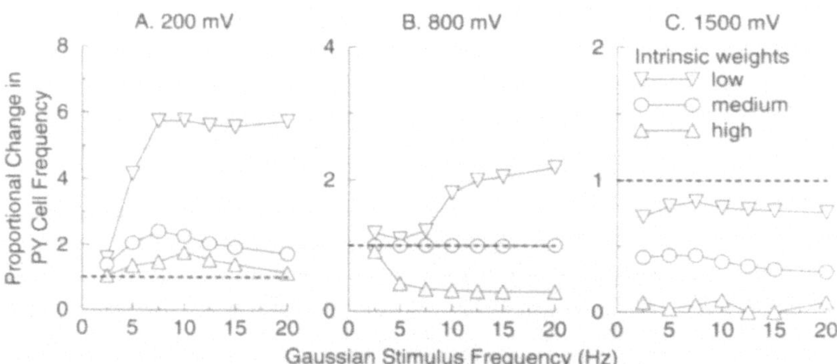

Figure 2. Effect of synaptic weights on pyramidal cell discharge. Proportional change in pyramidal cell discharge when each simulation was normalized by the equivalent discharge in the control circuit. Afferent EPSP amplitude in the interneurones is given above each panel. The dotted line in each panel indicates no change from the control circuit in which the mean EPSP amplitude in the pyramidal cells was 200 mV, the interneurone EPSP was 800 mV, and intrinsic PSPs were *medium.*

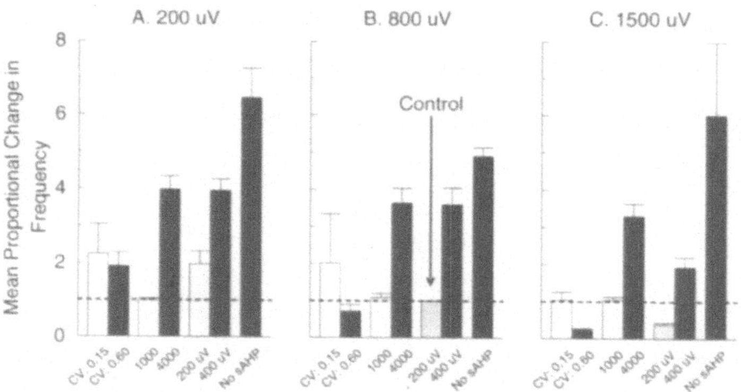

Figure 3. The influence of stimulation and circuit parameters on pyramidal cell discharge. The amplitude of EPSPs in the interneurones and the key to the parameter changes are given at the top and bottom of each panel respectively (see text for details). Stippled bars show pyramidal cell discharge when intrinsic PSPs were *medium*. Error bars give the standard deviation of cell discharge and indicate its sensitivity to the range of stimulation frequencies tested (2.5–20 Hz). The dotted line in each panel indicates no change from the control circuit (indicated by the arrow).

of inhibition on the frequency of PY cell discharge, an effect that increasing inhibition had little influence on (Fig. 3). PY cell discharge was more sensitive to changes in inhibition when the PY cell mean EPSP was doubled than when the extent of afferent convergence was altered. PY cell discharge was least effected, and most sensitive to inhibitory changes, when the CV of the distribution of afferent inpulses was altered.

Cell ablation studies showed removal of either the OA or BA cell had little influence on PY cell discharge, and that at *medium* to *high* levels of inhibition the LM cell typically controlled the inhibitory influence of the OA and BA cells on PY cell discharge (Fig. 4).

CONCLUSIONS

The two modes of PY cell discharge suggested a correlation with the two functions traditionally associated with hippocampal activity, learning, and memory or recall. The scaling of PY cell discharge with the frequency of stimulation may represent a learning mode, whereas, a low frequency of PY cell discharge in the presence of *high* inhibition suggests a highly reliable impulse transmission representative of a memory or recall function for the circuit.

Simulations showed the frequency of PY cell discharge to be much less influenced by increased inhibition at increased levels of convergence than when either the mean amplitude of the PY cell EPSP or the frequency of afferent stimulation was increased.

Unlike the OA and BA cells, the LM cell was not reciprocally connected to PY cells. Consequently, LM cell discharge did not scale with an increase in the discharge frequency of PY cells. A sufficiently high LM cell discharge frequency could reverse the decoupling of PY cell discharge from the frequency of stimulation seen at high levels of inhibition. By reducing the frequency of OA and BA cell discharge and converting their influence on PY cells from feed-foward to feed-back inhibition, the LM cell may act to gate the flow of information through area CA1.

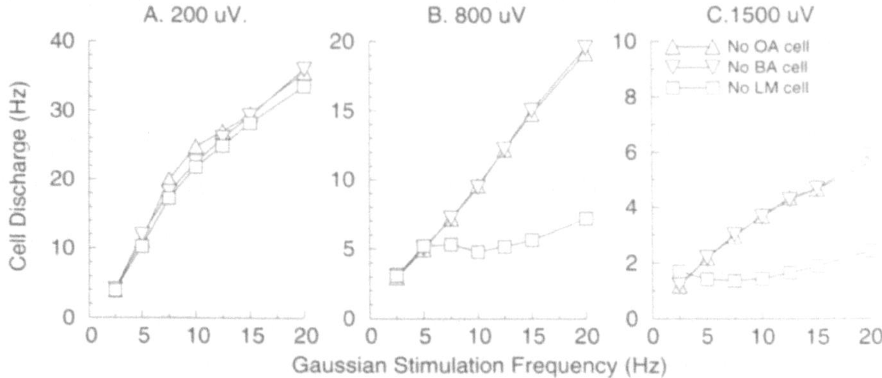

Figure 4. Frequency of pyramidal cell discharge after cell ablation. The effect on pyramidal cell discharge of removal of the *oriens-alveus* (OA), the basket (BA), and the *lacunosum-moleculare* (LM) interneurone at three interneurone EPSP amplitudes and a range of stimulation frequencies (indicated above and below each panel respectively).

REFERENCES

1. Douglas, R. J. and Martin, K. A. C. (1991) A functional microcircuit for cat visual cortex. *J. Physiol. Lond.* **440**:735–769.

2. Bernard, C. and Wheal, H. V. (1994) Model of local connectivity patterns in CA3 and CA1 areas of the hippocampus. *Hippocampus.* **4**:497–529.

3. Patton, P. E. and McNaughton, B. (1995) Connection matrix of the hippocampal formation: I. The dentate gyrus. *Hippocampus.* **5**:245–286.

4. Lacaille, J.-C. and Schwartzkroin, P. A. (1988) Stratum lacunosum-moleculare interneurons of hippocampal CA1 region. II. Intrasomatic and intradendritic recordings of local circuit synaptic interactions. *J. Neurosci.* **7**:1411–1424.

5. Olbrich, H.-G. and Braak, H. (1985) Ratio of pyramidal cells versus non-pyramidal cells in sector CA1 of the human Ammon's horn. *Anat. Embryol.* **173**:105–110.

6. Thomson, A. M. and Radpour, S. (1991) Excitatory connections between CA1 pyramidal cells revealed by spike triggered averaging in slices of rat hippocampus are partially NMDA receptor mediated. *Eur. J. Neurosci.* **3**:587–601.

7. Sayer, R. J., Friedlander, M. J. and Redman, S. J. (1990) The time course and amplitude of EPSPs evoked at synapses between pairs of CA3/CA1 neurons in the hippocampal slice. *J. Neurosci.* **10**:826–836.

8. Gulyás, A. I., Miles, R., Sik, A., Tóth, K., Tamamaki, N. and Freund, T. F. (1993) Hippocampal pyramidal cells excite inhibitory neurons through a single release site. *Nature.* **366**:683–687.

9. Williams, S., Vachon, P. and Lacaille, J.-C. (1993) Monosynaptic GABA-mediated inhibitory postsynaptic potentials in CA1 pyramidal cells of hyperexcitable hippocampal slices from kainic acid-treated rats. *Neuroscience.* **52**:541–554.

10. Storm, J. F. (1989) An after-hyperpolarization of medium duration in rat hippocampal pyramidal cells. *J. Physiol. Lond.* **409**:171–190.

11. Lacaille, J.-C. and Williams, S. (1990) Membrane properties of interneurons in stratum oriens-alveus of the CA1 region of rat hippocampus *in vitro. Neuroscience.* **36**:349–359.

12. Lacaille, J.-C. (1991) Postsynaptic potentials mediated by excitatory and inhibitory amino acids in interneurons of stratum pyramidale of the CA1 region of rat hippocampal slices *in vitro. J. Neurophysiol.* **66**:1441–1454.

13. Lacaille, J.-C. and Schwartzkroin, P. A. (1988) Stratum lacunosum-moleculare interneurons of hippocampal CA1 region. I. Intracellular response characteristics, synaptic responses and morphology. *J. Neurosci.* **7**:1400–1410.

A BIOLOGICAL MECHANISM FOR SYNAPTIC STABILITY IN DEVELOPING NEOCORTICAL CIRCUITS

Niraj S. Desai, Kenneth R. Leslie, Sacha B. Nelson,
and Gina G. Turrigiano

Department of Biology and Center for Complex Systems
Brandeis University
Waltham, MA 02254-9110

INTRODUCTION

A common feature of both learning and development is that the number and strengths of synaptic connections can change dramatically in a relatively short period of time. Precisely how such changes give rise to functioning and flexible circuits remains an outstanding, unanswered question in neuroscience. One idea, which enjoys widespread support, is that correlation-based rules of synaptic modification are responsible for at least some forms of learning and activity-dependent development in the nervous system.[1,2] By these rules, correlated activity in presynaptic and postsynaptic cells leads to an increase in synaptic strength, while uncorrelated activity leads to a decrease in strength. This sort of Hebbian synaptic modification is enormously attractive for a number of reasons. It gives neural circuits great flexibility, and suggests a way for them to wire themselves without extensive genetic instructions. It is also backed by experiments in the hippocampus and neocortex that demonstrate that real synapses can actually follow such rules — namely, in long-term potentiation (LTP) and depression (LTD).[2]

Correlation-based synaptic modification, however, is not sufficient for understanding much of what goes on in learning and development. In particular, there is the question of how neurons maintain their firing rates and synaptic properties within suitable operating ranges despite large fluctuations in the amount of input they receive. During development, synaptic connections are being formed and reconfigured substantially and continuously, thus dramatically altering the synaptic drive received by each neuron.[1] Some means of tuning a neuron's sensitivity to fit its ever-changing environment would seem to be needed. In addition, Hebbian-based learning can lead to instability in neural networks because of positive feedback[3,4]: the strengthening of the synapse between two cells causes the presynaptic cell to drive the postsynaptic one more strongly; this in turn strengthens the synapse even more; this leads to more driving; and so on. (One can also imagine a feedback mechanism that steadily erodes synaptic strength.) Moreover, an unconstrained Hebbian process may not be sufficiently specific for either learning or development because every cell that synapses onto a postsynaptic neuron may be potentiated (or depressed) to some extent because of an increase (or decrease) in the postsynaptic firing rate. Simple correlation-based modification rules can lead to situations in which all synapses eventually saturate or fall to zero.[4]

The problem of achieving stability in neural circuits that obey correlation-based rules has been recognized for some time, and several solutions have been proposed.[3,5] Many of these have been based on the idea that LTD can compensate for LTP. This idea has considerable merit, but it requires that synaptic strength lost through LTD roughly equal strength gained through LTP. A rather exquisite balance between the two processes must be maintained, leading to questions about how robust such a mechanism would be. A different solution would be to allow neurons to scale all their synaptic strengths up or down as a function of average activity, increasing them when activity is low, decreasing them when activity is high. Such a mechanism is attractive from a theoretical standpoint because of its simplicity and its manifest robustness. Here we present experimental evidence from cultured pyramidal neocortical neurons that such an activity-dependent regulatory process does in fact exist. We have found that blocking all activity for an extended period (days) causes a pronounced increase in the average amplitude of miniature excitatory postsynaptic currents (mEPSCs), while chronically increasing activity causes a significant decrease in mEPSC amplitude. This novel bidirectional form of synaptic plasticity may well serve to allow neurons to respond to dramatic long-lasting changes in synaptic excitation.

METHODS

Experiments were performed on pyramidal neurons kept in primary cell culture. These cultures were prepared (as previously described[6]) by removing visual cortex from postnatal (P4-6) Long-Evans rat pups, dissociating cells using a papain procedure, and plating neurons in a monolayer on top of astrocyte feeder layers. Cultures were maintained in an incubator for 7-9 days before use. During this time neurons in these cultures formed synaptic connections and developed spontaneous excitatory and inhibitory synaptic activity. We used pharmacological manipulations to increase or decrease spontaneous activity, and asked how this affected the average amplitude of excitatory postsynaptic currents. Activity levels were usually manipulated by treating cultures for up to two days before each experiment with 0.1 μM tetrodotoxin (TTX) or 20 μM bicuculline. The TTX served to block all activity, while the bicuculline increased activity substantially. After the treatment period, cultures were removed from the incubator and placed on the stage of an inverted microscope. Whole cell patch recordings were then obtained from pyramidal neurons. In most experiments, neurons were held in voltage clamp at -70 mV in the presence of TTX (to block action potentials), bicuculline (to block inhibitory currents), and 50 μM APV (to block NMDA-mediated currents). Under these conditions, miniature excitatory postsynaptic currents mediated by AMPA receptors could be recorded from neurons in both control cultures and treated cultures. Our data for each neuron sampled currents due to a large number of synapses. Average current amplitudes and waveforms for each cell were determined and were then used to calculate group averages for all cells in each condition; the group averages are the numbers presented in this paper.

RESULTS

The effects of both activity blockade and activity enhancement on excitatory quantal amplitudes were pronounced (Figure 1). In control cultures, recordings from pyramidal neurons produced a broad distribution of peak mEPSC amplitudes, with an average of 13.6±0.7 pA (n=27). Blocking all activity for 48 hours prior to the experiment with TTX nearly doubled the average (26.1±1.7 pA, n=10). On the other hand, treatment for 48 hours with bicuculline, which increased pyramidal cell firing rates by more than a factor of two, significantly decreased mEPSC amplitude (9.5±0.3 pA, n=10). In fact, not only were TTX and bicuculline treatments able to change the average value of the quantal amplitude at AMPA synapses, but they shifted the entire distribution of miniature currents, toward larger values for TTX-treated cells, toward smaller ones for bicuculline-treated cells (Figure 1C). It is important to point out that these pharmacological manipulations did not alter other measured neuronal properties (membrane resting potential, input resistance, and whole cell capacitance); nor did they appear to affect the rise time and kinetics of miniature currents. Figure 1B demonstrates that the average current waveforms recorded in the three conditions can be largely superimposed if scaled to the same peak, indicating that the activity-dependent regulation of mEPSC amplitude is not a result of differential electrotonic filtering. In addition, manipulating activity levels did not affect mEPSC frequency.

The regulation by activity of the amplitude of miniature currents was mirrored by a similar regulation of spike-mediated transmission (Figure 2). We obtained dual whole cell

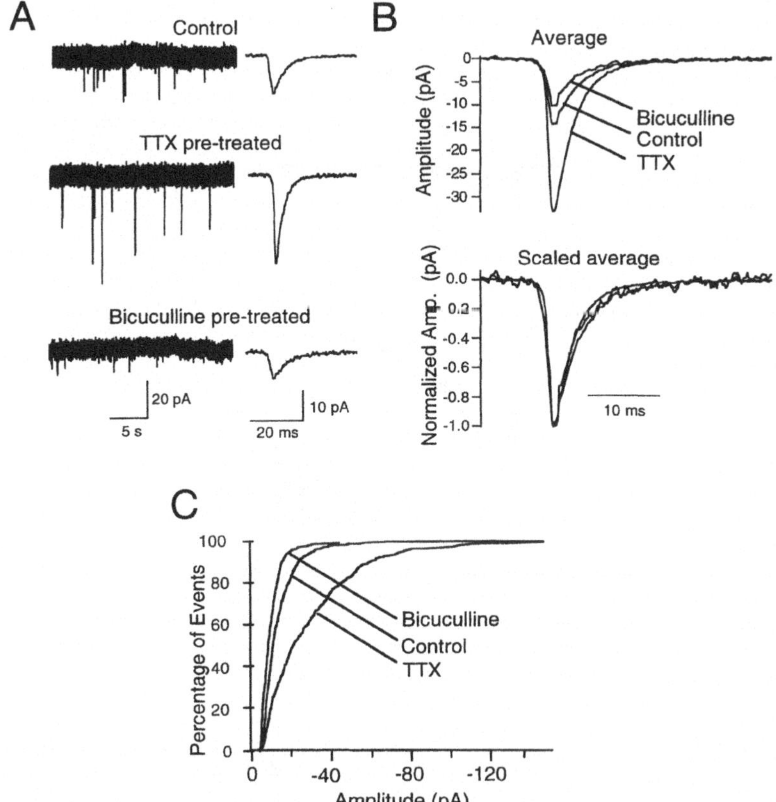

Figure 1. Regulation by activity of pyramidal neuron mEPSC amplitudes. (A) Recordings of mEPSCs from three representative neurons from sister cultures in which activity was normal (control), blocked (TTX), or enhanced (bicuculline), for 48 hours prior to the experiment. On the left are stretches of voltage clamp recordings which illustrate rapid, AMPA-mediated mEPSCs due to spontaneous single vesicle release. On the right are the average mEPSC waveforms for all mEPSCs recorded from the neurons on the left. (B) Average mEPSC waveforms for all cells in the each of the three conditions. Average amplitudes for TTX- and bicuculline-treated cells differed significantly from the control average (p<0.004 for TTX, p<0.01 for bicuculline, Student t-test with Bonferoni correction). The average waveforms could be superimposed by scaling their peaks. (C) Cumulative amplitude histograms showing the distribution of mEPSC amplitudes for all events recorded from all cells in the three conditions.

patch recordings from pairs of monosynaptically connected pyramidal neurons, held one neuron (the postsynaptic cell) in voltage clamp, and measured the current response elicited in it by firing an action potential in the other neuron (the presynaptic cell); here TTX was omitted from the recording solution. Excitatory postsynaptic currents (EPSCs) measured in this way, at a holding potential of -70 mV, increased in average amplitude from 89±65 pA (n=13) in control cultures to 219±52 pA (n=9) in cultures that had been treated with TTX for 48 hours. (We have not made similar measurements in bicuculline-treated cultures.) It is useful to note that the twofold increase in EPSC amplitude is comparable in size to the increase that activity blockade produced in mEPSC amplitude, suggesting (though not proving) that the change in quantal size alone is sufficient to account for the effect of TTX treatment on spike-mediated currents.

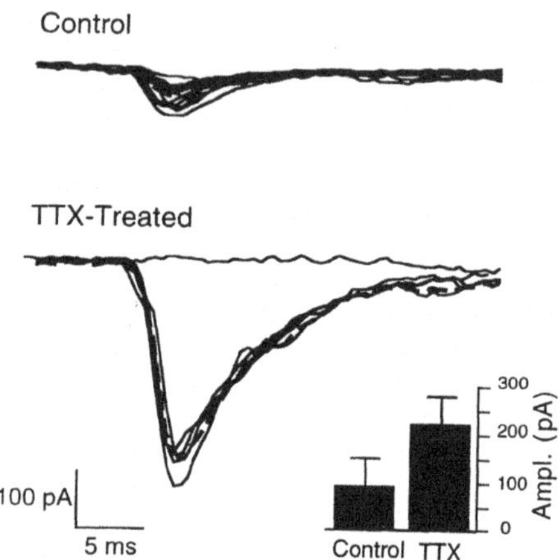

Figure 2. Spike-mediated excitatory postsynaptic currents (EPSCs) were also regulated by activity. Shown here are representative monosynaptic responses in postsynaptic cells to single action potentials fired in presynaptic cells. The inset shows average EPSC amplitudes for control and TTX-treated cultures. The differences are significant (p<0.005, Wilcoxon rank sum test).

We wished to understand more about the nature of the measured change in quantal amplitude. One question we sought to answer was whether the regulation of mEPSC amplitude was being produced by a postsynaptic change in responsiveness to glutamate or by a presynaptic change in the glutamate content of synaptic vesicles. We addressed this question by blocking all synaptic transmission, applying controlled amounts of glutamate directly to pyramidal neurons (at the soma and along the main dendrite), and measuring the amplitudes of the resulting currents. What we found was that, regardless of where on the neuron the glutamate was applied, the average amplitude of glutamate-induced currents was roughly twice as large in TTX-treated neurons as in control neurons. Again this increase with activity deprivation was comparable in size to the measured increase in mEPSC amplitude. These data suggest strongly that activity is regulating mEPSC amplitude through a postsynaptic change in receptor number or function.

Another question we addressed was about the time course of this form of synaptic plasticity. While in most of our experiments cultures were treated for 48 hours prior to the experiment, here activity was blocked with TTX for 15, 26, or 48 (±4) hours, and then mEPSCs were recorded. As Figure 3 shows, there was a progressive increase in mEPSC amplitude with treatment time; that is, this process is both slow (compared, for example, to the induction of LTP) and cumulative. The fact that this phenomenon is well separated in time from other, better known kinds of plasticity may mean that it is well placed to serve as a stabilizing mechanism. Our hypothesis, as described at the end of the introduction, was that stability in neural circuits could be maintained if synaptic strengths were scaled up or down as a function of *average* activity, with the average taken over a time long compared

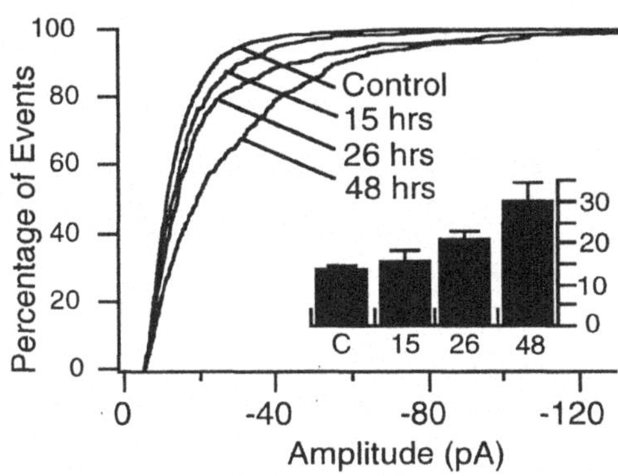

Figure 3. Cumulative amplitude histograms showing the progressive increase in mEPSC amplitude with time in TTX. Inset shows the mean mEPSC amplitudes for cells from sister cultures grown under control conditions (n=7), 15 hours in TTX (n=8), 26 hours (n=11), and 48 hours (n=10). The differences were significant (p<0.002, one way ANOVA).

to correlation-based synaptic modification. That the activity-dependent regulation described here is in fact quite slow may be valuable in this regard.

This regulation also differs from LTP in that it does not appear to depend on the activation of NMDA receptors. Blocking transmission through NMDA receptors with APV for 48 hours had no effect on average mEPSC amplitude. By contrast, blocking AMPA receptors with CNQX, which (unlike APV) reduced neuronal firing rates, produced increases in mEPSCs similar to those elicited by TTX treatment.

Finally, there is the difficult question of the relationship between mEPSC amplitude distributions produced by different activity levels. We can envision several ways in which the average excitatory quantal amplitude might be modified by activity: by increasing the strength of each synapse by the same amount (an additive model), by increasing strengths by random amounts (a random additive model), or by scaling each synaptic strength by the same multiplicative factor (a multiplicative model). Figure 4 explains our efforts to distinguish between these three possibilities. We were unable to fit the experimental data with either additive or random additive functions, but did obtain very good fits for data from TTX-treated and bicuculline-treated cultures by employing a multiplicative function. Multiplying mEPSCs from TTX-treated cells by a factor of 0.36 and mEPSCs from bicuculline-treated cells by a factor of 1.59 allowed us to collapse the cumulative amplitude distributions for both TTX and bicuculline data onto the control distribution (Figure 4B).

These data suggest an exciting idea: activity may be globally scaling the quantal amplitude of each of a neuron's synapses by the same multiplicative factor. Multiplicative scaling of synaptic strengths is valuable because it allows a neuron to adjust the total amount of synaptic excitation it receives, while preserving relative differences between synapses (such as those produced by LTP or LTD).[4] Put another way, multiplicative scaling offers a robust means of maintaining stability, while preserving flexibility.

SUMMARY

Using cultured visual cortical neurons, we have discovered a novel form of plasticity in which synaptic strengths, as represented by quantal amplitudes, are globally scaled by a neuron's history of activity. When overall activity is low, synaptic strengths between pyramidal neurons are scaled up; when activity is high, strengths are scaled down. This plasticity in mEPSC amplitude modifies spike-mediated transmission, is most likely postsynaptic, and does not depend on NMDA receptor activation. It also appears to have a slow time course and to be multiplicative in form. These last attributes especially make it an attractive candidate to serve as a stabilizing force in neural circuits in which correlation-based synaptic modification processes are present.

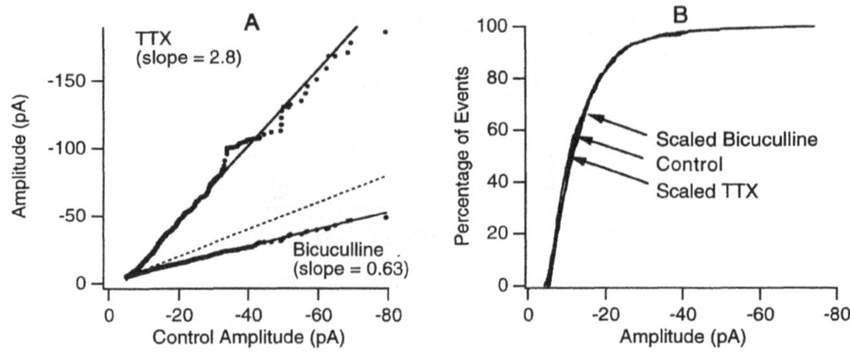

Figure 4. Tests to determine how activity scales mEPSC amplitude. (A) The relationship between control and experimental amplitudes was determined in the following manner. All mEPSC amplitudes recorded from control, TTX-treated, and bicuculline-treated cultures were ranked in ascending order. These amplitude distributions were linearly interpolated to produce an equal number of observations in each condition, and the experimental amplitudes were then plotted against control amplitudes (solid circles). On this type of plot, manipulations that have no effect on amplitude produce lines with slope one (as shown by plotting control against control, dashed line). Addition or subtraction of a constant value to each control mEPSC value leaves the slope unchanged but shifts the line up or down, while multiplication of each control value by a factor different from one changes the slope. The best fits of three different models of activity regulation of mEPSC amplitudes were calculated. A purely additive model provides very poor fits to the data. A random additive model, generated by first calculating differences between test and control amplitudes and then adding these values to control amplitudes in random order, also produces poor fits. Data are well fitted, however, by a multiplicative model, as the solid lines show. (B) TTX and bicuculline data can be collapsed onto control data by multiplicative scaling.

ACKNOWLEDGMENTS

Financial support was provided by the NIH grants R01 NS36853 and K02 NS01893 (GGT), the Whitehall Foundation, and the Sloan Foundation. NSD is a Sloan Center for Theoretical Neurobiology postdoctoral fellow. KRL is a HHMI predoctoral fellow.

REFERENCES

1. L.C. Katz and C.J. Shatz, Synaptic activity and the construction of cortical circuits, *Science* 274:1133 (1996).
2. M.F. Bear and R.C. Malenka, Synaptic plasticity: LTP and LTD, *Curr. Opn. Neurobiol.* 4:389 (1994).
3. K.D. Miller, Synaptic economics: competition and cooperation in synaptic plasticity, *Neuron* 17:371 (1996).
4. K.D. Miller and J.D.C. McKay, The role of constraints in Hebbian learning, *Neural Comp.* 6:100 (1994).
5. E.L. Bienenstock, L.N. Cooper, and P.W. Munro, Theory for the development of neuron selectivity: orientation specificity and binocular interaction in visual cortex, *J. Neurosci.* 2:32 (1982).
6. L.C. Rutherford, A. DeWan, H.M. Lauer, and G.G. Turrigiano, Brain-derived neurotrophic factor mediates the activity-dependent regulation of inhibition in neocortical cultures, *J. Neurosci.* 17:4527 (1997).

EDGE DETECTORS AND TEXTURE DETECTORS DIFFER IN THEIR LATERAL CONNECTIVITY

Alexander Dimitrov and Jack D. Cowan

Department of Mathematics
The University of Chicago
5734 S University Ave
Chicago IL 60637
{dim2,cowan}@math.uchicago.edu

INTRODUCTION

The structure of the early sensory pathways in most animals is determined in part by the statistical structure of signals they perceive from the surrounding environment. Because of evolutionary pressure, their sensory systems have developed so that they optimally process the incoming sensory data. Various theoretical and experimental studies of the early visual system suggest that a good optimal measure is given by Barlow's redundancy reduction hypothesis (Barlow, 1961). It correctly describes the spatial characteristics of the retino-ganglion filter (Atick and Redlich, 1992) and the temporal response function of LGN neurons (Dong and Atick, 1995) based on studies of the statistical structure of natural scenes. So far Information Theory has been use mostly to characterize forward processing in the sensory systems. We applied it to a recurrently connected network of orientation selective units to describe the structure of lateral connectivity in the striate cortex (Dimitrov and Cowan, 1997).

A well known fact about the visual system is the existence of signal representations on several spatial scales. We study the statistical properties of edges on two distinct scales - "objects" and "texture". In our study, an object has spatial structure on the scale of 0.25 to 1.00 degrees of visual angle, while a texture element is on the scale of 0.01 to 0.10 degrees of visual angle, repeated quasi-periodically on a scale of up to 1.00 degrees. As expected, they have a large scale structure which is scale-invariant, with a power law autocorrelator. The texture sample also has fine structure which is not present in the object sample and is caused by the quasi-periodic structure of textures.

The structure of the network we use for processing this signal reflects what is known about the structure of V1. We model a layer of orientation sensitive cells which are laterally connected to one another, each receiving oriented input from the previous layer. We assume that each unit receives as input the magnitude of the directional derivative of the luminosity signal along the preferred visuotopic axis of the cell. The natural development of units with oriented receptive fields, approximating directional derivatives, has been proposed in many theoretical works (Linsker, 1986; Hancock et al.. 1992; Olshausen and Field, 1996; Bell and Sejnowski, 1996) and of course there is ample experimental evidence for such units (Hubel and Wiesel, 1961). Our assumption implies that locally the input to a cell is proportional to $|\cos|$ of the angle between the unit's preferred orientation and the local gradient (edge). Thus each unit receives a broadly tuned signal, with a half-width at half-height of approximately 60°. With this feed-forward structure, the assumption that the system acts to decorrelate such inputs suggests a way to calculate the lateral connections that will perform this task. This calculation, and a further study of the statistical properties of the input is the topic of the paper.

MATHEMATICAL MODEL

We introduce a general model for a laterally connected network with a feed-forward input $V(x)$, which is attributed to signal from the environment. We shall consider particular forms of V later. The input of the network represents the information redundantly due both to its structure and the statistical structure of natural images.

We formulate the problem in terms of a recurrent kernel W, so that

$$O = V + W * O \tag{1}$$

This equation describes the linearized steady state of the dynamical system $\dot{O} = -O + \sigma[W * O + V]$, which is one form of the Wilson-Cowan equations (Wilson and Cowan, 1973).

For this linear system, Barlow's redundancy reduction hypothesis (Barlow, 1961; Barlow, 1989) can be formulated as a variational problem with a cost functional (Atick and Redlich, 1992)

$$E\{W\} = Tr(K(W)R_V K^T(W)) - \rho \log det(K^T(W)K(W)) \tag{2}$$

where $K = (\delta - W)^{-1}$ is the effective forward kernel of the system and $R_V = < V^T V >$ is the autocorrelation matrix of the input. The minimum of this functional with respect to W provides an output with minimal redundancy of the signal representation. As a first approximation, we consider an upper bound of it, given by the best Gaussian approximation of the input signals. The assumption that V is a Gaussian signal reduces the minimization of $E\{W\}$ to simply pairwise decorrelating the outputs of units:

$$\delta(x_1 - x_2) \sim < O(x_1) \circ O(x_2) > = < (K \cdot V)(x_1) \circ (K \cdot V)(x_2) > \sim K \cdot R \cdot K^T \tag{3}$$

A solution is $K \sim R^{-\frac{1}{2}}$, unique up to a unitary transformation. Following (Atick and Redlich, 1992), we fix a solution by requiring that it has the reflection symmetry found in the input signal. The corresponding recurrent filter is then

$$W = \delta - K^{-1} = \delta - \rho\, R^{\frac{1}{2}} \tag{4}$$

In order to compare our calculations to existing cortical structures, we need to explicitly include the effects of noise on the system. We need to recognize that this is a different system, which is described by

$$O_1 = V + N_v + W * M * (O_1 + N_o) \tag{5}$$

where N_v is the input noise and N_o is the individual unit noise in the recurrently connected layer. Similarly to (Atick and Redlich, 1992) (Atick and Redlich, 1990; Atick and Redlich, 1992), we can modify the decorrelation kernel W derived from (1) to $W * M$. The form of the correction M, which minimizes the effects of noise in the system, is obtained by minimizing the distance between the states of (5) and the noise-free system (1). If we define $\chi^2(M) = < (O - O_1)^2 > = Tr < (O - O_1)(O - O_1)^T >$ as the distance function, the solution to $\frac{\partial \chi^2(M)}{\partial M} = 0$ will give us the appropriate kernel:

$$W * M = \delta - (R + N_v^2 + N_o^2) * (\rho \, R^{1/2} + N_o^2)^{-1} \tag{6}$$

It is advantageous to write this expression in "measurable" variables. When working with the filter, we have at our disposal $Vm = V + N_v$, which generates $Rm = R + N_v^2$, since V and N_v are not correlated. In these terms,

$$W * M = \delta - (Rm + N_o^2) * ((Rm - N_v^2)^{\frac{1}{2}} + N_o^2)^{-1} \tag{7}$$

Equation (5) may be used to model redundancy reduction with lateral interactions for any sensory modality. Here we apply it to describe a layer of the visual cortex. In the equation O models the activity of a cortical layer with complex orientation-selective cells (e.g. V1 layer 2 in primates), W specifies its lateral connectivity and V is the effective input to this layer from the previous layer (V1 layer 4). In particular, based on empirical evidence, throughout the rest of the paper we shall use

$$V(\theta, x) = |\frac{d}{dn_\theta} L(x)| \tag{8}$$

RESULTS

Correlation Structure of the Input Signal

Let us consider the correlation structure of edges in natural images as represented by the input signal (Eq. 8). In all quantities we use the "log-contrast" function $L(x) = ln(I(x)/I_o)$ and its derivatives (Ruderman, 1994), where $I(x)$ is the luminosity intensity. We use two datasets - one with only few large objects per image and one with texture.

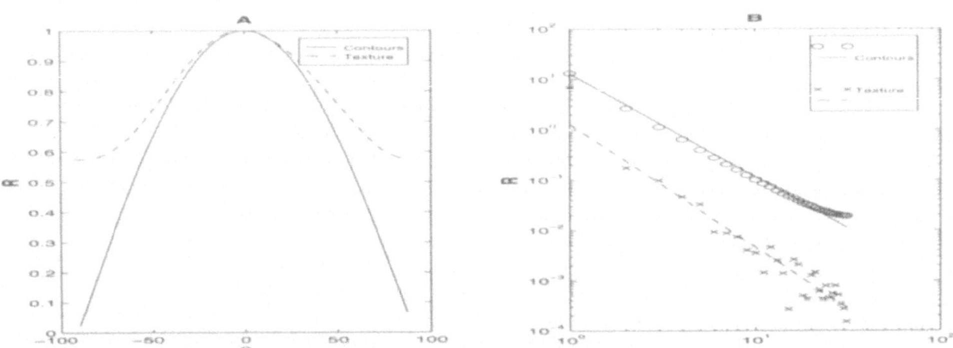

Figure 1. Local autocorrelation function (A) and local autocorrelation power spectrum (B) between orientation-selective units in object contours (–) and texture (- -). Both have similar scaling behavior of their power spectrum, proportional to $\eta^{-\lambda}$, with $\lambda \approx 2.05 \pm 0.01$ for contours and $\lambda \approx 2.31 \pm 0.02$ for texture. Texture has higher correlation for disparate orientations (A.) because of its repetitive nature.

We first consider the local correlation structure of edges. The correlation function $R(\theta - \theta') \equiv\, <V_\theta(x)V_{\theta'}(x)>$ depends only on the relative angle between preferred orientations (Fig. 1A). This allows us to define the autocorrelation power spectrum from its Fourier series (Fig. 1B) and study its spatial and scaling properties. Differences between the statistical structures of texture and objects can be seen in both representations of the autocorrelation function.

Next we consider the full correlation structure of edges. The correlation function $R(\theta_1, \theta_2, x - y) =\, <V(x, \theta_1)V(y, \theta_2)>$ is spatially translationally invariant, but there is interaction between the spatial and angular parts, so it is not translation invariant in angles anymore. A discrete version of this structure can be seen in Fig. 2. The distinction between objects and texture is even more evident here. Object contours have rather specific anisotropic spatial correlations, while texture has mostly isotropic correlation, which it is on average much higher than for objects (note the different scales on the color bars).

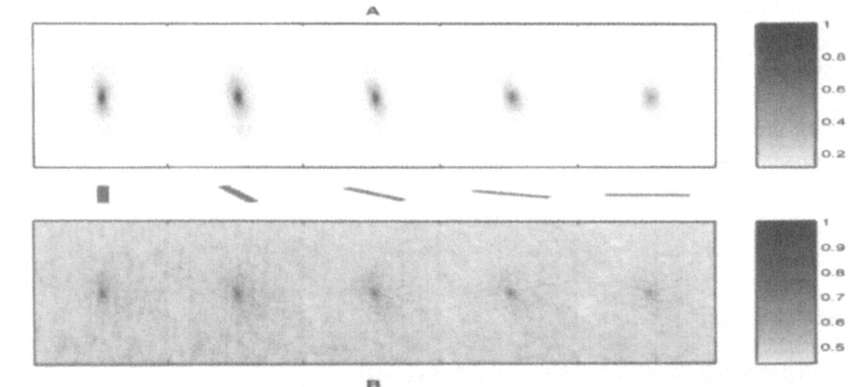

Figure 2. The autocorrelation function of a set with 8 orientations for contours (A) and texture (B). Only 5 of the 64 pairs of correlations in angle are shown, since the rest can be recovered by symmetry. The vertical label denote the preferred angle of the central unit and the horizontal labels – the preferred angle of the unit to which the central one is correlated. Dark represents high correlation, light – low correlation.

Connectivity Structure

Let us consider the implications of the model for the connectivity between units, processing the two different signals. Knowing the correlation structure of the signals from the previous section, we can find a linear filter that maximally reduces the redundancy of the output (Eq. 6). The connectivity within a single hypercolumn suggested by these calculations and the differences between object and texture processing units can be seen on Figure 3A.

We can perform the same calculation on the spatially extended case (Figure 3B,C). Again we notice the general inhibitory nature and the spatial isotropy of the connectivity between texture processing units. Contour-selective units have targeted anisotropic connections along their preferred visuotopic axis and in general no isotropic connections.

In related work (Dimitrov and Cowan, 1997; Dimitrov and Cowan, 1998) we compare our calculations for contour processing cells with physiological measurements and find good agreement with the experimental data. The current results suggest that tex-

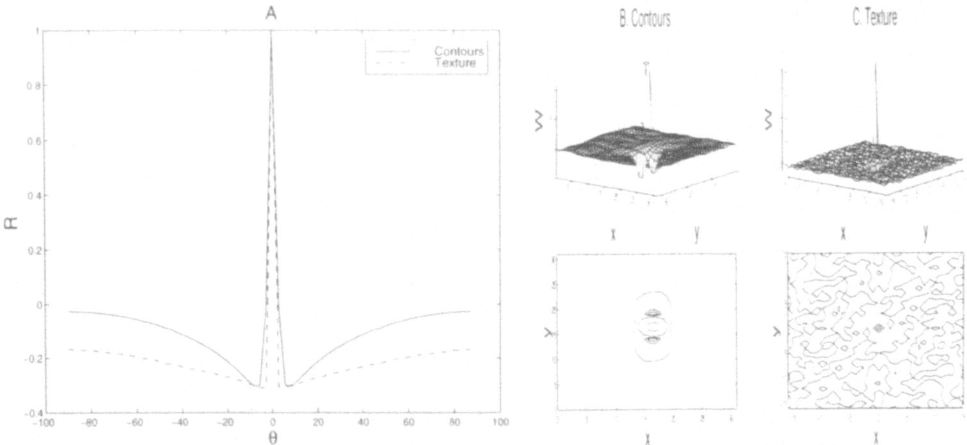

Figure 3. The local (A) and spatially extended (B,C) connectivity structure of contour and texture-selective units. Locally (A), the texture processing units have stronger inhibitory connections, which are also more uniform than the object units. Nonlocaly we observe the same effect, as well as the spatial isotropy of connectivity between texture-selective units (C) as contrasted to the anisotropic connection between object units (B). Only the connectivity between units of the same orientation preference is shown on the spatially extended plots, but the pattern persists for other relative orientations.

ture processing units have connectivity which differs substantially from that of edge detectors. Because of the higher overall correlation, fewer units are necessary to process texture than contours and they interact mostly inhibitory in a spatially uniform fashion. All this evidence points towards another structure in V1 as the candidate for texture processing tasks – those cells which are located in the cytochrome oxidase blobs (Livingstone and Hubel, 1988). Further studies, both theoretic and experimental, are necessary to elucidate this proposed functionality.

CONCLUSIONS

We show in this paper that properties of orientation selective cells in the visual cortex can be partially described by linear systems analysis. We also suggest a possible function for the cytochrome oxidase blobs. We achieve this by using a recurrent network as the underlying model. This was chosen for several reasons, mostly because we tried to give the model biological plausibility and recurrency is well established on the cortical level.

Our work is based on previous suggestions relating the second order statistics of the visual environment to the structure of the visual pathway. We studied two different parts of the visual environment – "objects" and "texture" and found that their statistical structure differs substantially, so that they need two different cortical systems to process them.

Further studies are needed to clarify the proposed texture-selective cortical subsystem. The attempt to relate it to the cells in the cytochrome oxidase blobs suggests that color may be of importance in processing texture. The measured statistical structure of textures suggests an isotropic inhibitory lateral interaction between texture processing units, which has to be verified experimentally.

359

REFERENCES

Atick, J. J. and Redlich, N. N. (1990). Towards a theory of early visual processing. *Neural Computation*, 2:308.

Atick, J. J. and Redlich, N. N. (1992). What does the retina know about natural scenes? *Neural Computation*, 4:196.

Barlow, H. B. (1961). Possible princilples underlying the transformation of sensory messages. In Rosenblith, W. A., editor, *Sensory Communications*. MIT Press. Cambridge, MA.

Barlow, H. B. (1989). Unsupervised learning. *Neural Computation*, 1:295.

Bell, A. T. and Sejnowski, T. J. (1996). The "Independent Components" of natural scences are edge filters. *Vision Research*, (submitted).

Dimitrov, A. and Cowan, J. D. (1997). Spatial decorrelation in orientation tuned cortical cells. In Mozer, M., Jordan, M., and Petsche, T., editors, *Advances in NIPS*, volume 9. MIT Press, Cambridge, MA.

Dimitrov, A. and Cowan, J. D. (1998). Spatial decorrelation in orientation selective cortical cells. *Neural Computation*, (under review).

Dong, D. W. and Atick, J. J. (1995). Temporal decorrelation: a theory of lagged and nonlagged responses in the lateral geniculate nucleus. *Network*, 6(2):159.

Hancock, P., Baddeley, R., and Smith, L. S. (1992). The principal components of natural images. *Network*, 3:61.

Hubel, D. and Wiesel, T. (1961). Receptive fields, binocular interaction and functional architecture in the cat's visual cortex. *J. Physio.*, 195:215.

Linsker, R. (1986). From basic network principles to neural architecture: Emergence of orientations-selective cells. *Proc. Natl. Acad. Sci. USA, Neurobiology*, 83:8390.

Livingstone, M. and Hubel, D. (1988). Segregation of form, color, movement. and depth: anatomy, physiology, and perception. *Science*, 240:740.

Olshausen, B. A. and Field, D. J. (1996). Emergence of simple-cell receptive field properties by learning a sparse code for natural images. *Nature*, 381:607.

Ruderman, D. L. (1994). The statistics of natural images. *Network*, 5:517.

Wilson, H. R. and Cowan, J. D. (1973). A mathematical theory of the functional dynamics of cortical and thalamic nervous tissue. *Kybernetik*, 13:55.

ORIENTATION CONTRAST ENHANCEMENT MODULATED BY DIFFERENTIAL LONG-RANGE INTERACTIONS IN VISUAL CORTEX

Udo Ernst, Klaus Pawelzik, Fred Wolf, Theo Geisel

MPI für Strömungsforschung
D-37073 Göttingen, Germany
Email: {udo,klaus,fred,geisel}@chaos.uni-frankfurt.de

INTRODUCTION

Recently Sillito and coworkers[18, 12] demonstrated that stimulation beyond the classical receptive field (RF) can not only modulate, but radically change a neuron's response to oriented stimuli. They revealed that patch-suppressed cells when stimulated with contrasting orientations inside and outside their classical RF can strongly respond to stimuli oriented orthogonal to their nominal preferred orientation.

Here we analyze the emergence of such complex response patterns[7, 8, 11, 16, 22] in a simple model of primary visual cortex. We show that the observed sensitivity for orientation contrast can be explained by the differential interaction between the local lateral microcircuitry[5, 20, 21, 23] and the long-range lateral connections between iso-oriented domains[6, 15]. The simplicity of our model allows for a rigorous analytical investigation of the mechanisms underlying cross-orientation enhancement in a one-dimensional model. In particular we demonstrate that the observed properties might arise without specific connections between sites with cross-oriented classical RFs.

To be more realistic, we also simulate the visual cortex as a two-dimensional rather than a one-dimensional system, thus gaining an additional degree of freedom. Surprisingly, the existence of strongly contrast-sensitive cells seems now to be restricted to specific regions in this artificial cortex. We interpret this phenomenon in terms of a conflict between the pattern formation process of the neuronal dynamics and the boundary conditions induced by the orientation preference and lateral connection scheme of each cell population.

MODEL

Abstracting from biophysical details, our cortex model consists of a layer of $N \cdot N$ sites, each of them representing a pool of biological neurons (Fig.1) in an orientation column. The stimulus consists of a center patch and a surround annulus of oriented moving gratings. Input from the stimulus is projected onto the cortex via the LGN.

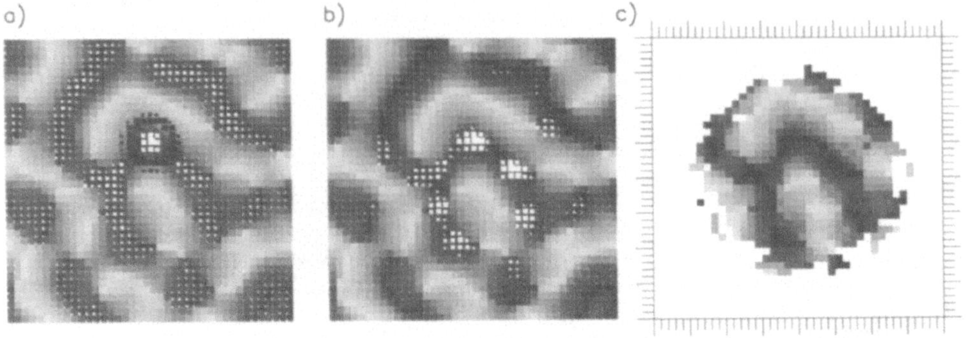

Figure 1. Structure and response properties of the model network. a) Coupling structure from one neuron on a grid of $N \times N = 1600$ elements projected on the orientation preference map that was used for stimulation. Inhibitory and excitatory interactions are marked with black and white squares, respectively, the sizes of which represent the interaction strength. b) Activation patterns of the network driven by a central stimulus of radius $r_c = 11$ and horizontal orientation. c) Self-consistent orientation map calculated from the activation patterns for all stimulus orientations. Note that this map matches the input orientation preference map shown in a) and b).

The divergence of inputs from one specific location on the retina, and the orientation tuning of one cortical unit, are modeled by a convolution with Gaussian functions in cortical and orientation space, respectively.

$$
w(\vec{r_j}, \vec{R_i}) \propto e^{-\frac{1}{2}\left(\frac{\vec{r_j}-\vec{R_i}}{\sigma_{recp,r}}\right)^2} \cdot \left[1 + \epsilon \cdot e^{-\frac{1}{2}\left(\frac{\phi_j-\Phi_i}{\sigma_{recp,\phi}}\right)^2}\right] \tag{1}
$$

$\vec{R}$ is a vector in visual space, $\vec{r}$ in cortical space, Φ the orientation of the stimulus at position $\vec{R}$, and ϕ the preferred orientation of the column at position $\vec{r}$. Note that apart from the divergence of afferent inputs, we always assume strict retinotopy such that $\vec{R_i} = \vec{r_j}$ for $i = j$. The orientation tuning ϕ_j of column j is assigned from an orientation preference map obtained by optical imaging from T. Bonhoeffer. This orientation preference shows up as a weak tuning ϵ of the LGN input. Due to the broad divergence $\sigma_{recp,r}$ of the afferent input from one location on the retina, the column which has its classical RF within the center patch gets its strongest afferent input from the center stimulus, but also weak subthreshold input from the surround stimulus. This assumption is in agreement with the definition of the classical RF that is not the whole region in the visual field providing input to one neuron, but only the subregion which elicits a response from this neuron (suprathreshold input).

The cortical columns are built up by excitatory and inhibitory cell populations, which are reciprocally connected together with weights $w_{ex}^{ex,in}$ and $w_{in}^{ex,in}$ (Fig.2 left). Each population receives afferent input from LGN and is connected to other columns via lateral connections. Dendrites originating from excitatory cells (pyramidal cells) extend over distances $\sigma_{ex,r}$ up to several hypercolumns, but are strongly orientation selective[17] ($\sigma_{ex,\phi} < \pi/4$). Therefore only columns having the same or a similar orientation make excitatory synapses onto each other. Inhibitory couplings made by interneurons and smooth cells act locally and do not extend over more than one hypercolumn ($\sigma_{in,r}$ small), but they are unspecific and connect columns with arbitrary orientation preferences ($\sigma_{in,\phi} >> 1$). Together with the pre-determined orientation preference map, this coupling scheme leads locally to a mexican-hat shaped coupling structure, and globally to a patchy connection pattern (Fig.1 a)) that has also been

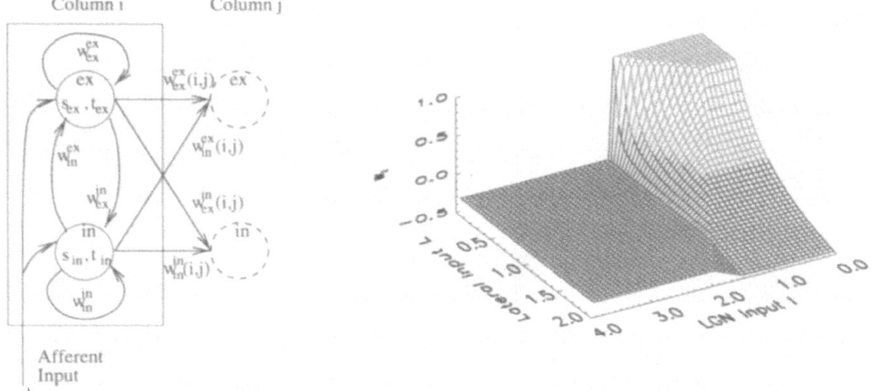

Figure 2. Cortical coupling structure induces differential interaction. Left: Afferent input projects from LGN onto each column i which consists of an excitatory (ex) and an inhibitory (in) cell population. Each population has a piecewise linear gain function which is $g(I) := s \cdot (I - t)$ with slope $s_{ex,in}$ and threshold $t_{ex,in}$ for $I > t$, and $g(I) := 0$ otherwise. The populations in the same column i (solid line) and neighboring columns j (dotted lines) are coupled with weights $w_{ex,in}^{ex,in}$. Right: Representation of the differential interaction ω_l of the cortical microcircuit. If the input I from the LGN and the input L from the lateral connections is low, $\omega_l > 0$ (light grey). If either I or L is large, $\omega_l < 0$ (dark grey).

found anatomically[3, 9, 10, 13, 14, 17].

$$w_{ex}^{ex,in}(\vec{r_i}, \vec{r_j}) \propto e^{-\frac{1}{2}\left(\frac{\vec{r_i}-\vec{r_j}}{\sigma_{ex,r}}\right)^2} \cdot e^{-\frac{1}{2}\left(\frac{\vec{\phi_i}-\vec{\phi_j}}{\sigma_{ex,\phi}}\right)^2}$$

$$w_{in}^{ex,in}(\vec{r_i}, \vec{r_j}) \propto e^{-\frac{1}{2}\left(\frac{\vec{r_i}-\vec{r_j}}{\sigma_{in,r}}\right)^2} \cdot e^{-\frac{1}{2}\left(\frac{\vec{\phi_i}-\vec{\phi_j}}{\sigma_{in,\phi}}\right)^2} \qquad (2)$$

Subscripts indicate the origin, and superscripts the destination of the interaction, respectively. Each cell population is modeled by a single differential equation which describes the dynamics of the firing rate in dependence of the synaptic input I to that population[24].

$$\tau_{ex}\dot{a}_{ex} = -a_{ex}^{ex} + g(I_{ex}^{ex} + I_{ex}^{in} + I_{ex}^{aff})$$

$$\tau_{in}\dot{a}_{in} = -a_{in}^{ex} + g(I_{in}^{ex} + I_{in}^{in} + I_{in}^{aff}) \qquad (3)$$

If both populations of one column are activated, additional lateral excitatory input changes the activation of the excitatory population in two counteracting ways. First, the population is excited directly, and second, the population is inhibited indirectly by the increased activation of the inhibitory interneurons. We assume that the combination of these inputs is negative, which can be achieved by a higher threshold and higher gain of the inhibitory population[21, 22], but more generally holds true if the following equation is fulfilled for each column in a simplified two-columnar model

$$1 + s_{in}(w_{in}^{in} - \epsilon w_{in}^{ex}) < 0 \qquad (4)$$

The factor ϵ describes the relation between the long-range lateral input on the excitatory population vs. the long-range input to the inhibitory population.

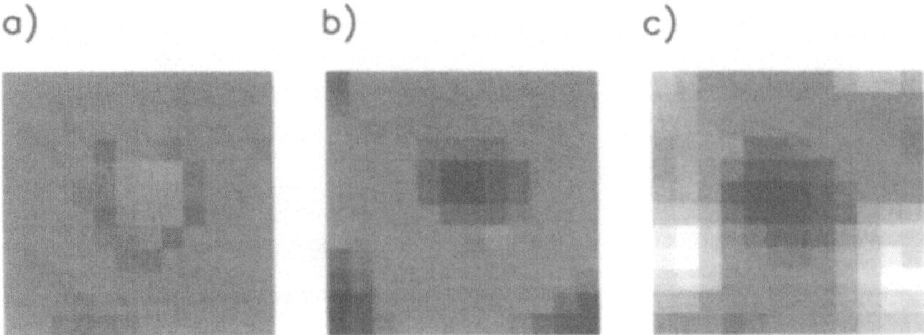

Figure 3. Changes in pattern of activity on the surface of the cortex induced by the additional presentation of a surround stimulus. Grey levels encode increase (darker grey) or decrease (lighter grey) in activation. a) center and surround parallel, b) and c) center and surround orthogonal. While in b), the center is stimulated with the preferred orientation, in c), the center is stimulated with the non-preferred orientation.

Contrarily, we find that as long as the inhibitory subpopulation remains inactivated, additional lateral excitatory input will excite only the excitatory subpopulation and increase its activity. Conclusively, long-range interaction may be effectively positive or negative, depending on the target column's activation. This differential interaction can be expressed in terms of an "effective" weight ω_L that is positive or negative if the effective interaction acts excitatorily or inhibitorily, respectively (Fig.2 right).

MECHANISMS

We assumed a rather weak selectivity of the afferent connections and a restricted contrast, which implies that every stimulus provides some input also to orthogonally tuned cells. This means that long-range excitatory connections, while not effective when only the surround is stimulated, can very well be sufficient for driving cells if the stimulus to the center is orthogonal to their preferred orientation. We want to present three examples of nonclassical RF phenomena reproduced in our model, and explain their mechanisms.

a) Surround suppression

First, we stimulate a cell with its preferred orientation in its classical RF. Adding an iso-oriented surround stimulus now activates many of the orientation columns which make long-range lateral connections to that cell. This interaction is negative, because the target neuron is highly activated, resulting in a suppression of activity (Fig.3 a)).

b) Orientation contrast enhancement

Now, we change the surround stimulus to the orthogonal orientation. The surrounding columns with the preferred orientation of our target cell now become inactivated. Therefore the long-range interaction and suppression is removed and the cell's activation gets effectively enhanced. The activation of the surrounding columns with orthogonal orientation preference does not contribute because long-range interactions are, as assumed, orientation selective (Fig.3 b)).

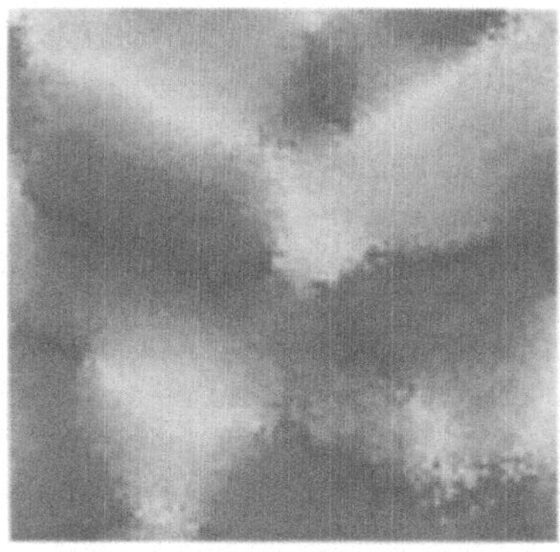 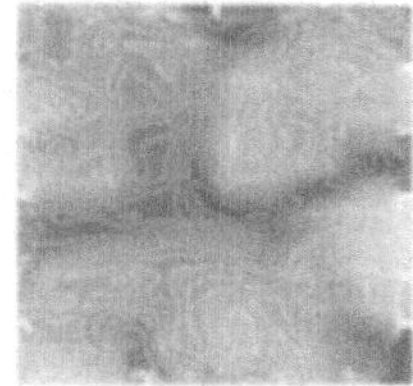

Figure 4. Strength of nonclassical receptive field properties (right) at different positions in the cortex. Light regions indicate strong orientation contrast sensitivity. Comparing this map with the orientation preference topography (left), we find that orientation contrast sensitivity is weakest (dark regions) at pinwheel positions or fractures.

c) Orientation contrast facilitation

Finally, we stimulate the cell with its non-preferred orientation in its classical RF. As expected, the cell remains inactivated. Adding a surround stimulus with preferred orientation activates the long-range lateral connections to this cell. This additional input can lead to an activation of the target cell, thus facilitating its response to non-optimal stimuli (Fig.3 c)).

In b), it is even possible that the activation of the cell gets even higher than stimulated with the center stimulus alone, because the combined stimulus provides an additional small afferent input from the surround. This phenomenon has already seen by Sillito et. al.[18], where the firing rate sometimes increases by a factor of two and more. These results show a surprising agreement with previous findings on nonclassical receptive field properties that culminated in the paper by Sillito et. al.[18], and they clearly demonstrate that the known intracortical interactions might lead to surprising effects on receptive fields.

GEOMETRICAL CONSTRAINTS

The mexican-hat shaped coupling structure induces blob-like patches of firing activity[19, 2]. Two blobs can not coexist within a radius being roughly of the same size as the inhibitory surround of one cell population. In one spatial dimension, this interaction leads to a strong competition between blobs, while in two spatial dimensions, the blobs have a much higher ability to shift away from each other.

Investigating cells at different positions in the cortex, we find that the strength of nonclassical receptive field phenomena varies with the location of these cells. Generally, we find numerous orientation contrast sensitive cells located at homogeneous patches of orientation preference. This sensitivity decreases as we approach orientation discontinuities as pinwheels or fractures (Fig.4). We assume that these differences are induced by the local coupling structure[4]. At pinwheels, we have both a smaller direct afferent input, and also a much weaker excitatory feedback due to the specificity of the excitatory couplings. This might be the reason for the observation that not all cells showing patch suppression are orientation selective.

Apart from these considerations, we also observe that there exist regions where columns with orthogonal orientation preferences are closer to each other. Here, activation of one orientation may suppress the activation of the orthogonal orientation by means of the inhibitory surround interaction, even if there is suprathreshold input to the orthogonal orientation column.

In typical one dimensional studies[1, 21], the distance between cells having a fixed orientation difference is always the same. Nevertheless, we should be aware of the differences that in reality are introduced by the typical geometry of orientation preference maps. Response properties can significantly be altered and even the mechanisms working well in one dimension may be set in question. Taking this into consideration, the analytical consequences in a one-dimensional simplified model of coupled orientation columns are in investigation and will be published elsewhere.

REFERENCES

1. H. Bartsch, M. Stetter, and K. Obermayer, Lecture Notes in Computer Science **1327**: Artifical Neural Networks - ICANN '97, ed. W. Gerstner, A. Germond, M. Hasler, and J.-D. Nicoud, Springer-Verlag Berlin/Heidelberg, 237-242 (1997).
2. R. Ben-Yishai, R.L. Bar-Or, and H. Sompolinsky, Proc. Nat. Acad. Sci. **92**, 3844-3848 (1995).
3. A. Das and C.D. Gilbert, Nature **375**, 780-784 (1995).
4. A. Das and C.D. Gilbert, Nature **387**, 594-598 (1997).
5. R.J. Douglas, C. Koch, M. Mahowald, K.A.C. Martin, and H.H. Suarez, Science **269**, 981-985 (1995).
6. U. Ernst, K. Pawelzik, F. Wolf, and T. Geisel, Lecture Notes in Computer Science **1327**: Artifical Neural Networks - ICANN '97, ed. W. Gerstner, A. Germond, M. Hasler, and J.-D. Nicoud, Springer-Verlag Berlin/Heidelberg, 231-236 (1997).
7. C.D. Gilbert and T.N. Wiesel, Vision Res. **30**, 1689-1701 (1990).
8. C.D. Gilbert, Curr. Op. Neurobiol. **6**, 269-274 (1996).
9. A. Grinvald, E.E. Lieke, R.D. Frostig, and R. Hildesheim, J. Neurosci. **14**, 2545-2568 (1994).
10. J.A. Hirsch and C.D. Gilbert, J. Neurosci. **11**, 1800-1809 (1991).
11. J.J. Knierim and D.C. van Essen, J. Neurophys. **67**, 961-980 (1992).
12. J.B. Levitt and J.S. Lund, Nature **387**, 73-76 (1997).
13. R. Malach, Y. Amir, M. Harel, and A. Grinvald, Proc. Nat. Acad. Sci. **90**, 10469-10473 (1993).
14. G. Mitchison and F. Crick, Proc. Nat. Acad. Sci. **79**, 3661-3665 (1982).
15. K.R. Pawelzik, U. Ernst, F. Wolf, and T. Geisel, Advances in Neural Information Processing Systems 9, ed. M.C. Moser, M.I. Jordan, and T. Petsche, MIT Press, 90-96 (1996).
16. U. Polat and D. Sagi, Vision Res. **33**, 993-999 (1993).
17. K.E. Schmidt, D.-S. Kim, W. Singer, T. Bonhoeffer, and S. Löwel, J. Neurosci. **17**, 5480-5492 (1997).
18. A.M. Sillito, K.L. Grieve, H.E. Jones, J. Cudeiro, and J. Davis, Nature **378**, 492-496 (1995).
19. D.C. Somers, S.B. Nelson, and M. Sur, J. Neurosci. **15**, 5448-5465 (1995).
20. H. Suarez, C. Koch, and R. Douglas, J. Neurosci. **15**, 6700-6719 (1995).
21. E. Todorov, A. Siapas, and D. Somers, Advances in Neural Information Processing Systems 9, ed. M.C. Moser, M.I. Jordan, and T. Petsche, MIT Press, 118-124 (1996).
22. L.J. Toth, S.C. Rao, D.-S. Kim, D. Somers, and M. Sur PNAS **93**, 9869-9874 (1996).
23. M.V. Tsodyks, W.E. Skaggs, T.J. Sejnowski, and B.L. McNaughton, J. Neurosci. **17**, 4382-4388 (1997).
24. H.R. Wilson and J. Cowan, Biol. Cyb. **13**, 55-80 (1973).

CARBACHOL-INDUCED RHYTHMS IN THE HIPPOCAMPAL SLICE: SLOW (.5-2HZ), THETA (4-10HZ) AND GAMMA (80-100HZ) OSCILLATIONS.

Jean-Marc Fellous, Taylor Jonhston, Michele Segal and John Lisman

Volen Center for Complex Systems
Brandeis University
Waltham, MA 02254

INTRODUCTION

The hippocampus is the locus of various in vivo brain rhythms. Early work in the freely moving rat showed that three types of hippocampal oscillatory activity could be detected, depending on the behavior of the animal[1]. Low frequency (.5-2 Hz) irregular oscillations predominate during slow wave sleep, and are completely absent during walking. A medium frequency (5-10 Hz) rhythmical component predominates during walking behavior or REM sleep, and is absent during slow wave sleep. Finally, a fast oscillatory component (40-100 Hz) can be observed during REM sleep or walking. The neuronal circuits involved in each of these oscillations are still largely unknown, and are likely to involve the complex interplay between intrinsic cellular and synaptic hippocampal properties, and external rhythmic inputs from subcortical areas. In particular, the septal cholinergic projection to the hippocampus has been shown to significantly contribute to some of these rhythms. Here we focus on the effects of carbachol (CCH, a cholinergic agonist) on the intrinsic hippocampal circuitry. Using field recordings, we show that rhythms in these three frequency ranges may be observed in vitro, and are therefore likely to be the result of synchronized population activity. Further work in this system will allow for a detailed exploration of the neural circuitry involved and its computational role in learning and memory.

METHODS

Long-Evans rats (20-30 days) were used to obtain 400 μm thick transversal hippocampal slices. Slices were submerged in ACSF (mM: NACl, 124; NaH2CO3, 26; D-glucose, 10; KCl, 5; CaCl2, 2; MgSO4, 2; NaH2PO4, 1.2) at 31-32 °C and perfused at constant flow (2ml/min). Electrophysiological recordings in CA1 were achieved using glass microelectrodes (ACSF filled, 300-400 KΩ). All drugs were freshly prepared in ACSF and bath applied. Stimulation was administered through a unipolar glass electrode, filled with ACSF, and placed in the Stratum Radiatum, close to CA3. Stimulation and recording were monitored by oscilloscope and computer, and saved on disk for off-line analysis. Data analyses were performed by programs written in C, and using Matlab.

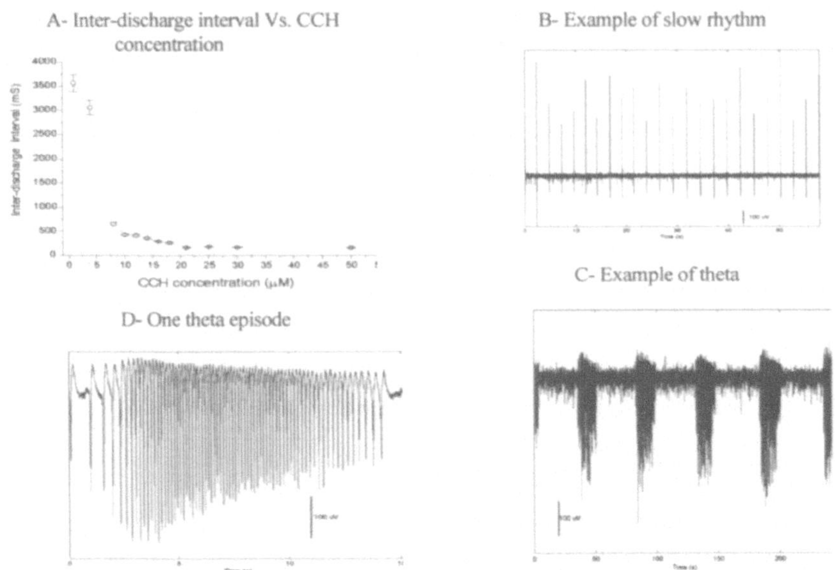

Fig1: Spontaneous carbachol-induced rhythms in the transversal hippocampal slice. A - Concentration-response curve: Each point is the average of at least 3 slices. B - Example of the slow rhythm observed at low CCH concentrations. C- Example of theta-like rhythm observed at higher CCH concentrations. D- Only one theta episode is represented. Note the marked difference of patterns and rhythm around 10 μM CCH.

RESULTS

We used standard field recording methods to record population activity in CA1. In standard ACSF no population activity was recorded. We then added carbachol to the solution. After about 1 minute, large spontaneous field events were recorded as shown in Figs. 1B, 1C. These events were typically in the 80-300 μV range. The events were rhythmic and we characterized their frequency by averaging their inter-discharge time interval over more than 10 minutes of recording. When carbachol was removed, or atropine sulfate added (data not shown), these discharges disappeared.

To study the effect of carbachol concentration on this rhythmicity, different doses were added. Each slice was first bathed in low CCH (<10 μM), then washed with ACSF only, and then bathed in high CCH (>20 μM). Low and high concentrations were paired so that the difference in the 2 concentrations used for each slice was maximal. Concentrations between 10 and 20 μM were not paired with any other concentrations.

At low concentrations (< 10μM), CCH induces slow rhythmic population discharges of frequencies between .5 and 2 Hz and of large irregular amplitudes (Figs. 1A, 1B). At higher concentrations (between 20 and 60 μM), the pattern of discharge changes to bouts of 10-15sec rhythmical discharges at theta-like frequencies (4-10Hz), followed by 15-25sec 'silence' (Figs 1A, 1C). Finally, higher CCH concentrations (> 80μM) increase background cellular activity, and with the occasional exception of one or two initial theta-like episodes, are not followed by consistent population discharges.

Because of the technique used here, it is not possible to observe the behavior of individual cells during each of the slow or theta-like oscillations. However, an analysis of the amplitude, half-width and area of the waveform generated by population discharge provide indirect evidence about the size of the population involved, and its synchrony.

Six episodes of spontaneous slow rhythm (8µM CCH) and six episodes of spontaneous theta-like rhythm (25µM CCH) were recorded from each of three slices. Fig. 2 shows plots of the amplitude, area and half width of each of the population discharges recorded. Slow oscillation amplitudes proved to be about twice as large as theta amplitudes. This result suggests that either more cells contribute to the slow rhythm than the theta rhythm, or that both rhythms involve the same number of cells, but that discharges are more synchronous in the case of slow rhythmic activity. The analysis of the surface area of these discharges evidences the same differences as seen for amplitudes. Slow rhythmic discharge area is about twice that of theta. This finding indicates that, on average, the synchrony of the discharges is similar between the two rhythms, leading to the conclusion that the slow rhythmic activity involves twice more cells than theta. However, plots of the individual discharge half- width shows two different subsets of discharges within theta episodes. Most of the population events appear to be more synchronous (n=339) than slow rhythmic discharges, with small half-widths (M= 0.5237), while there is a significant group of events (n=126) that are 2.8 times less synchronized than slow rhythmic discharges (M= 2.816). The first group indicates that, during these discharges, fewer cells are involved in theta than in the slow rhythm (more synchrony, but half amplitude) while the second group indicates that, during these discharges, the same number of cells might be active in both rhythms.

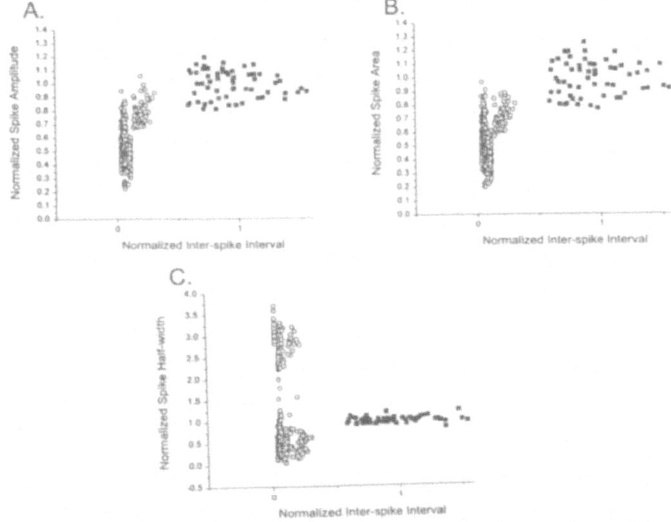

Fig2: Sow rhythm and theta-like population discharge (spike) amplitude, half-width and area. The discharge parameters (amplitude, half width and area) were quantified and normalized across slices as a percentage of the mean slow rhythm parameters for a particular slice (i.e. amplitude of episode 1a is given as percentage of the average slow rhythm amplitude for slice 1). Data were normalized with respect to the slow rhythm rather than the CCH-theta oscillation because the slow rhythm variability is significantly less than theta variability. Filled symbols: Slow rhythm, empty symbols theta-like rhythm.

We next assessed whether theta-like episodes can be elicited by stimulation of the network, in condition in which they would not occur spontaneously. Slices were bathed in 8µM CCH, and stimulation was given randomly between slow activity discharges. Several stimulus protocols were tried. We found that a theta-like pattern of four shocks that were able to elicit population discharge were successful in initiating theta-like episodes. This initiation did not however systematically occur, and in most cases failed. Moreover, the amplitude and frequency of the elicited episodes did not depend on stimulation strength (data not shown). Interestingly, in most cases where initiation did not occur, gamma-like oscillations could be observed as a result of some of the four shocks (Fig 3A). Single shocks delivered randomly during a spontaneous CCH-theta episodes (with 25 μM CCH) were however systematically successful in resetting the rhythm. With larger stimulus strength, this stimulation terminated the ongoing oscillatory episode (data not shown).

Finally, a closer inspection of the spontaneous rhythms induced by low CCH concentrations (8μM, Fig 3B), and higher concentrations (25 μM, Fig 3C) reveals that these two rhythms may be coupled with a spontaneous 'trailing' gamma-like oscillation of a high frequency (80-100Hz). The nature of the coupling was not investigated further here, and appeared to occur randomly.

DISCUSSION

Recent studies have shown that a slow rhythm (<1Hz) can be found in intracellular recordings in cortical association areas 5 and 7, motor areas 4 and 6 and visual areas 17 and 18 in cats[2] and in the thalamus[3]. It is also found in the EEG of naturally sleeping cats and humans[2]. This rhythm is generated by a complex interaction between pyramidal cells and local inhibitory neurons in the cortex and, unlike delta and spindle rhythms can exist without the thalamus[4,5]. Interestingly, the excitation of the pedunculopontine tegmental cholinergic nucleus abolishes the slow rhythm in single cells, shifts the EEG toward higher frequencies, and involves muscarinic receptor activation. The same effect has been described with stimulation of the locus coeruleus[6]. These results suggest that ACh (and NE) may participate in the regulation of this rhythm. Because it is found in so many brain areas, this rhythm must be an essential property of neocortical tissue, including the hippocampus, and might depend on cholinergic receptor activation.

In the hippocampus of urethane-anesthetized rats, oscillatory activity of a frequency as low as .5 Hz and large amplitude may also be detected[7]. This large irregular activity (LIA) appears in alternation with another oscillatory activity, theta, occurring at higher frequency and which will be discussed below. Of interest to the present study, is the finding that some pyramidal cells are sensitive to ACh levels, such that a reduction of ACh induces LIA activity, while an increase in ACh induces theta activity.

In the hippocampal slice, we showed that a slow rhythm can be found in low concentrations of CCH, and that it was the result of the synchronous activity of a large population of cells (Fig 1). In higher CCH concentrations, this rhythm gives way to a theta-like pattern of activity (see below). The technique used here does not allow for a detailed comparison with in vivo rhythms such as the slow cortical rhythm or hippocampal LIA mentioned above.

The theta rhythm has been implicated in several brain functions including sensory processing, memory and the control of voluntary movement[7-9]. In vivo studies suggest that the hippocampal theta rhythm depends on GABAergic and cholinergic inputs from the septum and have distinct physiological effects[10-13]. In vivo stimulations of the septum show that this rhythm can be initiated and reset by co-activation of both of the GABAergic and cholinergic septohippocampal pathways.

In awake animals, stimulation of the septum, as well as external sensory stimulation[14, 15], are capable of resetting the theta rhythm[16]. Cortical stimulations, on the other hand, are prevented from reaching the hippocampus if they occur during a natural or elicited theta episode[17, 18]. Other studies have shown that theta episodes recorded in the hippocampus can be elicited by stimulation of hypothalamo-septal fibers[19], stimulation of the reticular formation[20, 21], or by hippocampal infusion of CCH[22].

Previous studies have demonstrated that the hippocampus is capable of 'resonating' at theta frequency when activated by muscarinic receptor agonists, in slice preparations where septal inputs are absent[23-27]. Unlike in vivo, this rhythm is essentially generated in the CA3 region, and is mainly mediated by non-NMDA excitatory synaptic transmission. Together, these findings suggest that in vitro theta oscillations are the result of the complex interplay between synaptic and intrinsic phenomena that involve some (but not all) of the same mechanisms implicated in the generation and maintenance of in vivo theta.

Here we found that the carbachol-induced theta generated in CA3, and measured at the output of the network, in CA1, can be initiated, and reset by external stimulation (data not shown). We also showed that, in comparison with the slower rhythm described above (Fig 2), CCH-induced theta-like activity may involve a large fraction (if not all) of CA1 cells, and theta no relations between discharge amplitude and stimulus strength were found when the theta episodes were elicited. Because all CA1 cells are probably involved in each discharge, their firing does not represent an informational code. Within CA3, however, this oscillation does not involve all cells[24], and what we observe here in CA1 may be an indication that coactive subsets of CA3 cells might successfully drive nearly all of CA1 cells.

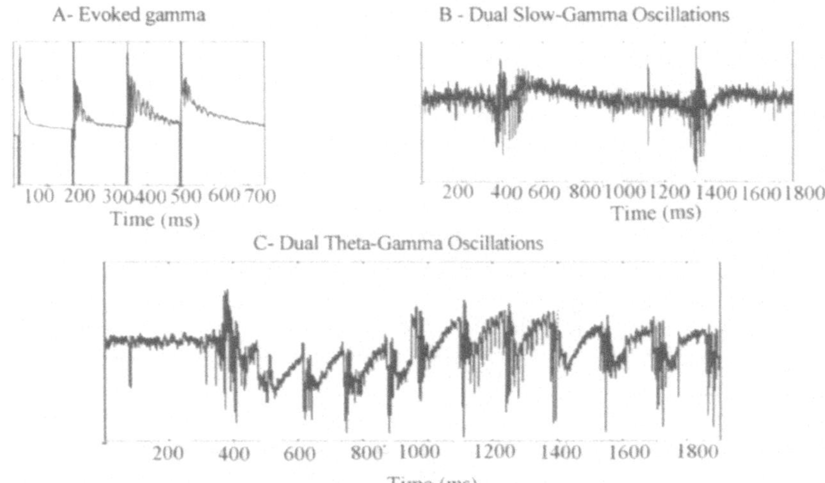

Fig 3: Gamma oscillations.
A- evoked in 8uMCCH, theta pattern of stimulation. B and C- spontaneously found mixed with slow episodes (in 8μM CCH) or theta (in 25 μM CCH) in some cases.

Synchronous gamma (40-100 Hz) activity has been observed in numerous brain structures[28-30]. In the hippocampus, this oscillation appears to be controlled by the dentate gyrus and forwarded to CA3 and CA1 in a less coherent manner[31]. However, lesions of the entorhinal cortex shifts the site of gamma oscillation generation from the dentate, to the CA3-CA1 system, indicating that, by itself, CA3-CA1 is able to oscillate at this frequency. Interestingly, there is also a strong correlation of amplitude and frequency between theta and gamma activity, even when the entorhinal cortex is removed. Recent work has also shown that coupled gamma and theta oscillations can be recorded in the entorhinal cortex, and that individual cells discharge in temporally defined gamma windows, within a theta cycle[32].

Gamma can be obtained in vitro by repeated stimulations of CA1 afferents, in the absence of any drugs[33]. Pressure ejection of glutamate, in spite of blockade of AMPA, NMDA and GABAb transmission, may also elicit this oscillation, while GABAa antagonists or metabotropic glutamate receptors antagonists block it. These results support the hypothesis that in the slice, gamma oscillations can be generated by GABAa synaptic transmission[34, 35], coupled with cellular mechanisms downstream to metabotropic receptors activation.

We showed here that gamma-like oscillations could be found in vitro spontaneously coupled with either of the slow or theta-like rhythms. This oscillation could also be observed after repeated stimulation. Since metabotropic glutamate receptor activation can produce gamma oscillations, and since both muscarinic and metabotropic receptors activate inositol triphosphate intracellular pathways, this result suggests that gamma discharges might depend on the activation of a common inositol triphosphate pathway.

ACKNOWLEDGMENT

We thank Chong-Hyun Kim, Nick and Nona Otmakhov and Ed Richard for their support throughout this study. Supported by the Sloan Foundation and by the Alzheimer's association RG3-96-015.

REFERENCES

1. Leung, L.W., Lopes da Silva, F.H. & Wadman, W.J. *Electroencephalogr Clin Neurophysiol* **54**, 203-19 (1982).
2. Steriade, M., Nunez, A. & Amzica, F. *J Neurosci* **13**, 3252-65 (1993).
3. Steriade, M., Contreras, D., Curro Dossi, R. & Nunez, A. *J Neurosci* **13**, 3284-99 (1993).
4. Steriade, M., Nunez, A. & Amzica, F. *J Neurosci* **13**, 3266-83 (1993).
5. Timofeev, I. & Steriade, M. *J Neurophysiol* **76**, 4152-68 (1996).
6. Steriade, M., Amzica, F. & Nunez, A. *J Neurophysiol* **70**, 1385-400 (1993).
7. Bland, B.H. & Colom, L.V. *Prog Neurobiol* **41**, 157-208 (1993).
8. Bland, B.H. *Prog Neurobiol* **26**, 1-54 (1986).
9. Vinogradova, O.S. *Prog Neurobiol* **45**, 523-83 (1995).
10. Stewart, M. & Fox, S.E. *Trends Neurosci* **13**, 163-8 (1990).
11. Brazhnik, E.S. & Fox, S.E. *Exp Brain Res* **114**, 442-53 (1997).
12. Ylinen, A., *et al. Hippocampus* **5**, 78-90 (1995).
13. Soltesz, I. & Deschenes, M. *J Neurophysiol* **70**, 97-116 (1993).
14. Vinogradova, O.S., Brazhnik, E.S., Kitchigina, V.F. & Stafekhina, V.S. *Neuroscience* **53**, 993-1007 (1993).
15. Buno, W., Jr., Garcia-Sanchez, J.L. & Garcia-Austt, E. *Brain Res Bull* **3**, 21-8 (1978).
16. Buzsaki, G., Grastyan, E., Czopf, J., Kellenyi, L. & Prohaska, O. *Brain Res* **225**, 235-47 (1981).
17. Vinogradova, O.S., Brazhnik, E.S., Stafekhina, V.S. & Kitchigina, V.F. *Neuroscience* **53**, 981-91 (1993).
18. Vinogradova, O.S., Brazhnik, E.S., Kichigina, V.F. & Stafekhina, V.S. *Neur. Behav Phys.* **26**, 113-24 (1996).
19. Smythe, J.W., Christie, B.R., Colom, L.V., Lawson, V.H. & Bland, B.H. *J Neurosci* **11**, 2241-8 (1991).
20. Vertes, R.P. *Prog Neurobiol* **19**, 159-86 (1982).
21. Kirk, I.J. & McNaughton, N. *Hippocampus* **3**, 517-25 (1993).
22. Oddie, S.D., Bland, B.H., Colom, L.V. & Vertes, R.P. *Hippocampus* **4**, 454-73 (1994).
23. Konopacki, J., MacIver, M.B., Bland, B.H. & Roth, S.H. *Brain Res* **405**, 196-8 (1987).
24. MacVicar, B.A. & Tse, F.W. *J Physiol (Lond)* **417**, 197-212 (1989).
25. Bland, B.H., Colom, L.V., Konopacki, J. & Roth, S.H. *Brain Res* **447**, 364-8 (1988).
26. Williams, J.H. & Kauer, J.A. *J Neurophysiol* **78**, 2631-40 (1997).
27. Traub, R.D., Miles, R. & Wong, R.K. *Science* **243**, 1319-25 (1989).
28. Demiralp, T., Basar-Eroglu, C. & Basar, E. *Int J Neurosci* **84**, 1-13 (1996).
29. Gray, C.M. *J Comput Neurosci* **1**, 11-38 (1994).
30. Bressler, S.L. *Trends Neurosci* **13**, 161-2 (1990).
31. Bragin, A., *et al. J Neurosci* **15**, 47-60 (1995).
32. Chrobak, J.J. & Buzsáki, G. *The Journal of Neuroscience* **18**, 388-398 (1998).
33. Whittington, M.A., Traub, R.D. & Jefferys, J.G. *Nature* **373**, 612-5 (1995).
34. Traub, R.D., Whittington, M.A., Colling, S.B., Buzsaki, G. & Jefferys, J.G. *J Physiol* **493**, 471-84 (1996).
35. Wang, X.J. & Buzsaki, G. *J Neurosci* **16**, 6402-13 (1996).

CHEMOTAXIS CONTROL BY LINEAR RECURRENT NETWORKS

Thomas C. Ferrée and Shawn R. Lockery

Institute of Neuroscience
University of Oregon
Eugene, OR 97403

INTRODUCTION

The nematode *Caenorhabditis elegans* provides an excellent opportunity to study biological computation. This is mainly because it has a very simple neuromuscular system, consisting of only 302 neurons and 95 muscle cells. Anatomical studies have revealed the morphology of every neuron and the location of nearly every electrical and chemical synapse[1], and it has recently become possible to make electrophysiological recordings from identified neurons in *C. elegans*[2].

Among its behaviors is chemotaxis[3], the ability to orient up gradients of chemical attractants (Fig. 1a). Laser ablation experiments have identified which neurons are important for chemotaxis in *C. elegans*[4]. The chemotaxis control circuit appears to involve a small number of chemosensory neurons, interneurons and motor neurons, with feedforward and feedback connections among them. In this paper, we seek to understand how small networks of this type might control chemotaxis in a nematode-like body.

LINEAR MODEL OF THE NEMATODE *C. ELEGANS*

In a recent paper[5], we derived a simple model of the nematode body and chemotaxis control circuit. The body model was based on the observation that, during much of their movement on a flat surface, nematodes move forward at constant speed[6]. Thus

$$\frac{dx}{dt} = v \, \cos\theta \tag{1}$$

Computational Neuroscience
edited by Bower, Plenum Press, New York, 1998

$$\frac{dy}{dt} = v \sin \theta \tag{2}$$

where v is the speed, and θ is the direction of movement. The rate of turning of the model nematode was determined by the degree of bend in the neck, which in turn was determined by the relative voltage of dorsal (D) and ventral (V) motor neurons. Thus

$$\frac{d\theta}{dt} = \gamma \left(V_{\mathrm{D}} - V_{\mathrm{V}}\right) \tag{3}$$

where γ is a constant parameter. In the present model, however, lateral movement of the head during ordinary, sinusoidal locomotion was ignored. This model focuses, therefore, on chemotaxis control strategies which can be characterized by movement of the body center-of-mass alone.

The model network included a single chemosensory neuron (V_1), three interneurons (V_2–V_4) and two motor neurons (V_{D} and V_{V}). Each neuron was connected to each other neuron. Electrophysiological experiments suggest that many *C. elegans* neurons are effectively isopotential, and do not fire classical, all-or-none action potentials[2]. Simultaneous recordings from pairs of neurons in *Ascaris suum*, a related species of nematode, suggest that nematode chemical synapses release neurotransmitter tonically, and that postsynaptic voltage is a sigmoidal function of presynaptic voltage at steady state[7]. We therefore modeled the voltage of each neuron as

$$\tau_i \frac{dV_i}{dt} = -V_i + \tanh\left(\sum_{j=1}^{N} w_{ij} \left(V_j - \overline{V}_j\right)\right) + \delta_{i1} C(t) \tag{4}$$

where w_{ij} represents the strength of the synaptic connection from cell j to cell i, and $\overline{V}_j$ represents the presynaptic voltage at which no transmitter is released. Since *C. elegans* is believed to sense the chemical concentration at only a single point in space[3], we took $C(t) \equiv C(x(t), y(t))$, the instantaneous chemical concentration at the tip of the nose. We then used simulated annealing to find sets of network parameters: τ_i, w_{ij} and $\overline{V}_j$, which produced chemotaxis in the model. Many such sets were found. Because the network model was nonlinear, however, the neural computations underlying chemotaxis control in the model remained elusive.

While the chemotaxis network in *C. elegans* is likely to be nonlinear, chemotaxis control by this network may be effectively linear. To demonstrate that small linear networks of graded-potential neurons are capable of controlling nematode-like chemotaxis, here we linearized the voltage dependence of (4). Because recordings from *Ascaris suum* suggest that nematode neurons tend to rest near the middle of their voltage-sensitive range for neurotransmitter release, this linearization was carried out about the voltage $\overline{V}_j$ for each neuron. The resulting linear network can be written in the form

$$\frac{dV_i}{dt} = \sum_{j=1}^{N} A_{ij} V_j + b_i + c_i(t) \tag{5}$$

where the matrix $A_{ij} = (-\delta_{ij} + w_{ij})/\tau_i$, the vector $b_i = -\sum_{j=1}^{N} w_{ij}\overline{V}_j/\tau_i$, and the time-dependent input vector $c_i(t) = \delta_{i1}C(t)/\tau_1$. We again used simulated annealing to find parameter sets which produced chemotaxis. Figure 1b shows the result of one such optimization, for one initial condition. Like the real nematode in Fig. 1a, the model nematode oriented almost directly up the gradient and dwelled at the peak.

Similar chemotaxis behavior was observed for a variety of other initial conditions. Similar behavior was also observed for other linear networks obtained by this optimization procedure.

COMPUTATIONAL RULES FOR CHEMOTAXIS

Because equations (5) are linear in voltage, they can be solved analytically. With a convenient choice of initial conditions $V_i(t_0)$, the solution can be written in matrix form

$$\mathbf{V}(t) = -\mathbf{A}^{-1}\,\mathbf{b} + \int_{t_0}^{t} \mathbf{K}(t - t')\,\mathbf{c}(t')\,dt' \tag{6}$$

where the kernel $\mathbf{K}(t - t')$ is an $N \times N$ matrix involving the eigenvalues λ_i and eigenvectors of the matrix $\mathbf{A}$. Inserting (6) into (3) and using the fact that our model has only a single chemosensory neuron leads to

$$\frac{d\theta}{dt} = \Omega_{\text{bias}} + \int_{t_0}^{t} k(t - t')\,C(t')\,dt' \tag{7}$$

where Ω_{bias} is a constant turning bias, and the scalar kernel $k(t - t')$ is the impulse response of the turning rate to chemosensory input at the nose.

The second term in (7) indicates that, formally, the rate of turning at time t depends on the entire history of the stimulus $C(t')$. Previous studies of chemotaxis, however, have shown that many animals use time-derivatives of the stimulus to direct oriented movement[8]. We therefore sought an expansion of (7) in terms of the instantaneous stimulus $C(t)$ and its time-derivatives. Mathematically, this expansion can be justified as follows: The kernel $k(t - t')$ is a sum of exponentials of the form $\exp(\lambda_i(t - t'))$. For the system to be stable, i.e., for the voltages to remain finite at all times t, each eigenvalue λ_i must have negative real part. This condition is indeed satisfied by the network which produced Fig. 1b, and must be satisfied by any network optimized to control chemotaxis. Consequently, in practice the second term in (7) accumulates contributions from the stimulus $C(t')$ for times t' only in the *recent* past. We therefore expanded $C(t')$ in a Taylor series about the time t:

$$C(t') = C(t) + \frac{dC}{dt}(t' - t) + \ldots \tag{8}$$

Substituting (8) into (7) and regrouping terms gives

$$\frac{d\theta}{dt} = \Omega_{\text{bias}} + Z_0(t - t_0)C(t) + Z_1(t - t_0)\frac{dC}{dt} + \ldots \tag{9}$$

where the expansion coefficients are defined

$$Z_n(t - t_0) = \int_{t_0}^{t} k(t - t')\,(t' - t)^n\,dt' \tag{10}$$

Note that the functions $Z_n(t - t_0)$ do not depend on the stimulus, but are intrinsic to the network. Equation (9) constitutes a simple computational rule, which can be used in place of the network (5) to control chemotaxis in the model nematode.

To demonstrate that (9) controls chemotaxis effectively, we computed the expansion coefficients $Z_n(t - t_0)$ from the network parameters which generated Fig. 1b. Figure

1c shows the result of keeping only terms through $C(t)$. The initial transient behavior results from the explicit time dependence in $Z_0(t - t_0)$. This rule failed to produce chemotaxis, but instead led to circular movement whose center traveled along a line of equal concentration. Figure 1d shows the result of keeping all terms through dC/dt. This rule led to oriented movement up the gradient and dwelling at the peak, much like the network behavior shown in Fig. 1b. Similar results were obtained for other initial positions and angles. This suggests that in order to control chemotaxis, the linear network computes the first time-derivative of the chemical stimulus.

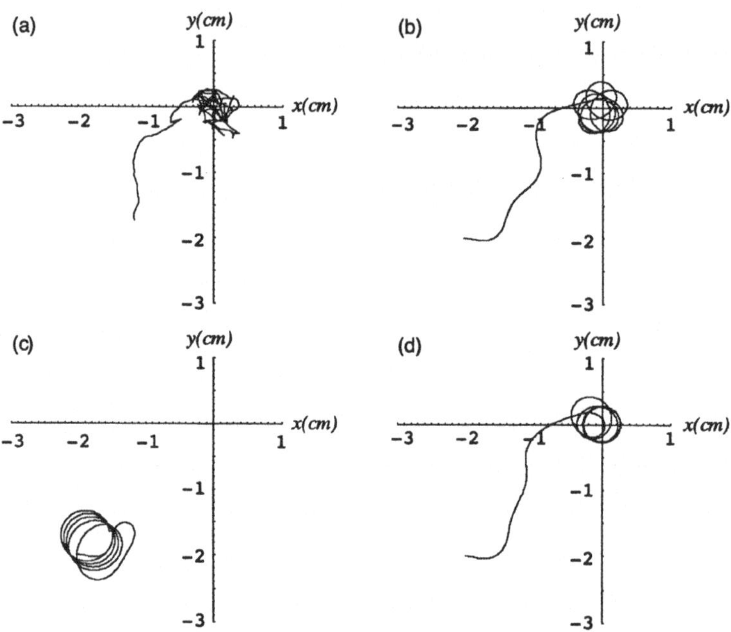

Figure 1: Nematodes chemotaxing in a Gaussian chemical gradient. (a) A real nematode[6]. (b) The model nematode controlled by the linear neural network (5). (c) The model nematode controlled by the computational rule (9), keeping terms only through $C(t)$. (d) Same as (c), but keeping terms through dC/dt.

DISCUSSION

We have shown that small networks of graded-potential neurons with linear voltage dependence are capable of controlling nematode-like chemotaxis. An expansion of the analytic solution in time derivatives of the stimulus led to simple, intuitive rules, which elucidate computations being performed implicitly by the network. The first time-derivative of the stimulus was shown to be necessary to control chemotaxis. These results have three interesting consequences. First, since the model network was linear, these results suggest that much of chemotaxis control in real nematodes may arise from linear computations only. Second, for a fully-connected recurrent network, each eigenvalue and eigenvector depends on *all* of the time constants τ_i and synaptic weights w_{ij}, and each voltage V_i depends on *all* of the eigenvalues and eigenvectors. Information processing in such a system is, in general, highly distributed. Third, since the analytic methods employed here made no assumption about the pattern of synaptic connectivity in the network, these results suggest that linear network models based more closely on the anatomical data are likely to evolve similar computational strategies for chemotaxis.

ACKNOWLEDGEMENTS

The authors thank C. I. Bargmann for providing unpublished data, and J. T. Pierce and T. M. Morse for helpful discussions. This work was supported by NIMH MH11373, NIMH MH51383, NSF IBN 9458102, ONR N00014-94-1-0642, the Sloan Foundation, and the Searle Scholars Program.

REFERENCES

[1] White, J. G., E. Southgate, J. N. Thompson and S. Brenner (1986). The structure of the nervous system of *C. elegans*, *Phil. Trans. R. Soc. London* **314**:1-340.

[2] Goodman, M. B., D. H. Hall, L. Avery and S. R. Lockery (1998). Active currents regulate sensitivity and dynamic range in *C. elegans* neurons. Submitted to *Neuron*.

[3] Ward, S. (1973). Chemotaxis by the nematode *Caenorhabditis elegans*: Identification of attractants and analysis of the response by use of mutants. *Proc. Nat. Acad. Sci. USA* **70**: 817-821.

[4] Bargmann, C. I., unpublished.

[5] Ferrée, T. C., B. A. Marcotte and S. R. Lockery (1997). Neural network models of chemotaxis in the nematode *Caenorhabditis elegans*. In: Mozer, M. C., M. I. Jordan and T. Petsche (Eds.) Advances in Neural Information Processing Systems 9. MIT Press. pp. 55-61.

[6] Pierce, J. T., and S. R. Lockery, unpublished.

[7] Davis, R. E., and A. O. W. Stretton (1989). Signaling properties of *Ascaris* motor neurons: Graded active responses, graded synaptic transmission, and tonic transmitter release, *J. Neurosci.* **9**:415-425.

[8] Dunn, G. A. (1990). Conceptual problems with kinesis and taxis. In: Armitage, J. P. and J. M. Lackie (Eds.) *Biology of the chemotactic response*. Cambridge University Press. pp. 1-13.

A VISUALLY DRIVEN HIPPOCAMPAL PLACE CELL MODEL

Mark C. Fuhs[1], A. David Redish,[2] and David S. Touretzky[1]

[1]Computer Science Dept. and Center for the Neural Basis of Cognition
Carnegie Mellon University
Pittsburgh, PA 15213
[2]Neural Systems, Memory and Aging
Life Science North Bldg, P.O. Box 24-5115
University of Arizona
Tucson, AZ 85724

INTRODUCTION

The firing of place cells in the rodent hippocampus is partly under the control of visual landmarks in the environment (O'Keefe and Conway, 1978). Most place cell models incorporating "visual" input assume noise-free bearing and/or distance information from idealized point objects (e.g., Burgess, Recce, and O'Keefe, 1994; Touretzky and Redish, 1996), rather than attempting to extract landmark information from real-world scenes. This leaves open the question of what kind of visual information is necessary for the hippocampal system to maintain place fields that are stable across trials yet sensitive to landmark position.

In this paper we first describe an algorithm that operates on real images taken from various viewing locations and returns "blob" descriptions: regions of roughly uniform intensity having a rectangular or ovoid shape. We then construct simulated place cells using radial basis functions tuned to blob parameters, and train them by competitive learning to develop realistic place fields. The result is a model that takes real-world scenes as input and produces a distributed activity pattern over a set of place cells as output, from which the current viewing location can be estimated with good accuracy. The implication of this work is that the visual pathway to the rodent hippocampus could involve a relatively simple representation of the local view; rodents may not require object recognition to navigate visually.

VISUAL PROCESSING

Collecting Images of a Scene

We collected twenty-five greyscale images of a computer laboratory using a digital camera atop a mobile robot. The pictures were taken from multiple viewpoints on one side of the room. The twenty-five viewpoints were organized as a 5x5 grid measuring 102 cm on a side.

Characterizing Images Using Blobs

Growing Blobs. Blobs are detected in images using a region growing algorithm. We define a blob B as a set of pixels which must satisfy a set of conditions. We define the blob's mean intensity I_B and mean intensity variation ΔI_B as:

$$I_B = \frac{1}{|B|} \sum_{(x,y)\in B} I(x,y) \tag{1}$$

$$\Delta I_B = \frac{1}{|B|} \sum_{(x,y)\in B} \|\nabla I(x,y)\| \tag{2}$$

Initially, thousands of pixels are randomly chosen from each image, each of which is the starting pixel for a blob. Neighboring pixels to each blob are then added if their intensity is "reasonably" close to the blob's mean intensity; specifically, the difference $|I(x,y) - I_B|$ must be less than a thresholding function T.

The thresholding function T must be defined carefully to allow for differences in the sharpness of object boundaries, and for the effects of non-uniform lighting in the scene. We start by requiring that all pixels simply be within a constant τ of the mean, in which case $T(\tau, ...) = \tau$. We can refine this constraint by scaling T based on local curvature, so that when there is a sharp edge nearby, the threshold is reduced, tightening the criterion for membership. We approximate the local curvature with a five-point discrete Laplacian:

$$\nabla^2 I(x,y) \approx \frac{1}{4}\left[I(x-1,y) + I(x,y-1) + I(x+1,y) + I(x,y+1)\right] - I(x,y) \tag{3}$$

We modulate T by an inverse sigmoidal function s of the Laplacian, so that for regions of high curvature the sigmoid is nearly zero. μ is the mean curvature value for which s should be 0.5, and γ is a gain factor:

$$\sigma\left(\nabla^2 I(x,y), \mu, \gamma\right) = \left(1 + \exp\frac{\nabla^2 I(x,y) - \mu}{\gamma}\right)^{-1} \tag{4}$$

Substituting in the experimentally determined values used in our simulations for μ and γ, the threshold equation becomes:

$$T(\tau, \nabla^2 I(x,y), ...) = \tau \cdot \sigma\left(\nabla^2 I(x,y), 15, 0.5\right) \tag{5}$$

Finally, to handle contiguous surfaces with non-uniform lighting we increase T based on the mean intensity variation ΔI_B. This allows individual blobs to grow over shallow intensity gradients in images, typically due to smooth surfaces that are lit at an angle, over which the threshold would otherwise be too small for blobs to be able to grow. Including this final constraint yields the following threshold equation:

$$T(\tau, \nabla^2 I(x,y), \Delta I_B) = (\tau + \Delta I_B) \cdot \sigma\left(\nabla^2 I(x,y), 15, 0.5\right) \tag{6}$$

In addition to pixels being added to blobs, a pixel that is a member of a blob B_1 may also switch to a neighboring blob B_2 if it satisfies the constraints for pixel addition to B_2 and its intensity more closely matches that of B_2 than B_1.

During the region growing algorithm, the vast majority of pixels were acquired by some blob within the first several iterations. The majority of iterations were therefore dedicated to switching pixels among blobs.

Merging Blobs. Though the region growing algorithm starts with thousands of blobs in order to insure that the entire image can be represented, the images themselves are comprised of far fewer regions of approximately uniform intensity. Without merging blobs, multiple blobs occupying part of the area of a particular region of uniform intensity would compete, and, with few exceptions, none of the blobs would be able to acquire all of the pixels of a particular region. Therefore, neighboring blobs of similar intensity must be merged.

Unfortunately, there is no single threshold of intensity which clearly distinguishes between neighboring blobs in a single region and neighboring blobs in distinct regions with similar intensity; this threshold depends upon the coarseness of features that we wish to detect. Therefore, the threshold was initially set to a low value (15 shades of gray) but was increased as region growing progressed. By intermittently saving the state of each blob at several points during region growing, we were able to create a database of blobs at varying levels of sensitivity to intensity changes. We then compared overlapping blobs collected at different points during region growing; for each set of overlapping blobs, only the blob obtained at the coarsest threshold whose shape was still highly elliptical or rectangular was retained.

Quantitatively Describing Blobs

Describing a blob in terms of the pixels that comprise it is useful for region growing; however, a description of such high dimensionality is far too complex for the purposes of self-localization. Therefore, after blobs were "grown," they went through a series of processing steps which allowed us to describe them using a few simple features. We chose features consistent with the description of an ellipse or rectangle: average intensity, elevation, azimuth, size, eccentricity (the ratio of the lengths of the major and minor axes) and angle of orientation of the major axis.

First, we smoothed the edges and filled in any small holes in the body of the blob, since such high frequency information tended to interfere with categorizing gross features of the blob.

Once blurred, the first moment, or center of mass, of the blob was calculated. The vertical and horizontal positions of the blob center in the image correspond to the elevation and azimuth of the blob, respectively.

A medial-axis transform was then performed on each blob. The medial-axis transform calculates the distance from each pixel to the nearest edge of the blob. The set of pixels whose distances are relative maxima – those pixels whose distance to an edge is greater than all of their neighbors – specifies the "skeleton" of a shape. We fit this skeleton to a line, the slope of which indicated the orientation of the blob's major axis.

Knowing the axis of orientation, we calculated the length of the major and minor axes by calculating the average distances of each pixel, D_x and D_y, from the major and minor axes, respectively. $4D_x$ yields a good approximation of the length of the minor axis; $4D_y$ yields a good approximateion of the length of the major axis. We then calculated the size of the blob as the product of the lengths of the major and minor axes.

Several template superellipses, whose shapes vary between an ellipse and a rectangle depending upon their parameterization, were compared to each blob, and the percentage of overlap of the two shapes was calculated. The highest percentage achieved

Figure 1. Rectangular blobs extracted from a real-world scene.

among the possible templates was then used to select out only those blobs with greater than 90% overlap. (See Figure 1.) This enforces our criterion that blobs be of approximately elliptical or rectangular shape, which ensures that the features extracted from the blob do, in fact, represent its shape accurately.

Once these features had been calculated, the 5x5 grid of blob features was interpolated into a 17x17 grid. This interpolation provided a more continuous view of the environment than the original 5x5 grid.

TRAINING PLACE CELLS

The Place Cell Model

We modeled place cells as radial basis functions. Each place cell was tuned to two blobs that were each described using four features: average intensity, elevation, azimuth, and size. Each place cell was therefore represented by two sets of feature values, M_1 and M_2; this pair of feature values described the pair of blobs that would maximize the place cell's activation. The similarity of a place cell's ideal feature value μ_f to the corresponding feature value of a blob x_f is measured using a Gaussian function:

$$G\left(x_f, \mu_f, \sigma_f\right) = \exp - \left(\frac{x_f - \mu_f}{\sigma_f}\right)^2 \qquad (7)$$

When x_f and μ_f are close, this function will be close to 1; as their difference increases, the function approaches zero. The magnitude of σ_f determines the sensitivity of the function to the difference between x_f and μ_f.

The response of a place cell to a particular blob B_i is determined by the product of feature similarities over the set of features F:

$$A_{B_i} = \prod_{f \in F} G\left(x_f, \mu_f, \sigma_f\right) \qquad (8)$$

All features of B_i must be similar to the place cell's maximal response value in order for the activation to be high; thus, the place cell responds to conjunctions of features.

There is no predetermination of which blobs are to be associated with which place cells. Those blobs B_i and B_j that generated the strongest response to the two sets of place cell feature values M_1 and M_2 determined the activation of the place cell:

$$A = \max_{i \in B} A_{B_i} \cdot \max_{j \in B, j \neq i} A_{B_j} \qquad (9)$$

Adapting Place Field Centers

Place cells were initially tuned to random feature values, which were unlikely to correspond to any particular blob in the environment. Therefore, place fields were initially quite broad and weak, conveying little spatial information. In order to increase the spatial information content, a competitive learning algorithm was used to adapt each place cell feature value μ_f to the corresponding blob feature value x_f. For each iteration t of the competitive algorithm, the place cell whose activation was the highest at a particular grid position was determined to be the "winner" of that grid position. For each position in the grid, the pair of blobs that maximized the activation of the winning place cell were used to train the feature values of the winning place cell. Each feature value of the winning place cell was adapted to the correspoonding blob feature value according to the following equation:

$$\mu_f^{(t+1)} = \alpha x_f + (1 - \alpha) \mu_f^{(t)} \qquad (10)$$

The coefficient α controls the rate at which feature values are adapted; for our simulations the α value of 0.5 was used.

Adapting Place Field Sizes

Overly large place fields lack spatial information content useful for tasks such as navigation; conversely, overly small place fields require too many place cells to represent the entire area of an environment. Therefore, it was important to control the size of the sensitivity parameters σ_f to normalize the sizes of place fields. Unfortunately, there is no single value for σ_f that is appropriate in all conditions, as features such as position and size vary nonlinearly through space.

Therefore, in parallel with the competitive learning algorithm, we adapted the sizes of the place fields by modifying the sensitivity parameters σ_f of the place cells. We defined an "ideal" place field size – in this case, the area around a 5x5 grid position –and used a simple gradient descent technique to adjust the σ_f values to produce the desired place field size.

This algorithm, in conjunction with the aforementioned competitive learning algorithm, was able to quintuple the information content of the place code, using the measure of Skaggs et al. (1993). The total information of the place code was sufficient to distinguish which of the twenty-five images were being presented to the place cells. (See Figure 2.)

DISCUSSION

This work demonstrates that a simple paradigm of feature extraction can provide sufficient information to derive place fields. Classically, one associates landmarks with objects; however, this paradigm makes the task of self-localization quite complex, as

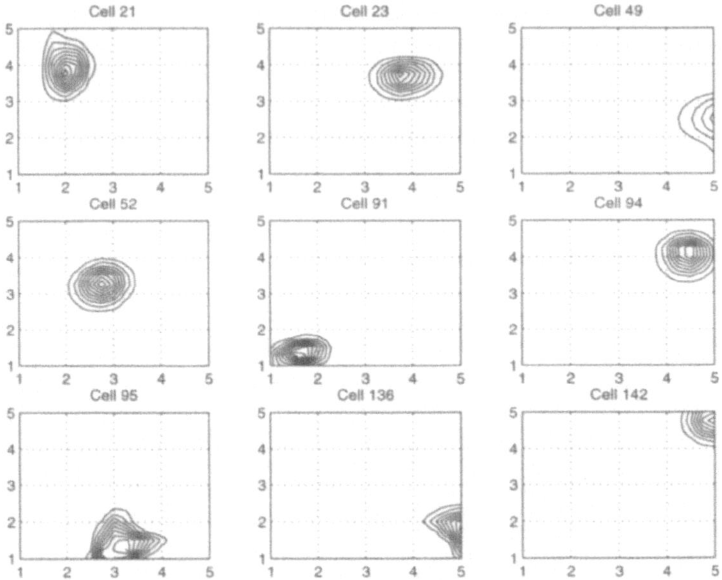

Figure 2. Sample place fields constructed by tuning a product-of-Gaussians function to blob parameters extracted from local views.

it requires that the abstract object be recognized through complex visual processing and that one's location relative to that object be calculated based on one's previous knowledge about the object's shape, size, etc. However, the wealth of information available in a single image, even at low resolution, provides enough information that such complicated cognitive processing is unnecessary.

In the theoretical sense, a landmark is any abstraction that may be characterized by features that possess both position sensitive and position invariant qualities. In order to recognize the same landmark from different viewpoints, there must be some function of the feature space that changes among landmarks, but does not change with different viewpoints of the same landmark. In order to determine one's location relative to the landmark, there must be another function of the feature space that changes as a function of one's relative position to the landmark. So long as these conditions are satisfied by the features extracted from sensory information, any visual processing system would likely be as successful as the system presented here.

Acknowledgments. Supported by National Science Foundation grant IBN-9631336.

REFERENCES

Burgess, N., Reccce, M., and O'Keefe, J., 1994, A model of hippocampal function, *Neural Networks*, **7**(6/7):1065-1081.

O'Keefe, J. and Conway, D. H., 1978, Hippocampal place units in the freely moving rat: Why the fire where they fire. *Experimental Brain Research*, **31**:573-590.

Skaggs, W. E., McNaughton, B. L., Gothard, K. M., and Markus, E. J., 1993, An information-theoretic approach to deciphering the hippocampal code, in: *Advances in Neural Information Processing Systems 5*, pp. 1030-1037, S. J. Hanson, J. D. Cowan, and C. L. Giles, eds., Morgan Kaufmann, San Mateo, CA.

Touretzky, D. S. and Redish, A. D., 1996, Theory of rodent navigation based on interacting representations of space, *Hippocampus*, **6**(3):247-270.

TRANSIENT SYNCHRONIZATION OF PROPAGATING DISCHARGES IN NEOCORTICAL SLICES

D. Golomb, E. Kozhinsky, I.A. Fleidervish and M.J. Gutnick

Zlotowski Center for Neuroscience and Dept. of Physiology,
Ben-Gurion University, Beer-Sheva 84105, Israel

INTRODUCTION

When cortical $GABA_A$ inhibition is blocked, a brief local stimulation above a certain intensity elicits a population discharge. In cortical slices[1,2,3], this discharge consists of spatially and temporally localized neuronal activity, which may be viewed as an *in vitro* experimental model for focal epileptic discharge. The cellular equivalent of this traveling activity is characterized by a depolarization envelope, upon which a train of action potential rides. Studying pulse propagation is expected to uncover the relationship between cortical underlying circuitry and collective dynamical phenomena.

During discharge propagation, neurons are synchronized in the discharge time scale. Is there coherence in the spike time scale? In a homogeneous network, the spikes are fully synchronized except for a time shift[4]. We use modeling to investigate the effects of sparseness, heterogeneity and synaptic noise on the transient synchrony. The model's predictions are tested experimentally using dual single-unit recording *in vitro*.

THE MODEL

A single compartment conductance based model includes: a sodium current I_{Na}, a persistent sodium current I_{NaP}, a delayed rectifier current I_{K-dr}, and a leak current I_L. The current I_{K-slow} represents several slow voltage- and Ca-activated potassium currents. Neurons are coupled by fast AMPA synapses and by slow NMDA synapses. The current-balance equation is

$$C\frac{dV(x,t)}{dt} = -I_{Na} - I_{NaP} - I_{K-dr} - I_{K-slow} - I_L - I_{AMPA} - I_{NMDA} \qquad (1)$$

A synapse is represented by a variable $s(x,t)$, $0 < s \leq 1$, that is the ratio of open synaptic channels. The variable s increases during a presynaptic action potential and decays following it. Synaptic depression is mimicked by decreasing the normalized number of synaptic vesicles T during an action potential with a rate k_t, and by slow a recover afterwards. The phenomenological equations[4] are

$$\frac{ds}{dt} = k_f s_\infty (V_{pre}) T(1-s) - k_r s \qquad (2)$$

$$\frac{dT}{dt} = -k_t s_\infty \left(V_{\text{pre}}\right) T + k_v \left(1 - T\right) \tag{3}$$

where $s_\infty(V) = \{1 + \exp\left[-\left(V - \theta_s\right)/\sigma_s\right]\}^{-1}$, $\theta_s = -20$ mV, $\sigma_s = 2$ mV, $k_f = 1$ ms^{-1}, $k_r = 0.2$ ms^{-1} for AMPA and 0.0067 ms^{-1} for NMDA, $k_v = 0.001$ ms^{-1}, and $k_t = 1$ ms^{-1}.

The network has a one-dimensional and sparse architecture[5]. The AMPA current a cell receives is

$$I_{\text{AMPA}}\left(x, t\right) = g_{\text{AMPA}} \left(V - V_{\text{Glu}}\right) \sum_j J_{i \leftarrow j} s\left(x', t\right) \tag{4}$$

where V_{Glu} is the reversal potential of glutamate. The probability that two cells are coupled decays with the distance between them.

$$P\left(J_{i \leftarrow j} = 1/M\right) = \frac{ML}{2N\lambda} \exp\left(-|x_i - x_j|/\lambda\right) \tag{5}$$

where N is the number of neuron, M is the average number of inputs to a cell, λ is the footprint (coupling) length, and L is the slice length; $\lambda << L$. The NMDA current I_{NMDA} is calculated in a similar manner, but is multiplied by a sigmoid function of the post-synaptic voltage.

Synaptic noise (spontaneous excitatory postsynaptic conductances (EPSCs)) is modeled by a linear synapse on each neuron (without saturation or depression) receiving a Poisson spike train at a rate $\nu = 50$ Hz. Each pre-synaptic spike at time t_0 evokes an EPSC with an amplitude $g_{\text{noise}} f\left(t - t_0\right)$ described by an extended α-function with a rise time of $\tau_1 = 1$ ms and a decay time of $\tau_2 = 5$ ms, corresponding to the AMPA kinetics.

The spiking function of the ith neuron, which fires at times $t_1 \ldots t_\mu \ldots$, is

$$\Lambda_i\left(t\right) = \sum_\mu \delta\left(t - t_\mu\right) \tag{6}$$

We define two types of correlation function. The trial-average correlation between two distinct cells

$$\bar{c}_{i,j}\left(\tau\right) = \left\langle \int dt\, \Lambda_i\left(t\right) \Lambda_j\left(t + \tau\right) \right\rangle_{\text{trials}} \tag{7}$$

and the system-average correlation between cells with distance k:

$$C_k\left(\tau\right) = \frac{1}{N - k} \sum_{i=1}^{N-k} \int dt\, \Lambda_i\left(t\right) \Lambda_{i+k}\left(t + \tau\right) \tag{8}$$

RESULTS

Rastergrams of neuronal firing during the discharge of network with $M = 128$ and $M = 16$ are shown in Fig. 1A. With $M = 128$ (left panel), the firing pattern looks more ordered than with $M = 16$ (right panel), but there is no full coherence as occurs without sparseness. The reduction of order with increasing sparseness is demonstrated by the system-average correlation function $C_k\left(t\right)$ (Fig. 1B). At $M = 16$, the nearest-neighbors cross-correlation C_1 has a peak near $\tau = 0$ and it decays monotonically as $|\tau|$ increases. Therefore, neurons are synchronized at the discharge time scale, but there is no global spike-to-spike synchrony. At $M = 128$, C_1 has wiggles with a time period of a few ms, reflecting some level of spike synchronization. The cross-correlation C_{1024} between neurons 8λ apart is similar to C_1 for both $M = 16$ and $M = 128$, but with a

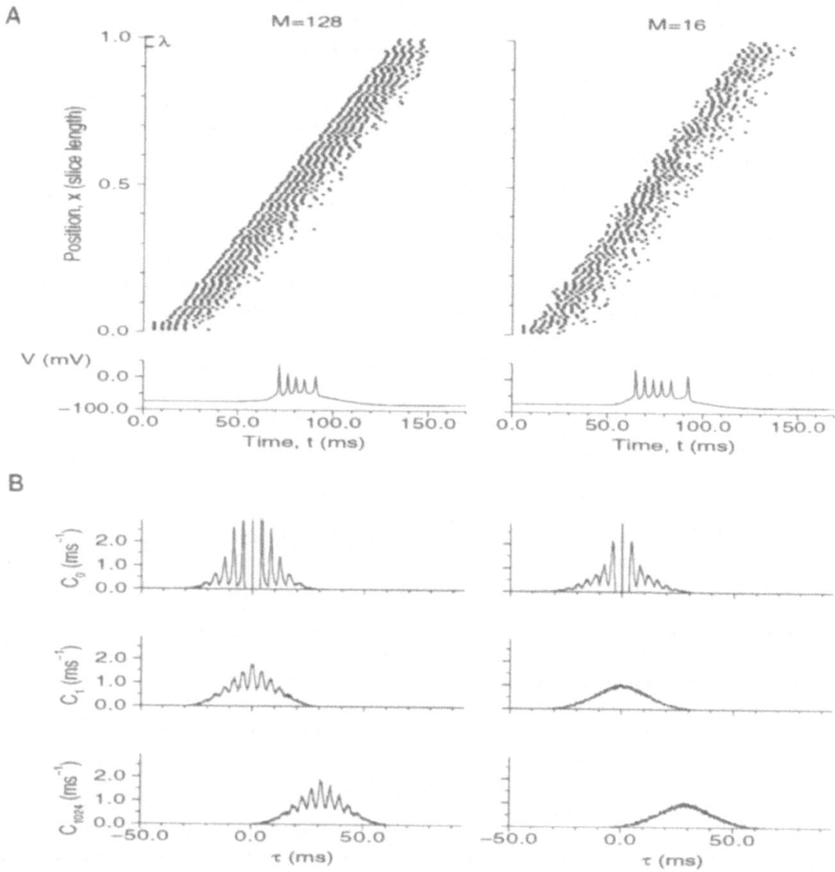

Figure 1. Discharge propagation in sparse networks with average number of synapses $M = 128$ (left) or $M = 16$ (right); $N = 4096$. A, *top*: Rastergrams. The spike times from every 32nd cell are shown. *Bottom*: Voltage time trace of one neuron located at $x = 0.5$. B. System-average auto- correlations C_0 (top), nearest-neighbors cross-correlations C_1 (middle), and average cross-correlation between neurons 8λ apart C_{1024} (bottom).

time shift. Generally, the time shift of C_k is $\Delta x / v_D$, where $\Delta x = kL/N$ and v_D is the discharge propagation velocity.

We define the synchrony measure χ_k to be the modulation depth of the wiggles in the cross-correlation function C_k. The measure χ_k is larger when there is more global spike synchronization. With a uniform architecture (no sparseness), it is infinity, because the C_k is a sum of δ peaks. With low M (high sparseness), it is very small and is governed by the numerical noise. Except for numerical fluctuations, χ_k remains constant with the distance ($k \geq 0$), *i.e.*, the global correlation between neurons does not depend on distance, except for a phase shift. Using the measure χ for quantifying the results of numerical simulations, we find[5] that spike synchronization occurs only when the average number of synapses M a cell receives is large enough. As M increases, there is a cross-over from an asynchronized to a synchronized discharge. With the parameter set chosen for the model, M at the cross-over varies between 30–100. This number is comparable with the estimated number of inputs a cell receives in a cortical slice[5].

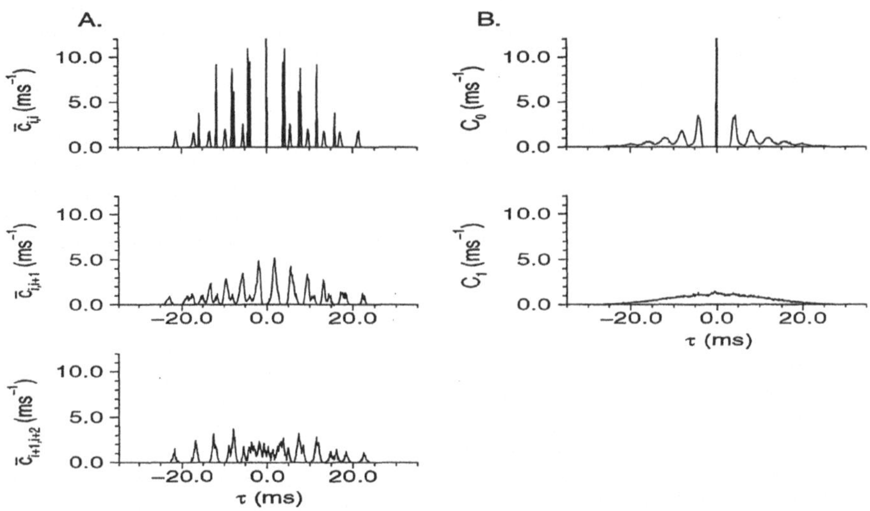

Figure 2. Correlations in a system with both sparseness (M=48) and synaptic noise ($\nu = 50s^{-1}$, $g_{\text{noise}} = 0.1\text{mS/cm}^2$). A. Trial-average correlations: Auto-correlation $\bar{c}_{i,i}$ of one neuron at $x = 0.5$ ($i = 2048$, upper panel), cross-correlation $\bar{c}_{i,i+1}$ between that neuron and its neighbor (middle panel), and cross-correlation $\bar{c}_{i+1,i+2}$ (lower panel). Averaging is done over 200 trials. B. System-average correlation: Auto-correlation C_0 (upper panel) and nearest-neighbor cross-correlation C_1 (lower panel). While the trial-average cross-correlation shows strong spike synchrony, the system-average spike synchrony is very small (note the small wiggles around $\tau = 0$).

Spike synchronization emerges because of the propagating and transient nature of the discharge. In the case of persistent activity, there is no synchrony even between adjacent neurons, and both trail-average and system-average cross-correlations are flat. Synaptic depression appears to help transient synchrony; it reduces the M value at the cross-over. The strengths of AMPA and NMDA conductances affect synchrony only weakly. Spike synchronization is robust even with large levels of heterogeneity in the intrinsic properties of the cells.

Synaptic noise has two effects. When it is weak, it reduces synchrony, and if it is strong enough, it also induces spontaneous discharges. We simulated the system with two values of ν, 20 and 50 s^{-1}. Increasing ν or g_{noise} reduces the synchrony measure χ. However, in the regime where no spontaneous discharges are observed, χ is much higher than the χ values obtained with sparseness for $M < 200$. Hence, synaptic noise contributes to synchrony reduction much less than sparseness.

Sparseness and heterogeneity affect the global (system-average) synchrony of the traveling discharge. They do not, however, affect the local (trial-average) synchrony between two neurons, because they are not stochastic in time. If the network model is stimulated again and again exactly in the same manner, the two neuron will respond exactly the same, and the trial-average cross-correlation will be a sum of δ-functions. Synaptic noise, which is stochastic, affects both global and local synchrony. In a cortical slice there are sparseness, heterogeneity and noise. Here the trial-average and system-average correlations are calculated for a system that has both sparseness and synaptic noise.

Examples of trial-average correlation functions between two adjacent neurons for a noisy and sparse network ($g_{\text{noise}} = 0.1\text{mS/cm}^2$, $\nu = 50s^{-1}$, $M = 48$) are shown in Fig. 2A, and the corresponding system-average correlations are shown in Fig. 2B.

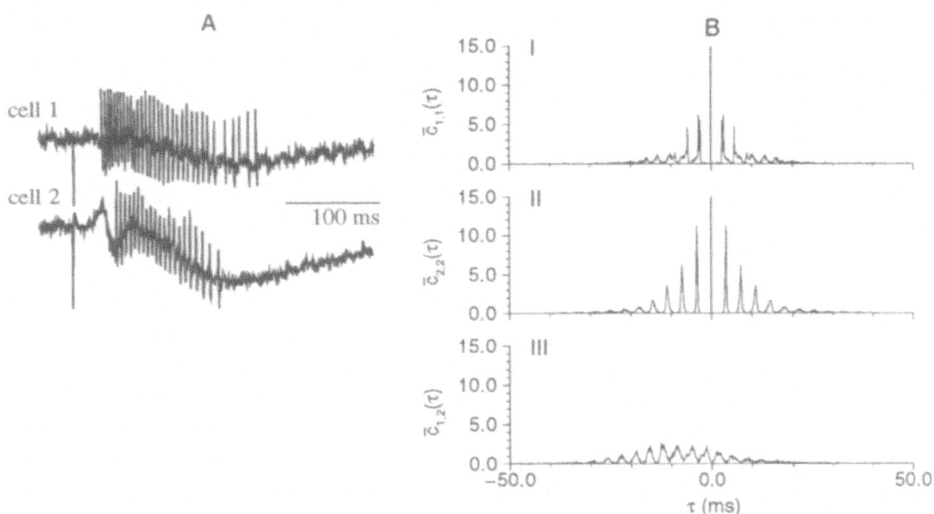

Figure 3. Experimental results. A. Typical voltage time traces of two cells. B. Auto-correlations $\bar{c}_{1,1}$ and $\bar{c}_{2,2}$ of the two neurons (I and II), and cross-correlation $\bar{c}_{1,2}$ between the two neurons (III).

Synaptic noise reduces synchrony only moderately, and hence trial-average auto- correlations show very sharp peaks, and cross-correlations reveal a decaying oscillatory pattern. The relative phase of firing varies from one pair of neurons to another. For example, the cross-correlation presented in middle panel of Fig. 2A indicates that those particular two cells fire almost in anti-synchrony. The central peak of the cross-correlation presented in lower panel of Fig. 2A is broader and the phase relationship between those two cell is less well-defined. In general, even if the phase relationship between two neurons in a pair is well-defined, it varies considerably from pair to pair. With a relatively low value of M (high sparseness), it is distributed almost uniformly. As a result, the system-average cross-correlation exhibits almost no wiggles in the time scale of the inter-spike interval, indicating that there is only a very small global (system-average) spike synchronization in the network. As a consequence of these results we predict that in cortical coronal or sagittal slices, strong local, trial-average spike synchronization will be measured, but that global, system-average spike synchrony will be low or absent.

To test the prediction regarding local synchrony, we record extracellularly from two cells in a coronal cortical slice during the discharge propagation. $GABA_A$ inhibition is blocked by bicuculline, and the slice is stimulated every 15 s at a distance of 4–5 mm from the recording electrodes, in the border between the white and gray matter. A typical dual recording from two adjacent (distance < 0.3 mm) neurons is shown in Fig. 3A. Auto-correlations of cells showed high peaks (Fig. 3B). The cross-correlation was significantly oscillatory in 4 out of 5 pairs, revealing a strong trial-average synchrony between the two cells.

DISCUSSION

As long as discharges propagate, they are always synchronous at the discharge time scale, and this synchrony does not decay with distance between neurons for any level of sparseness, heterogeneity and noise. Our modeling results show that global spike

synchronization can emerge during propagating discharges because of their propagating
and transient nature, if the average number of input a cell receives is large enough. The
global correlation between neurons does not depend on the distance between them,
except for a phase shift. The noise level in the network is not expected to abolish spike
synchronization between two specific neurons, but the sparseness (and heterogeneity)
in the system cause the global spike synchrony to be very small or zero.

Experimental results confirm the prediction that the trial-average cross-correlation
in cortical slices during propagating activity is significant, implying that the level of
noise in the preparation is weak.

Acknowledgments

This research was supported by a grant from the Israel Science Foundation.

REFERENCES

1.　　M.J. Gutnick, B.W. Connors, and D.A. Prince, Mechanisms of neocortical epileptogenesis in
vitro, *J. Neurophysiol.* 48: 1321-1335 (1982).
2.　　W.J. Wadman, and M.J. Gutnick, Non-uniform propagation of epileptiform discharge in brain
slices of rat neocortex, *Neuroscience* 52: 255-262 (1993).
3.　　R.D. Chervin, P.A. Pierce, and B.W. Connors, Periodicity and directionality in the propagation
of epileptiform discharges across neocortex, *J. Neurophysiol.* 60: 1695-1713 (1988).
4.　　D. Golomb, and Y. Amitai, Propagating neuronal discharges in neocortical slices: Computational
and experimental study, *J. Neurophysiol.*, 78: 1199-1211 (1997).
5.　　D. Golomb, Models of neuronal transient synchrony during propagation of activity through
neocortical circuitry, *J. Neurophysiol.*, 79: in press (1998).

A HEBBIAN ALGORITHM THAT BALANCES INFORMATION
RATE AND NEURAL RESOURCE CONSUMPTION

Allan Gottschalk

Department of Anesthesia and Institute of Neurological Sciences
University of Pennsylvania
Philadelphia, Pennsylvania 19104
ag@network3.entropy.upenn.edu

INTRODUCTION

Hebb's seminal concept that synaptic strength varies as a function of correlated pre- and post-synaptic activity[1] has impacted pharmacologic, physiologic, and computational models of the nervous system. Specifically, he proposed that correlated pre- and post-synaptic activity leads to an increase in synaptic strength. When computational models incorporate Hebbian synapses, some limit must be placed on synaptic growth if well-defined solutions are to be produced, and it may be necessary to incorporate some means by which synaptic strength can also decrease. Strategies to limit synaptic growth have included some type of limit on individual or total synaptic size.[2] Another approach incorporates a physiologically and computationally motivated "sliding threshold" for synaptic modification where the sliding threshold is stimulus dependant.[3] We previously demonstrated how a sliding threshold for synaptic modification naturally emerges out of the process of adjusting synaptic size so that the resulting flow of information justifies the neural resources consumed in its generation.[4] A Hebbian algorithm implementing this principle is now introduced and applied to the formation of the spatial receptive fields (RFs) of early vision and their functional organization. The properties of this solution are comparable to results previously obtained by direct numerical solution of the equations which govern the optimal solution to this problem.[5,6]

MODEL/ALGORITHM

Consider the simple visual system model of Fig. 1. Members of the stimulus ensemble are detected by an array of sensors whose outputs $\mathbf{g}$ are transformed by $\Phi=[\phi_1|\phi_2|...|\phi_N]^T$ and corrupted by noise $\mathbf{n}$ to produce $\mathbf{h}$. We now seek to adjust Φ to maximize the mutual information $I(\mathbf{g};\mathbf{h})$ between $\mathbf{g}$ and $\mathbf{h}$. When $I(\mathbf{g};\mathbf{h})$ is optimized with respect to Φ, it is necessary to modify the cost function $I(\mathbf{g};\mathbf{h})$ in order to obtain well-defined solutions. Examples of the types of terms which can be incorporated include total synaptic use $\text{tr}[\Phi\Phi^T]$ or total neural output power $\text{tr}[\Phi C\Phi^T]$, where $\text{tr}[\bullet]$ denotes the matrix trace, and C is the correlation matrix of sensor outputs ($C = E[\mathbf{g}\mathbf{g}^T]$, where $E[\bullet]$ is the expectation operator). Thus, we seek the Φ which maximizes the following expression

$$I(\mathbf{g};\mathbf{h}) - \alpha(\text{tr}[\Phi\Phi^T]) - \beta(\text{tr}[\Phi C\Phi^T]), \tag{1}$$

where α is a meta-parameter specifying the tradeoff between synaptic size and information

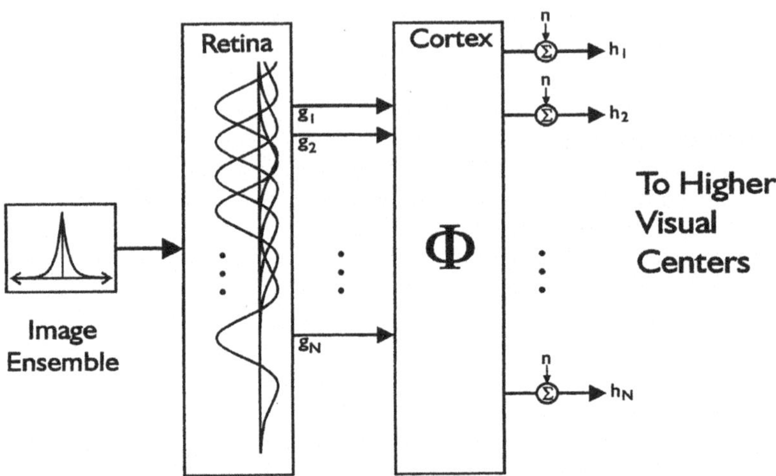

Figure 1. Basic model. See text for details.

flow, and β is a meta-parameter indicating the tradeoff between neural output power and information flow. As previously shown,[5] if the signal and noise are Gaussian and independent and Φ is linear, an expression for the optimal value of Φ can be obtained and the mutual information at the optimum can be analytically determined. This result suggests the following synaptic modification scheme, depicted schematically in Fig. 2 for a scaler synapse,

$$\Phi_{i+1}^T = \Phi_i^T + \gamma\{[\mathbf{g}\mathbf{h}^T]_i(I - 2\beta[\mathbf{h}\mathbf{h}^T]_i) - 2\alpha\Phi^T[\mathbf{h}\mathbf{h}^T]_i\}, \tag{2}$$

where $[\bullet]_i$ indicates the current value of the moving average of the enclosed product taken over some suitable interval, and γ $(0<\gamma\leq1)$ is a constant modulating the size of the synaptic modification at each step.

This algorithm was implemented for an array of sensors with a 2-dimensional (2D) difference-of-Gaussians (DOG) profile and a single cortical neuron. The input was a sequence of random stimuli with a circularly symmetric exponential correlation profile. Inputs were generated using independent outputs of a random number generator which were passed through a filter to produce stimuli with the desired statistical properties. These results are then compared with those obtained analytically for the corresponding problem.

When multiple cortical neurons are present, the optimal solution to the above problem lacks uniqueness for a number of reasons.[5] This problem can be overcome by modifying the above cost function so that the term $\text{tr}[\Phi\Phi^T]$ is replaced by $\Sigma\phi_i^T W_i\phi_i$, where W_i is a diagonal matrix of suitable dimension which is defined for each cortical neuron.[5] As conceived here, this matrix penalizes retinotopically distant connections so that each

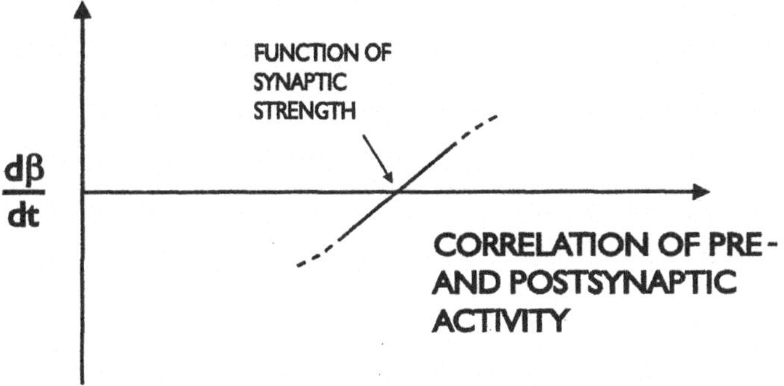

Figure 2. Schematic of synaptic modification algorithm.

matrix element factors the cost of connecting the given cortical neuron with each retinal output. The synaptic modification algorithm of Eq. 2 becomes

$$\Phi_{i+1}^{T} = \Phi_i^{T} + \gamma\{[\mathbf{gh}^{T}]_i(I - 2\beta[\mathbf{hh}^{T}]_i) - 2\alpha(W_1\phi_1|W_2\phi_2|...|W_N\phi_N)[\mathbf{hh}^{T}]_i\}, \tag{3}$$

where the only difference is that $(W_1\phi_1|W_2\phi_2|...|W_N\phi_N)$ replaces Φ^{T}. Simulations for the case of multiple cortical neurons were carried out as described above, where the elements of the W_i were chosen using a 2D Gaussian profile. Thus, the elements of each W_i are given by $Exp(r_{ij}^2/\tau)$, where r_{ij} is the retinotopic distance from the i-th cortical neuron to the j-th retinal neuron, and τ is the corresponding space constant. Results from the simulation are contrasted with those obtained by direct numerical solution of the equations governing the optimal solution of the same system.

RESULTS

For a single cortical neuron, simulations with multiple correlated input neurons and a single output neuron converged to a synaptic weight vector proportional to the eigenvector of the matrix of input correlations associated with the largest eigenvalue of the matrix. The

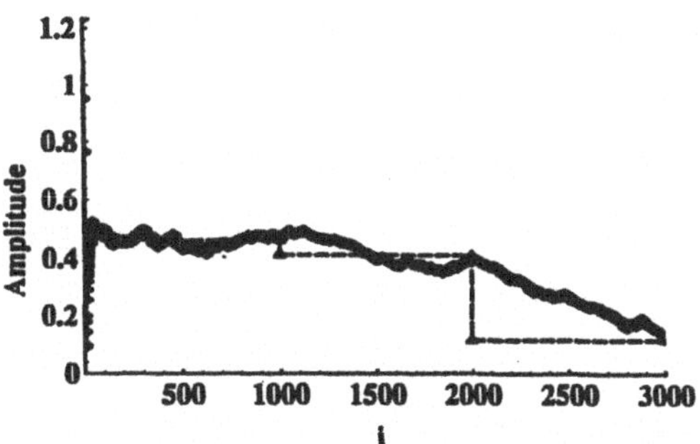

Figure 3. Amplitude of synaptic weight vector as a function of stimulus presentation (solid line) and its analytically determined optimum (dashed line).

constant of proportion converged to its theoretically determined optimal value. This is illustrated in Fig. 3, where the value of this constant is displayed as a function of stimulus number for three different stimulus conditions along with the theoretically determined value of the amplitude. The test stimulus decreases in power and becomes more highly correlated over space in going from left to right in Fig. 3.

When multiple output neurons are present, the algorithm of Eq. 3 led to localized oriented cortical RFs, samples of which are shown in Fig. 4 along with their corresponding neural output power. The functional organization of the ensemble of cortical RFs in the solution is demonstrated in Fig. 5 where the orientation of the preferred spatial wave vector of each cortical neuron is plotted as a function of cortical position. The properties of the solution obtained with the above algorithm are comparable to solutions to the same problem obtained by direct numerical solution of the equations which govern the optimal solution.

DISCUSSION:

These results demonstrate the feasibility of a synaptic modification algorithm to balance information rate against neural resource consumption. Such an algorithm is Hebbian in character and, as depicted in Fig. 2, can be conceived of as comparing the level of pre- and postsynaptic correlation with a sliding threshold. This sliding threshold is a

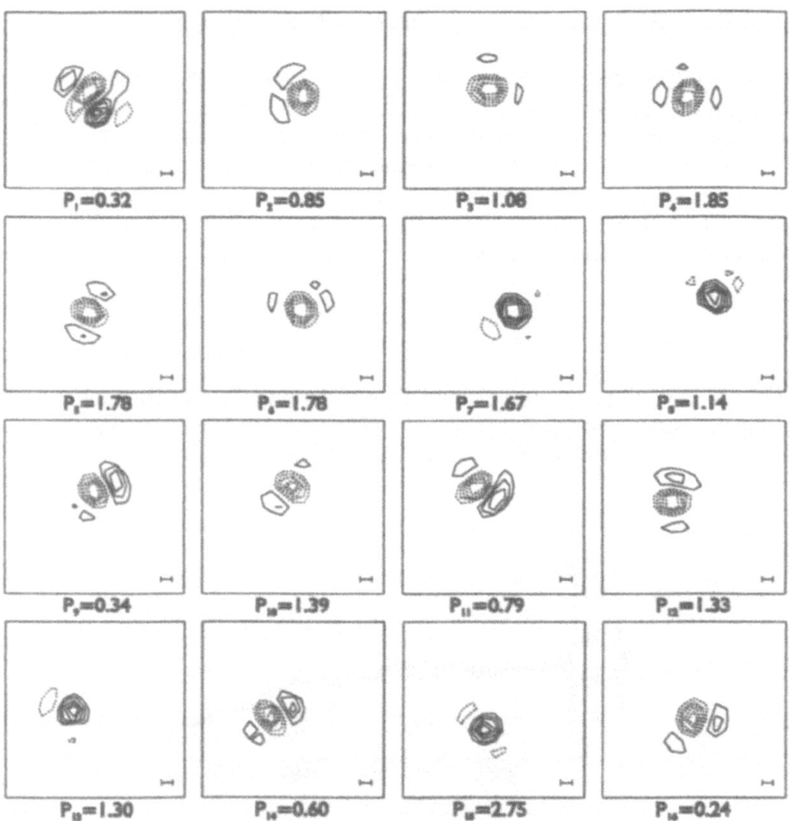

Figure 4. Sample of consecutive cortical RFs produced when multiple cortical neurons are present. The corresponding neural output power is shown below each RF.

function of the prevailing synaptic weights and stimulus history. Synaptic weights are modified so that the information rate they permit is balanced against their size and the resulting neural output power. When the information rate does not justify the expenditure of neural resources, the corresponding synapse is set to zero. This balance is determined *a priori*, and encoded in the meta-parameters α and β of the above equations. In the case of a fixed pool of neural resources available for representing a given stimulus ensemble, α and β would become the Lagrange multipliers that indicate the balance between neural resource consumption and information rate achieved by the optimal solution.

The actual distribution of synaptic weights obtained in the simulation is, for a single output neuron, proportional to the eigenvector of matrix of stimulus correlations associated

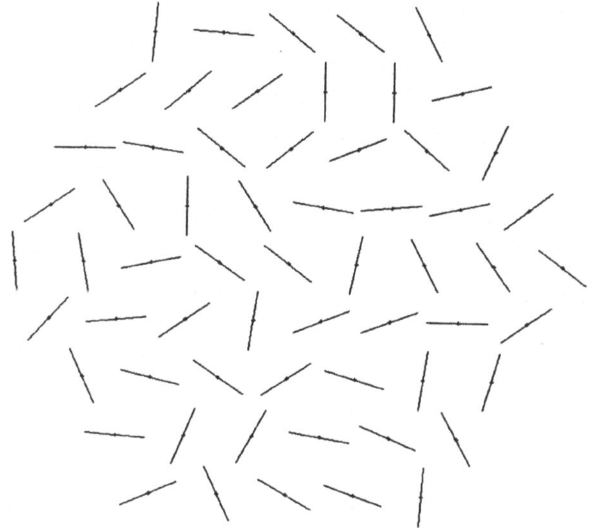

Figure 5. Orientation of optimal spatial wave vector as a function of cortical position.

with the largest eigenvalue, further demonstrating the role of Hebbian algorithms as principal component analyzers.[7] This eigenvector is altered as the properties of the stimulus ensemble are varied. The amplitude of the corresponding neuron depends upon the size of the associated eigenvalue as well as the power of the neural noise and the relative costs of the neural resources quantified by the meta-parameters α and β. It is conceivable that the optimal amplitude should be zero, a situation where the resulting information is not justified by the costs of obtaining it.

When multiple output neurons are present and a penalty is placed on retinotopically distant connections, a physiologically plausible solution emerges. This consists of an ensemble of localized orientation selective RFs whose functional architecture is pinwheel in nature. Since the properties of this RF ensemble do not differ from those obtained by numerical solution of the corresponding equations of optimality, these results indicate the neural feasibility of balancing information flow with the neural costs of providing that information.

This information theory based Hebbian algorithm for synaptic modification also indicates why synaptic strength should not necessary increase in the face of correlated pre- and post-synaptic activity, and why some synapses should simply fade away.

ACKNOWLEDGEMENTS

Supported by EY10915.

REFERENCES

1 Hebb, D.O. *The Organization of Behavior*, Wiley, New York (1949).
2 Brown, T.H. and Chattarji, S. Hebbian synaptic plasticity. In: *The Handbook of Brain Function and Neural Networks*, M. A. Arbid ed.. MIT Press, Cambridge, MA (1995).
3 Bienenstock, E.L., Cooper, L.N., and Munro P.W. Theory for the development of neuron selectivity: orientation specificity and binocular interaction in visual cortex. *J. Neurosci.* 2:32-48 (1982).
4 Gottschalk, A. Information based limits on synaptic growth in Hebbian models. In:. *Computational Neuroscience: Trends in Research, 1997*, J. Bower, ed., Plenum, New York (1997).
5 Gottschalk, A. Neural image representation in early vision. Submitted for publication (1997).
6 Gottschalk, A. Limits on image representation in early vision. In: *Computation in Neurons and Neural Systems*, f.H. Eeckman, ed., Kluwer, Boston (1994).
7 Oja, E. A simplified neuron model as a principal component analyzer. *J. Math. Biol.* 15:267-273 (1982).

A MODEL OF MONOCULAR CELL DEVELOPMENT BY COMPETITION FOR NEUROTROPHIC FACTOR: EFFECTS OF EXCESS NT WITH MONOCULAR DEPRIVATION AND EFFECTS OF NT ANTAGONIST

Anthony E. Harris,[1,2,5] G. Bard Ermentrout,[3,5] and Steve L. Small[4,6]

[1]Intelligent Systems Program
[2]Center for the Neural Basis of Cognition
[3]Dept of Mathematics and Statistics
[4]Dept of Neurology
[5]Univ. of Pittsburgh
Pittsburgh, Pa 15213
[6]Univ. of Maryland
Baltimore, MD 21201-1595

INTRODUCTION

Several lines of experimental data indicate that neurotrophic factors (NTs) such as brain-derived neurotrophic factor (BDNF) and NT-4/5 may play a role in the activity-dependent competition between thalamocortical afferents from each eye during ocular dominance (OD) column segregation. Application of excess NT can prevent OD segregation by allowing inputs from both eyes to remain[2], and we have previously presented a model which accounts for both normal OD column development and the prevention of that segregation with excess NT[4, 5].

More recent experimental evidence further implicates NT in the developmental process. When initiated early in development, monocular deprivation causes a shift to the columns subserving the open eye at the expense of columns from the closed eye. Recent experimental work suggests that excess NT can prevent this shift by allowing connections from both eyes to remain[7]. Furthermore, a recent experiment using trkB-IgG protein, an NT antagonist, has suggested that this prevents OD column segregation in normally sighted animals. However, while application of excess NT prevents OD column development by allowing connections from both eyes to *remain*, NT antagonist prevents segregation by causing connections from both eyes to be *lost*.

We present here an extension of previous results[4, 5]. In particular, we demonstrate through network simulations that our model can account for the two new experimental results outlined above. We first present the model's equations. Next we present simulation results modeling the two experimental manipulations mentioned, for a single cortical cell receiving inputs from two eyes. We have also simulated a full cortical network elsewhere[6], and a condensed manuscript is in preparation.

METHODS

Our model consists of a sheet of cortical units with linear activation functions receiving inputs from each eye. Cortical units interact via the standard Mexican hat function. The ith cortical unit has a fixed supply of NT for which incoming afferents compete (N_i). Each connection from the right eye to the ith cortical unit has associated with it two dynamic variables: a weight (w_i^r), or synaptic strength, and the amount of trophic factor currently taken up at that synapse (n_i^r). Each connection from the left eye has two similar variables. These variables change according to the following differential equations:

$$\dot{w}_i^r = n_i^r \overbrace{(\sum_{i'} I_{ii'}(C_{rr}w_{i'}^r + C_{rl}w_{i'}^l))}^{\text{'}LTP\text{'}-like\ rule}(1 - w_i^r) - \beta_1 \overbrace{(\sum_{i'} I_{ii'}(w_{i'}^r + w_{i'}^l))}^{\text{'}LTD\text{'}-like\ rule} w_i^r \qquad (1)$$

$$\dot{n}_i^r = (N_i - (n_i^r + n_i^l))w_i^r - \beta_2 n_i^r \qquad (2)$$

(with two identical equations for the connection from the left eye but with $l's$ and $r's$ interchanged). These equations can be understood as follows (derived previously[4, 5]). The rate at which weight increases is given by the first term in eqn.1. The "LTP" term represents the simple Hebbian learning rule of presynaptic times postsynaptic activity, and is obtained by averaging the weight change under the Hebbian rule over the entire input ensemble[10, 9], where $I_{ii'}$ is the intracortical interaction between the ith and the i'^{th} units, and C_{rr} (C_{lr}) is the intraeye (intereye) correlation of input activity. This Hebbian term is modulated by the amount of trophic factor currently taken up at that synapse, n_i^r. Thus NT positively affects LTP, consistent with experimental data[8]. The $(1 - w_i^r)$ term simply bounds the weight below 1. The rate at which weight decreases is given by the second term in eqn.1. It consists of an "LTD" term multiplied by w_i^r, which bounds the weights from below by zero (β_1 is a constant). Thus synaptic strength always remains nonnegative and bounded. The rate of increase of trophic factor at a synapse is given by the first term in eqn.2. As the amount of trophic factor taken up by both synapses ($n_i^r + n_i^l$) approaches the total amount available N_i, the rate of uptake goes to zero, since the available NT is depleted. The rate of trophic factor increase at a synapse is also positively modulated by the weight of that synapse (w_i^r), consistent with recent evidence[1]. Finally, the amount of NT at a synapse decays proportional to n_i^r, which keeps trophic factor bounded below by zero. The positive feedback between rate of connection strength increase (or LTP) and rate of trophic factor uptake drives the competitive process, and is eliminated when the amount of postsynaptic trophic factor available (N_i) is above a critical value.

RESULTS

Fig. 1 shows phase plane trajectories for a single cell developing under three different experimental paradigms: the control condition ($C_{ll} = C_{rr} = 0.9$, $C_{lr} = C_{rl} = 0.3$, $N = 1.0$), the MD alone condition ($C_{rr} = 0.9$, $C_{ll} = 0.6$, $C_{lr} = C_{rl} = 0.0$, $N = 1.0$), and the MD plus excess NT condition ($C_{rr} = 0.9$, $C_{ll} = 0.6$, $C_{lr} = C_{rl} = 0.0$, $N = 15.0$). Note that because we are modeling only one cortical cell, $i' = 1$ in the above equations and is dropped. In the control situation, the left eye wins the competition for this set

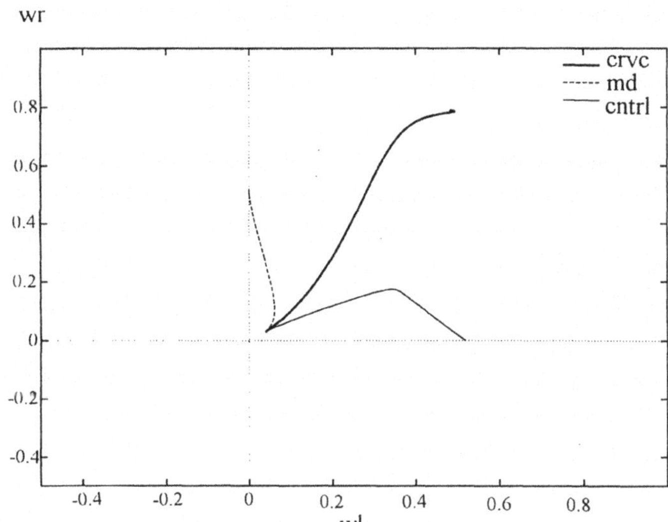

Figure 1 Phase plane trajectories of the weights from each eye under three different experimental conditions: control (cntrl), md alone (md), and md plus excess nt (crvc). See text for parameters and explanation.

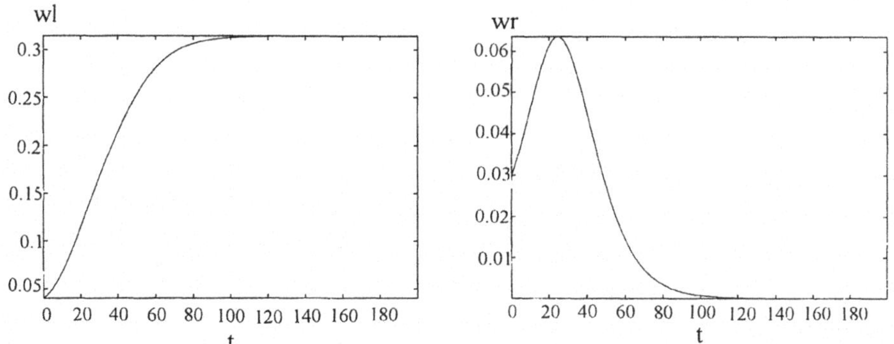

Figure 2 Trajectories for weights from the right and left eyes versus time. NT is at an intermediate level, so that one eye wins the competition (the "control" experiment). Parameters are:
$C_{ll} = C_{rr} = 0.9$, $C_{lr} = C_{rl} = 0.3$, $\beta_1 = 1.2$, $\beta_2 = 0.2$, $N = 1.0$.

of initial conditions. After MD is performed, the right eye now wins the competition (initial conditions are exactly the same as before), as shown by the 'md' curve in Fig. 1. The third simulation shows that when MD is simulated and excess NT is applied, a binocular cell results (the 'md+nt' curve). Thus the model suggests a mechanism by which excess NT can mitigate the effects of MD, as suggested by recent experiments[7]. In single cell simulations, when available NT is low enough, inputs from both eyes decay to zero. This can first be seen in the trajectories of Fig. 2, which shows the trajectories through time of the weights from the right and left eyes when NT is set to an intermediate value (here set to 1), the "control" situation. One eye wins the competition, and a monocular cell results, as the weight from one eye goes to an intermediate value while the weight from the other decays to zero. Fig. 3 shows the results after the available NT has been set to a very low value ($N = 0.1$), simulating the application of NT antagonist. Now the weights from both eyes decay to zero. Because there is not enough NT available to fuel the weight increase of either eye, synapses from both eyes decay to zero. This result is suggested by a previously published bifurcation diagram[5]. The top branch is the weight of the winning eye; the bottom branch is the weight from the losing eye (which is zero). The top branch crosses the zero branch and becomes negative before

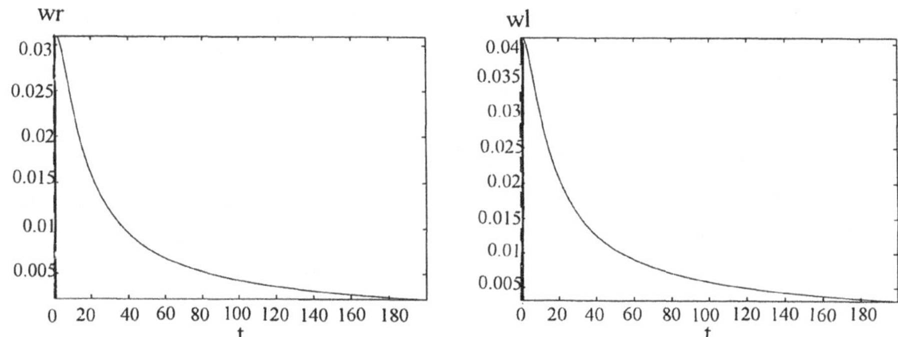

Figure 3 Same trajectories as Fig. 2, but now NT has been removed from the system, simulating the application of antagonist. Parameters same as before, except $N = 0.1$.

available NT reaches zero. In that region, therefore, the only stable solution is the zero solution (i.e. $w^r = w^l = n^r = n^l = 0$), since the non-negative branch is the only one that can be reached for non-negative initial conditions.

CONCLUSION

We have previously presented a model accounting for normal OD column development and the prevention of that development with excess NT[4, 5]. The current results suggest that the model can also account for the mitigation in the shift to the open eye with MD plus excess trophic factor[7] and the prevention of OD development with NT antagonist[3]. Further work will examine possible mechanisms controlling the length of the critical period after which MD is no longer effective.

REFERENCES

1.   Blochl, A. and Thoenen, H.,1995,Characterization of nerve growth factor (NGF) release from hippocampal neurons: evidence for a constitutive and an unconventional sodium-dependent regulated pathway,*European Journal of Neuroscience* 7:1220.

2.   Cabelli, R. J. and Hohn, A. and Shatz, C.J., 1995, Inhibition of ocular dominance column formation by infusion of NT-4/5 or BDNF,*Science* 267:1662-1666.

3.   Cabelli,R.J. and Shatz,C.J.,1996, An endogenous ligand of TrkB is required for ocular dominance segregation,*in* "Society for Neuroscience Abstracts, 1996".

4.   Harris, A.E., Ermentrout, G.B. and Small,S.L.,1996,A model of monocular cell development by competition for trophic factor,*in* "Computation and Neural Systems 1996 Proceedings".

5.   Harris,A.E,Ermentrout, G.B. and Small,S.L.,1997,A model of ocular dominance column development by competition for trophic factor,*Proceedings of the National Academy of Sciences, USA* 94:9944.

6.   Harris, A.E., 1997, A computational model of the effects of neurotrophic factor on ocular dominance column development,*PhD Thesis*, U. Pittsburgh.

7.   Hata,Y., Katsuyama,N., Fukuda,M., Ohshima, M., Tsumoto,T. and Hatanaka,H.,1996,"Brain-derived neurotrophic factor disrupts effects of monocular deprivation in kitten visual cortex,*in* "Society for Neuroscience Abstracts, 1996".

8.   Korte, M., Carroll, P., Wolf, E., Brem., G., Thoenen, H. and Bonhoeffer, T.,1995,Hippocampal long-term potentiation is impaired in mice lacking brain-drived neurotrophic factor,*Proceedings of the National Academy of Sciences, USA* 92:8856.

9.   Linsker, R., 1986,From basic network principles to neural architecture:emergence of spatial-opponent cells,*Proceedings of the National Academy of Sciences, USA* 83:7508-7512.

10.   Miller, K.D., Keller,J.B. and Stryker, M.P.,1989,Ocular dominance column development: analysis and simulation,*Science* 245:605.

SYNCHRONIZATION OF RANDOMLY DRIVEN NONLINEAR OSCILLATORS AND THE RELIABLE FIRING OF CORTICAL NEURONS

R. V. Jensen, L. Jones, and D. H. Gartner

Department of Physics
Wesleyan University
Middletown, CT 06457

INTRODUCTION

There has been a recent resurgence of interest in the idea that information may be encoded in the precise timing of neuronal spike trains rather than the average firing frequency alone. For example, Bialek and de Ruyter have clearly demonstrated that changes in the motion of the visual field of a housefly can be decoded from the timing of the firing sequence of a single, high level neuron in the fly visual system.[1] Moreover, a recent re-examination by Bair and Koch[2] of single neuron recordings in the monkey visual cortex, measured while viewing repeated presentations of random dot patterns on a TV screen[3], revealed that the complex firing patterns of these high level neurons are extraordinarily reliable and precise to a few milliseconds!

However, it has become common practice to view biological neurons as unreliable stochastic processors. How then does the nervous system preserve this temporal precision in single neurons at high levels in sensory systems? A necessary condition for the observed precision is the requirement that single neurons must respond to natural, *dynamic* inputs with much greater reliability than previously thought. In fact one recent electrophysiological study of isolated cortical neurons in slice preparations by Mainen and Sejnowski[4] has clearly demonstrated that cortical neurons may exhibit inconsistent firing patterns when driven by approximately uniform injection currents but will show extremely reliable, precise, and repeatable (though irregular) firing patterns when driven by more realistic, randomly fluctuating injection currents.

From a mathematical point of view the key question raised by the Mainen and Sejnowski experiments is how a *nonlinear oscillator* driven by an (approximately) constant input can exhibit an unstable, non-reproducible firing pattern while the same nonlinear oscillator with a fluctuating driving force can exhibit a highly stable and reproducible pattern of activity?

Computational Neuroscience
edited by Bower, Plenum Press, New York, 1998

RANDOMLY DRIVEN NONLINEAR OSCILLATORS

Mathematical models of nonlinear oscillators have been introduced to describe a wide variety of biological and physical phenomena that exhibit self-sustained oscillatory behavior, from the Hodgkin-Huxley model[5] for spiking neurons to the van der Pol equations for nonlinear electrical circuits[6]. When strongly driven by forces that are periodic in time, these oscillators often exhibit a remarkable "phase- locking" that synchronizes the nonlinear oscillations to the driving force. We have recently found that a similar phenomena occurs when a nonlinear oscillator, like the famous van der Pol oscillator,[6]

$$x''(t) = \epsilon(1 - x(t)^2)x'(t) - x(t) + F_R(t) \tag{1}$$

is strongly driven by a force, $F_R(t)$, that is randomly fluctuating in time[7].

In the randomly driven case the synchronization is less obvious for a single oscillator because there is no well defined "phase" to lock on to, but when several similar oscillators with different initial conditions or phases are driven with the same random force, their fluctuating behavior may reliably converge to an identical response. Using a variety of different examples of nonlinear oscillators from the biological and physical sciences, we find that the reliable synchronization of randomly driven nonlinear oscillators is a common, but largely overlooked, feature of these nonlinear dynamical systems. In particular, the nonlinear oscillators used to model the spiking, voltage dynamics of neurons, like the Fitzhugh-Nagumo, Morris-Lecar, Hindmarsh-Rose, and Hodgkin-Huxley equations[5], and even simple phase models[8] all exhibit asymptotically stable, synchronized behavior when strongly driven by randomly fluctuating currents.

Consider for example the Hodgkin-Huxley (HH) model described by the coupled, nonlinear differential equations for the membrane voltage, $V(t)$, and the sodium and potassium gating variables, $m(t)$, $h(t)$, and $n(t)$:

$$CV'(t) = -g_{leak}(V - E_{leak}) - g_{Na}m^3h(V - E_{Na})$$
$$-g_K n^4(V - E_K) + I_0 + I_R(t), \tag{2}$$
$$m'(t) = \alpha_m(V)(1 - m) - \beta_m(V)m, \tag{3}$$
$$h'(t) = \alpha_h(V)(1 - h) - \beta_h(V)h, \tag{4}$$
$$n'(t) = \alpha_n(V)(1 - n) - \beta_n(V)n. \tag{5}$$

Equations (2)-(5) can be easily solved for different initial conditions, parameters, and driving currents, I_0 and $I_R(t)$, and the solutions plotted using a neuronal modeling program like GENESIS[9] or a general purpose mathematical analysis program like *Mathematica*. (In this paper we examined the solution of Eqs. (2)-(5) for a 30 micron diameter spherical neuron with standard "squid" parameters[9].) With constant input current, I_0, (above firing threshold), the HH model exhibits self-sustained oscillations at a constant firing rate. In general when two identical oscillators are started out with different initial conditions, the spike timing of the two oscillators will not coincide. However, when a randomly fluctuating injection current, $I_R(t)$, that simulates realistic, summed synaptic currents, is applied to the two oscillators, the spike timing can be driven into synchrony, as shown in Fig. 1.

RELIABLE SPIKE TIMING

The synchronization of the randomly driven HH models shown in Fig. 1 is an illustration of the more general phenomena of the possibility of asymptotically stable[10]

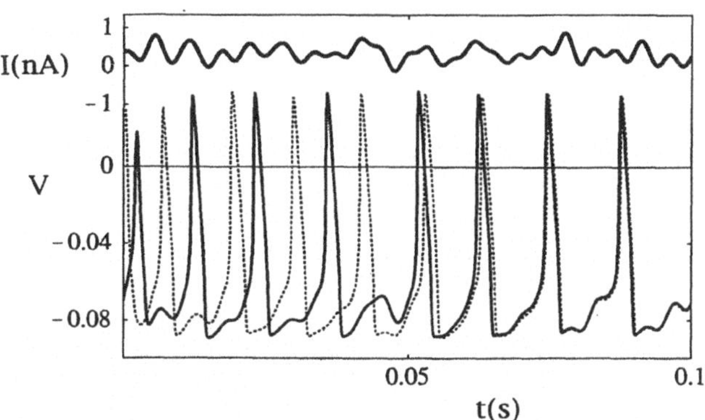

Figure 1. The voltage traces for two spiking HH neurons with different initial conditions (solid and dashed curves) are driven into synchrony by the randomly fluctuating injection current shown by the top curve with $I_0 = 0.3$ nA and rms fluctuations of ± 0.2 nA.

solutions for aperiodcally driven nonlinear systems[7]. When the fluctuations in the random drive are sufficiently strong and vary at frequencies comparable to the natural frequencies of the driven oscillator, then the long-time (asymptotic) solution converges exponentially to a unique (stable) solution. When two identical oscillators with different initial conditions (like those in Fig. 1) are strongly driven by the same force, then both solutions converge to the same asymptotic solution and therefore toward synchronization[11]. (Note that a comparison of the driving current in Fig. 1 with the spike timing of the Hodgkin-Huxley oscillators shows that this synchronization is not a trivial consequence of current threshold crossing but is a more subtle feature of the driven nonlinear oscillator.)

A very important property of this effect is that the synchronization of randomly driven nonlinear oscillators is structurally stable. This means that approximate synchronization is realized even in the presence of small variations (or errors) in the parameters of the systems and in small levels of additional noise in the driving signal. In particular, the structural stability provides a mathematical explanation for the remarkable reliability of the spike timing recently observed in the response of neocortical neurons to fluctuating input currents that resemble real synaptically generated currents[4]. These neurons exhibit highly irregular firing patterns, that can nevertheless be reliably reproduced when a fluctuating stimulus is repeatedly applied, even though the same neurons exhibit unreliable spike timing when conventional, constant current inputs are applied.

COMPARISON OF THEORY AND EXPERIMENT

How does the mathematical model reconcile the experimental observations of unreliable spiking patterns when neurons are driven by (ostensibly) constant currents and near perfect synchrony with random driving?

First, we find that just as in the case of periodically driven nonlinear oscillators, no synchronization occurs when the fluctuations in the driving force are small or constant. In these case the phases of the nonlinear oscillations are very sensitive to small (few percent) changes in the initial conditions, in the system parameters and in the

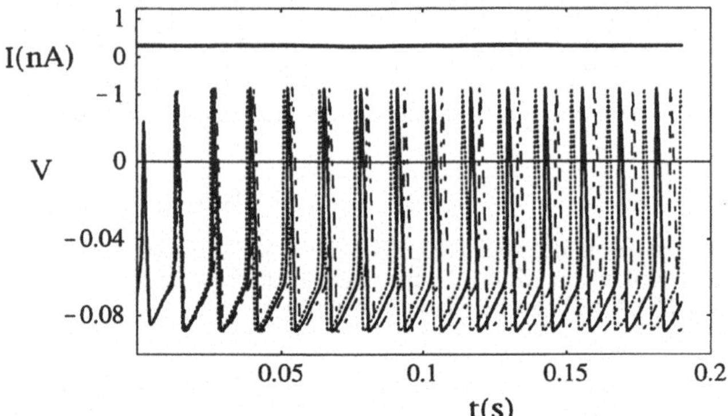

Figure 2. The voltage traces are plotted for three spiking HH neurons (solid, dashed, and dotted curves) with constant current injection of 0.3 nA, 0.33 nA, and 0.27 nA (respectively) indicated by the heavy line at the top of the Figure. These small variations in injection current result in the loss of spike timing synchrony as the phases of the HH oscillators drift apart. This theoretical picture should be compared with Fig. 1A of Ref. 4.

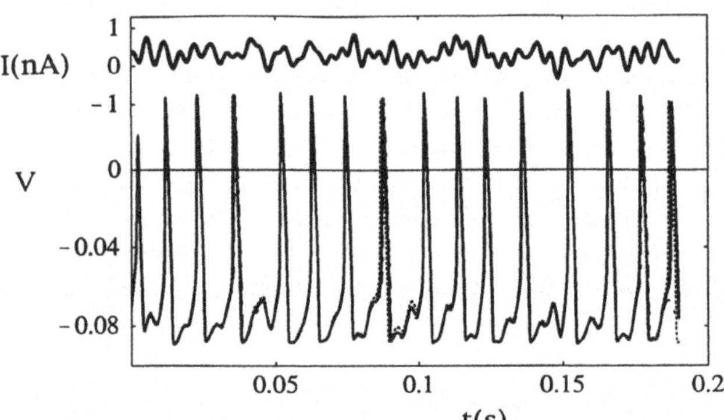

Figure 3. The voltage traces are plotted for three spiking HH neurons (solid, dashed, and dotted curves) with random current injection, with $I_0 = 0.3$ nA, 0.33 nA, and 0.27 nA (respectively) and rms fluctuations of $\pm$ 0.2 nA, indicated by the heavy curve at the top of the Figure. In this case precise spike timing synchrony is maintained in spite of the 10 % variations in the average current. This theoretical picture should be compared with Fig. 1B of Ref. 4.

amplitude of the driving force. With constant driving forces these small changes in parameters cause a steady accumulation of errors in the firing patterns as observed in the experiments[4].

Consequently, the experimental variation in neuronal firing patterns with the constant current injection could result if the neuronal membrane potentials, or the magnitudes of the intrinsic ion currents or even the external injection current were to drift by a only a few percent during the experiments. For example, Fig. 2 shows the voltage traces for three HH oscillators with constant injection currents, I_0, that vary by $\pm$ 10%. Because of the variations in the injection current the spike timing drifts steadily apart in relative phase for the three oscillators. The same results are obtained for small (5-10%)variations in the membrane reversal potentials as well as the kinetic parameters in the HH equations.

Second, we find that with strongly fluctuating inputs the same levels of spike timing reliability observed in the experiments can be achieved in the mathematical models, even with small changes in initial conditions, cell parameters, and driving currents! For example, Fig. 3 shows the voltage traces for three randomly driven HH oscillators with mean injection currents, I_0, that vary by $\pm$ 10% like those in Fig. 2. The random injection current, $I_R(t)$, has the same amplitude and frequency spectrum as the synchronizing current in Fig. 1. Despite the variations in the average injection current the spike timing is reliably maintained to millisecond accuracy. Again similar results are obtained for small (5-10%)variations in other parameters in the HH equations.

CONCLUSION

Reliable spike timing is essential if single neurons are expected to faithfully encode temporal information in the timing of successive spikes. We find that stable, reliable (though highly irregular) oscillations are a natural feature of nonlinear oscillators strongly driven by aperiodically fluctuating inputs. More detailed studies[7,12] show that synchronization does not occur if the amplitudes of the fluctuations are too small or if the typical frequencies of the driving currents are much larger or much smaller that the natural frequency of oscillator. (Synchronization fails for constant drives on both accounts.) Consequently, for neurons that rely on precise spike timing for the communication of information, our results suggest that the only messages that can be reliably conveyed are those that have a large enough amplitude and a frequency spectrum that matches the membrane properties of the cell at the time of transmission.

Useful discussions with L. Abbott, W. Bialek, B. Ermentrout, and Anne Williamson and support from the National Science Foundation (IBN-9634409 and PHY-9507574) are gratefully acknowledged.

REFERENCES

1. F. Rieke, D. Warland, R. de Ruyter van Steveninck, and W. Bialek, *Spikes, Exploring the Neural Code*, (MIT Press, Cambridge MA, 1997).
2. W. Bair and C. Koch, Neural Comp. **8**, 1184 (1996).
3. W.T. Newsome, K.H. Britten, and J.A. Movshon, Nature **341**, 52 (1989).
4. Z.F. Mainen and T.J., Sejnowski, Science **268**, 1503 (1995).
5. See for example C. Koch and I. Segev (eds.), *Methods in Neuronal Modeling*, (MIT Press, Cambridge, MA, 1989)
6. See for example S.H. Strogatz, *Nonlinear Dynamics and Chaos*, (Addison-Wesley, New York, 1994).

7.   R.V. Jensen, "Synchronization of Randomly Driven Nonlinear Oscillators", submitted to Physical Review Letters, December, 1997.

8.   G.B. Ermentrout and J. Rinzel, Am. J. Physiol. **246**, R102 (1984).

9.   J.M. Bower and D. Beeman, *The Book of Genesis*, (Springer-Verlag, New York, 1994).

10.   D.W. Jordan and P. Smith, *Nonlinear Ordinary Differential Equations*, (Clarendon Press, Oxford, 1987).

11.   R. He and P.G. Vaidya, Phys. Rev. A **46**, 7387 (1992).

12.   R.V. Jensen, in preparation.

NEURAL ENSEMBLE PROCESSING WITH TYPES

Don H. Johnson and Charlotte M. Gruner

Computer and Information Technology Institute
Department of Electrical and Computer Engineering
Rice University, MS 366
Houston, TX 77005–1892

INTRODUCTION

Neural ensembles — functionally uniform collections of neurons — process their inputs, which are both ascending and descending, and produce discharge patterns that encode those aspects of the stimulus enhanced by the ensemble. The "neural code" means how groups of neurons, responding individually and collectively, represent sensory information with their discharge patterns[1]; knowing the code would unlock the secrets of how neurons, working in concert, process and represent information. From the viewpoint of point process theory, the code is equivalent to the intensity of an accurate vector-channel, point-process model[2] for the data. The intensity summarizes the (probably complex) dependence of discharge probability on discharges occurring in the same neuron (temporal dependence), in other neurons at the same time (spatial dependence), and in other neurons at different times (spatiotemporal dependence). Unfortunately, traditional optimal estimation techniques depend heavily on the intensity's intrinsic structure (how one event depends on the timing of others)[1], which is part of the neural code we seek. Psychophysical and physiological evidence indicates that sensory systems are most responsive to stimulus changes. These changes induce time-varying (nonstationary) responses, which confound many techniques for quantifying the neural code: Mutual information calculations[3, 4], cross- and autocorrelation techniques[5], and artificial neural networks[6] all rely on stationary response patterns. Furthermore, these methods don't generalize easily to neural ensembles.

We have developed a technique based on recent results from universal source coding and empirical classification theory, problem areas in what is termed Shannon Theory by information theorists. To quantify the degree to which pairs of responses differ, we compute an information theoretic distance measure, the Kullback-Leibler distance. In this paper, we outline how this distance can be computed from ensemble recordings that are *cyclo-stationary*[7]: In the context of sensory systems, the response to any one stimulus presentation may vary with post-stimulus time, but that this time-variation is consistent from one presentation to another. Results from universal coding — the theory of lossless compression when no *a priori* statistics are available — place bounds on how much data are needed for an adequate analysis. Using our approach, ensembles can be studied in a systematic way, and we can assess

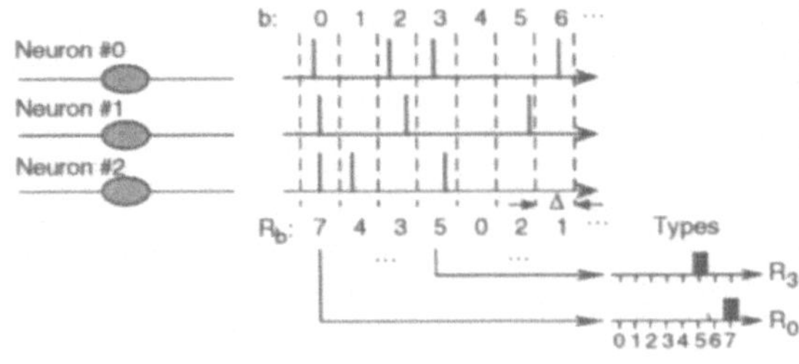

Figure 1. Individual neurons are given an arbitrary identification number. The discharge pattern for each is measured, and individual discharges placed in the bth bin (each bin has width Δ). Using the neuron identification number and the presence of discharges in a binary code, a number denoted by R_b is assigned to each bin to represent which neurons fired during that bin. Because the intensity corresponding to each neuron typically varies with time, we estimate types for each bin separately using multiple stimulus presentations.

the efficacy of stimulus feature coding in the ensemble vis-à-vis coding by the collection of individual neural activity. A companion paper[8] in this volume describes a direct application of this approach to auditory localization. This paper describes the technique's properties, focusing on determining how information is coded in ensembles.

TYPE-BASED ENSEMBLE RESPONSE MEASURE

To develop a measure of the ensemble's response, we first convert the ensemble's discharge pattern into a convenient representation as depicted in figure 1. Here, a neural population's response during the bth bin is summarized by a single number R_b that equals a binary coding for which neurons, if any, discharged during the bin. The response vector $R_m = \{R_{m,1}, \ldots, R_{m,b}, \ldots, R_{m,B}\}$ collects the responses obtained during the mth stimulus presentation from the N-neuron population, and $R = \{R_1, \ldots, R_m, \ldots, R_M\}$ represents the entire dataset, M equaling the number of stimulus presentations. The assumption is that the probability distribution of $R_{m,b}$ does not depend on stimulus presentation, but could vary from bin to bin: The response is cyclo-stationary. By examining the distribution of values of R_b at each bin, we generalize the PST histogram to the case of several neurons.* This new histogram would portray which combination of discharges occurred within the ensemble during a bin, and how this combination changes with post-stimulus time. However, this histogram would not reveal any sequential effects and whether temporal correlation, especially among neurons, varied with time. The quantity containing all response properties would be the *joint* distribution of R_m, which we could estimate from the observation collected from the repeated stimulus presentations. Unfortunately, the number of stimulus repetitions required to estimate this joint distribution is prohibitively large: The number of histogram bins would be 2^{BN}, each of which should be filled with enough events to obtain a reasonable estimate of occurrence probability.

Even if we had sufficient data, having the joint response distribution would not directly reveal the quality of sensory coding and when it was taking place. Modern information theory provides the tools not only to make this assessment, but also suggests how to cope with our inability to estimate accurately the joint response distribution. In information theory parlance, the histogram we seek is known as a *type*[9], which we denote by $\widehat{P}(R)$. What we seek is a measure that quantifies the differences between two responses. Modern classification theory suggests the use of the *Kullback-Leibler distance*. Letting $R^{(1)}$, $R^{(2)}$ denote the responses to

---

*The usual PST histogram would be obtained by considering average of the distribution of R_b in the single neuron case, then dividing by the binwidth.

two stimuli, the Kullback-Leibler (KL) distance is defined to be

$$\mathcal{D}(P(R^{(1)})\|P(R^{(2)})) = \sum P(R^{(1)}) \log \frac{P(R^{(1)})}{P(R^{(2)})} . \tag{1}$$

In this paper, the logarithm is base two, which means that the KL distance has units of bits. This quantity determines how easily responses can be distinguished that are generated according to their respective probability distributions: The larger this quantity, the easier the discrimination, as the probability of misclassification decreases exponentially with the KL distance $(2^{-\mathcal{D}(P(R^{(1)})\|P(R^{(2)}))})$. If the responses to both stimuli were statistically independent from bin to bin, the KL distance equals the sum of the individual response distances across bins. If each bin's response depended only on the previous bin, known as a first-order Markovian dependence, the distance calculation becomes more complicated.

$$\mathcal{D}(P(R^{(1)})\|P(R^{(2)})) = \sum_b \mathcal{D}(P(R_b^{(1)})\|P(R_b^{(2)})) \quad \text{Independent case}$$

$$\mathcal{D}(P(R^{(1)})\|P(R^{(2)})) = \mathcal{D}(P(R_1^{(1)})\|P(R_1^{(2)}))$$
$$+ \sum_b \mathcal{D}(P(R_{b+1}^{(1)}|R_b^{(1)})\|P(R_{b+1}^{(2)}|R_b^{(2)})) \quad \text{Markov case}$$

These formulas suggest that if the data had a simple statistical structure such as zeroth-order (independent) or first-order Markov, we would need much less data to estimate the types corresponding to the underlying distributions. If D denotes the Markovian order of the data, we would need to fill $(B - D)2^{(D+1)N}$ bins.

As we use types instead of the true distributions, we must be concerned with the statistical properties of the types as well as the properties of the KL distance computed between them. Because the KL distance must be non-negative, its estimate must be biased. We use the bootstrap procedure[10] to not only remove bias, but also to obtain confidence intervals for distance estimates without requiring knowledge of the estimate's detailed properties. In our experience, these procedures remove the inherent bias, but at the expense of increased variance, an increase that does not greatly influence the results. Thus, the only assumption type-based analysis makes is the maximal time (expressed in bins) over which one discharge influences the occurrence of another, regardless of whether these discharges occur in the same neuron or in different ones. Recent results in information theory[11] provide methods for finding the minimal joint probability description that, depending the amount of data available, most succinctly describes the data. From this minimal description, the exact nature of temporal dependence within the ensemble can be exposed. This theory details how much data are needed to measure a given degree of dependence (the number of bins spanned by the discharge correlation). The Markovian order of the data that can be adequately described by M stimulus representations for N is

$$D \leq \frac{\log(M + 1)}{\log(2^N + 1)} . \tag{2}$$

Approximately, this constraint means that $M \geq 2^{D\,N}$. It should be noted that type based techniques have been mathematically demonstrated to be the most data-efficient way of quantifying a response[12]. Thus, any measure that promises similar capabilities to those provided by types would need at least this amount of data.

Because the Markov order for the data is usually unknown, we must *assume* an order in computing our KL distances. Once the analysis order equals or exceeds the order of the data, the KL distance will not change, but for analysis orders smaller than the true order, the

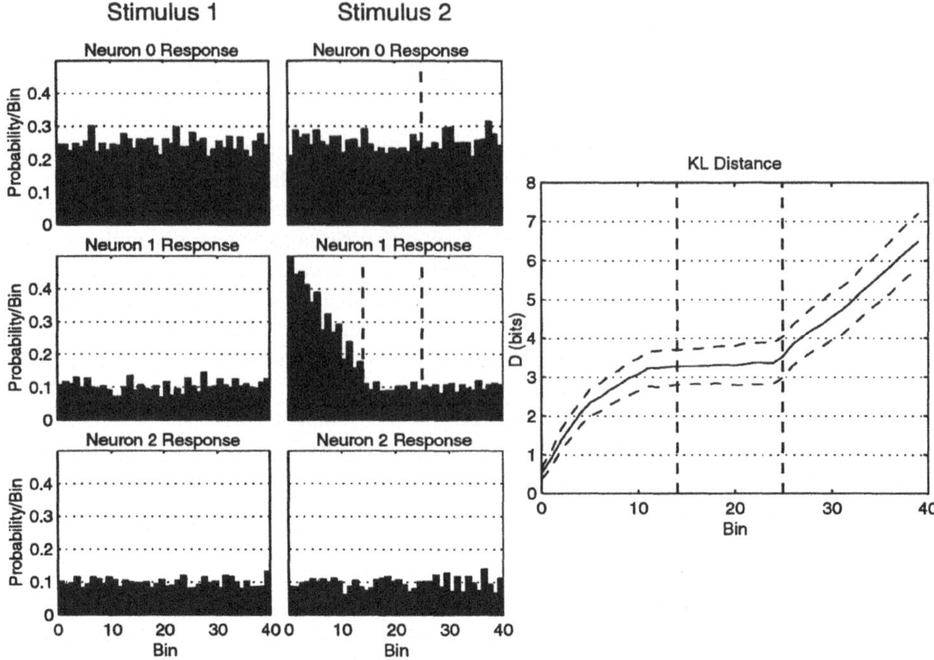

Figure 2. We simulated a three-neuron ensemble responding to two stimulus conditions in which 400 stimulus repetitions were presented. The left portion of the display shows PST histograms of each neuron, and these indicate that neuron 1 had a transient rate response to stimulus 2 (ending at the first vertical dashed line). The right panel shows the result of computing the KL distance to measure the difference between the two responses. The solid line is the debiased KL distance (debiased with the bootstrap technique) and the dashed lines are the 90% confidence intervals (also computed using the bootstrap). We plot how the KL distance increases with post-stimulus time. The second increasing portion occurred because of a stimulus-induced correlation ($\rho = 0.5$) between neurons 0 and 1 in the bins after the second dashed line in the second neuron's response to the second stimulus. Throughout all responses, the correlation between discharges in successive bins was first-order equaled -0.1. First-order Markovian analysis was used in computing the KL distance.

distance will not represent how different the two responses are. Unfortunately, simple examples show that these intermediate distances can be greater or smaller than the true distance. In applying our technique to data, we use a sequence of analysis orders, seeking constancy unless the bound of (2) is exceeded.

EXAMPLE

By computing the distance between the responses to two different stimuli, we can probe how the ensemble processes the stimulus, reveal what portion of the response encodes the stimulus change, and also determine which portion encodes the stimulus change most vigorously. Perhaps the most powerful aspect of type-based analysis is that it makes no *a priori* assumption about the nature of neural encoding. The KL distance quantifies how well the code expresses stimulus changes regardless of its form, whether it be a timing code, a rate code, or some combination of these. Figure 2 illustrates the application of this approach to a simple population of three neurons. Both a stimulus-induced rate response and a transneural correlation can be detected, and the relative contribution of each response component to sensory discrimination quantified. Interestingly, the correlation response is nearly as significant

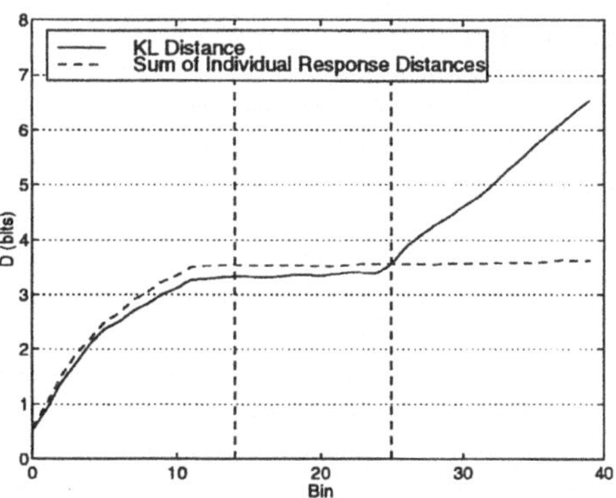

Figure 3. The KL distance between the ensemble response and the sum of the distances computed from the individual neural responses is shown for the two stimulus conditions used in figure 2. The statistically significant difference between the two in the latter portion indicates that the ensemble is representing the stimulus difference more than the components.

for distinguishing the stimuli as the transient rate response: The contribution of each to the total KL distance is about the same. In figure 3, we see that indeed this simulated ensemble is encoding more about the stimulus than the individual neurons. In this plot, the distance between the ensemble's response is compared with the sum of distances computed from the individual neural responses. The latter computation would mimic that of the ensemble if the neural discharges were statistically independent of each other.

Type-based calculations can also be used to determine the presence of correlation in an ensemble's response. This correlation can be stimulus- or connectivity-induced. In the former case, spike trains can be correlated merely because neurons are responding to the same stimulus. In the latter, the neurons receive common inputs or are interconnected. We compute the type of the measured response and derive from this quantity the type that would have been produced by the ensemble if it had statistically independent members (spatial dependence) or had no temporal dependence. Referring to figure 1 for an example, the probability of each neuron discharging in each bin can be calculated from the joint probability of various response patterns occurring in a bin (e.g., $\Pr[\text{discharge in neuron } \#0] = \Pr[R_n = 1] + \Pr[R_n = 3] + \Pr[R_n = 5] + \Pr[R_n = 7]$). From these component probabilities, we estimate the probability of all possible ensemble response patterns by multiplying according to the ensemble response the probabilities of each neuron discharging or not ($\Pr[R_n = 1] = \Pr[\text{discharge in neuron } \#0] \cdot \Pr[\text{no discharge in neuron } \#1] \cdot \Pr[\text{no discharge in neuron } \#2]$). By calculating the distance between these two types, we can infer when correlated responses are present; figure 4 illustrates an example wherein we tested for spatial dependence. Even though the PST histograms give no hint of the sudden spatial correlation, the KL distance clearly indicates its presence. Further analysis using the same technique can be used to narrow the nature of the correlation; for example, pairs could be considered to determine if the entire population contributes to the correlation or not.

SUMMARY

Type-based analysis is the only known technique that can measure the responsiveness of an ensemble, quantify the various contributions to this responsiveness, and provide some insight into the nature of interneuronal response correlation. We can quantify the degree of response detail warranted by the amount of data without making the usual implicit assumptions other variability measures make that the response measure is Gaussian or that the data are stationary.

Unfortunately, fundamental bounds on information representation indicate that the amount of data needed to quantify an ensemble's response grows exponentially with the

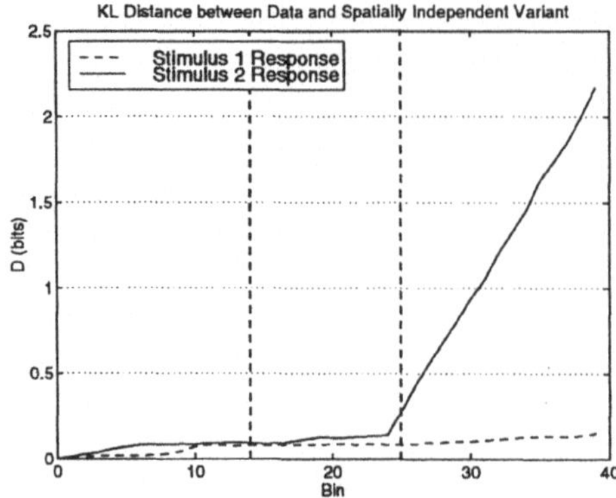

Figure 4. The KL distance between the type computed from response and the type computed derived from it that forces a spatially independent ensemble response structure is shown for the two stimulus conditions used in figure 2. As was the case in actuality, the responses to stimulus 1 demonstrate no trans-neural correlation. The second stimulus did induce a correlation in the latter portion of the response, and the KL distance clearly indicates the presence of such correlation.

number of neurons. Because the technique uses the amount of available data as efficiently as is possible with any statistical technique, all other procedures face the same data limitation. In applications, the biased nature of the KL distance estimates require the use of resampling techniques like the bootstrap to computationally remove bias so that response differences can be quantified. The computational load required by the bootstrap is large, requiring hundreds of times longer to compute than the raw KL distance calculation. Although the computations are time-consuming, the type-based technique provides a great deal of insight into neural ensemble coding.

Acknowledgments

CMG was supported in part by the Keck Center for Computational Biology, Houston, Texas, and grant 1T15 LM07093 from the Library of Medicine to the Keck Center. This paper's contents are solely the responsibility of the authors and do not necessarily represent the official views of the National Library of Medicine.

REFERENCES

1. W. Bialek, F. Rieke, R. R. de Ruyter van Steveninck, and D. Warland. Reading a neural code. *Science*, 252:1852–1856, 1991.

2. M. I. Miller and D. L. Snyder. *Random Point Processes in Space and Time*. Springer-Verlag, New York, second edition, 1991.

3. F. Gabbiani and C. Koch. Coding of time-varying signals in spike trains of integrate-and-fire neurons with random threshold. *Neural Computation*, 8:44–66, 1996.

4. F. Rieke, D.A. Bodnar, and W. Bialek. Naturalistic stimuli increase the rate and efficiency of information transmission by primary auditory efferents. *Proc. R. Soc. Lond. B*, 262:259–265, 1995.

5. M. Abeles and M. H. Goldstein, Jr. Multispike train analysis. *Proc. IEEE*, 65:762–773, 1977.

6. J. C. Middlebrooks, A. E. Clock, L. Xu, and D. M. Green. A panoramic code for sound location by cortical neurons. *Science*, 264:842–844, 6 May 1994.

7. W. A. Gardner. *Cyclostationarity in Communications and Signal Processing*. IEEE Press, New York, 1994.

8. C.M. Gruner and D.H. Johnson. *Analysis of sensory coding in the lateral superior olive*. Plenum, New York, 1998.

9. T. M. Cover and J .A. Thomas. *Elements of Information Theory*. Wiley, New York, 1991.

10. B. Efron and R. J. Tibshirani. *An Introduction to the Bootstrap*. Chapman & Hall, New York, 1993.

11. M. J. Wienberger, J. J. Rissanen, and M. Feder. A universal finite memory source. *IEEE Trans. Info. Th.*, 41:643–652, 1995.

12. M. Gutman. Asymptotically optimal classification for multiple tests with empirically observed statistics. *IEEE Trans. Info. Th.*, 35:401–408, 1989.

RESPONSE TO PERTURBATIONS OF A NEURAL NETWORK MODEL OF LOCOMOTOR CONTROL IN THE LAMPREY

Ranu Jung[1] and Suzanne Generazzo.[2]

[1]Center for Biomedical Engineering and Dept. of Electrical Engineering
21 WGRL
University of Kentucky
Lexington, KY 40506-0070

[2]Dept. of Mathematical Sciences
University of Massachusetts
1 University Ave.
Lowell, MA 01854

INTRODUCTION

In several vertebrates, stable rhythmic locomotor activity can be obtained from neural networks in the spinal cord without external periodic forcing. In an active animal, the spinal central pattern generator (CPG) is dynamically affected by descending input from the brain as well as afferent sensory inputs.[1] While certain aspects of spinal CPG mechanisms for locomotor control have been well investigated,[3,8] the interactions of the CPG with the descending supraspinal inputs and sensory inputs are not well understood. Changes in the topology of the system caused by modifying the supraspinal-spinal CPG interactions are likely to influence the response of the system to external perturbations.

Thus, changes that alter the boundary between stable states could alter the likelihood that a perturbation would cause a state transition. It may be advantageous to a biological system to exist near such bifurcation boundaries. For example, if the boundary existed between a stable non-oscillatory state and a stable oscillatory state, a small perturbation would allow the oscillations to turn on or off. If the supraspinal-spinal interactions move the system into a stable limit-cycle state then external perturbations should be capable of causing phase and strength dependent resetting.[2,7,10] Modifications of the supraspinal-spinal CPG interactions may also change the topology of the state space around the stable limit cycle, which would alter the phase dependent responses to external perturbations.

In *in-vitro* preparations in which the brain and the spinal cord are bathed with physiological saline (non-oscillatory state), we have found that brief trigeminal stimulation causes short latency polysynaptic EPSPs (with occasional action potentials) and IPSPs in Müller neurons.[4] During fictive locomotion induced by applying excitatory amino-acids to the spinal cord, brief trigeminal nerve stimulation is capable of inducing a prolonged excitatory afterdischarge in Müller neuron activity following the short latency responses. In addition, a weak or a strong resetting of the fictive locomotor rhythm with aftereffects (short-term memory) lasting several cycles can occur.[4] Thus, it appears that the feedforward-feedback (FF-FB) loop that exists between the reticular neurons and the CPG[6,9] alters the response of the system to afferent sensory stimulation. Also, we observed that in the brain - spinal cord preparation periodic perturbation of the spinal cord during fictive locomotion can cause a 1:1 entrainment of the rhythm.

To better understand aspects of the dynamical interaction of the spinal CPG with supraspinal and afferent sensory input, we investigated the response to external perturbations of a neural network model of locomotor control in the lamprey. In the model, we find both phase and strength dependent resetting of the rhythm with phase dependent aftereffects. The nature of the sensory input (excitatory versus inhibitory) alters the phase response characteristics. The FF-FB ratio can completely change the perturbation responses, with the altered topology resulting in termination of the oscillation at some phases. We also find that periodic perturbation can cause entrainment below the default period as well as aperiodic behavior which nonetheless has an underlying structure.

METHODS

The Experimental Preparation

Resetting of the fictive locomtor rhythm was examined in a split bath *in-vitro* brain spinal cord (50-70 segments long) preparations of the adult lamprey (Ichtyomyzon Unicuspis or Petromyzon Marinus). The brain was bathed with physiological saline and the spinal cord with a 0.5 mM D-glutamate or 0.2 mM N-methyl-DL-aspartate saline solution. Motor activity was recorded en-passant from the ventral roots using suction electrodes and reticular activity recorded intracellularly from Müller neurons using sharp electrodes. The left trigeminal nerve was stimulated with a brief electrical stimulation (300 μsec pulse). The caudal end of the left spinal cord surface (dorsal afferents) was stimulated with a 20 msec pulse at 1Hz.

The Neuron Model

The model (based on a model previously used [5]) is a left-right symmetric network in which the spinal CPG has excitatory (E) and lateral (L) inhibitory interneurons with ipsilateral projections, and crossed (C) inhibitory interneurons with contralateral projections. Motoneurons (MN) receive input from the E and C interneurons and are output units only.[2] Reticular neurons are modeled as two pools, tonically active (RT) and phasically active (RP) with excitatory ipsilateral projections to all CPG neurons. RP neurons receive ipsilateral excitatory input from RT neurons and feedback from ipsilateral E and contraleral C interneurons. Each class is represented by one neuron.

Each neuron has basic passive properties akin to biophysical neurons and receives tonic synaptic drive from reticular neurons and weighted synaptic input from other connecting neurons. The reticular neurons receive external pulse perturbation input as an excitatory synaptic input:

$$C_M^i \frac{dv_i}{dt} = G_R^i \left(V_R^i - v_i\right) + G_T^i \left(V_T^i - v_i\right) + \sum_j G_{ji} h(v_j)\left(V_{syn}^j - v_i\right) + G_{pulse}^i \left(V_{pulse}^i - v_i\right)$$

where v_i is the voltage of neuron i, G *is* the maximal conductance for a given ionic channel; the subscript indicates the type: R=resting, T=tonic, ji=synaptic from j to i, and pulse. The reversal potentials are given by V with the appropriate subscripts. $V_R = 0$, $V_T = 1$. V_{syn}^j, V_{syn}^i, V_{pulse}^i = +1 for excitatory connections and -1 for inhibitory connections. The neuron's output as a function of voltage is given by a piecewise seventh order polynomial with a strict threshold (0), saturation level (1), and a continuous third derivative.[5]

Model Analysis

Pulse inputs mimicking brief sensory stimulation are applied to both RT and RP neurons in the following configurations: symmetric single excitatory (inhibitory) pulse to the left and right sides, single excitatory pulse to the left and inhibitory pulse to the right sides; periodic excitatory pulse to the left and right sides. For single pulse perturbations transient and steady state cophase plots[2,7,10] are used to assess the phase dependence and strength dependence of the response as well as the aftereffects (qualitatively and as # of cycles to reach eventual phase shift). 100 stimuli are applied per cycle. MN outputs are used as the rhythm index. The FF-FB ratio between the reticular and CPG neurons is altered and the cophase plots reexamined. Responses to periodic pulse perturbation with periods above and below the default cycle period are obtained for 4000 or more cycles. First order delay plots are examined to assess the presence of entrainment and the structure in aperiodic rhythms.

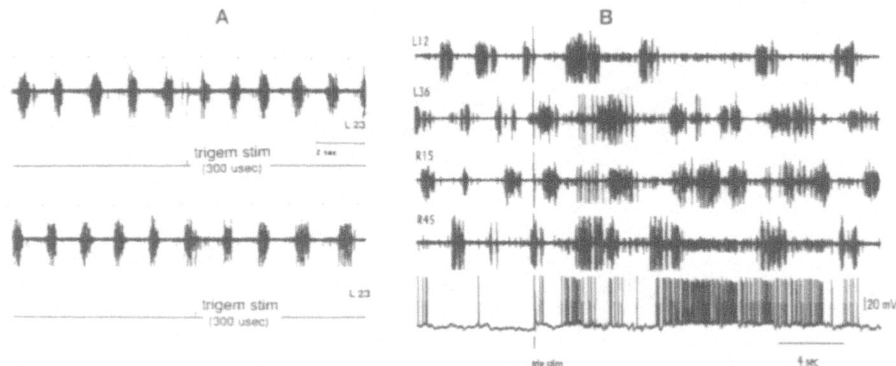

Figure 1. A) Resetting effect of trigeminal nerve perturbation. Dots indicate burst initiation before stimulation and in a hypothetical periodic rhythm that would exist in the absence of stimulation. Top: Phase advance, Bottom: Phase delay. B) Strong resetting effects of trigeminal nerve stimulation. Inter and intrasegmental coordination is affected by the stimulation. The contralateral Müller neuron is activated and shows strong afterdischarge concomitant to the effects on the ventral root activity. L-left, R- right. Numbers indicate segment at which ventral root was recorded.

Simulations were conducted on models developed in Matlab and Simulink (Mathworks Inc.) using a Runga-Kutta 4-5 order integration routine. XPP and AUTO were used to do conduct bifurcation analysis to determine the boundary between the oscillatory and non-oscillatory regimes obtained on altering the FF-FB connection strengths.

RESULTS

In the experimental preparations, resetting of the fictive locomtor rhythm on brief trigeminal nerve stimulation is observed. As shown in Figure 1A both phase advances as well as phase delay can occur. Figure 1B shows strong effects of brief trigeminal nerve stimulation on the fictive locomtor rhythm accompanied by an afterdischarge in the Müller neuron. Figure 2 shows 1:1 entrainment of the fictive locomotor rhythm with periodic perturbation which takes several cycles to establish itself on initiation of the perturbation as well as return to the control period on cessation of the perturbation.

Under the default parameters the CPG-reticular network yielded stable symmetric oscillatory output with appropriate phase relationships amongst the different neurons. A two-parameter

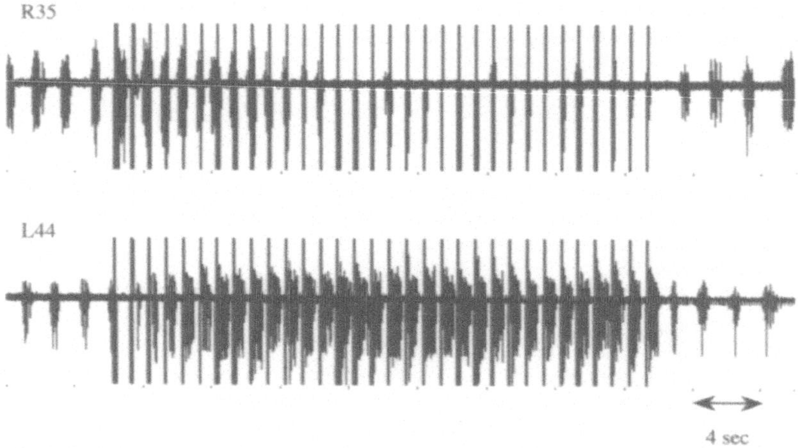

Figure 2. Effect of periodic stimulation of the spinal cord. The contralateral ventral root is excited and shows 1:1 entrainment, the ipsilateral root is strongly inhibited and silenced. R=right. L=Left ventral root

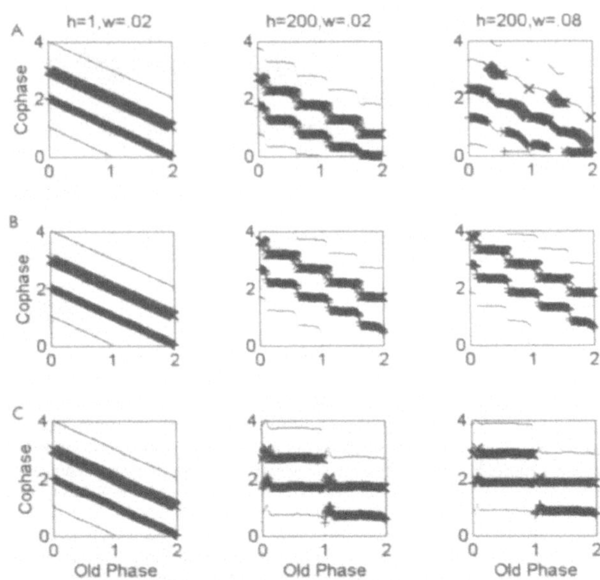

Figure 3. Cophase plots. Made by plotting old phase ϕ versus cophase θ_i for i=1 to 4 where i indexes the cycle number. Old phase ϕ, phase of stimulation, is the delay between the locomotor burst onset and the time of stimulation, divided by the control cycle period (T_0). The cycle preceding the stimulated cycle is the control cycle. The ith cophase, θ_i, is the time from the end of the stimulus to the onset of the ith cycle after the stimulus, divided by T_0. For short stimuli the "no effect" line is given by a slope of -1 and $\theta_i \cong i - \phi$. A slope close to -1 indicates weak Type 1 resetting, while a slope close to 0 indicates strong Type 0 resetting. Row A- Symmetric excitatory input; Row B- Symmetric inhibitory input, Row C- Excitatory input to left, inhibitory input to right. h= arbitrary height of pulse, w= width of pulse in seconds.

bifurcation diagram obtained on altering the FF-FB connection strengths showed regions of stable oscillatory and non-oscillatory states (see figures in 5).

The effect of the single pulse perturbations was both phase and strength dependent. With excitatory pulse inputs a Type 1 resetting with phase advance was observed (Figure 3). The cophase plots were highly nonlinear, with higher sensitivity at phases spanning burst onset and burst termination. In between these transition zones, the phase advance varied very little. Perturbations applied during bursting caused premature burst termination and reduced the off-time. The burst duration and off-time for the consecutive cycle were also slightly reduced. With longer stimuli perturbations applied just prior to burst termination there was an apparent phase delay (Figure 4). This occurred because of immediate initiation of a new cycle thereby effectively prolonging the burst in which the perturbation was given. Perturbations applied during the off-time also initiated a

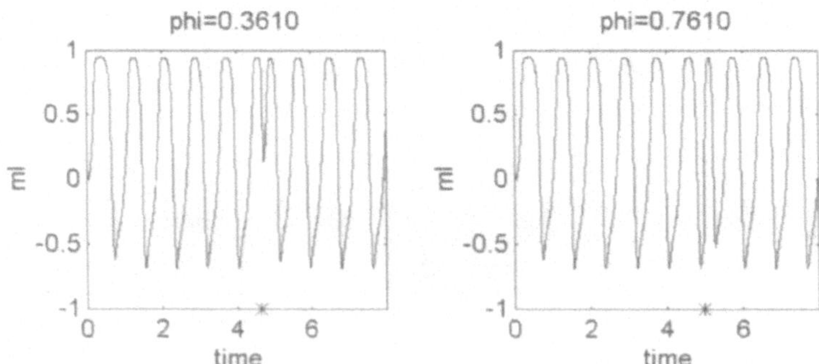

Figure 4. Effect of single excitatory pulse input. At both old phases (phi) a new burst is initiated, but in the left panel the burst initiation gives an apparent increase in burst duration. This would appear as a phase delay in the cophase plot, pulse height=200, pulse width=0.08 seconds, ml= left motoneuron output. time in seconds.

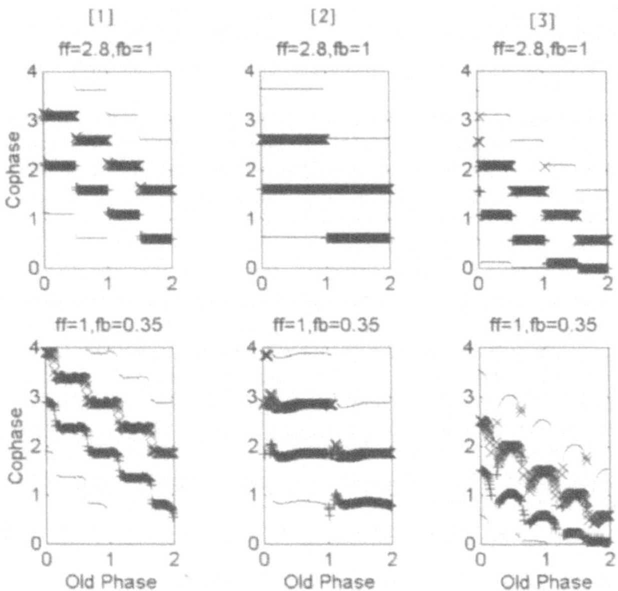

Figure 5. Effects of altering the FF-FB ratio on the cophase plots. Column [1]= Symmetric inhibitory input, Column [2]=Left inhibitory, Right Excitatory input, Column [3]=Symmetric excitatory input. ff= feedforward gain, fb= feedback gain (Default ff=fb=1, compare plots with figure 4). Pulse height = 200, pulse width= 0.02 seconds

new cycle but with a sharply reduced burst duration and off-time. Aftereffects were seen in second cycle following the perturbation. Similarly, with a symmetric inhibitory perturbation Type 1 resetting with highly nonlinear cophase plots were observed (Figure 3). However, now a phase delay was observed and longer stimuli applied near the end of the cycle affected the following

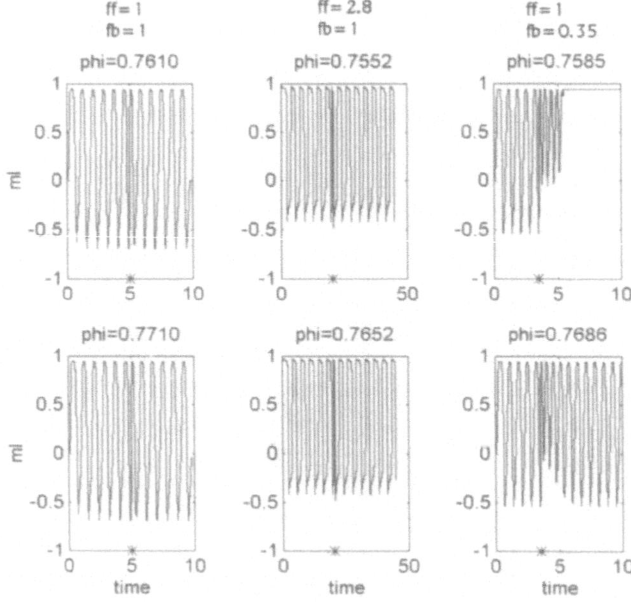

Figure 6. Effects of altering FF-FB ratio on the perturbation response. Single symmetric excitatory pulse. phi=old phase. Each column indicates the FF-FB values. Pulse height=200, pulse width =0.02 seconds, time in seconds.

cycle. Asymmetric pulse perturbation resulted in Type 1 resetting with a phase advance for perturbations early in the cycle and phase delays for perturbations late in the cycle. With longer stimuli this resetting was converted to a Type 0. Very small affects were observed on the consecutive cycle (Figure 3).

Altering the FF-FB gains changed the topology of the system and affected the phase and strength dependence of the responses (Figure 5). In all configurations, increases in FF gain made the cophase plots more non-linear. The range of phases for increased sensitivity sharply reduced while the degree of phase advance or delay observed under default conditions increased. With decreased FB gain, perturbations applied at some phases could push the limit cycle off it's attractor state to a stable fixed state (Figure 6). With the initial state of the system set to be in an asymmetric bistable fixed state, brief pulse perturbations switched the system between the bistable states.

Periodic symmetric excitatory trains of pulses were capable of entraining the rhythm. Figure 7 shows the first order delay plots. Stimulation was begun after 5 control cycles. The delay plots show the periods of cycles 10 through 225 only. 1:1 entrainment is seen for a stimulation period of 0.498. As predicted by the cophase plots 1:1 entrainment occurred only below the default period. Even at stimulation periods greater than control there occurred a phase dependent advance and no cycles period was extended. Hence 1: 1 entrainment could not be obtained. 1:2 and 2:2 entrainment were also observed. Interspersed with the periodic rhythms were aperiodic rhythms which showed definite structure in the delay plots. Stimulation at approximatley one-tenth the default period (0.083) caused cessation of oscillation.

CONCLUSIONS

Phase and strength dependent resetting of biological limit cycle oscillators can be caused by a brief perturbing stimulus.[1,6,8] Our model shows a nonlinear phase and strength (intensity and duration) dependent resetting of the locomotor rhythm. Both advances and delays can be obtained depending on the nature of the stimulus: excitatory or inhibitory; symmetric or asymmetric. Comparison of the model results to the experimental data suggests that the trigeminal stimulation acts either as an excitatory or as an asymmetric sensory input. We observe both phase resetting and aftereffects in experimental investigations. In the model, perturbations at some phases lead to aftereffects such that the limit cycle does not relax back to it's attractor within the perturbed cycle. Aftereffects can last for more than one cycle. Thus, the network architecture itself is conducive to

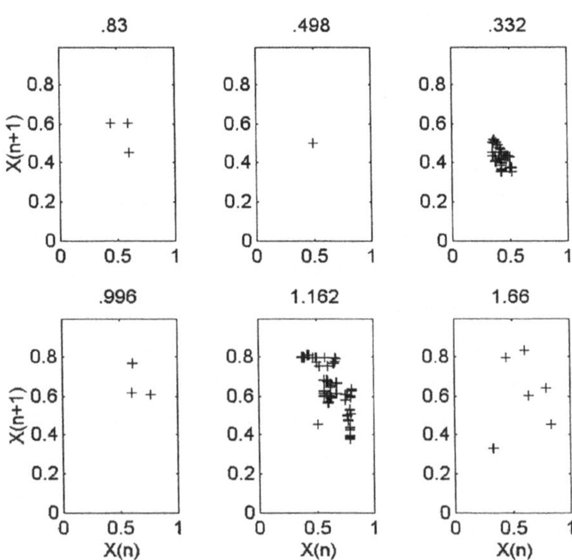

Figure 7. First order delay plots. X(n) and X(n+1) are consecutive cycle periods. Default cycle period was 0.834 secs. Title on each panel gives the period of the stimulating train. Pulse height=200, pulse width=0.02, ff=fb=1.

short term memory effects. The apparent increase in burst duration, on perturbations applied just prior to burst termination in the model, are also observed experimentally (unpublished). However, in experiments it is not possible to distinguish initiation of a new cycle from a true burst elongation. The model exhibits both n:m entrainment and aperiodic responses with underlying structure. The aperiodicity observed may partly be due to the aftereffects of the perturbations. The reticulo-spino-reticulo loop can also alter the response of the system to perturbing stimuli by changing the underlying topology of the system. The model analysis thus suggests mechanisms for some of the experimental observations. It suggests that the reticular-spinal loop can affect short and long term locomotor control in the lamprey by a) modulating the rate at which the system returns to a given basin of attraction after brief perturbing stimuli and, b) altering the phase resetting characteristics of the rhythm.

ACKNOWLEDGMENTS

This work was supported in part by grants from NSF (IBN9601345) and The Whitaker Foundation to Ranu Jung. Suzanne Generazzo was supported by an NSF REU to University of Kentucky.

REFERENCES

1. F. Delcomyn, Neural basis of rhythmic behavior in animals, *Science* 210:492-498 (1980).
2. L.Glass and M.C. Mackey. *From Clocks to Chaos*, Princeton University Press, New Jersey (1988).
3. S. Grillner, T. Deliagina, Ö. Ekeberg, A. El. Manira, R.H. Hill, A. Lansner, G.N. Orlovsky, and P. Wallén, Neural networks that coordinate locomotion and body orientation in lamprey, *Trends Neurosci.* 18:270-279 (1995).
4. R. Jung and A.H. Cohen, Effects of trigeminal input on locomotor pattern and reticular neural activity in the lamprey, *Soc. Neurosci. Abstr.* 21:277.3 (1995).
5. R. Jung, T. Kiemel, and A.H. Cohen, Dynamical behavior of a neural network model of locomotor control in the lamprey, *J. Neurophysiol.* 75(3):1074-1086 (1996).
6. S. Kasicki, S.Grillner, Y. Ohta, R. Dubuc, and L. Brodin, Phasic modulation of reticulospinal neurones during fictive locomotion and other types of spinal activity in lamprey, Brain Res. 484:203-216 (1989).
7. M. Kawato, Transient and steady state phase response curves of limit cycle oscillators, *J. Math. Biol.*, 12:13-30 (1981).
8. S. Rossignol and R. Dubuc, Spinal pattern generation, *Curr. Opin. Neurobiol.* 4:894-902 (1994).
9. L.Vinay and S. Grillner, Spino-bulbar neurons convey information to the brainstem about different phases of the locomotor cycle in the lamprey, Brain Res. 582:134-138 (1992).
10. A.T. Winfree, *The Geometry of Biological Time*, Springer-Verlag, New York (1980).

MODELING DYNAMIC RECEPTIVE FIELD CHANGES PRODUCED BY INTRACORTICAL MICROSTIMULATION

George J. Kalarickal and Jonathan A. Marshall

Department of Computer Science, CB 3175, Sitterson Hall
University of North Carolina, Chapel Hill, NC 27599-3175, U.S.A.

INTRODUCTION

Receptive field topography changes during intracortical microstimulation

Intracortical microstimulation (ICMS) of a localized site in the somatosensory cortex of rats and monkeys produces reorganization of receptive field (RF) topography over a large region of the cortex.[1] ICMS excites nearly all afferents and excitatory and inhibitory cortical neurons within a few microns of the stimulating electrode. This stimulation produces nearly simultaneous activation of all pre- and postsynaptic elements and modulatory inputs close to the ICMS site. In addition, some of the afferent terminals receive ortho- and antidromic excitation. However, not all anti- and orthodromically excited afferents succeed in driving their target neurons above threshold.[1] ICMS of the cortex for 2–6 hours produces a large (2-fold to over 20-fold) increase in the cortical representation of the skin region represented by the ICMS-site neurons before ICMS.[1]

EXIN model of dynamic RFs changes during ICMS

We have formulated and tested a neural network model that exhibits similar RF topographical changes. The model, which uses the EXIN (excitatory+inhibitory) learning rules,[2] predicts some experimentally testable RF characteristics. The EXIN model uses modifications in the synaptic efficacy of feedforward excitatory and lateral inhibitory connection pathways to describe the effects of ICMS.

A novel feature of the EXIN model is the role of lateral inhibitory learning rule. We have used the EXIN lateral inhibitory learning to model dynamic RF changes produced by artificial scotoma conditioning and retinal lesions.[3,4,5] Recanzone et al.[1] realized a need for different learning in excitatory and inhibitory synapses during ICMS to account for all the results. However, they did not propose any specific rule for modifying inhibitory synapses.

METHODS

We simulated a patch of somatosensory cortical neurons, arranged in a 30×30 grid of spatial positions. The position of each neuron's RF corresponded to the neuron's position in the grid. Adjacent RFs initially had more than 50% spatial overlap.

During ICMS excitatory and inhibitory neurons and synaptic terminals are directly activated by the electrical current, and others are activated/receive input indirectly through synaptic activation and via anti- and orthodromic excitation.[1] Therefore, in the simulations, ICMS was modeled by: (1) directly activating model cortical neurons close to the ICMS site, and (2) activating thalamocortical excitatory and lateral inhibitory pathways.

The strength of the direct excitation to model cortical neurons is maximum at the ICMS site and decreases with distance from the ICMS site according to a Gaussian function. The strength of excitation received by the presynaptic excitatory and inhibitory terminals decreased according to a Gaussian function of the distance of the pathway from the stimulation site and the distance of the target neuron from the stimulation site. The Gaussian distributions of the direct excitation to model cortical neurons and excitation to presynaptic excitatory and inhibitory synaptic terminals are based on the observation that the strength of electrical stimulation decreases with distance from the stimulation site.[1]

The spread of direct excitation was small, and excitation to presynaptic excitatory and inhibitory synaptic terminals spread over a large distance. In the model as well as in the experimental data, the presynaptic excitations spread over large distances, but most were initially ineffective in driving cortical neurons.

Simulation procedure

To simulate ICMS, direct excitation to model cortical neurons, and excitation to presynaptic excitatory and inhibitory presynaptic terminals are held constant. The indirect excitation and inhibition via synapses change during ICMS because the strength of synapses change according to the EXIN learning rules.

The EXIN learning rules

The EXIN lateral inhibitory learning rule. The EXIN lateral inhibitory learning rule[2] is an anti-Hebbian outstar learning rule. The weight Z_{ij}^{-} of the lateral inhibitory pathway from neuron i to neuron j is governed by the following equation:

$$\frac{d}{dt} Z_{ij}^{-} = \delta g(x_i) \left(-Z_{ij}^{-} + q(x_j) \right), \tag{1}$$

where $\delta > 0$ is a small learning rate constant, x_i and x_j are the activation levels of neurons i and j, respectively, and g and q are half-rectified increasing functions. Thus, *whenever a neuron is active, its output inhibitory connections to other active neurons tend to become slightly stronger (i.e., more inhibitory), while its output inhibitory connections to inactive neurons tend to become slightly weaker.*

In an outstar learning rule,[6] *presynaptic* activity "enables" the learning at a synapse; when the learning is enabled, the weight tends to become proportional to the postsynaptic activity. In an instar learning rule, *post*synaptic activity enables the learning; when the learning is enabled, the weight tends to become proportional to the presynaptic activity. Thus, in an instar rule, the subscripts for x_i and x_j in the equation above would be interchanged.

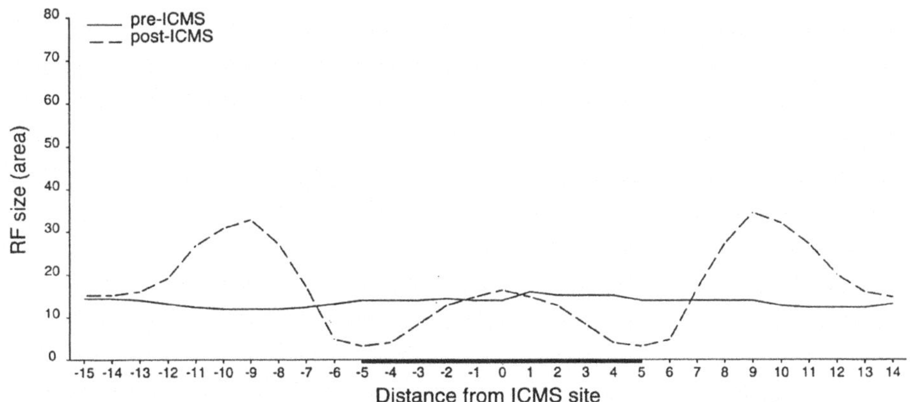

Figure 1. The figure shows the RF area of neurons in a cross-section of cortex passing through the ICMS site before and after ICMS. The thick line segment on the abscissa represents the neurons active during the initial phase of ICMS.

The EXIN feedforward excitatory learning rule. The afferent excitatory pathway weight changes are governed by the EXIN excitatory learning rule, which is an instar Hebbian rule. The rule can be expressed [2,7] as

$$\frac{d}{dt}Z_{ij}^{+} = \epsilon f(x_j)\left(-Z_{ij}^{+} + h(x_i)\right), \tag{2}$$

where i is a modeled thalamic neuron and j is a modeled cortical neuron, x_i and x_j are the activation levels of neurons i and j, respectively, Z_{ij}^{+} is the feedforward excitatory weight from neuron i to neuron j, $\epsilon > 0$ is a small learning rate constant, and f and h are half-rectified increasing functions. Thus, *whenever a neuron is active, its input excitatory connections from active neurons become slightly stronger, while its input excitatory connections from inactive neurons become slightly weaker.*

The activation equation

The activation level x_j of each modeled cortical neuron is governed by a shunting equation[6] based on the Hodgkin-Huxley model:

$$\frac{d}{dt}x_q = -Ax_q + \beta(B - x_q)E_q - \gamma(C + x_q)I_q, \tag{3}$$

where A, B, C, β, and γ are positive constants, and E_j and I_j represent respectively the neuron's total excitatory and inhibitory input signals. Because Equation 3 is a shunting equation, neuron activation levels remain within a bounded range, between $-C$ and B.

RESULTS
Synaptic reorganization during ICMS in the EXIN model

During ICMS the following synaptic modifications occur in the EXIN model. The EXIN inhibitory learning rule decreases the weights of the active lateral inhibitory

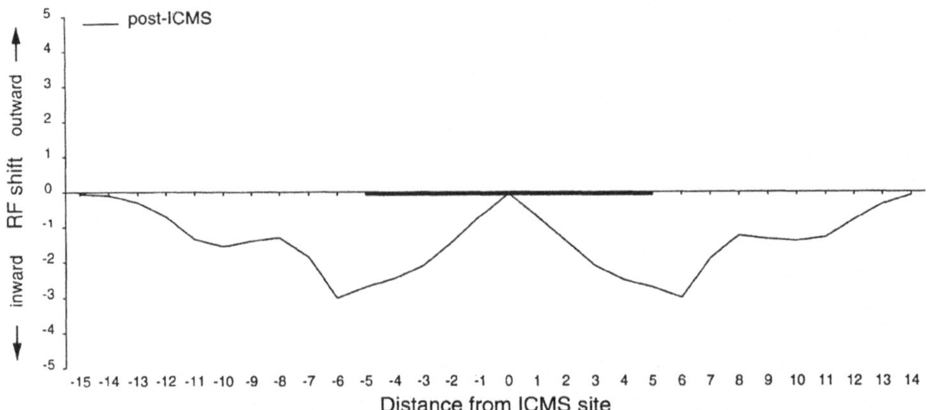

Figure 2. The figure shows the RF shift of neurons in a cross-section of cortex passing through the ICMS site after ICMS. The thick line segment on the abscissa represents the neurons active during the initial phase of ICMS.

pathways to the surrounding inactive or very weakly active neurons. This decrease in inhibitory weights to the initially inactive neurons allows these neurons to respond to the excitation they receive via anti- and orthodromically activated afferents. In addition, the decreased inhibition causes an expansion in the RF size of these neurons and thereby an increase in RF overlap with the ICMS-site RF. However, inactive lateral inhibitory weights to active or inactive neurons do not change. Thus, as the RFs of the previously inactive neurons from the ICMS site expand toward the ICMS-site RF, the neighboring active neurons receive more inhibition, and their RFs contract (Figure 1).

According to the EXIN feedforward excitatory learning rule, the active cortical neurons during ICMS strengthen excitatory synapses from the strongly active afferents at the ICMS site or the branches of the afferents at the ICMS site and weaken excitatory synapses from weakly active and inactive afferents. Thus, during ICMS the RFs of the active neurons shift to overlap with the ICMS-site RF. Neurons closest to the ICMS site eventually have RFs almost identical to the ICMS-site RF (Figures 1 and 2). Some neurons lose responsiveness to parts of their original RF and gain responsiveness to the ICMS-site RF. This RF change is termed "RF substitution."[1]

In response to the relatively weak activation of neurons close to the ICMS site during ICMS, the learning rule in the simulation causes the lateral inhibitory weights between the active neurons to weaken. This keeps the RF size and responsiveness of the ICMS-site neurons roughly constant (Figures 1 and 3).

Comparison of the EXIN model with neurophysiological data

During ICMS, the computational simulation of the EXIN model produced the following effects, corresponding closely to the experimental neurophysiological results reported by Recanzone et al.[1]: (1) increase in the cortical representation of the ICMS-site RF (Figure 4); (2) almost no change in RF size and responsiveness of ICMS-site neuron (Figures 1 and 3); (3) substitution of parts of the ICMS-site RF for the former RF of surrounding neurons (Figures 1 and 2); (4) RF expansion in some neurons and contraction in others (Figure 1); and (5) RF shift toward the ICMS-site RF (Figure 2).

426

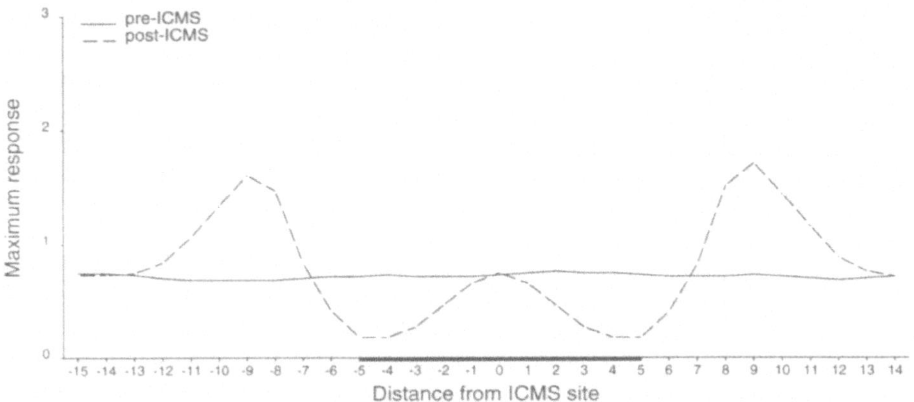

Figure 3. The figure shows the maximum responsiveness of neurons in a cross-section of cortex passing through the ICMS site before and after ICMS. The thick line segment on the abscissa represents the neurons active during the initial phase of ICMS.

With no learning in excitatory synapses and learning in inhibitory synapses, the model predicts an increase in the number of cortical neurons with RFs overlapping with the ICMS-site RF. An experimental idea is to abolish learning in excitatory synapses without affecting inhibitory learning by injecting an NMDA antagonist into the cortex during ICMS. NMDA antagonists have been shown to abolish changes in excitatory synaptic efficacy without completely eliminating postsynaptic responses to electrical and visual stimulation.[8,9] To draw firm conclusions, controls would be required to estimate changes in excitatory and inhibitory synapses.

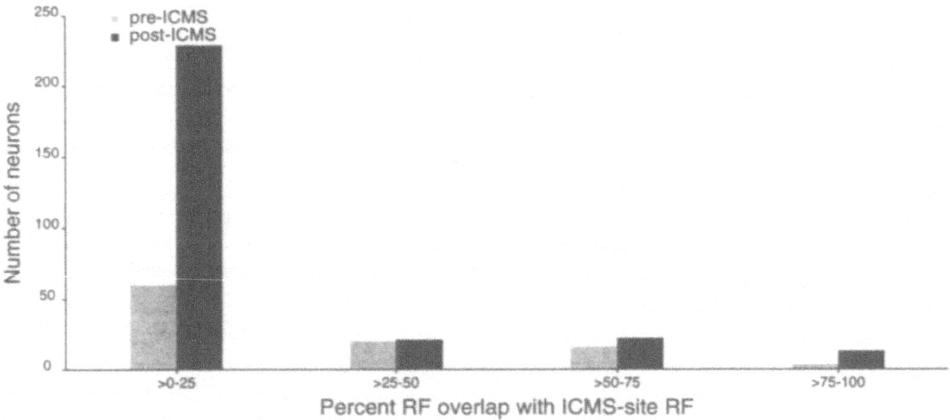

Figure 4. The figure shows the number of simulated cortical neurons with RF overlapping with the ICMS-site RF before and after ICMS.

DISCUSSION

The EXIN model reproduces the main effects of ICMS on cortical RF location, size, and responsiveness. The combination of the EXIN afferent excitatory and lateral inhibitory learning produces an increase in the cortical representation of the skin region corresponding to the ICMS site, without much effect on ICMS-site RF size and responsiveness.

The EXIN model predicts an increase in the cortical representation of the skin region corresponding to the ICMS site, even with no afferent excitatory learning. A neurophysiological experiment abolishing learning in excitatory synapses without shutting down learning in inhibitory synapses would test the predictions of the EXIN lateral inhibitory learning rule. Other experiments to verify the predictions of the EXIN learning rule have been proposed by Marshall and Kalarickal.[4]

This work, in conjunction with the EXIN model of artificial scotoma conditioning and retinal lesions,[4] suggests the hypothesis that outstar lateral inhibitory plasticity is an important and general principle in cortical organization.

ACKNOWLEDGMENT

Supported by the Office of Naval Research (ONR N00014-93-1-0208) and the Whitaker Foundation (Special Opportunity Grant).

REFERENCES

1. G.H. Recanzone, M.M. Merzenich, and H.R. Dinse, Expansion of the cortical representation of a specific skin field in primary somatosensory cortex by intracortical microstimulation, *Cereb. Cortex* 2:181–196 (1992).
2. J.A. Marshall, Adaptive perceptual pattern recognition by self-organizing neural networks: context, uncertainty, multiplicity, and scale, *Neural Networks* 8: 335–362 (1995).
3. C. Darian-Smith and C.D. Gilbert, Topographic reorganization in the striate cortex of the adult cat and monkey is cortically mediated, *J. Neurosci.* 15(3):1631–1647 (1995).
4. J.A. Marshall and G.J. Kalarickal, Modeling dynamic receptive field changes in primary visual cortex using inhibitory learning, *in*: "Computational Neuroscience: Trends in Research, 1997," J.M. Bower, ed., Plenum Press, New York (1997).
5. M.W. Pettet and C.D. Gilbert, Dynamic changes in receptive-field size in cat primary visual cortex, *Proc. Natl. Acad. Sci. USA* 89(7):8366–8370 (1992).
6. S. Grossberg, Neural expectation: cerebellar and retinal analogs of cells fired by learnable or unlearned pattern classes, *Kybernetik* 10:49–57 (1972).
7. S. Grossberg. "Studies of Mind and Brain: Neural Principles of Learning, Perception, Development, Cognition, and Motor Control," Reidel Press, Boston (1982).
8. M.F. Bear, W.A. Press, and B.W. Connors, Long-term potentiation in slices of kitten visual cortex and the effects of NMDA receptor blockade, *J. Neurophysiol.* 67(4):841–851 (1992).
9. S.M. Dudek and M.F. Bear, Homosynaptic long-term depression in area CA1 of hippocampus and effects of N-methyl-D-aspartate receptor blockade, *Proc. Natl. Acad. Sci. USA* 89:4363–4367 (1992).

LOCAL SPINAL MODULATION OF THE K_{Ca} CHANNEL UNDERLYING SLOW ADAPTATION IN A MODEL OF THE LAMPREY CPG

Anders Lansner[1], Jeanette Hellgren Kotaleski[1,2], Maria Ullström[2], and Sten Grillner[2]

[1]Dept. of Numerical Analysis and Computing Science, Kungl. Tekniska Högskolan, Stockholm

[2]Dept. of Neuroscience, Karolinska Institutet, Stockholm

INTRODUCTION

The lamprey is a primitive water-living vertebrate that moves by means of undulatory swimming. It is of particular interest as an experimental model for the neural generation of locomotion [1]. A major advantage of this system is that a motor pattern similar to that seen during swimming can be elicited in an isolated piece of spinal cord. Although many important details are still unknown, the lamprey spinal central pattern generator (CPG) is one of the best characterized vertebrate neuronal systems, and it has been the subject of a number of modeling and simulation studies.

We have previously demonstrated the effects on rhythm generation and intersegmental co-ordination, of incorporating modulation of the slow after-hyperpolarization (AHP) in a neural network model of the spinal burst generating circuitry [2,3]. This model was capable of replicating many important characterisitics of the lamprey swimming motor pattern, such as a bursting frequency range of about 1 to 10 Hz, a burst proportion around 35-40 %, and intersegmental co-ordination with a phaselag around 1% per segment, independent of bursting frequency. However, such motor activity resulted only with end-compensated excitatory connectivity and a slight tonic excitatory head bias. Further, neuromodulation was assumed to be subject exclusively to brainstem control, causing some complication when an isolated piece of spinal cord with the brainstem removed is considered. Contrary to experimental observation, lack of AHP modulation in the model distorts the rhythm generation and intersegmental co-ordinaton quite severely.

This model could thus not account very well for the bursting seen in a cut spinal cord piece maintained under *in vitro* conditions. Our aim in the study reported here has been to investigate further the possibility that neuromodulatory control to a significant degree may be mediated via the local neuronal circuitry of the spinal cord itself. This would explain the relatively minor experimental effects of removing brainstem neuromodulatory input. Local control is a definite possibility, since, the dense midline plexus of the lamprey spinal cord, in addition to descending glutamate and 5-HT/DA projections that provide brainstem input to the swimming pattern generator, also contain local 5-HT/DA neurons that may be a target for such regulation [4].

The effects on locomotor bursting patterns of neuromodulators like e.g. 5-HT (serotonin) are well established from invertebrates to primitive vertebrates and mammals, but the precise underlying mechanisms are still unclear [5-7]. One target of these substances is the AHP, on which they act directly or indirectly, via calcium dependent potassium channels (K_{Ca}). For example, experiments show that 5-HT directly modulates (decreases) K_{Ca} conductances in lamprey neurons [8].

There is evidence for at least two sub-types of K_{Ca} channels, one which is activated by Ca^{2+} influx during action potentials (K_{CaAP}), and the other by Ca^{2+} entering via NMDA channels (K_{CaNMDA}). The characteristic time constants of these are different, the former being about one order of magnitude faster than the latter [9,10].

Assuming a neuromodulator that reduces the AHP, as e.g. 5-HT, the model predicts a reduction of neuromodulator release with bursting frequency. Thus, an inhibitory input to the neuromodulatory neurons must be assumed. An intriguing possibly is that the contralaterally and caudally projecting inhibitory interneurons (C) provide such a control. Anatomically this is quite plausible, since the axons of these neurons pass nearby or through the midline plexus while crossing over to the contralateral side. Details of how endogenous neuromodulator levels change with activation of the spinal locomotor network are, however, not yet known but is presently under investigation.

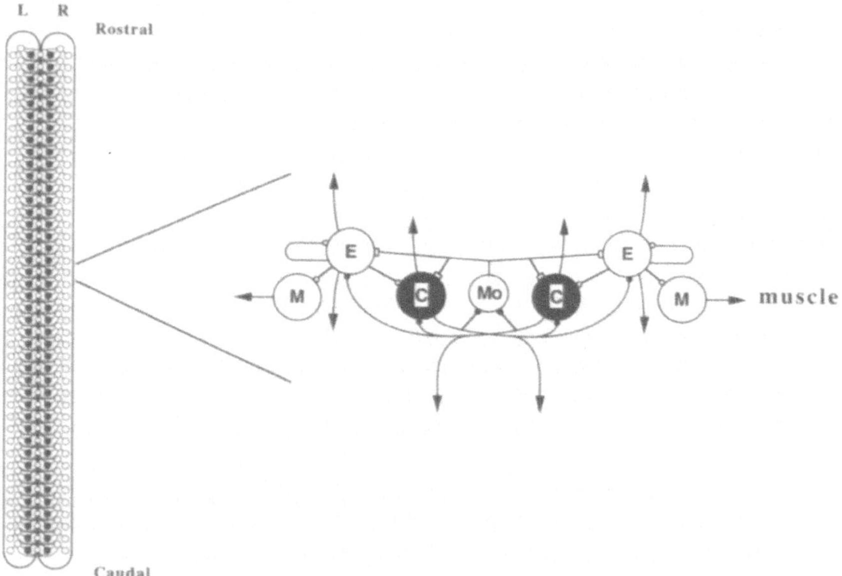

Figure 1. The distributed CPG (left) can be seen as a chain of local CPG:s (enlarged, right) connected by "intersegmental" connections. These connections extend longitudinally and contact target neurons within a certain distance from their point of origin. Units shown as circles represent populations of neurons.

The CPG network

The local, "segmental", rhythm generating network can be seen as two reciprocally inhibiting halves, each one comprised of premotor interneurons, i.e. excitatory interneurons (E), contralaterally and caudally projecting inhibitory interneurons and inhibitory lateral interneurons. Recent studies have indicated that the lateral interneurons may not be of primary importance for burst generation [11]. Thus, in the model presented here we have included E, C and, in addition, neuromodulator releasing neurons (Mo). E neurons excite other E neurons as well as C neurons ipsilaterally, whereas C neurons provide reciprocal inhibition to E and C and inhibit neuromodulatory neurons. The latter release neuromodulators bilaterally onto E and C neurons (Figure 1). The rhythm generating network is assumed to be activated by an excitatory amino acid, either in the surrounding bath solution or via reticulo-spinal projections.

A distributed CPG of a certain size can be modelled by coupling together local oscillators by means of "intersegmental" connections in a way that is compatible with what is known about this connectivity in the lamprey. The principle of "synaptic spread" is used [12]. Moreover, the same kind and strength of connections is used independent of distance within the projection range of the unit [13], implying that the premotor network lacks segmental boundaries. E connections extend ipsilaterally two segments rostrally and caudally. C contralateral connections extend two segments rostrally and twenty segments caudally. The inhibitory connections from C to Mo extend two segments rostrally and caudally, whereas the neuromodulator release is assumed to lack significant rostro-caudal extent.

DESCRIPTION OF THE UNIT AND ADAPTATION MODEL

The simulation model used in this study is comprised of simplified continuous output adapting model units, corresponding to a population of neurons, of a similar kind as was first used in a neuro-mechanical model of lamprey swimming [13]. This model has a degree of detail intermediate between mathematical coupled oscillator models [14,15] and simple connectionist models [16,17] on one side and biophysically detailed network models [18] on the other.

The units used here are similar to those in our previous model [3], but differs, in particular, with respect to the slow NMDA dependent adaptation. Further, all model units have linear transfer functions. It was attempted to make E and C as similar in their properties as possible. However, due to the different roles of the two types of neurons the parameters ended up slightly different. The adaptation of C units was significantly smaller than that of E units.

The adaptation dynamics is described by the following set of equations:

$$\frac{d\vartheta}{dt} = \frac{1}{\tau_\vartheta} (\mu u - \vartheta)$$

$$\frac{d\mu}{dt} = \frac{1}{\tau_\mu} (\mu(f) - \mu)$$

$$\mu(f) = \max(\mu_0 - \gamma f, 0)$$

Here, u represents unit activity, ϑ represents adaptation and μ corresponds to, for example, the conductance of the K_{Ca} channels. Subscripted τ represents time constants, f is level of neuromodulator reaching a unit, μ_0 and γ are parameters describing the modulatory effect on the steady state value of the K_{Ca} conductance. Values used for τ_ϑ and τ_μ in the simulations were 0.800 s.

Synaptic as well as bath application of neuromodulator could be simulated. Antagonism was simulated by a negative value of f.

SIMULATION RESULTS

The main focus of this first study has been to show that a network of the above described type can explain experimental data relating to phase constancy seen during *in vitro* experiments on isolated lamprey spinal cord. When nothing else is stated, results are from simulation of a piece which is seventy segments long. Burst frequency, burst proportion and phase lag where calculated on five independent runs for 5 or 10 seconds of stable bursting. Measurements were made at half-amplitude of the output of E units. The first and last 15 segments were not included to avoid end effects.

The distributed CPG and intersegmental co-ordination

A simulated network was activated by tonic stimulation of the E and C units. Our model of the local CPG produced a stable bursting with an adequate frequency range. As activation of C increased, the modulatory units were gradually more inhibited thus producing a more pronounced adaptation building up quicker to terminate the burst.

Figure 2 (left) demonstrates the burst frequency range and intersegmental co-ordination of the simulated distributed CPG during ten seconds of activity produced by initial weak and stepwise increasing stimulation. The frequency range of the simulated network was at least 0.5 to 8 Hz. Burst proportion and phase lag was approximately constant for frequencies from 2 to 9 Hz (Figure 2, right). Furthermore, when introduced into a previous neuromechanical simulation of the entire swimming behavior [13] this modified CPG model demonstrated significantly improved swimming speed range and efficiency.

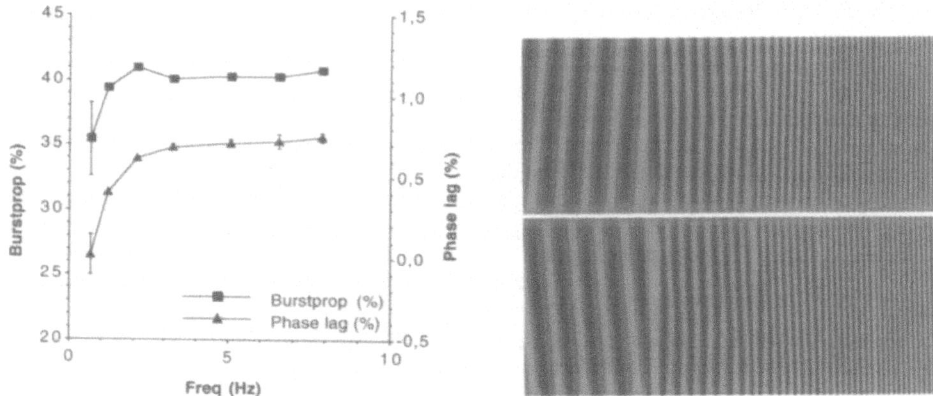

Figure 2. <u>Left:</u> The activity of C-units over ten simulated seconds, starting with a stimulation corresponding to 0.125 arbitrary units and increasing to 2.0 a.u. Bursting frequency goes from less than 2 Hz up to about 9 Hz. Activity is shown as light and inactivity/inhibition as dark. Time is along the horizontal dimension. Activity in the rostral part is shown centrally with the right and left sides unfolded. <u>Right:</u> Burst proportion and phase lag at different bursting frequencies.

The intersegmental co-ordination of locomotor activity was also highly flexible. A backward swimming motor pattern could easily be produced by addition of an inhibitory input to the most rostral segments together with an optional excitation of the caudal segments. Figure 3 shows ten seconds of activity during which co-ordination is changed. As can be seen the transitions are quite fast.

Occationally, at very low frequencies the network spontaneously produced backward co-ordination. In such cases, simulated bath application of neuromodulator agonist could increase burst proportion and produce a more stable forward swimming motor pattern. Reciprocal inhibition prevented the burst proportion to increase beyond 50 %.

Bursting of short spinal cord pieces and bath neuromodulator

Experimentally, even short pieces of a few segments are capable of bursting, but they also sometimes display tonic or othewise disturbed activity. We simulated acitivity in shorter pieces down to three segments at a stimulation of 0.25 (Figure 3, right). Bursting was maintained down to four segments. A three segment piece displayed tonic activity which could be turned into bursting by simulated bath application of neuromodulator antagonist. This result conforms qualitatitively to what can sometimes be seen experimentally.

DISCUSSION

The model of the lamprey spinal swimming CPG presented here employs a partly new kind of burst generating mechanism in which local spinal modulation of the slow adaptation plays a key role. The distributed CPG structure conforms with biology in that it lacks segmental boundaries and has a functionally strong long-range longitudinal connectivity [19]. Simulation of this pattern generating network demonstrates a good bursting frequency range as well as an adequate control of burst proportion.

The model of an isolated piece of spinal cord presented here is capable, without any end compensation of connectivity or extra excitation to head or tail regions, of generating intersegmental co-ordination with a phase lag independent of bursting frequency over an adequate bursting frequence range. The CPG simulated is flexible enough to allow additional brainstem control in the intact animal to modulate many aspects of the swimming motor pattern by simple tonic commands. In these respects, the present model replicates real system behavior quite well. However, lack of experimental quantitative data on neuromodulatory effects leaves considerable freedom in model architecture and parameters, and so far only a small subset of possibilities has been explored. More experimental data at the cellular level are expected to provide additional constraints for the model. The adaptation included and modulated here corresponds to the slower K_{CaNMDA} according to the magnitude of the time constant. In other systems, K_{Ca} channels with a similar time constant are modulated [20]. In the lamprey, the faster K_{CaAP} conductance is modulated by 5-HT/DA[5] and it is presently under investigation to what degree this modulation may contribute.

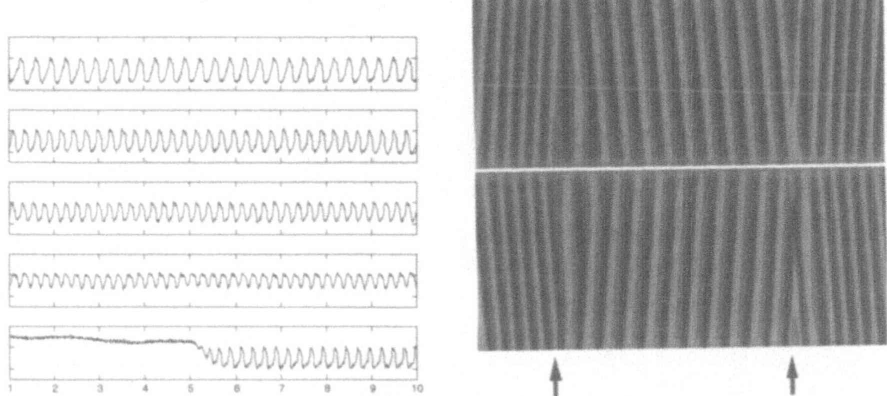

Figure 3. <u>Left:</u> The simulated network showed good flexibility with respect to forward-backward co-ordination of the swimming motor pattern. By adding a small amount of extra excitation/inhibition to the head and tail (changes marked by arrows) co-ordination could be changed between forward and backward swimming. <u>Right:</u> Simulated activity of cut spinal cord pieces. From top to bottom nine seconds of activity in simulated pieces with a length of 3, 4, 5, 7, and 10 segments. In the three segment piece, simulated bath application of a low level of neuromodulator antagonist was introduced after five seconds.

Further, some important facts are not directly accounted for in this model, such as the NMDA oscillating properties of lamprey premotor interneurons. In addition, the brainstem neuromodulatory input, which was not included in the present model, is also likely to be important. Moreover, possibly other local inhibitory interneurons serve the purpose attributed to C in this model [21].

Our model is currently being further developed and analyzed in order to characterize the burst generating mechanism in more detail and to investigate the significance of different parameters in determining different aspects of burst generation and intersegmental co-ordination. Further experimental studies of the candidate neuromodulatory systems have been intensified.

CONCLUSIONS

The modified network model presented here matches many important aspects of the experimental data on burst generation and intersegmental co-ordination in an isolated piece of lamprey spinal cord. The local spinal modulatory control of the slow adaptation of premotor interneurons provides phase constancy at medium to fast bursting frequencies in a simulated spinal cord piece. The model remains hypothetical, however, until further experimental results are available on the possible involvement of neuromodulators like e.g. 5-HT/DA in the spinal locomotor network.

REFERENCES

1. S. Grillner, T. Deliagina, Ö. Ekeberg, A. El Manira, R.H. Hill, A. Lansner, G.N. Orlovsky, and P. Wallén. (1995). "Neural networks that co-ordinate locomotion and body orientation in lamprey", *Trends Neurosci.*, 18(6), 270-279.
2. A. Lansner, J. Hellgren Kotaleski, M. Ullström, and S. Grillner (1996). "Modulating the Calcium Dependent Potassium Conductance in a Model of the Lamprey CPG" *Computation and Neural Systems*, Boston, J.M. Bower, ed., Plenum Press.
3. M. Ullström, A. Lansner, J. Hellgren Kotaleski, and S. Grillner. (1997). "Significance of modulated adaptation for rhythm generation and intersegmental co-ordination in Lamprey", this volume.
4. J. Schotland, O. Shupliakov, M. Wikström, L. Brodin, M. Srinvasan, Z.B. You, M. Herrera-Marschitz, W. Zhang, T. Hökfelt, and S. Grillner. (1995). "Control of lamprey locomotor neurons by co-localized monoamine transmitters", *Nature Lond.*, 374, 266-268.
5. P. Wallén, J. Buchanan, S. Grillner, J. Christenson, and T. Hökfelt. (1989). "The effects of 5-hydroxytryptamine on the afterhyperpolarisation, spike frequency regulation and oscillatory membrane properties in lamprey spinal cord neurons.", *J. Neurophysiol.*, 61, 759-768.
6. R.M. Harris-Warrick, and A.H. Cohen. (1985). "Serotonin modulates the central pattern generator for locomotion in the isolated lamprey spinal cord", *J. Exp. Biol.*, 116, 27-46.
7. J. Christenson, P. Wallén, L. Brodin, and S. Grillner. (1991). "5-HT Systems in a Lower Vertebrate Model: Ultrastructure, Distribution, and Synaptic and Cellular Mechanisms" *In* Volume Transmission in the Brain: Novel Mechanisms for Neural Transmission, K. Fuxe and L.F. Agnati, eds., Raven Press, New York, 159-170.
8. J. Schotland, and S. Grillner. (1993). "Effects of serotonin on ficitve locomotion coordinated by a neural network deprived of NMDA receptor-mediated cellular properties", *Brain Res.*, 93, 391-398.
9. J. Tegnér. (1997). "Modulation of Cellular Mechanisms in a Spinal Locomotor Network" PhD thesis, Karolinska institutet.
10. L. Brodin, H. Tråvén, A. Lansner, P. Wallén, Ö. Ekeberg, and S. Grillner. (1991). "Computer simulations of N-methyl-D-aspartate (NMDA) receptor induced membrane properties in a neuron model", *J. Neurophysiol.*, 66, 473-484.
11. P. Fagerstedt, P. Wallén, and S. Grillner (1995). "Activity of interneurons during fictive swimming in the lamprey" *4th IBRO Word Congress of Neuroscience*, Kyoto, Japan, 346.
12. T.L. Williams. (1992). "Phase coupling by synaptic spread in chains of coupled neuronal oscillators.", *Science Wash. DC*, 258, 662-665.
13. Ö. Ekeberg. (1993). "A combined neuronal and mechanical model of fish swimming", *Biol. Cybern.*, 69, 363-374.
14. A.H. Cohen, P.J. Holmes, and R.H. Rand. (1982). "The nature of the coupling between segmental oscillators of the lamprey spinal generator for locomotion", *J. Math. Biol.*, 13, 345-369.
15. N. Kopell, and G.B. Ermentrout. (1988). "Coupled oscillators and the design of central pattern generators", *Math. Biosci.*, 90, 87-109.
16. J.T. Buchanan. (1992). "Neural Network Simulations of Coupled Locomotor Oscillators in the Lamprey Spinal Cord", *Biol. Cybern.*, 66, 367-374.
17. T.L. Williams. (1992). "Phase Coupling in Simulated Chains of Coupled Oscillators Representing the Lamprey Spinal Cord", *Neural Comput.*, 4, 546-558.
18. T. Wadden, J.H. Kotaleski, A. Lansner, and S. Grillner. (1997). "Intersegmental coordination in the lamprey: simulations using a network model without segmental boundaries", *Biol. Cybern.*, 76(1), 1-9.
19. N. Mellen, T. Kiemel, and A.H. Cohen. (1995). "Correlational analysis of fictive swimming in the lamprey reveals strong functional intersegmental coupling", *J. Neurophysiology*, 73(3), 1020-1030.
20. P. Sah. (1996). "Ca^{2+}-activated K^+ currents in neurones: types, physiological roles and modulation", *Trends Neurosci.*, 19, 150-154.
21. Y. Otha, R. Dubuc, and S. Grillner. (1991). "A new population of neurons with crossed axons in the lamprey spinal cord", *Brain Res.*, 546, 143-148.

SEQUENCE COMPRESSION BY A HIPPOCAMPAL MODEL: A FUNCTIONAL DISSECTION

William B Levy, Per B. Sederberg & David August

Department of Neurological Surgery
University of Virginia Health Sciences Center
Charlottesville, Virginia 22908 USA
E-mail: wbl@virginia.edu, pbs5u@virginia.edu, daafis@virginia.edu

INTRODUCTION

The hippocampus is hypothesized to store memories of experiences during waking hours and then teach the cerebral cortex by broadcasting back this stored information during slow-wave sleep or during moments of wakeful inactivity[1,2,3]. Skaggs and McNaughton[4] identified this rebroadcast in rats that experienced a repetitively sequenced experience. We have presented a hippocampal model of this phenomenon[5,6] which spontaneously rebroadcasts previously learned sequential information. In addition to spontaneous recall, the model also replays the learned sequence information much faster than the original experience, a phenomenon known as temporal compression. The increases produced by the model, ranging from ten to forty-fold faster than the learned rate, are quite comparable to the reported compression observed experimentally by Skaggs and McNaughton[4]. In the model, two parameters are important for controlling the amount of temporal compression. The first is the overall level of activity (e.g., number of cell firings per timestep) and the second is the timespan of the rule governing associative synaptic modification. Increasing either of these parameters can produce more compression.

The hippocampal model, essentially region CA3, contained a moderate amount of biological detail including pyramidal cells of the capacitative, integrate-and-fire type, an inhibitory interneuron with a faster time-constant than the pyramidal cells, and a spectrum of axonal conduction lags on the order of what is seen in the hippocampus. This model encapsulated several different timescales, ranging from the relatively long off-time constant of the activated NMDA receptor (> 100 ms) and the theta-cycle matched duration of individual input patterns (100 ms) all the way down to the short (1-2 ms) time-constant of AMPA class synaptic events and the axonal conduction lags that varied by tenths of milliseconds. The question we now ask is whether all this detail was necessary to simulate temporal sequence compression. That is, can this compression be understood with simpler models?

Here we proceed by simplifying and dissecting the more complicated model; in particular, creating compromises between the integrate-and-fire model and a very simple recurrent McCulloch-Pitts model of area CA3. Specifically, we study four models of increasing levels of complexity. Our goal is to understand the necessity of some of the features included in the integrate-and-fire model.

The four models investigated here use McCulloch-Pitts neurons. That is, the neurons do not have capacitance; they are reset every timestep. Moreover, there is an equivalent time lag (one timestep) of axonal conduction between all connected neurons of this sparsely connected network, rather than a range of axonal conduction delays. Also, in all four of these models, activity is controlled by a simple competetive scheme in which a fixed number of neurons are active at each timestep, rather than being controlled by an interneuron. As always, a local Hebbian associative modification rule spanning at least one timestep is used. The four models are:

- Model 1: The standard McCulloch-Pitts model that this laboratory has been studying as a model of hippocampal function[7,8,9], with an associative modification rule that spans one timestep, and in which individual elements of the input sequence (patterns) also last for just one timestep.

- Model 2: The same model as in (1), but in which input patterns last for multiple timesteps.

- Model 3: The same model as in (1), but with a time-constant in the associative modification rule that extends over several timesteps. This time-constant, denoted τ, is implemented simply as a running averager of presynaptic activity (see, e.g., Levy and Sederberg[10]).

- Model 4: A combination of models (2) and (3), in which input patterns are active for several timesteps and the associative synaptic modification spans several timesteps.

In previous studies, the average network activity level (e.g., the number of cells firing per timestep) was shown to be important in controlling the speed of recall. Therefore, we again study the effect of activity levels on each of the four models described above. In addition, models (2) and (4) allow us to study the effect of input pattern duration on temporal compression. Finally, for models (3) and (4), we investigate the effect of varying the time-constant of associativity, τ, on compression.

Each simulation has two phases, training and testing. During training, the external sequence that drives learning is presented several times to the network, and the recurrent synaptic weights change according to the associative modification rule. After the synaptic weights have stabilized, the testing phase begins. During testing, the network is presented with the first pattern of the sequence as a prompt and is then allowed to run for a period of time without any external input. If learning has been successful, the network will recall an approximate version of the entire sequence on its own.

The input sequence used in these simulations is circular (see Figure 1a). That is, some overlap exists between those cells activated by the first pattern of the sequence and those activated by the (nominal) final pattern. Because of this circularity, the network typically recalls the sequence many times during testing, providing a better opportunity to study the speed and quality of recall. For ease of explicating the simulation, each timestep, Δt, can be thought of as 20 ms.

Figure 1 shows the difference in network firings before and after training for the first and simplest model. During learning, the initially random neuronal firing of the recurrently activated cells (Figure 1a) gives way to a more patterned appearance (Figure 1b), in which individual cells fire in short bursts marking different subsequences analogous to hippocampal place cells.

RESULTS AND DISCUSSION

As shown in Figure 2, model (1) does not perform temporal compression and exhibits poorly selective recall as the average activity level is increased.

Model (2), which includes patterns that last for multiple timesteps, also does not perform temporal compression. In addition, this model has difficulty learning at all when the patterns become too long. For example, when each pattern lasts for 5 timesteps, the network forms an attractor within the sequence, and repeatedly recalls this pattern during testing (data not shown).

In contrast to model (2), model (3) – with its time-spanning associative modification rule – is capable of temporal compression. However, in contrast to the model of August and Levy and as shown in Figure 3a, activity has little effect on the amount of compression, although it again decreases the selectivity of recall. This lower quality of recall makes intuitive sense because, in a situation where each input pattern lasts for only one timestep, compression only comes about by skipping some of the patterns (see Figure 3b). Figure 4 (dashed line) shows the effect of τ, the associative time-constant, on the amount of temporal compression. Increasing τ produces compression. However, this effect appears to plateau.

As implied from the previous model, model (4) also exhibits temporal compression. Moreover, activity increases lead to increased compression, but the additional increase is only 20% when the average activity level is increased from 5% to 50%. Notably, the selectivity of recall at high activity levels is better in this model than model 3 because it is now possible to compress subsequences of individual patterns without skipping them altogether (data not shown). Finally, as shown in Figure 4, higher values of τ produce more compression. Overall, more compression is possible in this model than in model (3), but the five-fold compression shown here is still far from the 20-40-fold compression observed experimentally and observed in the August and Levy model.

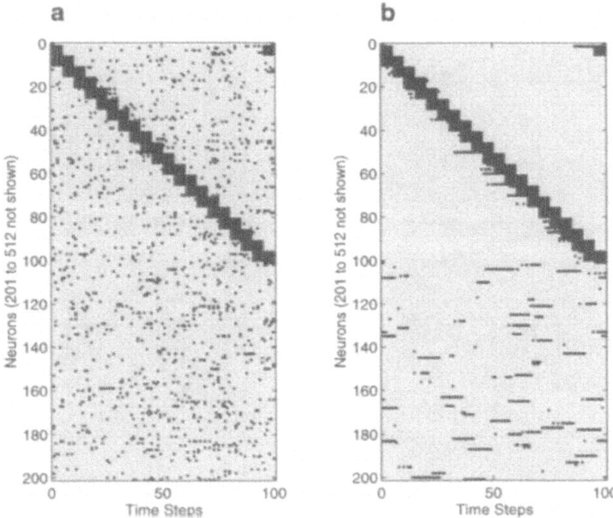

Figure 1: *Network states before and after training.* (a) This rastergram depicts the state of the network in model (1) in response to external input before synaptic modification. Each large dot is a firing event, thus each row is the firing pattern of one cell over time. Think of the network in two parts (although there is no such explicit distinction). The first part contains cells directly activated by the external input; the second part contains the neurons that are only fired, in this example, by recurrent excitation. Here, the first part corresponds to the first 100 cells and the second part is the remaining 900 cells of the 1000-cell network. The diagonal band of large dots starting in the upper left corner of each figure is the firing of those cells that are directly activated by the 20-pattern input sequence. The remaining, recurrently-activated cells, appear to fire randomly in Fig. (a). Note that to save space, only the first 200 cells of the network are shown. (b) After 100 presentations of the sequence, the recurrent synaptic weights have changed and stabilized, cells no longer fire randomly, instead, cells fire over brief periods of time spanning a few consecutive input patterns analogous to hippocampal place cells.

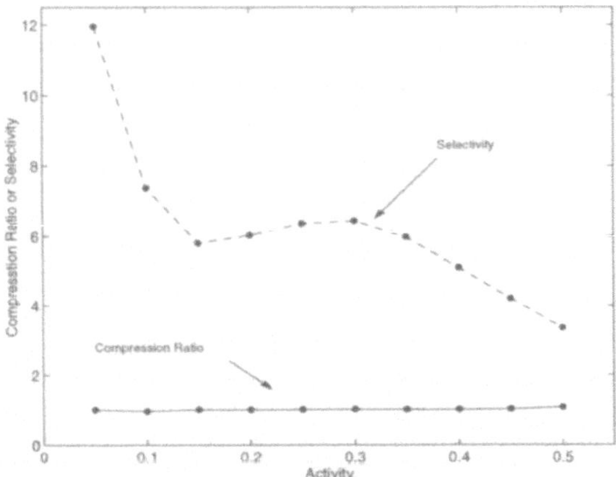

Figure 2: *Effect of activity on model (1).* As activity levels during recall increase, the selectivity decreases, while the compression ratio remains the same. The compression ratio, CR, is the ratio of the original to the recalled sequence duration. As shown here, $CR \approx 1$ no matter what activity level at which the network is tested. The selectivity is the ratio of the highest similarity value at any timepoint during recall to the average of all the other similarity values at that time, where similarity is measured by a normalized product of the network state matrix at the beginning and ending of training.

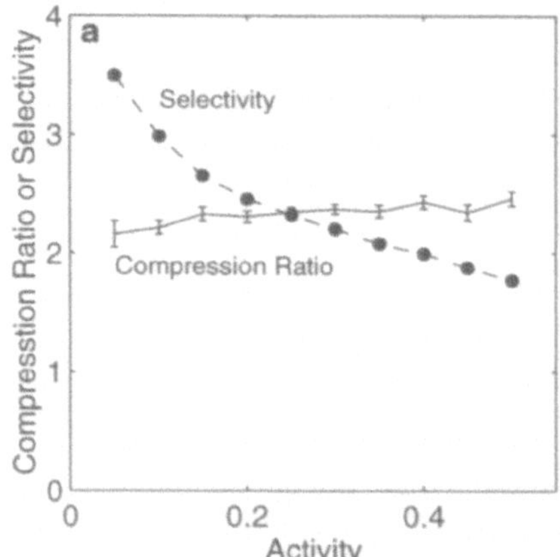
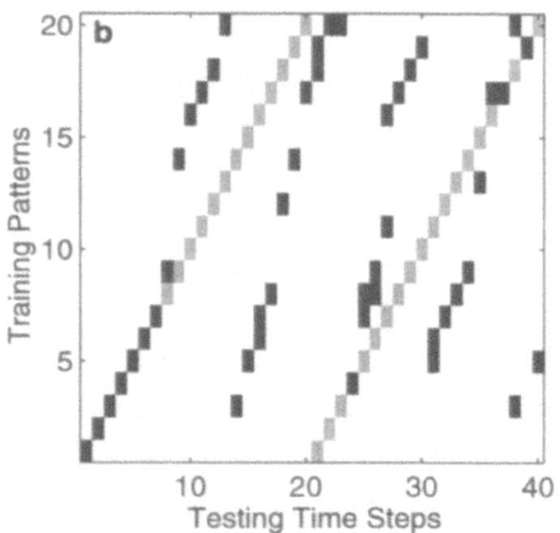

Figure 3: *Effect of activity on recall for Model (3).* (a) Unlike the two previous models, this model does compress the sequence by about 2-fold. Increasing the average activity level during testing from 5% to 50% increases compression somewhat, but still decreases the selectivity of recall substantially. (b) This panel shows the "winners" as a function of time during recall. The winner at any given timestep is the pattern, of the last learning trial, that has the highest similarity value to the network state as one timestep during testing. Thus, this depression represents a crude decoding of the CA3 network state. Here, the winners are shown in black and the original sequence is shown for comparison in gray. Compression is apparent based on the increased slope of the sequence of winners compared to the original sequence. However, also apparent, during each recall of the sequence, is the tendency of this network to skip patterns which lowers the selectivity of recall. Here, the network is running at 5% activity and $\tau = 100$ ms.

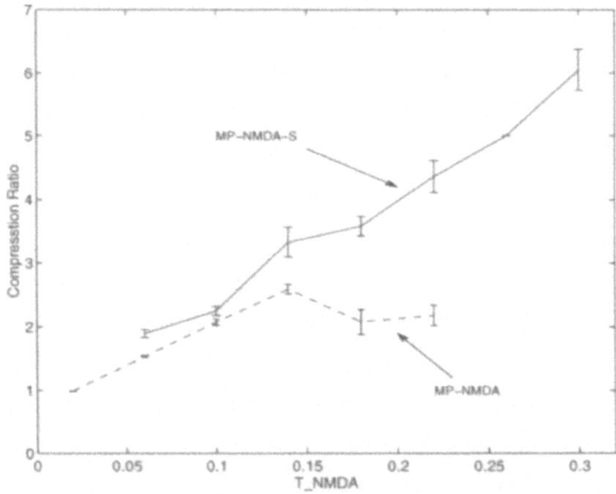

Figure 4: *Increasing τ increases temporal compression.* The compression ratio increases with τ, the time-constant of the running averager of presynaptic activity in the synaptic modification rule. Shown here are simulations using model (3) in which input patterns last for 20 ms (dashed line), and model (4) in which input patterns last for 100 ms (solid line). Two effects are apparent. First, as τ increases, so does compression. There is a range, for relatively low values of τ, for which increasing the pattern duration makes little difference in the amount of compression. However, using longer duration patterns allows us to extend the range over which τ can increase CR to produce up to 5-fold compression. Increasing τ much beyond the range of values illustrated here leads to failure of recall.

In conclusion, temporal compression can be achieved with a McCulloch-Pitts based recurrent network that lacks the complexity of a multi-timescale model with integrate-and-fire, capacitative neurons. A model studied previously by our laboratory similar to model (1) does perform a type of compression called jump-ahead, in which the network simply skips from early in the sequence to the final pattern during recall[11]. Although model (1) studied here can reproduce this behavior (data not shown), it is not our primary interest here because it does not resemble the type of sequence compression observed experimentally by Skaggs and McNaughton[4].

In our simulations, the crucial feature for obtaining compression is the presence of some time-spanning capability in the rule governing associative synaptic modification. However, although simpler models qualitatively work, they fail to produce quantitative predictions of compression seen in behaving animals and are not robust to parameter settings. The earlier, more physiological model, with its more detailed appreciation of the timescales of various biophysical events, is much more appropriate for predicting quantitative aspects of temporal compression.

ACKNOWLEDGMENTS

This work was supported by NIH MH48161 and MH00622, by Pittsburgh Supercomputing Center Grant BNS950001 to WBL, and by the Department of Neurosurgery, Dr. John A. Jane, Chairman.

REFERENCES

1. Buzsaki, G., Horvath, Z, Urioste, R., Hetke, J., and Wise, K., 1992, High-frequency network oscillation in the hippocampus, *Science* 256:1025-1027.
2. Pavlides, C. and Winson, J., 1989, Influences of hippocampal place cell firing in the awake state on the activity of these cells during subsequent sleep episodes, *J. Neurosci.* 9(8):2907-2918.
3. Wilson, M.A. and McNaughton, B.L., 1994, Reactivation of hippocampal ensemble memories during sleep, *Science* 265:676-679.
4. Skaggs, W.E. and McNaughton, B.L., 1996, Replay of neuronal firing sequences in rat hippocampus during sleep following spatial experience, *Science* 271:1870-1873.
5. August, D.A., 1996, Sequence learing with an integrate-and-fire neural network model of hippocampal area CA3. Dissertation thesis.
6. August, D.A. and Levy, W.B, 1997, Spontaneous replay of temporally compressed sequences by a hippocampal network model, in: Computational Neuroscience: Trends in Research 1997, J.M. Bower, ed., Plenum, New York, 231-236.
7. Levy, W.B, Wu, X.B., and Baxter, R.A., 1995, Unification of hippocampal function via computational/encoding considerations, *Int. J. Neural Sys.* 6 (Supp.):71-80.
8. Levy, W.B and Wu, X.B., 1996, The relationship of local context codes to sequence length memory capacity, *Network* 7:371-384.
9. Wu, X.B., Baxter, R.A., and Levy, W.B, 1996, Context codes and the effect of noisy learning on a simplified CA3 model, *Biol. Cyber.* 74:159-165.
10. Levy, W.B and Sederberg, P.B., 1997, A neural network model of hippocampally mediated trace conditioning, *IEEE International Conference on Neural Networks* I-372-376.
11. Prepscius, C. and Levy, W.B, 1994, Sequence prediction and cognitive mapping by a biologically plausible neural network, *INNS World Congress on Neural Networks* IV-164-169.

FROM TOUCH LOCALIZATION TO DIRECTED MOTOR OUTPUT IN THE LEECH LOCAL BEND NETWORK

John E. Lewis[*] and William B. Kristan, Jr.

Department of Biology, University of California San Diego
La Jolla, CA 92093-0357
[*]current address: Cellular and Molecular Medicine, University of Ottawa,
Ottawa, Ontario, Canada, K1H 8M5

ABSTRACT

Local bending in the leech is a body positioning response directed away from a touch to the body wall. We have used several approaches toward understanding this transformation between touch location and directed behavioral output. Here, we summarize our results and describe a model of the local bend network that is based on known physiological and anatomical properties.

INTRODUCTION

Some classes of behaviors are particularly amenable to study due to their easily defined relationships between sensory input and behavioral output. For instance, in some behaviors, a directional input produces a corresponding directional output. Examples of such *directed behaviors* are orienting toward an auditory stimulus in the barn owl[1] and insect escape turns directed away from an approaching predator[2]. The general problem in producing directed behaviors involves accurately localizing the sensory stimulus and translating this information into the appropriate motor command. The local bend of the medicinal leech is a simple type of directed behavior that is elicited by a touch to the body wall and results in a body positioning response directed away from the touched location (fig. 1A). This behavior is elicited in a single segment of the leech segmented body. The neuronal network underlying the local bend consists of three layers of neurons in a single ganglion (fig. 1B): mechanosensory P neurons, interneurons, and motorneurons. The four P neurons show a slowly adapting response to a touch stimulus and provide the major input to the local bend network [3,4]. These P cells activate a layer of interneurons that in turn activate a motorneuron layer[5]. These motorneurons control the length of longitudinal muscle and therefore determine body positioning — shortening near the stimulus site and lengthening opposite the stimulus site results in a movement away from the stimulus.

Our studies have involved various aspects of the sensorimotor transformation performed by the local bend network, from the encoding of touch location by the P neurons to the organization of motor output. This system is one of the few examples where it may be possible to describe such a transformation completely.

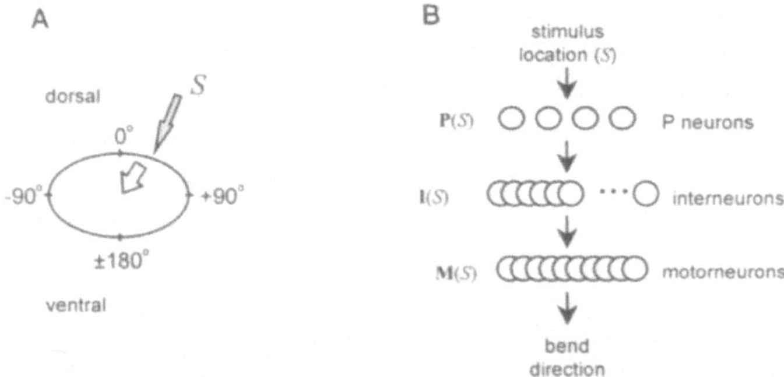

Figure 1. (A) The local bend network translates the location of touch stimulus, S, on the body perimeter to a motor command that results in a bend directed away from the touched site. Shown also is the coordinate system used to describe location on the body cross-sectional perimeter. (B) The three neuronal layers of the local bend network.

On the behavioral side, video and electromyographic analyses have revealed that the local bend is directed away from a touched site on the body perimeter with a root-mean-squared (RMS) error of 8%[6]. On the sensory side, we have investigated the representation of touch location by the mechanosensory P neurons[4]. We quantified the P neuron spike trains elicited by stimuli at different locations on the body wall perimeter. Such tuning curves are cosine-shaped, with the peaks distributed around the body perimeter (fig. 2A). Using a previously described spike train decoding method, the population vector method[7,8], we found that a code involving only the P neuron spike count can be decoded with a RMS error of less that 3%. In other words, a more complicated code involving the details of spike timing is not necessary to account for the accuracy achieved by the behavior.

In another study, we investigated how the P neuron representation of touch location is transferred to the motorneurons to produce the appropriate behavioral output. In a set of experiments in which we delivered touch stimuli, recorded the activity in all the active P neurons, and simultaneously measured bend direction, we were able to show that the population vector formed by the P neuron spike counts was well-correlated with bend direction on a trial-by-trial basis[9]. We proposed a physiologically-based model to illustrate how the local bend network extracts the information encoded in the P neuron population vector, and the following section describes this model in more detail.

A SIMPLE MODEL OF THE LOCAL BEND NETWORK

Our model of the local bend network consists of three neuronal layers (fig. 1B). The input layer comprises four units (P_i) to model the four P neurons. The firing properties of the P units (i.e. spike count for given stimulus location, S) are given by the P neurons' experimentally derived tuning curves (fig. 2A) and response variance[4].

The output layer consists of 10 units (M_n) , to model the five bilateral classes of longitudinal motorneurons: dorsal, dorsolateral, lateral, ventrolateral, and ventral classes[4]. Each of these motorneurons innervates longitudinal body wall muscle with overlapping fields, the centers of which are distributed evenly around the body perimeter, extending about 15-25% of the perimeter in size[4,10]. The *preferred location* for a given neuron is the stimulus location that produces the maximum response in that neuron. The preferred locations of two of the motorneurons are known and they correlate well with the center of their innervation fields[4]. Thus, we assign the preferred locations for the motorneuron units, M_n^*, to be uniformly distributed, on average, over the interval $[-180°, +180°]$.

The middle layer of units (I_j) represents the local bending interneurons, the exact number (N_I) of which is not known. Several studies have characterized a subset of the local bending interneurons[11,12]. Using electrical stimulation of P neurons, the connections to seventeen interneurons (eight bilateral pairs and one unpaired) of the local bend network were identified and characterized by the size of the resulting post-synaptic potential (PSP). We reanalyzed these data for each identified interneuron[12] in the present context by plotting the PSP size versus the preferred location of the P cell that was stimulated. Electrical stimulation of a single P neuron mimics an actual touch at the preferred location of each P neuron, because at these locations only one P neuron is active on average (fig. 2A). Plotting the data in this way provides an estimate of the tuning curve for each interneuron. We found that these tuning curves were well described by cosine functions, with preferred locations ranging from 18° to 115° (fig. 2B). Six of nine interneurons were fit very well by a cosine function ($R^2 > 0.9$; two others with $R^2 > 0.7$; and cell 157 with $R^2 = 0.5$). An implication of cosine tuning of the interneurons is that the connection strength from a P neuron to a given interneuron is related to the cosine of the difference in their preferred locations. This is exactly the type of connectivity that results in the accurate transfer of information encoded in a population vector[13]. Figure 2C shows the normalized PSP produced in four different interneurons by stimulation of each P neuron. These data fit very well to a cosine function. The preferred locations of the identified interneurons are distributed over about two-thirds of the body perimeter (note that fig. 2B shows the right interneurons only). There are no interneurons with preferred locations near ±180° because of the selection criteria for identifying the known interneurons[12]. Presumably, with different selection criteria, several more interneurons will be identified with preferred locations that fill out the entire perimeter. Thus, for the model we assume that the I_j^- are evenly distributed, on average.

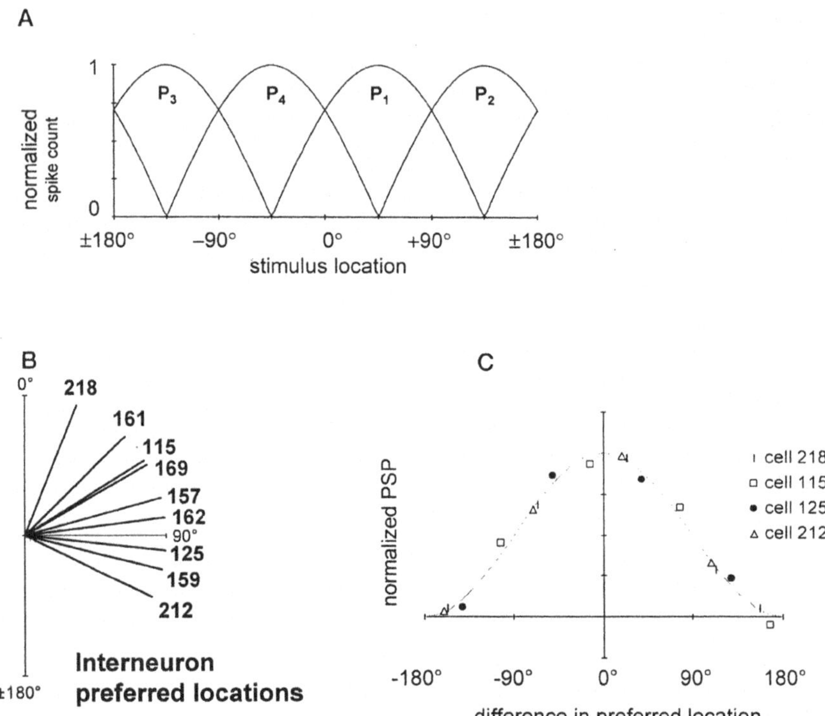

Figure 2. (A) Tuning curves for the four mechanosensory P neurons showing normalized spike count versus stimulus location. Shown are the curve fits to the data. (B) Preferred stimulus locations of the identified interneurons. Only the right interneurons are shown. The bold labels refer to the numbers used to identify each interneuron[12]. (C) Demonstration that the connection strength between P neurons and interneurons is related to the cosine of the difference in the respective preferred locations. Data from the four interneurons that best illustrate this relationship are shown.

The I and M units are characterized only by their preferred locations (I_j^- and M_n^*). There is certainly variability in these preferred locations between animals and presumably between different ganglia in the same animal. These variations can be easily included in the model. The strength of the connectivity between units is proportional to the cosine of the difference between their respective preferred locations (fig. 2C). The measure of neural activity for the I and M units can be thought of as the size of the post-synaptic potential. The model network is summarized in Eq. 1, giving the activity levels for each P, I, and M unit as a function of stimulus location (S):

$$P_i(S) = \eta + \cos(S - P_i^*)$$

$$I_j(S) = \eta + \sum_{i=1}^{4} \left[w \cos(P_i^* - I_j^*) \right] F\left(P_i(S) \right) \qquad \text{(Eq. 1.)}$$

$$M_n(S) = \eta + \sum_{j=1}^{N_I} \left[w \cos(M_n^* - I_j^*) \right] F\left(I_j(S) \right)$$

where $i=(1,\ldots,4)$, $j=(1,\ldots,N_I)$, $n=(1,\ldots,10)$, η is a noise term, w is a scaling constant for the synaptic strength, and F is a threshold function defined as $F(x)=0$ if $x \leq 0$ and $F(x)=x$ if $x>0$. The ratio of post-synaptic to pre-synaptic voltage is about 0.6 for the physiological range of synaptic potentials[14], so the parameter w is fixed at $w=0.6$. We introduce variability in a unit's response by adding a gaussian random number η with zero mean, and standard deviation such that the standard deviation σ of the response of a given unit with mean response, r, is $\sigma=kr$ (k = coefficient of variation). The same value of k is used for both I and M units. To decode the activity of the M units and obtain an estimate of the bend direction, we use the population vector, i.e. the vector sum of the M_n^* weighted by their respective activity levels. A model error is then calculated in the same way as for the behavioral experiments.

One issue that we have investigated using this model involves the number of interneurons required in the local bend network. We varied the number of interneurons and found that only 16 were required for the model error to match the behavioral error[9]. Considering the criteria used to identify the known interneurons, there are probably 25 to 30 interneurons in the local bend network[12]. One possible explanation for this is that the interneurons are involved in the production of other behaviors, and a greater number allows greater flexibility when choices are made between behaviors.

CONCLUSION

We outline a model of the leech local bend network that is both physiologically and anatomically based. Although the model is relatively simple, it demonstrates the mechanisms used by this system to extract information encoded in a population vector.

REFERENCES

1. Knudsen EI, Blasdel GG, Konishi M (1979) Sound localization by the barn owl (*Tyto alba*) measured with the search coil technique. J Comp Physiol 133:1-11.

2. Miller JP, Jacobs GA, Theunissen FE (1991) Representation of sensory information in the cricket cercal sensory system. I. Response properties of the primary interneurons. J Neurophysiol 66:1680-1689.

3. Kristan WB Jr (1982) Sensory and motor neurones responsible for the local bending response in leeches. J Exp Biol 96:161-180.

4. Lewis JE. From touch localization to directed behavior: Neural computation in the leech [PhD Thesis]. San Diego, CA: University of California, San Diego, 1997.

5. Lockery SR, Kristan WB Jr (1990) Distributed processing of sensory information in the leech. I. Input-output relations of the local bending reflex. J Neurosci 10:1811-1815.

6. Lewis JE, Kristan WB Jr (1998) Quantitative analysis of a directed behavior in the medicinal leech: implications for organizing motor output. J Neurosci 18 (in press).

7. Georgopoulos AP, Schwartz AB, Kettner RE (1986) Neuronal population coding of movement direction. Science 233:1416-9.

8. Salinas E, Abbott LF (1994) Vector reconstruction from firing rates. J Comput Neurosci 1:89-107.

9. Lewis JE, Kristan WB Jr (1998) A neuronal network for computing population vectors in the leech. Nature 391:76-79.

10. Stuart AE (1970) Physiological and morphological properties of motoneurones in the central nervous system of the leech. J Physiol (Lond) 209:627-646.

11. Lockery SR, Wittenberg G, Kristan WB Jr, Cottrell GW (1989) Function of identified interneurons in the leech elucidated using neural networks trained by back-propagation. Nature 340:468-71.

12. Lockery SR, Kristan WB Jr (1990) Distributed processing of sensory information in the leech. II. Identification of interneurons contributing to the local bending reflex. J Neurosci 10:1816-1829.

13. Salinas E, Abbott LF (1995) Transfer of coded information from sensory to motor networks. J Neurosci 15:6461-6474.

14. Lockery SR, Sejnowski TJ (1992) Distributed processing of sensory information in the leech. III. A dynamical neural network model of the local bending reflex. J Neurosci 12:3877-95.

INFORMATION EXCHANGE BETWEEN PAIRS OF SPIKE TRAINS IN THE MAMMALIAN VISUAL SYSTEM

Steven B. Lowen,[1] Tsuyoshi Ozaki,[2] Ehud Kaplan,[3] Malvin C. Teich[4]

[1] Department of Electrical & Computer Engineering
Boston University
8 Saint Mary's St., Boston, MA 02215
Email: lowen@bu.edu
[2] The Rockefeller University
1230 York Ave., New York, NY, 10021
Email: yoshi@camelot.mssm.edu
[3] Department of Ophthalmology
Mt. Sinai School of Medicine
One Gustave Levy Pl., New York, NY, 10029
Email: kaplane@rockvax.rockefeller.edu
[4] Departments of Electrical & Computer Engineering
and Biomedical Engineering
Boston University
8 Saint Mary's St., Boston, MA 02215
Email: teich@bu.edu

ABSTRACT

We have studied the neural firing patterns of retinal ganglion cells (RGCs) and their target lateral geniculate nucleus (LGN) cells. Reliable information transmission coexists with fractal fluctuations which appear in RGC and LGN firing patterns. Unexpectedly, these fluctuations appear not to be independent across LGN cells; information is also shared among pairs of LGN spike trains. Over short time scales, we find that clusters of spikes in the RGC neural firing pattern appear at the LGN output essentially unchanged, while isolated RGC firing events are more likely to be eliminated; thus the LGN action-potential sequence is not simply a *randomly* deleted version of the RGC spike train. Employing information-theoretic techniques, we estimate the information efficiency of the LGN neuronal output — the proportion of the variation in the LGN firing pattern that carries information about its associated RGC input — to be in the vicinity of 50% over counting windows below about 10 ms. We develop a new information-theoretic measure which helps determine at what time scale a neural spike train changes from a time code to a rate code.

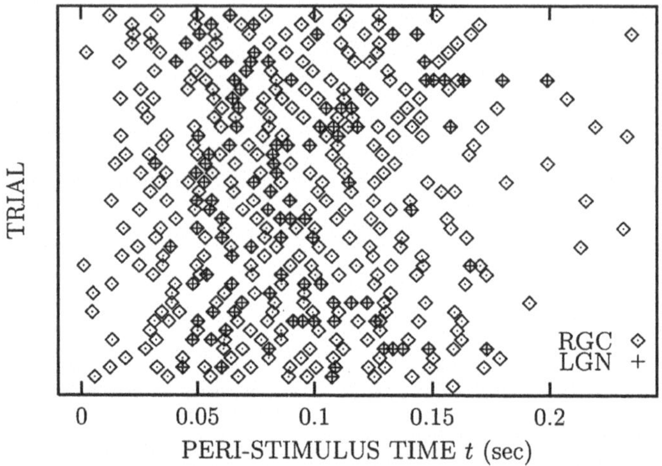

FIGURE 1: RGC AND LGN JOINT RASTER PLOT

INTRODUCTION

The relationship between spike trains at the RGC and its target LGN cells have yielded to increased understanding over the course of many years' research (see the excellent review article by Funke and Wörgötter, 1997). One observation that repeatedly emerges is that conventional measures of noisiness, such as interspike-interval coefficient of variation and signal-to-noise ratio, generally increase at the LGN relative to the RGC (see, e.g., Tables 1 and 2 in Teich et al., 1997). This observation indicates that other important characteristics of the transformation should be sought and investigated; indeed many researchers, typically using a linear-systems approach, have carried out studies along these lines (Kaplan and Shapley, 1984; Levine and Troy, 1986; Funke and Wörgötter, 1997, to cite only a few). However, a thorough exploration of the inhibition and shunting mechanisms used by the LGN in its role as gatekeeper requires an analysis in which the joint statistical features of RGC and LGN spike trains are examined.

SELECTIVE PASSAGE OF RGC SPIKES BY THE LGN

The LGN spike train derives from the RGC spike train by a deletion mechanism. Aside from occasional clusters of calcium spikes, every LGN spike occurs simultaneously with an RGC spike at the input to the LGN. How are these spikes selected and why? The joint raster plot in Fig. 1 shows 40 stimulus-cycle trials for an ON/X cat RGC/LGN pair driven by a 4.2 Hz drifting grating at 100% contrast with a mean luminance level of 50 cd/m^2. Spikes from the RGC are denoted by diamonds, while crosses signify the associated LGN action potentials. Rather than forming a randomly deleted version of the RGC spikes, in which each RGC spike undergoes a Bernoulli trial (coin flip) to determine whether it survives as a spike at the output of the LGN, the LGN action potentials appear to occur either at the end of or during clusters of RGC action potentials. Figure 1 reveals several LGN spike triplets closely spaced in time. Other LGN firings tend to occur after a sequence of closely spaced RGC spikes. The result is that the LGN neuron tends to fire when the RGC is most active, near the peak of the

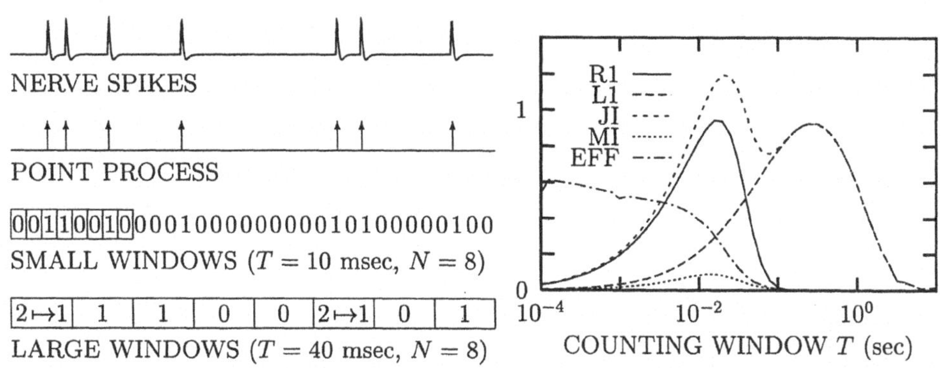

FIGURE 2: STANDARD INFORMATION-THEORETIC MEASURES FIXED NUMBER OF WINDOWS N

RGC peri-stimulus time (PST) histogram, and when the signal is strongest. This can actually sharpen timing reliability in the LGN, by some measures, relative to that of the RGC (Funke and Wörgötter, 1997).

INFORMATION EXCHANGE BETWEEN RGC/LGN SPIKE-TRAIN PAIRS: STANDARD APPROACH

The signal-processing benefit that an LGN cell might afford an input spike train can be evaluated by examining several measures of the input and output spike trains. We consider the mutual information which provides an indication about the nature of the code used by the LGN and why clusters might be favored. We subsequently consider information exchange across pairs of LGN spike trains.

We first use the standard information-theoretic approach for determining the information content of a neural spike train as a measure of the relative reliability of these signals (Meister, 1996; Rieke *et al.*, 1997; Saleh and Teich, 1987). This method treats segments of spike trains as neural codes, and indicates how information is transmitted in a quantifiable manner without making any assumptions about the nature of these codes. A spike train, for example RGC activity transmitted to an LGN cell by S potentials (Kaplan and Shapley, 1984), is first divided into a sequence of counting windows of the same duration T. As shown in Fig. 2, the code is generated by writing a 1 if at least one point-process event (idealized nerve spike) occurs in a window, and a 0 if not, and this process is repeated for all counting windows. The binary stream thus produced is grouped into overlapping segments of $N = 8$ windows each. Each such segment forms one particular codeword out of a total of $M = 2^N$ possibilities. A histogram is then formed from the codewords, as an estimate of their probabilities of occurrence. The entropies are then calculated (Rieke *et al.*, 1997) for this RGC spike train, as well as for the corresponding LGN sequence of action potentials, and also for the two spike trains jointly (JI), from which the mutual information (MI) may be obtained.

An efficiency (EFF) can be defined (Rieke *et al.*, 1997) as the ratio of the MI to the LGN spike-train entropy; this serves to quantify the proportion of information available in the LGN spike train that is shared with the RGC input. A higher efficiency does not necessarily mean improved performance, however, since an ideal axon that transmits action potentials perfectly, but without any processing, has a theoretical efficiency of unity. The LGN apparently selects part of the signal transmitted from the RGC, discarding some entropy which presumably does not correspond to useful information for the task at hand. In the process, the relative spike-time reliability can increase, since the LGN preferentially transmits clusters of spikes. The procedure described above is carried out for a variety of counting times T. To examine the implementation of this approach, Fig. 2 presents curves of the three entropies and the mutual information as functions of the counting window, for an RGC/LGN spike train pair recorded under maintained-discharge conditions (ON/X cat RGC/LGN pair driven by a blank screen of luminance 50 cd/m^2) and with a data segment length of $N = 8$ windows per codeword. This choice of N yields the most details about RGC/LGN spike train behavior given the data size available.

For larger values of N the number of possible joint RGC/LGN codewords 2^{2N} becomes comparable in magnitude with the number of such codes available in the data, and spuriously large values of the efficiency result. The values are normalized by dividing by N, to give entropies and mutual information in bits per window. As seen in Fig. 2, both the RGC and LGN curves approach the theoretical maximum of 1 bit per window, but at different times; the times correspond to approximately half the average interevent intervals of each spike train as expected. For large window durations, all windows contain a spike, and thus the information is identically zero; this limitation is fundamental to the counting method used. The joint entropy lies below the theoretical maximum of two bits per window (one for each spike train) and above the entropies of both the RGC and LGN spike trains, as it must. The mutual information remains below 0.2 bits per window, but the efficiency exceeds one-half for counting windows less than 10 msec, indicating that most of the variation in the LGN spike train represents information transmitted from the RGC at these window sizes.

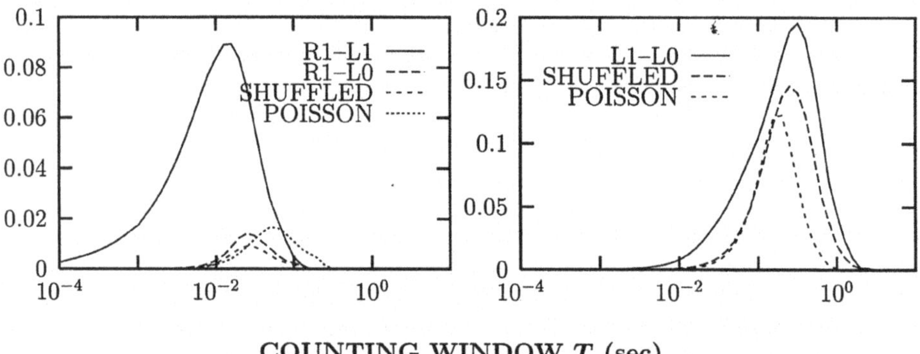

COUNTING WINDOW T (sec)

FIGURE 3   **FIGURE 4**

Figure 3 shows the mutual information as a function of the counting time, with a data segment length of $N = 8$, between 1) an RGC/LGN spike train pair (R1-L1) recorded under maintained discharge conditions (the same MI curve as shown in Fig. 2), 2) the same RGC spike train and an unrelated LGN spike train recorded at the same time (R1-L0), 3) a shuffled version of the spike trains used in 1) (same interspike intervals, but in a random order), and 4) homogeneous Poisson processes with the same mean rate as those in 1). The mutual information for the associated RGC/LGN spike-train pair significantly exceeds that of the two surrogates and of the same RGC with an alternate LGN cell for counting times less than about 40 msec, so that the RGC/LGN spike trains likely exhibit reliable timing precision over these time scales. Indeed, the similar shapes of the shuffled surrogate and the unrelated RGC/LGN spike trains shows that this surrogate successfully captures the behavior of these spike trains when studied with this measure, while removing the effects of precision timing. For counting times greater than 80 msec, however, the Poisson surrogate exhibits a larger apparent mutual information than the original data, indicating that timing precision and reliability reported by this estimation procedure at this time scale likely arises from limitations in the procedure itself.

INFORMATION EXCHANGE ACROSS PAIRS OF LGN SPIKE TRAINS

As shown in Fig. 3, significant mutual information exists between RGC/LGN pairs as expected, but not in unrelated RGC/LGN cells nor between two RGCs (not shown), for which the mutual information is small, roughly equal to that of the surrogates. However, evidence exists for mutual information between two simultaneously recorded LGN cells, although their inputs, the associated RGCs, do not appear to be related. Figure 4 shows the mutual information as a function of the counting time, with a data segment length of $N = 8$, between two LGN cells recorded under maintained discharge conditions, and for two surrogate data sets. The two LGN cells show a small but significant increase in the mutual information over that of the surrogates, especially for counting windows less than 30 msec. Evidently some process is acting jointly on the two LGN cells which introduces dependencies and thus joint reliability between the two spike trains, since no such dependency exists in the RGC spike trains to any significant degree. For driven cells, in contrast, the mutual information calculated between any combination of spike trains resembles that of the solid curve in Fig. 3. In that case, results for RGC/LGN pairs resemble those for RGC/RGC, LGN/LGN, and unrelated RGC/LGN pairs; all exhibit mutual information values which exceed those of the surrogates by significant amounts (not shown). We see that such relationships do not exist under maintained discharge conditions, so that it will be useful to investigate whether the close relation among all four of the spike trains under driven conditions arises from common fluctuations imposed by the stimulus or its presence, and how different stimulus conditions will affect these relationships.

INFORMATION EXCHANGE BETWEEN RGC/LGN SPIKE-TRAIN PAIRS: NEW APPROACH

We have developed a new information-theoretic approach, shown in Fig. 5, in which the overall length of the data segment remains fixed ($N \cdot T = 0.3$ sec in this example), while the number of windows within the segment changes. Figure 5 presents the same quantities shown in Fig. 2. In the limit of a small window the relative precision attains a high value, and the spike train is analyzed as a precise *time code*. At the other limit the entire segment is a single window, and the spike train is analyzed as a *rate code*. This method has the advantage of keeping the data segment length fixed, rendering the effects of including varying amounts of data less important. The knee in the efficiency curve at 37.5 msec suggests a crossover time near which the time code can be said to change to a rate code. The neural system itself can, of course, function simultaneously at all scales, thereby transmitting time-code and rate-code information simultaneously.

Larger windows have higher counting resolution (rather than registering any positive number of spikes with a 1 symbol), extending the range of usefulness to longer

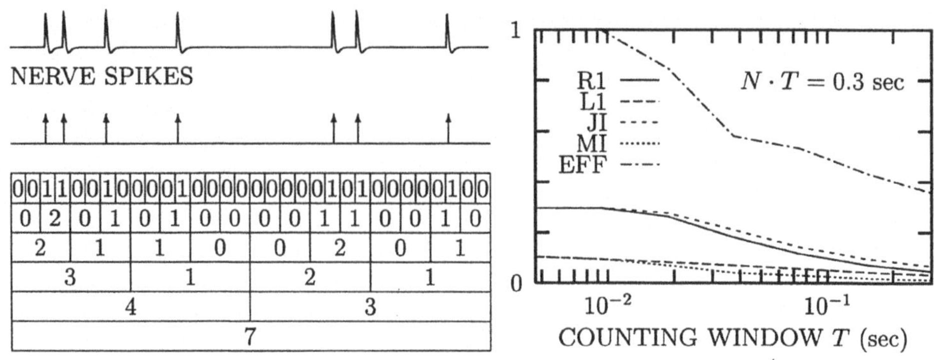

**FIGURE 5: NEW INFORMATION-THEORETIC MEASURES
FIXED DATA-SEGMENT DURATION $N \cdot T$**

window sizes. In particular, the RGC counts need not all assume a value of unity, and thus yield no useful information, when the LGN has appreciable information, as in Fig. 2 near 0.2 seconds.

Further, large numbers of windows are associated with small window sizes, alleviating the finite-data problems mentioned above. Indeed, we were able to extend the number of windows N to 64 before encountering the limitations of our data size. For this large value of N and the data sets we have used, each RGC code becomes unique; however, the LGN, having a much lower rate, does not generate unique codes. Since the LGN does not add any information to the RGC/LGN code pairs already present in the unique RGC codes, the joint information is equal to that in the RGC, and the MI assumes a value equal to that of the LGN. Thus in this case, the efficiency becomes 100%. This illustrates another limitation of the use of efficiency calculations, which require care in their interpretation; the spuriously high efficiency values shown here result from data limitations, and not from neural processing at the LGN.

REFERENCES

Funke, K., and Wörgötter, F., 1997, On the significance of temporally structured activity in the dorsal lateral geniculate nucleus (LGN), *Prog. in Neurobiol.* 53:67.

Kaplan, E., and Shapley, R. M., 1984, The origin of the S (slow) potential in the mammalian lateral geniculate nucleus, *Exp. Brain Res.* 55:111.

Levine, M. M., and Troy, J. B., 1986, The variability of the maintained discharge of cat dorsal lateral geniculate cells, *J. Physiol. (London)* 330:339.

Meister, M., 1996, Multineuronal codes in retinal signaling, *Proc. Natl. Acad. Sci. USA* 93:609.

Rieke, F., Warland, D., de Ruyter van Steveninck, R., and Bialek, W., 1997, *Spikes: Exploring the Neural Code*, MIT Press, Cambridge, MA.

Saleh, B. E. A., and Teich, M. C., 1987, Can the channel capacity of a lightwave communication system be increased by the use of photon-number-squeezed light?, *Phys. Rev. Lett.* 58:2656.

Teich, M. C., Heneghan, C., Lowen, S. B., Ozaki, T., and Kaplan, E., 1997, Fractal character of the neural spike train in the visual system of the cat, *J. Opt. Soc. Am. A* 14:529.

THE ROLE OF FEEDFORWARD AND FEEDBACK INHIBITION ON FREQUENCY-DEPENDENT INFORMATION PROCESSING IN A CEREBELLAR GRANULE CELL

Huo Lu, [1] F. W. Prior, [2] and L. J. Larson-Prior [1]

[1] Department of Neuroscience and Anatomy
Pennsylvania State University
College of Medicine
M.S. Hershey Medical Center
Hershey, PA 17033, USA
[2] Philips Multimedia Center
1070 Arastradero Rd
Palo Alto, CA 94304

ABSTRACT

Multi-modal sensory information entering the cerebellum via mossy fibers is processed through the granule cell (GC) network, the major cellular elements of which are the GC and an inhibitory interneuron, the Golgi cell. A GC model supporting both feedforward (FF) and feedback (FB) inhibition to its dendritic arbor was constructed. This model was used to examine the influence of Golgi cell inhibition on GC responses to mossy fiber inputs ranging from 10 - 100 Hz. Both FF and FB inhibitory signals reduced GC output. When both inhibitory loops accessed the same GC dendrite, the greatest decrement in GC output was produced by FB inhibition alone. However, if each inhibitory loop accessed a different GC dendrite, both inhibitory inputs were required to produce the greatest decrement in GC output.

INTRODUCTION

In the cerebellar cortex, both mossy fiber (MF) and parallel fiber (PF) systems contact Golgi cell interneurons, which provide strong FF and FB inhibitory control over GC output. While climbing fiber input to the cerebellar cortex is quite uniform and of low frequency, MF inputs vary over a wide range of firing frequencies. The architecture of the GC network is such that both Golgi cell interneurons and GC are co-activated by MF afferents. These Golgi cells then act to inhibit the same GC in a FF manner. The FB limb consists of Golgi cells activated by GC axons constituting the PF system. Thus, the FF loop samples MF inputs while the FB limb samples GC output processed through the PF axon.

As MF firing frequencies are often strongly influenced by the animal's state, the information transmitted by the granule cell network is likely to rely on the mechanisms by which the GC handles multi-rate data streams. To test the effect of Golgi cell inhibition on GC output, a biologically realistic model of the GC under both FF and FB inhibitory control was presented with a range of MF input patterns which varied in their rate from 10 - 100 Hz.

METHODS

Electrophysiological   Recording

Whole-cell patch clamp recordings were made from GCs in the granule cell layer of rat cerebellar slices. Cerebellar slices were prepared from Sprague-Dawley rats (either gender, 10 to 25 days old). Animals were anaesthetized with isoflurane (Ohmeda) and sacrificed by decapitation. The vermis of he cerebellum was isolated and placed in oxygenated Kreb's solution at 5-10°C. Slices were prepared as 150 mm thick sagittal sections on a Vibratome (Pelco 101, Series 1000) and were maintained at room temperature in oxygenated Kreb's solution. Recording commenced 1 hour after slice preparation.

Patch clamp recordings were obtained with micropipettes (A-M Systems, Corning #7052, 1.65 mm OD, 1.20 mm ID), prepared on a horizontal puller (Flaming/Brown Model P-97). To reduce their electrical capacitance, all pipettes were coated to within 100-200 mm of their tips with Sylgard 184 resin (Dow Corning). The tips of the pipettes were polished using a heated platinum filament. Final resistances ranged between 8 and 14 MΩ. An Axoclamp 200A amplifier (Axon Instruments) was used for both current and voltage clamp experiments. Data acquisition and analysis were performed with the aid of a personal computer running pClamp 6.0 (Axon Instruments).

Slices were cut in a Kreb's solution composed of (mM): NaCl, 120; KCl, 2; $MgSO_4$, 1.2; $NaHCO_3$, 26; NaH_2PO_4, 1.2; $CaCl_2$, 2; glucose, 11; saturated with 95% O_2, 5% CO_2. Recording pipettes were filled with an internal solution of the following composition (mM): potassium gluconate, 122; KCl, 4; NaCl, 4; $MgCl_2$, 1; $CaCl_2$, 0.02; BAPTA, 1; glucose, 15; ATP, 3; HEPES, 5; GTP, 0.3; pH was adjusted to 7.2 with KOH, and the final osmolarity was adjusted to 290-300 mmol/kg.

The white matter was stimulated using a concentric bipolar electrode (0.25-4.0 mA). IPSCs were elicited by electrical stimulation in the presence of (-)D-2-amino-5-phosphonopentanoic acid (AP-5, 50 mM) and 6-cyano-7-nitroquinoxaline-2,3-dione (CNQX, 10 mM) to depress excitatory transmission.

Computational   Neuronal   Modeling

A 59 compartment GC model was developed using GENESIS simulation software (v. 2.1, Bower and Beeman 1994) and included a somal compartment, 4 dendrites and their dendritic bulbs, and the complete axonal arbor. To a previously described GC model (Lu et al. 1995), a 14 compartment ascending axon, a bifurcation site, and two 15 compartment PFs were added. Based on anatomical studies from the literature (Palay and Chan-Palay 1974), morphological features of the axon were chosen. The length of the PF on either side of the bifurcation was initially set at 600 mm, which is approximately the size of a single Golgi cell dendritic arbor. Each axonal compartment was 40 mm in length with the non-beaded portion of the axon exhibiting a diameter of 0.1 - 0.2 mm and beaded portions (40% of the total length) defined as 0.3 mm in diameter. To obtain a conduction velocity through the axonal arbor of 0.2 -0.3 m/s (Eccles et al. 1966; Llinás et al. 1969; Vranesic et al. 1994), the axial resistance (RA) was set in a range of 0.1 to 0.4 Ω*m. Each compartment of the ascending and PF axon was modeled with fast sodium and delayed rectifier potassium conductances to ensure efficient signal transmission.

Synaptic conductances were modeled on the dendritic bulb compartment and consisted of α-amino-3-hydroxy-5-methyl-4-isoxazolepropionic acid (AMPA), N-methyl-D-aspartate (NMDA) and γ-aminobutyric acid (GABA$_A$) gated channels of the A-subtype. AMPA and NMDA conductances were as previously reported (Lu et al. 1995). Based on our experimental data (Fig. 1A), the GABA$_A$ conductance was modeled with a maximum conductance of 50 nS and a rising phase time constant of 0.2 ms. This resulted in an IPSP of 0.7-0.8 mV amplitude (Fig. 1C) which is 20% of the unit EPSP amplitude generated by the modeled excitatory conductances in response to a single pulse synaptic input at one dendritic bulb. As demonstrated experimentally, the reversal potential of this conductance was set at -65 mV (Fig. 1B).

FF and FB loops were simulated as 5 ms time delays. The FF loop was triggered simultaneously with activation of the GC dendrite by incoming MFs, resulting in the triggering of inhibitory inputs to the same dendrite with a 5 ms time delay. For FB inhibition, the 5 ms loop time was added to the transmission time from GC synaptic activation to the firing of the PF distal axon, producing a final delay time of about 8.5 ms, which is consistent with delay times reported for disynaptic inhibitory currents evoked by whole cell stimulation of GC somata (Barbour 1993).

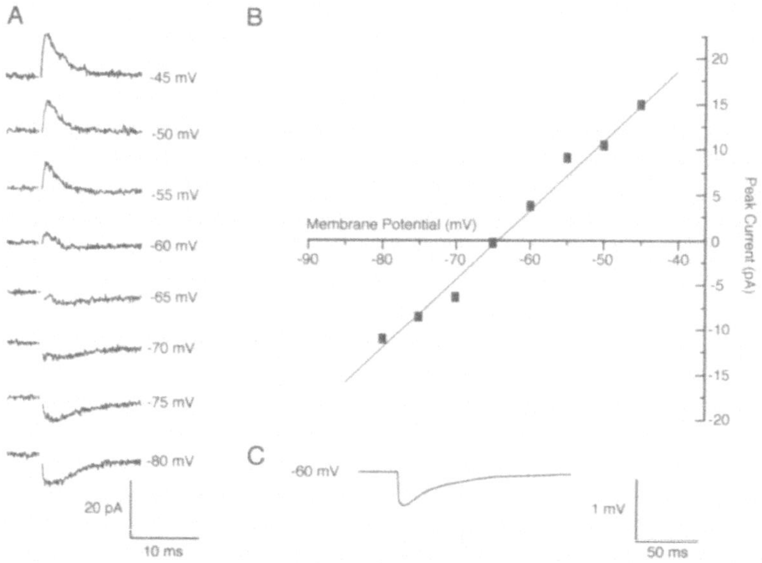

Figure 1. A GABA$_A$ conductance was modeled using parameters from whole-cell patch clamp recordings. A. IPSCs were elicited by electrical stimulation in the presence of AP-5 (50 mM) and CNQX (10 mM). B. The reversal potential of this conductance is at -65 mV. C. In simulation experiments, this modeled conductance produced a fast IPSP in response to dendritic stimulation.

RESULTS

Synaptic inputs varying from 10-100 Hz were delivered to one dendritic bulb to test the influence of FF and FB inhibition on GC responses with the assumption that each input signal was capable of evoking the FF loop and each action potential transduced to the distal axon was able to activate the FB loop.

Both FF and FB inhibition strongly modulated GC output frequency. When both systems accessed the same GC dendrite, FF inhibitory inputs reduced GC output rates by 30% for MF input frequencies of 40 - 100 Hz. FB inhibition, which sampled PF output,

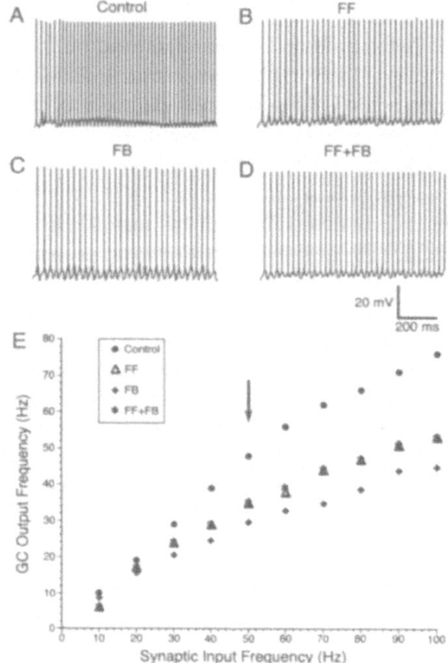

Figure 2. Effect of FF and FB inhibition on GC output frequency. Synaptic inputs varying from 10-100 Hz were delivered to one dendritic bulb. GC responses to a 50 Hz (arrow in E) input are illustrated in A-D. **A**. Control responses were elicited in the absence of inhibition. **B**. FF inhibition reduced GC firing rate. **C**. FB inhibition alone resulted in the greatest reduction in response frequency. **D**. The addition of FB to FF inhibition did not strengthen the effect of FF inhibition alone. **E**. Summary of data across all tested input frequencies illustrates several points. Note that both FF and FB inhibitory influences are greatest at input frequencies of $\geq$ 30 Hz and that the addition of FF to FB inhibition reduces the efficacy of FB inhibition in isolation.

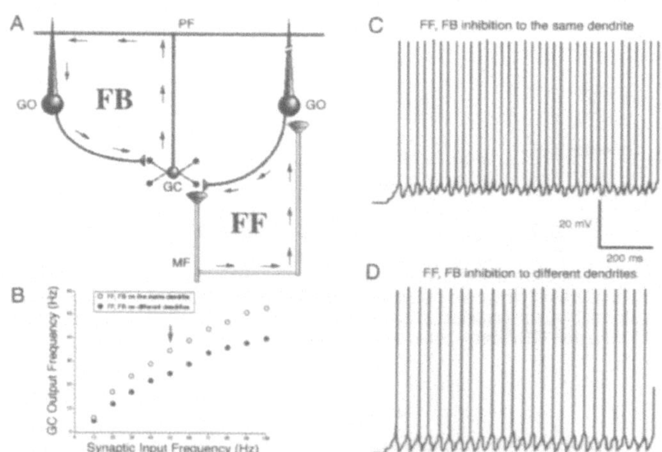

Figure 3. Effect of FF and FB inhibition applied at different dendrites. **A**. Schematic diagram illustrating relationship of FF and FB loops which terminate on different GC dendrites. **B**. GC output frequency across a range of MF input frequencies with both FF and FB inhibition active at the same (O) or at different (●) dendrites. **C**. GC output to a 50 Hz input (arrow in B) with both FF and FB inhibition active at the same dendrite. **D**. GC output to a 50 Hz input (arrow in B) with both FF and FB inhibition active at different dendrites.

resulted in a 40% reduction of GC output frequency over the same range of MF input rates. The reduction in GC output produced by activation of the FF inhibitory loop was not enhanced by the addition of FB inhibition to the same dendrite. Interestingly, the effect of FB inhibition alone was attenuated by the addition of the FF inhibitory loop (Fig. 2).

Activation of both FF and FB inhibition applied to different GC dendrites resulted in a greater reduction of GC output frequency (17-22%) than produced by application of both loops to the same GC dendrite (Fig. 3). In this case, activation of both FF and FB inhibitory loops resulted in the greatest reduction in GC output rates, enhancing the effect of either inhibitory loop applied separately.

The delay time of the FB loop (Tfb) is likely to vary with differences in the location of activated PF-Golgi cell synapses. To test the effect of such differences, FB loop times of 8.5, 11 and 16 ms were tested. Loop times of 8.5 and 11 ms produced equivalent changes in GC outputs. At the longest loop time tested, GC responses to input frequencies of $\geq$ 80 Hz were less effected by FB inhibition than at shorter loop times. Within this range of input frequency, an irregular firing pattern was exhibited which may indicate that FB inhibition with long time delays modulates not only GC firing rate but also GC firing pattern.

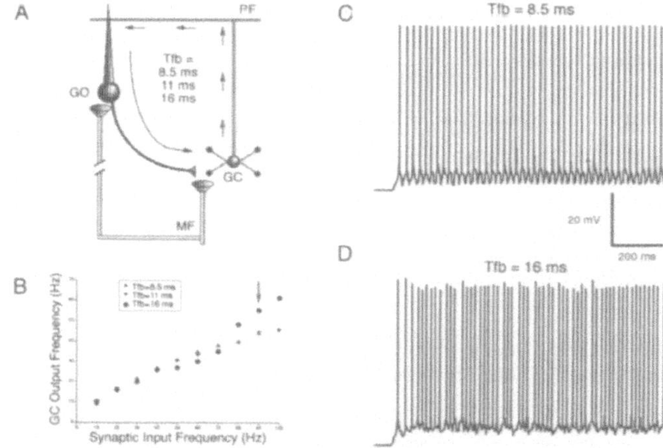

Figure 4. Effect of longer loop times on the modulatory influence of the FB loop.   **A**. Schematic of the experimental design in which Tfb, the time delay of the FB loop, was varied (8.5, 11 and 16 ms).   **B**. GC output responses to lengthened FB loop times. A major change in GC output responses was noted at the longest loop times. Simulated output responses to 90 Hz inputs (1 sec duration, arrow in B) with Tfbs of 8.5 (**C**.) and 16 ms (**D**.) are illustrated.

DISCUSSION

The simulation studies reported here indicate that Golgi cell activation not only modulates GC firing rate, but also shapes temporal patterning in PF outputs. Marr (1969) hypothesized that the Golgi cell might sample both MF and PF activity to adjust the firing threshold of the GCs. In our studies, The GC output is abolished by FF inhibition when inputs were $\leq$ 10 Hz. It suggests that Golgi cell inhibition acts to shift the GC sensitivity to the right, such that higher frequency inputs are required to elicit a given output rate.

Our simulation results indicate that FB inhibition has a stronger influence on GC output frequency than FF inhibition (Fig. 2E). When FF and FB input reaches the same dendrite of the GC, the output frequency falls back to that exhibited when only the FF loop existed, i.e., FF inhibition becomes dominant when two inhibitory inputs are applied at the same dendritic bulb. It is possible that this effect results from saturation of the GABA$_A$ channels on the dendritic bulb. Consistent with this hypothesis, inhibitory inputs reaching two different dendritic bulbs result in a further reduction of the GC output frequency (Fig. 3B). This result also indicates that FF and FB inhibition can be either additive or non-additive depending on whether or not both inhibitory inputs are applied to the same PC dendrite.

Relative timing differences between FF and FB inhibitory signalling may represent one mechanism for plastic remodeling of GC output. Loop transit times create a situation in which FF inhibitory inputs access GC dendrites prior to the arrival of FB signals. As FB loop times are increased their influence on GC output rate is lessened, as might be expected due to the loss of summation of IPSPs. At the longest loop time delays, however, FB inhibitory inputs appear to influence the temporal pattern of GC output. These effects are most apparent at higher MF input rates and may represent one mechanism by which the GC network segregates MF inputs based on their firing rate.

The Golgi cell interneuron samples both MF and PF responses in order to regulate GC output. Our simulation studies suggest that this regulation may be extremely plastic, relying on differences in loop transit times, dendritic location, and relative timing to produce a varied effect of GC output rate and pattern. The ability of Golgi cell inhibition to regulate PF input to the Purkinje cell network in what may be a highly plastic manner is critical to the output of the cerebellar cortical network.

ACKNOWLEDGMENTS

The authors gratefully acknowledge the financial support of the National Institutes of Health (NS 30759) and the National Science Foundation (IBN 9514844) to LLP.

REFERENCES

Barbour B (1993) Synaptic currents evoked in Purkinje cells by stimulating individual granule cells. Neuron, 11:759-769

Bower JM and Beeman D (1994) The book of GENESIS : exploring realistic neural models with the GEneral NEural SImulation System. TELOS, Springer-Verlag, New York.

Eccles JC, Llinás R and Sasaki K (1966) Parallel fibre stimulation and the responses induced thereby in the Purkinje cells of the cerebellum. Exp. Brain Res., 1:17-39

Llinás R, Bloedel JR and Hillman DE (1969) Functional characterization of neuronal circuitry of frog cerebellar cortex. J. Neurophysiol., 32:847-870

Lu H, Prior FW and Larson-Prior LJ (1995) Signal transduction in a cerebellar granule cell: a modeling approach. Neurosci Abstr 21:916

Marr D (1969) A theory of cerebellar cortex. J Physiol 202:437-470

Palay SL and Chan-Palay V (1974) Cerebellar cortex cytology and organization. Springer-Verlag, New York

Vranesic I, Iijima T, Ichikawa M, Matsumoto G and Knöpfel T (1994) Signal transmission in the parallel fiber-Purkinje cell system visualized by high-resolution imaging. Proc. Natl. Acad. Sci., 91:13014-13017

USING THE DYNAMIC CLAMP TECHNIQUE TO STUDY FREQUENCY REGULATION OF THE PYLORIC RHYTHM

Yair Manor, Farzan Nadim and Eve Marder

Volen Center For Complex Systems
Brandeis University
415 South Street
Waltham, MA 02254

INTRODUCTION

The stomatogastric ganglion (STG) of the spiny lobster *Panulirus interruptus* produces a triphasic pyloric rhythm which, in the behaving animal, is subject to changes in frequency, duty cycle and phase relationships. The pyloric rhythm is driven by a pacemaker ensemble that includes two Pyloric Dilator (PD) neurons. The inhibitory synaptic connection from the Lateral Pyloric (LP) neuron to the PD neurons provides the sole feedback from the rest of the pyloric network to the pacemaker ensemble. This synaptic connection is therefore potentially involved in the regulation of the pyloric rhythm.

The chemical synaptic connections among the neurons of the STG are inhibitory. Moreover, the threshold for transmitter release is close to the resting potential, and synaptic release is both spike-mediated and graded. At some synapses the graded component of transmitter release is significantly larger than that evoked by the action potentials, and is the major contributor to network dynamics (Graubard et al., 1983). When spike-mediated transmission is blocked by tetrodotoxin, an alternating pattern of slow wave oscillation characteristic of the pyloric rhythm persists in the presence of various activating modulatory substances (Raper, 1979). We have previously characterized the time-dependent dynamics of the graded component of the LP to PD synapses (Manor et al., 1997). We found that the strength and phasing of this depressing synapse is sensitive to changes in frequency and duty cycle of the presynaptic waveform.

Even though the synaptic connection from the LP neuron to the PD neurons clearly affects the pyloric rhythm, it is not known how, or whether, the *dynamical properties* of this synaptic connection affect the overall properties of the pyloric rhythm. It is not possible to eliminate or manipulate this synaptic connection pharmacologically without modifying other components of the pyloric circuit. An alternative is to remove LP from the network, either by hyperpolarization or by photoinactivation, and replace the biological synapse with an artificial

connection, implemented with the dynamic clamp technique (Sharp et al., 1993).
In the present work we set up the framework for investigating how the dynamical properties of the LP to PD synaptic connection influence the frequency of, and the phase relationships within, the pyloric network.

METHODS

The experimental setup consists of one intracellular electrode in the LP neuron and two intracellular electrodes in one of the PD neurons (Fig. 1). An intracellular electrode in the second PD neuron was sometimes employed. The electrode in the LP neuron is used to eliminate the biological LP to PD synapse, by injecting a large hyperpolarizing current in the LP neuron. The two electrodes in PD neuron are used to monitor the membrane potential and inject the dynamic-clamp current in real time (on-line). To calculate this current on-line, we solve a set of differential equations that describe the synaptic dynamics, as explained below.

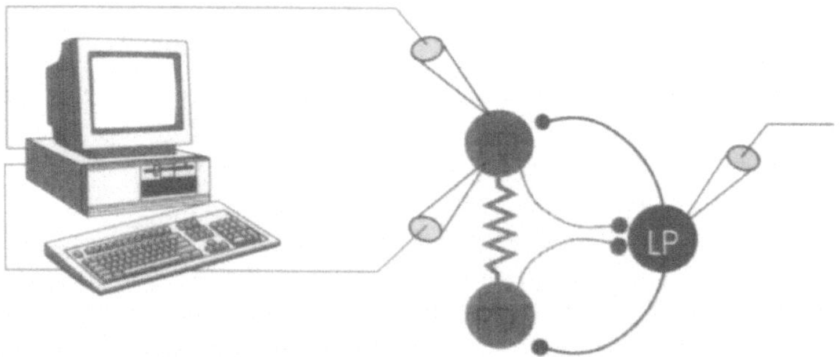

Figure 1. The experimental setup. Resistor symbol indicates electrical synapse. The filled circles show inhibitory chemical synapses.

The artificial synapse is created by injecting, in the postsynaptic neuron PD, a current that mimics the normal function of the synapse. This current is given by

$$I = g_{syn}(V_{LP},t)\ (V_{PD} - E_{syn}) \tag{1}$$

where g_{syn} is the change in conductance of the postsynaptic cell produced by the synapse, V_{PD} is the membrane potential of PD, V_{LP} is the presynaptic potential and E_{syn} is the equilibrium potential for this synapse set at -80 mV.

We have previously constructed a computational model that captures the dynamics of the LP to PD synapse (Nadim et al., 1997). In this model, $g_{syn}(V_{LP},t)$ consists of a sustained component, P, and two transient components, T_1 and T_2:

$$g_{syn}(V_{LP},t) = g_P m + g_{T1} m h_1 + g_{T2} m h_2 \tag{2}$$

where m, h_1 and h_2 are state variables that obey the following differential equations:

$$\tau_m(V_{LP})\frac{dm}{dt} = m - m_\infty(V_{LP})$$

$$\tau_{h_1}(V_{LP})\frac{dh_1}{dt} = h_1 - h_{1,\infty}(V_{LP}) \tag{3}$$

$$\tau_{h_2}(V_{LP})\frac{dh_2}{dt} = h_2 - h_{2,\infty}(V_{LP})$$

460

and m_∞, $h_{1,\infty}$, $h_{2,\infty}$, τ_m, τ_{h1} and τ_{h2} are sigmoidal functions of the LP membrane potential.
To demonstrate the predictive value of this model, we compared the response of the biological postsynaptic neuron with the model output (Equation (1)), to the same presynaptic waveform. Figure 2 shows a comparison between the biological data and the model output. The LP neuron (bottom traces) was voltage clamped with a single pulse (A), a train of pulses (B) and a realistic waveform (C). The postsynaptic (PD) responses are shown in the top traces. In each of these three cases, the model output (thick traces) fits reasonably well the current clamp response of the biological PD neuron (thin traces).

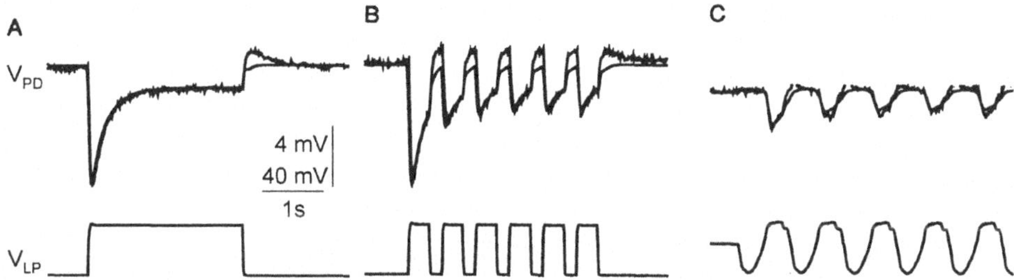

Figure 2. Comparison of biological data to model output. The LP neuron (bottom traces) was voltage clamped with: *A.* a single pulse; *B.* a train of pulses; and *C.* a realistic waveform. The postsynaptic (PD) biological (gray traces) and model (black traces) responses are shown in the top traces.

We used the model described in Eqs. (1)-(3) to compute the synaptic conductance $g_{syn}(V_{LP},t)$ on-line. Equation (3) was integrated numerically using a fourth order Runge-Kutta method. The membrane potential of the PD neuron, V_{PD}, was continuously monitored to update the driving force for the synaptic current.

Because in our experimental setup the presynaptic LP neuron was hyperpolarized, we used a representative LP waveform as the presynaptic input for the artificial synapse. This representative waveform was calculated by recording and averaging several cycles of the normal activity of an LP cell. The presynaptic signal was generated by playing back the representative waveform repeatedly into the model.

The period and shape of the presynaptic input (V_{LP}) was therefore completely determined by us. In contrast, the period of the PD cells was determined by the biological pacemaker and the artificial synaptic input (I_{dyn} in Fig. 3). Because the pyloric rhythm fluctuated in period, and was influenced by the dynamic clamp synaptic input, we could not use a fixed period for the presynaptic waveform. Therefore, to imitate the biological effect of the LP neuron on the PD neuron, we adjusted the period of each cycle according to the period of the postsynaptic cell, thereby "closing the loop."

Figure 3 illustrates the algorithm used to adjust the period of the presynaptic cycle. We continuously monitor the postsynaptic voltage and define the points where this voltage crosses some threshold with negative slope as t_i. After several crossings, we measure the biological period $T_{bio,i} = t_i - t_{i-1}$. Each time $T_{bio,i}$ is measured, $T_{sim,i}$, the period of our simulated presynaptic cell, is updated to $T_{bio,i}$. The time at which the next cycle of the simulated synaptic event is triggered is given by $s_i = t_i + \phi T_{bio,i}$, where ϕ is a parameter defining the injection phase, measured from the end of the PD burst. The synaptic current injection may prolong or shorten T_{bio}, the period of the postsynaptic cell. If T_{bio} is prolonged, the presynaptic cycle will be completed before the postsynaptic cell reaches the phase ϕ again ($s_i + T_{sim} < s_{i+1}$). Hence, the injection of the next presynaptic cycle must be delayed until s_{i+1}. If T_{bio} is shortened, the postsynaptic cell reaches the phase ϕ before the presynaptic cycle is completed ($s_i + T_{sim} > s_{i+1}$). In this case, the old presynaptic cycle is terminated at s_{i+1} and a new cycle is initiated.

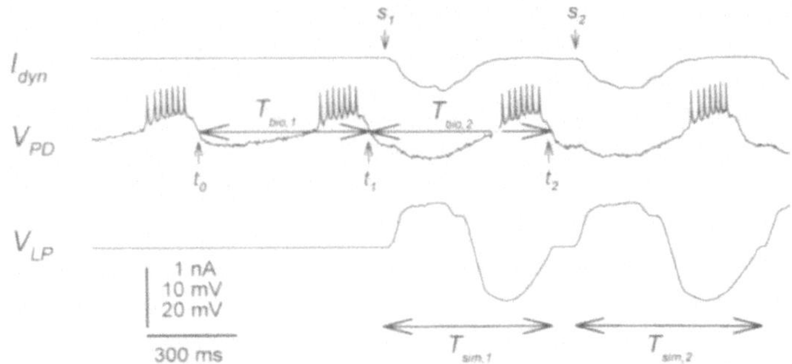

Figure 3. On-line adjustment of the period of the artificially-generated presynaptic waveform. See text for additional explanation.

RESULTS

The effect of the injection phase of the artificial synapse on the pyloric period.

The LP neuron is inhibited by the PD neurons and bursts on rebound from this inhibition, that is, between the PD bursts. The phase ϕ of the LP burst relative to the end of the PD burst is not constant (Hooper, 1997a; Hooper, 1997b). The parameter ϕ may be important for the regulation of the pyloric period. We therefore examined the effect of phase of the injected artificial synapse on the pyloric period.

We used injection phases ϕ ranging from 0 to 0.3, where 0 denotes the end of the PD burst. This range is compatible with the physiological range as described in (Hooper, 1997b).

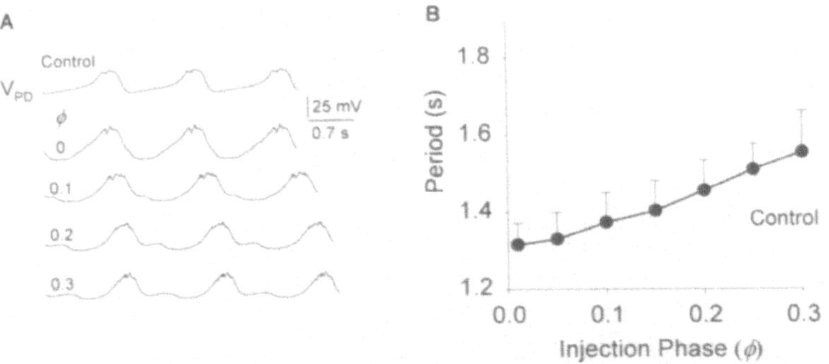

Figure 4. The effect of phase of the artificial synapse on the pyloric period. *A.* Three cycles of the PD voltage at different injection phases (ϕ) of the artificial synapse. *B.* Pyloric period as function of ϕ. Error bars indicate standard deviation across 15 trials. Dotted line shows the average control period.

In Figure 4A, each trace represents 3 PD cycles when the artificial synapse is injected at a fixed phase ϕ. Each trace is an average over 15 trials. The action potentials are eliminated as a result of this averaging. The top trace is control (no current injection). The consequent traces show the averaged runs for ϕ values of 0, 0.1, 0.2 and 0.3, respectively. This figure indicates that the period increases as the injection phase increases. This result is shown in Figure 6B, where the period is plotted against ϕ. The dotted line represents the control period (no current injection).

A comparison between the artificial synapse and constant current injection.

The pyloric period can also be modified by injection of *constant* current in PD. The pyloric period increases as the amplitude of the injected negative constant current is increased (inset of Fig. 5A). As noted above, the pyloric period also increases as the phase of the artificial LP to PD synapse is increased (inset of Fig. 5B). We therefore examined how these two types of stimuli affect the burst and interburst durations of PD. In both cases we measured the pyloric periods and the corresponding PD interburst and burst durations. We then plotted the burst durations (filled circles) and interburst durations (open squares) against the pyloric periods. In the case of constant current injection, both the interburst and burst durations were larger with longer pyloric periods. As the pyloric period increased (by increasing the constant hyperpolarizing current), the interburst durations increased 3 times as much as the burst durations (slopes: 0.73 and 0.27, respectively).

We repeated the same procedure using the artificial synapse at different injection phases ϕ to modify the pyloric period. In this case, the interburst duration was larger with longer pyloric periods (slope: 1.28). However, the burst duration *decreased* with pyloric period (slope: -0.28). Longer periods correspond to later phases shown in Figure 4.

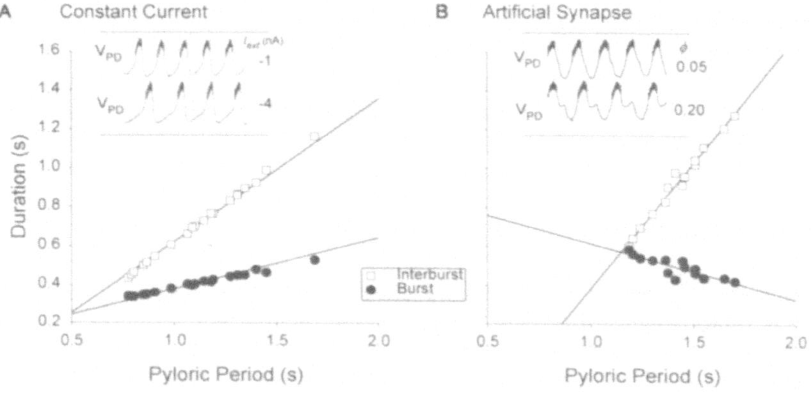

Figure 5. The effect of constant and phasic current on burst (filled circles) and interburst (open squares) durations. Duration of PD burst and interburst shown as function of pyloric period when pyloric period was modified by *A*. injection of constant current with different amplitudes; and *B*. injection of an artificial synapse at different injection phases.

SUMMARY

We investigate the role of the LP to PD synaptic connection in controlling the frequency of the rhythm. We have previously constructed a model that described the dynamics of the biological synapse (Nadim et al., 1997). The model consisted of a persistent, a slow transient, and a fast transient component. To study the effect of this synapse, we eliminate the biological LP to PD synapse (by injecting current in LP) and we replace it with an artificial synapse, using the model. We implement the artificial synapse by using a modification of the dynamic clamp technique (Sharp et al., 1993). During the experiment the PD membrane potential is monitored and used for computing the driving force of the synapse as well as the pyloric period during each cycle. A computer-generated presynaptic LP waveform is used as input for the model synapse. The shape of this waveform is taken from a previous recording of the LP neuron. The frequency is adjusted on-line to reflect the change in the pyloric period.

Using this method we find that the LP to PD synapse, depending on its input phase, is capable of slowing down the pyloric rhythm. Constant negative current injection in the PD cells can also slow down the rhythm. However, unlike constant current injection which increases both the PD burst and interburst durations, this synaptic input increases the interburst duration but moderately decreases the burst duration. A similar effect is obtained if the input phase is kept constant and the synaptic *strength* is increased (not shown).

There are many different ways to affect the pyloric rhythm by stimulating the pacemaker ensemble. These include injection of constant current, injection of constant conductance and phasic injection of conductance. Each of these methods could have different effects on the phase relationship between the pacemaker ensemble and the follower neurons. Previous studies of phase and frequency regulation of the pyloric rhythm have used constant current injection to modify the rhythm (Abbott et al., 1991; Hooper, 1997a; Hooper, 1997b). The method described in this paper (phasic injection of conductance) affects the pyloric pattern in a considerably different manner than constant current injection (as shown in Fig. 5). The difference between these two methods could be due to the nature of the input (that is, current versus conductance) or its temporal pattern (constant versus phasic). We are currently investigating this question by comparing injection of constant conductance with phasic conductance. However, our preliminary results indicate that the method described here is closest to the mechanism employed by the biological LP to PD synapse in regulating the pyloric rhythm. These experiments shed light on the role of the feedback from the LP neuron to the PD neurons in controlling the frequency and pattern of the pyloric rhythm.

Different LP waveforms may have different effects on the regulation of pyloric network frequency. The LP to PD synapse may act to keep the PD burst duration approximately constant, as frequency varies.

ACKNOWLEDGMENTS

This research was supported by NS17813, MH46742, and the Sloan Center for Theoretical Neurobiology at Brandeis University.

REFERENCES

Abbott, L. F., S. L. Hooper and E. Marder, 1991, Oscillating networks: control of burst duration by electrically coupled neurons. *Neural Comp.* **3:** 487-497.

Graubard, K., J. A. Raper and D. K. Hartline, 1983, Graded synaptic transmission between identified spiking neurons. *J. Neurophysiol.* **50:** 508-521.

Hooper, S. L., 1997a, Phase maintenance in the pyloric pattern of the lobster (*Panulirus interruptus*) stomatogastric ganglion. *J. Comput. Neurosci.* **4:** 191-205.

Hooper, S. L., 1997b, The pyloric pattern of the lobster (*Panulirus interruptus*) stomatogastric ganglion comprises two phase maintaining subsets. *J. Comput. Neurosci.* **4:** 207-219.

Manor, Y., F. Nadim, L. F. Abbott and E. Marder, 1997, Temporal dynamics of graded synaptic transmission in the lobster stomatogastric ganglion. *J. Neurosci.* **17:** 5610-5621.

Nadim, F., Y. Manor, L. F. Abbott and E. Marder, 1997, Strength and timing of graded synaptic transmission depend on frequency and shape of the presynaptic waveform., pp. 391-394 in *Computational Neuroscience: Trends in Research, 1997*, edited by J. Bower. Plenum Press, New-York.

Raper, J. A., 1979, Nonimpulse-mediated synaptic transmission during the generation of a cyclic motor program. *Science* **205:** 304-306.

Sharp, A. A., M. B. O'Neil, L. F. Abbott and E. Marder, 1993, The dynamic clamp: artificial conductances in biological neurons. *Trends Neurosci.* **16:** 389-394.

ATTRACTOR DYNAMICS IN REALISTIC HIPPOCAMPAL NETWORKS

Elliot D. Menschik, Shih-Cheng Yen, and Leif H. Finkel

Institute of Neurological Sciences and Department of Bioengineering
University of Pennsylvania
3320 Smith Walk
Philadelphia, PA 19104

INTRODUCTION

Autoassociative attractor neural networks[1,2] provide a powerful paradigm for the storage and recall of memories, however, their biological plausibility has always remained in question. Given the complexity and variability of biological networks, it seems likely that a "biological" attractor network must be regulated by some control structure. We describe a functional architecture for implementing the Hopfield attractor paradigm in a model of the CA3 region of the hippocampus. In this model the control structure is provided by the intrinsic and extrinsic rhythms (gamma and theta) of the hippocampus and neuromodulatory input. Cellular-level simulations are shown that serve as a "proof-of-concept."

The function of the model is demonstrated in the left-hand panel of Figure 1. Shown are hypothetical spike traces of CA3 pyramidal cells with idealized gamma and theta population rhythms shown for reference. The state of the network is defined by the spatial pattern of temporally-precise spikes during the time window defined by a single gamma cycle. A memory is reached when the network converges to a fixed-point attractor. The gamma-band synchronization is induced by a network of mutually inhibitory interneurons[3-5]. Theta-band oscillations, induced by septal interneurons, act to clock new perforant input to the network from entorhinal cortex, terminate each attractor state, and reset the network for the next set of entorhinal inputs. Lastly, cholinergic input from the medial septum is responsible for maintaining otherwise bursting pyramidal cells in a single spike firing mode. As such, this model provides some putative roles for hippocampal interneurons, synchronous oscillations, and cholinergic neuromodulation.

The CA3 region provides an ideal neural substrate for an autoassociative network[6]. In fact, numerous investigators have used the autoassociative model to create functional hippocampal networks for memory and spatial navigation, but, to date, nearly all have relied upon simplified models of individual neurons. Such studies have provided both insight and novel theories of hippocampal function. However, more detailed and realistic, compartmental models are necessary for studying the effects of normal and pathological cellular mechanisms on network function. Our study is directed to the following questions: Can networks of biological neurons perform the same or analogous computations as networks of artificial (i.e. formal, mathematical) neurons? What are the functional consequences of neuromodulation upon network dynamics? How might pathological perturbations at the cellular and subcellular levels translate into network dysfunction at a functional level?

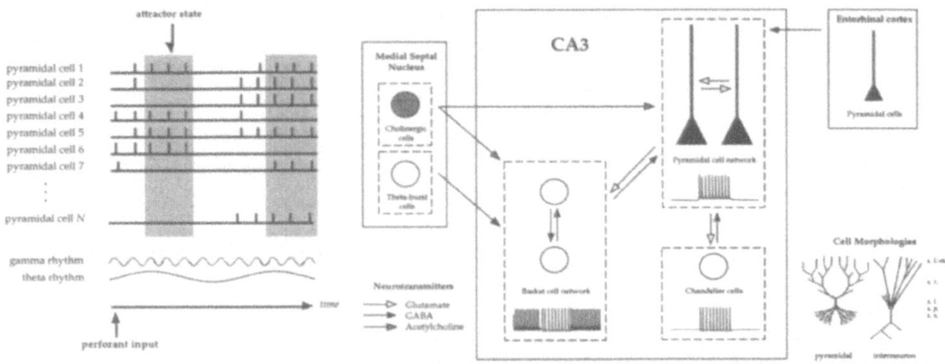

Figure 1. (*Left*) Operation of the biological analog of an autoassociative attractor neural network. (*Right*) The architecture and behavior of the model is consistent with known hippocampal anatomy and physiology. Shown is a network schematic and sample simulated somatic voltage traces from the three CA3 cell populations. *s. l.-m.*, stratum lacunosum-moleculare; *s. r.*, stratum radiatum; *s. l.*, stratum lucidum; *s. p.*, stratum pyramidale; *s. o.*, stratum oriens.

METHODS

Compartmental simulations were constructed using PGENESIS[7], the recent parallel implementation of the GENESIS development package. Simulations were performed on a network of 18 Silicon Graphics Indy workstations. Differential equations were solved using Crank-Nicholson implicit integration with a step size of 25µs.

Pyramidal cells, interneurons, and synapses

The cellular models chosen for the simulations are the most highly detailed and realistic hippocampal cells developed to date: the 66-compartment hippocampal CA3 pyramidal and the 51-compartment hippocampal interneuron developed by Traub and colleagues[8,9]. For the most part, AMPA, NMDA, and $GABA_A$ synapses were implemented as described by Traub and colleagues[10].

Network connectivity

This preliminary network consists of 24 neurons (8 pyramidal cells, 8 basket cells, and 8 chandelier cells). Its design is inspired by the known anatomy of CA3 and is sketched in the right-hand panel of Figure 1. Mutually inhibitory basket cells are depolarized by septal cholinergic and local pyramidal cell glutamatergic input while being simultaneously inhibited by septal perisomatic GABAergic input oscillating at theta frequencies. The inhibition between basket cells creates synchronous gamma oscillations[3-5] which are themselves modulated at theta frequencies by the septal inhibition. CA3 pyramidal cells are depolarized in the *s. p.* by septal cholinergic input, in the *s. r.* by recurrent glutamatergic input, and in the *s. l.-m.* by perforant glutamatergic input from entorhinal cortex. Synchronous oscillatory inhibition from the basket cells in the perisomatic region constrains pyramidal cell firing while recurrent inhibition from chandelier cells at the axonal initial segment balances recurrent excitation. Two random 8-bit patterns (10011100 and 10101011) were chosen and stored in the recurrent synaptic matrix using Hopfield's original algorithm[1] to scale maximal synaptic conductances.

Cholinergic neuromodulation

Neuromodulation via acetylcholine (ACh) is implemented at the cellular level inhibiting intrinsic membrane currents and diffusely depolarizing cells. For ionic current inhibition we derived dose-response curves based on a Michaelis-Menten model. Nonlinear curve fits matched very closely the data of Madison *et al.*[11] for I_{AHP} and Toselli and Lux[12] for I_{Ca}. Diffuse cholinergic depolarization of pyramidal and basket cells is modeled indirectly using depolarizing somatic current injection. Additional detail is provided in a recent, larger-scale study[13].

RESULTS

Network function

Before examining the behavior of the compartmental network model, our first step was to establish control conditions by exploring the trajectories of all 256 initial states of an 8-cell artificial Hopfield attractor network storing the randomly chosen binary patterns 10011100 and 10101011 (9C and AB in hexadecimal notation). For each initial state we recorded the subsequent network states at each time step using synchronous updating. A map of these trajectories is shown in Figure 2.

With the control established, each of the 256 initial states was presented to the biologically-based network as simulated entorhinal input on the perforant pathway (i.e. transient glutamatergic excitation of the distal apical arbor). After the initial spike pattern, the network state was allowed to evolve according to its recurrent connectivity. A sample spike trace for the pyramidal cell network is shown in Figure 3 with the inputs 00011000 (18 hex) and 00011001 (19 hex). The attractors reached are in fact the same stored patterns as found in the artificial network. While it is not clear from this data that the final state reached is an attractor, additional simulations show that if the theta rhythm is interrupted (i.e. the pyramidal cells are not inhibited) at the end of a theta cycle, the firing pattern is indeed stable (data not

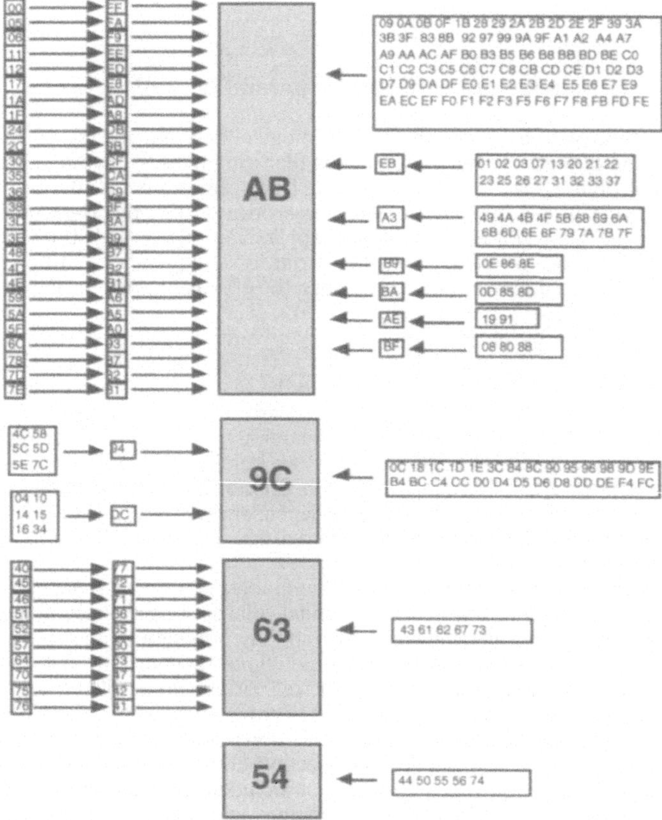

Figure 2. Trajectories of the 256 initial states of the artificial attractor network storing the binary patterns 10101011 and 10011100 (AB and 9C, respectively, in hexadecimal notation). By symmetry the inverse patterns are also attractors. All initial states move to one of the four attractors within 2 time steps.

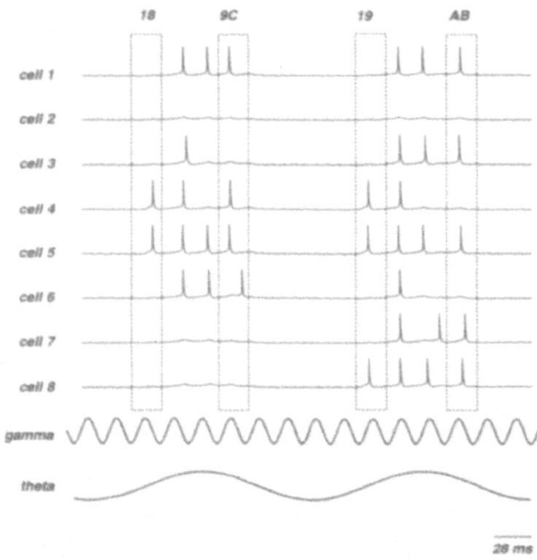

Figure 3. Performance of the biological attractor network. Shown are spike traces from the 8 pyramidal cells following two entorhinal inputs at the onset of each theta cycle. Idealized gamma and theta population oscillations are shown for reference. A comparison with Figure 2 shows that the correct attractors are reached by the end of the theta cycle, but that different trajectories are taken than in the artificial attractor network. The hexadecimal equivalents of the network inputs and outputs are shown at the top.

shown).

Cholinergic neuromodulation of single pyramidal cells

Acetylcholine plays a critical role in the biological network model not only by providing diffuse excitation of the cells, but by regulating the firing mode of pyramidal cells. Shown in the top panel of Figure 4 is the effect of varying [ACh] for the pyramidal cell model using the dose-response curves discussed in the Methods section. Somatically-recorded voltage traces are shown for the same cell using simulated application of 0.1 to 100 µM ACh. As [ACh] rises, the cell undergoes a marked transition from low-frequency bursting to high-frequency spiking. In contrast to the pyramidal cell model, the effects of ACh on the fast spiking of the interneuron model is negligible (data not shown).

The transition of the intrinsically bursting pyramidal cell to a spiking regime can be qualitatively understood by recognizing that I_{Ca} is responsible for the slow depolarization underlying the burst and also the interplay with the fast sodium current that causes the rapid series of action potentials riding the slow depolarization. In contrast, I_{AHP} is the current responsible for terminating the burst and maintaining a long hyperpolarization following it. The inhibition of I_{Ca} by ACh removes the slow calcium-dependent wave, diminishes the reverberating depolarization between soma and adjacent dendrites that create the rapid series of overlying spikes, and affects the calcium dependency of I_{AHP}. The inhibition of I_{AHP} by ACh markedly reduces the afterhyperpolarization that terminates the burst and maintains the low inter-burst interval.

It should also be noted that under these conditions, a very small depolarizing current (0.1 nA) results in a very high spike rate for pyramidal cells as the simulated concentration of ACh rises. This rise in spike frequency can be (and in our simulations is) held in check by interneuronal control. Finally, while our simulations demonstrate that cholinergic input is sufficient to induce a transition in pyramidal cell firing mode, Traub and colleagues have shown that tonic somatic current injections[8,14] or tonic stimulation of slow dendritic $GABA_A$ receptors[8] may be responsible for a similar functional shift. Our results are independent of these two factors as we have held current injection constant at a level that does not induce spiking in the absence of ACh, and we have not included slow dendritic $GABA_A$ receptors.

Our simulations also suggest that bursting and spiking behavior in hippocampal pyramidal cells have distinct advantages that may be exploited by switching between these two firing modes. The bottom panels of Figure 4 plot the calcium concentration in the pyramidal cell model as a function of time and space for a spiking cell and a bursting cell.

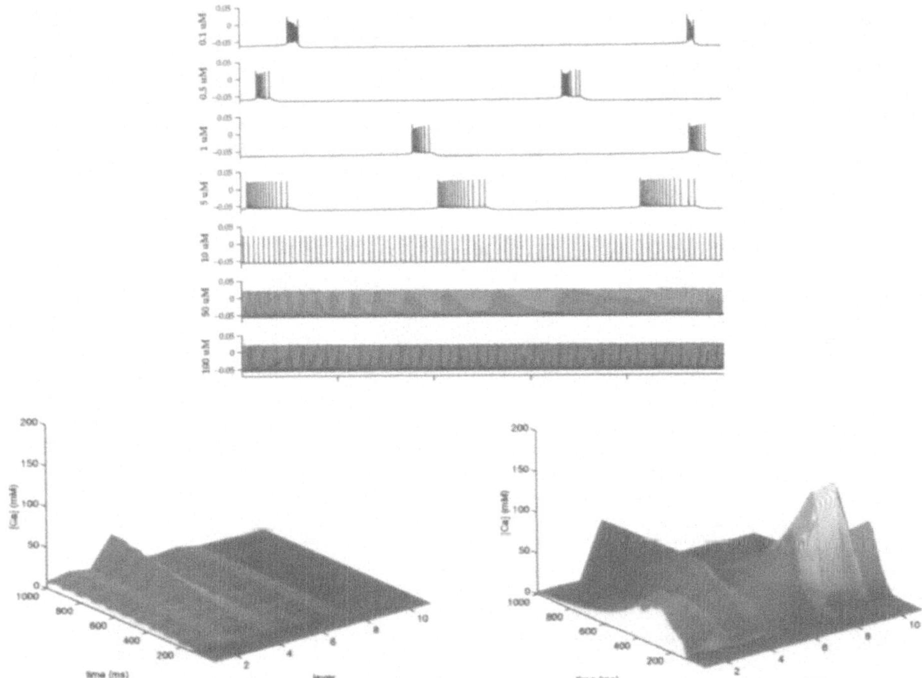

Figure 4. Muscarinic neuromodulation of a pyramidal cell. (*Top*) Shown are 2.5 seconds of simulated somatic recordings from the model pyramidal cell at various levels of cholinergic input. The cell received 0.1 nA current injection in all traces. (*Bottom*) Compartmental distribution of calcium concentration in mM as a function of time in spiking (*left*) and bursting (*right*) hippocampal pyramidal cells. Layers 1-3 are the basal dendrites, layer 4 is the soma, and layers 5-11 are the apical dendritic arbor.

The data shows that backpropagating single spikes are quite poor at inducing calcium influx in passive distal dendrites in contrast to backpropagating bursts.

DISCUSSION

The biological attractor network

It should be noted that the results presented here are of a preliminary nature and our primary goal was to present a concept for biological attractor networks, describe the structure of such a network, and demonstrate its practical feasibility. Nevertheless, use of only eight pyramidal cells, while facilitating the comparison with an artificial attractor network as well as making practical the simulation on a local network of workstations, is a serious limitation. Storing just two patterns heavily overloads the storage capacity of such a small network. A much larger version of this network has since demonstrated the scalability of the model[13].

Another problem encountered by the network is the relatively frequent (approximately 40%) failure to converge to the appropriate attractor in the time window provided by the theta rhythm. This flaw is due to the small size of the network and the relatively slow gamma-band rhythm produced by the basket cells. Again, a larger network with more gamma cycles per theta cycle has shown nearly perfect convergence to any of several stored patterns even in the presence of significant noise[13].

Acetylcholine and the functional role of spikes and bursts

Physiological recordings show that hippocampal pyramidal cells are capable of either bursting or spiking, and that these different modes are correlated with the behavioral state of the rat[15]. From a theoretical perspective, these two firing modes have their respective advantages. Spiking is rapid and can have a temporal precision of a millisecond or so, allowing for efficient representation of information and a well-defined network activity state. However, our simulations have demonstrated that spiking may be poorly suited for inducing and/or maintaining synaptic plasticity in the distal dendritic arbor. Rather, this function appears to be better fulfilled by backpropagating bursts which are capable of causing significant alterations of calcium levels in the dendrites. The drawback for hippocampal bursts lies in their typical low-frequency and variable length which make the representation of information difficult and inefficient at best.

Together with the behavioral correlations, our findings suggest that spiking behavior is necessary for the initial processing of novel information and its later recall, while bursting is necessary for more permanent storage of patterns via the induction of LTP and LTD. This view is wholly consistent with Buzsáki's "two-stage" memory model[16]. Our model indicates that ACh acts on at least two levels to initiate and manage a transition from bursting to spiking behavior, at least in intrinsically bursting hippocampal pyramidal cells. At the cellular level, the transition is due to a reduction in the afterhyperpolarizing calcium-dependent potassium current and the high-threshold calcium current. At the network level, the diffuse depolarizing action of ACh on interneurons can serve as the driving force for mutually inhibitory interneuronal networks which can generate gamma-band rhythmicity and thereby control the timing of pyramidal cell spiking.

ACKNOWLEDGEMENTS

We thank Nigel Goddard and Greg Hood at the Pittsburgh Supercomputing Center for providing a beta-test version of PGENESIS and help with its implementation. Supported by grants from Mrs. Patricia Kind, The Whitaker Foundation, and the Office of Naval Research.

REFERENCES

1. J.J. Hopfield, Neural networks and physical systems with emergent collective computational abilities, *Proceedings of the National Academy of Sciences of the United States of America*, 79:2554-8 (1982).
2. D.J. Amit, *Modeling Brain Function: The world of attractor neural networks*, Cambridge University Press, New York, 1989.
3. X.J. Wang and G. Buzsáki, Gamma oscillation by synaptic inhibition in a hippocampal interneuronal network model, *Journal of Neuroscience*, 16:6402-13 (1996).
4. R.D. Traub, M.A. Whittington, I.M. Stanford and J.G.R. Jefferys, A mechanism for generation of long-range synchronous fast oscillations in the cortex, *Nature*, 383:621-4 (1996).
5. P. Bush and T.J. Sejnowski, Inhibition synchronizes sparsely connected cortical neurons within and between columns in realistic network models, *Journal of Computational Neuroscience*, 3:91-110 (1996).
6. A. Treves and E.T. Rolls, Computational analysis of the role of the hippocampus in memory, *Hippocampus*, 4:374-391 (1994).
7. N.H. Goddard and G. Hood, Large Scale Simulation with PGENESIS. In J. M. Bower and D. Beeman (Eds.), *The Book of GENESIS: Exploring Realistic Neural Models with the GEneral NEural SImulation System*, Springer-Verlag, in press.
8. R.D. Traub, J.G.R. Jefferys, R. Miles, M.A. Whittington and K. Tóth, A branching dendritic model of a rodent CA3 pyramidal neurone, *Journal of Physiology*, 481:79-95 (1994).
9. R.D. Traub and R. Miles, Pyramidal cell-to-inhibitory cell spike transduction explicable by active dendritic conductances in inhibitory cell, *Journal of Computational Neuroscience*, 2:291-8 (1995).
10. R.D. Traub, M.A. Whittington, S.B. Colling, G. Buzsáki and J.G.R. Jefferys, Analysis of gamma rhythms in the rat hippocampus *in vitro* and *in vivo*, *Journal of Physiology*, 493:471-484 (1996).
11. D.V. Madison, B. Lancaster and R.A. Nicoll, Voltage clamp analysis of cholinergic action in the hippocampus, *Journal of Neuroscience*, 7:733-41 (1987).
12. M. Toselli and H.D. Lux, GTP-binding proteins mediate acetylcholine inhibition of voltage dependent calcium channels in hippocampal neurons, *Pflugers Archiv - European Journal of Physiology*, 413:319-21 (1989).
13. E.D. Menschik and L.H. Finkel, Neuromodulatory control of hippocampal function: Towards a model of Alzheimer's disease, *Artificial Intelligence in Medicine, Special Issue: Computational Modeling of Brain Disorders* (in press).
14. R.D. Traub, R.K. Wong, R. Miles and H. Michelson, A model of a CA3 hippocampal pyramidal neuron incorporating voltage-clamp data on intrinsic conductances, *Journal of Neurophysiology*, 66:635-50 (1991).
15. R.D. Traub and R. Miles, *Neuronal Networks of the Hippocampus*, Cambridge University Press, New York, 1991.
16. G. Buzsáki, Two-stage model of memory trace formation: a role for "noisy" brain states, *Neuroscience*, 31:551-70 (1989).

ENTRAINMENT OF A SLOW NEURONAL OSCILLATOR BY A FAST ONE

Farzan Nadim,[1] Yair Manor,[1] Steve Epstein,[2] and Eve Marder[1]

[1] Volen Center For Complex Systems
Brandeis University
415 South Street
Waltham, MA 02254
[2] Department of Mathematical Sciences
Rensselaer Polytechnic Institute
Troy, NY 12180

INTRODUCTION

Networks of neurons that produce rhythmic activity often need to operate in a coordinated way. Although there is extensive theory of coupled oscillators that are close in period, much less is known on the interplay of oscillators with considerably different time scale periods.

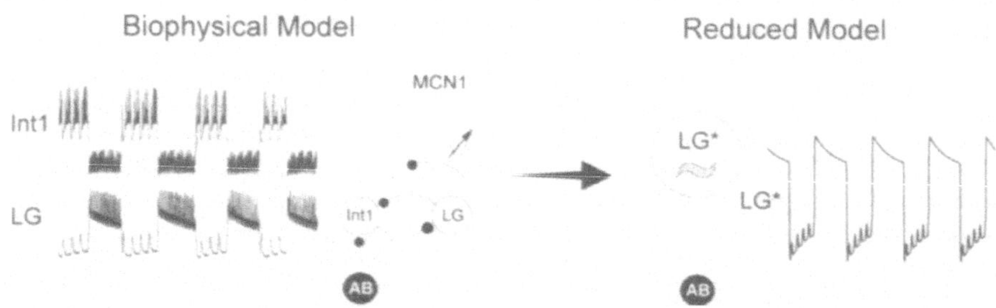

Figure 1. Stimulation of the modulatory neuron MCN1 generates a slow rhythm in the pair of reciprocally inhibitory neurons, Int1 and LG. The pacemaker of a faster rhythm, AB, makes a synaptic connection with Int1. *A.* Biophysical model of the circuit. *B.* Reduced model of the system. The reduced model is built by collapsing MCN1, Int1 and LG into a single relaxation oscillator.

We study the interactions between two oscillators that differ in period by an order of magnitude. The gastric mill rhythm of the crab *Cancer borealis* is a network-generated rhythm with a period of 5 to 20 s. At the center of this network is a pair of neurons, the lateral gastric neuron (LG) and interneuron 1 (Int1), that make reciprocal inhibitory connections. Stimulation of the modulatory neuron MCN1 results in a gastric mill rhythm in which LG bursts in antiphase with Int1. The pacemaker of the fast (0.5 to 2 Hz) pyloric rhythm, the Anterior Burster neuron (AB), makes an inhibitory connection with Int1. The LG neuron presynaptically inhibits the axonal terminals of MCN1 and is also electrically coupled to MCN1 (Figure 1A, schematic diagram). Based on the physiological findings, Coleman et al. (1995) suggested that the period of the gastric mill rhythm is determined by the time course of the MCN1 to LG chemical excitation.

We previously built a conductance-based model of this network (Manor et al., 1997). Our model suggested that factors other than the time course of the MCN1 to LG chemical excitation are important in the generation and frequency-control of the gastric mill rhythm. The model showed that the gastric mill rhythm may emerge from the interplay of several subcomponents of this circuit. These subcomponents include the LG/Int1 reciprocally inhibitory pair of neurons, the MCN1/LG excitatory/inhibitory interaction, and the fast periodic inhibition of Int1 by the AB neuron.

Surprisingly, we found that the most important factor in determining the model gastric mill period was the fast pyloric input from the AB neuron. As the strength of the AB to Int1 synapse was decreased, the model gastric mill rhythm became extremely slow and was eventually disrupted. The model gastric mill period was also dependent on the pyloric period. In general, as the pyloric period increased, the gastric mill period increased linearly. However, occasionally the gastric mill period decreased sharply. The relationship between the gastric mill and pyloric periods can be described with a sawtooth function. We found that the gastric mill rhythm was coupled to the pyloric rhythm through a time-locking mechanism in which the LG burst always started with a fixed delay after the AB input. Thus, each gastric cycle contained an integer number of pyloric cycles. Interestingly, within the same run, the gastric mill period could vary between n, n−1 and n+1 pyloric periods.

The conductance-based model was instrumental in suggesting a possible role for the pyloric input in the generation of the gastric mill rhythm. It was also useful to make several predictions, which are now under experimental investigation (Bartos and Nusbaum, 1997). However, because of the complexity of the biophysical model, our understanding of the circuit was incomplete, and several questions remained unanswered. Foremost among these questions is the following: how does a fast biological oscillator control the period of a rhythm that is 10 times slower?

In this paper we present a reduced version of the conductance-based model, built by collapsing the MCN1, Int1 and LG neurons into one model cell. This phenomenological model consists of a relaxation oscillator, with a fast variable V that represents the membrane potential of the reduced LG, and a slow variable w that represents the MCN1 chemical excitation to LG. The periodic AB inhibition of Int1, which results in a periodic disinhibition of LG, is modeled as a direct excitation of LG. Figure 1B shows a cartoon representation of the reduced model. The phenomenological LG is labeled as LG*, to emphasize its difference from the conductance-based LG.

RESULTS

Phase Plane Analysis of the Phenomenological Model

The two-dimensional relaxation oscillator is given by

$$\frac{dV}{dt} = f(V,w) \tag{1a}$$

$$\frac{dw}{dt} = \varepsilon g(V,w) \tag{1b}$$

where $0 \leq \varepsilon \ll 1$,

$$f(V,w) = -F(V) + w + I \tag{2a}$$
$$g(V,w) = G(V) - w \tag{2b}$$

and I represents the input currents to the reduced LG cell.

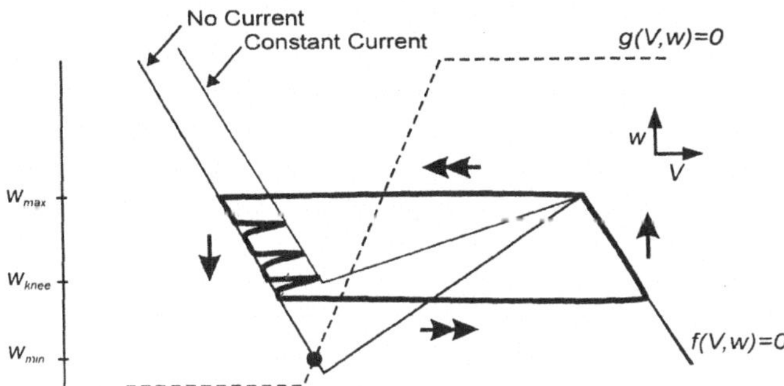

Figure 2. V-w phase plane for the relaxation oscillator representing LG*. Dashed line represents the w-nullcline, $g(V,w) = 0$. Thin solid lines represent two limits of the V-nullcline, $f(V,w) = 0$: with no current (lower curve) and with constant current (upper curve). Thick line represent the trajectory. w_{max} is the maximal w value (equal to the w value at the right knee point on the V-nullcline). w_{knee} is the minimal value of w for which the trajectory can switch from the left branch to the right (equal to the w value at the left knee point on the upper V-nullcline). w_{min} is the minimal possible value of w (equal to the w value at the intersection of the w-nullcline and the lower V-nullcline.

The graphs of $f(V,w) = 0$ and $g(V,w) = 0$ are the V- and w-nullclines. To simplify the analysis, we use piecewise linear curves to describe these nullclines. In Figure 2, the nullclines are represented as thin lines (solid and dashed, respectively) in the phase plane of V and w. In the biophysical model and the biological circuit the periodic inhibition from AB to Int1 results in intermittent disinhibition of LG. In the phenomenological model, we introduce the periodic AB input as a direct excitation of LG. This input is given by setting $I = I_{pyl}(t, V)$ in Equation (2a), as a periodic half-sine depolarization of LG:

$$I_{pyl}(t,V) = i_{max} \, H(-V) \sin\left(\frac{\pi}{d}(t \bmod P_{pyl})\right) H(d - (t \bmod P_{pyl})) \tag{3}$$

where i_{max} is the maximal current, P_{pyl} is the period of the pyloric input, d is the duration of the pyloric input, and H is the Heaviside function. We make the experimentally valid assumption that the pyloric input affects LG only when LG is in its trough phase. Because I in Equation (2a) is not constant, the V-nullcline swings between two limits. These two limits are shown as solid lines in Figure 2: the lower V-nullcline represents the case of no current injection ($I = 0$); the upper V-nullcline denotes the case of constant maximal current injection ($I = i_{max}$). With no current injection, the equilibrium point (solid circle) is stable.

The periodic shifts in the V-nullcline result in a large horizontal velocity vector, and transient depolarizations. So long as w is above the knee point of the left branch of the upper V-nullcline (w_{knee}), the trajectory is trapped in the trough phase and moves back and forth between the left branches of the two V-nullclines. As w decreases below w_{knee}, at the next pyloric cycle the left branch of the V-nullcline shifts above the current value of w. As a result, the large horizontal vector field now moves the trajectory rapidly to the right branch of the V-nullcline. This transition defines the onset of the LG burst. Because the shift in the V-nullcline occurs during a pyloric input, the transition into burst is time-locked to the pyloric rhythm.

Figure 3 shows the time course of the LG membrane potential in the biological system (Coleman et al., 1995), the biophysical model (Manor et al., 1997) and the reduced model. This figure demonstrates that the phenomenological model captures the essential time course behavior of both the biophysical model and the biological system.

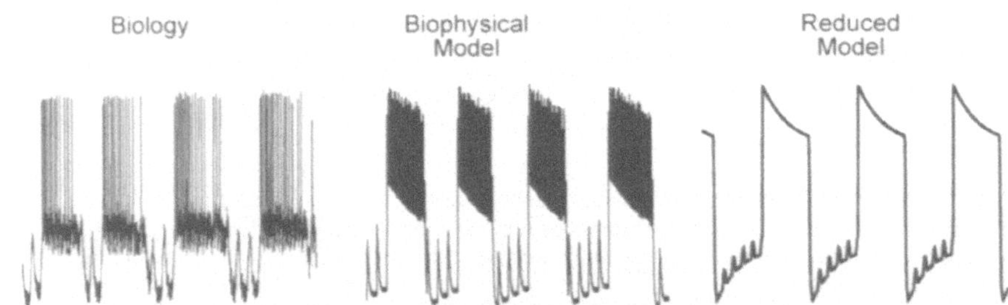

Figure 3. Time course of LG membrane potential. Left: Intracellular recording of LG in the biological preparation (from Coleman et al., 1995). Middle: LG membrane potential in the biophysical model (from Manor et al.,1997). Right: LG* membrane potential in the reduced model.

Dependence of the Model Gastric Mill Period on the Period of the Pyloric Input.

The fact that the LG transition from trough to plateau occurs exclusively during a pyloric input implies that $P_{gastric}$, the period of the gastric mill oscillation, is an integer multiple of P_{pyl}, the pyloric period. To examine the dependence of $P_{gastric}$ on P_{pyl}, we integrated Equations (1)-(2) numerically, and measured $P_{gastric}$. We repeated this procedure for various values of P_{pyl} as shown in Figure 4A. The dotted lines denote integer multiples of P_{pyl}. All data points fall on these lines ($P_{gastric} = nP_{pyl}$) and are restricted between some minimal value, P_{min} (horizontal dashed line) and $P_{min}+P_{pyl}$ (diagonal dashed line). Along the dotted lines, n is constant. Consequently, as P_{pyl} increases, $P_{gastric}$ increases. At the limit $P_{min}+P_{pyl}$, n decrements by one. At these values of P_{pyl}, $P_{gastric}$ drops sharply to P_{min}.

In the reduced model, we can derive an analytical formula for the dependence of $P_{gastric}$ on P_{pyl}. Since $P_{gastric} = nP_{pyl}$ we only need to find the dependence of n on P_{pyl}. The following term describes $n(P_{pyl})$ (a formal proof is beyond the scope of this paper):

$$n(P_{pyl}) = \text{floor}(P_{min}/P_{pyl}) + 1 \qquad (4)$$

where P_{min} is the minimal possible gastric period. P_{min} can be calculated as follows: $P_{min} = A_{min}+B_{min}$, where A_{min} and B_{min} are the minimal possible trough and plateau durations, respectively. These two intervals can be calculated exactly. A_{min} is the time it takes for w to decrease from w_{max} to w_{knee} on the left branch of the lower V-nullcline. B_{min} is the time it takes

w to increase from w_{knee} to w_{max} on the right branch of the V-nullcline. During the trough and also during the plateau, $G(V)$ is constant (the w-nullcline in Figure 3 is constant on both the left and right branches of the V-nullcline). As a consequence, during those time intervals dw/dt is a linear function of w. Therefore, during the trough phase w decays exponentially; during the plateau, it increases exponentially. Using these facts, we can calculate P_{min}:

$$P_{min} = (1/\varepsilon)\ (\log\frac{w_{knee}}{w_{max}} + \log\frac{w_{min} - w_{max}}{w_{min} - w_{knee}}) \qquad (5)$$

We now derive an analytical expression for $P_{gastric}$ as function of P_{pyl}. Figure 4B shows a plot of this analytical solution. The analytical solution matches the numerical solution exactly (compare solid lines in Fig. 4B with circles in Fig. 4A).

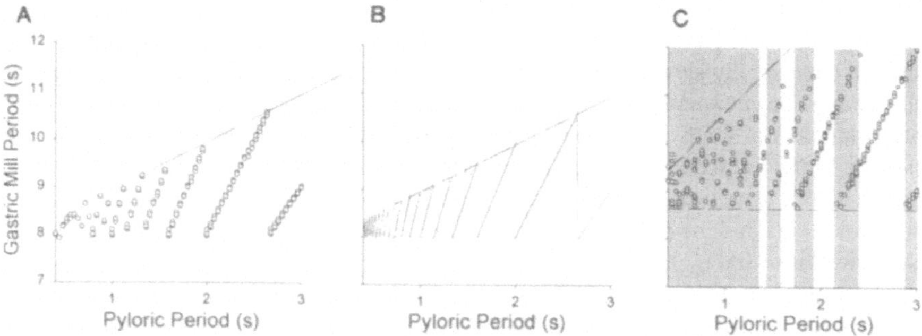

Figure 4. Dependence of the period of the slow (gastric mill) rhythm on the period of the fast (pyloric) rhythm. In **A-C**, dotted lines are integer multiples of the pyloric period (the x-axis). Horizontal and diagonal dashed lines are the curves P_{min} and $P_{min}+P_{pyl}$ (see text) **A**. Open circles are results of numerical integration of Eqs. (1)-(3). **B**. Solid lines are the analytical solution of Eqs. (4) and (5). **C**. Open circles are results of numerical integration of Eqs. (1)-(3) when high frequency negative current was added to the right hand side of Equation (2). Gray areas represent pyloric periods for which there were 2 or more distinct gastric mill periods.

Fast inhibitory input (noise) in LG is sufficient for producing variability in the gastric mill period.

In the model presented here, the period of the slow rhythm is regulated by the fast input through an N:1 time-locking mechanism. Indeed, the analytical solution developed above demonstrates that, for any value of P_{pyl}, $n(P_{pyl})$ is unique and therefore for each P_{pyl}, there is a unique value of $P_{gastric}$. This result is also supported from the numerical simulation of Eqs. (1)-(2) (Figure 4A). However, both in the biophysical model and in the biological system, the gastric mill period alternates between several integral multiples of the pyloric period (Manor et al., 1997). We postulate that this variability could be due to high frequency input to LG, such as fast IPSPs from Int1 or fast electrical EPSPs from MCN1.

We examine the effect of a high frequency current on the period of the gastric mill oscillation. We add a negative sinusoidal waveform, with a cycle period 5 times smaller than the smallest P_{pyl} used, to Equation (2a). We then integrate Eqs. (1)-(2) numerically, with various values of P_{pyl}. For each case, we measure $P_{gastric}$. Figure 4C shows a plot of the measured values of $P_{gastric}$ versus P_{pyl}. The introduction of a high frequency negative current has two major effects. First, the gastric mill period is slowed down. In particular, the lower bound

P_{min} increases from 8 s (dashed horizontal line in Figure 4A) to 8.7 s (dashed horizontal line in Figure 4C). Second, for each P_{pyl} value, $P_{gastric}$ assumes 1 or 2 (and in some cases, even 3) distinct values. To emphasize this point, at P_{pyl} values for which $P_{gastric}$ has more than one distinct value, the background is colored in gray. At P_{pyl} values where $P_{gastric}$ is unique, the background is white.

SUMMARY

We have developed a conductance-based model that describes the interaction between the slow MCN1-activated gastric mill rhythm and the fast pyloric rhythm in the stomatogastric nervous system of the crab. The conductance-based model yielded an interesting set of predictions. However, some of the results were difficult to understand because of the complexity of the model. In this paper, we reduce the conductance-based model to a phenomenological model, while retaining its essential features. The phenomenological model provides a simple framework to capture the mechanism underlying the interaction between the slow and fast rhythms. The simplicity of the phenomenological model allows us to develop analytical solutions to describe the dependence of the gastric mill period on parameters of the pyloric input, such as period (Fig. 4B) and strength (not shown in this paper). In the reduced model, the period of the slow rhythm is regulated by the fast input through an N:1 time-locking mechanism. This result contrasts with the behavior of both the biophysical model and the biological system, where more complex patterns are observed. In this work, we show that high frequency input to LG could result in patterns with both N:1 and (N+1):1 locking. We suggest that high frequency input to LG, such as provided by IPSPs from Int1 and electrical EPSPs from MCN1, are responsible for the complex pattern of bursting observed in the biophysical model and the biological system.

ACKNOWLEDGMENTS

This research was supported by NS17813, MH46742, the Sloan Center for Theoretical Neurobiology at Brandeis University, and the W.M. Keck Foundation. We wish to thank Drs. N. Kopell, M.P. Nusbaum and M. Bartos for useful comments and discussions.

REFERENCES

M.J. Coleman, P. Meyrand and M.P. Nusbaum, 1995, A switch between two modes of synaptic transmission mediated by presynaptic inhibition, *Nature* **378**:502-505.

B. Bartos and M.P. Nusbaum, 1997, Frequency regulation of a slow rhythm by periodic inputs from a fast rhythm in the crab stomatogasric ganglion, *Soc Neurosci Ab* **23**:476.

Y. Manor, F. Nadim, M.P. Nusbaum and E. Marder, Modeling the MCN1-activated gastric mill rhythm: the interaction between fast and slow oscillators, in: *Computational Neuroscience: Trends in Research, 1997*, J. Bower, ed., Plenum Pub. Corp., New-York (1997).

ENTRAINMENT OF A RECIPROCAL INHIBITION NEURAL NETWORK MODEL TO A PERIODIC PULSE TRAIN

Hirofumi Nagashino, Kazumi Achi and Yohsuke Kinouchi

Department of Electrical and Electronic Engineering,
Faculty of Engineering, The University of Tokushima,
Tokushima 770-8506 Japan

INTRODUCTION

It is important to elucidate the basic mechanisms of how the impulse trains are transformed through neuronal processing. The characteristics of firing of a neuron has been investigated by physiological experiments and theoretical modeling when a periodic impulse train activates a neuron.

It has been reported that in multifrequency coordinated rhythmic movements in humans, the frequency of voluntary rhythm is entrained by periodic external force so that the frequency ratios of the voluntary rhythm to the external force are rational numbers in low order .[1] In this behavior two input pulse trains are considered to be incorporated in the neural network level, from the intrinsic activity and external signal. To clarify its underlying mechanism it is necessary to study the integration property of the two periodic rhythm in neuronal networks.

The results of analysis of the response of oscillators or neuron models to one periodic pulse train have been reported in several cases .[2,3,4,5,6] The analysis of the neuronal network models with two periodic inputs has been open to investigation. Analysis of the response of a neuron model to two pulse trains has also been reported .[7]

It has been observed that in physiological studies of the decerebrated cat electrical continuous stimulation to mid-brain causes rhythmic activities of the spinal neural networks without periodic input .[8] In the present paper we propose a conceptual model in the neural network level that generates the synchronized output to a periodic pulse train that consists of two periodic impulse trains which are anti-phase to each other. We formulate a mathematical model of reciprocal inhibition network of neurons that have integration and firing characteristics and sustained impulse train input from the central nervous system. Numerical analysis of the entrainment characteristics have been made in the model which approximates the central pulse train input with constant input.

A RECIPROCAL INHIBITION NEURONAL NETWORK MODEL

We proposed a conceptual model in which a neuron receives two input pulse trains, one from the central nervous system and the other as external force .[7] We analyzed the integration property of the two inputs in a neuron.

Here we propose another model from a different point of view which incorporates reciprocal inhibition between extensor and flexor drivers. For the extensor and flexor of a joint one excitatory neuron each, e and f, is assumed to represent the driver. In the model the neurons receive common input that is temporarily constant. They are reciprocally coupled to each other through inhibitory interneurons. This network produces the intrinsic rhythmic activities. The excitatory neurons receive a periodic impulse train whose frequency is different from that of the intrinsic rhythm.

A number of theoretical models for a neuron that incorporate the integration and firing .[9,10,11,12] Here we intend to make as simple formulation as possible for the first step of the analysis. We formulate the neuronal network model in the following way. Let u_k denote the membrane potential of neuron k. The dynamics of u_k is expressed as

$$\tau_u \frac{du_k}{dt} = -u_k - qy_{ik} + E_c \sum_{i=0}^{\infty} \delta[t - (i - \frac{j}{2})T_c] + P \sum_{n=0}^{\infty} \delta[t - (n - \frac{j}{2})T_p], \qquad (1)$$

where $k = e$ or f and e is associated with the driver of the extensor, f with that of the flexor. In Eq.(1) τ_u is the time constant of the membrane potential, E_c is the intensity of the central input, T_c is the period of the central input, y_{ik} denotes the output of the inhibitory interneuron that inhibits the neuron k, q is a coupling constant, T_p and P are the period and amplitude of the external force respectively and $\delta[\cdot]$ is Dirac's impulse function. The value of j is set to 0 when $k = e$, while $j = 1$ when $k = f$. For simplicity of analysis, y_{ik} is assumed to be equal to y_m, where $m = f$ when $k = e$ and $m = e$ when $k = f$.

The relative refractory period is modeled by the change of the threshold v_k driven by the output of the neuron y_k as

$$\tau_v \frac{dv_k}{dt} = -v_k + ry_k, \qquad (2)$$

where τ_v is the time constant of the threshold and r ia a constant.

The output of the neuron y_k is expressed as

$$y_k = \delta[1(u_k - v_k) - 1], \qquad (3)$$

where $1(\cdot)$ is the unit step function. Eq. (3) means that an output impulse is elicited when the membrane potential exceeds the threshold.

The central input is expressed by a pulse train. However, it can be approximated with a constant value when the period is short and the intensity is small effectively. The numerical analysis was made employing the following equation for the membrane potentials instead of Eq. (1).

$$\tau_u \frac{du_k}{dt} = -u_k + E - qy_{ik} + P \sum_{n=0}^{\infty} \delta[t - (n - \frac{j}{2})T_p], \qquad (4)$$

where E denotes the intensity of the constant central input. This model is referred to as constant input model and The model that is expressed using Eq. (1) is referred to as pulse input model. .

Employing these neuronal network models, the coordinated rhythmic movement system is modeled conceptually as follows. The angle of the joint of one index finger and that of the other are denoted by θ_1 and θ_2, respectively. The angle θ_1 is forced to oscillate by external force, which generates the external input to the neurons that control the other joint. For simplicity of the model, it is assumed that this input is converted to a single impulse for every cycle of oscillation by some mechanism. The impulse train for the extensor and that for the flexor have the phase difference of half the period in the external force.

Parameters are set to the following values.

$$\tau_u = \tau_v = 1, \, E = 1, \, r = 5, \, q = 2.$$

NUMERICAL ANALYSIS

Let f_p denote the frequency of external input and f_o denote output frequency. An example of the time series of the output, membrane potential and threshold of each neuron of pulse input model are illustrated with the external input p_e and p_f in Figure 1. In the example the model is synchronized with the external input pulse train in $f_p : f_o = 2 : 1$.

The characteristics of the constant input model have been analyzed numerically. First the timing relationship of a periodic solution with n/m synchronization in low order is determined from the conditions that the membrane potential is equal to the threshold at the instants when the neuron fires. And then from the conditions that at some critical instants in which the neuron does not fire, it is required that the membrane potential is lower than the threshold, the range of the frequency of external force is obtained in which the periodic solution can exist.

The example of the calculation is shown below for $f_p : f_o = 2 : 1$. Figure 2 illustrates the timing relationship of the membrane potentials and threshold. In the periodic solution the membrane potential of neuron e, $u_e(t)$, satisfies the following equations.

$$u_e(T) = u_e(0)y + E(1 - y) \tag{5}$$
$$u_e(T_p) = u_e(T) - q)x^2/y + E(1 - x^2y) + P \tag{6}$$
$$u_e(2T_p) = u_e(T_p)x^2 + E(1 - x^2) \tag{7}$$
$$u_e(0) = u_e(2T_p) + P \tag{8}$$

where x and y are expressed respectively as

$$x = e^{-T_p/2} \tag{9}$$
$$y = e^{T/2} \tag{10}$$

The threshold of neuron e, $v_e(t)$, satisfies the following equations.

$$v_e(T) = (v_e(0) + r)y \tag{11}$$
$$v_e(T_p) = v_e(T)x^2/y \tag{12}$$
$$v_e(2T_p) = v_e(T_p)x^2 \tag{13}$$
$$v_e(0) = v_e(2T_p) \tag{14}$$

For the neuron f similar equations for the periodic solution with respect to the membrane potential $u_f(t)$ and threshold $v_f(t)$ can be obtained.

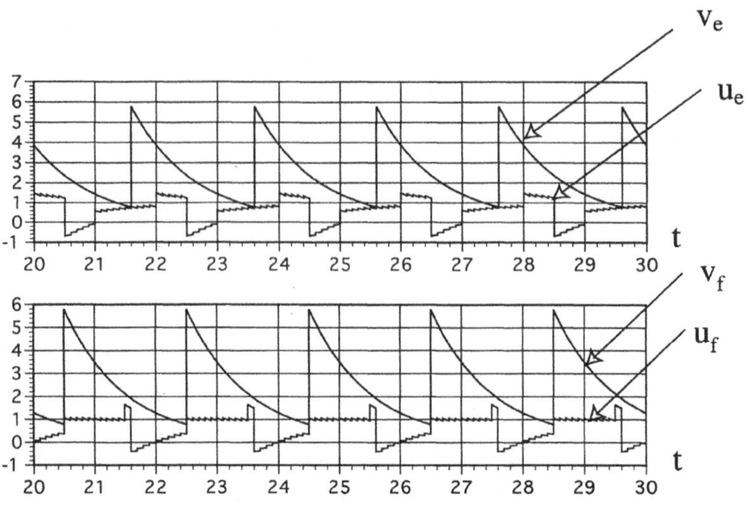

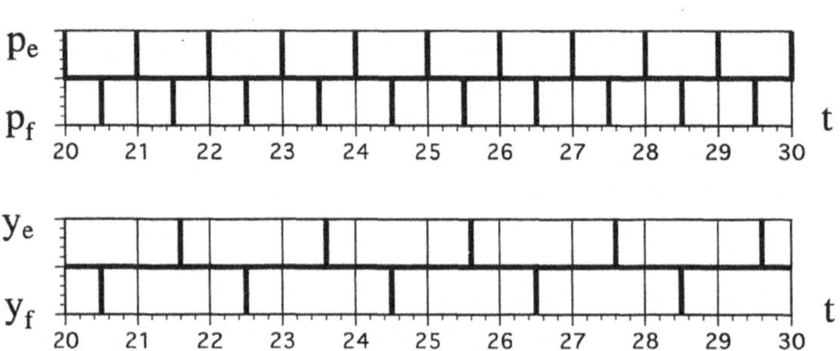

Figure 1. An example of simulation results of pulse input model, $E_c = 0.1$, $T_c = 0.1$, $f_p : f_o = 2 : 1$.

For firing at the instants shown in Figure 2, the following inequalities should hold.

$$u_e(0) > v_e(0) \tag{15}$$

$$u_e(T) < v_e(T) \tag{16}$$

$$u_e(T_p) < v_e(T_p) \tag{17}$$

$$u_e(2T_p) < v_e(2T_p) \tag{18}$$

$$u_f(\frac{3T_p}{2} - T) < v_f(\frac{3T_p}{2} - T) \tag{19}$$

$$u_f(2T_p - T) < v_f(2T_p - T) \tag{20}$$

$$u_f(\frac{5T_p}{2} - T) < v_f(\frac{5T_p}{2} - T) \tag{21}$$

The result of the numerical analysis is shown in Figure 3. When P is too small, the output of the neurons is not affected by the external force. When the intensity of the impulses of external force P is too large, the neurons are entrained completely to the external force, so that the ratio of the external input frequency and the output frequency of the neurons e and f is 1:1. For the appropriate range of P, the range of the

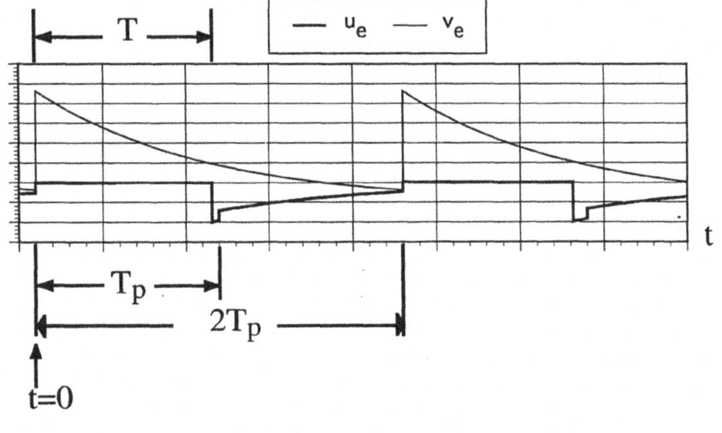

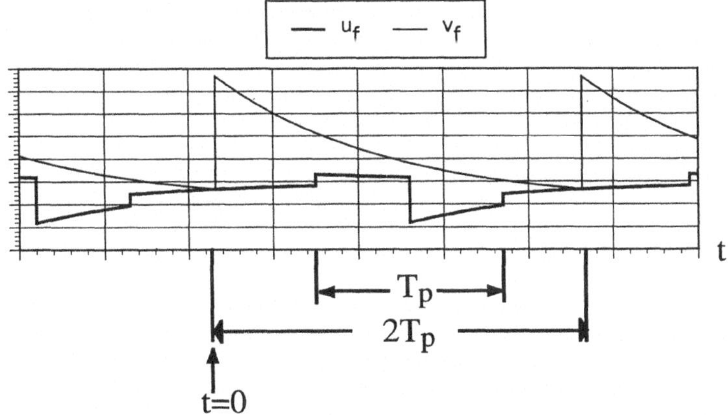

Figure 2. Determination of periodic solution for $f_p : f_o = 2 : 1$.

frequency of the external force in the periodic solutions appears in which the ratios of output frequency of the neuron to the external frequency are rational numbers in low order such as 2:1, 3:1, 3:2 and so on.

Such relationship is accomplished with more complicated phenomina caused by the mixed influence of the intrinsic output and the external input pulse trains than the case for a single pulse train. There are synchronization only to the external force, only to the central input and complex entrainment to the external force and the central input.

CONCLUSIONS

We have proposed a mathematical model for the neuronal network that generates $n : m$ synchronization to a pulse train which is applied to the network as a pair of anti-phase pulse trains. The regions in which the ratio of the frequency of the output to that of the external force are rational numbers in low order have been obtained by numerical analysis in a reciprocal inhibition neuronal network model.

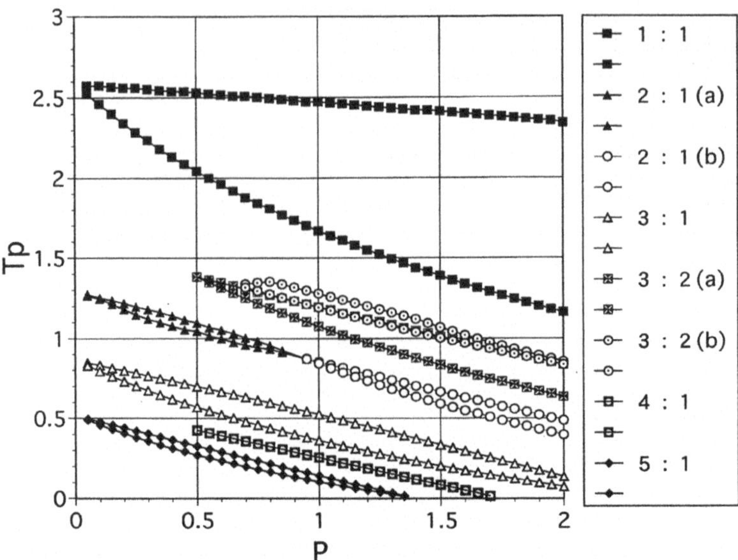

Figure 3. Region of $n : m$ synchronaiztion in $P - T_p$ plane.

The model needs to be improved in such a way that a lower order region is wider. Modeling the change to the simpler ratio output is open to future work.

REFERENCES

1. G. C. deGuzman and J. A. S. Kelso. Multifrequency behavioral patterns and the phase attractive circle map. *Biological Cybernetics*, 64:485–495, 1991.

2. J. Nagumo and S. Sato. On a response characteristic of a mathematical neuron model. *Kybernetik*, 10:155–164, 1972.

3. G. B. Ermentrout. n : m phase locking of weakly coupled oscillation. *J. Math. Biol.*, 12:327–342, 1981.

4. T. Kiemel and P. Holmes. A model for the periodic synaptic inhibition of a neuronal oscillator. *IMA J. Math. Appl. Med. Biol.*, 4:145–169, 1987.

5. T. Nomura, S. Sato, and S. Doi. Global bifurcation structure of a bonhoeffer van der pol oscillator driven by periodic pulse trains. *Biol. Cybern.*, 72:55–67, 1994.

6. S. Doi and S. Sato. The global bifurcation structure of the bvp neuronal model driven by periodic pulse trains. *Math. Biosci.*, 125:229–250, 1995.

7. H. Nagashino and Y. Kinouchi. Entrainment of a spiking neuron model to two periodic pulse trains. *Computational Neuroscience: Trends in Research, 1997*, pages 147–152, 1997.

8. M. L. Shik, F. V. Severin, and G. H. Orlovsky. Control of walking and running by means of electrical stimulation of the mid-brain. *Biofizyka*, 11:659–666, 1966.

9. R. F. Reiss. A theory and simulation of rhythmic behavior due to reciprocal inhibition in small nerve nets. *Proc. AFIPS Spring Joint Computer Conference*, 171-194, 1962.

10. J. P. Segundo, D. H. Perkel, H. Wyman, H. Hegstad, and G. P. Moore. Input-output relations in computer simulated nerve cells. *Kybernetik*, 4:157–171, 1968.

11. M. Tsukada and R. Sato. Statistical analysis and functional interpretation of input-output relation in computer-simulated nerve cells (in japanese). *Japanese J. of Med. Electr. and Biol. Engnr.*, 10:370–378, 1997.

12. H. Nagashino, H. Tamura, and T. Ushita. Firing modes in reciprocal inhibition neural networks and their analysis (in japanese). *Trans. of IECE of Japan*, 61-A:588–595, 1978.

EXTRACELLULAR RECORDING FROM MULTIPLE NEIGHBORING CELLS: RESPONSE PROPERTIES IN PARIETAL CORTEX

John S. Pezaris[1], Maneesh Sahani[1,2], and Richard Andersen[1,2]

[1]Computation and Neural Systems
[2]Sloan Center for Theoretical Neurobiology
California Institute of Technology
Mail Code 216-76,
Pasadena, CA, 91125

ABSTRACT

Multiple single unit extracellular recordings were made using tetrodes in macaque posterior parietal cortex while the animal was performing a visual memory saccade task. Recordings were made over a 2×2 mm area at both superficial and deep locations in one hemisphere. Signals were analyzed using an Expectation-Maximization algorithm for spike separation based on spike peak height or the first two principle components of spike shape. 27 sites were selected for analysis based on task response and clarity of separation, yielding 85 total neurons with a mode of 3 cells per site. The response criteria and stereotaxic location used were consistent with identifying neurons within the lateral intraparietal area (LIP).

For cells within the set of selected sites, responses to the task were further categorized based upon spatial characteristics (preferred direction) and temporal characteristics (time of maximal response). Neighboring cells were found to be very likely to have similar tuned direction (75% within one octant), but not comparatively likely to have similar temporal characteristics. We take this as evidence that area LIP is heterogeneous at the local level.

INTRODUCTION

Neighboring cortical cells are known to have similar inputs, often synapse locally onto each other, and yet do not necessarily compute identical outputs. By examining simultaneously recorded single-unit activity from neighboring cells, we are able to observe their response to stimuli, while also examining the level of intercell interaction, and therefore can explore the circuitry present at the local scale. Specifically, we set out to measure the spatial and temporal tuning differences between neighboring cells in area LIP.

EXPERIMENTAL METHODS

We trained a rhesus monkey (*Maccaca Mulatta*) to perform the memory saccade task, monitoring his eye position using the scleral coil technique. The task, performed in a darkened room, requires the animal to fixate a centrally-presented light while a peripheral target is flashed in one of eight equally-spaced positions on a 15° radius circle about the fixation point. The animal must remember the location of the target as long as the fixation light is illuminated, delaying the instructed saccade until the fixation light is extinguished. If the saccade is performed accurately, the target is briefly reilluminated, and the animal rewarded with a drop of juice. LIP neural response to this task is characterized by three phases: a sensory response to the target flash, an elevated baseline during the delay period, and a perisaccadic burst during the cued eye motion, often combined with a second sensory response during target reillumination.

To collect neural responses, tetrodes[1] were inserted daily into cortical tissue near the intraparietal sulcus through a chronically maintained craniotomy over the lateral intraparietal sulcus (stereotaxic coordinates 6 mm anterior, 12 mm lateral), using techniques previously described[2,3]. The four tetrode voltages were filtered, digitized, and streamed to digital media while the animal performed the task. These recordings were then analyzed off-line using statistical techniques described by Sahani, *et al*[4,5], and the resulting spike trains pairwise examined for spatial tuning and temporal response profile as described in the following sections. Neurons were considered *neighbors* if they were recorded simultaneously at the same site.

Penetrations were made between 3000 and 10000 μm down from the putative cortical surface, to arbitrary depths (usually every 500 μm along a single penetration). At the sites where an on-line isolation could be made using any single channel of the four tetrode signals with traditional (single-channel) equipment, the animal was run on a block of 100 trials, and the tetrode voltages recorded for later analysis. Sites reported below were further selected for clarity of signal, multiplicity of cells, and either brisk response to the task during the memory phase, or proximity to a previously recorded site with a strong response.

RESULTS

Twenty-seven recordings were selected, with a range of 2 to 6 cells found per recording. Nine (9) sites, or 33%, carried evidence of three cells, the most commonly found number of cells per site. The median was also 3 cells per site. A histogram of the number of sites with a given cell count is shown in figure 1.

Example Spatial Profile

A typical example of spatial tuning over target location is given in figure 2. For each cell, the response is taken as the average number of spikes for the experimentally relevant period of all successful trials for each target direction, and plotted on a polar graph. Consistent with previous reports[6], we find many of the cells recorded express target selectivity within the task response.

Example Temporal Profile

A more detailed example of task response at a single site where three neurons were identifiable is shown in figure 3. The graph depicts the peristimulus time histogram (PSTH) for the target location which evoked the strongest response for this site. The

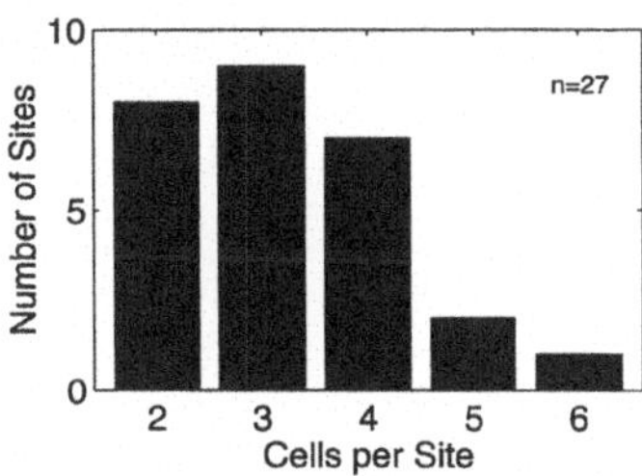

Figure 1. Tetrode recordings in posterior parietal cortex currently yield an average of three cells per tip location in our lab, when selected for task response, multiplicity of cells, and ease of signal separation.

two largest profiles (solid and dot-dashed lines) show typical LIP responses for the memory saccade task: a brief burst of activity in response to the target flash followed by a period of sustained memory activity, and a lesser perisaccadic burst. The third cell (dashed line) has a slower initial response, perhaps even lacking a distinct visual burst, followed by a decaying memory response and no perisaccadic burst.

Spatial Tuning Differences

As introduced above, we extracted the preferred direction for each cell, based on the maximum response elicited for each target location, and the time of maximum response, computed over all targets. Then, comparisons were made between these values for neighboring cells (those which were simultaneously detected in a single recording). The scattergram of preferred direction among pairs of neighboring cells is shown in figure 4. The data cluster around the unit-slope line, suggesting a tendency for neighboring cells to have similar preferred direction. Considering previous single-unit findings on the patchwork nature of LIP[7], outliers on this figure may represent recordings made when the tetrode was situated at the transition between two groups of neurons with distinct clustered preferred direction.

To quantify the similarity of tuning direction, we compute the histogram of differences between directions for neighboring cells, as shown in figure 5. The preferred direction was computed for each cell by considering the response for each target location as a vector in that direction and calculating the vector average of the eight responses. The large majority of pairs of neighboring neurons (75%) have preferred

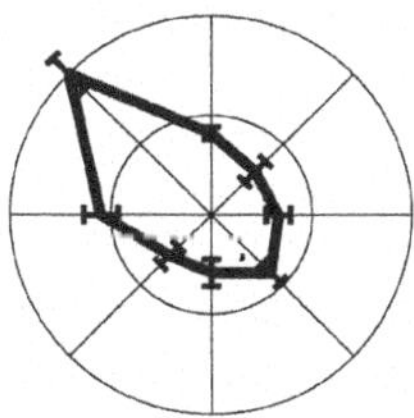

Figure 2. Example polar tuning curve. The task response is computed for a given cell for each target location and plotted in polar form where the radial excursion of the curve represents the magnitude of the response for the corresponding target direction, and error bars denote the standard error. The computed preferred direction is marked with an asterisk.

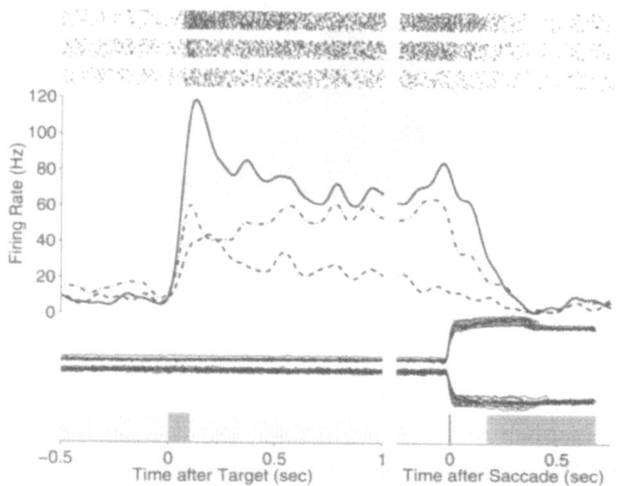

Figure 3. Example neural response to the memory saccade task. The left half of the figure is for a portion of the experimental task aligned to target presentation; the right half is aligned to the saccade. There is a small but varying amount of overlap between the two halves. At the top of the diagram are three rastergrams, separated out by cell; the corresponding rows from each group of rasters are from the same trial. In the middle are the peristimulus time histograms (PSTHs) for the same three cells, binned to 10 ms and smoothed to three bins. The solid line corresponds to the uppermost rastergram, the dash-dotted line to the middle rastergram, and the dashed line to the lower rastergram. Below the PSTHs are eye position traces, vertical position above horizontal position, with the multiple trials overlaid. At the bottom are the behavioral events. The stippled band represents the fixation point illumination, the solid band the represents the target presentation (although not apparent from this diagram, the fixation point continues to be illuminated while the target is flashed), and the sharp vertical line at $t = 0$ on the right half of the diagram represents the time of saccade. (The behavioral bands actually consist of a set of horizontal lines, one for each trial).

directions within $\pi/4$ radians, corresponding to the central cluster of points around the unit-slope line in figure 4.

Temporal Tuning Differences

The response profile of each cell is computed by measuring the firing rate to the preferred target during each of five arbitrary but experimentally relevant epochs: *background* (250 ms before target flash), *visual* (0–250 ms after target flash), *early memory* (250–500 ms post flash), *late memory* (500–1000 ms), and *perisaccadic* (-250–250 ms about the saccade). Plotting the response versus direction for each epoch, we find that neighboring cells vary considerably in the epoch of maximal response. In the example pair shown in figure 6, the upper cell fires most vigorously in the perisaccadic epoch while the lower cell fires most vigorously in the visual epoch, despite their preferred directions (during the epochs of highest response) matching quite well.

We compared the distance in time between epochs of maximal response for neighboring cells and plot the histogram, as shown in figure 7. The temporal response for each cell was computed in five epochs as described above, and the enumerated value of the epoch with maximal response used as the index of temporal tuning. Thus, a cell

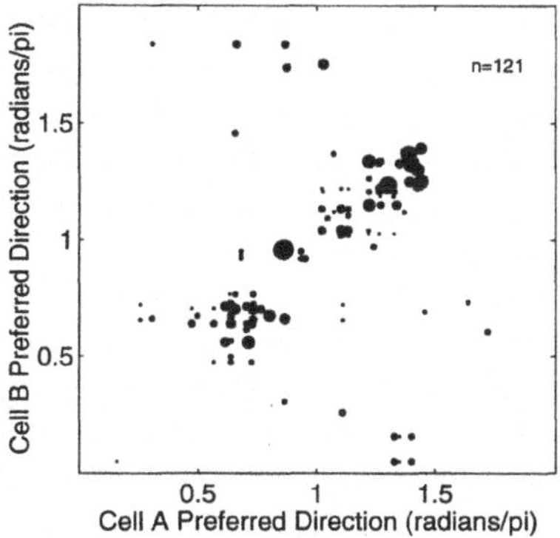

Figure 4. Scattergram of preferred directions for neighboring pairs of cells (A, B). The preferred direction was computed for each cell by taking the vector average of the response for each target direction, and pairwise plotted for neighboring cells. The diameter of each point is proportional to the square root of the product of the magnitude of the maximum responses in each neighboring pair.

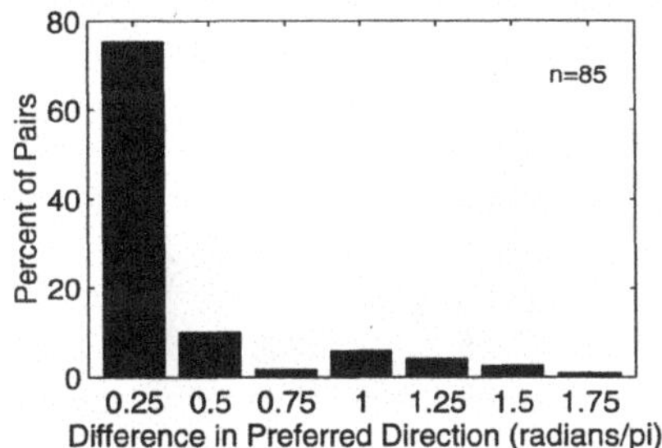

Figure 5. Histogram of differences in preferred direction. The differences in neighboring preferred direction has been binned to octants. The large majority of pairs of cells have similar preferred direction, as indicated by the large leftmost bin.

with strongest response in the visual period would have a temporal tuning of 2, one in the late memory period would have a tuning of 4. The temporal distance between tunings then was computed as the difference in tuning index. While not a rigorous metric, it does serve to show the wide distribution of temporal response peaks between neighboring cells, suggesting the cells have different computational tasks. Fully determining the exact nature of the differences between response types will require significant additional work. The metric used here intended merely to assist a gross determination of similarity among response types: as only 40% of the pairs peak in the same epoch, we conclude that the response characteristics have a broad tendency to be different among neighboring cells.

SUMMARY

We have presented a count of cells recovered per tetrode recording made at arbitrary penetration depths in parietal cortex. Our findings suggest that the tetrode

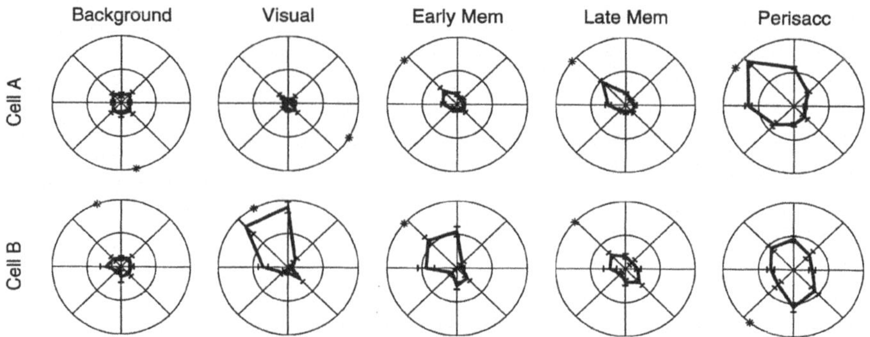

Figure 6. Neighboring cells tend to have less similar response dynamics. Here are normalized tuning curves versus target location for each of the five arbitrarily defined experimental epochs (see main text) for two simultaneously recorded cells. The vertical direction on the polar plots corresponds to the upper target location; the value along each radial arm indicates the neural response to the corresponding target location. Asterisks indicate the computed preferred direction for each tuning curve.

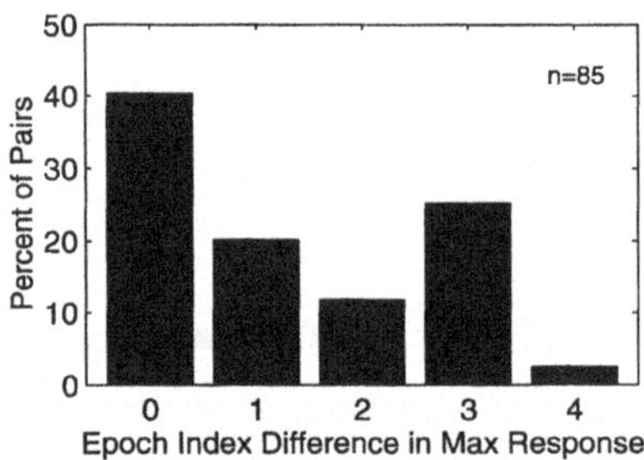

Figure 7. Using the arbitrary segments defined in the text, we determined the epoch of maximal response for each cell, and then measured the difference in epoch index for neighboring cells. The wide distribution of index differences indicates the responses for neighboring cells are not typically identical.

recording technology makes it easy to detect multiple cells per site in monkey cortex. We further examined the spatial and temporal tuning for simultaneously recorded cells and found that such neighboring neurons tended to have similarity in spatial tuning, but dissimilarity in temporal response. We take these results to be evidence for locally heterogeneous circuitry within parietal cortex, specifically area LIP, that carries locally homogeneous spatial selectivity.

ACKNOWLEDGMENTS

This work was generously supported by the Keck Foundation, the Sloan Center for Theoretical Neuroscience at Caltech, the Center for Neuromorphic Systems Engineering at Caltech, the Office of Naval Research, and the National Institutes of Health. The experiments would not have been possible without the assistance of B. Gillikin.

REFERENCES

1. M.L. Recce, and J. O'Keefe, The tetrode: an improved technique for multi-unit extracellular recording, *Society for Neuroscience Abstracts*, 15(2):1250 (1989).
2. J.S. Pezaris, M. Sahani, and R.A. Andersen, Tetrodes for monkeys, *Computational Neuroscience: Trends in Research 1997*, J.M. Bower, ed., Plenum Press, New York (1997).
3. J.S. Pezaris, M. Sahani, and R.A. Andersen, Multiple single unit recording using tetrodes in macaque visual cortex: electrode design and spike identification, *Society for Neuroscience Abstracts*, 21(2):905 (1995).
4. M. Sahani, J.S. Pezaris, and R.A. Andersen, Extracellular recording from multiple neighboring cells in primate cortex, *Neural Networks: from Biology to Hardware Implementations*, Chia, Italy, September 23–27, 1996.
5. M. Sahani, J.S. Pezaris, and R.A. Andersen, Extracellular recording from multiple neighboring cells: a maximum-likelihood solution to the spike-separation problem, *Computational Neuroscience: Trends in Research 1998*, J.M. Bower, ed., Plenum Press, New York (1998).
6. S. Barash, R.M. Bracewell, L. Fogassi, J.W. Gnadt, and R.A. Andersen, Saccade-Related Activity in the Lateral Intraparietal Area II. Spatial Properties, *J. Neurophysiology*, 66(3), (1991).
7. G.J. Blatt, R.A. Andersen, and G.R. Stoner, Visual receptive field organization and cortico-cortical connections of the lateral intraparietal area (area LIP) in the macaque, *J. Comparative Neurology*, 299:421-445 (1990).

ANALYSIS OF TETRODE RECORDINGS IN CAT VISUAL SYSTEM

Sergei Rebrik[1], Svilen Tzonev[1], and Ken Miller[1,2]

[1]Keck Center for Integrative Neuroscience and Department of Physiology,
[2]Department of Otolaryngology,
University of California at San Francisco, CA 94143
rebrik@phy.ucsf.edu, svilen@phy.ucsf.edu, ken@phy.ucsf.edu
http://mccoy.ucsf.edu/

INTRODUCTION

From simultaneous recording of closely located neurons one can obtain information about local inter-neuronal connectivity and acquire fine-scale neuronal response mappings. A multitrode (by which we mean a tight bundle of traditional extracellular electrodes) (McNaughton et al., 1983) is a natural choice for this kind of recording, because it allows discrimination of multiple neurons at a single recording site. Separation of spikes from different neurons is based on the assumption that spike amplitude decreases with the distance between the neuron and the tip of the electrode. Thus, the ratio of spike amplitudes measured on different wires provides reliable information for spike discrimination. The tetrode, which is made of four very thin wires (typical wire diameter with the insulation is 20 μm) is the minimal set of electrodes that provides "triangulation" in the three-dimensional space. If tetrode tips are not located in a plane, and under the idealization that neurons are point sources in a homogenous medium, the ratios of spike amplitudes will be unique for each neuron. In contrast to the tetrode, a single electrode is "spatially blind", and its use for multi-neuron recordings can easily lead to mixing of signals from different cells whose spike amplitudes as measured at the electrode site are equal.

The multi-unit nature of tetrode recordings imposes additional requirements on spike-sorting procedures. While in traditional extracellular recordings an experimentalist can position the electrode to enhance the signal from the neuron of interest, tetrodes are best positioned to record from as many cells as possible. This means that spike sorting can no longer be based on simple thresholding or on window discrimination (often implemented in hardware) - sorting of tetrode data is essentially software-based. As the tetrode technique is relatively new, and the number of labs using it is still small, the choice of software for spike-sorting is very limited. Existing packages are often system- and hardware-dependent, which makes it hard to port them to different lab environments.

In this paper we describe our ongoing development of spike sorting software for tetrode recordings. We discuss a natural spike-parameter representation that improves spike clustering, because noise common to all four electrodes is suppressed. We conclude by discussing future development of spike-sorting algorithms.

SPIKER: SOFTWARE FOR OFFLINE SORTING OF TETRODE DATA

The main goal behind the design of this software was to provide a user with a simple and universal program that could be used "out of the box". Program requirements included an intuitive graphical user interface, openness to modifications by the end user, and cross-platform portability.

Spiker is written in C using the commercially available "Galaxy C" environment. "Galaxy C" is essentially a C library that allows one to write system-independent programs that have a "native" graphical user interface.

The input file format of Spiker is a continuous binary stream of signed 16-bit ADC samples (any byte order). The total number of recording channels in the input file can be in the 1 to 16 range. Data samples are assumed to be in the following format: for every ADC time tick, samples for all channels are recorded in the channel order, and this structure is repeated for the duration of the recording. No information on number of channels, ADC sampling rate or timing is stored in the file. This allows use of Spiker with any data acquisition system that can continuously stream data to the disk (i.e. raw data format). An alternative to this format is to record only a preset number of points surrounding a spike peak, thus reducing the size of the data file. We chose not to use this approach because it severely limits the ability to reconstruct spike overlaps, an issue we intend to address in future work (see also Lewicki, 1994; Sahani et al., 1997). In addition, such extraction requires setting of thresholds at recording time. As we discuss later, threshold settings are very important for proper spike clustering, and it is easy to introduce irrecoverable losses of data if thresholds are set before data acquisition. Finally, in conditions in which many threshold-crossing events occur (many neurons and/or reasonable sustained firing rates), data compression obtained by extracting 1.5 msec spikes may not be substantial.

The spike sorting procedure implemented in Spiker is outlined below. First, spikes are detected in traces by a thresholding procedure. By default the threshold is set at 4 standard deviations from the channel mean, measured using data points taken from 10 random chunks. If needed, the threshold can be interactively adjusted by the user. After the spike detection, the amplitudes of spikes at all channels are measured. Peaks at different channels do not always precisely coincide in time, so the search for an extremum is performed on each channel individually. To reduce digitization error, data points are 10 times oversampled around the detected peak using Fourier interpolation over 80 neighboring points, and the amplitudes are corrected accordingly. This step is especially important for low ADC rates (20 kHz and below).

After the previous step, each spike (event) has a set of associated amplitudes; for a tetrode, there are four such amplitudes. Then each event can be treated as a point in this 4-dimensional space. The task of a clustering procedure is to determine the number of clusters and to assign events to clusters. Currently this task is performed manually. To visualize the distribution of points in four-dimensional space, we, like many others, project this space onto 6 two-dimensional planes (the number 6 is given by the number of different pairs of amplitudes). In other words, for the plane defined by channels X and Y, for each event a point is plotted at coordinates A_x and A_y, the amplitudes on the two channels (the remaining two amplitudes are disregarded for this plane). Thus, each event generates 6 points, one point per plane. Figure 1a shows a snapshot of a Spiker window used for clustering. The user can draw an ellipse around the points and mark them as belonging to a particular cluster, and this marking will be seen simultaneously in all displayed planes.

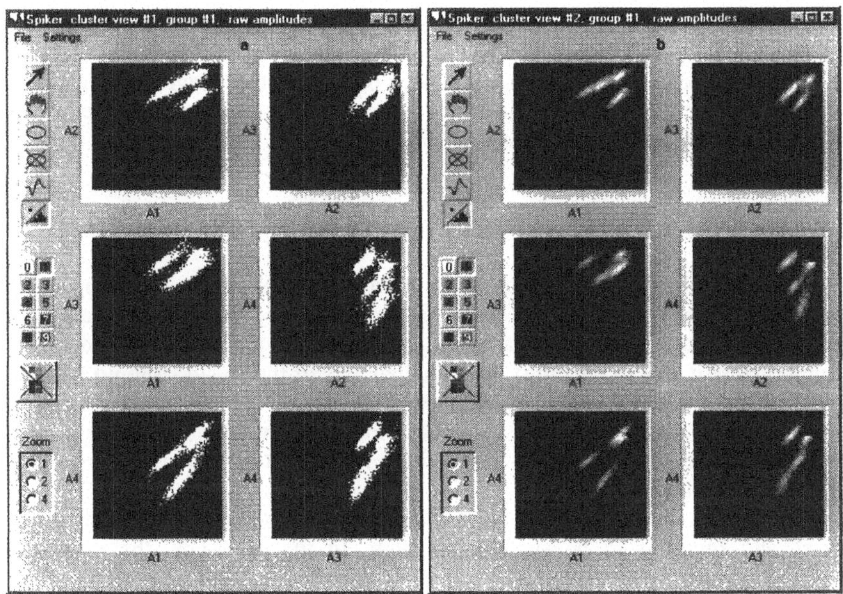

Figure 1. Clustering in raw amplitude coordinates. The point with (0,0) coordinates is located in the upper right corner of the quadrant. Spike amplitudes are negative and the absolute value of the spike amplitude grows in the down-left direction. Panel **a** shows events in the "point-per-event" representation, while panel **b** presents density of events coded by brightness. Note better cluster separation in panel **b**.

The total number of events in the recording affects perceived cluster size. As the total number of events grows, clusters increase in size because the number of outliers also grows. This can lead to overlap of clusters and make classification difficult or impossible. To avoid this kind of problem, a density distribution of events within the plane can be calculated and displayed. The portion of the plane containing spikes is divided by a rectangular grid into equal cells (in Spiker a grid of 128 by 128 is used) and the number of events in every cell is calculated. Each cell is displayed as a pixel, with brightness increasing with the number of events in the cell. To extend the dynamic range of displayed densities, a slower than linear dependency of brightness on the density is used. This prevents masking of low-density clusters by those with high-density. Clustering in the density representation is illustrated in Figure 1b.

In many multi-cell studies, neuronal interactions revealed by correlated firing patterns are of primary interest. This leads to an additional requirement on the "purity" of clustering. The thresholding procedure being used for spike detection can miss certain spikes in the clusters that lie in the low-amplitude zone. One can detect a bright-dark rectangular boundary in the upper-right corner of the quadrants in Figure 1, which is formed by this phenomenon. In other words, clusters crossing this boundary (representing the value of the threshold) can be missing a substantial number of spikes. Measures of reliability and synchronization based on such an incomplete cluster will be misleading, even though the incomplete cluster may be "clean" in the sense that it contains spikes from one cell only. Furthermore, if the missing part of the cluster (smaller-amplitude spikes) represents a non-random sample of the cell's output, cross-correlations and other measures based on these spikes will also be tainted.

If a relatively small piece (10-20%) of the cluster is cut off, one can adjust the threshold to recover the missing points. However, it is generally impossible to recover clusters that are missing a larger number of spikes, since the low-amplitude part of the cluster is usually inseparable from the noise. Such partial clusters should be discarded. The ability to recover some clusters by adjusting threshold illustrates one advantage of keeping the full voltage traces, rather than extracting spikes online

CROSS-CHANNEL CORRELATION IN THE TETRODE RECORDINGS.

The clusters in Figure 1, which come from cat visual cortex, are elongated. This indicates that amplitudes on different channels tend to co-vary. Two different mechanisms can lead to this. If the observed covariance was caused by the intrinsic variation of the spike amplitude (e.g. due to bursting), the principal axis of the clusters should point to the origin of the coordinate system. Another possible source of the covariance is noise correlated between channels. In this case the principal axes would all have a common slope, i.e. be parallel to one another. Inspection of the clusters in Fig. 1 shows that the second source is probably dominant.

To test this hypothesis we have calculated cross-channel correlation in our recordings and found it to be surprisingly high. In the cat LGN, the cross-channel coefficient of correlation (averaged over pairs of wires) was around 0.94, while in the cat visual cortex it was about 0.78. These values were calculated as an average across five random chunks of data (each 10,000 samples long) taken from six different files of each type. To exclude possible influence of spikes on these calculations, all points exceeding the level of four standard deviations were removed from the dataset; however, this makes very little difference.

The common noise can be suppressed by the use of channels in a differential mode, in which the signal at one channel is subtracted from the signal at another one. We, however, use a slightly different approach. The Hadamard transformation of the measured amplitudes allows us to reveal the differential part of the signal and to suppress the common source. The Hadamard transformation is given by the equation:

$$\begin{pmatrix} h_1 \\ h_2 \\ h_3 \\ h_4 \end{pmatrix} = \begin{bmatrix} 1 & -1 & -1 & 1 \\ -1 & 1 & -1 & 1 \\ -1 & -1 & 1 & 1 \\ 1 & 1 & 1 & 1 \end{bmatrix} \bullet \begin{pmatrix} a_1 \\ a_2 \\ a_3 \\ a_4 \end{pmatrix},$$

where a_1-a_4 are the original spike amplitudes which are used to calculate a new vector of amplitudes shown on the left side of the equation. As can be seen from the matrix, the first three new components represent differences in the amplitudes, while the fourth is just the sum of all four amplitudes. The Hadamard transform is linear and can be viewed as a scaling and rotation in the four-dimensional space. Common noise is suppressed in coordinates h_1 through h_3, but not in h_4, the sum of all four amplitudes. The projections in the $h_1/h_2/h_3$ subspace produce less elongated clusters, which are in most cases better separated than clusters in the raw amplitude coordinates. This is illustrated by Figure 2. Nevertheless, in certain cases raw amplitudes still provide better separation.

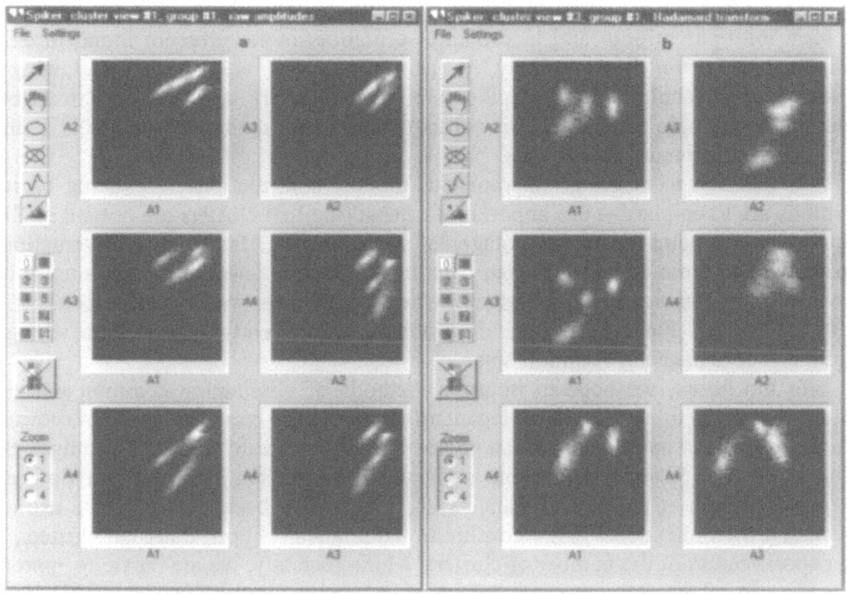

Figure 2. Comparison of clusters in the raw amplitude representation (**a**) and in the Hadamard representation (**b**). Note that in b clusters are less elongated and better separated, particularly in the three projections not involving h_4.

Geometrically, the Hadamard transform is a local linear approximation (about the point $(1,1,1,1)$) to a transformation into polar coordinates. That is, axis h_4 is the radial $(1,1,1,1)$ coordinate, and the other three axes locally approximate the remaining 3 angular coordinates. It may seem tempting to ignore the h_4 coordinate and work in the remaining 3D space, on the assumption that cells from different spatial locations will have different angular coordinates (Zhang et al., 1997). However, it is not infrequently the case that the h_4 coordinate is critical to successful sorting. Geometrically, two clusters may have centers with angular coordinates sufficiently close that, given the finite size of clusters, they will be hopelessly entangled without the absolute amplitude information in h_4. Near-threshold noise, which is distributed across all angular directions, also presents a problem. (Note, the Hadamard transformation handles these problems better than truly working in angular coordinates, because amplitude information is retained in the difference coordinates, but not in angular coordinates). In addition, it is always vital to keep the full amplitude information in order to check for clusters that have been cut off by the threshold, as discussed above; this checking must be done in the raw amplitude representation. Spiker allows simultaneous clustering in both representations.

The observed common source of noise is presumably due to the activity of distant neurons whose spikes are too small at the tetrode to be discriminated. One can probably significantly improve clustering if the common source suppression is performed before the spike detection. This will reduce the number of missing spikes in the low-amplitude clusters.

CONCLUSIONS

Spiker is available in the public domain (http://mccoy.ucsf.edu). It allows sorting of data from multiple tetrodes and incorporates a number of features not found in existing commercial software. Most notably, it uses continuous voltage data, allowing user adjustment of threshold and other spike-defining parameters; and it allows simultaneous clustering in multiple projections, in particular projections onto axis pairs in both raw amplitude and Hadamard coordinates.

In addition to creating software, our modest contributions to existing clustering algorithms are to emphasize the importance of ensuring that clusters are not cut off by the threshold, and to point out the advantages of sorting in the Hadamard representation. We have shown that these advantages stem from the large component of common noise on the four wires in tetrode recording, at least in the neural structures we have examined (cat LGN and visual cortex). Future algorithm development may greatly benefit from taking into account this dominance of common noise.

In the future, we hope to improve methods of eliminating common noise. One obvious means is to replace the Hadamard matrix with one that uses the covariance between channels to optimize common noise suppression; hardware solutions might also be explored. We also plan to implement automatic spike-sorting algorithms in Spiker. We, like others (Sahani et al., 1997; Zhang et al., 1997) have been exploring the use of the Expectation-Maximization (EM) algorithm to automatically form Gaussian clusters, given a user specification of the number of clusters. More generally, we are pursuing approaches that make use of the full waveform information, analogous to Lewicki (1994) (see also Sahani et al., 1997) but (we hope) with a number of improvements as well as adaptation to the tetrode situation. As these algorithms develop to the point of reliability, we will implement them in Spiker.

References

Lewicki. M.S. (1994). Bayesian modeling and classification of neural signals.*Neural Comput.* 6:1005-1030.

McNaughton, B.L., J. O'Keefe, and C.A. Barnes, (1983). The stereotrode: A new technique for simultaneous isolation of several single units in the central nervous system from multiple unit records. *J. Neurosci. Methods*, 8:391-397.

Sahani, M., J. S. Pezaris, and R. A. Andersen, (1997).Extracellular recording from multiple neighboring cells: A maximum-likelihood solution to the spike-separation problem. I*n Bower, J.M. (Ed.) Computational Neuroscience: Trends in Research 1996. Plenum Press. New York.*

Zhang, K., T.J. Sejnowski and B.L. McNaughton (1997). Automatic separation of spike parameter clusters from tetrode recordings. *Soc. Neuro. Abstr.* 23:504.

CORRELATION CODING IN STOCHASTIC NEURAL NETWORKS

Raphael Ritz and Terrence J. Sejnowski

Computational Neurobiology Laboratory
The Salk Institute for Biological Studies
10010 North Torrey Pines Road, La Jolla, CA 92037, USA

INTRODUCTION

It is commonly believed that the neural code used by nerve cells to transmit information in the cerebral cortex is the mean firing rate of action potentials. Whereas there is solid evidence for this coding scheme at the neuromuscular junction, where this concept originated, the temporal averaging involved in the decoding process causes problems at the cortical level, where neurons usually fire at rates too low to allow for a sufficiently long decoding time. As a possible solution to this problem it has been proposed that cells could also perform a spatial average instead of, or in addition to, temporal averaging. But this form of population code also assumes that information is coded in a firing *rate*—whether spatial or temporal—and that a neuron simply reflects changes in its input firing rates by modulating its output firing rate. This is the underlying assumption allowing the common reduction to a transfer function used by most artificial neural network models to describe single neuron processing.

Recently, deCharms and Merzenich[1] presented evidence for a different form of coding in the primary auditory cortex of marmosets. They showed that rapidly adapting cells responded to elongated tone stimuli with a fast transient onset response returning quickly to spontaneous firing rates. Thus, these cells cannot convey information about a steady–state stimulus by their firing rate. However, these cells do show an increase in their tendency to fire *simultaneously* as revealed by correlation analysis if they are tuned to the presented stimulus frequency. Nevertheless, each spike train looked almost like it was randomly generated and there was no stimulus–locked component as shown by a flat shift predictor.

Most characteristics of these experimental findings can be reproduced in a simple neuronal model using leaky integrate–and–fire units with Poisson–distributed, balanced input, as shown below.

THE RANDOM WALK MODEL

Assume that the generation of action potentials relies on the membrane potential $u_i(t)$ of cell i ($1 \leq i \leq N$) at time t crossing a firing threshold θ and that deviations from the resting potential (set to 0 here) are due to an input current $C_i(t)$ and given these deviations decay exponentially with the membrane time constant τ_m. The following equation governs the temporal evolution of the membrane potential:

$$\frac{d}{dt}u_i(t) = -\frac{1}{\tau_m}u_i(t) + C_i(t) \ . \tag{1}$$

A spike occurs when $u_i(t) = \theta$, and u_i is reset to its resting level. To avoid unrealistically large hyperpolarizations, we also introduce a negative saturation limit θ^{inh}, i.e., we assure $u_i(t) \geq \theta^{inh}$ for all t. To specify the input current, $C_i(t)$, assume that this input can be subdivided into a background and a stimulus component, $C_i^{bg}(t)$ and $C_i^{stim}(t)$ respectively

$$C_i(t) = C_i^{bg}(t) + C_i^{stim}(t) \ , \tag{2}$$

and that each of these components consists of excitatory as well as inhibitory parts

$$C_i^{bg,stim}(t) = E_i^{bg,stim}(t) - b^{bg,stim}I_i^{bg,stim}(t - \Delta^{inh}) \ , \tag{3}$$

where b denotes a balancing factor indicating the relative strength of the inhibition with respect to the excitation and Δ^{inh} represents a delay.

To introduce noise in the model, assume that all excitatory and inhibitory signal components are realizations of an ideal Poisson process, i.e.,

$$E_i^{bg}(t) = k \ \text{ with probability } \ p(k) = \frac{\lambda^k}{k!}e^{-\lambda} \tag{4}$$

where k is drawn at each time step for every component independently. The parameter λ denotes the mean and the variance of the distribution. Here, it can be interpreted as $\lambda = n_{aff} \cdot p_f$ the product of the number of afferents times the probability of firing in a single time step, thus fixing the input firing rate. For $\lambda = 10$ and a basic time step of 1 ms, this might correspond to 100 afferents each firing at a rate of 100 Hz.

Due to the randomness in the input the membrane potential undergoes a sort of a random walk with renewal[2].

SIMULATION RESULTS

For simulations, we used the following set of parameters. The thresholds were set to $\theta = 15$ and $\theta^{inh} = -30$, both in units of single EPSP amplitudes. The time scale was fixed by $\tau_m = 10$ ms and $\Delta^{inh} = 20$ ms. We solved (1) using a simple forward Euler method with a time step size of 1 ms. The input was specified by $\lambda = 10$ for all four components and the balancing factors are $b^{bg} = 1$ and $b^{stim} = 1.1$. A typical simulation run lasted for three seconds where a stimulus was switched on after the first second and turned off after the second.

Single Cell Properties

Consider first the firing of a single cell. The total input current fluctuates vigorously. The resulting membrane potential is smoother due to the temporal integration. Threshold crossings of the membrane potential resulting in spike emission were only driven by fluctuations except for the stimulus onset period, where there was an excess of excitatory input due to the delayed arrival of the balancing inhibitory input. This can clearly be seen in a spike histogram obtained by averaging over many repetitions of the same experiment (but using a different seed for initializing the random number

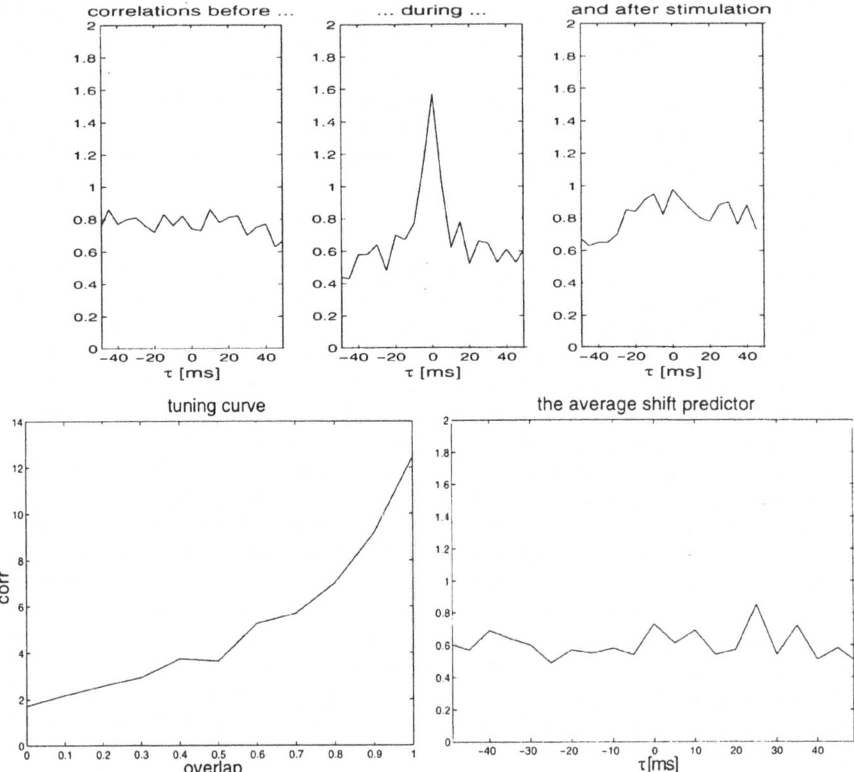

Figure 1. A pair of cells receiving a common input. (Top row) Average correlations from three different time periods: (Left) before stimulation; (Middle) during stimulation but after the onset response and (Right) after stimulation. There is a clear peak at $\tau = 0$ for the stimulation period indicating that these two neurons have a tendency to fire in synchrony during presence of the stimulus. Correlations were calculated for every trial using 5 ms bins and averaged afterwards. (Bottom, left) Tuning curve: Height of the central peak in the correlation during stimulation as a function of the fraction of identical input. The peak height increases with the overlap. (Bottom, right) Due to the overall noisy structure of the observed response, the shift predictor, correlating responses from different trials, is flat. Thus, there was no stimulus–locked activity during the tonic phase of the response.

generator each time). The mean firing rate stayed constant throughout the whole run except for a pronounced burst at stimulus onset and a reduction of firing after stimulus offset. The large trial–to–trial variability in firing is also shown by the model.

Multiple Cell Properties

In the experiments of deCharms and Merzenich[1] simultaneous recordings of spike trains from pairs of cells were analyzed. We simulated two cells getting independent background signals but sharing identical stimulus components in their input ($C_1^{\mathrm{stim}}(t) = C_2^{\mathrm{stim}}(t)$ for all t). The top row of Fig. 1 shows correlations between the firing times of the two cells calculated for every single trial for three different periods of time (before, during, and after stimulus presentation) and averaged afterwards. There was a strong peak at zero time shift only during common stimulation indicating an increased tendency of the two cells to fire simultaneously.

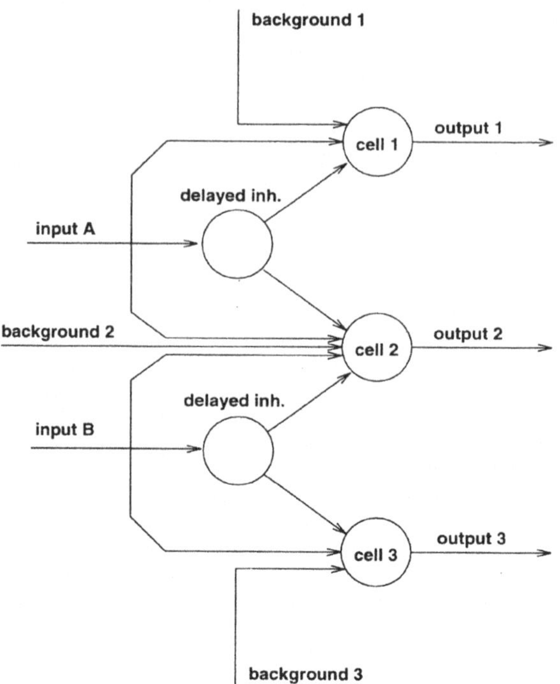

Figure 2. Basic structure for the three cell network. Now, cells 1 and 2 as well as 2 and 3 share some common input.

This is not a surprising result, because one might expect the common input to drive both cells to firing threshold simultaneously, but it is worth noticing since only a fraction of the emitted spikes are affected. These synchronous spikes happen to occur at random times and are not stimulus–locked, as indicated by the flat shift predictor in the lower–right part of Fig. 1. The height of the central peak in the correlation depends mainly on the amount of common input relative to the total input to both cells as shown in Fig. 1 (bottom left). The overlap here is defined as the ratio of common versus total input, ranging from zero (no common input) to one (absolutely identical input).

During the entire stimulation period, there was a pronounced increase in the correlations, which disappeared when the stimulus was turned off.

A FUNCTIONAL ROLE?

Finally, we suggest further computational implications of this mode of operation. It has been argued that the temporal structure of neuronal signals might be used for solving the binding problem[3]. Some time ago, there seemed to be experimetal evidence for this concept through the discovery of stimulus–related, collective oscillations, first found in the primary visual cortex of cats[4, 5, 6]. Similar observations have also been made in monkeys[7] but it is not clear whether the observed oscillations really have a crucial role in perception.

Here, we stress that the underlying mechanism for solving the binding problem is *simultaneaous* activity, not necessarily involving oscillations at all. Consider the neuronal network in Fig. 2. Two pairs of cells receive common input as before, so their

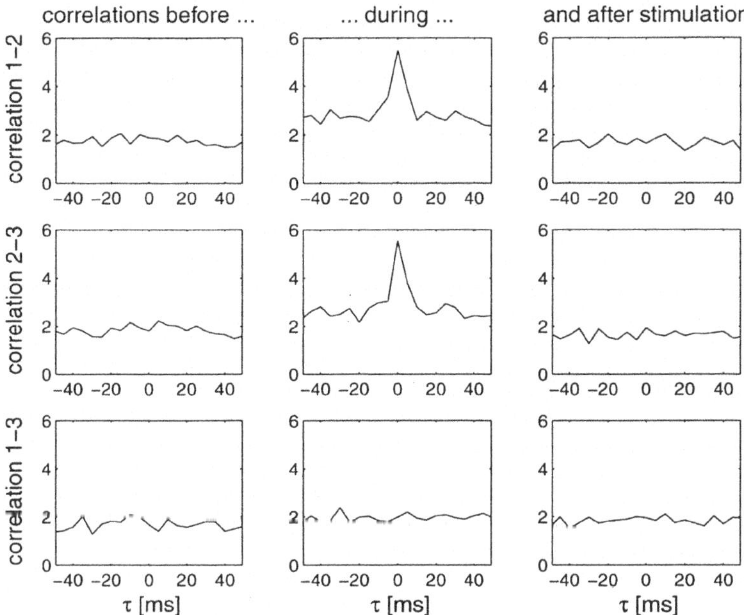

Figure 3. Demonstration of non–transitivity in a three neuron network. During stimulation, the responses of cells 1 and 2 as well as 2 and 3 show a tendency to fire in synchrony, but not cells 1 and 3. This is remarkable, since both are correlated with cell 2. This non–transitivity could be a useful property in avoiding the superposition catastrophe that can occur in binding the cell assemblies that represent multiple objects.

output spikes show an increased tendency to appear simultaneously, as reflected in the correlations shown in Fig. 3. It is remarkable, however, that cells 1 and 3 are not correlated, despite the fact that these two cells both are correlated with cell 2.

DISCUSSION

In contrast to the common belief that neurons code information only in their mean firing rate, deCharms and Merzenich[1] have shown that there is another possibility of coding, based on the relative timing of spikes from different neurons. We have replicated their results in a neural model. Conceptually, this idea is not new, and the underlying firing pattern may be even more complicated than just synchronous firing, as in synfire chains[8] or arbitrary firing patterns[9] or with respect to an internal neuronal clock[10].

What is new here is the observation that relative spike timing might be used in a noisy mode of operation. For this regime, it has commonly been assumed that the only way to get at reliable information transmission should be based on a rate code[11]. But there is increasing evidence for the possibility of temporal codes. First, it has been shown by Mainen and Sejnowski[12] that neocortical neurons fire very reliably if driven mainly by input fluctuations instead of a constant current. Therefore, the well known high variability in cortical spike firing times might reflect a high variability in the input to a neuron instead of intrinsic noise due to the spike generation process. Second, correlations in firing times between neurons tuned to similar stimulus features are omnipresent, but they have usually been interpreted as an artifact of common stimulation causing redundancy and having no use. Recently, this interpretation has been questioned. In the visual system, correlations seem to improve stimulus representation on the level of the retina[13] as well as the LGN[14]. In the auditory system, deCharms and Merzenich[1] provided evidence for the crucial role of correlations in stimulus representations. Their study was the starting point for the model presented here. We do not claim to have reproduced every single detail of their data. For this, a biophysically more realistic model should be appropriate. But we have shown here how such a code might work naturally and reliably even in a noisy environment.

The final question, however, whether this type of coding is really used in the brain (i.e., read out at the next level) remains to be experimentally examined. Correlations are easily read out by neurons and they play a central role in learning, so there is every reason to continue along this line of investigation.

Acknowledgments. Supported by DFG (grant Ri 821/1-1) and The Howard Hughes Medical Institute.

REFERENCES

1. R. C. deCharms and M. Merzenich, Primary cortical representation of sounds by the coordination of action–potential timing, *Nature.* 610 (1996).
2. G. L. Gerstein and B. Mandelbrot, Random walk models for the spike activity of a single neuron, *Biophyhs. J.* 4:41 (1964).
3. C. von der Malsburg, The correlation theory of brain function. Internal Report 81-2, MPI für Biophysikalische Chemie, Göttingen (1981). *Reprinted in:* "Models of Neural Networks II," E. Domany, J.L. van Hemmen, and K. Schulten, eds., Springer, Berlin, Heidelberg, New York (1994).
4. R. Eckhorn, R. Bauer, W. Jordan, M. Brosch, W. Kruse, M. Munk, and H. J. Reitboeck, Coherent oscillations: A mechanism of feature linking in the visual cortex? *Biol. Cybern.* 60:121 (1988).
5. C. M. Gray, and W. Singer, Stimulus-specific neuronal oscillations in orientation columns of cat visual cortex, *Proc. Natl. Acad. Sci. USA.* 86:1698 (1989).
6. C. M. Gray, P. König, A. K. Engel, and W. Singer, Oscillatory responses in cat visual cortex exhibit inter-columnar synchronization which reflects global stimulus properties, *Nature.* 338:334 (1989).
7. A. K. Kreiter and W. Singer, Stimulus–dependent synchronization of neural responses in the visual cortex of the awake macaque monkey, *J. Neurosci.* 16:2381 (1996).
8. M. Abeles, H. Bergman, E. Margalit, and E. Vaadia, Spatiotemporal firing patterns in the frontal cortex of behaving monkeys, *J. Neurophysiol.* 70:1629 (1993).
9. W. Gerstner, R. Ritz, and J. L. van Hemmen, Why spikes? Hebbian learning and retrieval of time–resolved excitation patterns, *Biol. Cybern.* 69:503 (1993).
10. J. J. Hopfield, Pattern recognition computation using action potential timing for stimulus representation, *Nature.* 376:33 (1995).
11. M. N. Shadlen and W. T. Newsome, Noise, neural codes and cortical organization, *Curr. Opin. Neurobiol.* 4:569 (1994).
12. Z. F. Mainen and T. J. Sejnowski, Reliability of spike timing in neocortical neurons, *Science.* 268:1503 (1995).
13. M. Meister, L. Lagnado, and D. A. Baylor, Concerted signaling by retinal ganglion cells, *Science.* 270:1207 (1995).
14. Y. Dan, J. J. Atick, and R. C. Reid, Efficent coding of natural scenes in the lateral geniculate nucleus: experimental test of a computational theory, *J. Neurosci.* 16:3351 (1996).

A MODEL OF THE EFFECTS OF LAMINATION AND CELLTYPE SPECIALIZATION IN THE NEOCORTEX

Adrian Robert

Department of Cognitive Science
University of California, San Diego
La Jolla, CA 92093-0515

INTRODUCTION

Each hemisphere of the neocortex consists of a relatively flat, laminated sheet in which several anatomically distinguishable layers (six, by convention) of cells are juxtaposed. In addition to local connections, cells within one localized region, or *area*, project to and receive from a limited set of other areas. These projections originate from and terminate in subsets of the layers, often following particular patterns[1] (figure 1, left). The existence of the patterns suggests a functional correlate, and various speculations have been put forth[2,3,4].

The nature of the anatomy, in which most cortical celltypes possess dendritic and axonal arborizations extending over multiple layers[5] (figure 1, right), and technical constraints on physiology experiments render it difficult to constrain these speculations by determining the influences that activity in and inputs to particular layers have on other layers. In this note, I investigate these influences using an anatomically realistic neural simulation model.

The next section describes the construction of a model of a small patch (4 x 4 mm) of neocortex, consisting of a stack of two-dimensional lattices of units representing

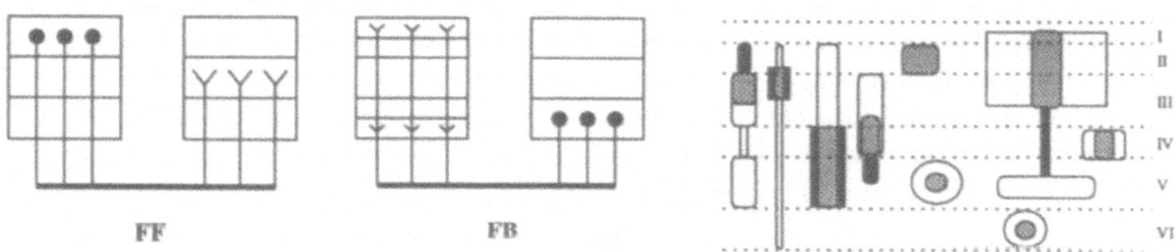

1. As noted by Felleman and Van Essen, connections between areas fall largely into one of the two forms above. The connection from V1 to V2 in the visual system (among others) is of the type on the left ('feedforward'), while the reverse projection from V2 to V1 (among others) is of the type in the middle ('feedback'). The right figure illustrates typical within-area cortical cell arborization patterns. Black, white, and grey indicate dendritic, axonal, and overlapping arborizations.

cells of particular types located in particular layers. The aim was not to model a specific cortical area, but to capture general architectural characteristics common to most association cortical areas. I also describe the construction of a simplified model consisting of a single layer with two cell populations. The behavior of both models is compared with a set of recordings from neocortical slices, allowing determination of which aspects of the experimental results are due to general properties of laterally-connected networks of excitatory and inhibitory cells and which depend on the more specific architectural characteristics of neocortex.

CONSTRUCTION OF THE LAMINATED MODEL

The main goal guiding the model's construction was accurate representation of those aspects of neocortical structure likely to be important in determining interlayer influence. Because the amount of time for transmission between layers can determine dynamical structure such as local feedback circuits that could magnify or shrink influences, a single-cell model capable of representing this temporal element was chosen. Because the arborization characteristics of the different celltypes determine routes of communication between layers, the distribution and arborization of the different cell-types (figure 1) was used to determine local connectivity.

Single Cell Model

Cortical cells were modeled by single compartment conductance-based units analogous to those described in [6]. The equation for the evolution of membrane potential in a cell with membrane capacitance C_m, membrane conductance g_{leak}, resting potential E_m, and membrane potential V_m is:

$$C_m\frac{dV_m}{dt} = (E_m - V_m)g_{leak} + \sum_{k=1}^{n_{syn}}[(E_k - V_m)g_k] + \sum_{l=1}^{n_{chan}}[(E_l - V_m)g_l]$$

If the membrane potential reaches a threshold in a unit, it is reset to the resting potential and a spike event is propagated with a delay to synapses on other cells, the effects of which are modeled by alpha-function conductance changes. That is, the conductance g of a synapse after being activated at $t = 0$ is:

$$g(t) = g_{peak}\frac{t}{\tau}e^{(1-\frac{t}{\tau})}$$

The time constants and peak conductances for synapses were determined according to what portion of the cell the particular class of input involved is known from anatomical data to contact.

Cortical cells fall into three classes based on intrinsic response properties[7,8]. Additional conducting channels were included in the cell models to mimic the effects of currents differentiating these classes (figure 2).

Cell Distribution and Interconnection

The neocortex is a regular latticework of more than a dozen celltypes distinguished by axonal and dendritic arborization pattern, distribution across the layers, and connection preferences among the other types[5,9]. 70–75% of the total cell population consists of pyramidal cells which are excitatory and split into 3-4 arborization subtypes per layer. The remaining dozen or so celltypes are inhibitory and may appear in multiple

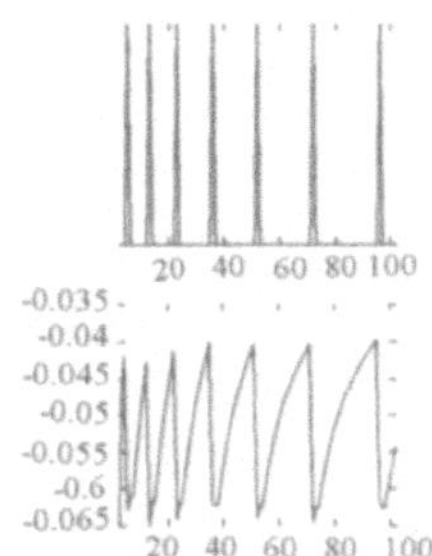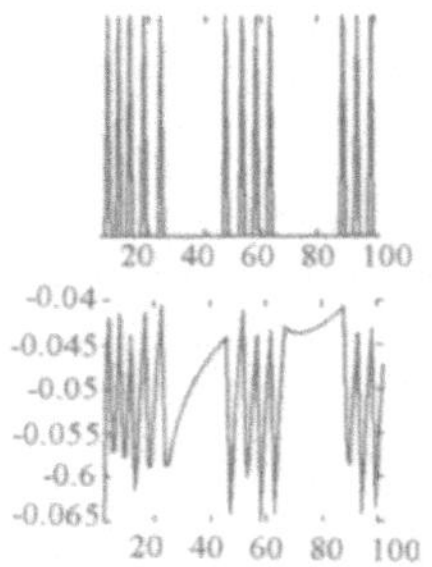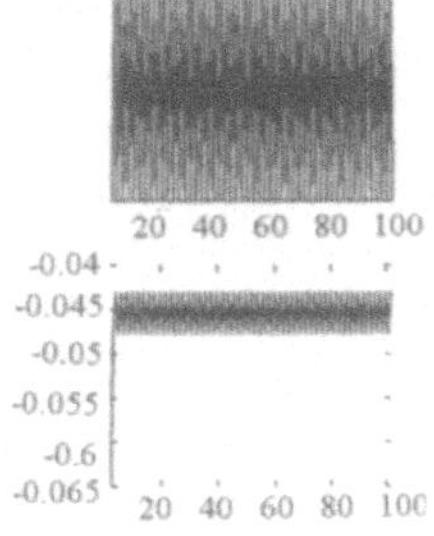

2. Model cell responses to current injection (0.5 nA). Top: firing; Bottom: membrane potential; Left-to-right: regular spiking pyramidal, intrinsic burst pyramidal, fast spiking inhibitory.

layers. Essentially complete data on arborization characteristics was gathered from the literature on macaque monkey cortical microanatomy, along with nearly complete data on the distribution of different celltypes across the layers and partial information on the numerical densities and type-type contact preferences[10].

Starting from a quantitative specification of inputs to pyramidal cells[11], a process of progressive constraint satisfaction was applied to determine celltype densities and contact preferences incompletely specified by the literature (details in [10]). These estimated values were then employed in conjunction with the arborization data to specify connections between cells in the model network. The results are provisional and subject to refinement as more data becomes available, and the experimental comparisons below were designed to determine the faithfulness of the model's connections to biology.

The neocortex of most mammals in most areas contains 10^5 neurons and 10^9 synapses per mm^2 surface area, however many of the cell arbors extend for substantial fractions of a millimeter or more. Accurately representing the effects of these arbors requires modeling several mm^2, out of the reach of present computational resources. The approach adopted here was to reduce the total density of cells and the number of connections by a factor (usually 30) while leaving arbor sizes unchanged, as done in other neural modeling work (e.g., [6]); synaptic weights were increased in compensation.

CONSTRUCTION OF THE SINGLE-LAYER MODEL

A single layer model was constructed with two cell populations, regular-spiking pyramidal and inhibitory. Connection arborizations were set to the average of the entire classes of connections they replaced in the full model, and synaptic weights were set similarly. The aim was to construct a network with roughly the same spatial scales of interaction but without lamination or celltype specialization. One version of the network contained only $GABA_A$ inhibitory synapses, while the other version contained inhibitory connections using $GABA_B$ in similar proportion to the laminated network.

EXPERIMENTAL RESULTS FROM NEOCORTICAL SLICES

To assess the faithfulness of the laminated model to cortical dynamical functioning, its behavior in replicated experimental conditions was compared to data on the lateral spread of activity in rat Sm-I neocortical slices under varying low levels of bicuculline-induced disinhibition following electrical stimulation at a single point (Chagnac-Amitai and Connors[12,13] – henceforth CC). This allowed assessment of whether the model's excitability and characteristic lengths of activity spread accurately matched the real cor-

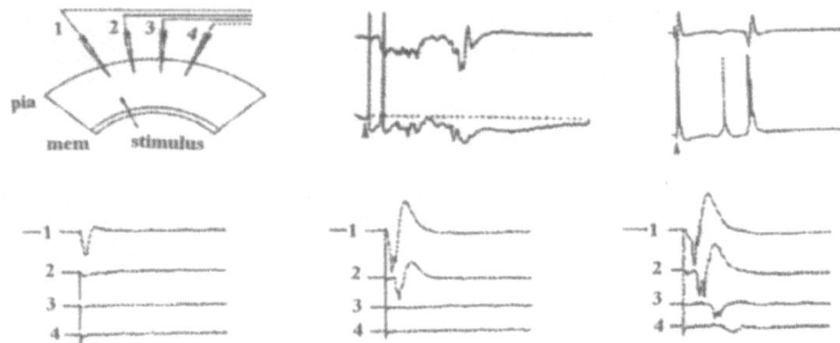

3. Slice recordings of Chagnac-Amitai and Connors[12]. Upper left: experimental setup; bottom row: extracellular field potentials recorded at four points for three increasing levels of disinhibition; upper right: dual extracellular (top) and intracellular (bottom) traces for superficial regular spiking pyramidal (left) and deep bursting pyramidal (right). Reprinted with permission.

tex. The single-layer model's behavior was also compared, and this allowed determination of which aspects of the experimental results depended on the special characteristics of cortical architecture.

CC recorded local field potentials from a series of 4 electrodes in layers II/III laterally spaced at 0, 600, 1400, and 2200 μm from a brief shock stimulation point located in layer VI (figure 3). The results reflected the following main features:

1. Wave propagation occurs with reliability dependent on disinhibition level; about 20% is sufficient to generate completely reliable epileptiform propagation.

2. Propagation rates are in the range of 80-100 μm/msec.

3. Field potential event: width of positive fp = 10 msec, width of positive-negative fp = 20 msec.

4. Regeneration/reflection sites are observed at higher disinhibition levels.

5. Concurrent intracellular recordings revealed that superficial pyramidal cells are dominated by IPSPs during wavefront passage, and rarely fire. Deep layer bursting cells on the other hand receive greater excitation and usually fire (figure 3); deep regular spiking cells are variable.

COMPARISON OF SINGLE-LAYER MODEL WITH EXPERIMENT

A 1 msec input pulse was applied to the central excitatory and inhibitory cells in the network, to imitate the effects of both dendritic and antidromic activation of cells with processes within a sufficiently small radius of the stimulation point. The strength of GABA$_A$ synapses onto both pyramidal and inhibitory cells was varied to imitate the effects of bicuculline disinhibition.

In the network employing only GABA$_A$ inhibition, two regimes of behavior were observed (figure 4): a spreading activity lump that lasted up to 20 msec and spread out to at most double the initial stimulation radius before collapsing to quiescence, or an indefinitely growing lump of activity at maximal firing rate that spread through the entire network. The transition on the disinhibition scale between these two regimes was sharp, but variation was smooth (in terms of length of time to quiescence/radius of spread and explosion spread rate) within them.

The network additionally employing GABA$_B$ inhibition exhibited the above two regimes plus a third of wave propagation (figure 4), situated between them. The thickness of the wavefront varied smoothly with disinhibition level (GABA$_A$ only) but the velocity remained constant.

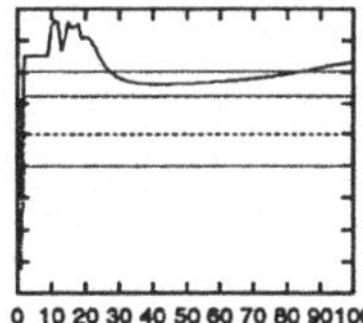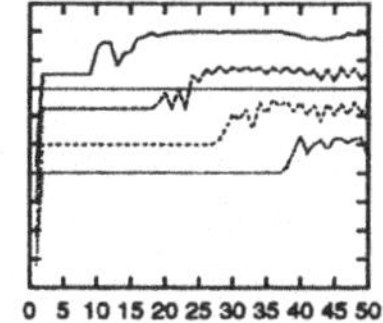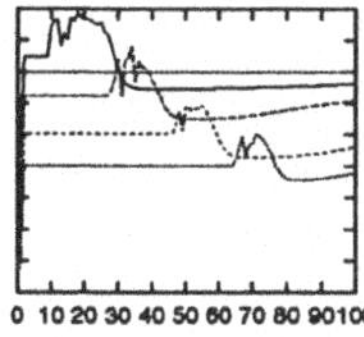

4. Modes of response of single-layer model to point-shock stimulation. Traces represent local average membrane potentials from pyramidal cells at four points spaced as in experiment. From left to right, quiescence regime, explosion regime, wave propagation regime.

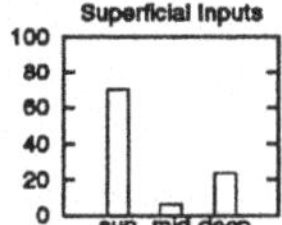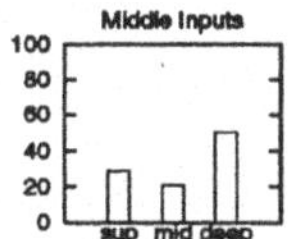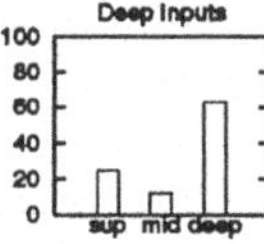

5. Proportions of input that pyramidals in superficial, middle, and deep layers receive from pyramidals in each layer class (note compartmentalization between the superficial and deep populations, responsible for effect illustrated in figure 6; see text).

Several parameters were varied in the model besides $GABA_A$ strength: various connectivity lengths and $GABA_B$ strength. Variations affected mainly the details of behavior within regimes and their transition points; we summarize only the overall qualitative characteristics below (cf. 1–5 above, italics indicates nonreplication):

1. Adjoining wave propagation and quiescence regimes were found, when $GABA_B$ was included in the model; reliability depended on disinhibition level.

2. *Propagation velocity was slower in the model by 50%; this may reflect the absence of the longer distance connections in the real cortex and laminated model that were averaged out here.*

3. Wavefront width was 10–20 msec.

4. *Regeneration sites were not observed.*

5. *Differences between excitatory cell subpopulations were not observed.*

COMPARISON OF LAMINATED MODEL WITH EXPERIMENT

In the laminated model, the 1 msec pulse was applied to central cells in layer VI, as well as, to a lesser radius, those cells in other layers extending processes there. Relative strengths of $GABA_A$ synapses were varied.

The same three regimes as in the $GABA_B$ single-layer network were observed, with essentially the same dependency on the relative strengths of the inhibitory synapses. There were three main qualitative differences in the details of the behavior:

- Wave propagation velocity was higher, comparable to that observed experimentally.

- Wave instantaneous thickness was greater and propagation omnidirectional and messier.

- Compartmentalization existed such that activity propagated primarily in deep layers with only occasional spilling through to the upper layers.

Figure 6 illustrates both the wave propagation and compartmentalization. Regarding replication of the slice results, the replication was improved over that of the single-layer model in the respect that the wave propagation velocity was accurately reproduced and the cell population differences between superficial and deep layers were observed (figure 5 shows part of the basis of this in the model).

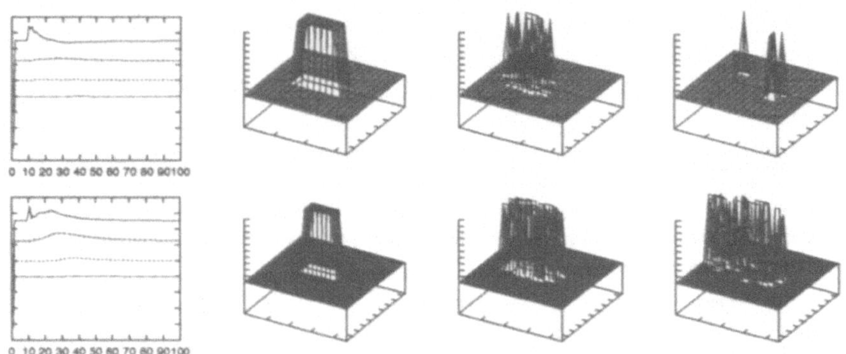

6. Activity on layer II pyramidals (top) and layer VI pyramidals (bottom). Left: membrane potentials of layer II and layer V pyramidals as in figure 4; right: spike activity on layer II and layer VI at 5,10,15 msec in a different, more disinhibited simulation. Note greater propagation on deep layers.

DISCUSSION AND CONCLUSIONS

The experimental results reported suggested that activity could propagate amongst deep layer pyramidals with only minimal correlate in superficial pyramidal activity. The laminated model also displayed this phenomenon and provided a clearer visualization of the underlying dynamics, along with a connection to an anatomical basis in the connectional compartmentalization of the pyramidal populations. In the future, the model can be used to investigate this structure-function relation in more detail.

REFERENCES

1. D.J. Felleman and D.C. Van Essen. Distributed hierarchical processing in the primate cerebral cortex. *Cerebral Cortex*, 1:1–47, 1991.
2. B.A. Vogt. The role of layer I in cortical function. In A. Peters and E.G. Jones, editors, *Cerebral Cortex, Volume IX: Normal and Altered States of Function*. New York: Plenum, 1991.
3. C.M. Fair. *Cortical Memory Functions*. Boston: Birkhauser, 1992.
4. R.P.N. Rao and D.H. Ballard. Dynamic model of visual recognition predicts neural response properties in the visual cortex. *Neural Computation*, 9:805–47, 1997.
5. E.G. Jones and A. Peters, editors. *Cerebral Cortex, Vol. I. Cellular Components of the Cerebral Cortex*. New York: Plenum, 1984.
6. M.A. Wilson and J.M. Bower. The simulation of large-scale neural networks. In C. Koch and I. Segev, editors, *Methods in Neuronal Modeling*. Cambridge, Mass.: MIT Press, 1989.
7. D.A. McCormick, R.W. Connors, J.W. Lighthall, and D.A. Prince. Comparative electrophysiology of pyramidal and sparsely spiny stellate neurons of the neocortex. *J. Neurophysiol.*, 59:782–806, 1985.
8. D.A. Prince and J.R. Huguenard. Functional properties of neocortical neurons. In W. Singer P. Rakic, editor, *Neurobiology of Neocortex*. New York: Wiley, 1988.
9. E.L. White. *Cortical Circuits*. Boston: Birkhauser, 1989.
10. A. Robert. The neocortex: A summary of anatomical and physiological information for modeling purposes. manuscript, 1995.
11. J. DeFelipe and I. Farinas. The pyramidal neuron of the cerebral cortex: Morphological and chemical characteristics of the synaptic inputs. *Progr. Neurobiol.*, 39:563–607, 1992.
12. Y. Chagnac-Amitai and B.W. Connors. Horizontal spread of synchronized activity in neocortex and its control by GABA-mediated inhibition. *J. Neurophysiol.*, 61:747–58, 1989a.
13. Y. Chagnac-Amitai and B.W. Connors. Synchronized excitation and inhibition driven by intrinsically bursting neurons in neocortex. *J. Neurophysiol.*, 62:1149–62, 1989b.

SELF-ORGANIZING MAPS OF SPIKING NEURONS USING TEMPORAL CODING

Berthold Ruf and Michael Schmitt

Institute for Theoretical Computer Science, Technische Universität Graz,
Klosterwiesgasse 32/2, A-8010 Graz, Austria
E-mail: {bruf, mschmitt}@igi.tu-graz.ac.at

ABSTRACT

Kohonen's self-organizing map has been thoroughly investigated for artificial neural networks. There have been several approaches for biologically more realistic neural networks which all rely on rate coding. Here we show that a topology preserving behavior can be also achieved by networks of spiking neurons using temporal coding. Besides being generally faster during learning and application, our approach has the additional advantage that the winner among competing neurons can be determined fast and locally. Our model is a further step towards a more realistic description of unsupervised learning in biological neural systems. Furthermore, it may provide a basis for fast implementations of neural networks in pulsed VLSI.

INTRODUCTION

Biological neural systems are known to be extremely powerful information processing devices. It is not clear which essentials of these networks are relevant for an understanding and a simulation of these abilities[1, 2, 3, 4]. Discrete models, such as threshold gates or McCulloch-Pitts neurons, are undoubtedly very simplistic descriptions of biological neurons. Models with real-valued output, such as the sigmoidal gate, where analogue values are interpreted as firing rates of biological neurons, are more suitable for the modelling of neural processes in terms of analogue computations. However, both types of models do not capture a phenomenon which is widely believed to be the basis of fast analogue computations in biological neural networks: the timing of single action potentials. Models of spiking neurons that use temporal coding are a first attempt to explore the computational significance of this phenomenon[5, 6, 3]. Furthermore, spiking neuron networks (SNNs), where the computations are based on this coding scheme, have recently been shown to be computationally more powerful than networks consisting of threshold or sigmoidal gates with respect to time and network complexity[4]. The issue of learning was raised by Ruf and Schmitt[7] where it was shown that supervised learning is possible in such SNNs.

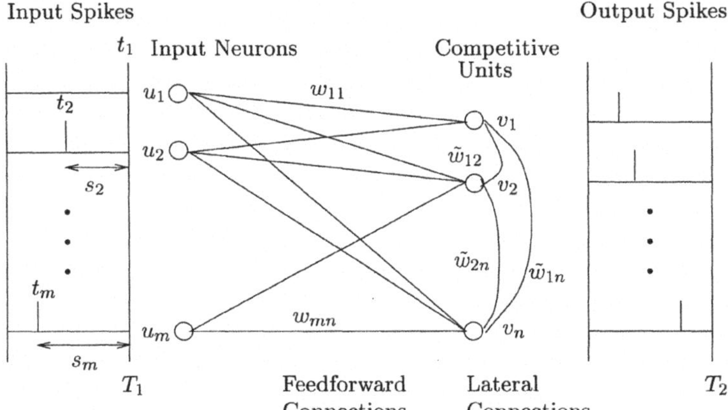

Figure 1. The basic architecture. Each input-unit u_i fires exactly once at time t_i. The ith input value is given by $T_1 - t_i$. Each competitive unit fires exactly once before some constant T_2.

Under certain weak assumptions (basically the initial rising segments of the excitatory postsynaptic potentials have to be linear) one can guarantee that v fires at a time determined by $T_2 - \sum w_i s_i$ with T_2 being some constant.

This computation can also be performed on the basis of "competitive temporal coding", such that no explicit reference times T_1 and T_2 are necessary (see Maass[8] for details).

If one considers now a set $V = \{v_1, \ldots, v_n\}$ of neurons with various weight vectors, all receiving the same input, the $v_j \in V$ can compute in parallel their weighted sums. If we assume that the input vector and the weight vector for each neuron are normalized, then this weighted sum represents the similarity between the two vectors with respect to the Euclidean distance. Hence the earlier a $v_j \in V$ fires, the more similar is its weight vector to the input vector. V forms a competitive layer where the neuron in V which fires first wins the competition. The next section shows how this idea can be used to achieve self-organizing behavior.

SELF-ORGANIZATION

It is now possible to use the construction described above for implementing a variation of Kohonen's SOM algorithm in the context of SNNs as follows: Given a set S of m-dimensional input vectors $\mathbf{s}^l = (s_1^l, \ldots, s_m^l)$ and an SNN with input neurons $U = \{u_1, \ldots, u_m\}$ and competitive neurons $V = \{v_1, \ldots, v_n\}$, where each competitive neuron $v_j \in V$ receives synaptic "feedforward" input from each input neuron $u_i \in U$ with weight w_{ij} and "lateral" synaptic input from each competitive neuron $v_k, k \neq j$, with weight $\tilde{w}_{kj}$. At each cycle of the learning procedure one $\mathbf{s}^l \in S$ is randomly chosen and the input neurons are made fire such that they temporally encode $\mathbf{s}^l$. Each $v_j \in V$ then starts to compute $\sum_i w_{ij} s_i^l$ as described in the previous section. The firing time of the winner neuron, say v_k, actually indicates its weighted sum. Its firing influences however the firing times of the other neurons in V through the lateral connections (see Figure 1).

If the lateral connections are strongly inhibitory, such that the firing of the winner neuron v_k inhibits all other neurons in the competitive layer from firing, one can implement in a straightforward way competitive learning: One simply has to apply the standard competitive learning rule (also known as instar learning rule) to the winner neuron v_k. The postsynaptic spike of v_k, propagating backwards to the synapses of v_k, may serve as a "switch" for the application of the learning rule. For this type of unsupervised learning, the weight update for the synapse between u_i and v_k is given by

$$\Delta w_{ik} = \eta(s_i^l - w_{ik}), \qquad i \in \{1, \ldots, m\}$$

where $\mathbf{s}^l$ is the current pattern and η the learning rate.

This way of realizing competitive learning can be extended to a self-organizing behavior as follows: essentially one has to find a way to implement a given neighborhood matrix $(m_{kj})_{1 \leq k, j \leq n}$ on the basis of locally available information, where m_{kj} describes the distance between the kth and jth competitive neuron. We use a monotonously increasing function $m_{kj} \mapsto \tilde{w}_{kj}$ such that the lateral connections $\tilde{w}_{kj}$ among the competitive neurons reflect the structure of the neighborhood: initially neurons which are topologically close together have strong excitatory lateral connections whereas remote neurons have strong inhibitory connections. This means that the firing of the winner neuron v_k at time t_k drives the firing times of neurons in the neighborhood of v_k towards t_k, thus increasing the values they encode. The firing of remote neurons is delayed by the lateral inhibition. In the example of Figure 1, the firing of the winner v_1 at t_1 shifts the firing time of a topologically close neuron v_2 towards t_1 and postpones the firing of a topologically remote neuron v_n. We suggest the following learning rule:

$$\Delta w_{ij} = \eta \frac{T_{out} - t_j}{T_{out}}(s_i^l - w_{ij}) \tag{1}$$

where t_j is the firing time of the jth competitive neuron. The learning rule applies only to neurons that have fired before a certain time T_{out} (which has to be chosen sufficiently large). The factor $(T_{out} - t_j)/T_{out}$ realizes the neighborhood function, which is largest for the winner neuron and decreases for neurons which fire at later times. After each learning iteration the weight vectors for each neuron have to be normalized again. However, in Ruf and Schmitt[14] it is shown that the assumption that the weight vectors have to be normalized can be dropped without considerably deteriorating the quality of learning.

The lateral weights $\tilde{w}_{kj}$ are decreased during the learning process, thus reducing the size of the neighborhood.* As in the standard formulation of the SOM, η is slowly decreased during learning.

SIMULATIONS

We performed computer simulations for leaky integrate-and-fire neurons where the initial segment of the postsynaptic potentials is linear. We tested our approach with one-dimensional input patterns. 10 input patterns, uniformly distributed over $[0, 1]$, were presented to a layer of 10 competitive units, which had initially random weights of values around the midpoint of the patterns. The lateral weights for the immediate neighbors were chosen slightly positive, for the second neighbors zero and for all other neurons negative. The goal was the formation of a linear map. Figure 2 shows that

---

*This requires that the lateral weights change their sign during learning, which is not very realistic in the context of biological networks. However, one can consider variations of our simplifying architecture, e.g. one may use instead two connections, one excitatory, which decreases, and one inhibitory, which increases during learning.

initially nearly all neurons react strongly, i.e. fire early on each input, whereas after learning only few neurons, being topologically close, react on a certain input pattern. Furthermore, the neurons became sorted according to the ordering of the input patterns. Hence the topology of the neurons reflects the topology of the input space.

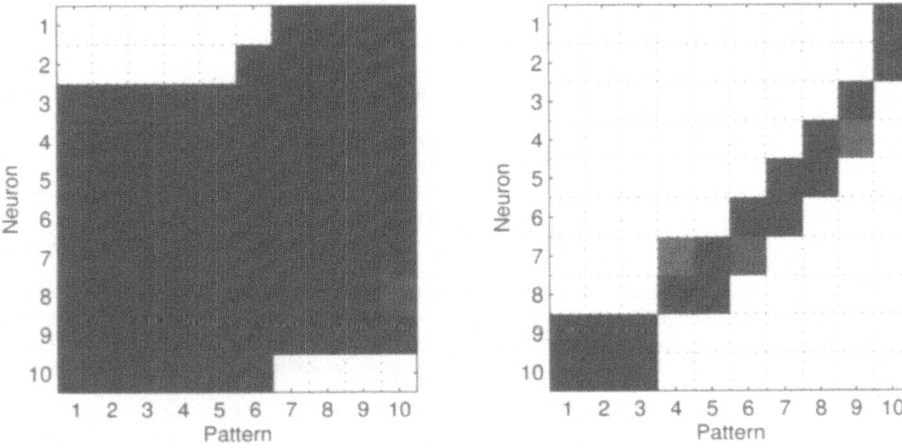

Figure 2. Learning process for one-dimensional input patterns after 10 iterations (left) and 4000 iterations (right). Each column represents the reaction of the network to a particular input pattern. The colors of the squares indicate the firing times: the darker a square, the earlier the corresponding firing time. The non-firing of a neuron is represented by a white square.

We were also able to show that our method performs well on one of the standard examples for the SOM, where two-dimensional input patterns are chosen randomly from a square and the competitive units are expected to organize themselves in a topology preserving grid. The detailed results using a measure for quantifying the neighborhood preservation known as "metric multidimensional scaling" (see e.g. Goodhill and Sejnowski[15]) are presented in Ruf and Schmitt[14].

CONCLUSIONS

To the best of our knowledge, this is the first implementation of Kohonen's learning algorithm for SNNs using temporal coding. This work is a further step towards showing that biological neurons can indeed achieve a topology preserving behavior using a similar learning procedure like the one suggested by Kohonen. Our approach also may give rise to fast hardware implementations of self-organizing networks in pulse coded VLSI.

As mentioned above we have assumed as in Maass[8] that the neurons are of the leaky integrate-and-fire type, where the initial segment of the postsynaptic potentials rises linearly. Recent simulations[16] have shown that even when using a more detailed neural model which includes non-linear effects and more realistic shapes for the post-synaptic potentials (e.g. α-functions), spiking neurons can still compute weighted sums in temporal coding. This indicates that the above-mentioned simplifying assumptions can be dropped.

REFERENCES

1. W. Gerstner and L. van Hemmen. How to describe neuronal activity: spikes, rates or assemblies? In *Advances in Neural Information Processing Systems*, volume 6. Morgan Kaufmann, 1994.
2. T. Sejnowski. Time for a new neural code? *Nature*, 376:21 – 22, 1995.
3. F. Rieke, D. Warland, W. Bialek, and R. de Ruyter van Steveninck. *SPIKES: Exploring the Neural Code*. MIT-Press, Cambridge, 1996.
4. W. Maass. Networks of spiking neurons: The third generation of neural network models. To appear in: *Neural Networks*.
5. M. A. Arbib, editor. *The Handbook of Brain Theory and Neural Networks*. MIT Press, Cambridge, 1995.
6. W. Maass. Lower bounds for the computational power of networks of spiking neurons. *Neural Computation*, 8:1–40, 1996.
7. B. Ruf and M. Schmitt. Learning temporally encoded patterns in networks of spiking neurons. *Neural Processing Letters*, 5(1):9–18, 1997.
8. W. Maass. Fast sigmoidal networks via spiking neurons. *Neural Computation*, 9:279–304, 1997.
9. T. Kohonen. *Self-Organizing Maps*. Springer, Berlin, 1995.
10. T. Kohonen. Physiological interpretation of the self-organizing map algorithm. *Neural Networks*, 6:895 – 905, 1993.
11. J. Sirosh and R. Miikkulainen. Topographic receptive fields and patterned lateral interaction in a self-organizing model of the primary visual cortex. *Neural Computation*, 9:577 – 594, 1997.
12. Y. Choe and R. Miikkulainen. Self-organization and segmentation with laterally connected spiking neurons. Technical Report AI TR 96-251, Department of Computer Science, University of Texas at Austin, 1996.
13. A. Murray and L. Tarassenko. *Analogue Neural VLSI: A Pulse Stream Approach*. Chapman & Hall, 1994.
14. B. Ruf and M. Schmitt. Unsupervised learning in networks of spiking neurons using temporal coding. In W. Gerstner, A. Germond, M. Hasler, and J.-D. Nicoud, editors, *Proceedings of the 7th International Conference on Artificial Neural Networks - ICANN'97*, volume 1327 of *Lecture Notes in Computer Science*, pages 361-366, Springer, Berlin, 1997.
15. G. J. Goodhill and T. J. Sejnowski. A unifying objective function for topographic mappings. *Neural Computation*, 9:1291 – 1303, 1997.
16. B. Ruf. Computing functions with spiking neurons in temporal coding. In J. Mira, R. Moreno-Díaz, and J. Cabestany, editors, *Biological and Artificial Computation: From Neuroscience to Technology*, volume 1240 of *Lecture Note in Computer Science*, pages 265–272. Springer, Berlin, 1997.

A MODEL FOR DEVELOPMENT OF CORTICAL LATERAL CONNECTIVITIES USING MOTION INFORMATION

Ladan Shams and József Fiser

Univ. of Southern California
Dept. of Computer Science
Los Angeles, CA 90089
lshams@selforg.usc.edu, fiser@bcs.rochester.edu

ABSTRACT

We present a prediction optimization model inspired by classical conditioning for learning to predict the visual stimuli in the next time frame. The learning process operates on V1-simple cell type filter outputs of gray scale real image sequences capturing moving objects as retinal input. The learning results in development of an iso-orientation connectivity pattern as well as a gating mechanism based on motion information. The network learns to predict the response of the orientation selective cells for both the moving portion and the background of the scene. We argue that learning object independent visual transformations, such as translation in our model, can occur prior to learning of complex object features in the visual system.

INTRODUCTION

The essential problem of object recognition in higher level vision is how to discard certain aspects of a visual input while detecting other features in order to identify that the input belongs to a given category among many possible ones. One traditional computational approach to this problem is to represent the object under a given condition (such as pose, size, illumination), and then expand this representation by attaching additional information about the object under different conditions. This approach leads to the problem of how identification of an object is possible from a condition/view which has never been experienced before [1]. In this paper we suggest an alternative approach to object recognition based on data on visual development. The basic idea is that the developing visual system first detects typical transformations of the visual inputs (such as translation, looming, occlusion), and their *effects* on the visual input. Since these transformations are objet independent they can apply to any new input, moreover, they can be hardwired in the nervous system prior to learning specific objects. Applying a transformation might involve developing a representation for new "latent" dimensions, such as speed in 3 dimensions for transforming a looming image. Nevertheless, the result is that an object representation has to be developed in this transformed input space facilitating invariant object perception. We present a scheme which develops circuitry for such a transformation based on real image sequences, and sets the stage for learning complex features of objects from lower dimensional input.

TRANSLATION AS AN EXAMPLE

To demonstrate the basic concept we selected the transformation of *translation*. When an image of an object is shifting on the retina the visual input is changing in every moment. Extracting features from this input for object representation is difficult, unless the effects of changes in the input due to shifting can be discarded. This can be done if a predictory scheme can tell where a given feature will be and what changes might be anticipated based on its present condition. Such a predictory scheme for shift can be developed based on motion detection. If an area of the visual field is coherently moving in one direction with a given speed, the position of feature A in the next time frame can be anticipated with high certainty. Thus information unrelated to shift can be extracted from the input.

Two points about this scheme are worth noting. First, learning a transformation does not require identification of the object in advance. In our example the only requirement is existence of a circuitry that can detect common motion. There is evidence in the physiological, psychophysical and child development literature that the circuitry and the functionality of motion detection emerges very early preceding complex object recognition abilities [2-4].

The second point is that our learning strategy is different from the traditional scheme of learning shift invariant object recognition [5-7]. In the traditional scheme every object has to appear in each position during training where it to be identified later. This scheme still requires for every new type of transformation to show each object in all possible condition (subsampled). In our case only the transformation has to be learned with a sufficient number of examples. Thus we break down the learning process into learning first some "meta-knowledge" which can be applied across the set of inputs, and only then we attempt to encode individual inputs. This paper presents an example for the first step.

THE LEARNING MODEL

We simulate a recurrent network of V1 type cells with lateral connections via a one-layer feedforward network where the number of input and output units are equal, and each unit in the output layer is an abstraction of the same unit in the input layer. The forward connections from an input unit to its corresponding output unit represents a feedback loops, and connections from an input unit to other output units represent lateral connections (Fig. 1A, B). The input to the network is a small patch in the visual field represented by a grid of 3x3 "hypercolumns" (set of detectors selective for a given scale and orientation). The model illustrates how learning is performed for each hypercolumn in the output layer. We use a set of simple cell-like filters selective for three orientations and one spatial frequency at every sampling locus of the visual field (a Gabor-jet) to represent the hypercolumns in V1. We assume that these cells are fully connected to all of their retinotopic neighbors initially. The input to the network consists of nine jets along with motion information for each of the nine coordinates. As retinal input we used sequence of frames taken by a camera, containing various objects moving horizontally to the right at a fixed rate. Each gray scale frame was complemented by a binary motion map. For every jet at (x,y) in

516

In this paper we investigate unsupervised learning processes in SNNs. On the basis of a construction provided by Maass[8] we show how competitive learning can be performed by SNNs using temporal coding. We extend this idea to a learning mechanism that is closely related to one of the most successful paradigms of unsupervised learning: the self-organizing map (SOM) by Kohonen[9].

Topology preserving maps have been found in many regions of the brain, e.g. in the visual, auditory, or somatosensory cortex[5]. The SOM provides a possible explanation how such maps can develop. However, its biological relevance depends strongly on its implementation. Previous versions of the SOM assume that the output of a neuron is characterized by its firing rate and not by the timing of single firing events. Using lateral connections the procedure for detecting the so-called winner neuron is usually implemented as a recurrent network. This has the consequence that the winner neuron is detected not before the network has settled down into an equilibrium state. However, since the computation relies on the convergence of the network, this approach disregards the benefits that temporal coding offers to fast information processing in SNNs.

In addition to these conventional implementations there has also been some research on biologically more realistic models of self-organizing map algorithms, e.g. by Kohonen[10], Sirosh and Miikkulainen[11], Choe and Miikkulainen[12]. Also in these approaches the output of a neuron is assumed to correspond to its firing rate, and learning takes place in terms of this rate after the network has reached a stable state of firing.

In this article we propose a mechanism for achieving self-organization in SNNs where computing and learning are both based on the timing of single firing events. In contrast to the standard formulation of the SOM, our construction has the additional advantage that the winner among the competing neurons can be determined fast and locally by using lateral excitation and inhibition. These lateral connections also constitute the neighborhood relationship among the neurons. We assume that initially neurons which are topologically close together have strong excitatory lateral connections whereas remote neurons have strong inhibitory connections. During the learning process the lateral weights are decreased, thus reducing the size of the neighborhood.

We have investigated the capability of the model to form topology preserving mappings in a series of computer simulations. Our results show that it exhibits the same characteristic behavior as the SOM. The typical emergence of topology preserving behavior could be observed for a wide range of parameters. The model of unsupervised learning that we propose in this paper is therefore a candidate for a more realistic description of fast analogue computation in biological neural systems. Moreover, it also provides a link to possible industrial applications via silicon implementations in pulse coded VLSI[13].

COMPUTING WITH SPIKING NEURONS

The investigation and analysis of computing with spiking neurons is easier if it is restricted to the model of leaky integrate-and-fire neurons which disregards non-linearities, e.g. the spatio-temporal integration of postsynaptic potentials in the dendritic tree. Recently Maass has shown how such neurons can compute weighted sums in temporal coding, where the firing time of a neuron encodes a value in the sense that an early firing of the neuron represents a large value[8]. More precisely, one considers a neuron v which receives excitatory input from m neurons $u_1, \ldots, u_m$; the corresponding weights are denoted by $w_1, \ldots, w_m$. Each u_i fires exactly once within a sufficiently small time interval $[0, T_1]$ at a time t_i with $t_i = T_1 - s_i$, where s_i is the ith input to v and $T_1 > 0$ some constant.

frame i of a sequence the goal of learning is to predict the values of the same jet in the next time unit (captured in frame i+1). Thus, we used grids extracted from frame i as input and the jets corresponding to their center extracted from frame i+1 as target for all frames of each sequence. The network was trained using the Widrow-Hoff gradient method.

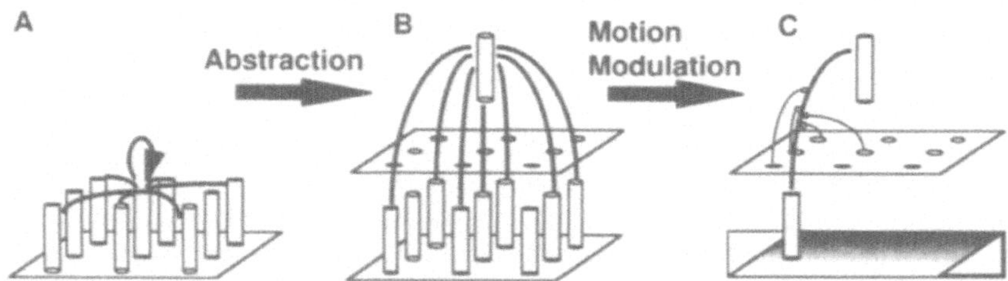

Figure 1. Deriving the model architecture. A) Schematic representation of a V1 patch - with full connectivity. Cylinders symbolize hypercolumns, links represent recurrent and horizontal interactions (shown only for the central hypercolumn). B) Abstraction of the recurrent network by a feedforward network (only for the central hypercolumn). The top cylinder represents the central hypercolumn in time t+1. A separate layer of motion detectors (shown in the middle) interacts with the feedforward connections. C) Multiplicative modulation of each feedforward connection by local motion inputs.

We did not build in any a priori knowledge enforcing any binding among the units of the same jet or between jets. The only a priori constraint was the multiplicative interaction among the Gabor units and the motion units transforming the network into a sigma-pi network [8, 9](Fig. 1C). Each feedforward connection is gated by 5 motion inputs (4 neighbors' and its own) leading to a total of 2^5 possible motion configurations. The weights of the motion synapses are all fixed to one, and only the weights of connections going to the output units are adaptive. The output of unit j is computed as follows:

$$o_j = \sum_{i,k} w_{ijk} * x_{ik} = \sum_{i,k} w_{ijk} * x_i \prod_{v=1}^{5} \mu_{vk}$$

where k is one of the 32 possible combinations of 5 binary motion variables and

$$\mu_{vk} = \begin{cases} m_v & if \ k_v = 1 \\ n_v & if \ k_v = 0 \end{cases}$$

m and n are motion and no-motion detectors, respectively, and w_{ijk} is the weight of the connection going from ith Gabor input unit which is multiplied by kth motion input aggregate.

RESULTS

Out of the 1152 connection weights corresponding to the 3 orientation filters at 9 coordinates and several motion configurations for each only 60 connections survived. From the neighboring coordinates only connections coming from the left neighbors with motion configuration indicating that the left neighbor is moving (therefore, replacing the center jet in the next time unit) were remained. The rest of the surviving connections belonged to the same coordinate with motion configuration signaling no motion neither at the location nor at the left neighbor. In both cases only connections linking the same orientation in input and output survived, showing an iso-orientation connection pattern. In short, motion in one direction and one speed led to development of lateral connections, and a proper gating based on spatio-temporal information

DISCUSSION

Our scheme readily extends to all four motion directions in plain, and more velocities (sufficient for linear extrapolation of the diagonal directions) by simply introducing a more elaborate layer of motion detectors, implicating that this scheme can clearly be used in predicting translation. In our model, we assumed the motion information accurately computed and ready to be used. We acknowledge that the computation of the global motion direction involves addressing the local apperture problem, and may not be locally available with the level of the accuracy which we have used. In order to extend the paradigm to 3D motion, depth information should be incorporated into the modulatory part of the network. Processing sequences containing looming or receding stimuli can lead to development of connections among cells with same orientation tuning but of varying spatial frequency, potentially useful for achieving size invariant prediction and/or recognition. The predictory paradigm is not limited to predicting motion-based changes, but can be applied for any visual transformation, such as illumination or edge based transformations.

The fact that in case of moving objects a prediction mechanism leads to iso-orientation connectivity offers an alternative or extension to earlier explanations for existence of the significant iso-oriented axonal patterns in the primary visual cortex [10, 11].

REFERENCES

1. Logothetis, N.K. and D.L. Sheinberg, *Visual object recognition.* Annual Review of Neuroscience, 1996. **19**: p. 577-621.
2. Katz, L.C. and E.M. Callaway, *Development of local circuits in mammalian visual cortex.,* in *Annual Review of Neuroscience,* W.M. Cowan, *et al.*, Editor. 1992, Annual Reviews, Inc.: Palo Alto. p. 31-56.
3. Burkhalter, A., K.L. Bernardo, and V. Charles, *Development Of Local Circuits in Human Visual Cortex.* The Journal Of Neuroscience, 1993. **13**(5): p. 1916-1931.
4. Spelke, E.S., *Principles of Object Perception.* Cognitive Science, 1994. **14**: p. 29-56.
5. Fukushima, K., *Neocognitron: A self-organizing neural network model for a mechanism of patern recognition unaffected by shift in position.* Biological Cybernetics, 1980. **36**: p. 193-202.
6. Foldiak, P., *Learning Invariance from Transformation Sequences.* Neural Computataion, 1991. **3**: p. 194-200.
7. Hinton, G.E., *Learning translation invariant recognition in a massively parllel network.,* in *PARLE: Parallel Architectures and Languages Europe,* G.G.&.G. Hartmanis, Editor. 1987, Springer-Verlag: Berlin. p. 1-13.
8. Mel, B.W. and C. Koch, *Sigma-pi learning: On radial basis functions and cortical associative learning.* Advances in Neural Information Systems, ed. T.S. Touretzky. 1990, San Mateo: Morgan Kaufmann.
9. Rumelhart, D.E., G.E. Hinton, and J.L. McClelland, *A general framework for parallel distributed processing.* J. L. McClelland, ed. D.E. Rumelhart. 1986, Cambridge: Bradford.
10. Lund, J.S., Y. Takashi, and J.B. Levitt, *Comparison of intrinsic connectivity of macaque monkey cerebral cortex.* Cerebral Cortex, 1993. **3**: p. 148-162.
11. Malach, R., *et al.*, *Relationship between intrinsic connections and functional architecture revealed by optical imaging and in vivo targeted biocytin injections in the primate striate cortex.* Proceedings of the National Academy of Sciences, 1993. **90**: p. 10469-10473.

ANALOG VLSI MODEL OF THE LEECH HEARTBEAT ELEMENTAL OSCILLATOR

Mario F. Simoni[1], Girish N. Patel[1], Stephen P. DeWeerth[1], and
Ron L. Calabrese[2]

mario@ece.gatech.edu, girish@ece.gatech.edu, steved@ece.gatech.edu, and
rcalabrese@biology.emory.edu

[1]Electrical and Computer Engineering   [2]Department of Biology
Georgia Institute of Technology   Emory University
Atlanta, GA 30332-0250   Atlanta, GA 30322

INTRODUCTION

Although significant complexity has been demonstrated in software-based neural system models, it is very difficult to simulate these models near real time. Such real time operation is critical both in the extraction of the maximal information from the models and in the utilization of these models in the creation of artificial systems. Analog very large-scale integrated (aVLSI) circuits have been shown to be a useful medium for implementing real-time neural system models[1]. Additionally, aVLSI circuits are compact and dissipate little power, facilitating the engineering of artificial systems based on biological principles. Much research has been performed in the aVLSI modeling of early sensory (e.g., visual and auditory) processing[2,3,4]. Little application has been made, however, in the modeling of biological motor systems, even though the technology has significant potential in this area.

For example, pattern generating circuits can be created using aVLSI model neurons that have nonlinear membrane properties with both fast and slow dynamics. Although some aVLSI modeling of CPGs has been performed[5,6], it has been based on very simple neuron models. When silicon neurons are designed at the level of ionic channels, modelers can investigate, in real time, the effects of membrane and synaptic properties on patterns produced by CPG circuits. We summarize our efforts in the design of silicon neurons for the modeling of a two-cell, reciprocally-inhibitory network that forms the elemental oscillator in the leech heartbeat system. To validate our model we compare and contrast the data obtained from the aVLSI system with data obtained from a detailed computer model of the leech heartbeat elemental oscillator[7,8].

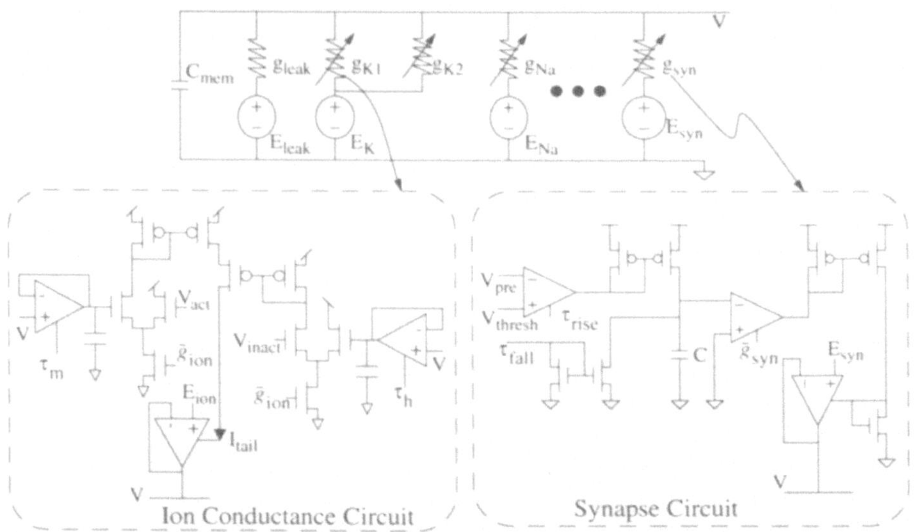

Figure 1. A schematic representation of the aVLSI model neuron showing the single–compartment, electrical equivalent circuit model architecture, where V is the membrane potential. The bottom two schematics show a more detailed transistor level schematic of the circuits that model the voltage gated conductances and the spike mediated synapse. For a voltage gated conductance that has only activation only half of the shown circuit is used. The triangles represent a standard, five–transistor operational transconductance amplifier (OTA) where the third input is to set the bias, or tail, current which ultimately sets a conductance.

AVLSI NEURON MODEL

In order to implement pattern generating circuits, we first developed a single-compartment, aVLSI model neuron as shown in Figure 1. This model neuron, similar to those developed by Mahowald and Douglas[9], is the electrical equivalent circuit of a biological neuron with ionic conductances that have Hodgkin–Huxley type dynamics. Unlike their predecessors, our model neurons include both fast and slow ionic currents and spike-mediated synaptic transmission. The programmable channel parameters include the reversal potentials, E_{ion}, time constants for activation and inactivation of channel conductances, τ_m and τ_h, maximum channel conductances, $\bar{g}_{ion}$, and the potential at which activation and inactivation

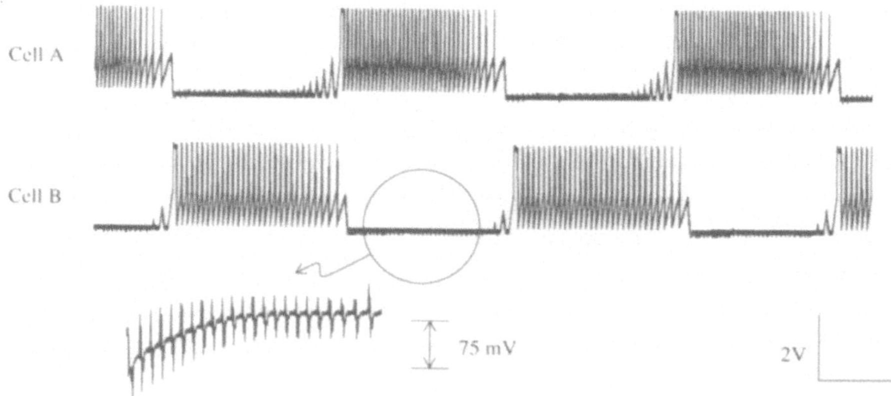

Figure 2. Typical bursting patterns produced from two silicon neurons connected in a reciprocally inhibitory network. Inset shows magnified membrane potential showing sag potential due to hyperpolarization-activated inward currents.

curves reach 50 percent, V_{act} and V_{inact}. The synapses are implemented such that an action potential in the presynaptic cell causes a change in the postsynaptic conductance that is characterized by a quasi–alpha function. The change in postsynaptic conductance is not a true alpha function because the conductance continuously increases in magnitude while the presynaptic cell's membrane potential is above V_{thresh}. The control parameters for synaptic transmission are the rising and falling time constants of the alpha function, τ_{fall} and τ_{rise}, maximal synaptic conductance, $\bar{g}_{syn}$, synaptic reversal potential, E_{syn}, and threshold at which the presynaptic potential will elicit a postsynaptic potentiation, V_{thresh}.

The behavior of the aVLSI model is determined by the combination of the different channel types and their respective parameters. Extensive biological experimentation and thorough computer modeling of the leech heartbeat system have provided the design guidelines for the choice of channel types and their parameter spaces that are necessary for bursting type behavior[7,8]. The channel types contained in the aVLSI model include: a leak current I_l, a fast sodium current I_{Na}, a delayed rectifying potassium current I_{K2}, a fast activating/slow inactivating potassium current I_{K1}, a persistent sodium current I_P, a hyperpolarization activated inward current I_h, and a slow calcium current I_{CaS}. Three currents implemented in the software model were shown to have less significance for bursting behavior and were thus omitted from the aVLSI model to help simplify the circuits. The three omitted currents include a rapidly activating and inactivating potassium current, I_A, a rapidly activating and inactivating calcium current, I_{CaF}, and a graded synaptic current that is dependent on the presynaptic calcium concentration, I_{SynG}. The membrane currents, I_A and

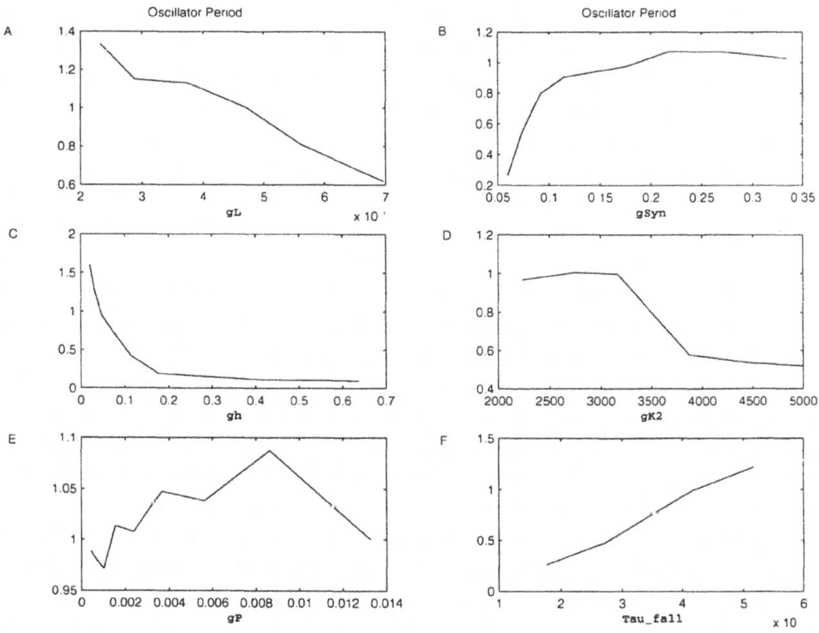

Figure 3. Sensitivity of bursting period to maximal conductances (A–E) and fall-time of spike-mediated synaptic conductance (F). Burst periods and Tau_fall are normalized to oscillator period at operating point. All conductances are in μS.

I_{CaF}, were shown to have minor effects on bursting period when varied in the software model except when their parameters vary significantly from the standard[8]. The graded synaptic transmission as mediated by I_{SynG}, was shown in voltage clamp experiments to play a minor functional role in the biological neurons under normal conditions[10].

ANALYSIS OF DATA

To validate our model we compared and contrasted the behavior of a pair of reciprocally-inhibited aVLSI neurons to the computer simulations and biological data. We first found an operating point in the parameter space that produced bursting behavior similar to biological neurons, where the ratio between the bursting period and spiking period is on the order of 1:100, this is shown in Figure 2. From this operating point, we varied the maximal conductances individually while fixing all other parameters. The resulting changes in bursting period were recorded and compared to changes from similar parameter shifts in the software model. The parameters that have an effect on the bursting period are shown in plots A through F of Figure 3.

One of the major features we observed was that the "fast" currents can have dramatic effects on "slow" behaviors. A good example of this is plot D which shows that $\bar{g}_{K2}$ can alter the bursting period by almost 50%. As $\bar{g}_{K2}$ is increased, it decreases the pulse width of action potentials. This decrease in pulse width corresponds to less synaptic effect on the postsynaptic cell because the synaptic current is dependent upon the amount of time that the membrane potential is above V_{thresh}. The decreased synaptic input allows the postsynaptic cell to escape from the inhibition sooner than it would normally, thus decreasing the bursting period. This same parameter has a similar effect on the bursting period in the software model but for different reasons. In the software model, an increase in $\bar{g}_{K2}$ decreases the level of the plateau potentials which causes a decrease in action potential frequency, and thus a decrease in synaptic inhibition[8].

There are two reasons why the I_{K2} current has different mechanisms for altering the bursting period in the two models. The primary reason is that the I_{K2} current plays different roles in the two models. In the software model this current has a time constant between that of the action potentials and the bursting period. This time constant allows the current to gradually build up during the plateau phase of the bursting period and, as it builds up, it decreases the level of the plateau potential as mentioned above. However, in the aVLSI model this current is used to repolarize the membrane potential during an action potential because the sodium conductance inactivates differently from the software model. In the aVLSI model, the inactivation of the sodium conductance has less of an effect than in the software model because of the manner in which the activation and inactivation dynamics interact. The circuits used to implement the sigmoidal activation and inactivation curves, in the aVLSI model, have very narrow sloped regions, and thus do not allow much overlap in these regions of the curves, which causes an "all or nothing" effect on the current. However, to use I_{K2} in this manner, the time constant of activation had to be much faster than in the software model and, as such, it was unable to play the same role. The second reason is because the role played by I_{K2} in the aVLSI model means that it has a direct effect on the width of the action potentials which in turn directly affects the magnitude of the synaptic current. This is not an issue in the software model because the synaptic current is independent of the width of the action potentials.

Shifts in other parameters such as the leak current (Figure 3A) and the hyperpolarization activated inward current (Figure 3C) produce similar effects for similar reasons in both the aVLSI model and the software model. The I_{K1} current was the only current for which the aVLSI model showed little sensitivity to bursting period while the full software model showed a relatively large sensitivity. This lack of sensitivity is because the I_{K1} current has the same problem with the circuits implementing the activation and inactivation curves as mentioned above for the sodium conductance. This problem is made even greater for I_{K1} because the time constant of inactivation is even faster than that for sodium so the magnitude of the instantaneous conductance is constantly close to zero.

CONCLUSIONS AND FUTURE WORK

We have described aVLSI model neurons with detailed Hodgkin–Huxley-type dynamics, and have shown that a reciprocally inhibited pair of these neurons are capable of producing qualitatively similar behavior to the leech elemental oscillator. These circuits provide a real-time facility, which greatly accelerates the exploration of the parameter space of the modeled systems. In this paper, we have explored only a small portion of the parameter space for the leech elemental oscillator. We are presently conducting a more thorough examination of this parameter space to study the effects of other individual parameters and to test the interdependence among the parameters.

We are also addressing the aVLSI issues that affect the model to create dynamics that are closer to biological neurons. The major problem facing this particular model is the circuits that model the activation and inactivation curves. We have already designed a sigmoid producing circuit that has a programmable sloped region[11] and have used this circuit in the fabrication of another neuron model. Preliminary testing results of this improved aVLSI model show that this new circuit corrects many of the issues that were presented here. We are also exploring ways to cut down on the number of parameters and improve the model's stability without affecting the underlying dynamics[12]. As a longer range goal, we are exploring circuit technologies to make even slower time constants and ways of storing analog values within the model to implement forms of short and long term learning.

This aVLSI neuron model has a lot of potential to create complex systems that span multiple levels of computation from sub-cellular dynamics to emergent, system–level behavior. Furthermore, all of this computation is performed in real-time so that we can use this model to create autonomous systems. We are currently working on a system of coupled pattern generators using an improved version of this aVLSI model neuron to simulate lamprey swimming. We then plan to combine this aVLSI network with a mechanical system to close the sensory feedback loop, and, thus, study various ways that motor output can alter the sub-cellular dynamics. Then, with a closed–loop system, we can explore how alterations in the sub-cellular dynamics can be used to achieve a desired form of behavior through various learning algorithms. The ability to work with a complete, closed-loop system in real-time provides an tremendous advantage over working with isolated parts of a system under unnatural conditions.

ACKNOWLEDGMENTS

This research was funded by NSF grant #IBN-9511721. The authors would also like to thank Farzan Nadim and Øystein Olsen for their discussions during the preliminary design stage and analysis and testing stage of this project. We would also like to thank the Georgia Tech Analog Consortium for their funding of travel expenses.

REFERENCES

1. C. Mead. *Analog VLSI and Neural Systems*. Addison–Wesley, New York, NY (1989).
2. L.Watts, R. Lyon, and C. Mead, A bidirectional analog VLSI cochlear model, *Advanced Research in VLSI, Proceedings of the 1991 Santa Cruz Conference*, C. Sequin, ed., MIT Press, Cambridge, MA 153–163, (1991).
3. R. Sarpeshkar, R. Lyon, and C. Mead, An analog VLSI cochlea with new transconductance amplifiers and nonlinear gain control, *Proceedings of the IEEE Symposium on Circuits and Systems, Vol. 3*, 292–295, (1996).
4. K. Boahen and A. Andreou, A contrast sensitive silicon retina with reciprocal synapses, *Advances in Neural Information Processing Systems*, 4:764-772, (1991).
5. S. Ryckebusch, J. Bower, and C. Mead, Modeling small oscillating biological networks in analog VLSI. in: *Advances in Neural Information Processing Systems 1*, J. D. Cowan, G. Tesauro, and J. Alspector, eds., Morgan Kaufman Publishers, Palo Alto, CA 384–393 (1989).
6. S. Wolpert, J. Masi, R. Davis, M. Fox, R. Peura, Recreating neural oscillators in VLSI, *Proceedings of the 1994 20th Annual Northeast Bioengineering Conference*, Springfield, MA, 64-66, (1994).
7. F. Nadim, Ø. Olsen, E. De Schutter, and R. Calabrese, Modeling the leech heartbeat elemental oscillator: I. Interactions of intrinsic and synaptic currents, *J. Comp. Neurosci.* 2(3): 215–236 (1995).
8. Ø. Olsen, F. Nadim, E. De Schutter, and R. Calabrese, Modeling the leech heartbeat elemental oscillator: II. Exploring the parameter space,. *J. Comp. Neurosci.* 2(3): 215–236, (1995).
9. M. Mahowald and R. Douglas, A silicon neuron, *Nature*, Vol. 354. 19(26): 515–518, (1991).
10.Ø. Olsen and R. Calabrese, Activation of intrinsic and synaptic currents in leech heart interneurons by realistic waveforms, *J. Neurosci.* 16(16):4958–4970, (1996).
11.S. DeWeerth, G. Patel, M. Simoni, Variable linear-range subthreshold OTA, Electronics Letters, 33(15):1309–1311, (1997).
12.M. Simoni and S. DeWeerth, Adaptation in an aVLSI model of a neuron, *Proceedings of the IEEE Symposium on Circuits and Systems*, in press, (1998).

A MATHEMATICAL DESCRIPTION FOR GABAERGIC MODULATION OF SEQUENCE DISAMBIGUATION IN HIPPOCAMPAL REGION CA3

Vikaas S. Sohal and Michael E. Hasselmo*

Department of Psychology and Program in Neuroscience
33 Kirkland St., Harvard University, Cambridge, MA 02138
sohal@katla.harvard.edu, hasselmo@katla.harvard.edu

ABSTRACT

Changes in $GABA_B$ modulation may underlie experimentally observed changes in the strength of synaptic transmission at different phases of the theta rhythm. Analysis demonstrates that these changes improve sequence disambiguation by a model of CA3, and that changes in the level of GABAergic modulation correspond to changes in the temperature of an analog network. These results suggest that $GABA_B$ modulation in CA3 may produce dynamics which resemble annealing. These dynamics may underlie a role for the theta cycle in improving sequence retrieval for spatial navigation.

INTRODUCTION

Manipulations that abolish the hippocampal theta rhythm impair spatial navigation (Winson, 1978), but the computational function of theta is not known. Previous experiments have found that the amplitude of evoked field potentials due to neuronal spiking activity changes during different phases of the theta rhythm (Rudell et al., 1980; Buzsaki et al., 1981; Rudell and Fox, 1984). This could result from changes in pre and/or postsynaptic inhibition during a theta cycle. More recent evidence suggests that the strength of synaptic transmission also depends on the phase of theta rhythm (Wyble et al., 1997), which may also reflect phasic changes in presynaptic inhibition.

Phasic changes in presynaptic inhibition could result from the rhythmic activation of presynaptic $GABA_B$ receptors, which selectively suppress recurrent synaptic transmission in piriform cortex (Tang and Hasselmo, 1994) and hippocampus (Ault and Nadler, 1982; Colbert and Levy, 1992). Stewart and Fox have hypothesized (1990), and *in vivo* recordings have shown (Buszaki and Eidelberg, 1983; Fox et al., 1986; Skaggs et al., 1996), that some interneurons tend to fire near a preferred phase of theta. Phasic firing by even a subset of interneurons that activate presynaptic $GABA_B$ receptors could rhythmically activate

* Author to whom correspondence should be addressed

presynaptic GABA$_B$ receptors, producing the changes in synaptic strength observed *in vivo* (Wyble et al., 1997). Paired pulse depression, thought to result from activation of presynaptic GABA$_B$ receptors on interneurons, has a rise time and fast component of decay (Otis et al., 1993) that are compatible with theta-frequency oscillations in GABA$_B$-mediated synaptic suppression.

These rhythmic changes could be important for retrieval of sequences of activity in the hippocampus. Evidence suggests that the hippocampus may represent a series of adjacent locations by a sequence of successively firing place cells (Tsodyks et al., 1996; Jensen and Lisman, 1996; Wallenstein and Hasselmo, 1997). In a neural network model of hippocampal region CA3, we find that changes in GABA$_B$ suppression during a theta cycle improves sequence disambiguation: the retrieval of sequences with identical starting points. These dynamics may underlie a role for the theta cycle in improving sequence retrieval for spatial navigation.

OPTIMAL NETWORK PARAMETERS DURING SEQUENCE DISAMBIGUATION

When two sequences, $\{X_1, ..., X_m\}$ and $\{X_1, ..., X_1, Y_{i+1}, ..., Y_m\}$ are both stored in a network, it is ambiguous whether X_i should be followed by X_{i+1} or Y_{i+1}. The only information that can resolve this ambiguity is knowledge of the desired endpoint, X_m or Y_m. If the patterns of activity are neural representations of locations, then disambiguating forked sequences using knowledge of the desired endpoint corresponds to the everyday problem of deciding which way to turn at an intersection based on one's destination.

The simplest neural realization of this problem, shown in Figure 1, contains two sequences: $\{a_1, a_2\}$ and $\{a_1, a_3\}$. When a_1 becomes active, the pattern of activity that should follow a_1 is ambiguous. We represent disambiguation of the sequences $\{a_1, a_2\}$ and $\{a_1, a_3\}$ by assuming that a_2 receives a small amount of afferent input when a_1 is active. Afferent input to a_2 represents the knowledge that "a_2 is the desired goal," and should bias the network so that it completes the sequence $\{a_1, a_2\}$ rather than $\{a_1, a_3\}$. We are not studying whether the network *can* disambiguate $\{a_1, a_2\}$ from $\{a_1, a_3\}$ when a_2 is given a biasing input. Instead we describe some *optimal* characteristics of a network that does perform this disambiguation.

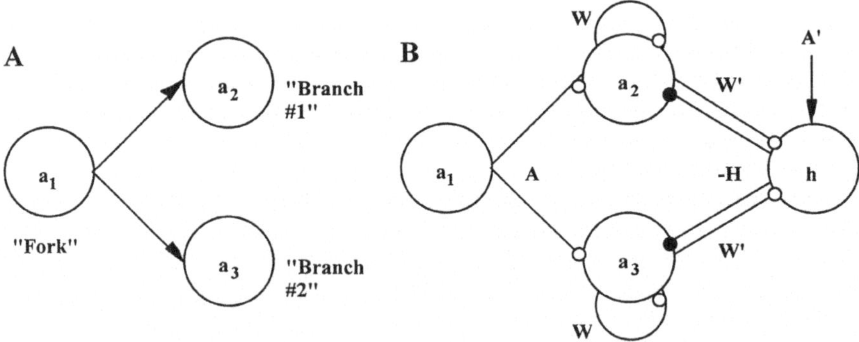

Figure 1. A. A simple sequence disambiguation problem. B. A neural network realization of the sequence disambiguation problem diagrammed in A.

The network parameters, shown in Figure 1B, are: A, the strength of excitation from a_1 to a_2 and a_3; A_{bias}, afferent input to a_2; A', afferent input to the inhibitory interneuron; W, the strength of recurrent excitatory connections from a_2 to a_2 and a_3 to a_3; W', the strength of excitatory connections from a_2 and a_3 to the interneuron; -H, the strength of the inhibitory connections from this interneuron to a_2 and a_3; h, activity of the model interneuron; η, the rate of passive decay for the activities, a_2, a_3, and h; θ, the threshold of the neurons represented by a_2, a_3, and h.

To represent the rhythmic changes in the strength of recurrent synaptic transmission that have recently been observed experimentally (Wyble et al., 1997), we introduce the parameters k(t) and k'(t). The strengths of all recurrent excitatory connections in the network (the parameters A, W, and W') are multiplied by k(t). So [1-k(t)] represents the suppression of recurrent excitation at time t. Similarly, k' multiplies the strength of the recurrent inhibition in the network (the parameter H), so [1-k'(t)] represents the suppression of recurrent inhibition at time t.

The activity of each model neuron is a continuous variable that represents mean firing rate of a pool of real neurons. The equations for the evolution of a_2, a_3, and h, have been derived by averaging over short timescales and many neurons, and are similar to those used elsewhere (Wilson and Cowan, 1972):

$$\dot{a}_2 = -\eta a_2 + kA + A_{bias} + kW(a_2 - \theta)_+ - k'H(h - \theta)_+$$
$$\dot{a}_3 = -\eta a_3 + kA + kW(a_3 - \theta)_+ - k'H(h - \theta)_+ \qquad (1)$$
$$\dot{h} = -\eta h + A' + kW'(a_2 - \theta)_+ + kW'(a_3 - \theta)_+$$

where $(x)_+ = x$ if $x > 0$ and 0 otherwise. As long as $A_{bias} > 0$, a_2 will cross threshold before a_3, giving $a_2 > \theta$ and $a_3 < \theta$. Assume that the interneuron will be activated before $a_3 > \theta$. Then the activities are in the regime $a_2 > \theta$, $a_3 < \theta$, and $h > \theta$, in which network dynamics reduce to a linear system. We measured the performance of sequence disambiguation by this system as $a_2(T) - a_3(T)$, where T is a fixed time. This quantity measures the differential completion of the two sequences, $\{a_1, a_2\}$ and $\{a_1, a_3\}$ at the end of the time interval during which the network processes this sequence disambiguation problem.

We found the k(t) and k'(t) that maximized the quantity $a_2(T) - a_3(T)$ using the *maximum principle* (Pontryagin et al., 1962). The maximum principle determines necessary conditions on the functions k(t) and k'(t) to maximize the difference $a_2(T) - a_3(T)$. These conditions uniquely specify k(t) and k'(t):

$$k = \begin{cases} k_{min} \ \mathrm{if}\ t < t_{final} - t_1, \\ k_{max} \ \mathrm{if}\ t > t_{final} - t_1 \end{cases}, \ \mathrm{where}\ t_1 = \frac{\pi}{W k_{max}} \sqrt{\frac{W^2 k_{max}}{4HW'k'_{max} - W^2 k_{max}}} \qquad (2)$$

$$k' = \begin{cases} k'_{min} \ \mathrm{if}\ t < t_{final} - t_1 - t_2, \\ k'_{max} \ \mathrm{if}\ t > t_{final} - t_1 - t_2 \end{cases}, \ \mathrm{where}\ t_2 = \frac{\pi}{W k_{min}} \sqrt{\frac{W^2 k_{min}}{4HW'k'_{max} - W^2 k_{min}}} \qquad (3)$$

This indicates that optimal sequence disambiguation occurs when the level of suppression of recurrent synaptic transmission decreases from its maximum values ($1 - k_{min}$, $1 - k'_{min}$) to its minimal values ($1 - k_{max}$, $1 - k'_{max}$) during the sequence disambiguation task. In contrast, application of the maximum principal to the biasing input, A_{bias}, shows that optimal sequence disambiguation occurs when A_{bias} remains at its maximum value throughout the sequence disambiguation task.

According to these results, $a_2(T) - a_3(T)$ should increase if the suppression of recurrent synaptic transmission decreases while the network described above is performing sequence disambiguation. However, these results do not directly show that a decreasing suppression of recurrent synaptic transmission allows this network to successful disambiguate two sequences, nor do they indicate whether changes in the performance of this network are significant. To answer these questions, we solved for the final state of the system described above and studied how that final state changed if the suppression of recurrent synaptic transmission decreased while that system disambiguated two sequences.
$k(t)$ and $k'(t)$ took the form:

$$k(t) = k'(t) = \alpha t + \beta$$
$$\alpha(100 \, \text{m sec}) + \beta = 1$$

(4)

The choice of 100 msec corresponds to one-half of a theta cycle. Thus, α measures the amount by which the suppression of recurrent synaptic transmission decreases during the sequence disambiguation task.

If a_1 is active, and a_2 is receiving biasing input, then the system has only two possible final states: for low A_{bias}, the system reaches a final state in which both a_2 and a_3 are active, whereas for A_{bias} sufficiently large, only a_2 is active in the final state, so sequence disambiguation is successful. We found the minimum A_{bias} necessary for successful sequence disambiguation as a function of α. The results, shown in Figure 2, demonstrate that increasing α, corresponding to a *greater change* in the suppression of recurrent synaptic transmission, allows successful sequence disambiguation using *weaker biasing inputs*, i.e. using less information about the desired goal.

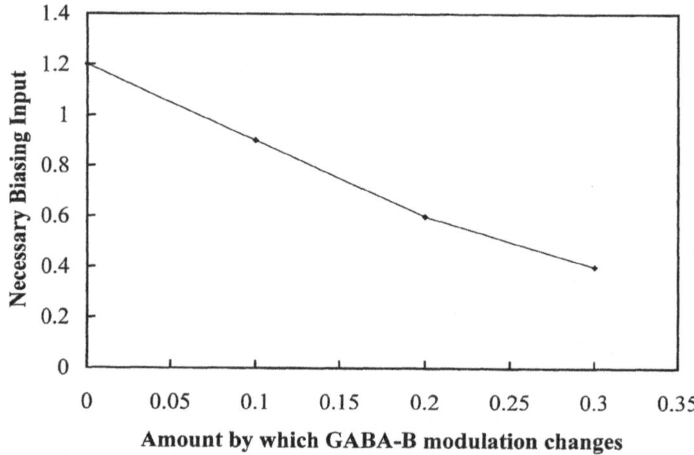

Figure 2. Shows the biasing input necessary for successful sequence disambiguation, as a function of the amount by which $GABA_B$ modulation decreases during the task.

SIMULATIONS

We have also studied sequence disambiguation in continuous-firing-rate and sparsely-connected integrate-and-fire models of hippocampal region CA3 (Sohal and Hasselmo, 1998b). In both models, theta-frequency inputs from the septum entrained oscillations in the $GABA_B$ suppression of recurrent synapses, producing phasic changes in the strength of synaptic transmission during a theta cycle similar to those observed experimentally (Wyble et al., 1997). These oscillations caused the $GABA_B$ suppression of recurrent synapses to fall during sequence disambiguation. Consistent with the preceding analysis, this fall improved sequence disambiguation by these networks.

RELATION TO ANNEALING

During successful sequence disambiguation GABAergic suppression is initially high. As a result, recurrent excitation and inhibition are weak, so many neurons, representing multiple sequences, are active. However, as GABAergic suppression falls, recurrent excitation and inhibition become stronger, so that a few neurons representing one sequence become active while the remaining neurons are inhibited. Thus, as the level of GABAergic modulation falls, the network shifts from sampling many possible states to selecting the optimal state. A shift from sampling to selecting states also occurs during annealing as the temperature of a system gradually decreases. Thus the role of GABAergic modulation in sequence disambiguation may be analogous to that of temperature in annealing, and a falling level of GABAergic modulation may improve sequence disambiguation just as a decreasing temperature drives annealing. Hopfield and Tank (1985) showed that the equilibrium solution in an analog network equals the effective field solution for a Boltzmann machine (Hinton and Sejnowski, 1986) at an effective temperature determined by the gain width. Similarly, we have shown that the equilibrium solution to our network is the same as the effective field solution in a Boltzmann machine, and the fall in GABAergic modulation which occurs in our network is equivalent to decreasing both the effective temperature and the energy of data terms relative to constraints (Sohal and Hasselmo, 1998a).

CONCLUSIONS

As explained in the Introduction, the storage and disambiguation of sequences by CA3 may be relevant to spatial navigation. We have shown that the dynamics of sequence disambiguation may resemble those of annealing, and other studies have used simulated annealing by a Boltzmann machine for robot navigation (Lin and Lee, 1995). Storage of temporal sequences in the hippocampus may also be important for non-spatial tasks such as transitive inference (Sohal and Hasselmo, 1998a).

These results suggest that one function of the theta rhythm may be to cause rhythmic changes in $GABA_B$ suppression of synaptic transmission. Our analysis demonstrates that such changes in $GABA_B$ suppression may produce dynamics that resemble annealing and improve sequence disambiguation in the hippocampus.

REFERENCES

Ault, B. and Nadler, J.V., 1982, Baclofen selectively inhibits transmission at synapses made by axons of CA3 pyramidal cells in the hippocampal slice. *J Pharmacol Exp Ther.* 223: 291-297.

Barkai, E. and Hasselmo, M.E., 1994, Modulation of the input/output function of rat piriform cortex pyramidal cells. *J Neurophys.* 72: 644-658.

Buszaki, G. and Eidelberg, E., 1983, Phase relations of hippocampal projection cells and interneurons to theta activity in the anesthetized rat. *Brain Res.* 266: 334-339.

Buszaki, G., Grastyan, E., Czopf, J., Kellenyi, L., and Prohaska, O., 1981, Changes in neuronal transmission in the rat hippocampus during behavior. *Brain Res.* 225: 235-247.

Colbert, C.M. and Levy, W.B., 1992, Electrophysiological and pharmacological characterization of perforant path synapses on CA1: mediation by glutamate receptors. *J Neurophys.* 68: 1-8.

Hinton, G.E. and Sejnowski, T.J., 1986, Learning in Boltzmann machines. In: *Paralell Distributed Processing: Explorations in the Microstructure of Cognition. Volume 1: Foundations,* D.E. Rummelhart and J.L. McClelland, eds., The MIT Press, Cambridge, MA.

Hopfield, J.J. and Tank, D.W., 1985, Neural computation of decisions in optimization problems. *Biol Cybern.* 52: 141-152.

Jensen, O. and Lisman, J.E., 1996, Theta/gamma networks with slow NMDA channels learn sequences and encode episodic memory: role of NMDA channels in recall. *Learning and Memory.* 3: 264-278.

Levy, W.B., 1996, A sequence predicting CA3 is a flexible associator that learns and uses context to solve hippocampal-like tasks. *Hippocampus.* 6: 579-590.

Lin, C.T. and Lee, C.G.S., 1995, A multi-valued Boltzmann machine. *IEEE Trans Syst Man Cyber.* 25: 660-669.

Morris, R.G.M., Garrud, P., Rawlins, J.N.P., and O'Keefe, J., 1982, Place navigation impaired in rats with hippocampal lesions. *Nature.* 297: 681-683.

Otis, T.S., Dekoninck, Y., and Mody, I., 1993, Characterization of synaptically elicited $GABA_B$ responses using patch-clamp recordings in rat hippocampal slices. *J Physiol.* 463: 391-407.

Rudell, A.P. and Fox, S.E., 1984, Hippocampal excitability related to the phase of the theta rhythm in urethanized rats. *Brain Res.* 294: 350-353.

Rudell, A.P., Fox, S.E. and Rank, J.B., 1980, Hippocampal excitability phase-locked to the theta rhythm in walking rats. *Exp Neurology.* 68: 87-96.

Pitler, T.A. and Alger, B.E., 1994, Differences between presynaptic and postsynaptic $GABA_B$ receptors in rat hippocampal pyramidal cells. *J Neurophys.* 72: 2317-2327.

Pontryagin, L.S., Boltyanskii, V.G., Gamkrelidze, R.V., and Mishchenko, E.F., 1962, *The Mathematical Theory of Optimal Processes*, Neustadt LW, ed., Interscience Publishers, New York.

Sohal, V.S. and Hasselmo, M.E., 1998a, The role of $GABA_B$ modulation in hippocampal region CA3 may be analogous to that of temperature in annealing. *Neural Comp.* In press.

Sohal, V.S. and Hasselmo, M.E., 1998b, $GABA_B$ modulation improves sequence disambiguation in computational models of hippocampal region CA3. Submitted.

Stewart, M. and Fox, S.E., 1990, Do septal neurons pace the hippocampal theta rhythm? *Trends Neurosci.* 13: 163-168.

Tang, A.C. and Hasselmo, M.E., 1994, Selective suppression of intrinsic but not afferent fiber synaptic transmission by baclofen in the piriform (olfactory) cortex. *Brain Res.* 659: 75-81.

Tsodyks, M.V., Skaggs, W.E., Sejnowskii, T.J., and McNaughton, B.L., 1996, Population dynamics and theta rhythm phase precession of hippocampal place cell firing:a spiking neuron model. *Hippocampus.* 6:271-280.

Wallenstein, G.W. and Hasselmo, M.E., 1997 GABAergic modulation of hippocampal activity.: sequence learning, place field development, and the phase precession effect. *J Neurophys.* (in press).

Wyble, B.P., Linster, C. and Hasselmo, M.E., 1997 Evoked synaptic potential size depends on the phase of theta rhythm in rat hippocampus. *Soc Neurosci Abstr* 23: 508 (197.7).

BIDIRECTIONAL COMPLETION OF CELL ASSEMBLIES IN THE CORTEX

Friedrich T. Sommer, Thomas Wennekers and Günther Palm

Department of Neural Information Processing
University of Ulm, 89069 Ulm, Germany

INTRODUCTION

Reciprocal pathways are presumedly the dominant wiring organization for cortico-cortical long range projections[5]. This paper examines the hypothesis that synaptic modification and activation flow in a reciprocal cortico-cortical pathway correspond to learning and retrieval in a bidirectional associative memory (BAM): Unidirectional activation flow may provide the fast estimation of stored information, whereas bidirectional activation flow might establish an improved recall mode. The idea is tested in a network of binary neurons where pairs of sparse memory patterns have been stored in bidirectional synapses by fast Hebbian learning (Willshaw model). We assume that cortical long-range connections shall be efficiently used, i.e., in many different hetero-associative projections corresponding in technical terms to a high memory load. While the straight-forward BAM extension of the Willshaw model does not improve the performance at high memory load, a new bidirectional recall method (CB-retrieval) is proposed accessing patterns with highly improved fault tolerance and also allowing segmentation of ambiguous input. The improved performance is demonstrated in simulations. The consequences and predictions of such a cortico-cortical pathway model are discussed. A brief outline of the relations between a theory of modular BAM operation and common ideas about cell assemblies is given.

THE MODEL

Cortical circuitries with strong feedback are the most probable structures realizing Hopfield like associative memory modules[8]. Where anatomy ensures strong direct feedback, Hebbian synaptic learning[7], as verified experimentally[6], enhances the symmetry of the synaptic connections, since reciprocal synapses between a pair of a small group of simultaneously active neurons should be strengthened by roughly the same amount. Columnar networks of strongly interconnected pyramidal cells formed by local intra-cortical connections[3] have been ascribed to this kind of auto-associative memory function[1]. Recent systematic evaluations of anatomical tracer studies in animals shed light on the organization of cortico-cortical long-range connections: Strong reciprocal pathways between two cortical areas x and y – independent from their distance – have turned out to be a dominating connection scheme[5]. These pathways terminate in a patchy way and, strikingly, the patch size coincides with column size[12].

We model a reciprocal cortico-cortical pathway as a densely connected bidirectional hetero-associative memory (BAM), and do not consider explicitly the auto-associative intra-columnar connections (but see result section). Two pools of neurons x and y are fully bidirectionally connected. Synaptic modification is modeled by Hebbian learning of M pattern pairs: $\{(x^\nu, y^\nu) : x^\nu \in \{0,1\}^n, y^\nu \in \{0,1\}^m, \nu = 1, ..., M\}$. We use randomly generated memory patterns with constant numbers of '1'-entries, i.e., $|x^\nu| = \sum_i x_i^\nu = a, |y^\nu| = b \; \forall \; \nu$. Since in the cortex low activation is observed, the memory patterns are sparse, i.e., $a, b << n, m$. For this case the Willshaw model[16, 18] is efficient, where only minimum synaptic depth is required by binary clipping of the outer product sum:

$$C_{ij} = \sup_\nu x_i^\nu y_j^\nu. \tag{1}$$

During retrieval an associative memory maps an initial pattern $x(0)$ closest to a memory pattern x^ν onto a pattern close to y^ν. In the Willshaw model each neuron j forms the dendritic potential

$$d(0)_j := [Cx(0)]_j = \sum_i C_{ij}x(0)_i, \tag{2}$$

and determines its activity value by threshold comparison

$$y(1)_j = H(d(0)_j - \theta) \;\; \forall j, \tag{3}$$

with the global threshold value θ and $H(x)$ denoting the Heaviside function. In the following eqs. (2) and (3) are denoted as *simple retrieval.*

In auto-associative memories usually iterative retrieval is employed[8]. In parallel update schemes simple retrieval is iterated until a stable pattern is reached which is the retrieval result. Iterative retrieval improves the Willshaw model, if combined with an activity dependent threshold strategy or with a Boolean ANDing ($\wedge$) in each component between the old and the new input pattern applied after the second iteration step[13], i.e., $r \geq 1$:

$$x(r+1)_j = x(r)_j \wedge H(d(r)_j - \Theta) \; \forall j. \tag{4}$$

Biologically, Boolean ANDing could be realized by after-depolarization effects, where the cell response is fascilitated by its recent previous activation *(cell fascilitation)*[4, 9, 2].

Crosswise Bidirectional Retrieval

Iterative retrieval can be understood as a kind of high precision recall mode[1] which will now be extended to the hetero-associative case. At higher memory load the input error tolerance of simple retrieval is small, see Fig. 3. For bidirectional iteration of the retrieval as in the Kosko BAM[10] the modification (4) can be first applied after the third step which turns out to be too late to achieve significant retrieval error reduction[14]. Since in the hetero-associative matrix row and column sums are different, retrieval methods combining both in each step might be superior to a sequence of simple retrieval steps (2): In a decision whether a y-neuron is active in a particular memory pattern, we include lateral constraints between different y-units – also stored in the matrix – by the use of second order quantities, denoted as *conditioned links*, see Fig. 1.

A retrieval process based on this idea should return a pattern where '1'-components are connected by the highest conditioned links. We approximate such a process by using

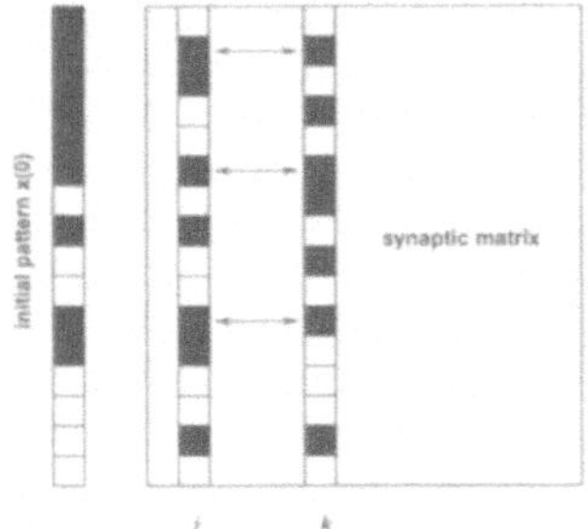

Figure 1. Conditioned link: Black squares correspond to '1'-entries in the pattern and columns. The link between y-units j and k conditioned by the pattern $x(0)$ is defined as overlap between matrix columns restricted on the rows $\{i : x(0)_i = 1\}$:

$$w_{jk}^{x(0)} := \sum_i C_{ij} C_{ik} x(0)_i$$

Provided $x(0)$ is a fair estimation of a learning pattern x^ν then a high conditioned link indicates high probability that $y_j^\nu = y_k^\nu = 1$. In the example $w_{jk}^{x(0)} = 3$.

conditioned links as auto-associative weights for the y-units. The resulting dendritic potentials can be transformed by $\sum_k w_{jk}^{x(r)} y(r{-}1) = \sum_{i \in x(r)} C_{ij} [C^T y(r{-}1)]_i$ (and similiarly in the x-layer). Obviously, they can be computed by a forth and back propagation through the ordinary synaptic matrix. The use of conditioned links in updates of both layers yields a new retrieval scheme. We call it *crosswise bidirectional (CB) retrieval* since matrix rows and columns are evaluated simultaneously:

$$y(r{+}1)_j = H\Big(\sum_{i \in x(r)} C_{ij} [C^T y(r{-}1)]_i - \Theta \Big) \tag{5}$$

$$x(r{+}1)_i = H\Big(\sum_{j \in y(r)} C_{ij} [C x(r{-}1)]_j - \Theta' \Big) \tag{6}$$

For $r = 0$ pattern $y(r{-}1)$ has to be replaced by the simple retrieval result $H([Cx(0)] - \theta)$, for $r > 2$ Boolean ANDing with results from timestep $r - 1$ can be applied.

RESULTS

Memory Performance

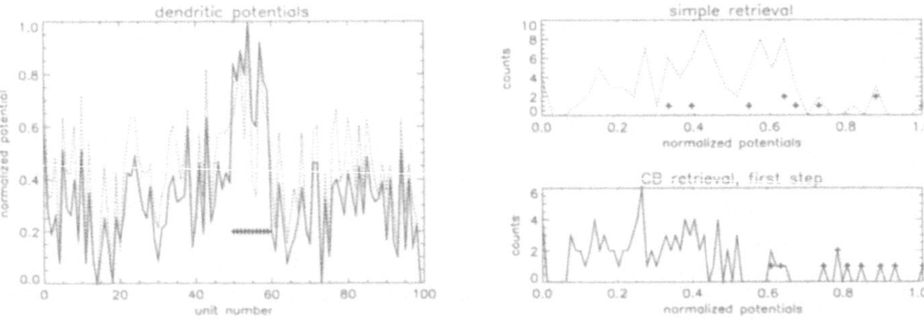

Figure 2. 36 stored pattern pairs with $n = m = 100$ and activities $a = b = 10$ resulting in a memory load of $p(C_{ij} = 1) = 0.3$: Dendritic potentials in the y units and their histograms obtained with an initial pattern corrupted by 35 wrong one-components. Crosses denote one-components in the addressed learning pattern.

Fig. 2 demonstrates in a small network with 100 neurons in each layer that already the first CB retrieval step produces highest dendritic potentials at sites of '1'-components in the addressed learning pattern whereas for simple retrieval a threshold detection of the '1'-components is impossible. Fig. 3 shows at a memory load

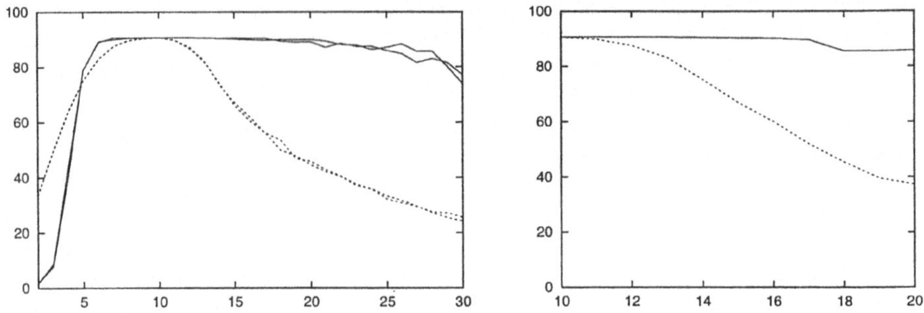

Figure 3. Mean output transinformation about a memory pattern for $n = m = 2000$, $M = 15000$, and again $a = b = 10$. Simple retrieval: dotted, CB-retrieval: solid. The x-axes display the address activity: $|x(0)| = 10$ corresponds to an errorfree memory pattern, lower activities are due to "miss" errors, higher activities are due to "add" errors. Left: Retrieval from addresses with random noise, two trials using different sets of randomly generated memory patterns. The output activities are adjusted near $|y| = b$ by threshold setting. Right: Retrieval from addresses where add errors are '1'-components of a second learning pattern.

$p(C_{ij} = 1) = 0.3$ retrieval results for addresses distorted either by random noise (left), or by a second learning pattern (right) where the address contains one learning pattern and '1'-components of a second learning pattern are successively added with increasing abscissa. The information in a memory pattern with $m = 2000$ and $b = 10$ is 90.8 bits. Lower transinformation values indicate output patterns disrupted by a nonvanishing retrieval error rate. For errorfree addresses, i.e., $|x(0)| = 10$, both retrieval methods work errorfree. However, with increasing input noise the transinformation in the simple retrieval results drops much faster than in the CB retrieval outputs*. With superimposed inputs simple retrieval behaves similiarly poor as with random noise. CB-retrieval does quite well, even if a complete superposition of two learning patterns is presented and yields almost perfectly one of the learning pattern pairs. Subsequently, the other pattern can be retrieved by the original address where the '1'-components from the first pattern pair have been deleted.

Thus, CB-retrieval exhibits much more fault tolerance than simple retrieval and can also be used for a segmentation of ambiguous inputs. The achieved experimental information capacity in the parameter range of Fig. 3 is strikingly high, reaching the order of the theoretical expectations of 0.52 bit/synapse. The higher performance is due to better exploitation of information stored in the Hebbian learning matrix and does not require more complex learning procedures[11] or dummy augmentation in the pattern coding[17] as has been proposed earlier to improve BAM. For a theoretical treatment and more detailed simulation results, see[15].

Implications and questions raised by the model

We have shown that information stored by a rough (binary clipped) hetero-associative Hebbian learning process can be recalled with significantly increased precision by bidirectional retrieval. It provides the fault tolerance and the ability of segmentation one would expect from associative memory realized in a biological system.

Our model of a reciprocal pathway assumes the existence of short closed loops where symmetric weights can be formed by Hebbian learning. This would be provided by a patchwise correspondence of both directions of a pathway, which has not

---

*The range of very small input activities is irrelevant since there retrieval errors are already high with both methods.

yet been examined experimentally. The time-delays in long reciprocal pathways can be significantly higher than in local connections. Therefore, the onset of the predicted hetero-associative high precision recall should be accompanied by a considerable slow-down in performance, although only a few iterations are required in the model. It may be related to cognitive or perceptual tasks, such as those where γ-oscillations where observed in behaving monkeys.

High precision recall can be realized by combinations of cell fascilitation and the use of quantities like conditioned links. How conditioned links can computed in detail by biological mechanisms in cortical pyramidal cells requires more detailed model assumptions. The present model, however, provides some hints which of its parts have to be refined: Eq. (5), for instance, prescribes the population of active neurons at one side of the pathway $x(t)$ for a short time to be fixed and to echo the signals from the other side (with linear dependence on the input). This could be realized by the interplay between fast and slow dendritic-neuronal processes: Increased slow components in the dendritic potentials would determine the population $x(t)$. Patterns of coinciding single spikes $y(t-1)$ affect the spike-rates in this cell population. The spike-rates influence the slow components in the dendritic potential and by this the new population of firing cells in $y(t+1)$, see Fig. 4. Another possible realization of eq. (5) is that the columnar auto-associative short-range connections stabilize the population $x(t)$.

Figure 4. Formation of conditioned links in a cortico-cortical pathway: Increased slow components in the dendritic potentials of neurons in x determine the population of firing neurons. Coinciding activities on a fast timescale in area y affect the spikerates (with linear dependence on the input) in the x-cell population. The spikerates influence the slow components in the dendritic potential and by this the new population of firing cells in area y.

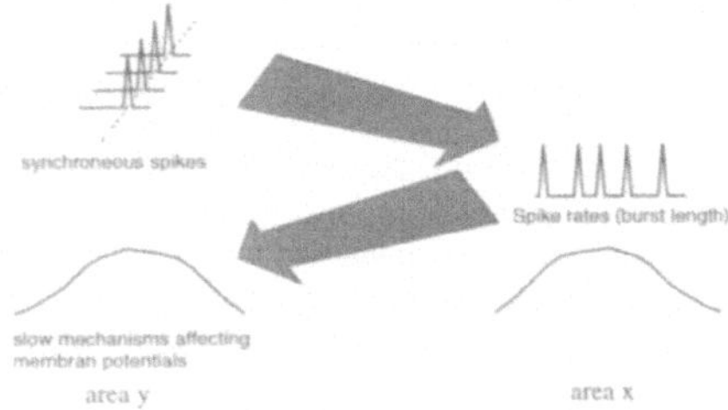

CONCLUSIONS

Three types of associative memory operations might play an important role in the cortex: auto-associative memory (AAM) using intra-cortical short range connections, one-way hetero-associative memory (HAM) on unidirectional cortico-cortical pathways, and bidirectional associative memories (BAM) on reciprocal cortico-cortical pathways.

Like in AAMs the feedback in a reciprocal pathway may allow recall procedures involving iterative retrieval providing higher precision at reduced speed. The idea is tested in a BAM model using the synaptic modification of the Willshaw model to store sparse binary random patterns. Simple bidirectional extension of the Willshaw model does not improve the performance. Indroducing the concept of conditioned links in BAM a new and very efficient BAM model for sparse memory patterns can be proposed. The high performance of the proposed CB-memory model and its segmantation ability for superimposed memory patterns is shown experimentally. We argue how the basic assumptions of our BAM model, namely symmetry of connections and high connectivity may be realized in a cortical pathway. Possible candidates for biological mechanisms corresponding to the computation of conditioned links might be the interplay of fast and slow dendritic processes and/or the influence of cortical short-range connections.

The proposed role of reciprocal cortico-cortical pathways as effective BAM devices supports Hebb's theory of concept representation by persistent distributed acticity patterns (cell assemblies): Fault tolerance and the capability to separate superpositions of several memory patterns – as provided by the CB-model – is crucial in a cell assembly theory[14]. Moreover, the proposed bidirectional retrieval in cortical pathways breaks up the monolithic cell assemblies of previous theories. Cell assemblies including many cortical areas can naturally emerge, if the areas are connected by reciprocal pathways. However, the cell assembly can be used in parts by restricting the recall on one or a few pathways involved (representing different facettes of a learned concept).

Acknowledgement: Two of the authors where supported by grants of the Deutsche Forschungsgemeinschaft: (F.T.S.) SO352/3-1 and (T.W.) PA268/8-1.

REFERENCES

1. D. A. Amit. The Hebbian paradigm reintegrated: Local reverberations as internal representations. *Behavioural and Brain Sciences*, 18:617 – 657, 1995.
2. A. Bibbig and T. Wennekers. Hippocamal two-stage learning and memory consolidation. In *Proceedings of the 13th European Meeting on Cybernetics and Systems Research*, 1996.
3. V. Braitenberg and A. Schüz. *The anatomy of the cortex. Statistics and Geometry*. Springer, Berlin Heidelberg New York, 1991.
4. M. Caesar, D. A. Brown, B. H. Gahwiler, and T. Knopfel. Characterization of a calcium-dependent current generating a slow afterdepolarization of CA3 pyramidal cells in rat hippocampal slice cultures. *European Journal of Neuroscience*, 5:560–569, 1993.
5. D. J. Felleman and D. C. Van Essen. Distributed hierarchical processing in the primate cerebral cortex. *Cerebral Cortex*, 1:1 – 47, 1991.
6. B. Gustafsson and H. Wigström. Physiological mechanisms underlying long-term potentiation. *Trends Neurosci.*, 11:156–162, 1988.
7. D. O. Hebb. *The Organization of Behaviour*. Wiley, New York, 1949.
8. J.J. Hopfield. Neural networks and physical systems with emergent collective computational abilities. *Proceedings of the National Academy of Sciences, USA*, 79, 1982.
9. O. Jensen and J. E. Lisman. Novel lists 7 ± 2 known items can be reliable stored in an oscillatory short-term memory network: Interaction with long-term memory. *Learning & Memory*, 3:257–263, 1996.
10. B. Kosko. Adaptive bidirectional associative memories. *Applied Optics*, 26(23):4947–4971, 1987.
11. C.-S. Leung, L.-W. Chan, and E. Lai. Stability, capacity and statistical dynamics of second-order bidirectional associative memory. *IEEE Trans. Syst, Man Cybern.*, 25(10):1414–1424, 1995.
12. A. Schüz. Patchiness as a means to get a message across. *Trends in Neuroscience*, 17(9):365, 1994.
13. F. Schwenker, F. T. Sommer, and G. Palm. Iterative retrieval of sparsely coded associative memory patterns. *Neural Networks*, 9(3):445 – 455, 1996.
14. F. T. Sommer and G. Palm. Improved bidirectional retrieval of sparse patterns stored by Hebbian learning. *Submitted to Neural Networks*, 1997.
15. F. T. Sommer and G. Palm. Bidirectional retrieval from associative memory. In *Neural Information Processing Systems 10*. Lawrence Erlbaum Ass. Inc., 1998.
16. K. Steinbuch. Die Lernmatrix. *Kybernetik*, 1:36–45, 1961.
17. Y. F. Wang, J. B. Cruz, and J. H. Mulligan. Two coding stragegies for bidirectional associative memory. *IEEE Trans. Neural Networks*, 1(1):81–92, 1990.
18. D. J. Willshaw, O. P. Buneman, and H. C. Longuet-Higgins. Nonholographic associative memory. *Nature*, 222:960–962, 1969.

MODEL OF HIPPOCAMPAL LTP INDUCED BY TIME-STRUCTURED STIMULI

Masami Tatsuno, and Yoji Aizawa

Department of Applied Physics
Waseda University
Shinjuku, Tokyo 169, JAPAN

INTRODUCTION

Information representation in the brain is one of the most challenging issues in neuroscience, and it is currently surmised that the rate-coding mechanism plays a significant role in the peripheral nervous systems[1,2], and that the timing process of spike trains may play an important role in the central nervous systems[3,4,5]. However, the mechanism of information representation in the brain has not yet been elucidated in detail.

To comprehend the entire process of information representation in the brain, it is necessary to understand memory information representation, since memory is the elementary function for all higher order functions in the brain. For this purpose, Tsukada et al. studied the effect of time-structured stimuli on long-term potentiation (LTP) in area CA1 of the hippocampus, and observed a strong relation between the amplitude of induced LTP and the serial correlation of Markov stimuli[6,7]. In this paper, we submitted an explanation of their experimental data by proposing a simple biological neuron model by taking into account both N-methyl-D-aspartate (NMDA) and non-NMDA receptors, together with the new rule for the synaptic efficacy change. Furthermore, we obtained the notable result that the chaotic stimuli in the nonstationary regime have a significant effect on LTP, which seems to suggest that the nonstationarity in spike trains emphasizes LTP in the hippocampus.

MODEL

The neuron model of the hippocampal CA1 pyramidal cell was proposed in our previous paper[8]. The basic time evolution of the activity of a single neuron was given by

$$O(t+1) = \begin{cases} 1, & \text{with prob. } f(V(t)) \\ 0, & \text{with prob. } 1 - f(V(t)) \end{cases},\qquad(1)$$

where $O(t+1)$ is the output of the neuron at time $t+1$, $V(t)$ is the membrane potential at time t, and $f(\cdot)$ is a probability function which describes the relation between the membrane potential and the output. Here, we adopted the following form for $f(\cdot)$:

$$f(V(t)) = \frac{1}{1 + e^{-(V(t)+|V_{th}|)}}, \tag{2}$$

where V_{th} is the threshold membrane potential ($V_{th} < 0$) and $f(V(t)) = 1/2$ at $V(t) = V_{th}$. The membrane potential $V(t)$ comprised of three parts, i.e.,

$$V(t) = \sum_{n=-\infty}^{t} E(n)\sigma(n)e^{-\frac{(t-n)}{\alpha}} + V_r + \delta(t), \tag{3}$$

where the first term of the right hand side represents the post-synaptic potential induced by the input current at time t, $E(n)$ is the synaptic efficacy at time n, $\sigma(n)$ is the spike train of the input stimuli which represents the existence of the input spike at time n ($\sigma(n) = 1$ with input, $\sigma(n) = 0$ without input), α is the decay constant for non-NMDA receptors, V_r is the constant resting membrane potential, and $\delta(t)$ is a small fluctuation of the membrane potential at time t where the average value of $\delta(t)$ is zero.

Next, we considered LTP of this model in terms of the change of $E(t)$. Physiologically, both depolarization of the membrane and calcium entry into the post-synaptic neuron are necessary for LTP to occur[9]. Therefore, we introduced the variable $S(t)$, which represents the concentration of calcium ions in the post-synaptic neuron and is expressed as:

$$S(t) = C \sum_{n=-\infty}^{t} \{V(n) + |V_L|\}\theta(V(n) + |V_L|)e^{-\frac{(t-n)}{\beta}}, \tag{4}$$

where C is a constant, $\theta(\cdot)$ is a step function, V_L is the membrane potential required for removal of magnesium ions from NMDA receptors ($V_L < 0$), and β is the decay constant for NMDA receptors.

Several experimental observations reveal strong correlation between the LTP amplitude and the frequency or the time structure of the input stimuli[6, 7, 10, 11]. By also considering the saturation of $E(t)$ due to the limited number of neurotransmitters and receptors, we proposed the following rule for the change of the synaptic efficacy $\Delta E(t)$:

$$\begin{aligned}
\Delta E(t) &= E(t+1) - E(t) \\
&= C_1(1 - e^{-C_2(S(t)-C_3)^2})\left(\frac{1}{1 + e^{-C_4(-(E(t)-C_5))}}\right) \\
&\quad \times \left(\frac{1}{1 + e^{-(S(t)-C_3)}}\right),
\end{aligned} \tag{5}$$

where $C_1 \sim C_5$ are all parameters. A schematic of $\Delta E(t)$ is shown in Fig. 1.

We considered a population of our neuron model by assuming that each neuron is independent and receives input signals from only Schaffer collaterals. Then, suppose we have N neurons which are stimulated independently, the averaged output at time t, $< O(t) >$ can be written as

$$< O(t) >= \frac{1}{N} \sum_{i=1}^{N} O_i(t) \sim f(V(t)), \tag{6}$$

in a large N limit according to the central limit theorem. Therefore, in this paper we considered $f(V(t))$ to represent the characteristic measure of the LTP amplitude. In Figs. 2 - 4, the amplitude variable of the vertical axis shows $<< f(V(t)) >_t >_e$ for comparison with the physiological data; the notation $< \cdot >_t$ represents the time average for a sample process and $< \cdot >_e$ represents the ensemble average for a large number of spike train sample processes.

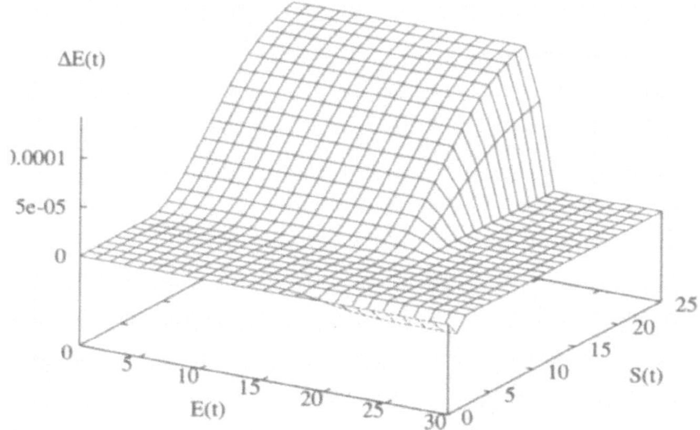

1. $\Delta E(t)$ as a function of $E(t)$ and $S(t)$.

NUMERICAL SIMULATIONS

To investigate the effect of the spike structure on synaptic modification, we consider three different spike stimuli: (i) regular, (ii) Markov, and (iii) chaotic. The detail of the setup of the numerical LTP experiments was explained in our previous paper[8].

Regular Stimuli

The regular periodic stimuli ranging from 10 Hz - 0.5 Hz were used in the simulation. The results are shown in Fig. 2. Figure 2 clearly shows that a higher-frequency stimulus produces a larger LTP than lower-frequency ones, and that no LTP occurs at frequencies of less than 2 Hz. These results are consistent with the results of experiments performed by Tsukada et al.[6].

Markov Stimuli

For Markov stimuli, the following transition matrix was used;

$$P = [P_{ij}] = \begin{bmatrix} WW & WS \\ SW & SS \end{bmatrix}, \tag{7}$$

where W, and S stand for the Markovian signal states which produce one tenth and a half of the maximum response, respectively. Here inter-spike interval was adjusted to 100 ms. The transition matrices used in the simulations are given as

$$P_{wo} = [P_{ij}] = \begin{bmatrix} 1 & 0 \\ 1 & 0 \end{bmatrix} \qquad P_{sp} = [P_{ij}] = \begin{bmatrix} 0.9 & 0.1 \\ 0.1 & 0.9 \end{bmatrix}$$

$$P_{sn} = [P_{ij}] = \begin{bmatrix} 0.1 & 0.9 \\ 0.9 & 0.1 \end{bmatrix} \qquad P_{so} = [P_{ij}] = \begin{bmatrix} 0 & 1 \\ 0 & 1 \end{bmatrix},$$

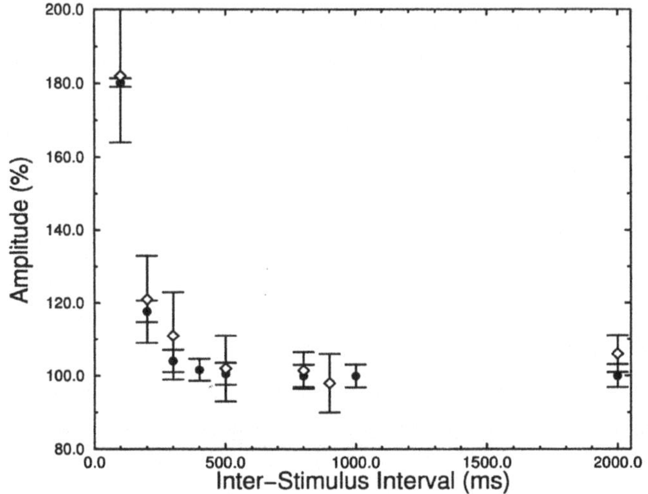

2. LTP amplitude induced by regular stimuli. The filled circles and the dotted diamonds represent the averages obtained by numerical simulations and experiments respectively, and the errorbars represent the standard deviations.

where P_{wo}, P_{sn}, P_{sp}, and P_{so} correspond to the weak only (Type 1), the strongly negative correlation (Type 2, $\rho = -0.8$), the strongly positive correlation (Type 3, $\rho = 0.8$), and the strong only (Type 4), respectively.

Figure 3 shows the simulation results. The Type 1 is the least effective and the Type 4 is the most effective because the Type 1 is the weakest stimulus and the Type 4 is the strongest. In the case of the Type 2 and the Type 3, the larger the serial correlation, the larger the LTP amplitude. Although the simulation results of the Type 1 and the Type 4 measured with the weak test stimulus do not fit with the experimental data, all others fit quantitatively well. Therefore, overall agreement in Figs. 2 and 3 shows that our neuron model is suitable for the investigation of the relation between the spike timing and the induced LTP in area CA1 of the hippocampus.

Chaotic Stimuli

Next, we further investigated information coding in the hippocampus by applying chaotic stimuli to our neuron model. Since there are several experimental evidences that the brain uses chaotic dynamics for information processing[12, 13], it is important to elucidate what type of chaotic stimulus is effective for LTP.

For this purpose, we used the modified Bernoulli map $x_n \rightarrow x_{n+1}$ (mod.1) to produce the chaotic input stimuli $\sigma(t)$,

$$x_{n+1} = \begin{cases} x_n + 2^{B-1}(1 - 2\epsilon)x_n^B + \epsilon, & (0 \leq x_n \leq 1/2) \\ x_n - 2^{B-1}(1 - 2\epsilon)(1 - x_n)^B - \epsilon, & (1/2 < x_n \leq 1), \end{cases} \tag{8}$$

where $1 \leq B \leq 3$ and $\epsilon = 10^{-13}$ are parameters. The properties of this map were thoroughly studied in terms of the symbolic dynamics[14], and we transformed the value

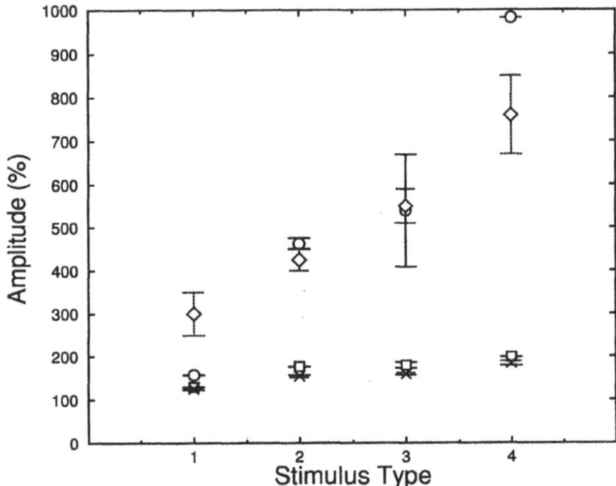

3. Amplitude of LTP induced by Markov stimuli. The circles and the squares represent the simulation averages measured with the weak and strong test stimuli, respectively. The diamonds and the crosses represent the experimental averages measured with the weak and strong test stimuli, respectively. The standard deviations are displayed by the errorbars.

of x_n to the spike strength in the following way:

$$\begin{cases} 0 \leq x_n < 1/2 & \rightarrow \quad \text{Strong Impulse} \\ 1/2 \leq x_n \leq 1 & \rightarrow \quad \text{Weak Impulse.} \end{cases} \tag{9}$$

The results are shown in Fig. 4. In the weak test-stimulus case, the LTP amplitude gradually increases as the value of B increases, and the deviation in the nonstationary regime is much larger than in the stationary regime. On the other hand, in the strong test stimulus case, the LTP amplitude does not change, but again the deviation becomes larger in the nonstationary regime.

SUMMARY

In this paper, we proposed a neuron model of area CA1 of the hippocampus taking into account the effects of NMDA and non-NMDA receptors, together with the rule for the synaptic efficacy change. To investigate the relation between the time structure of the input stimuli and the amplitude of LTP, we performed numerical simulations that reproduce most of the experimental data, and that reveal that LTP is strongly influenced by chaotic stimuli in the nonstationary regime.

For further understanding of memory information representation, we need to investigate in more detail the processes of the hippocampal network on a more complex scale. To this end, we must study the properties of the network of our neuron model. Further research towards the analysis of the network level will be reported in a forthcoming paper.

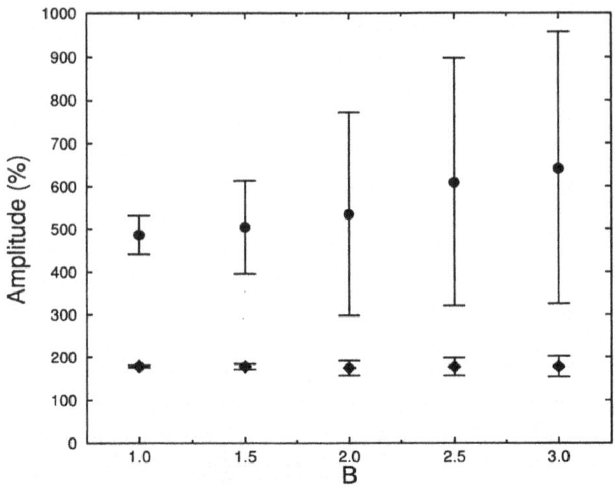

4. Amplitude of LTP induced by chaotic stimuli. The circles and the diamonds represent the simulation averages measured with the weak and strong test stimuli, respectively. The standard deviations are displayed by the errorbars.

REFERENCES

1. E. D. Adrian, "The Physical Background of Perception", Clarendon Press, Oxford (1946).
2. W. Calvin and C. Stevens, Synaptic noise and other sources of randomness in motoneuron interspike intervals, *J. Neurophysiol.* 31:574 (1968).
3. A. M. H. J. Aertsen and G. L. Gerstein, Dynamic aspects of neuronal cooperativity: fast stimulus-locked modulations of effective connectivity, *in*: "Neuronal Cooperativity", J. Krüger, ed., Springer-Verlag, Berlin (1991).
4. W. R. Softky and C. Koch, The highly irregular firing of cortical cells is inconsistent with temporal integration of random EPSPs, *J. Neurosci.* 13:334 (1993).
5. H. Fujii, H. Ito, K. Aihara, N. Ichinose and M. Tsukada, Dynamical cell assembly hypothesis – theoretical possibility of spatio-temporal coding in the cortex, *Neural Networks* 9:1303 (1996).
6. M. Tsukada, T. Aihara, M. Mizuno, H. Kato and K. Ito, Temporal pattern sensitivity of long term potentiation in hippocampal CA neurons, *Biol. Cybern.* 70:495 (1994).
7. M. Tsukada, T. Aihara, H. Saito and H. Kato, Hippocampal LTP depends on spatial and temporal correlation of inputs, *Neural Networks* 9:1357 (1996).
8. M. Tatsuno and Y. Aizawa, Theory for characterization of hippocampal long-term potentiation induced by time-structured stimuli, *J. Phys. Soc. Jpn.* 66:572 (1997).
9. P. S. Churchland and T. J. Sejnowski, "The Computational Brain", MIT Press, Cambridge (1992).
10. G. M. Rose and T. V. Dunwiddie, Induction of hippocampal long-term potentiation using physiologically patterned stimulation, *Neurosci. Lett.* 69:244 (1986).
11. J. Larson and G. Lynch, Theta pattern stimulation and the induction of LTP; the sequence in which synapses are stimulated determines the degree to which they potentiate, *Brain Res.* 489:49 (1989).
12. G. J. Mpitosos, R. M. Burton Jr., H. C. Creech and S. O. Soinila, Evidence for chaos in spike trains of neurons that generate rhythmic motor patterns, *Brain Res. Bull.* 21:529 (1988).
13. C. A. Skarda and W. J. Freeman, How brains make chaos in order to make sense of the world, *Behav. Brain Sci.* 10:161 (1987).
14. Y. Aizawa, Nonstationary chaos revisited from large deviation theory, *Prog. Theor. Phys.* 99:149 (1989).

MODULATION OF OSCILLATORY PROPERTIES, BURST RATES, INTERSEGMENTAL COORDINATION BY GABAB- RECEPTOR ACTIVATION IN THE LAMPREY

Jesper Tegnér,[1,2] Anders Lansner,[2] and Sten Grillner[1]

[1]Nobel institute for Neurophysiology, Dept. of Neuroscience, Karolinska Institutet, S–171 77 Stockholm, Sweden
[2]Dept. of Numerical Analysis and Computing Science, Kungliga Tekniska Högskolan, S-100 44 Stockholm, Sweden

INTRODUCTION

Transmitters can modulate both the synaptic connectivity and intrinsic membrane properties underlying the operation of neuronal networks. The abundance of modulatory mechanisms acting on different levels provides a large degree of flexibility in the operation of neural networks[1]. The GABAergic modulation of the locomotor circuits in the lamprey spinal cord is further investigated in this study

We have used the spinal cord preparation of the lamprey since it provides a good model system for a study on how different levels of organization are linked together[2]. The GABAergic modulation via $GABA_B$ receptors acts through several mechanisms at the cellular level which can modulate the overall swimming pattern in the lamprey.

The effects of $GABA_B$ receptor activation in the lamprey spinal cord includes: (i) a reduction of low–voltage–activated (LVA) calcium currents[3,4], leading to (ii) a decreased tendency of rebound depolarization[3,4], (iii) high–voltage–activated (HVA) calcium currents are reduced[3] which decreases (iv) the amplitude of the apamin sensitive slow afterhyperpolarization[3,4]. $GABA_B$ receptor activation also (v) decreases the IPSP and EPSP of premotor interneurons due to an increased degree of presynaptic inhibition[6,5]. On the network level (vi) the overall effect is a reduction of the burst rate[7] and (vii) the intersegmental coordination can be modified by a local activation of $GABA_B$ receptors[7].

The object of the present study was to determine whether the $GABA_B$ receptor activation exert its main action indirectly via the K_{Ca} channels or LVA calcium channels. In the analysis we have combined intracellular recordings with computer simulations. Furthermore, we have performed experiments to determine whether the $GABA_B$ receptor induced reduction of the burst rate could be accounted for by the action on K_{Ca} channels and/or a reduction of LVA calcium current.

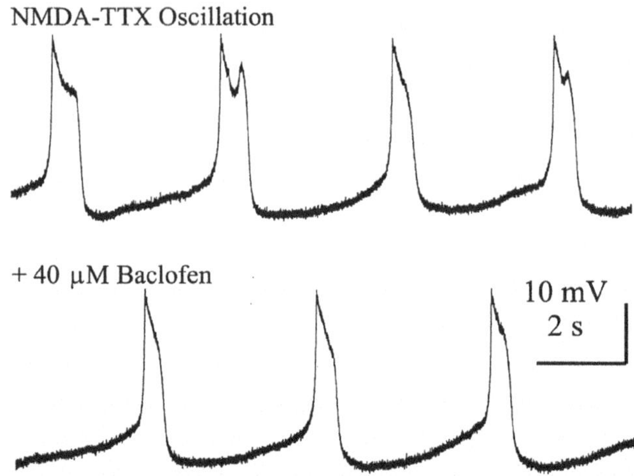

Figure 1. Effects of Baclofen (lower trace) on NMDA–induced TTX resistant membrane potential oscillations. Intracellular recording from a spinal neuron using discontinuous current clamp conditions allowing current injections in order to control the trough membrane potential. Bath application of 150 μM NMDA and 3 μM TTX induced membrane potential oscillations.

Methods–Experiments and Computer Simulations

Experiments were performed using adult lampreys and standard electrophysiological techniques. Intracellular recordings were performed using sharp microelectrodes. Discontinuous current clamp (DCC) was used to control the membrane potential as current was injected into the cell. The efferent motor activity induced by NMDA was recorded from the ventral roots at their exit from the spinal cord by extracellular recording. Simulations utilized a compartmentalized model neuron of Hodgkin-Huxley type with Na^+, K^+ and calcium channels of both an HVA[8, 9] and LVA type[4] and calcium–dependent potassium (K_{Ca}) channels[8, 9, 10]. The K_{Ca} channels are activated either by the calcium entering during the action potential (K_{CaAP}) or the calcium inflow through the NMDA channels, K_{CaNMDA} [8, 9]. Moreover, in this study we have used the extended cell model with an LVA calcium channel[4] with an intracellular calcium pool representing the calcium entering via LVA calcium channels thus activating a K_{Ca} channel referred to as K_{CaLVA}.

Results and Discussion

To examine the summed action of the LVA and K_{Ca} conductances, we have tested the effects of $GABA_B$ receptor activation, with baclofen, on the NMDA–induced TTX resistant membrane potential oscillations. These oscillations[11, 9] are characterized by a rapid depolarization followed by a plateau depolarization during which a gradual Ca^{2+}-dependent repolarization takes place, due to activation of K_{Ca}-channels. At a certain potential level the NMDA–channels close and a rapid repolarization follows. Thereafter a slow, gradual depolarization follows until the NMDA–channels open again. Intracellular recordings revealed that $GABA_B$ receptor activation increases the duration of the hyperpolarized phase, decreases the plateau proportion and reduces the peak amplitude for all trough potentials tested in all cells (n=8). The cycle duration was increased whereas the plateau duration was decreased.

As shown, in earlier experiments[12] and computer simulations[13], a reduction of the K_{Ca} –channel, increases the duration of both the plateau and the hyperpolarized phase. Thus, the plateau proportion under those conditions is approximately constant. The effects of baclofen on the NMDA–induced TTX resistant membrane potential oscillations could therefore not only be due to an effect on the K_{Ca} –channels since baclofen reduces the plateau proportion. It was therefore important to investigate how calcium conductances, could influence the NMDA–induced TTX resistant membrane potential oscillations. We used the earlier developed model of the LVA calcium channel[4]. By computer simulations it is possible to separate the effects of the calcium conductances from their respective activation of K_{Ca} –channels. The simulations showed that a reduction of the conductance of the LVA calcium channel, decreases the duration of the plateau. The simulations revealed that the duration of the hyperpolarized depend on the LVA calcium conductance and the level of NMDA calcium during the plateau. The slow decay dynamics of the NMDA calcium which control the K_{CaNMDA} conductance provide a depression during the hyperpolarized phase.

To further elucidate the role of the LVA conductance for the duration of the hyperpolarized phase the activation and deinactivation curves of the LVA calcium current were shifted between 1 and 5 mV (Fig. 2A) in the hyperpolarizing direction. Thus, as the "LVA calcium window" (the none 0 $m^3 h$ factor) is shifted towards more negative potentials, the duration of the plateau phase and the hyperpolarized phase decreases (Fig.2B). Figure.2C shows that a reduction of an LVA calcium conductance has a clear effect on the duration of the hyperpolarized phase when the $m^3 h$ factor is translated by -3 mV.

Thus, a simulated reduction of the LVA calcium conductance can account for the effects of $GABA_B$ receptor activation on the NMDA–induced TTX resisistant membrane potential oscillations .

To examine whether the calcium entering through the LVA calcium channels could activate K_{Ca} channels, we simulated the effects of an additional calcium pool which represents the LVA calcium (see also Methods), which activates a K_{CaLVA} -channel, on the NMDA–TTX oscillations. The main results are: (i) if the maximal conductance of the K_{CaLVA} –channel is comparable with K_{CaAP} and K_{CaNMDA} then a reduction of the LVA calcium leads to an increased duration of the plateau phase, (ii) if the decay time of the intracellular calcium is comparable to the calcium entering via NMDA channels (slow kinetics), then a reduction of the LVA calcium should lead to a decreased duration of the hyperpolarized phase. Baclofen decreases the duration of the plateau and increases the hyperpolarized phase in the experiments. We therefore conclude that if the LVA calcium activates a K_{CaLVA} –channel, then the intracellular calcium would be expected to have a fast kinetics comparable to the calcium entering during the action potential. Furthermore, the maximal conductance of K_{CaLVA} should be expected to be small as compared to the maximal conductance of K_{CaAP} .

In summary, experiments and computer simulations indicate that the modulation of the LVA calcium channel is sufficient to account for the modulation by $GABA_B$ receptors on the NMDA–induced TTX resisistant membrane potential oscillations . A dominant effect on K_{CaAP} is not consistent with the experimental data. Furthermore, the experiments are consistent with a low conductance of a K_{CaLVA} channel with a calcium kinetics of not the slow NMDA type.

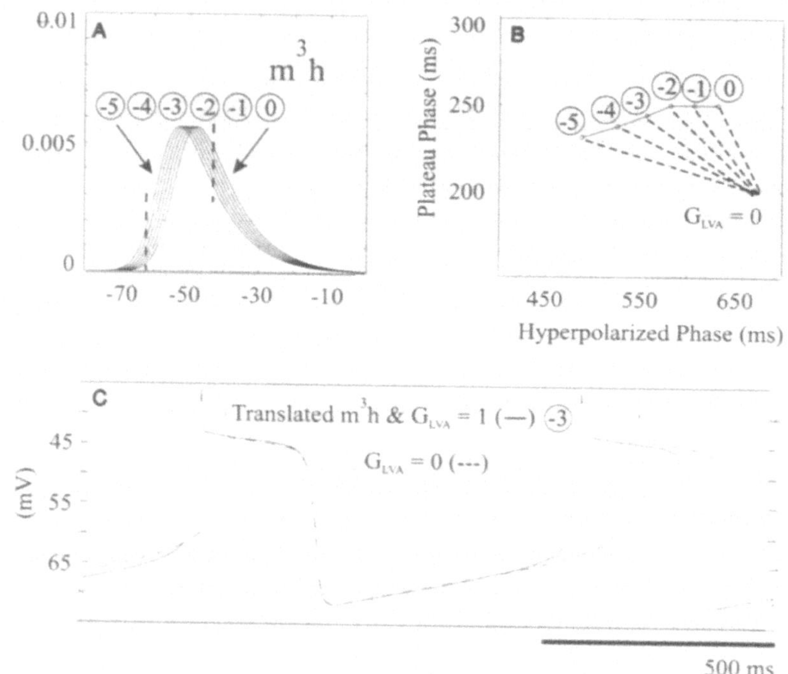

Figure 2. Effects of a voltage shift in the activation and deinactivation curves of the LVA calcium current on NMDA–induced TTX resisistant membrane potential oscillations . A The effect on the m^3h factor (A1) as the $B_{\alpha_m}, B_{\beta_m}, B_{\alpha_h}$ and B_{β_h} parameters were decreased between 1 and 5 mV (see methods in [4]). The dashed lines indicate the voltage level at which the rapid depolarization and hyperpolarization, occurs respectively. B shows the duration of the plateau and hyperpolarized phase as the m^3h factor was shifted. One example is shown in C. Note the longer duration of the hyperpolarized phase when the LVA conductance is turned off in all cases.

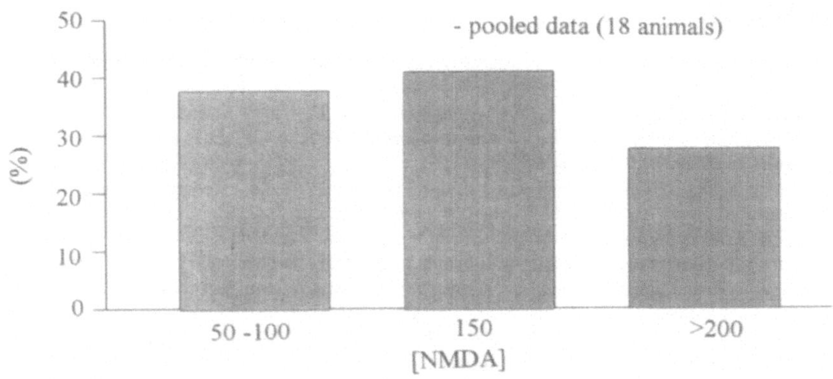

Figure 3. The reduction of the burst rate induced by an activation of $GABA_B$ receptors by baclofen. Note that the burst rate is reduced also at high levels of NMDA which would not be expected if the main action was exerted via the K_{Ca} channels.

The reduction of the burst rate by apamin, which reduces the K_{CaAP} conductance, was weaker at higher burst rates corresponding to the larger level of NMDA used to induce locomotion[12]. A simulation study detailed the sufficient conditions for this unexpected dependency of the NMDA level of the K_{CaAP} mechanism[13]. Since $GABA_B$ receptor reduces the K_{CaAP} indirectly through a reduction of the calcium entering through the HVA calcium channels it was therefore important to re-investigate the network effect of baclofen on the cycle duration and to test whether the effect of $GABA_B$ receptor activation is smaller at a larger level of NMDA as compared to a lower level of NMDA. If the main action of baclofen is via the K_{CaAP} mechanism then a small effect (in average 6 % [12]) would be expected at the higher levels of NMDA. That was not, however, observed in any preparation tested (Fig. 3). The effect of baclofen is clear even at the highest NMDA levels. This result further supports the notion that it is not the K_{CaAP} conductance which is major mechanism through which $GABA_B$ receptor activation acts.

An activation of $GABA_B$ receptors can modify the intersegmental coordination in lamprey[7]. To test whether a reduction of the K_{CaAP} or the LVA calcium conductance could account for this we simulated a local modulation, corresponding to the experimental "split–bath" configuration. It showed that either mechanism is sufficient to modify the intersegmental coordination. Thus, a forward intersegmental coordination could be decreased by a reduction of either LVA calcium or K_{CaAP} conductance in the rostral pool. The role of presynaptic inhibition for the intersegmental coordination remains, however, to be determined.

Acknowledgments

Dr.Örjan Ekeberg is acknowledged for the development of the SWIM simulation software (see http://www.nada.kth.se/sans/). This work was supported by the Medical Research Council (proj. no. 3026), the Swedish Natural Science Research Council (proj. no. B–AA/BU03531), the Swedish National Board for Industrial and Technical Development, NUTEK, (proj. no. 8425–5–03075) and the Swedish Society for Medical Research.

REFERENCES

1. E. Marder and R.L. Calabrese, Principles of rhythmic motor pattern generation, *Physiol. Rev.* 76:687–717 (1996).
2. S. Grillner, T. Deliagina, Ö. Ekeberg, A. El Manira, R.H. Hill, A. Lansner, G.N. Orlovsky, and P. Wallén, Neural networks that coordinate locomotion and body orientation in lamprey, *Trends Neurosci.* 18:270–279 (1995).
3. T. Matsushima, J. Tegnér, R.H Hill, and S. Grillner, $GABA_B$ receptor activation causes a depression of Low- and High-Voltage-Activated Ca^{2+} currents, Postinhibitory Rebound, and Postspike Afterhyperpolarization in lamprey neurons, *J. Neurophysiol.* 70:2606–2619 (1993).
4. J. Tegnér, J. Hellgren–Kotaleski, A. Lansner and S. Grillner, Low Voltage Activated Calcium Channels in the Lamprey Locomotor Network–Simulation and Experiment, *J. Neurophysiol.* 77:1795–1812 (1997).
5. S. Alford, J. Christenson, and S. Grillner, Presynaptic $GABA_A$ and $GABA_B$ receptor–mediated phasic modulation in axons of spinal motor interneuron, *Eur. J. Neurosci.* 3:107-117 (1991).
6. S. Alford and S. Grillner, The involvement of $GABA_B$ receptors and coupled G–proteins in spinal GABAergic presynaptic inhibition, *J. Neurosci.*
7. J. Tegnér, T. Matsushima, J. Tegnér, A. El Manira, and S. Grillner, The spinal GABA system modulates burst frequency and intersegmental coordination in the lamprey: differential effects of $GABA_A$ and $GABA_B$ receptors, *J. Neurophysiol.* 69:647–657 (1993).
8. Ö. Ekeberg. P. Wallén, A. Lansner, H. Tråvén, L. Brodin and S. Grillner, A Computer based Model for Realistic Simulations of Neural Networks. I: The Single Neuron and Synaptic Interaction, *Biol. Cybern.* 65:81–90 (1991).
9. L. Brodin, H. Tråvén, A. Lansner, P. Wallén, Ö. Ekeberg and S. Grillner, Computer Simulations of N-methyl-D-aspartate (NMDA) Receptor Induced Membrane Properties in a Neuron Model, *J. Neurophysiol.* 66:473–484 (1991).
10. H. Tråvén, L. Brodin, A. Lansner, Ö. Ekeberg, P. Wallén and S. Grillner, Computer Simulations of NMDA and non-NMDA Receptor-Mediated Synaptic Drive — Sensory and Supraspinal Modulation of Neurons and Small Networks, *J. Neurophysiol.* 70:695–709 (1993).
11. P. Wallén S. Grillner, N-methyl-D-aspartate Receptor-Induced, Inherent Oscillatory Activity in Neurons Active During Fictive Locomotion in the Lamprey, *J. Neurosci.* 9:2745–2755 (1987).
12. A. El Manira, J. Tegnér, and S. Grillner, Calcium-Dependent Potassium Channels Play a Critical Role for Burst Termination in the Locomotor Network in Lamprey, *J. Neurophysiol.* 72:1852–1861 (1994).
13. J. Tegnér, A. Lansner and S. Grillner, Modulation of Burst Frequency by Calcium–Dependent Potassium Channels in the Lamprey Locomotor System–Dependence of the Activity Level, *J. Comp. Neurosc.* in Press (1998).

ACTIVITY DEPENDENT MODULATION OF THE BURST RATE BY CALCIUM-DEPENDENT POTASSIUM CHANNELS IN LAMPREY

Jesper Tegnér,[1,2] Anders Lansner,[2] and Sten Grillner[1]

[1]Nobel institute for Neurophysiology, Dept. of Neuroscience, Karolinska Institutet, S–171 77 Stockholm, Sweden
[2]Dept. of Numerical Analysis and Computing Science, Kungliga Tekniska Högskolan, S–100 44 Stockholm, Sweden

INTRODUCTION

The operation of neural networks can be modified by a variety of neuromodulators which act on the intrinsic membrane and synaptic properties of network neurons[1, 2]. It is less clear how and to what extent these mechanisms and their effects depend on the level of network activity and therefore are state–dependent. Some insights have been obtained in studies on invertebrates. The purpose of this study is to further investigate the activity dependent modulation of the burst rate by calcium–dependent potassium channels, K_{Ca} channels in the locomotor circuits in lamprey[3].

In networks with reciprocal inhibition it is crucial to identify factors which either terminate the burst activity on one side thus releasing the inhibition from the other side or promote escape from inhibition[4]. The K_{Ca} channels in lamprey provides membrane properties which play an important role for the control of the termination in the spinal locomotor network [5]. This mechanism can be influenced by a variety of modulators such as 5–HT, dopamine, substance P and GABA, thus changing the burst rate during fictive locomotion. The afterhyperpolarization, which follows each action potential due to K_{Ca} channels is sensitive to apamin [6, 7]. Qualitatively, it could be expected that a prolonged burst duration induced by an inhibition of K_{Ca} –channels would lead to a prolonged cycle duration. Previous studies reported conflicting results on the role of K_{Ca} channels in the lamprey locomotor network [6, 7]. These apparently conflicting data were due to differences in the activity level [5]. Apamin reduced the burst frequency and disrupted the motor pattern, at low burst frequencies whereas at higher burst frequencies (larger levels of NMDA), the effects of apamin were considerably smaller [5]. That finding which appeared enigmatic could be accounted for in a computational study [3]. A puzzling observation was that apamin increased the duration of the hyperpolarized phase during intrinsic membrane oscillations (NMDA–TTX). One assumption that could account for both results was that there is a subdivision of the K_{Ca} –channels sensitive to apamin, activated by the calcium entering during the action potential (K_{CaAP}), and the K_{Ca} –channels activated by the calcium which enters through NMDA channels (K_{CaNMDA})

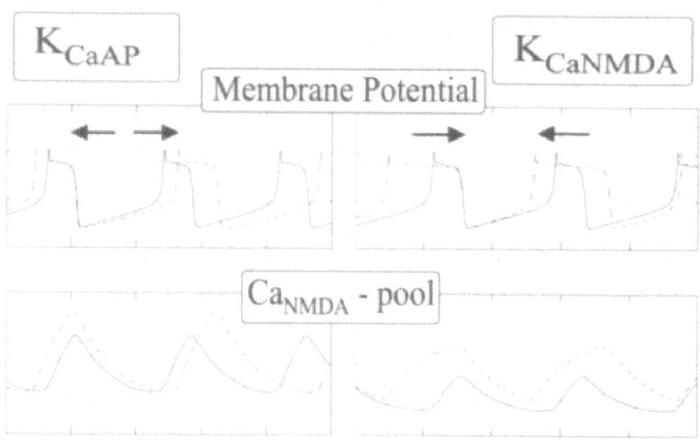

Figure 1. Simulated reduction of the K_{CaAP} (Left panel) and K_{CaNMDA} (Right panel) conductance during NMDA pacemaker oscillations . A reduction of the K_{CaNMDA} –conductance prolonged the plateau phase and shortened the duration of the hyperpolarized phase (dashed line). The level of calcium (arbitrary units) in the Ca_{NMDA} -pool is monitored in the lower panels.

which supposedly do not respond to apamin. Thus, the larger the NMDA drive the less relative influence the K_{CaAP} conductance have as compared to the K_{CaNMDA} , and thus a smaller modulatory effect[3]. Here we investigate the role of the K_{CaNMDA} conductance on NMDA–TTX induced membrane potential oscillations and its influence on the burst rate. We compare the role of K_{CaNMDA} conductance with the K_{CaAP} conductance.

Methods–Experiments and Computer Simulations

Intracellular recordings (DCC) on adult lampreys spinal neurons were performed using sharp microelectrodes. The simulations utilized a compartmentalized model neuron of the Hodgkin-Huxley type with Na^+, $K^+$and calcium channels and K_{Ca} channels [8, 9]. The single segment model of the spinal cord consisted of a population of neurons coupled through reciprocal inhibition[10, 11].

Results

Earlier experiments showed that addition of apamin increased the duration of the hyperpolarized phase[5]. This could be simulated provided that only the K_{CaAP} conductance was reduced (Fig.1 Left Panel,top)[3].

A simulated reduction of the K_{CaNMDA} during NMDA–TTX oscillations prolonged the plateau and decreased the duration of the hyperpolarized phase (Fig.1 Right Panel,top). This result is consistent with the notion that apamin does not affect the K_{CaNMDA} conductance. The mechanisms depend on the slow calcium dynamics in the Ca_{NMDA} -pool (Fig.1 lower). A reduced K_{CaAP} conductance prolongs the plateau thus increasing the Ca^{2+} level in the Ca_{NMDA} –pool during the plateau (Fig.1 lower left). The hyperpolarized phase is therefore prolonged due to the slow decay of the NMDA

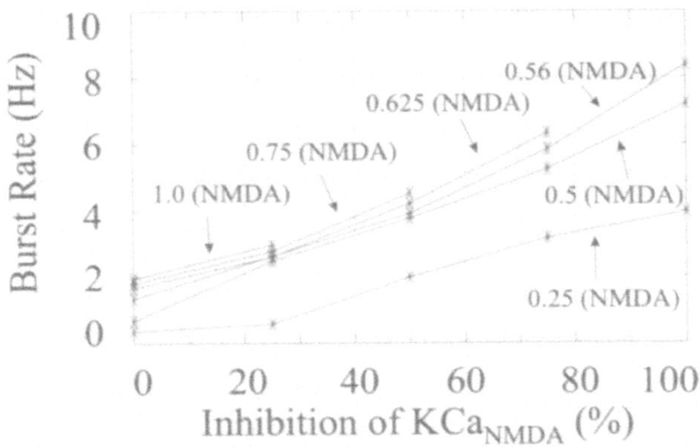

Figure 2. Computer simulation of the effects of K_{CaNMDA} –channels on the burst rate of the NMDA driven segmental network. A parameter plot of the burst rate versus the degree of inhibition of the K_{CaNMDA} –channels. A progressive inhibition of the K_{CaNMDA} conductance increases the burst frequency. The points in control (0 % inhibition of the K_{CaNMDA} conductance) with increasing frequency corresponds to increasing levels of NMDA. The lines corresponding to an NMDA level larger than 0.56 are terminated before an inhibition of 100 % was tested, because the alternating pattern broke down.

calcium. A reduced K_{CaNMDA} conductance on the other hand leads to a reduced degree of depression exerted by the K_{CaNMDA} conductance during the hyperpolarized phase (Fig.1 lower right). This gives a shorter duration of the hyperpolarized phase.

Furthermore, an increased degree of reduction of K_{CaNMDA} conductance disrupted the oscillations due to the decreased strength of the repolarization. Thus, in contrast to the experimental observations[5] apamin would be expected to abolish the membrane potentials if it acted on the K_{CaNMDA} –channels.

Intracellular recordings were performed to test whether the NMDA induced afterhyperpolarization was sensitive for apamin. NMDA was added to the perfusing solution and Cd^{2+} was used to reduce the Ca^{2+} inflow from calcium channels. Addition of apamin did not reduce the peak amplitude of the AHP when using long current pulses (50–300 ms).Thus, only the K_{CaAP} and not the K_{CaNMDA} conductance appears to be sensitive for apamin.

If the K_{CaNMDA} channels were apamin sensitive the burst rate would be expected to decrease with depression of the K_{CaNMDA} channels [5]. A simulated reduction of the K_{CaNMDA} channels increase the burst rate and disrupt the reciprocal burst pattern at large levels of simulated NMDA an effect more marked, the higher the level of NMDA drive (Fig.2). Both effects differ from that of apamin. This result further supports the notion that only the K_{CaAP} conductance is sensitive for apamin. The increase in burst rate by a reduced K_{CaNMDA} conductance is due to the fact that the K_{CaNMDA} channel effectively inhibits the neuron during the hyperpolarized phase due to the slow decay of the intracellular calcium.

The role of K_{CaAP} conductance was investigated in a simulated AMPA/kainate network. A reduction of the K_{CaAP} was found to either decrease or increase the burst rate in the segmental network depending on whether the AMPA/kainate level was low or high (Fig.3, top). The somewhat unexpected increase in burst rate is due to the

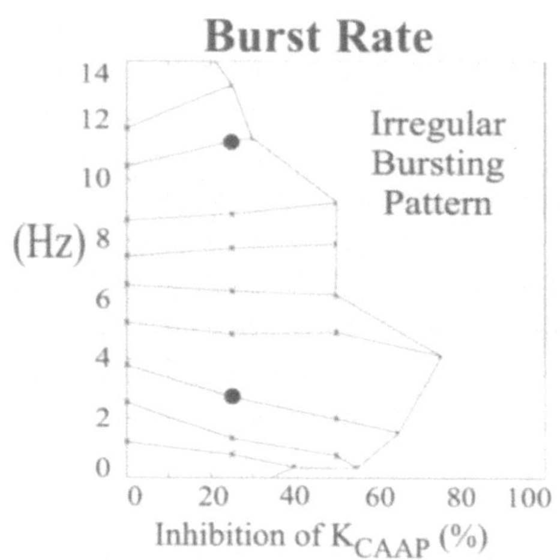

Calcium Levels

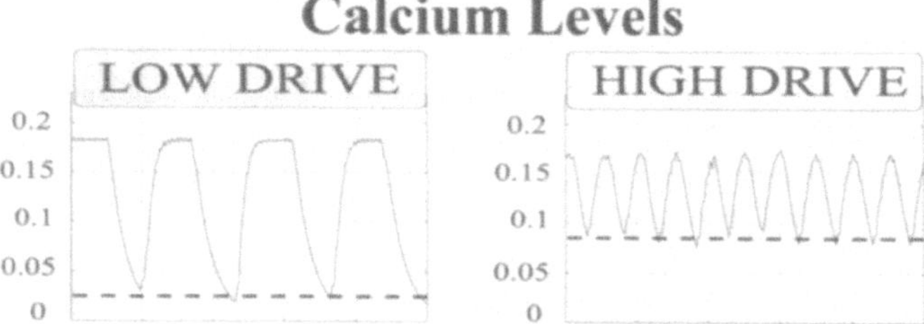

Figure 3. The effect of a reduction of K_{CaAP} depends on the initial burst rate. The top panel show a parameter plot of the effect of the K_{Ca} –conductance activated by the Ca_{AP} –pool (K_{CaAP}) on the burst rate as the segmental network was driven by AMPA/kainate. The filled circles indicated the parameter values (low and high drive) at which the calcium levels were measured. Note that a reduction of the K_{CaAP} conductance increase the burst rate if the initial burst rate is high. The trough level of the intracellular calcium at this activity level is clearly above zero (lower right). Thus, the K_{CaAP} conductance provide inhibition during the silent phase when the network has a high alternation rate. A reduced K_{CaAP} conductance leads to a reduced inhibition, thus a faster burst rate at this activity level.

fact that the K_{CaAP} channel effectively inhibits the neuron during the hyperpolarized phase when the network has a high burst rate since the time course of the intracellular calcium regulating the K_{CaAP} channels is on a similar time scale as the duration of the hyperpolarized phase itself (Fig.3, lower right). The role of K_{CaAP} channels in this range is thus similar to that of K_{CaNMDA} channel in an NMDA driven network.

Discussion

The observations that (i) the effect of K_{CaAP} on the burst rate is less, the higher the degree of NMDA activation and (ii) the hyperpolarized phase during NMDA–TTX oscillations is prolonged by K_{CaAP} suggested a selective action of apamin on K_{CaAP} channels [3]. The fact that this assumption could account for two unexpected experimental results could be interpreted as support for this notion.

Further evidence from the single cell level include: Intracellular recordings of the NMDA induced afterhyperpolarization suggests that the K_{CaNMDA} is not apamin sensitive. A simulated reduction of the K_{CaNMDA} increased the hyperpolarized phase during the NMDA–TTX oscillations. An increased degree of inhibition of the K_{CaNMDA} conductance disrupts the NMDA–TTX oscillations. At the network level we showed that an inhibition of the K_{CaNMDA} conductance increase the burst rate and disrupt the pattern in the higher NMDA range.

The activity dependence of the modulatory role of K_{CaAP} is not limited to an NMDA driven network. The simulations suggest that the K_{CaAP} conductance acts as a burst termination factor in the low activity range whereas at an increased degree of activity the K_{CaAP} provides inhibition during the hyperpolarized phase. This gives a similar activity dependence as during the NMDA driven condition but the mechanisms are different. In the NMDA case, the K_{CaNMDA} gradually takes over as a factor which regulate the burst rate whereas in the AMPA/kainate network the time course of the intracellular calcium related to the K_{CaAP} is closer to the duration of the hyperpolarized phase the faster the burst rate of the network. Thus a reduced afterhyperpolarization in a network without NMDA properties would be expected to increase the burst rate at a high activity level.

These results suggest that the effect (reduction of burst rate) of a modulation (reduction) of the K_{CaAP} conductance on the network, either in the NMDA or AMPA/kainat mode, is smaller, the faster the burst rate of the network. In conclusion, the influence of a single conductance such the K_{CaAP} or K_{CaNMDA} depend on the activity level of the network. Thus, whether the reciprocal network operates in an escape or release mode [4] may depend on the mode of activation (with or without NMDA properties), the relative balance between K_{CaAP} and K_{CaNMDA} as the drive increase, and the initial state of the network (alternation rate). These observations are of importance when considering the intersegmental coordination (see Hellgren et al, this volume), suggesting a difference whether the network is activated with NMDA as compared to AMPA/kainate or D–Glutamate since the contribution of different factors may depend on how the network is activated.

Acknowledgments

We are indebted to Dr.Örjan Ekeberg for the development of the SWIM simulation software (http://www.nada.kth.se/sans/). This work was supported by the Medical Research Council (proj. no. 3026), the Swedish Natural Science Research Council (proj. no. B–AA/BU03531), the Swedish National Board for Industrial and Technical Development, NUTEK, (proj. no. 8425–5–03075) and the Swedish Society for Medical Research.

REFERENCES

1. S. Grillner, T. Deliagina, Ö. Ekeberg, A. El Manira, R.H. Hill, A. Lansner, G.N. Orlovsky, and P. Wallén, Neural networks that coordinate locomotion and body orientation in lamprey, *Trends Neurosci.* 18:270–279 (1995).

2. E. Marder and R.L. Calabrese, Principles of rhythmic motor pattern generation, *Physiol. Rev.* 76:687–717 (1996).

3. J. Tegnér, A. Lansner and S. Grillner, Modulation of Burst Frequency by Calcium–Dependent Potassium Channels in the Lamprey Locomotor System–Dependence of the Activity Level, *J. Comp. Neurosc.* in Press (1998).

4. K. Skinner, N. Kopell and E. Marder, Mechanisms for oscillation and frequency control in reciprocally inhibitory model neural networks, *J. Comp. Neurosc.* 1:69–87 (1994).

5. A. El Manira, J. Tegnér, and S. Grillner, Calcium-Dependent Potassium Channels Play a Critical Role for Burst Termination in the Locomotor Network in Lamprey, *J. Neurophysiol.* 72:1852–1861 (1994).

6. R.H. Hill, T. Matsushima, J. Schotland and S. Grillner, Apamin blocks the slow AHP in lamprey and delays termination of locomotor bursts, *Neuroreport.* 10:943–945 (1992).

7. D.P. Meer, and J.T. Buchanan, Apamin reduces the late afterhyperpolarization of lamprey spinal neurons, with little effect on fictive swimming, *Neurosci. Lett.* 143:1–4 (1992).

8. Ö. Ekeberg. P. Wallén, A. Lansner, H. Tråvén, L. Brodin and S. Grillner, A Computer based Model for Realistic Simulations of Neural Networks. I: The Single Neuron and Synaptic Interaction, *Biol. Cybern.* 65:81–90 (1991).

9. L. Brodin, H. Tråvén, A. Lansner, P. Wallén, Ö. Ekeberg and S. Grillner, Computer Simulations of N-methyl-D-aspartate (NMDA) Receptor Induced Membrane Properties in a Neuron Model, *J. Neurophysiol.* 66:473–484 (1991).

10. J. Hellgren, S. Grillner, and A. Lansner, Computer Simulation of the Segmental Neural Network Generating Locomotion in Lamprey by using Populations of Network Interneurons, *Biol. Cybern.* 68:1–13 (1992).

11. J. Tegnér, J. Hellgren–Kotaleski, A. Lansner and S. Grillner, Low Voltage Activated Calcium Channels in the Lamprey Locomotor Network–Simulation and Experiment, *J. Neurophysiol.* 77:1795–1812 (1997).

SYNCHRONIZATION IN NETWORKS OF NOISY INTERNEURONS

P.H.E. Tiesinga, W-J Rappel, and Jorge V. José

Center for Interdisciplinary Research on Complex Systems,
and Department of Physics, Northeastern University,
Boston, Massachusetts 02115

INTRODUCTION

It is generally believed that rhythm synchronization of different neuronal regions
of the brain are of paramount importance to understand the perception problem.[1] A
particular type of synchronous neuronal oscillations, which may be linked to the binding of different brain regions, occurs in the 20 − 80 Hz frequency range, known as
gamma oscillations.[2] Recent advances in the understanding of this problem have indicated that random networks of GABA-ergic interneurons are capable of producing
the gamma oscillations under physiological conditions.[3] This synchronization by mutual inhibition was reproduced in computer simulations using physiologically realistic
models.[3] In the experimental preparation there is always some heterogeneity in neuronal properties, as well as synaptic noise. An important issue is therefore whether
the mechanism suggested by these simulations is robust. In a recent paper Wang and
Buzsáki (WB),[4] and also White et al,[5] have carried out computer simulations of a network of interconnected GABA-ergic interneurons. The synchronization was found to be
only moderately robust against heterogeneity in the distribution of driving currents. In
this work we consider the stability of this type of gamma oscillations against synaptic
noise. In addition, we consider the spatio-temporal coherence of a set of interneurons
with specific spatial connectivity properties. To analyze the results quantitatively we
calculate measures that characterize the network's temporal coherence as well as its
spatial synchronization.

Our aim in this work is to characterize the nature and stability against noise of the
synchronous properties of a simple interneuronal model system, in order to ascertain the
necessary conditions for binding vis á vis more physiologically realistic model systems
(see for instance the work by Traub et al[6]). We study the spatial binding properties
of a simple model of coupled interneuron clusters. Here we understand for binding the
ability of interneuron clusters to synchronize over long distances. The interneurons are
grouped in clusters. Within the clusters the neurons are connected all to all, whereas all
the neurons in each cluster are only directly connected to the neighboring cluster. We
find that the network of clusters can be in one of several coupled states. Among them

we find modes in which spatially separated (i.e. not directly connected) clusters, spike in synchrony at gamma frequencies between 20-80 Hz, while the in-between clusters (including those connected directly) spike with a phase delay. This mode may be seen as a simple realization corresponding to a network binding. Our results suggest that the occurrence of this binding-like state depends especially on the initial conditions, the intrinsic spiking frequencies of the neurons, and the noise-level. In this paper we present some of our preliminary results on a system of ten clusters each with ten interneurons.

METHODS

We model the neurons as a single compartment with Hodgkin-Huxley type channels with the neurons connected to each other by inhibitory GABA-ergic synapses. The equation for an individual neuron is (the index i of a given neuron is omitted)

$$C_m\frac{dV}{dt} = -I_{Na} - I_K - I_L - I_{syn} + I_{app} + C_m\xi. \tag{1}$$

Here I_{Na}, I_K, I_L, I_{syn}, I_{app} and ξ are the sodium, potassium, leak, synaptic, externally applied and noise currents, respectively. The currents are measured in $\mu A/cm^2$ units and $C_m = 1\mu F/cm^2$ is the membrane capacitance. We use a Hodgkin-Huxley type of voltage-gated sodium and potassium currents, with the rate functions and values for the maximum conductances as given in WB.[4] The equations (1) are integrated using an adapted second order Runge-Kutta method designed for stochastic equations,[7] with a step size $dt = 0.01$ms. The accuracy of the zero noise results was checked against results obtained with a smaller step size and using a 4th order Runge-Kutta algorithm. Noise in neurons derives from various sources,[8] including synaptic noise due to spontaneous inhibitory post synaptic potentials (IPSP) and excitatory post-synaptic potentials (EPSP), the stochastic nature of the number of vesicles released by a presynaptic action potential, and the stochastic nature of the opening of the voltage-gated channels. We have modeled the effect of noise as a Gaussian distributed, delta correlated, current *in the soma*,[9] i.e. $\langle \xi_i(t)\xi_j(t')\rangle = 2D\delta(t - t')\delta_{i,j}$ (D is expressed in mV^2/ms). Models for synaptic noise,[10] and many experimental results yield interspike intervals (ISI) distributed according to a gamma distribution.[11] The ISI obtained from our neuron model are also gamma-model distributed,[12] yielding support to our implementation of synaptic noise. We typically simulate $N = 100$ neurons on a DEC alpha workstation, and obtain the spike $X_i(t)$ and the voltage $V_i(t)$ traces at discrete times $t = n\tau$ (we take $\tau = 4 - 20\ dt$). The average spiking rate is $f_i = \langle X_i\rangle$ (here $\langle\cdot\rangle$ is a time average), and the network average is

$$f_\mu = (\sum_i f_i)/N, \tag{2}$$

The coherence function κ, that measures the synchronization of the network, is defined as (slightly modified from the expression in WB[4]):

$$\kappa(\tau_2) \equiv \sum_{i\neq j} \frac{\langle \hat{X}_i(t)\hat{X}_j(t)\rangle}{\sqrt{\langle \hat{X}_i^2(t)\rangle\langle \hat{X}_j^2(t)\rangle}}. \tag{3}$$

In this expression $\hat{X}_i(t)$ is $X_i(t)$ binned in bins of size $\tau_2 \sim 2\ ms$. In addition we calculate the histogram of the interspike intervals (ISIH). The interspike interval is the time between two consecutive action potentials. We average the ISIH of individual neurons over the whole network. We have also plotted our results as rastergrams, the firing times t of a neuron x are depicted as a circle on a x-t graph.

RESULTS

Random Networks

We started looking at a single neuron with noise. We find that weak noise causes a variability in the interspike intervals and consequently a finite width of the ISIH (Figure 1(c)), though the average firing rate remains constant. For stronger noise the fluctuations can give rise to additional action potentials, thereby increasing the average spiking rate. Next we considered a network of $N = 100$ neurons, connected all to all. In Figure 1(a) and (b) we plot both κ ($\tau_2 = 2\ ms$ in Eq. (3)), f_μ (Eq. (2)), respectively, as a function of D. Initially, for weak noise, the coherence is reduced and the average spiking rate decreases. For strong noise, synchronization is destroyed, and the firing rate starts to increase. The fluctuations are large enough to cause additional action

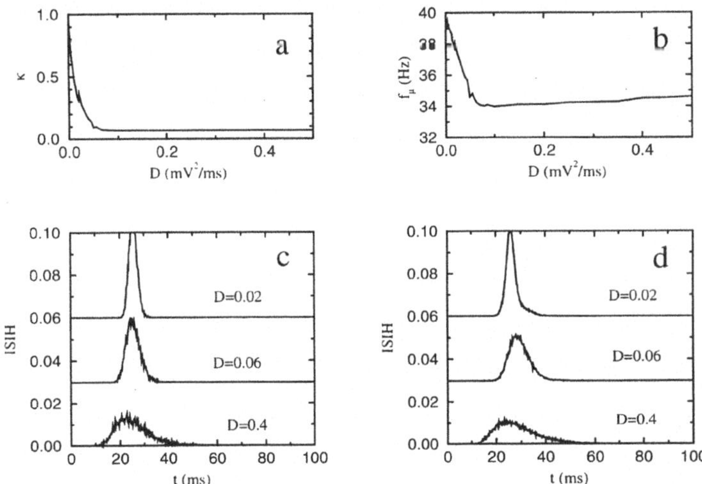

Figure 1. Here we show the coherence function κ (a) defined in Eq. (3), and the average frequency f_μ (b) defined in Eq. (2), versus D for a network of all to all connected interneurons ($N = 100$, $I_{app} = 1$). We plot the ISIH for an isolated (unconnected) neuron (c) and the network average (d), the first two curves are shifted over 0.06 and 0.03 units in the vertical direction, respectively. In the network results we discarded a transient of 1000 ms and averaged over 3000 ms for the measurements. For the isolated neuron the transient was 500 ms, and we used an averaging time of 20000 ms.

potentials themselves. The coherence can be studied using the network ISIH introduced earlier. The synchronization parameter κ measures the fraction of neurons that spike within a time τ_2, in essence this is determined by the width of the peak in the ISIH. From Figure 1(d) one notes that with the drop in κ, the width of the ISIH increases. Next we considered a network of $N = 100$ neurons, with an average of 60 connections to other neurons. The connections between any two neurons are chosen at random with a probability $p = 0.6$. We find that the robustness against noise decreases with p. We find that synchronization, and more specifically the value of κ, in the semi-ordered regime ($0.1 < \kappa < 1$) depends sensitively on the particular realization for the random connections (not shown).

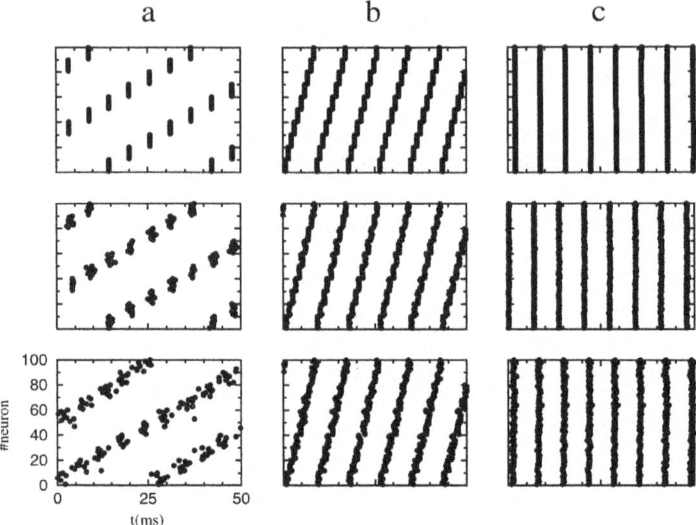

Figure 2. Rastergrams for different values of the applied current $I_{app} = 1.0$ (a), 3.4 (b), and 4.4 (c). The rastergrams are plotted for different noise strengths $D = 0$, 4×10^{-3}, and 0.02 from top to bottom. The data is obtained from numerical simulations of a network of 10 clusters with 10 neurons each. A transient of 2000 ms was discarded.

Clusters of Connected Neurons

In this subsection we study the synchronization of a network of 10 clusters with 10 neurons per cluster as a function of the current and noise-strength D. The neurons are labeled by their cluster number c and the index j within the cluster. All the neurons in a particular cluster c are connected among themselves and to the ones in the next cluster (i.e. $c + 1$). We start by determining the behavior of the connected network without noise, $D = 0$. Depending on the value of the current and the initial configuration we find three generic types of network behavior. (a) All the clusters spike at the same time; (b) two clusters spike at the same time; and (c) each cluster spikes at different times. This can be deduced from rastergrams in Figure 2. In the rastergrams the neuron number is given by $c \times 10 + j$. What is interesting in case (b) is that the clusters that are synchronized with a zero phase-difference are not connected directly to each other. This effect can be thought of as a possible example of the binding phenomenon.[6] The modes are stable against weak noise within the gamma-frequency range. Which mode is chosen depends on the value of the applied current and the initial phases of the neurons. Sometimes the $D = 0$ mode for a particular value of the current and initial configuration of phases, was replaced by a different one for $D \neq 0$. We also studied the effect of a time delay in the transmission of action potentials between the clusters. The network settled in different modes depending on the different amounts of time delay. In addition we have studied the dynamics of networks with a different connectivity between the clusters, for example in a system with both feedforward and backward connections or finite boundary conditions. These networks display qualitatively similar binding-like phenomena that will be reported elsewhere.[12]

CONCLUSION

In this paper we have reported on the results of a numerical study of a network of interneurons under physiological conditions. This system exhibits gamma oscillations that are of significant importance to neural behavior. We have considered the stability of these oscillations against synaptic noise. We found that the gamma oscillations are stable only for small amounts of noise. This result together with those of previous studies delimit the regions of stability and robustness of this mechanism for synchronization. One is in need of an experimental determination of the prevailing noise levels in hippocampal networks. Only then can one conclude whether synchronization by mutual inhibition is strong enough to account for the synchronization observed in both in vivo and in vitro experiments. We started by considering the properties of one interneuron and then extended our study to networks of interneurons with different connectivity properties. We have also considered interneuron networks coupled to excitatory neurons. We found that in the latter case the stability of the gamma oscillations is strengthened. There are still a number of questions to be answered, in particular to better understand the type of binding-like behavior exhibited by these neuronal networks. A more detailed presentation of these and other related results will appear elsewhere.[12]

Acknowledgments

This work was partially supported by the Center for Interdisciplinary Research on Complex Systems (CIRCS) funded by Northeastern University.

REFERENCES

1. W.J. Freeman, *Mass Action in the Nervous System*. Academic Press, New York (1975).
2. W. Singer and C.M. Gray, *Visual feature integration and the temporal correlation hypothesis*, Annu. Rev. Neurosci. 18:555 (1995).
3. M.A. Whittington, R.D. Traub, and J.G.R. Jeffreys, *Synchronized oscillations in interneuron networks driven by metabotropic glutamate receptor activation*, Nature 373:612 (1995).
4. X.J. Wang and G. Buzsáki, *Gamma oscillation by synaptic inhibition in a hippocampal interneuronal network model*, J. Neurosci. 16:6402 (1996).
5. J.A. White, C.C. Chow, J. Ritt, C. Soto-Treviño, and N. Kopell, *Synchronization and oscillatory dynamics in heterogeneous, mutually inhibited neurons*, Preprint (1997).
6. R.D. Traub, M.A. Whittington, I.M. Stanford, and J.G.R. Jeffreys, *A mechanism for generation of long-range synchronous fast oscillations in the cortex*, Nature 383:621(1996).
7. H.S. Greenside and E. Helfand, *Numerical integration of stochastic differential equations-II*, Bell Syst. Tech. J. 60:1927 (1981).
8. D. Johnston and S. Wu, *Foundations of Cellular Neurophysiology*, MIT Press, Cambridge (1995).
9. D. Golomb and J. Rinzel, *Synchronization properties of spindle oscillations in a thalamic reticular nucleus model*, J. Neurophys. 72:1109 (1994).
10. R.B. Stein, *A theoretical analysis of neuronal variability*, Biophys. J. 5:173 (1965).
11. P.F.M. Teunis, F. Bretschneider, J.J.M. Bedaux, and R.C. Peters, *Synaptic noise in spike trains of normal and denervated electroreceptor organs*, Neuroscience 41:809 (1991).
12. P.H.E. Tiesinga and Jorge V. José, (*to be published*).

SIGNIFICANCE OF MODULATED ADAPTATION FOR RHYTHM GENERATION AND INTER-SEGMENTAL CO-ORDINATION IN LAMPREY

Maria Ullström[1], Anders Lansner[2], Jeanette Hellgren Kotaleski[1,2], and Sten Grillner[1]

[1]Nobel Institute for Neurophysiology,
S-171 77 Stockholm, Sweden.
[2]SANS - Studies of Artificial Neural Systems,
Dept. of Numerical analysis and Computing science,
Royal Institute of Technology,
S-100 44 Stockholm, Sweden.

INTRODUCTION

The lamprey is a primitive eel-like water-living vertebrate that swims with an undulatory movement. Since detailed information about the spinal neural network generating swimming is available (Grillner et al. 1995) a number of modelling and simulation studies has been performed. The simulations presented here are based on a previous neuro-mechanical model of lamprey swimming (Ekeberg 1993) using a simplified, non-spiking, adapting neuron model. This model worked over a limited range of swimming frequencies and one goal with this study has been to extend the frequency range with maintained intersegmental co-ordination. We have modelled an active modulation of the afterhyperpolarisation (AHP) by means of 5-HT (serotonin) into the original model (Ekeberg 1993). 5-HT is known to depress the late phase of the AHP amplitude following the action potential in the spinal neurons and its effects are very powerful with an increased spike frequency and increased burst duration (Wallén et al. 1989). There is a dense medial plexus of 5-HT neurons in the lamprey spinal cord into which network interneuron extend their dendrites. It may produce such a modulation of the AHP, but the precise model of action is as yet unknown.

THE NETWORK MODEL

The simulated neurons are modelled as units representing a population of functionally similar neurons. These units are non-spiking and have a graded output that corresponds to the mean firing frequency of the modelled neurons. Spike frequency adaptation, due to current through Ca^{2+}-activated K^+-channels producing a slow AHP, is included in the model as a delayed negative feedback to the model units and is modulated actively. The units are stimulated by tonic excitatory input and at the same time the level of adaptation is increased linearly in proportion to the level of stimulation. This would correspond in the real lamprey to an increased AHP in spinal neurons at higher burst frequencies, which in turn would correspond to a decreased level of 5-HT release in the spinal cord at higher burst rates. This will produce a more pronounced AHP that can build up quicker and terminate the burst.

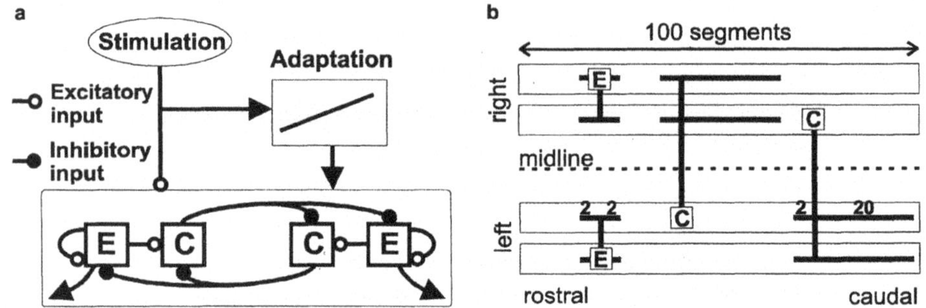

Fig. 1. a The model of a local CPG with an active modulation of the adaptation. The units represent populations of excitatory (E) and contra-lateral inhibitory (C) interneurons. The E units excite all ipsilateral units and the C units depress contra-lateral activity. In the lamprey spinal cord the E interneurons excite the motor neurons and are the output neurons. The network is stimulated by tonic excitatory input and the level of adaptation is increased linearly with increasing stimulation. b Schematic drawing of the rostrocaudal extent of connections in the simulated network for one E and one C unit on each side. The E units have ipsilateral projections extending over two segments both in caudal and rostral direction. The C units project over two segments in rostral direction and over twenty segments in caudal direction to unit on the contralateral side.

The simulated network contains model units of two different spinal interneurons: excitatory (E) interneurons and contra-lateral (C) inhibitory interneurons. The synaptic connections in the central pattern generator (CPG) network are compatible with what is known about synaptic connectivity in the lamprey spinal cord (Buchanan 1982; Buchanan and Grillner 1987; Grillner et al. 1995). In the model of a local CPG (Fig. 1 a) the E unit excites all ipsilateral units including itself and the C unit inhibits the contra-lateral units. The model of the whole spinal cord includes one hundred local CPG:s in a column with longitudinal connections (Fig. 1 b). The E units have projections reaching over two segments in both rostral and caudal direction, and the C units project two segments in rostral direction and over twenty segments in caudal direction.

SIMULATION RESULTS

A local CPG produces a stable burst activity alternating between the contra-lateral sides with a frequency range of 0.5-16 Hz when the adaptation level is increased with higher stimulation (Fig. 2 a). This active modulation of the adaptation is tuned to restrict the burst duration to be around 30-40% of the cycle duration (Fig. 2 b), as seen during fictive locomotion in isolated lamprey spinal cord (Wallén and Williams 1984). If the adaptation level is constant with respect to the degree of stimulation the frequency range becomes small and an increase in burst proportion is seen when increasing the burst frequency and does not remain constant. If the adaptation level is increased more at higher stimulation, the frequency range increases markedly and the burst proportion can remain constant and become lowered beyond that seen in the lamprey. A hemi-segment, i.e. a self-exciting E unit, can also produce a burst activity with the same frequency range as above or even more. This shows that the adaptation is the dominating burst terminating factor and not the reciprocal inhibition from the C units.

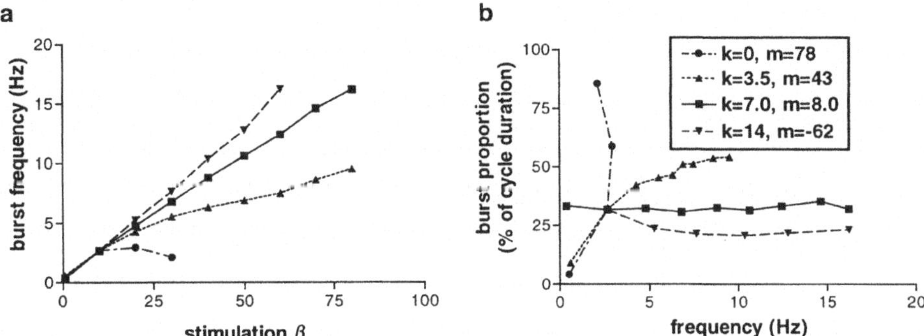

Fig. 2. Burst frequency and proportion when simulating a local CPG as a function of the stimulation (β) and the burst frequency with differently modulated adaptation according to $\mu = k\beta + m$ where μ is the level of adaptation. **a** The control ($k=7.0$, $m=8.0$; solid line) produces a burst frequency range of 0.5-16 Hz. If the adaptation level is kept constant ($k=0$, $m=78$) or increases less ($k=3.5$, $m=43$) the burst frequency range will become smaller. **b** The control ($k=7.0$, $m=8.0$; solid line) produces a constant burst proportion of around 35 % of the cycle duration, if the adaptation level is kept constant ($k=0$, $m=78$), or increases less ($k=3.5$, $m=43$) and the burst proportion will increase with burst frequency. If the adaptation level increases more quickly with β ($k=14$, $m=-62$) the burst proportion will become almost constant but at a lower level. Note that the free parameter is the stimulation level β.

Lamprey swims with a rostro-caudal time delay between the burst activity along the spinal cord, this phase lag is proportional to the cycle duration and around 1% of the cycle duration per segment producing one wave length along the body (Wallén and Williams 1984). Our model of a complete spinal cord consists of one hundred local CPG:s.

We started to simulate a column of hemi-segments containing only self-exciting E units with longitudinal connections. This network produces a stable phase lag along the column that is strongly influenced by how the ends of the network are treated. A network without compensation for the reduced excitatory input to the most rostral and caudal E units produces a phase lag that increases when increasing the stimulation, i.e. the burst rate (Fig. 3 a; dashed line). This lag is rostro-caudal in the rostral half, and a caudo-rostral in the caudal half of the column. If we instead provide a compensation by adding excitatory connections to the E units near the ends, a rostro-caudal phase lag can be produced provided that the rostral part is stimulated more than the remaining network (Fig. 3 a; solid line). In this a way a constant phase lag of around 1% of the cycle duration per segment can be produced over a burst frequency range of 1.5-15 Hz that fits nicely with what is seen during lamprey swimming.

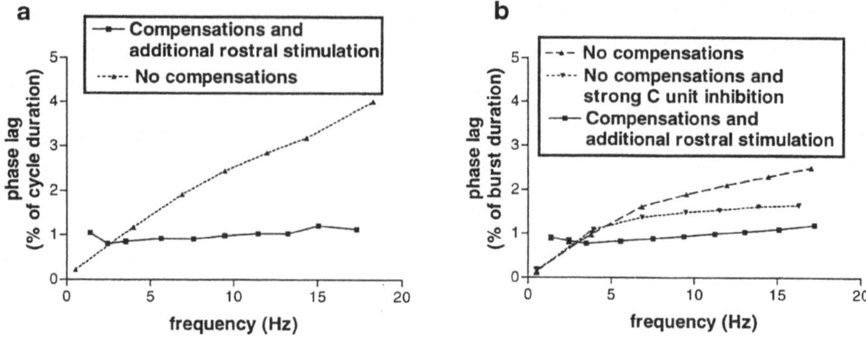

Fig. 3. a One hundred self-exciting E units distributed in a column with intersegmental connections projecting over two segments in both rostral and caudal direction. A phase lag increase is seen with increasing burst frequency when the ends are not compensated for reduced excitatory input (dotted line). When we do compensate with extra connections in the ends and the rostral part has a additional stimulation, the phase lag is constant and around 1% of the burst cycle per segment (solid line). **b** One hundred local CPG:s distributed in a column where the E units project two segments in both rostral and caudal direction and C units two segments in rostral and twenty segments in caudal direction. With compensations in the ends and an extra input to the rostral part the phase lag remain around 1% of cycle duration (solid line), but when we do not compensate an increasing phase lag with frequency is produced. The phase lag can however be constraint to an upper limit if the C unit projections are strong enough (dotted line).

When we introduce the inhibitory C units to two columns of E units, a network with one hundred local CPG:s is formed. The C units set the burst activity in the E unit columns out of phase, producing an alternation between contra-lateral sides. If this network is without "end compensation" the phase lag will be forced to a rostro-caudal direction along the whole network due to the long descending C unit projections. The increase in phase lag when increasing the burst rate can be limited to a upper bound of around 1% if the C unit projections are strong enough (Fig. 3 b; dotted line). With compensations in the ends and an additional stimulation of the rostral part a constant phase lag around 1 % of cycle duration per segment will however remain (Fig. 3 b; solid line).

CONCLUSION

By introducing AHP modulation as the dominating burst terminating mechanism (El Manira et al. 1994) in a semi-detailed model of the lamprey distributed spinal CPG, we have been able to simulate a stable swimming motor pattern that displays phase constancy over an adequate frequency.

ACKNOWLEDGEMENTS

Our thanks to Dr. Örjan Ekeberg for making his neuronal model of fish swimming available. This work was supported by the Medical Research Council (proj. no. 3026), the Natural Science Research Council and NUTEK.

REFERENCES

Buchanan, J.T., 1982, Identification of interneurons with contralateral, caudal axons in the lamprey spinal cord: synaptic interactions and morphology. *J. Neurophysiol.* 47:961-975.

Buchanan, J.T., and Grillner, S., 1987, Newly identified 'glutamate interneurons' and their role in locomotion in the lamprey spinal cord. *Science.* 236:312-314.

Ekeberg, Ö., 1993, A combined neuronal and mechanical model of fish swimming. *Biol Cybern.* 69:363-374.

El Manira, A., Tegnér, J., and Grillner, S., 1994, Calcium-dependent potassium channels play a critical role for burst termination in the locomotor network in lamprey. *J Neurophysiol.* 72:1852-1861.

Grillner, S., Deliagina, T., Ekeberg, Ö., El Manira, A., Hill, R.H., Lansner, A., Orlovsky, G.N., and Wallén, P., 1995, Neural networks that co-ordinate locomotion and body orientation in lamprey. *Trends Neurosci.* 18:270-279.

Wallén, P., Buchanan, J.T., Grillner, S., Hill, R.H., Christenson, J., and Hökfelt, T., 1989, Effects of 5-hydroxytryptamine on the afterhyperpolarization, spike frequency regulation, and oscillatory membrane properties in lamprey spinal cord neurons. *J. Neurophysiol.* 61:759 768.

Wallén, P., and Williams, T.L., 1984, Fictive locomotion in the lamprey spinal cord in vitro compared with swimming in the intact and spinal animal. *J. Physiol.* 374:278-290.

A HIPPOCAMPAL-LIKE NEURAL NETWORK MODEL SOLVES THE TRANSITIVE INFERENCE PROBLEM

Xiangbao Wu and William B Levy

Department of Neurological Surgery
University of Virginia Health Sciences Center
Charlottesville, Virginia 22908, USA
E-mail: xw3f@virginia.edu, wbl@virginia.edu

INTRODUCTION

Both rats and humans can solve configural learning problems. Based on lesion experiments in rats, configural learning is regarded as a hippocampally dependent function[1] when reconfigurability is critical. We have previously shown that a hippocampal-like neural network model[2,3] solves the configural problem of transitive inference (TI)[4]. Here we confirm this result and investigate the robustness of this demonstration as a function of network activity levels.

THE PROBLEM OF TRANSITIVE INFERENCE

Figure 1a describes the TI problem which is regarded as a logical problem in cognitive psychology (see references in Siemann and Delius,[5]). TI is based on a relationship such as "greater than". TI is said to develop if subjects can infer from A is greater than B and B is greater than C, that A is greater than C. In psychological experiments, at least 4 pairs of 5 atomic items as shown in Figure 1a are used. After learning which atomic stimulus (A, B, C, D) is the correct choice in each of these four pairs (A>B, B>C, C>D, and D>E), the subject gets the novel BD combination in which B should be the correct choice. Here the BD test is critical, but one can also test on the AE pair. However, combinations involving an end element (A or E) are quite easy because A has always been the right answer and E has always been the wrong answer.

Figure 1b suggests a control example for TI which is called non-transitive inference. Note that the atomic stimulus E is discarded, and in the fourth stimulus pair, this E is replaced by the atomic stimulus, A. One example of non-transitive inference is the game 'paper, rock, scissor', where paper defeats (covers) rock, rock defeats (breaks) scissor, and scissor defeats (cuts) paper.

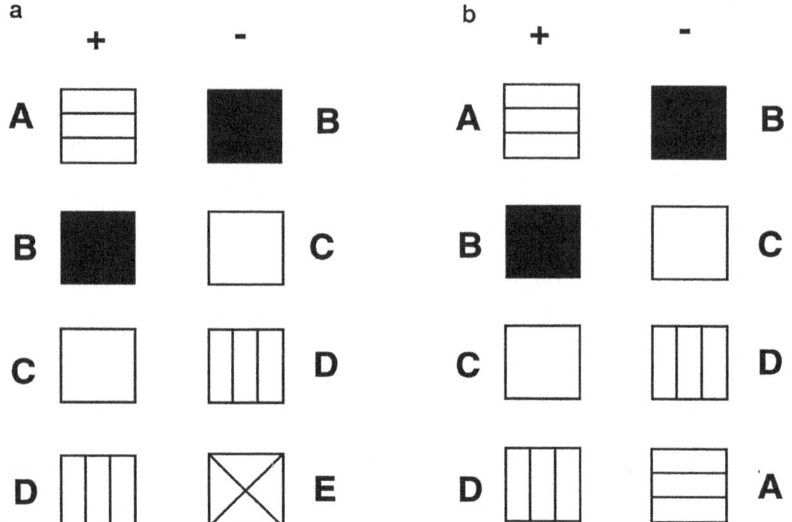

Figure 1: The two problem sets showing the input pairs (e.g., A+B-). Correct and incorrect stimuli are represented as + and - respectively. In both transitive inference (a.) and non-transitive inference (b.), the test input is the novel pair BD. For transitive inference B is the right answer, but there is no logically correct answer for the BD pair in the non-transitive inference situation.

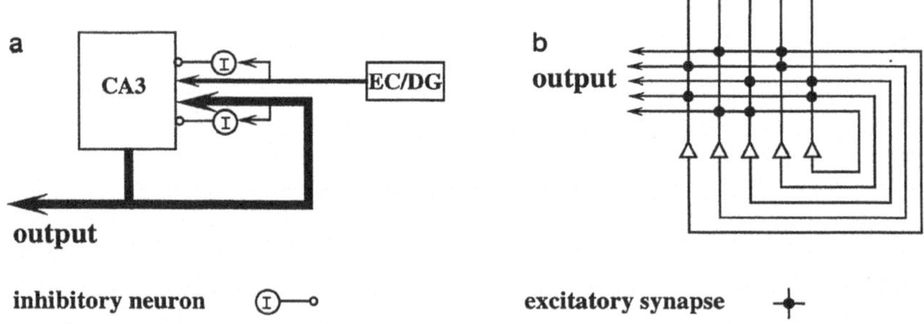

Figure 2: The Model. a. In the model the input layer is a combination of the entorhinal cortex and dentate gyrus. Accompanying this feedforward excitation is a proportional feedforward inhibition. The strong excitation of the network results from the recurrent connections, which is also accompanied by a feedback inhibition. The output of the network is the state of the excitatory CA3 cells themselves, and this is decoded by a simple cosine comparison. b. The recurrent excitatory synapses are sparse and randomly placed.

THE NETWORK

The hippocampal model is essentially a model of region CA3 (Figure 2). The input layer corresponds to a combination of the entorhinal cortex and dentate gyrus. To make the system's operation as transparent as possible, decoding is performed by similarity comparisons rather than a CA1-subiculum-entorhinal decoding system. The CA3 model is a sparsely (10%) interconnected feedback network of 512 neurons where all direct connections are excitatory and the network elements are McCulloch-Pitts neurons. There is an interneuron mediating feedforward inhibition, and one mediating feedback inhibition. Inhibition is of the divisive form, but the system is not purely competitive because of a slight delay. Synaptic modification develops over training. The process controlling synaptic modification is a local, self-adaptive postsynaptic rule that includes both potentiation and depression aspects[6,7]. The network computations are all local and are contained in three equations: spatial summation adjusted by inhibition; threshold to fire or not; and local Hebbian synaptic modification (see Levy and Wu,[8] Wu et al.,[9] Levy et al.,[10] for details).

In the simulations activity levels are varied by adjusting the inhibition constants K_r prior to learning. The range of activity level we tested goes from 10% to 15%, corresponding K_r from 0.054 to 0.047.

ENCODING TRANSITIVE INFERENCE

To study transitive inference, the same hippocampal-like network (but with different settings of parameters) that learned transverse patterning[11] is employed, and a similar set of input codings is constructed. That is, the inputs are sequences of stimulus, response, and reinforcement (e.g. (AB)(AB)(AB)aaa+++), where each pattern is repeated three times. A staged learning paradigm is followed as in the experiments of Dusek and Eichenbaum[1]. In this paradigm, there are five phases of training. Phase 1 consisted of the presentation of 10 trials of each pair in serial (i.e., ten trials of A+B- followed by ten trials of B+C-, then ten trials of C+D- and ten trials of D+E-). While in Phases 2, 3 and 4, trials were presented in blocks of 5, 3 and 1 trial respectively. The last phase of training is a random pair presentation of the four premise pairs.

RESULTS

Figure 3 is a comparison of experimental data of Dusek and Eichenbaum[1] and our results of network simulation. As noted, the BD test rather than the AE test is critical here because the AE comparison is not affected by the hippocampally inactivating lesions[1]. From Figure 3, one can see that B is the typically chosen answer (percent correct about 80%) when the BD pair is tested (which is not presented during learning). Likewise, A is the typical answer for the AE probe pair.

For the non-transitive inference problem (Figure 1b), where the novel test pair is essentially totally ambiguous, it turns out that the network failed to make consistent responses on the BD comparison. In fact, the networks randomly choose B or D without any noticeable preferences (percent correct about 50%, see Figure 4) when the BD pair is tested. In addition, their selection between A and C is random (Figure 4).

In the transitive inference problem, whether or not the network will succeed or fail is determined by the cell firing patterns the network constructs. That is, when BD

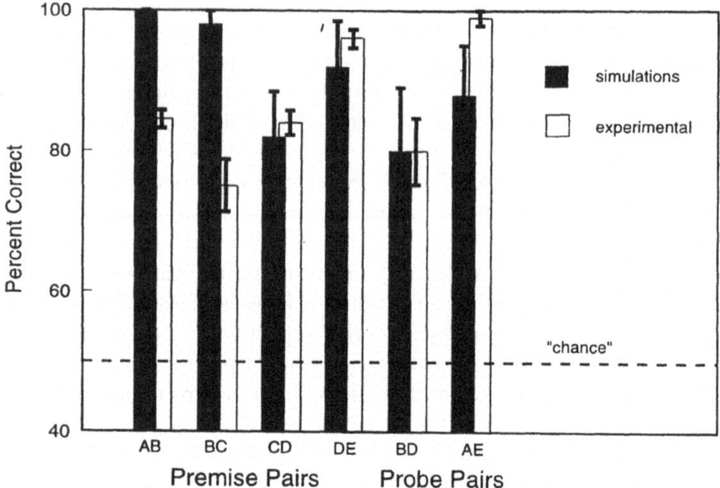

Figure 3: A comparison of experimental data and network simulations of the transitive inference problem. The open rectangles are the experimental results of Dusek and Eichenbaum[1]. The filled rectangles are the simulation results. Note that, as in the rat experiment, the model performs significantly above chance on the BD comparison. Error bars represent the standard error of the mean over 15 simulations. Dashed lines represent chance performance level. The activity level used here is 13% and K_r is 0.049.

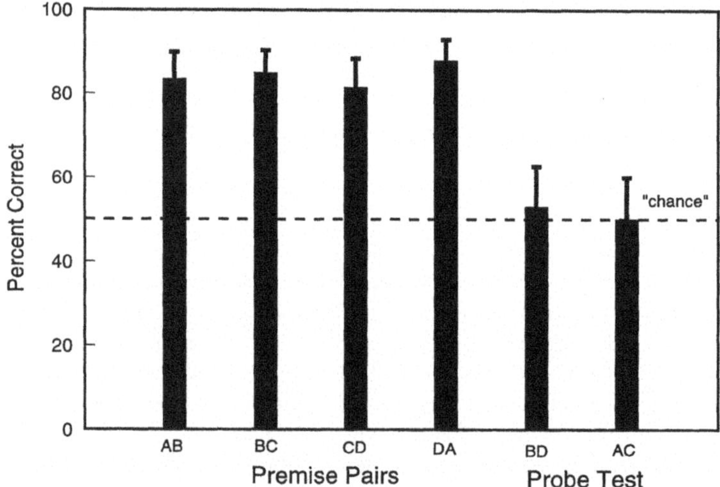

Figure 4: Network simulations of the non-transitive inference problem. Note that, the network randomly chooses B or D without noticeable preference on the BD comparison. Error bars represent the standard error of the mean over 20 simulations. Dashed lines represent chance performance level.

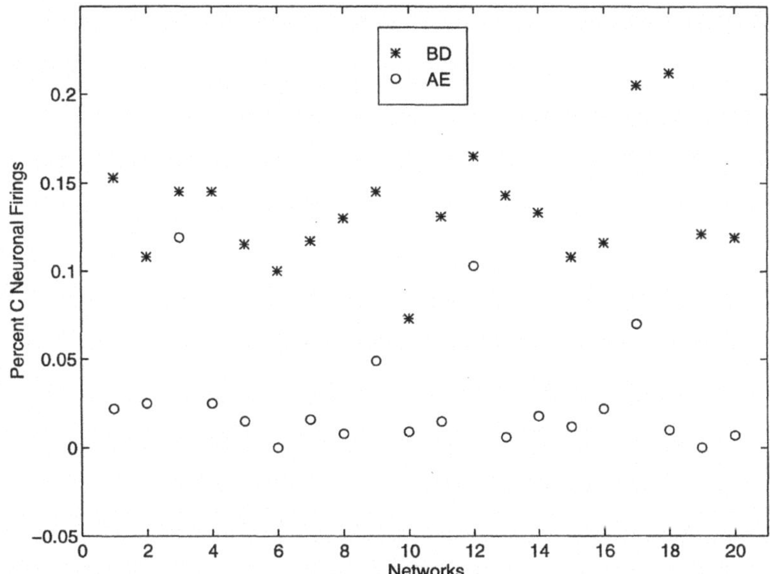

Figure 5: Percent C neuronal firings evoked by BD (*) and AE (o) for 20 different simulations. Note that BD fires many more C neurons than AE does. Moreover, when AE has relatively high C firings, the networks (4 out of 20) fail in performing either on BD or CD tests.

activates enough ($\sim 13\%$, see Figure 5, asterisks) C neuronal firings, which are defined by testing of C alone, the B answer is chosen over the D answer. This activation of so many C neurons by BD rarely occurs when the non-transitive inference problem is learned (Figure 5, open circles).

DISCUSSION

Configural learning problems, studied experimentally by psychologists, require a hippocampus for successful learning. In this report, we have again shown that a hippocampal-like network can solve a particular configural learning problem. Such results together with the network's ability to perform successful sequence prediction/goal finding[10,12], strengthen the qualifications of this hippocampal model.

However, not every simulation will succeed on the TI problem. A network's activity level plays an important role because it affects the learned patterns of neural firing. To test the robustness of the model, we have tried several activity levels. For example, the activity level used here (13%), gives a better network performance than the one used elsewhere (e.g. 15% in Levy and Wu[4]). However, if the activity level is too high (e.g. 25%) or too low (e.g. 4%), the network fails to produce good performance on the TI problem, although such statements are dependent on network size with lower activities working in larger networks.

ACKNOWLEDGMENTS

This work was supported by NIH MH48161 and MH00622, by Pittsburgh Super-computing Center Grant BNS950001 to WBL, and by the Department of Neurosurgery, Dr. John A. Jane, Chairman.

REFERENCES

1. Dusek, J.A. and Eichenbaum, H., 1997, The hippocampus and memory for orderly stimulus relations, *Proc. Natl. Acad. Sci. USA* 94:7109-7114.
2. Levy, W.B, 1989, A computational approach to hippocampal function, in: *Computational Models of Learning in Simple Neural Systems*, R.D. Hawkins and G.H. Bower, ed., pp.243-305, Academic, New York.
3. Minai, A.A. and Levy, W.B, 1993, Sequence learning in a single trial, *INNS World Congress on Neural Networks*, pp.505-508.
4. Levy, W.B and Wu, X.B. 1997, A simple, biologically motivated neural network solves the transitive inference problem, *Proceedings of the 1997 IEEE International Conference on Neural Networks*, pp.368-371.
5. Siemann, M. and Delius, J.D., 1994, Processing of hierarchic stimulus structures has advantages in humans and animals, *Biol. Cybern.* 71:531-536.
6. Levy, W.B and Steward, O., 1979, Synapses as associative memory elements in the hippocampal formation, *Brain Res.* 175:233-245.
7. Levy, W.B, 1982, Associative encoding at synapses, *Proceedings of the Fourth Annual Conference of Cognitive Science Society*, pp.135-136.
8. Levy, W.B and Wu, X.B., 1996, The relationship of local context codes to sequence length memory capacity, *Network* 7:371-384.
9. Wu, X.B., Baxter, R.A., and Levy, W.B, 1996, Context codes and the effect of noisy learning on a simplified hippocampal CA3 model, *Biol. Cybern.* 74:159-165.
10. Levy, W.B, Wu, X.B., and Baxter, R.A., 1995, Unification of hippocampal function via computational/encoding considerations, *Int. J. Neural Sys.* 6(Supp.):71-80.
11. Levy, W.B, Wu, X.B., and Tyrcha, J.M., 1996, Solving the transverse patterning problem by learning context present: A special role for input codes, *INNS World Congress on Neural Networks* pp.1305-1309.
12. Wu, X.B. and Levy, W.B, 1996, Goal finding in a simple, biologically inspired neural network, *INNS World Congress on Neural Networks* pp.1279-1282.

FINITE ELEMENT DECOMPOSITION OF HUMAN NEOCORTEX

David A. Batte, Travis S. Chow, and Bruce H. McCormick

Scientific Visualization Laboratory
Department of Computer Science
Texas A&M University
College Station, TX 77843-3112

IMPORTANCE OF FINITE ELEMENT BRAIN

Modeling brain morphology at both cellular and tissue levels brings richness to our understanding of brain organization that both complements and transcends knowledge derived exclusively from neuron tracing and brain atlases.

The neocortex of the human brain has been geometrically modeled and visualized using finite element decomposition and grid generation techniques. The work reported here extends modeling techniques reported previously (CNS*96) [1] to the whole human neocortex. Approximately 2500 finite elements (FEs) of the scale and complexity shown in Figure 1 are used to model the right hemisphere of the human neocortex. The finite elements of the human cortical shell are established in a manner consistent with developmental neurobiology.

Each neuron in a neuron morphology data repository, whether a traced biological neuron or a synthetically generated neuron, can be assigned to the FE which contains its soma. The cerebral cortex, so modeled, can be viewed as a giant "chest of drawers" where a "drawer" (any selected FE or cluster of neighboring FEs) can be "opened" as a file and its population of neurons visualized as illustrated in Figure 2. These FEs therefore define a file structure isomorphic to the neocortex as modeled and visualized at both cellular and tissue levels.

Neuron morphology data repositories for the human neocortex are in an early state of development. In future work, adding stochastic L-system modeling of the neuron populations will lead to a normative morphological model of the human neocortex. In principle this normative model would generate neurons statistically indistinguishable in morphology from the neuron forests of an actual brain as viewed at the limit of optical resolution. The required storage space for the normative model is estimated as six megabytes of data.

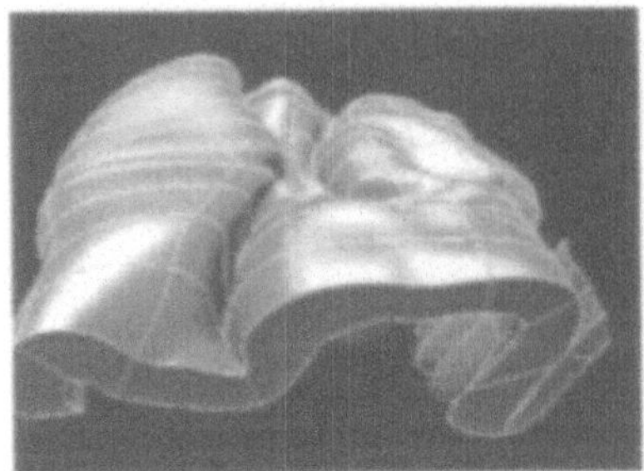

Figure 1. Finite element model for a piece of the neocortical shell.

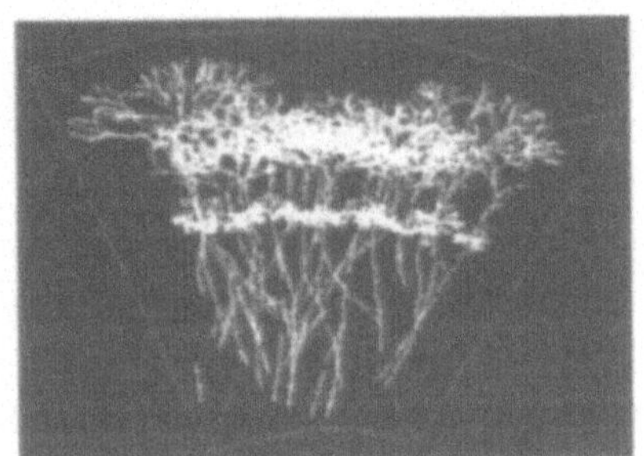

Figure 2. Finite element populated with synthetic neurons.

OBJECTIVES

The three objectives of the research are briefly described below.

Biologically Consistent Finite Element Decomposition of Human Cortical Shell

To establish finite elements in a manner consistent with developmental neurobiology, the elements are designed to follow the natural symmetries of the tissue. One coordinate axis is defined by the local orientation of its primary native neuron type (pyramidal cells). The other axes are chosen to simulate how a neuroanatomist would cut tissue locally to make successive sections look as similar as possible. Numerical grid generation algorithms establish a curvilinear coordinate system within each finite element and, hence, by extension throughout the entire solid model. This coordinate system provides a basis for orienting neurons and specifying physical barriers and chemical gradients within the tissue.

Mapping Neurons into Gridded Solid Model

Our second objective is to embed a population of neurons within the gridded solid model. Sparse populations of neurons can be drawn from databases of traced neurons, or generated stochastically using L-system modeling as described by McCormick and Mulchandani [2, 3, 4]. The grids provide a framework for positioning the neurons in the solid model of the reconstructed tissue and for specifying positional forces within the tissue. The embedded neurons are chosen to be statistically consistent with those found in the actual tissue.

Visualization of Neuron Populations within Gridded Solid Model

Our third objective is to visualize a population of neurons within the reconstructed tissue using computer graphics in conjunction with the grid generation and neuron mapping methods. An interactive approach, developed in [5], allows the user to freely explore the brain forest.

METHODOLOGY

Our methodology for the FE decomposition involves five steps, as shown in Figure 3, and the development of supplementary object-oriented software tools.

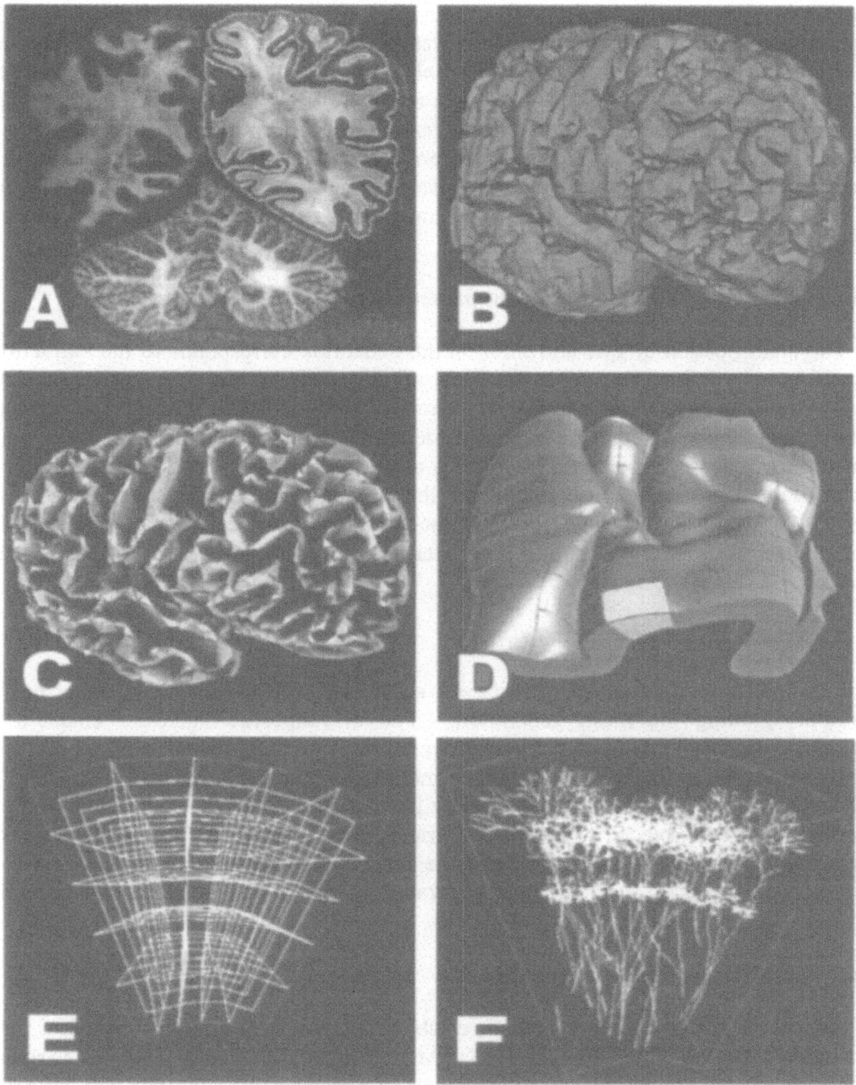

Figure 3. The five steps of our methodology. **A, B**: 3D reconstruction of the manually traced contours of the human neocortex yields a solid model. **C**: Feature extraction extricates the prominent *sulci* and *gyri* from the reconstructed solid model. **D**: Finite element decomposition of the model provides "drawers" for neuron populations. **E**: Curvilinear 3D grids are generated for the finite element "drawers." **F**: Neuron populations are implanted into the gridded finite elements.

3D Reconstruction of Neocortex

Serial reconstruction of cross-sectional slices of a *post mortem* human brain generates a solid model of the neocortex with minimal loss in morphological detail. The dataset for the reconstruction consists of 217 x 512 x 512 images of a 76-year-old normal female human cadaver brain cryosectioned through the horizontal plane [6]. Contours of the neocortex were manually traced using a third party application (*Elastic Reality* from Avid Technologies, Inc.). Tissue segmentation was performed manually with the subsequent evaluation by a neuroanatomist (Dr. Ian Russell). Because of the precision and consistency of the cryosectioning technique, the spatial alignment of the consecutive images along the cutting axis automatically insures the registration of the contours between adjacent sections. Then, a Delaunay-based surface reconstruction algorithm generates triangulated surfaces for both the exterior and interior side of the neocortical tissue [7]. Thus, the inner and outer surfaces jointly define the volume of the neocortex, producing a *boundary representation* (B-Rep) solid model.

Feature Extraction from Boundary Representation Model

The identification of *gyri* and *sulci* on the neocortex corresponds to the extraction of extremal points, defined loosely as the local minima and maxima of a surface, and other shape metrics, such as curvedness and shape index, from the B-Rep model. The extrication of these topological features rely on the determination of principal curvature values, defined loosely as the curvature of a point on a surface, for each vertex on the exterior and interior triangulated surfaces obtained from the previous step. Approximation techniques estimate the principal curvatures; then, various shape metrics are computed directly from these values, and extremal points are determined by solving a set of "extremality" equations [8].

Finite Element Decomposition

Extremal points and shape metrics extracted from the B-Rep model determine the constraints guiding the mesh generation. The decomposition takes a coarse-to-fine approach. First, the convoluted tissue is broken down into more manageable pieces, which roughly corresponds to the major *gyri* defined in anatomical atlases [9]. Second, 2D meshes, congruous with the ridge and valley lines, are determined for the inner and outer surfaces for each piece. Third, a graph correspondence algorithm associates complementary extremal points between the inner and outer mesh. In the absence of pyramidal cell axis data, coarse 3D grid generation is used to compute the hexahedral finite elements [10].

Grid Generation for Finite Element Model

Numerical grid generation methods establish a curvilinear coordinate system for the solid model. Batte [5] has applied 3D ITTM grid generators and several 3D variational grid generators [11] to parametric solid models of neocortical tissue segments. We apply these generators to the finite elements obtained in the finite element decomposition above.

First, we construct local 3D boundary-conforming grids within each finite element. Algorithms that iteratively optimize boundary and inter-element geometric continuity are available to generate unfolded, continuous grids for each finite element (as opposed to the convolutions of the entire neocortex). We have explored both geometric and variational grid generation methods to fill the solid model with locally-defined coordinate systems.

A global coordinate system minimizing discontinuities between grids of adjacent finite elements greatly facilitates neuron implantation. The continuity reduces irregularities and mitigates population shifts at finite element boundaries. We have investigated reparametrization techniques and related strategies for solving the continuity constraints to unify the disjoint grids into a continuous 3D grid for the entire solid model.

Neuron Implantation within Finite Elements

Computer graphics in conjunction with neuron mapping methods provides the facilities to visualize a neuron population within the gridded solid model. Sparse populations of neural cells can be drawn from databases of traced neurons or generated stochastically using L-system modeling as described by [2].

The finite element model provides a spatial indexing scheme for a neuron morphology data repository. First, neurons are assigned to the finite element containing their somas. This spatial decomposition gives the initial categorization criterion for the data repository. Subsequently, various relational ordering can be defined by creating different index-trees to the finite elements. One such relational ordering is a hierarchical breakdown through multiple levels of neocortical regions, represented as a group of finite elements, to the level of individual finite elements.

RESULTS

The resulting finite elements in B-Spline tensor product representation [12] follow the natural symmetries of the tissue as shown in Figures 4 and 5.

Figure 4. Band of reconstructed neocortex in triangulated mesh representation (shown in reduced resolution).

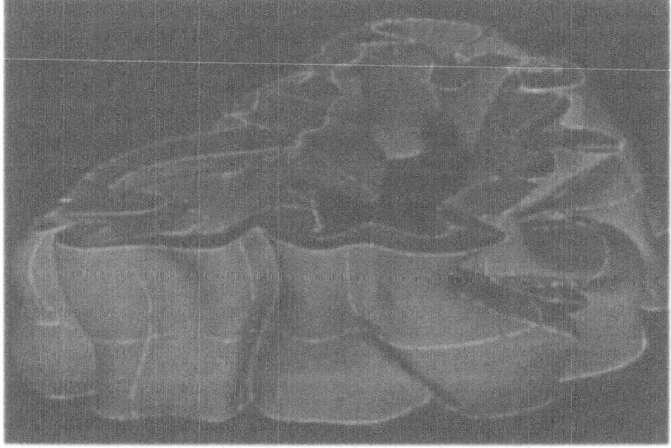

Figure 5. Finite element decomposition in B-Spline tensor product representation of above neocortical band.

One of the curvilinear coordinate axis, obtained by interpolating the cortical surface normals between corresponding extremal lines on the exterior and interior surfaces, approximates the local orientation of the pyramidal cell axes. The finite element boundaries geodesic to the exterior and interior surfaces correspond to the ridge and valley lines along gyral and sulcal folds.

SIGNIFICANCE

The methodology described above:
1. Constructs a parametric solid model of the human neocortex that captures the morphology hidden within its deep convolutions.
2. Builds a finite element mesh for the human neocortex.
3. Provides a global framework and coordinate system for a 3D atlas of the human brain and for a normative neocortical model.
4. Shows the viability of mapping neuron populations within a 3D graphical model as a means for structural and functional analyses.
5. Prototypes a hierarchical spatial data management system for neuron data sets.

The supplementary software tools are planned to allow the visualization and modeling of the neocortex through an exploratory environment accessible over the Internet.

ACKNOWLEDGMENTS

This work was supported by Texas Advanced Technology Program grant 999903-124 (McCormick) from the Texas Higher Education Coordinating Board.

REFERENCES

1. D. A. Batte and B. H. McCormick, Grid generation for brain visualization at the cellular and tissue level, in: *Computational Neuroscience: Trends in Research, 1997*, J. Bower (ed), Plenum Press, New York (1997).
2. B. McCormick and K. Mulchandani, L-system modeling of neurons, *Proc. Visualization in Biomedical Computing*, SPIE, 2359 (1994).
3. K. Mulchandani. *Morphological Modeling of Neurons*, Master's Thesis, Department of Computer Science, Texas A&M University, College Station, TX (1995).
4. K. Mulchandani and B. McCormick, A framework for modeling neuron morphology, in: *Computational Neuroscience: Research Trends for 1995*, J. Bower (ed), Academic Press, San Diego (1996).
5. D. A. Batte. *Finite Element Decomposition and Grid Generation for Brain Modeling and Visualization*, Master's Thesis, Department of Computer Science, Texas A&M University, College Station, TX (1997).
6. A. Toga, K. Ambach, and S. Schluender, High-resolution anatomy from in situ human brain, *NeuroImage*, 4 (1994).
7. B. Geiger, Three dimensional modeling of human organs and its application to diagnosis and surgical planning, *INRIA Rapports de Recherche–Sophia Antipolis*, 2105 (1993).
8. J. Thirion and S. Benayoun, Image surface extremal points, new feature points for image registration, *INRIA Rapports de Recherche–Sophia Antipolis* , 2003 (1993).
9. H. Duvernoy. *The Human Brain: Surface, Three-Dimensional Sectional Anatomy and MRI*, Spring-Verlag, Wien, New York (1991).
10. T. Chow. *Finite Element Decomposition of the Human Neocortex*, Master's Thesis, Department of Computer Science, Texas A&M University, College Station, TX (January 1998).
11. P. Knupp and S. Steinberg. *Fundamentals of Grid Generation*, CRC Press, Boca Raton (1994).
12. P. Dierckx. *Curve and Surface Fitting with Splines*, Oxford University Press (1995).

PATH INTEGRATION IN THE RAT HEAD-DIRECTION CIRCUIT

Hugh T. Blair[1], Patricia E. Sharp[1], Jeiwon Cho[1], Jeremy P. Goodridge[2], Robert W. Stackman[3], Edward J. Golob[3], and Jeffrey S. Taube[3]

[1]Department of Psychology
Yale University
New Haven, CT 06520

[2]Center for the Neural Basis of Cognition
Carnegie-Mellon University
Pittsburgh, PA 15213

[3]Department of Psychology
Dartmouth College
Hanover, NH 03755

INTRODUCTION

As a rat navigates through space, neurons called *head-direction (HD) cells* provide an ongoing signal of the animal's directional heading in the horizontal plane[1,2]. It is believed that the population of HD cells may function as a neural compass, providing the rat with its sense of direction during spatial navigation. The HD cell signal is thought to be generated, in part, by a path integration mechanism that uses angular motion information to compute the rat's directional heading[3-5]. Here we review empirical findings that suggest how different brain regions might participate in this process of angular path integration.

HD CELLS

A HD cell fires action potentials only when the rat's head is facing in a particular direction with respect to the surrounding environment, regardless of the animal's location within that environment (Figure 1). HD cells are *not* influenced by the position of the rat's head with respect to its body, they are only influenced by the direction of the head with respect to the surrounding environment. Each HD cell is tuned to have its own directional preference, referred to as the *preferred firing direction (PFD)* of the cell. Each HD cell has a different PFD, so the population of cells can represent any direction the animal faces.

Sensory Influences

Landmarks. Visual cues are known to influence HD cells, because the PFD of HD cells can be manipulated by relocating familiar visual landmarks in the environment[6-11]. HD cells can maintain accurate directional firing in the absence of visual landmarks[6-9], but

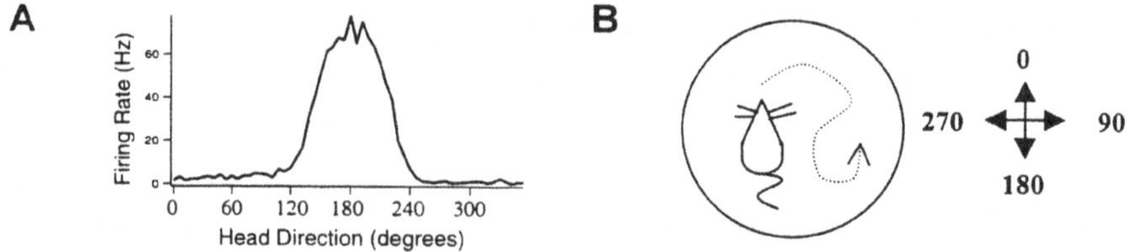

Figure 1. Tuning properties of a HD cell recorded in the anterior thalamus. A) Firing rate of the cell as a function of the rat's directional heading, with peak at PFD. B) Overhead view of the experimental chamber.

the directional signal is more stable when visual landmarks are present[11,12]. It is possible that when visual landmarks are absent, HD cells might rely on non-visual landmarks, such as odor and sound cues, to maintain the directional signal. However, evidence strongly suggests that HD cells also rely on "ideothetic" cues, which convey information about the angular motion of the head, to maintain the HD signal.

Ideothetic Cues. *Ideothetic cues* provide an animal with information about its own self motion. Evidence suggests that the HD signal is influenced by at least two types of ideothetic information: vestibular signals and efferent motor command signals.

Several experiments have demonstrated that HD cells are dependent upon vestibular signals. In one experiment[13], HD cells were recorded from AD as rats roamed freely in a cylindrical chamber. It was shown that if the chamber was rotated very slowly (below the rat's vestibular threshold), then the PFD of HD cells would rotate along with the chamber. But if the rotation was performed quickly, the PFD of HD cells did not follow the chamber, suggesting that vestibular cues allowed HD cells to track the rotation. Another experiment has shown that vestibular disorientation degrades the stability of the HD signal[14]. Finally, it has been shown that vestibular lesions abolish the HD signal[15] (see below).

When a rat is restrained and passively rotated to face different directions, HD cells often cease or reduce their normal directional firing[9,10,14,16]. This finding suggests that motor commands may be important for generating the HD signal. However, it is also possible that restraint alters the rat's attentional state, which has a deleterious effect on the processing of the directional signal. Another relevant finding is that when a rat turns its head to face a new direction, some HD cells in AD begin to change their firing rate about 20 ms *before* the rat's head starts turning[17]. Since vestibular motion cues cannot be generated until after the head starts to turn, this finding suggests that HD cells may be influenced by motor command signals that are initiated just before the head turn begins.

Angular Path Integration

HD cells are thought to rely, in part, on *angular path integration* to keep track of the rat's directional heading[3-5] (Figure 2). According to this hypothesis, HD cells compute the directional position of the rat's head by integrating the angular motion of the head over time. Like any neural integration mechanism, this "dead-reckoning" process would be prone to the accumulation of integration errors over time. It is thought that rats might use familiar visual landmarks to periodically "reset" the HD signal, correcting for path integration errors that have accumulated over time[18,19]. This suggestion is supported by observations[11,12] that HD cells can maintain directional firing in the absence of visual landmarks (presumably because the path integrator is intact), but they become less stable (perhaps because integration errors are harder to correct for without visual landmarks).

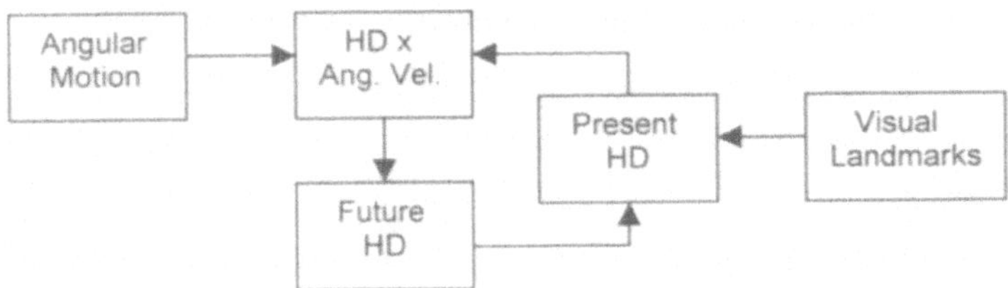

Figure 2. Steps in the angular path integration process. Information about the present head direction is combined with angular motion information, to compute a prediction of the future head direction. This prediction updates the representation of the present head direction. Visual landmarks can reset the representation of the present head direction.

ORGANIZATION OF THE HD CIRCUIT

HD cells have been observed in several regions of the rat brain (Figure 3). Some of these brain structures may be involved in generating the HD signal, while other regions may passively receive the HD signal from structures where it is generated.

Relating Structure and Function

The path integration hypothesis (Figure 2) provides a possible explanation for how the HD signal is generated, and may account for why the signal is influenced by both visual landmarks and ideothetic signals (e.g., motor and vestibular cues). If the path integration hypothesis is correct, then it should be possible to relate the functional steps of the path integration process (Figure 2) with the anatomical components of the HD circuit (Figure 3). However, it is important to note that the functions represented by each box in Figure 2 do not have to be performed in different parts of the brain. The path integration circuit may reside in a single brain region, or it may be distributed across several connected brain regions. Alternatively, there might be redundant copies of the circuit in different parts of the brain, each equally capable of angular path integration on its own.

The anatomy of the HD circuit (Figure 3) provides some preliminary clues about the possible functions of different brain structures[20]. For example, the visual system provides inputs to RsC, PoS, and LD, suggesting that these structures might be sites where visual landmark cues can influence the HD signal. LMN receives a major input from DTN, which might provide HD cells with ideothetic (vestibular and motor) information.

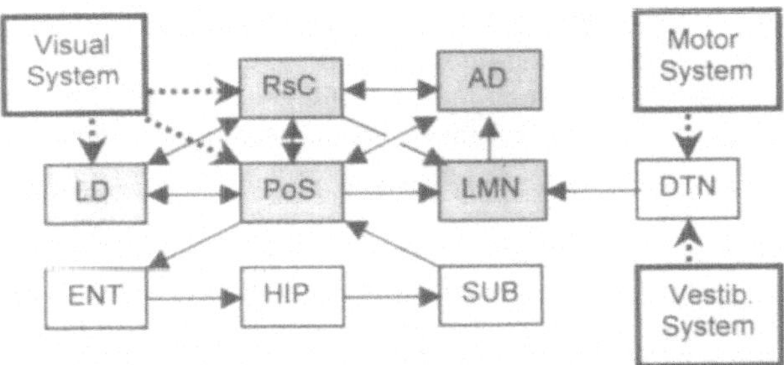

Figure 3. Anatomical organization of the HD circuit and primary sensory inputs. Shaded regions are the areas where HD cells have been observed. Abbreviations: retrosplenial cortex (RsC), anterodorsal thalamus (AD), laterodorsal thalamus (LD), postsubiculum (PoS), lateral mammillary nucleus (LMN), dorsal tegmental nucleus (DTN), entorhinal cortex (ENT), hippocampus (HIP), subiculum (SUB).

Properties of HD Cells in Different Brain Regions

Anterodorsal Thalamus (AD). The firing of HD cells in AD correlates best with the *future* direction of the rat's head[21,22]. On average, HD cells in AD anticipate the rat's future head direction by about 20 ms, but some AD cells anticipate by as much as 50 ms[23]. In addition, AD HD cells fire faster when the head is turning than when it is not turning[10,21]. These data suggest that AD might be involved in predicting the future direction of the rat's head, an operation that could be essential for path integration[21,23].

Postsubiculum (PoS). Unlike HD cells in AD, the firing of HD cells in PoS is best correlated with the direction that the rat's head is presently facing, and not with the future head direction[21,22]. Also in contrast with AD cells, the firing rate of most PoS HD cells does not increase when the rat's head is turning[21], although a few PoS HD cells are modulated by head turns[2]. These findings suggest that PoS may be involved in storing a representation of the present head direction, which could be important for path integration (see Figure 2). The suggestion that PoS is involved in path integration is supported by the finding that PoS contains neurons that fire in correlation with angular head motion[24].

Lateral Mammillary Nucleus (LMN). Like HD cells in AD, the firing of HD cells in LMN is best correlated with the rat's future head direction[25]. LMN cells anticipate the rat's future head direction by a greater amount than AD cells, which suggests that LMN might be the primary site for generating a prediction of the future head direction. LMN HD cells also have another interesting property: many of them seem to increase their firing rate when the rat's head is turning in one direction (clockwise or counterclockwise), but not the other[26]. By contrast, most AD HD cells increase their firing rate when the head turns in either direction[10,21]. This suggests that LMN might be the primary site where information about head direction (from PoS) is combined with information about angular head motion (perhaps from DTN)[20,23,27].

Laterodorsal Thalamus (LDN) and Retrosplenial Cortex (RsC). HD cells have been recorded in LDN[11] and RsC[6,28], but the detailed firing properties of these cells have not been investigated as intensively as HD cells in other areas. Further study is necessary to learn about how the HD cells in LDN and RsC are influenced by angular head velocity, and whether they are best correlated with the rat's past, present, or future head direction.

Lesion Studies

AD Lesions. Following lesions of AD, HD cells are no longer found in PoS[29]. One explanation for this could be that AD performs a function that is essential for computing the HD signal, such as predicting the future head direction (see above). Another possibility is that AD is not essential for generating the HD signal, but it provides the only route by which the HD signal can reach PoS from the location where it is generated. For example, the signal might be generated in LMN, but fail to reach PoS after AD lesions (Figure 3).

PoS Lesions. HD cells are still observed in AD after PoS lesions[29], suggesting that PoS is not essential for generating the HD signal. However, HD cells in AD exhibit three significant changes following PoS lesions: 1) their directional tuning width becomes broader, 2) their firing anticipates the rat's future head direction by a larger amount, and 3) they are less influenced by visual landmarks than AD cells in non-lesioned animals[29]. These findings suggest that PoS may be a site where visual cues exert an influence on the HD signal. This suggestion is supported by anatomical data (Figure 3), and also by the fact that PoS HD cells are normally correlated with the rat's present head direction, a property which would be expected of HD cells that rely mainly on visual cues.

Hippocampal Lesions. Hippocampal lesions (HL) do not abolish the HD signal in AD or PoS. However, when HL rats walk through a passageway from a familiar to a novel chamber, HD cells become unstable for several minutes before settling on a new PFD in the novel chamber[30] (by contrast, control rats keep the same PFD in both chambers[8]). In a novel setting, HD cells may have to rely on path integration alone (without help from visual landmarks) until the new chamber becomes familiar. If so, then HD cells in HL rats may be initially unstable in the novel environment because path integration is impaired by HL. Recent behavioral studies of path integration in animals with HL have reported conflicting findings[31,32]. Alternatively, HL might retard the process of learning to recognize landmarks in the new chamber. If so, HL rats would be forced to do without familiar cues for a longer time than control rats in the new chamber.

Vestibular Lesions. The directional firing of HD cells in AD is completely abolished following lesions of the vestibular system[15]. This strongly suggests that the HD signal is dependent on vestibular input for angular path integration. The effect of vestibular lesions may be analogous to removing the "Angular Velocity" box in Figure 2.

DISCUSSION

The experiments we have reviewed here suggest insights into the functional and anatomical organization of the rat HD circuit, but several open questions remain.

The core of the angular path integrator is likely to involve a feedback loop for updating the directional representation (see Figure 2). One tempting hypothesis is that this feedback loop might correspond to the classical circuit of Papez, which forms a loop between the hippocampal formation, mamillary nuclei, and anterior thalamus. The HD circuit is embedded within the Papez circuit, since it includes PoS/RsC, LMN, and AD (see Figure 3). If the Papez circuit contains the primary path integrator, then a lesion that severs the Papez loop should abolish the activity of HD cells throughout the brain. This might explain why lesions of AD abolish the HD signal in PoS. However, lesions of PoS do not abolish the HD signal in AD. One explanation for this could be that the function of PoS might be performed redundantly by RsC. Several facts support this idea: (1) PoS and RsC are closely interconnected, (2) they occupy a similar position within the Papez circuit, and (3) they share many of the same anatomical connections with other brain areas.

Another possibility is that the Papez circuit theory is wrong. Instead, the feedback loop for the path integrator might reside in another part of the brain, or it may be redundantly implemented in many different brain regions. It is particularly interesting to consider the possibility that angular path integration might be carried out by subcortical structures. It is well known that subcortical circuits are largely responsible for some other behaviors that involve neural integration of angular head motion, such as the vestibulo-ocular reflex and visual·orienting/tracking. Further studies are necessary to localize the angular path integration mechanism for the rat HD circuit.

REFERENCES

1. J.B. Ranck Jr., Head-direction cells in the deep layers of dorsal presubiculum in freely moving rats. *Soc. Neurosci. Abstr.* 10:599 (1984).
2. J.S. Taube, R.U. Muller, and J.B. Ranck Jr., Head-direction cells recorded from the postsubiculum in freely moving rats. I. Description and quantitative analysis. *J. Neurosci.* 10:420 (1990).
3. B.L. McNaughton, L.L. Chen, and E.J. Markus, "Dead-reckoning," landmark learning, and the sense of direction: A neurophysiological and computational hypothesis. *J. Cog. Neurosci.* 3:190 (1991).

4. A.D. Redish, A.N. Elga, and D.S. Touretzky, A coupled attractor model of the rodent head-direction system. *Network* 7:671 (1996).
5. K. Zhang, Representation of spatial orientation by the intrinsic dynamics of the head-direction cell ensemble: a theory. *J. Neurosci.* 16:2112 (1996).
6. L.L. Chen, L.H. Lin, E.J. Green, C.A. Barnes, and B.L. McNaughton, Head-direction cells in the rat posterior cortex. I Anatomical distribution and behavioral modulation. *Exp. Brain Res.* 101:8 (1994).
7. J.P. Goodridge and J.S. Taube, Preferential use of the landmark navigation system by head-direction cells. *Behav. Neurosci.* 109:49 (1995).
8. J.S. Taube and H.L. Burton, Head-direction cell activity monitored in a novel environment and during a cue-conflict situtation. *J. Neurophysiol.* 74:1953 (1995).
9. J.S. Taube, R.U. Muller, and J.B. Ranck Jr., Head-direction cells recorded from the postsubiculum in freely moving rats. II. Effects of environmental manipulations. *J. Neurosci.* 10:420 (1990).
10. J.S. Taube, Head-direction cells recorded in the anterior thalamic nuclei of freely moving rats. *J. Neurosci.* 15:70 (1995).
11. S.J.Y. Mizumori and J.D. Williams, Directionally selective mnemonic properties of neurons in the lateral dorsal nucleus of the thalamus of rats. *J. Neurosci.* 13:4015 (1993).
12. J.P. Goodridge, K.A. Worboys, P. Dudchenko, and J.S. Taube, The response of head-direction cells to non-visual cues. *Soc. Neurosci. Abstr.* 21:945 (1995).
13. H.T. Blair and P.E. Sharp, Visual and vestibular influences on head-direction cells in the anterior thalamus of the rat. *Behav. Neurosci.* 110:643 (1996).
14. J.J. Knierim, H.S. Kudrimoti, and B.L. McNaughton, Place cells, head-direction cells, and the learning of landmark stability. *J. Neurosci.* 15:1648 (1995).
15. R.U. Stackman and J.S. Taube, Firing properties of head-direction cells in the rat anterior thalamic nucleus: Dependence on vestibular input. *J. Neurosci.* 17:4349 (1997).
16. E.J. Markus, B.L. McNaughton, C.A. Barnes, J.C. Green, and J. Meltzer, Head-direction cells in the dorsal presubiculum integrate both visual and angular velocity information. *Soc. Neurosci. Abstr.* 16:441 (1990).
17. B.W. Lipscomb, H.T. Blair, and P.E. Sharp (1996). Evidence for motor command signal influences on anticipatory head-direction cells. *Soc. Neurosci. Abstr.* 22:913 (1996).
18. W.E. Skaggs, J.J. Knierim, H.S. Kudrimoti, and B.L. McNaughton, A model of the neural basis of the rat's sense of direction, in: *Advances in Neural Information Processing Systems 7*, G. Tesauro, D. Touretzky, and T. Leen, eds., MIT Press, Cambridge, MA (1995).
19. W.E. Skaggs, Influence of landmarks in a model of the head direction system. *Soc. Neurosci. Abstr.* 23:504 (1997).
20. J.S. Taube, J.P. Goodridge, E.J. Golob, P.A. Dudchenko and R.W. Stackman, Processing the head-direction signal: a review and commentary. *Brain Res. Bull.* 40:477 (1996).
21. H.T. Blair and P.E. Sharp, Anticipatory head-direction signals in anterior thalamus: Evidence for a thalamocortical circuit that integrates angular head motion to compute head direction. *J. Neurosci.* 15:6260 (1995).
22. J.S. Taube and R.U. Muller, Head-direction cell activity in the anterior thalamic nuclei, but not the postsubiculum, predicts the animal's future directional heading. *Soc. Neurosci. Abstr.* 21:945 (1995).
23. H.T. Blair and P.E. Sharp, Anticipatory time intervals of head-direction cells in the anterior thalamus of the rat: Implications for path integration in the head-direction circuit. *J. Neurophysiol.* 78:145 (1997).
24. P.E. Sharp, Multiple spatial/behavioral correlates for cells in the rat postsubiculum: multiple regression analysis and comparison to other hippocampal areas. *Cereb. Cortex* 6:238 (1996).
25. C.L. Leonhard, R.W. Stackman, and J.S. Taube (1996). Head-direction cells recorded from the lateral mammillary nucleus in rats. *Soc. Neurosci. Abstr.* 22:1873 (1996).
26. R.W. Stackman and J.S. Taube, unpublished observations.
27. J.P. Goodridge, A.D. Redish, D.S. Touretzky, H.T. Blair, and P.E. Sharp, Lateral mammillary input explains distortion in tuning curve shapes of anterior thalamic head-direction cells. *Soc. Neurosci. Abstr.* 23:503 (1997).
28. J. Cho and P.E. Sharp, unpublished observations.
29. J.P. Goodridge and J.S. Taube, Interaction between the postsubiculum and anterior thalamus in the generation of head-direction cell activity. *J. Neurosci.* 17:9315 (1997).
30. E.J. Golob and J.S. Taube, Head-direction cells are less responsive to ideothetic cues in rats with hippocampal lesions. *Soc. Neurosci. Abstr.* 22:1873 (1996).
31. S.H. Alyan, B.M. Paul, E. Ellsworth, R.D. White, and B.L. McNaughton, Is the hippocampus required for path integration? *Soc. Neursci. Abstr.* 23:504 (1997).
32. I.Q. Whishaw and H. Maaswinkel, Absence of dead-reckoning in hippocampal rats. *Soc. Neurosci. Abstr.* 23:1839 (1997).

Supported by NIMH grant 1 F31 MH-11102-01A1 to H.T.B., NIH grant R01 NS35191-01A1 to P.E.S., NIMH grants MH48924 and MH01286 to J.S.T., and NIDCD grant DC00236 to R.W.S.

WEIGHT-SPACE MAPPING OF fMRI LANGUAGE TASKS

Jeremy B. Caplan*, Randall R. Benson, James M. Hodgson, Kaaren
E. Bekken, Bruce R. Rosen and Jeffrey P. Sutton

Neural Systems Group and NMR Center, Massachusetts General
Hospital, Harvard University Medical School, Building 149— 9th
Floor, Thirteenth Street, Charlestown, MA 02129
*current address: Department of Biology, Brandeis University,
Waltham, MA 02254

INTRODUCTION

In the past several years, there has been tremendous growth in mapping cortical
activity using neuroimaging, including the ability to map multiple spatial scales
simultaneously. Beyond delineating functional regions and observing their sequencing in
time, one would hope to learn some of the fundamental computational principles governing
the dynamics of those regions. Essential components of this approach include
characterizing the spatial and temporal activity at different scales and comparing dynamic
network topologies *across* scales. These aspects are interconnected. In this report, we test
the hypothesis that the neocortex is comprised of nested functional clusters that emerge
simultaneously across spatial scales. We further posit that the rules of structure and
dynamics of these clusters are similar across levels (Sutton and Anderson, 1995).

The approach is based on the following intuition. Let us start with the idea that a
single task, T, such as those used in functional neuroimaging, involves the activity of a set
$\{\Lambda_k\}$ of brain regions. Λ_k are typically called "regions of interest" (ROIs) and they are
indexed by k. Each Λ_k is composed of the activation of voxels $\{\lambda_{ik}\bar{\eta}_i\}$ where λ_{ik} is the
magnitude of activation of the ith voxel within Λ_k and $\bar{\eta}_i$ is the position in 3-D space of
the ith voxel. The Λ_k typically range in size from 1 to 20 voxels and number from 1 to 15.
The exact ways in which the Λ_k are determined vary considerably. One can think of each
Λ_k as being associated with an aspect of the task which it subserves. The Λ_k are generally
not restricted to any characteristic spatial extent, nor are they comprised of any
characteristic number of neurons. Moreover, the Λ_k need not be independent.

Currently, most neuroimaging studies begin with a hypothesis driven by the experimenter's model and/or intuition relating to how to define the Λ_k for a single task. Subsequently, a control task is chosen such that it differs from the task in only one of its putative cognitive or behavioral components. A statistical test is applied to identify the regions corresponding to those components. For task T, the activation maps, $\{a_i \bar{\eta}_i\}$, are obtained from a statistical comparison, f, of the raw task activation, $\{a_i' \bar{\eta}_i\}$, with the raw baseline activation, $\{b_i \bar{\eta}_i\}$ [e.g., a Kolmogorov-Smirnoff (KS) comparison or cross-correlation of the complete signal timecourse with a paradigm-based reference function]:

$$\begin{pmatrix} \{a_i' \bar{\eta}_i\} \\ \{b_i \bar{\eta}_i\} \end{pmatrix} \xrightarrow{\ f\ } \{a_i \bar{\eta}_i\} \tag{1}$$

A clustering function, g, is then applied to $\{a_i \bar{\eta}_i\}$ to decompose it into the Λ_k:

$$\{a_i \bar{\eta}_i\} \xrightarrow{\ g\ } \{\Lambda_k\} \tag{2}$$

It is specifically the hypothesis that dictates the experimental approach.

We follow the inverse strategy. Instead of testing for the existence of some Λ_k corresponding to an individual task, we choose a set of tasks, $\{T^m\}$, where m=1,...,M, and set out to uncover some of the common networks, Θ_n, that are activated across this set of tasks. Instead of starting with the goal of looking for the neural substrate of some aspect of a single cognitive function, we start by mapping across several tasks, with the aim of probing the neural substrates subserving groups of tasks. In much the same way as multiple memories are stored in a distributed manner throughout small neural networks, we suggest that multiple tasks are associated with different distributed patterns of activity at levels detectable by functional neuroimaging.

In this scheme, each T^m has a corresponding activation pattern, $\{a_i^m \bar{\eta}_i\}$, obtained for each task relative to its own baseline (different tasks may be compared with different baselines; see Price and Friston, 1997). In a manner analogous to equation (1) in the single-task case, raw signal for each task T^m is mapped to a set of activity patterns:

$$\begin{pmatrix} \left\{ \left(a_i^m\right)' \bar{\eta}_i \right\} \\ \left\{ b_i^m \bar{\eta}_i \right\} \end{pmatrix} \xrightarrow{\ f\ } \left\{ a_i^m \bar{\eta}_i \right\} \tag{3}$$

The mapping, g, from $\{a_i^m \bar{\eta}_i\}$ to Θ_n, analogous to equation (2), involves a comparison between voxels across tasks:

$$\begin{pmatrix} \{a_i^1 \bar{\eta}_i\} \\ \vdots \\ \{a_i^M \bar{\eta}_i\} \end{pmatrix} \xrightarrow{\ g\ } \Theta_n \tag{4}$$

In this way, we interpret Θ_n as subserving some specific aspect of function in the very last step in the analysis, rather than in the first, as in the single-task case above.

One example of g in equation (4) includes the determination of a weight function across tasks for each pair of voxels. The weight can be thought of as a generalized tally or a count of the number of tasks over which two voxels are simultaneously activated or deactivated, relative to the baseline condition(s). Then, a fully-connected cluster of such voxels at a given weight threshold can be interpreted as a particular $\Theta_{n'}$ that is active during a subset of the tasks (the number of tasks roughly corresponds to the weight value). The Θ_n are analogous to Λ_k but they link multiple tasks and potentially, multiple scales of activity.

In recent work (Caplan et al., 1997), applying a weight-space transformation to functional magnetic resonance imaging (fMRI) data from six motor tasks, we found evidence for nested networks involved in motor function. We also observed the temporal course of the weights and found that the rules governing the temporal dynamics were similar across spatial scales. In this paper, we conduct the same weight-space transformation with language fMRI data and find the same type of nested functional organization implicated in the performance of these tasks.

METHODS

Functional Neuroimaging

Human language tasks were examined in six subjects, under informed consent, using fMRI at 1.5 T. The tasks consisted of ten different reading paradigms, involving words, pseudowords, nonwords, false-font strings, text and verb generation. All tasks were

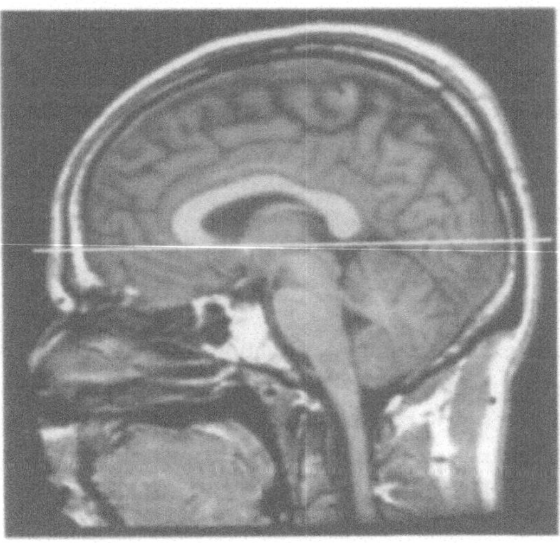

Figure 1. Mid-sagittal high-resolution anatomical MRI.
The white line illustrates the slice orientation.

performed covertly and followed a rest-task-rest-task-rest paradigm, where the rest state consisted in the subject visually fixating. Data were collected for 17 axial slices (thickness = 7 mm, gap = 1 mm) using an asymmetric spin-echo pulse sequence (TE = 20 ms, offset = 25 ms, TR = 2000-3000 ms) over a period of 4-6 minutes (fig. 1). The tasks were motion-corrected and analyzed using the KS statistic, and activation maps for task versus rest were generated. From the examination of these maps, many ROIs (i.e., the Λ_k) were observed; however, there was no compelling evidence at this stage of large-scale networks, nor was there any clue as to how such networks might be organized.

Weight-Space Mapping

Aiming to identify the functional organization of the brain regions involved in these tasks, we performed a weight-space transformation as previously described for motor tasks on functional maps (Sutton et al., 1996; Caplan et al., 1997). After obtaining the activation maps, $\left\{a_i^m \bar{\eta}_i\right\}$, for each task, we performed the mapping based on the results of the statistical comparison for each voxel within a task:

$$a_i^m \rightarrow A_i^m = \begin{cases} +1 & p < p_\tau, \text{ more activity than baseline} \\ 0 & p > p_\tau \\ -1 & p < p_\tau, \text{ less activity than baseline} \end{cases} \tag{5}$$

where p is the p-value from the KS test and p_τ is a significance threshold. Weights were calculated for all pairs of voxels (i,j) within each of the 17 slices (64 x 64 voxels per image, voxel size = 3.125 x 3.125 x 7 mm^3) by summing the inner products of the activation values across the ten tasks:

$$w_{ij} = \sum_m A_i^m A_j^m \tag{6}$$

The technique included no anatomical biasing. Pairs of voxels with a weight value above a threshold were then joined by lines, creating weight maps that were then overlaid upon high-resolution anatomical scans (fig. 2). This was a way to represent the functional connectivity in a clear visual manner.

Identifying Cluster Boundaries

To retrieve the Θ_n, the common activation patterns across tasks one must identify those voxel-pairs that are present at strong weight thresholds (i.e., in many tasks) and impose a threshold criterion for density of connection within a cluster. Then, a set of voxels would be interpreted as a network at some weight threshold, w_τ if a most (or all) of the weights between pairs of those voxels satisfied the condition

$$\left|w_{ij}\right| \geq w_\tau \tag{7}$$

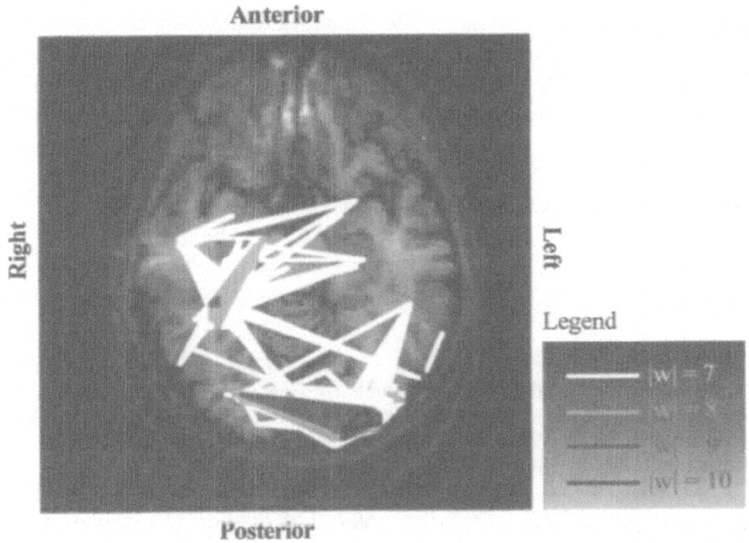

Fig. 2. Weight maps for one slice superimposed upon its corresponding anatomical MRI. Lines connect pairs of voxels with a given weight between them (see legend).

RESULTS

- Clusters of many different sizes were observed within 7 mm slices, with two-dimensional cluster sizes ranging from 80 mm^2 (single ROI) to the order of 9000 mm^2.
- Fully-connected networks were apparent at many different scales, suggesting that scaling might be continuous.
- The nodes of the networks were closely related to ROIs as identified from the original functional maps.
- Networks had similar connectivity patterns independent of scale.
- Networks were found to be nested within each other.

DISCUSSION

We have demonstrated the utility of the weightspace transformation as an exploratory procedure to reveal some of the common functional networks underlying a set of cognitive tasks which can shed insight on how the brain performs complex tasks. Extending this approach to time-varying experimental paradigms, such as learning studies or multicomponent tasks, may also provide clues about the dynamics of these networks. There is also the potential for application to the diagnosis and treatment monitoring of pathological conditions.

These findings support the notion that the neocortex may be organized into nested functional networks across multiple spatial levels. The presence of such nested clustering in language as well as in motor tasks reinforces the idea that dynamical cluster organization may be a fundamental mechanism of neocortical function. Further confirmation of this putative computational principle may be demonstrated by technologies which measure activity at different scales simultaneously (e.g., multiple unit recordings, fMRI, PET, optical imaging).

Supported by NIH grant MH01080 and the James S. McDonnell Foundation.

REFERENCES

Caplan, J. B., Bandettini, P. A. & Sutton, J. P. 1997. Weight-space mapping of fMRI motor tasks: evidence for nested neural networks. In: Bower, J. M. (ed.), *Computational Neuroscience: Trends in Research, 1997*. New York: Plenum Press. 585-589.

Price, C. J. & Friston, K. J. 1997. Cognitive conjunction: a new approach to brain activation experiments. *Neuroimage*. 5:261-270.

Sutton, J. P. & Anderson, J. A. 1995. Computational and neurobiological features of a network of networks. In: Bower, J. M. (ed.), *Neurobiology of Computation*. Boston: Kluwer Academic. 317-322.

Sutton, J. P., Caplan, J. B. & Bandettini, P. A. 1996. fMRI evidence of nested networks associated with motor tasks. *Human Brain Mapping*. 3:S370.

REPRESENTING ODOR QUALITY SPACE:
A PERCEPTUAL FRAMEWORK FOR OLFACTORY PROCESSING

Christine W.J. Chee-Ruiter and James M. Bower

Computation and Neural Systems Department
Division of Biology
California Institute of Technology
Pasadena CA 91125

INTRODUCTION

For most people, olfaction is a sense which is often ignored or used unconsciously during day-to-day activities. Yet it is a remarkably powerful sense, providing us with the means to remotely monitor our chemical environment. The main components of olfactory perception (odor quality, odor intensity, and hedonic value) are all active areas of research in olfactory behavior and psychophysics. Nevertheless, the nature and extent of olfactory experience remain elusive, and this especially hampers efforts to approach olfaction computationally.

One problem with addressing olfactory function from a computational standpoint is that it is difficult to conceptualize what the input (stimulus) space is, what the output (perceptual) space is, or what intermediate spaces might exist. The output space in particular is counterintuitive because odors are invisible, silent, amorphous, and hard to describe, while our analytical strengths are more suited to vision, hearing and touch.

It follows, then, that the mapping between the stimulus space (of chemicals) and the perceptual space (of odors) is not well understood. Currently we cannot randomly select a molecule, and reliably predict what odor it might elicit, or even whether it will elicit an odor at all. The answer to as basic a question as, "What makes a molecule capable of eliciting an odor?" is still unknown, with the result that a pair of molecules which, to a chemist, seem structurally similar, can elicit an intense odor on the one hand, and odorlessness on the other. Nevertheless, structure-function studies of chemicals and the odors which they elicit have revealed some interesting relationships which may shed light on the nature of the specificity of some olfactory receptors and how they may contribute to odor quality perception[1].

Olfactory researchers continue to face difficulties in selecting appropriate stimuli, and are often reduced to selecting an odorant on the basis of its prior use in research, on the basis of its chemical properties, or on an arbitrary basis. Consequently, a set of olfactory stimuli selected according to these methods may not span "odor space" appropriately (akin to testing color perception with only "blue" hues), with the result that our knowledge of olfactory perception is a widely ranging collection of disparate facts which are difficult to bring together to form any unifying theories.

More specifically, an olfactory space may be ordered chemically, biologically, or perceptually. An example of chemical ordering might be by the number of carbons in a molecule; an example of biological ordering might be molecules which are produced by certain species of animals; and an example of perceptual ordering might be odors which smelled "rose-like". It seems reasonable that an input space (the space of stimuli) might be best ordered by physical or chemical parameters, while the output space (the space of perception) might be best ordered by perceptual parameters. However, while it is likely that there may be a strong relationship between olfactory input spaces and output spaces (e.g., that sulfurous compounds probably smell like sulfur), we cannot assume that the principles used to order the former will also apply to the latter (in other words, the mapping may not be linear). However, for lack of a systematic, perceptually-relevant method of selecting stimuli, olfactory researchers often try to examine perceptual space by selecting stimuli using techniques which might more appropriately be applied to a study of the input space.

To address this issue, our technical objectives are twofold. First, we wish to provide a quantifiable basis for selecting olfactory stimuli according to a perceptually relevant parameter, odor quality. Second, we wish to propose a model of odor quality space based upon a close examination of olfactory function from a computational standpoint, derived from psychophysical results. Such an understanding of odor quality space may serve as a starting point for examining and testing the olfactory system.

APPROACH

A few notes on terminology may be useful at this time:
1. *Odorants* are chemicals which elicit odor perceptions (*odors*). In this study, only monomolecular odorants are discussed.
2. Odor perception includes several components: odor quality (e.g., "apple" vs. "orange"), intensity, and hedonic value (pleasantness or unpleasantness). Thus, every odor necessarily possesses an odor quality, an intensity and a hedonic value, much as every color has hue, saturation and lightness.
3. Odor *quality* is complex, and in detailed analysis is often described as comprising a set of one or more odor *notes*.
4. Odor notes are assigned names (odor quality *descriptors*), usually derived by analogy with a common, distinct and well-known odor quality (e.g., "apple" is the odor quality if you are smelling an apple, but may describe an odor note if it is a component of a more complex odor quality, say one which includes "apple" and "pear"). The concept of odor "notes" has some vagueness to it since, for example, there are several varieties of apples which may vary in odor quality, and also the odor of a given apple will vary as a function of time and of ripeness. Nevertheless, the concept of an odor note and its associated odor quality descriptor conveys essential information in language, and is used extensively in the fragrance and flavor professions[1].
5. Individual odor notes can be perceived at different relative intensities within an overall odor quality perception (e.g., a given chemical can elicit an odor which has a strong "citrus" note and a weaker "soapy" note).

Our approach focuses on the abstract problems of olfactory processing, independent of the constraints of a particular processing structure. This approach presumes that different kinds of olfactory structures (e.g., the antenna-based "noses" of arthropods, the nostril-based olfactory epithelia of vertebrates, or even "electronic" noses) share common problems in processing chemical information. Such an approach in color vision has proven useful both in aiding our understanding of how color is perceived by the eye, and in guiding experiments aimed at establishing the neural and perceptual pathways for color in the visual system. Likewise, this approach will allow us to study fundamental principles in the processing of chemical stimuli, which can be used to guide experiments in both biological and non-biological chemosensory systems.

There are three olfactory psychophysical phenomena upon which this model is based:

1. At a given concentration, a monomolecular stimulus frequently elicits multiple odor notes. For example, Isoamyl hexanoate ($C_{11}H_{22}O_2$) elicits the notes of apple, pineapple, fruity, and green, simultaneously[2].

2. For many odorants, the odor quality elicited by that odorant can vary slightly as a function of concentration. However, a small class of monomolecular stimuli can elicit not only vastly different odor notes, but notes of opposing hedonic quality (pleasant/unpleasant), with changes in concentration. For example, Indole (C_8H_7N) elicits a jasmine note at low concentrations, while at high concentrations elicits a fecal odor[2].

3. At a given concentration, the quality of an odor evoked by a monomolecular stimulus is context-dependent[3].

The first phenomenon shows a temporal relationship between specific odor notes: they are elicited simultaneously. Because these odor notes are elicited by the same stimulus molecule, there is an implied "closeness" in perceptual space (note that there is no reason *a priori* for these odor notes to have similar odor qualities—the "closeness" is determined temporally). The second phenomenon describes extreme changes in odor quality based on changes in stimulus concentration—in other words, concentration is an important variable which can cause the same chemical to be perceived as if it were a different stimulus. Finally, the last phenomenon indicates a complex relationship between the perceived odor quality of a stimulus, and the odorous background against which it is applied, suggesting that the "state" of the olfactory system may be an important factor in odor quality perception. Together, these phenomena provide a means of quantifying odor quality relationships.

METHODS & RESULTS

Odor Quality Clustering

We first asked the following question: What odor notes are simultaneously elicited in common by monomolecular stimuli? Data was obtained from two sources[4,5], which combined included approximately 1000 chemicals and 327 odor qualities. Each chemical and individual odor note was assigned a unique number, and a matrix of chemical vs. evoked odor notes was then created for each data source. For each data matrix A, an odor cross-correlation matrix was created:

$$C = A^T A = M \text{ X } M \text{ where } M = \text{number of odor notes}$$

Each element of C describes how often two odor notes were elicited together across all chemicals in our database, while the diagonal tells the frequency of occurrence of each odor

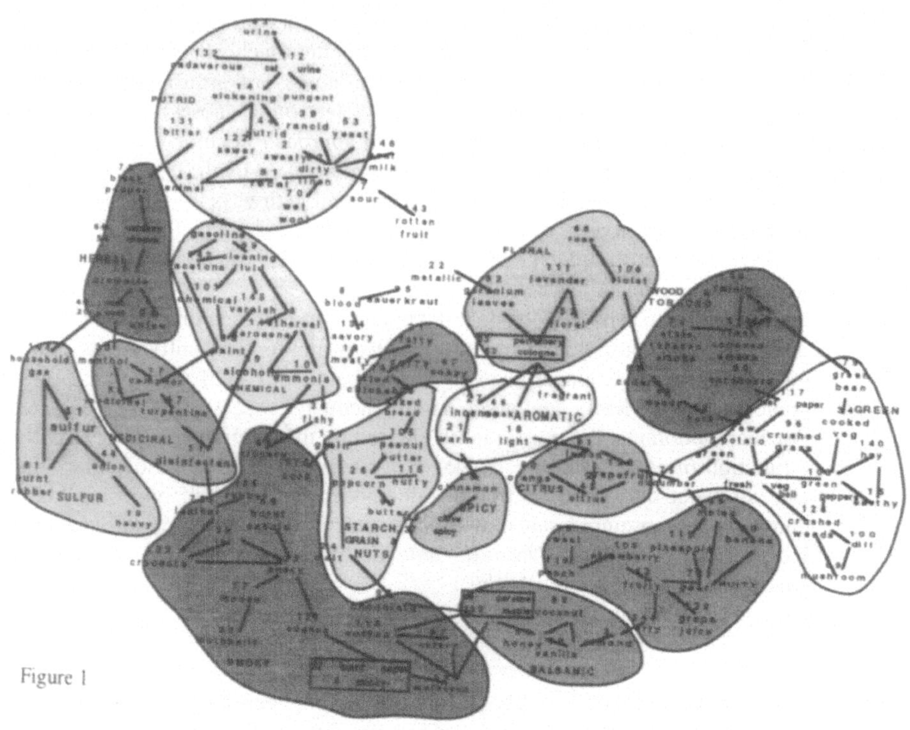

Figure 1

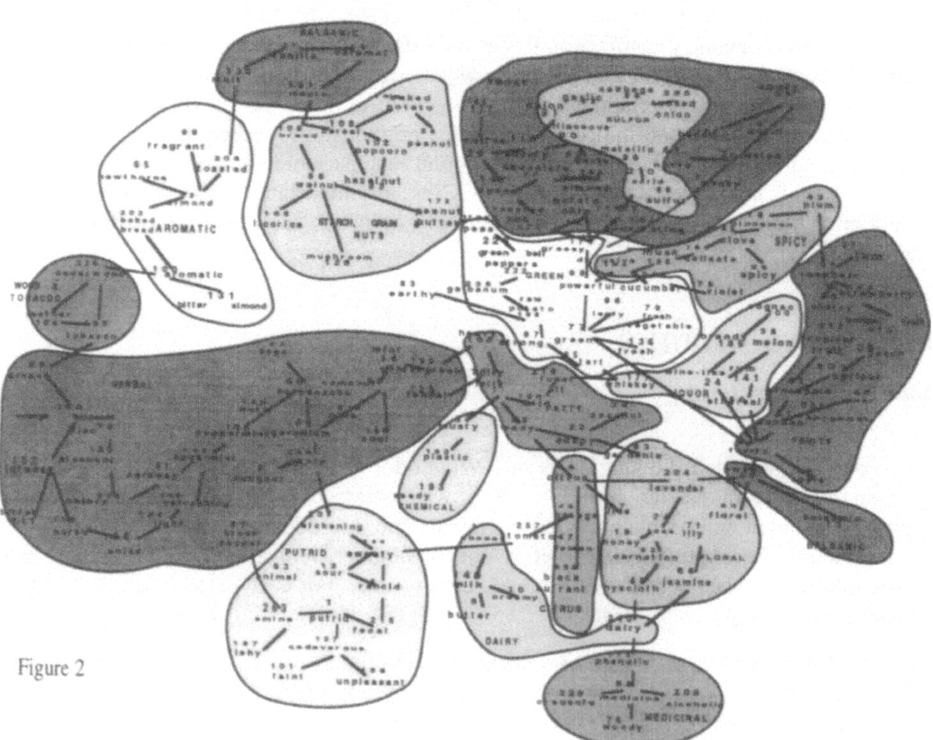

Figure 2

note. From this we can determine the probability that one note will appear given that another is present. Note that these conditional probabilities may be asymmetric (i.e., $P(B|A) \neq P(A|B)$). For example, odor A may sometimes be elicited with odor B, and other times appears with odors C and/or D, while odor B may only ever appear in combination with odor A. Thus, to capture the "importance" of the link between two notes, we use the metric $I = P(B|A) * P(A|B)$.

The following figures show the odor quality maps for each database. To be included on the map, an odor note must have been evoked separately by at least two different chemicals. Each eligible odor note is shown along with its identifying code number. Each link denotes the most significant connection between one odor note and another. In some cases secondary connections are shown, to allow small odor "islands" to be linked with larger groupings of odors. As noted earlier, while there is no *a priori* reason for simultaneously-elicited odor notes to smell similar (e.g., garlic and onion are temporally linked as well has having similar odor qualities), there does seem to be a high degree of correlation between clusters of odors on the map, and odor classification schemes based on other measures of odor "similarity" (described by Ohloff[1]).

There are three important caveats to these maps:

1.   To aid the viewer in intuitively grasping relationships between general odor "families," subjective groupings as denoted by shaded areas have been superimposed on the maps. Note that while the links between odor notes on the map were quantitatively determined by the method described above, the shaded areas and capitalized labels are somewhat subjective.

2.   The position of odor "groups" relative to one another as described by these map configurations is somewhat flexible. Odor notes were linked pairwise according to their most significant "I" value, and positioned near other odor notes which shared less significant "I" values. However, odors were often linked weakly to odors which are placed distantly on the map, and often to odors on the opposite edge; in other words, these map configurations could be viewed as "Mercator Projections" of the surface of a sphere. This strongly suggests that the odor quality space has three or more dimensions, and also that a much larger database of chemicals might allow more definite relationships to be made.

3.   There is a possibility that the databases may be inadvertently skewed in the selection of chemicals which evoked odors considered to be "pleasant" by the general population. However, inspection of the maps shows a healthy representation of "unpleasant" odors, and in any case it is hoped that this model will continue to serve as a repository for information on odor quality, and that time will serve to reconcile such bias if it exists.

Figures 1 & 2. Each figure illustrates relationships between odor qualities for a separate data source. Figure 1 shows odor quality profiles from Dravnieks, et. al. [4], while Figure 2 shows odor quality profiles obtained from Aldrich[5]. In each diagram, the largest connection between a pair of odors is shown. In some cases, two odors connect most strongly to each other, so the next largest connection of the pair is shown to allow the pair to be linked to another group. While secondary connections are not shown, odors are placed such that they are nearby odors to which they share connections. Each figure is just one configuration of many possible arrangements; in similar arrangements, however, odors near the top of the page have secondary links to odors placed near the bottom, and likewise between the left and right sides of the page. In other words, this map may be thought of as a 2D projection of a 3D "surface". Subjective groups of shaded regions superimposed on the maps make it easier to compare the two figures; the regions are not definitive and should not be considered as such.

Imposing a Structure on the Cluster Map

While it captures relationships between pairwise odor notes, this cluster representation does not necessarily show relationships between general odor families. Just as the color wheel illustrates color opponents of green/red and blue/yellow, it is desirable to determine if any "structure" exists for odor quality space. Such structure could both allow us to explain olfactory phenomena, and provide an intuitive framework for understanding how odor mixtures might be perceived. Furthermore, structure in the perceptual space might provide clues to possible analogous structure in the stimulus space. (The olfactory system may provide a biologically significant, as opposed to a chemically significant, means of classifying chemicals—a classification system which might be revealed by perceptual structure, but which might not otherwise be conceptualized by researchers). Olfactory phenomena such as odor metamerism (in which the same odor is elicited by different stimuli), cross-adaptation and odor cancellation suggest that the concept of an "orthogonal" odor may not be unreasonable. (An analogy would be color opponency in the color system.)

To seek an underlying structure to the odor quality map, we next consider that a small class of odorants can elicit not only vastly different odor qualities but odor qualities of opposing hedonic value (pleasant/unpleasant), with a change in concentration. For all odorants considered, surprisingly only three odor pairs were identified which met these criteria:

    (1)   fruity - sulfur
    (2)   floral - putrid
    (3)   green - fatty

This property of changing hedonic value (and therefore its property of exclusivity, since other odors do not have this property) with concentration lends itself to the establishment of "axes" through odor space. One representation of odor space might thus look like Figure 3. This representation suggests that as one were to move along a trajectory through odor space, one would perceive odors which varied smoothly from one odor family to another. However, comparison with the cluster map suggests that the links between odor families do not allow for smooth "trajectories" in any direction, and that the odor families are not contiguous as the representation suggests.

Thus, an alternate representation might look like Figure 4. This grouping suggests first of all that odor perceptions may "flip" between hedonic states, and then within a

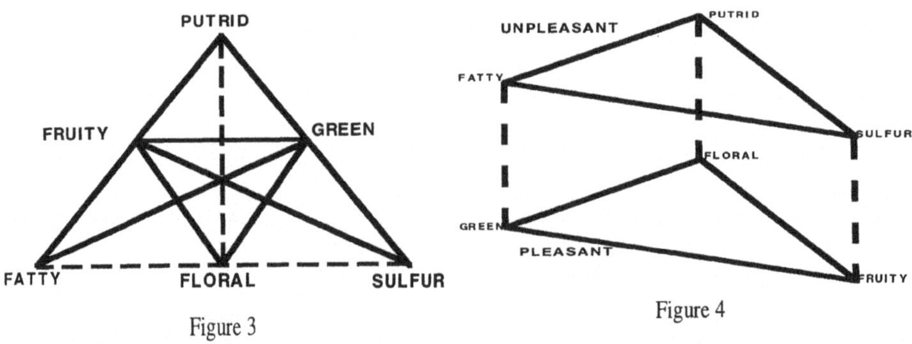

Figure 3

Figure 4

Figures 3 and 4. Candidate structures for odor quality space.

hedonic space odors vary smoothly from one odor family to another (this observation is often noted in everyday experience). Further, a given odorant may be represented twice: once in a pleasant context, once in an unpleasant context. The perceptual "flip" may depend on the state of the olfactory system, modifiable either centrally or through background adaptation. Of course, other representations besides these readily present themselves; however these are presented as examples of hypothetical frameworks from which subsequent psychophysical experiments may be designed.

It is important to note that individually, axes or clusters are interesting but also inconclusive. Together, however, they unite two aspects of olfactory perception (odors related by time and by concentration) which provide some explanations for observed phenomena, and therefore are worthy of further consideration.

Olfactory Function

What could this cluster map, with its structural underpinnings, tell us about olfactory function? We now call upon a third psychophysical phenomenon, that the quality of a perceived odor is context-dependent. If we consider that in the color system, color opponency forms the basis for processing of light into perceptual channels, it is likewise plausible that a similar "odor opponency" function in olfaction may be at work to process chemical stimuli via perceptual channels. In the most simple formulation, odors sharing a common perceptual channel may "interact" with one another (e.g., cause a perceptual shift in odor quality), while odors processed by different perceptual channels should not affect one another. For example, adaptation to an odorant eliciting an odor A should influence the perception of any subsequent stimulus eliciting an odor A' which shares that same perceptual channel, while adaptation to an orthogonal (contrasting) odor B should not. Lawless et. al.[3] demonstrated this phenomenon of perceptual shift with odorants eliciting woody and citrus odors.

Thus, this cluster map can be used as a guide for investigating possible olfactory perceptual channels, and to establish computational hypotheses for olfactory function.

CONCLUSIONS

We propose a model of odor-quality space which can be used as a guide to investigate possible olfactory perceptual channels, and to establish computational hypotheses for olfactory function. The model is based on quantitative measures of odor relationships using temporal and stimulus concentration parameters, and correlates well with qualitative odor classification models. We are currently performing psychophysical experiments on odor quality relationships with human subjects using this model as a tool for stimulus selection.

REFERENCES

1. Ohloff, G., *Scent and Fragrances: The Fascination of Odors and their Chemical Perspectives*, Springer-Verlag, NY (1994)
2. Arctander, S., *Perfume and Flavor Chemicals (Aroma Chemicals)*, published by the author, Montclair NJ (1969)
3. Lawless, H.T et.al., *Chem Sens*, 16(4):349-360 (1991)

4. Dravnieks, A., et. al., *Atlas of Odor Character Profiles*, ASTM Data Series DS 61, Philadelphia PA (1985)

5. ______, *Aldrich Flavor & Fragrances Catalog* (which used odor quality descriptors referenced from Arctander[2] and Furia & Bellanca[6]), Aldrich Chemical Company, Milwaukee WI (1996)

6. Furia, T.E. and Bellanca, N., *Fenaroli's Handbook of Flavor Ingredients*, CRC Press, Boca Raton FL (1994)

NEURONAL REPRESENTATIONS IN A CATEGORIZATION TASK: SENSORY TO MOTOR TRANSFORMATION

Emilio Salinas and Ranulfo Romo

Instituto de Fisiología Celular
Universidad Nacional Autónoma de México
04510 México D. F.
México

INTRODUCTION

When motor actions are triggered by sensory events, the neural networks that generate the motor commands must receive an adequate driving input, which, in general, is derived from the original sensory signals. We studied this conversion of sensory information into motor commands in a task in which monkeys categorized the speed of tactile stimuli as either high or low, reaching for one of two switches to indicate their choice. The input signal, the speed of probe movement, was varied systematically. The output of this process was an arm movement that did not depend directly on speed, but on a function of speed, its category, and the monkey had to compute it to obtain a reward. There are three essential quantities in this task: stimulus speed, speed category, and arm movement. Where and how are these variables represented in terms of neuronal activity? Are those representations independent? And, do they correlate with the monkey's behavior? Here we investigate these questions through analytical techniques applied to extracellular recordings from three cortical structures, primary somatosensory cortex[1] (S1), the supplementary motor area[2] (SMA, or medial premotor cortex), and primary motor cortex[3] (M1). We suggest that the primary sensory representation of the stimulus is used to compute two sets of responses that encode speed category and that participate in the generation of arm movements.

NEURONAL RESPONSES DURING CATEGORIZATION

Figure 1A illustrates the temporal sequence of events in the task[1,2,3]. First, a metal probe is lowered (PD) on the distal segment of one digit of the left, restrained hand, and touches the fingertip; the monkey reacts to this initial indentation by holding an immovable key (KH) with the free hand; then, after a variable delay interval (1–4.5 s), the stimulator tip starts moving (ON) at one of 10 speeds; after the probe stops moving (OFF) the monkey releases the key (KR) and projects its free hand to one of two pushbuttons (PB) to indicate whether the speed was low (12, 14, 16, 18 or 20 mm/s) or high (22, 24, 26, 28 or 30 mm/s). Low speeds correspond to a medial button and high speeds to a lateral one. Correct categorization is rewarded with a drop of water or juice. The time between OFF and KR is the reaction time, and the time between KR and PB is the movement time.

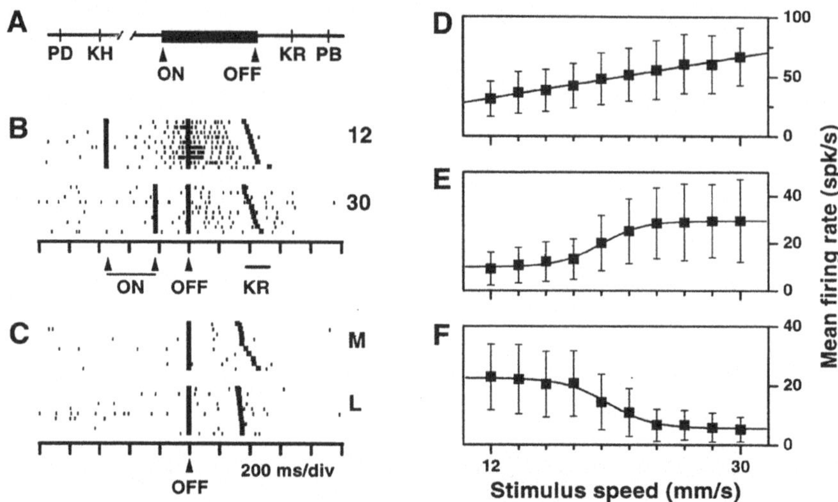

Figure 1. Cortical neuronal responses recorded during performance of the task. *A*, Sequence of events in the categorization paradigm; labels are explained in the text. *B*, Spike rasters from an M1 neuron. Stimulus speed is indicated on the right; 10 trials per speed are shown. *Small dots* correspond to action potentials and *large squares* correspond to behavioral events: stimulus onset (ON), stimulus offset (OFF) and key release (KR), in that order. Stimulation time varies with speed because the distance scanned by the probe was kept constant (6.5 mm). *C*, Spike rasters of the same neuron when the monkey made identical arm movements, medial (M) and lateral (L), in response to an LED light turning off (OFF). *D*, Average tuning curve for a population of 27 neurons in S1 that increased their activity with increasing speed. Error bars indicate ±1 SD with respect to the 27 individual means. *E*, Average tuning curve for a population of 40 SMA neurons selective for high speeds. *F*, Average tuning curve for a population of 20 M1 neurons selective for low speeds. Firing rates were computed by counting the spikes in one of three possible activation periods, during stimulation, during the reaction time or during arm movement. Continuous lines are fits to the data.

How do S1 neurons respond during stimulation? Two types of responses were observed[1]. About half of the S1 neurons increased their firing rates with increasing speed. As shown in Fig. 1D, on average this increase was linear, with the firing rate roughly doubling from 12 to 30 mm/s. The rest of the neurons also increased their activity during stimulation but did so identically for all speeds; they responded to stimulation itself, not to stimulus speed. No neurons were observed that decreased their firing rates with increasing speed. It is important to note that all S1 responses were identical during active categorization and when the same stimuli were applied passively. In the latter condition the monkey's right arm was restrained and no movements were performed.

In SMA and M1 most neurons responded during some part of the task but did so regardless of speed, ie., their activity was modulated by certain aspects of the task but not by the metrics of the stimulus[2,3]. However, a small fraction of the neurons in these areas (191/745 in SMA and 71/477 in M1) did respond differentially, not to speed itself, as observed in S1, but to the speed categories. These category-tuned neurons were extensively studied.

In both SMA and M1 two types of category-tuned neurons were observed. Ones fired more intensely for high versus low speeds, and the others did the opposite. While differential activity typically appeared during the reaction time, some neurons were differentially active also or exclusively during stimulation, and still others were active also or exclusively during arm motion. An example is shown in Fig. 1B. This neuron fired more action potentials during the reaction time when stimulus speed was low. When the responses of category-tuned neurons are plotted as functions of speed, the resulting traces typically have sigmoidal shapes. Figure 1E shows the average response of a population of neurons from SMA that were selective for high speeds; Fig. 1F shows a similar population average for M1 neurons that preferred low speeds.

Since low and high speeds are associated with medial and lateral movements, respectively, it is possible that the 'category-tuned' responses are actually related to the preparation or execution of impending arm movements or, in general, to motor behavior during the task. Are these neurons coding speed category or arm movement? Distinguishing these two possibilities is crucial to understand their functional role. Several lines of evidence suggest that they are related to both sensory and motor aspects of the task.

(1) When the animals performed identical arm movements towards the same pushbuttons but were guided by visual cues, more than two-thirds of the tested neurons (57/71 in SMA; 29/42 in M1) stopped responding differentially. They either modulated their firing rates identically for both movements or did not show any change at all with respect to baseline[2,3]. An example of this control experiment is shown in Fig. 1C. This neuron was strongly differential during categorization (Fig. 1B), but did not respond when the same movements were triggered by lights turning on and off. This experiment shows that most differential responses do not correlate unconditionally with motor behavior. (2) None of the category-tuned neurons (0/30 in SMA; 0/5 in M1) responded when the tactile stimuli were delivered passively and the animal did not make any responding movements[2,3], which indicates that these neurons are not purely sensory-driven. (3) Analyses of the timing of neuronal responses showed that the onsets and offsets of increased activity were not consistently timelocked to sensory (ON or OFF) or motor (KR) events[3]. (4) Other evidence makes it is quite unlikely that the differential activity is related to muscle precontraction, proprioceptive input or eye movements[2,3].

SENSORY OR MOTOR DIFFERENTIAL ACTIVITY?

The above points indicate that the majority of differential responses in SMA and M1 are conditional on the full categorization process. Now we apply two analytical approaches to further explore whether they are predominantly related to sensory input or motor behavior during actual performance of the task.

Neuronal versus Behavioral Performance

Decoding techniques were implemented to compare the psychophysical performance of the monkeys to the performance expected solely on the basis of a set of observed category-tuned responses. Neurometric performance curves directly comparable to the measured psychometric curves were constructed using decoding methods[3,4]. These curves correspond to the average accuracy with which an observer can estimate the speed category at any trial given only two pieces of information: a set of category-tuned responses at that trial—their firing rates, in our case—and a previous characterization of the response statistics of the neurons—their mean firing rates and SDs as functions of speed. The neurometric curves below were obtained using the maximum likelihood decoding method coupled to computer simulations of the recorded neurons[3,4]. The only free parameter of the decoding procedure was the number of neurons included, N.

Figure 2 shows the results of this analysis. When the observer estimates the speed category from the response of a single M1 neuron (right column, middle row), his performance is noticeably worse than the monkey's. However, the shapes of the neurometric and psychometric curves in this case are very similar, suggesting that a good match to the psychophysics can be obtained simply by looking at the responses of more neurons. Indeed, when four neurons are considered in each trial, the observer's performance is practically identical to the monkey's.* This excellent agreement is interpreted as meaning that the firing rates of category-tuned cells in M1 provide a neuronal representation of the speed category that is not subject to further processing. In contrast, when the category is estimated from a single S1 neuron (left column, middle row) the resulting neurometric curve is different from the corresponding psychometric curve. In this case increasing the number of decoded neurons again increases overall accuracy,

---

*Only four M1 neurons are needed to match the animal's performance because the decoding method used, maximum likelihood, is optimal. Other methods might require more neurons.

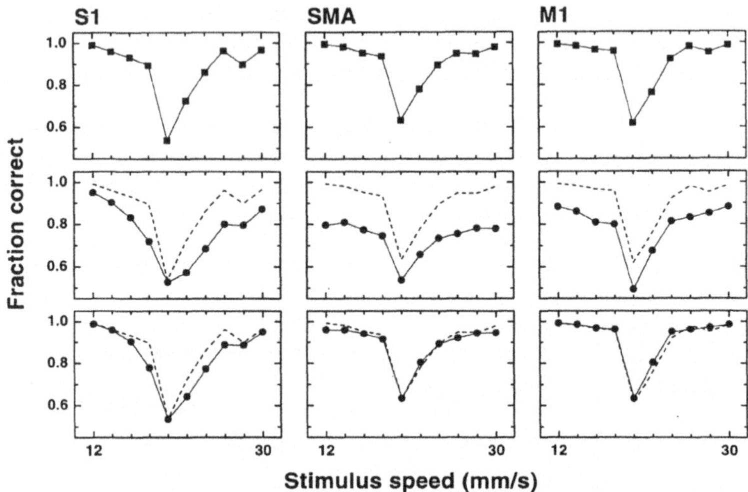

Figure 2. Psychometric versus neurometric performance in the categorization task. Each *column* corresponds to analysis of a different cortical area. The *top row* shows the psychophysical performance of the monkeys during the recording sessions for each of the areas indicated. For each *column*, this curve is reproduced with *dotted lines* on the lower *rows*. The rest of the curves represent the expected performance of an observer that, trial by trial, measures a set of neuronal responses and infers or decodes from them the speed category. The *middle rows* shows the average accuracy in categorization based on the response of $N=1$ neuron; in each iteration one firing rate was randomly selected from the corresponding population. The *lower row* shows the average accuracy in categorization based on the responses of $N=4$ neurons; in each iteration four firing rates were randomly chosen and were then combined to estimate speed category.

as expected, but the shapes of the curves are still noticeably different. When SMA neurons are used, the results are slightly less accurate than with M1 neurons.

On average, performance in categorization is identical for the monkey and for an observer that estimates the category according to the responses of a number of differential neurons. This suggests that the monkey's performance is based on a similar representation of speed category. This is not to say that the differential neurons in M1 and SMA directly participate in the decision-making process. Rather, they possibly constitute a copy of the output of the categorization process that is relayed to the motor networks only when such output needs to be indicated through an arm movement.

In Fig. 2, the mean decoding accuracy was roughly the same for the three neuronal populations considered. However, when simpler, non-optimal decoding methods are used, more S1 neurons than SMA or M1 neurons are required to achieve a given accuracy (to match the psychophysics, using the comparison method[3]: $N=40$ for S1, $N=27$ for SMA, $N=19$ for M1). This suggests that speed category can be read out more easily from a small number of category-tuned neurons than from the same number of S1 neurons with linearly increasing tuning curves. Hence, rather than containing more information per neuron, the sigmoidal tuning curves may provide a code for speed category that is more easily read out or 'understood' by the motor networks. Notice that, in contrast, speed itself is very poorly encoded by the differential neurons. For example, distinguishing between 12, 14 and 16 mm/s or between 30, 28 and 26 mm/s from their activity would be extremely difficult because their responses at these speeds are practically indistinguishable.

Analysis of Error Trials

We wish to determine whether, on an individual trial basis, the observed differential activity is a function of motor behavior only, of sensory input only, or of both. One way to do this is to sort the neuronal responses according to hit and error trials—that performance be

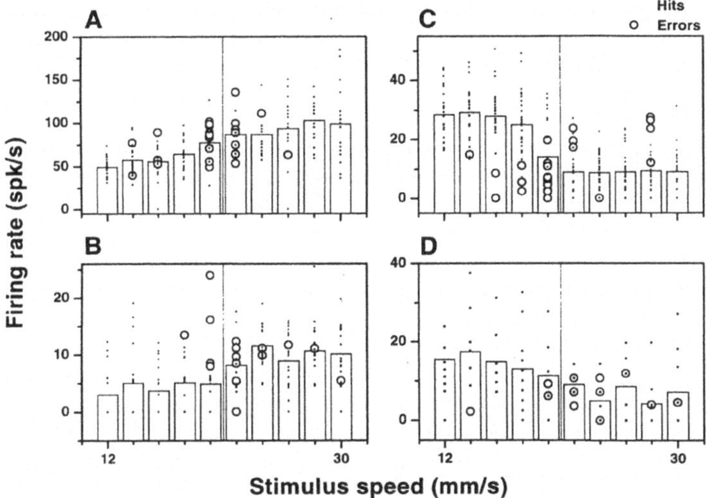

Figure 3. Responses of four neurons as functions of stimulus speed sorted according to trial outcome, hit or error. *Gray bars* indicate the mean firing rate—including responses for both hits and errors—at each speed. *Small dots* and *open circles* indicate firing rates at individual trials and correspond, respectively, to correct and incorrect categorizations. These neurons frequently fired the same number of spikes in different trials. Thus, some points represent more than one trial and contribute more to the mean rate averages. *A*, A typical tuned neuron from S1 that responds during stimulation. *B*, A neuron from SMA that responds during the arm motion and is selective for high speeds. *C*, A neuron from SMA that responds during the reaction time and is selective for low speeds. *D*, A neuron from M1 that is selective for low speeds during stimulation.

below 100% correct is crucial here—and look for significant differences between the resulting distributions. For example, for a given stimulus, an ideal sensory neuron should not show any significant differences between hit and error responses, because its activity should be determined exclusively by the input stimulus. Conversely, an ideal motor neuron would be expected to respond as a function of arm movement, regardless of the stimulus.

Thus, the key step in this analysis is to test whether pairs of response distributions (hit versus error responses) are significantly different.[†] Two statistical tests were applied to make the comparisons. A permutation test was used to determine whether the hit and error response distributions had significantly different means. The permutation technique is nonparametric—it does not make any assumptions about the underlying distributions—and can be applied to any statistic even when small numbers of data points are involved[5,6]. A minimum of 4 points were required in each distribution. As a check, the overlap between the two compared distributions was evaluated using the methods of signal detection theory[5,7], and the significance of this overlap was again determined through a permutation procedure. In general, the two tests gave similar results. Two distributions were considered as significantly different when both tests were positive at the $P<0.05$ level.

S1 neurons did behave as expected, in a purely sensory fashion. This is illustrated in Fig. 3A. For this neuron, the responses in error trials are distributed very much like those in hit trials. About the same numbers of errors fall above and below the overall means. The same was true for 64 of 66 S1 neurons analyzed (tuned and non-tuned); only 2 showed statistically significant differences between hits and errors ($P<0.05$) and were thus considered not sensory.

Category-tuned neurons in SMA and M1 behaved quite differently. The neuron in Fig. 3B responded as if related to arm motion when the speed was low; for low speeds the error responses are all above the means ($P<0.005$), as if they belonged to the hit response distribution for high speeds. However, the error responses for high speeds are close to their respective

---

[†]When hit and error trials for the same category are compared, each firing rate measured at speed x should be standardized, ie., expressed in units of SDs away from the overall mean rate at speed x.

means, and are significantly higher than the hit responses at low speeds ($P<0.005$). So for low speeds this neuron correlates with arm movement, whereas for high speeds it seems to ignore arm movement and react only to the stimulus. The neuron was not sensory because hit and error responses at low speeds were significantly different, and it was not motor because hit responses at low speeds were significantly different from error responses at high speeds, even though they corresponded to the same arm movement (medial). Figure 3C shows another SMA neuron. The hit and error responses at both low and high speeds were significantly different ($P<0.0005$), which made it not sensory. However, the error responses for high speeds and the hit responses for low speeds were also significantly different ($P<0.005$), which made it not motor. Figure 3D shows the same analysis for an M1 neuron that gave identical responses in correct and incorrect categorizations at high speeds, as expected from a *sensory* neuron. This unit was classified as not motor.

Of 68 SMA neurons analyzed, 35 were not sensory, 26 were not motor and 15 were neither sensory nor motor. And of 35 M1 neurons similarly analyzed, 19 were not sensory, 19 were not motor and 11 were neither sensory nor motor. Many neurons behaved differently across activation periods. For example, some neurons were not motor during the reaction time but not sensory during movement; these were classified as neither sensory nor motor. The differential neurons exhibit both sensory and motor characteristics in a wide variety of combinations.

CONCLUSIONS

The neural representation of the tactile motion signal goes through a sequence of recodings as it is transformed into the trigger for a motor reaction. First, speed motion is encoded by S1 neurons, and this representation is essential for all further processing[8]. Then, as suggested by the decoding analyses, the animal's choice, high or low, is encoded by the category-tuned neurons in SMA and M1. These cells probably act as an interface between the output of the categorization process and the generation of the motor command, because their responses are associated with both sensory and motor aspects of the task.

ACKNOWLEDGMENTS

The research of R. Romo was partially supported by an award from the Howard Hughes Medical Institute and grants from DGAPA-UNAM, CONACyT and Fundación Miguel Alemán. We thank W. Newsome and C. Brody for helpful comments and discussions and A. Hernández, H. Merchant, A. Zainos, and S. Méndez for their technical assistance.

REFERENCES

1. R. Romo, H. Merchant, A. Zainos, and A. Hernández, 1996, Categorization of somaesthetic stimuli: sensorimotor performance and neuronal activity in primary somatic sensory cortex of awake monkeys, *NeuroReport* 7:1273.
2. R. Romo, H. Merchant, A. Zainos, and A. Hernández, 1997, Categorical perception of somesthetic stimuli: psychophysical measurements correlated with neuronal events in primate medial premotor cortex, *Cereb. Cortex* 7:317.
3. E. Salinas and R. Romo, 1998, Conversion of sensory signals into motor commands in primary motor cortex, *J. Neurosci.* 18:499.
4. E. Salinas and L. F. Abbott, 1994, Vector reconstruction from firing rates, *J. Comput. Neurosci.* 1:89.
5. K. H. Britten, W. T. Newsome, M. N. Shadlen, S. Celebrini and J. A. Movshon, 1996, A relationship between behavioral choice and the visual responses of neurons in macaque MT, *Vis. Neurosci.* 13:87.
6. S. Siegel and N. J. Castellan Jr. "Nonparametric Statistics for the Behavioral Sciences," McGraw-Hill, New York (1988).
7. F. Rieke, D. Warland, R. de Ruyter van Steveninck and W. Bialek. "Spikes: Exploring the Neural Code," MIT Press, Cambridge MA (1996).
8. A. Zainos, H. Merchant, A. Hernández, E. Salinas and R. Romo, 1997, Role of primary somatic sensory cortex in the categorization of tactile stimuli: effects of lesions, *Exp. Brain Res.* 115:357.

THE PAPERLESS LABORATORY: AN INTEGRATED ENVIRONMENT FOR DATA ACQUISITION, ANALYSIS, ARCHIVING, AND COLLABORATION.

Thomas D. Coates, Jr.

Department of Neuroscience and Anatomy
Pennsylvania State University
Milton S. Hershey Medical Center
Hershey. PA 17033

THE PROBLEM

Collaboration with distant colleagues usually takes the form of sending facsimile copies of analyzed data or, if the colleague has the same analysis program, sending the data itself. Discussion takes place either by telephone or e-mail but, without the ability to interactively and simultaneously view the same data its difficult to discuss particular features. Similar problems are encountered when two distant colleagues coauthor papers and need to discuss changes to text, layout, and figures.

Organized long term storage of experimental data, notes, drawings, etc. poses a particularly challenging problem since they are often on different media (e.g. floppy disks, tapes, bound notebooks, large sheets of paper). In instances where part or all of the information for a project is stored electronically often several different incompatible formats are used. For example, notes may be stored in a particular word processor's format, experimental data in a different package–specific format, and graph/charts in yet another. Different types of media and non–standard storage formats make it difficult to efficiently index, archive, and retrieve information.

THE SOLUTION

A computing platform independent environment which provides:

1. A multimedia electronic notebook for the display of notes, experimental data, illustrations, video, audio, and other information in one consolidated format.
2. Standardized data acquisition, analysis, and storage tools accessible from the electronic notebook.
3. Support for the users existing platform dependent (external) applications.
4. A platform independent, compact, and permanent method of archiving all

information generated by both the paperless laboratory tools and external applications.
5. Fast tools for searching and retrieving the archived information.
6. Real–time network collaboration tools (audio/video conferencing, shared electronic whiteboard, and shared document editing).

Hypertext markup language (HTML), Java applets and applications, and virtual reality modeling language (VRML) are used to create the electronic notebook which is the heart of the paperless laboratory. The notebook is viewable on any system running Netscape 2.02 (or higher) with the appropriate plug–ins and helper applications.

An HTML editor such as Softquad's "HoTMetal" can be used to create pages in the notebook and insert various elements. Types of elements include (but are not limited to) Java applets, links to other notebook pages or web sites, and illustrations. HotMetal allows the user to create HTML documents in a user friendly environment which resembles a word processor. Additionally, HotMetal will import and convert documents created with many popular word processors and spreadsheets.

The tools discussed here are divided into two categories:

1. Tools used to create, manipulate, and archive notebook pages (and their associated data) on the local machine.
2. Tools used to exchange data over the network. This includes data import/export and collaboration tools.

LOCAL TOOLS

External Applications

The helper application utility in Netscape provides support for the user's existing applications. For example, the user may have a data acquisition/signal processing package which they would like to continue to use until comparable Java based tools are available. Netscape permits the user to associate their external application (helper application) with the file type(s) generated by the application. Clicking a link in the notebook page which references the desired file will result in the appropriate application being launched with the referenced file. The disadvantage of helper applications is they are machine specific. So, the user would only be able to view the referenced file on a machine that has the application installed and set up as a helper in its copy of Netscape.

Java Based Tools

Data Manager. The data manager application accepts data from a local or network source. File header information and/or information provided by the user allows the application to take the following actions:
1. Store the imported data in a user specified directory. If the data is a text document it can be converted to an HTML document. If the data type is binary then it is stored in its raw format unless the user specifies a filter method to re–map the data.
2. Add a reference to the data/document in the appropriate index.
3. Create a link in an existing HTML document to the newly created HTML document. If the imported data is binary the data manager can add an <APPLET>, <IMG>, or other tag to an existing HTML document to specify how to display the data on the page.

The data manager can also convert the incoming data into a new format. This is done by specifying a filter method to apply to the incoming data stream. Filters under development include ASCII text to HTML, ASCII Spreadsheet to VRML–2 elevation grid and a customizable filter to convert binary files generated by data acquisition packages to a common format useable by the notebook's signal processing and display applets.

Data Acquisition. Java based tools to acquire from supported data acquisition cards, sound and video cards, and peripherals attached to the PCs serial or parallel ports are not yet implemented. These tools will provide a standard way of capturing data from an attached card or peripheral regardless of its type. More work must be done to determine the best way to accomplish this task. The data acquisition tools, when implemented, will be part of the Data Manager.

Data Analysis. Tools for signal and image processing, time–frequency analysis, and statistics are likely future additions to the notebook's toolbox. Development of various Java based math tools can be followed by reading postings to **comp.lang.java.programmer** and browsing the Java repository at **http://java.wiwi.uni–frankfurt.de**

NETWORK TOOLS

The Java Media APIs (JMAPI) and the Java Telephony API (JTAPI) provide a platform independent programming environment for the development of collaboration tools.

Video Conferencing

Machine specific solutions have been developed in our lab. Cross platform solutions using JMAPI will be developed when the API is available.

Audio Conferencing and Shared Whiteboard

Cooltalk (free download http://www.insoft.com) and other Netscape plug–ins provide audio conferencing with a shared whiteboard. However, midday network congestion usually prevents use of the audio conferencing feature.

Electronic notebook specific JTAPI based applications are under development. These applications will have the option to use standard telephone service for the audio portion of the conference while using the network for exchanging whiteboard data. Additionally, the notebook whiteboard will permit interactive markup of an HTML document.

Data Exchange and Control

Development of a suite of Java applications which stream real time experimental data over the internet and permit the remote user to control peripherals attached to the local machine will provide a vehicle for direct collaboration. This utility, when fully developed, will allow the creation of a "virtual" laboratory where remote users can participate in experiments in real time and control associated laboratory equipment.

All files are archived on Compact Disk–Read Only Memory (CD–ROM) using a CD–ROM burner. There are currently no plans to develop Java based CD–ROM burner software. Most CD–ROM burners come bundled with software which can create ISO–9660 compliant disks. CD–ROMs created in ISO–9660 format are readable on systems running Unix, OS/2, DOS (ISO–9660 level 1), Windows, and MacOS (with foreign file extensions).

CD–ROM provides a standardized, high density storage solution for the archiving of all laboratory information. One standard ISO–9660 disk can store up to 640 megabytes of information. Digital Versatile Disk (DVD) is a new optical disk format that can store up to 17 gigabytes, enough for 9 hours of studio quality (MPEG–2) audio & video. DVD is rapidly replacing tape and "laser–disk" in high end video mastering and playback equipment. As DVD equipment prices fall it is likely to become the standard solution for high density, archival data storage.

The computing platform for the paperless laboratory is a networked multimedia PC with a 32 bit operating system or an Apple Macintosh. PC Operating systems which meet this criteria include Unix (e.g. Solaris, Linux, BSD), Windows–NT, OS/2, and Windows–95. Security is largely dependent on the chosen operating system (Unix is recommended) and can be as restrictive as the user desires. The total cost of the computer system and associated hardware to implement a paperless laboratory (average size university lab) is in the range of $3,000 to $5,000.

CLOSING REMARKS

The present version of the laboratory notebook fully supports archiving and retrieval of laboratory data, notes, and audio/visual files. The information in the notebook is accessible to the user both locally and over the World Wide Web with secure access. Files from the electronic notebook are permanently archived on CD–ROM thus providing some measure of protection against falsification (similar to keeping notes in a sewn, bound paper notebook). Java applets which allow the display of numerical data in a spreadsheet format, count down/up timers, playback of audio files, and panning of high resolution images are presently implemented. Applets for data acquisition, analysis, and collaboration are being developed. Support for the user's existing programs and the data generated by them is provided by Netscape's helper application utility.

The paperless environment described above has several advantages over traditional methods. First, archiving and retrieval of information is improved through the use of permanent storage media and search utilities. Second, the stored information is accessible to the user from anywhere in the world either via the web or on CD–ROM. Third, seamless integration of data, notes, and other information in a searchable environment provides a convenient way of discussing research with others. Finally, the amount of effort it takes to organize raw and analyzed data, notes, illustrations, and other information is significantly reduced since the notebook automates a large part of this process. Future plans for the electronic notebook include the migration of machine specific applications (e.g. data acquisition, analysis, and video conferencing) to platform independent Java applications and/or applets.

ACKNOWLEDGEMENTS

Matt Wyczalkowski

WSU IEEE Image Archive (http://www.eecs.wsu.edu/~ieee/gifs/image.html) for some of the buttons used in the electronic notebook

THE QUALITATIVE REASONING NEURON: A NEW APPROACH TO MODELING IN COMPUTATIONAL NEUROSCIENCE

Jeffrey L. Krichmar[1], Giorgio A. Ascoli[1], James L. Olds[1, 2], Lawrence Hunter[1, 3]

[1]Krasnow Institute for Advanced Study, Mail Stop 2A1, George Mason University, Fairfax, VA, 22030-4444
[2]Laboratory of Adaptive Systems, NINDS, NIH, Bethesda, MD
[3]National Library of Medicine, Bethesda, MD

INTRODUCTION

Modeling in computational neuroscience generally falls into one of two categories. In one approach, differential equations describing the state of a neuron are developed by studying the electrical properties of a neuron under laboratory conditions. The equations are then used to model the behavior of a neuron in an idealized situation[1,2]. Mathematical models tend to be computationally expensive and difficult to scale up to large neural networks. An alternative approach is artificial neural networks (ANN). ANNs sacrifice features of real brain networks, such as, synaptic transmission, temporal properties and architecture for the ability to build large assemblies of neuronal elements[3].

We have developed an approach, called the Qualitative Reasoning Neuron (QRN), that reproduces single neuron behavior, but is computationally simple enough to use in large scale neural networks. QRN is based on Kuiper's qualitative simulation methodology (QSIM)[4]. The technique is extremely efficient because precise quantitative values need not be calculated. Accurately detailed and predictive properties emerge from QRN simulations. QRN has simulated the response predicted by the Hodgkin-Huxley squid axon model[5] and adaptive motor control in a large network model of the cerebellar cortex[6]. This paper describes QRN methodology, outlines the development of a single QRN cerebellar Purkinje cell (PC) model, and presents simulation results.

METHODS

Qualitative reasoning uses *landmarks* to describe the critical values for a parameter. A landmark represents the upper or lower boundaries of a parameter and important events within the parameter's bounds. The set of landmark values associated with a parameter is defined as the *quantity space*. A qualitative reasoning system has two types of parameters: *Continuous* and *Discrete*. Continuous parameters have a value ordered quantity space. Discrete parameters have an unordered quantity space. The *qualitative state* of a parameter consists of a *value*, a *direction*, and a *weight*. The value of a continuous parameter can either be at or between landmarks in its quantity space. The value of a discrete parameter is always equal to a landmark in its quantity space. The direction can be *increasing, decreasing* or *steady*. The weight represents the relative significance of a parameter's effect in a qualitative system. *Constraints* ensure parameters stay within the range of the quantity space by containment rules that dictate changes between landmarks. Constraints can be mathematical functions or arbitrary functions. The multiplication function, $*(a,b,c)$, is described by "reasoning" rather than the traditional meaning: If a or b is zero, then c equals zero. If a is increasing and b is steady, c is increasing. Similarly, other mathematical functions can be "reasoned" instead of calculated. The QRN constraints are listed below:

1. M+(a,b): Monotonically increasing function. As a increases (decreases), b increases(decreases).
2. M-(a,b): Monotonically decreasing function. As a increases (decreases), b decreases(increases).
3. EXPD(a): Exponential decay function. a always decreases.
4. EXPI(a): Exponential increase function. a always increases.
5. TH(a,b): Threshold step function. If $a.value > b.value$, $a = a$. Otherwise, $a = \{0, steady\}$.
6. $+ (a,b,c)$: Addition function. $c = a + b$.
7. $* (a,b,c)$: Multiplication function. $c = a * b$.

Constraint resolution obtains a new qualitative state for a parameter. A *time step* in a QRN simulation refers to an iteration of constraint resolution for each parameter in the qualitative system. Equations (1) through (3) describe the calculation of a new qualitative state for any parameter Q. The new qualitative state is based on the direction and weight of the parameters, q_1 through q_n, that constrain parameter Q. The new value of Q is dependent on the qualitative direction given by equation (2) and the quantity space. If the qualitative direction is increasing (decreasing), the value of Q increments (decrements). The value of parameter Q is then checked to ensure it has not gone out of the quantity space bounds.

$$TotalWeight = \sum_{i=1}^{n} q_i.qdir \times q_i.qwgt \qquad (1)$$

$$Q.qdir = \begin{cases} decr. TotalWeight < 0, \\ incr. TotalWeight > 0, \\ stdy. TotalWeight = 0 \end{cases} \qquad (2)$$

$$Q.qwgt = ABS(TotalWeight) \qquad (3)$$

Where: qdir equals -1 for decreasing, 0 for steady, and +1 for increasing.

A QRN model of a PC is developed based on a recent GENESIS model of a cerebellar PC[7,8]. The morphology is based on a guinea pig PC (*cell 1*, Ref. 9). The QRN model contains 1 soma (SO), 9 main dendrite (MD), 60 thick dendrite (THD), and 1530 spiny dendrite (SPD) compartments. Each SPD has a spine compartment (SP) that receives parallel fiber input. A single climbing fiber contacts each THD and MD. A stellate cell synapses on each SPD and two stellate cells synapse on each THD. Each SO and MD receives basket cell inhibition.

The flux of Ca^{++}, K^+, Na^+ ions, and membrane potential, E_m, are described with continuous parameters. Channels are described by discrete parameters. QRN models a fast sodium current (NaF), a persistent sodium current (NaP), a P calcium current (CaP), a T calcium current (CaT), an anomalous rectifier (Kh), a delayed rectifier (Kdr), a persistent potassium current (KM), an A current (KA), a BK calcium-activated potassium current (KC), a K2 calcium-activated potassium current (K2), a glutamate channel for parallel fiber input (PF), a glutamate channel for climbing fiber input (CF), a GABA channel for basket cell input (BK), and a GABA channel for stellate cell input (ST). The kinetics of the channels in the QRN implementation of the PC is given in Table 1A. The landmark, t_d, describes the time delay before a channel opens. The landmark, t_o, describes the time a channel is open before inactivating. Logic was necessary to represent the faster time course of NaF and Kdr channels in response to large increases in E_m. The accelerated NaF and Kdr channel landmarks (denoted by * in Table 1A) are used. Table 1B describes the distribution of the channels and their relative weight, $\bar{g}$, by cell compartment type.

Table 1. A. Kinetics of the channels modeled by QRN. **B.** Distribution and relative weights of the channels modeled by QRN. A weight of N/A indicates that the channel is not present in that compartment type.

Name	t_d (ms)	t_o (ms)
NaF	0.2	0.8
		0.6*
NaP	0.1	N/A
CaP	0.0	1.0
CaT	0.0	1.0
Kh	0.0	N/A
Kdr	1.0	12.0
		2.0*
KM	0.0	N/A
KA	0.1	N/A
KC	0.9	N/A
K2	0.9	N/A
Glu	0	1.7
GABA	0	26.5

Name	SO	MD	THD	SPD
NaF	40	N/A	N/A	N/A
NaP	2	N/A	N/A	N/A
CaP	N/A	7	7	7
CaT	1	2	2	2
Kh	2	N/A	N/A	N/A
Kdr	10	1	N/A	N/A
KM	1	1	N/A	N/A
KA	2	2	N/A	N/A
KC	N/A	50	10	9
K2	N/A	1	1	1
PF	N/A	N/A	N/A	10
CF	N/A	50	50	N/A
BK	25	15	N/A	N/A
ST	N/A	N/A	20	20
Leak$_{Em}$	1	1	1	1
Leak$_{Ca}$	1	1	1	1
Leak$_K$	1	1	1	1
Leak$_{Na}$	1	N/A	N/A	N/A

The constraint model diagram for each compartment type is given in Figure 1. A simple compartment model of one SO, one MD, one THD and one SPD was developed for tuning open parameters. Different voltages were input to this simplified model. The parameters given by the GENESIS PC were used as a starting point[7]. The weights and landmarks of the various channels were adjusted until the simulated channels emulated the actual channels. These parameter values were used in the large-scale compartment model. Additional fine-tuning of NaF and Kdr channels was necessary in the large-scale model.

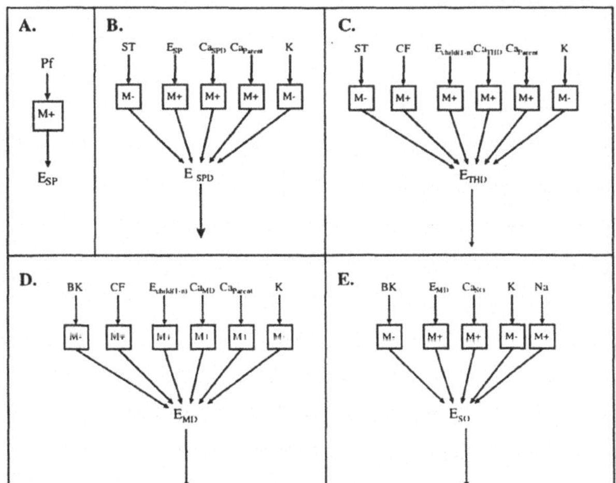

Figure 1. A. Constraints on the spine (SP). **B.** Constraints on the spiny dendrite (SPD). **C.** Constraints on the thick dendrite (THD). **D.** Constraints on the main dendrite (MD). **E.** Constraints on the Soma (SO).

The inputs to the QRN model are simulated, as in the GENESIS PC model, by varying PF excitation asynchronously from 1 to 100 Hz. Inhibitory stellate cells fired asynchronously from 1 to 30 Hz[10, 11]. CF activity was simulated as an ascending volley from the MDs to the THDs[12]. Inhibitory basket cells fired synchronously to the SO and MDs. QRN and GENESIS simulations of the PC were conducted on the same hardware with the same number of time steps. A time step represents a simulation time of 20 µs.

RESULTS

Figure 2 illustrates QRN's ability to model a complex spike in response to CF input. The GENESIS model's response to CF input is provided for comparison purposes.

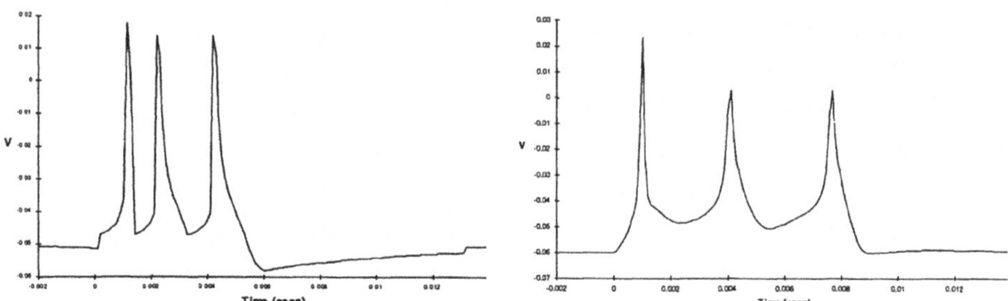

Figure 2. Complex spikes in response to climbing fiber input. The left chart is the QRN model of a Purkinje cell. The right chart is the GENESIS model of the Purkinje cell.

Different levels of PF (1-100 Hz) and ST activity were input during simulations. The mean (± S.D.) spike frequencies for QRN were 56±13, 38±10, 28±9 and 23±8 Hz with ST inhibition at 1, 10, 20 and 30 Hz respectively. The mean (± S.D) spike frequencies for GENESIS were 119±32 and 0.34±0.4 Hz with ST inhibition at 1 and 10 Hz respectively.

Figure 3 shows the interspike intervals (ISI), in response to 30 Hz PF excitation and varying ST inhibition levels. The solid line in Figure 3 represents ISI constructed from *in vivo* PC recordings in crus IIa of the rat. In Figure 4, simulated basket cell inhibition on the QRN PC model with 30 Hz PF and 1 Hz ST suppressed simple spike activity over 30 ms.

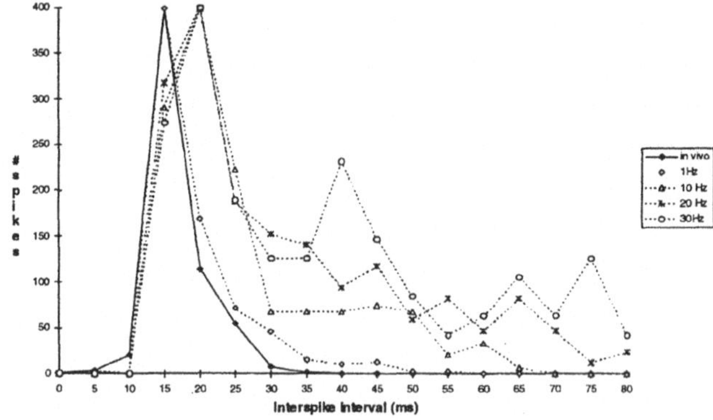

Figure 3. Interspike intervals (ISI) in response to inhibition. Parallel fibers asynchronous firing rates were set at 30 Hz. Stellate cell asynchronous firing rates were set at 1, 10, 20 and 30 Hz. The solid line is an ISI from crus IIa rat in vivo Purkinje cell recordings (adapted from Ref. 8).

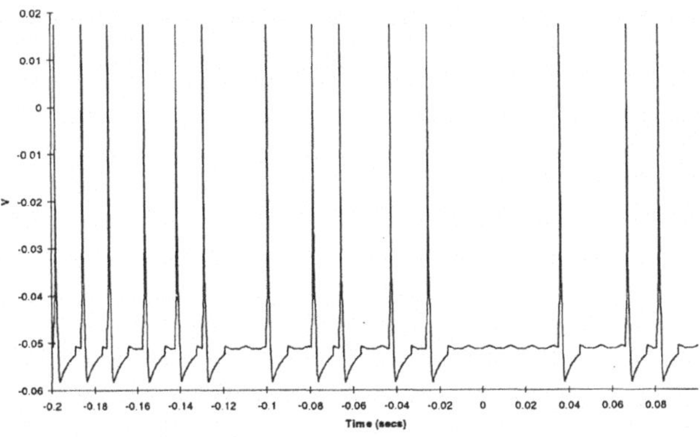

Figure 4. Basket cell inhibition. Model receives inhibition from the basket cells at time 0 with a background level of parallel fiber excitation set at 30 Hz and stellate cell inhibition set at 1 Hz.

The computer time needed to execute the PC simulations is given in Table 2. QRN is approximately 2.5 times faster than the GENESIS model.

Table 2. Comparison of computation times. Table entries represent computer time in minutes needed to complete a 5-second (250,000 time step) simulation. All tests were run on a 75Mhz Pentium PC running LINUX. The PF activity is asynchronous at 25 Hz and the stellate inhibition is asynchronous at 1 Hz. The GENESIS Purkinje cell model ran under version 2.01 of the GENESIS simulator. Models executed their simulations under LINUX on a 75MHz Pentium Personal Computer with 16Mb of RAM.

Model	Computer Time for 5 sec. Simulation
GENESIS	464 minutes
QRN	189 minutes

DISCUSSION

The QRN algorithm facilitated the creation of an efficient, yet detailed model of the cerebellar PC. The simulated complex spike output by the QRN models qualitatively matches the GENESIS simulation (see Figure 2) and electrophysiological data[12]. The complex spike was achieved by modeling active calcium channels in the dendritic tree sodium and potassium channels in the soma. In our model, a compartment received membrane potential from its children and calcium from its parent. However, this asymmetry is not necessary and could be improved upon by modifying compartments to receive membrane potential from its parent instead of calcium.

The ST inhibition levels used in this paper were 1, 10, 20 and 30 Hz. These inhibition levels are based on *in vitro* recordings of stellate cells in the turtle (mean frequency 13 Hz with a range from 5 to 20 Hz, Ref. 11) and *in vivo* recordings of Golgi cells in the cat (mean frequency 14.5 Hz with a standard deviation of 7.3 Hz, Ref. 10). The lowest level of inhibition was used for comparison with the GENESIS model[8]. The firing frequency of QRN was comparable to simple spike frequencies (30 to 100 Hz) reported *in vivo*[13]. The GENESIS model only fired within realistic ranges when ST inhibition was set at 1 Hz.

The simple spike ISI for QRN (Figure 3) is comparable to *in vivo* recordings. Increasing the level of inhibition resulted in lowering the firing frequency and shifting the ISIs toward longer intervals. Similar to the GENESIS model, different levels of inhibition set different firing threshold for the PC. However, the GENESIS PC did not exhibit spiking at ST inhibition levels above 10 Hz. The firing rates of the QRN model were lower than expected when PF activity was greater than 60Hz. This is due to long, inflexible refractory periods. Improvements could be accomplished by adjusting the weight and landmarks of the Kdr channel. Basket cell inhibition negated simple spike activity for approximately 30 ms in QRN simulations (see Figure 4). In the GENESIS model, basket inhibition interrupted simple spike activity for approximately 60 ms. Basket cell inhibition is thought to shunt PC activity, but the exact time course of the inhibition *in vivo* is not yet known.

QRN's modeling approach differs from mathematical models such as NEURON and GENESIS. The QRN method employs reasoning as opposed to calculation to get results. Numerical methods, no matter how optimized, are be more computationally intensive than reasoning. Qualitative reasoning does not require division, multiplication, exponential or floating-point arithmetic: It performs calculations through inequalities and the application of simple constraint resolution rules. The QRN model, which exactly matched the GENESIS PC model's compartments, channels and ions, was 2.5 times faster. In addition to greater efficiency, the QRN method also has the advantage of matching more closely the level of description generally used in neurobiological literature. For example, the Hodgkin-Huxley model uses differential equations to describe the sodium and potassium currents that induce an action potential. QRN models an action potential through reasoning. A large voltage increase opens sodium channels before potassium channels because the threshold for the sodium channel is less than the threshold of the potassium channel. As sodium ions enter the cell through the open channels, the membrane potential increases (i.e. $M+(Na, E_m)$. At a critical value (i.e. the sodium maximum landmark), the sodium channels begin to close and potassium channels in the soma are open causing the cell to repolarize (i.e. $M-(K, E_m)$). In this intuitive manner, QRN allows the construction of models.

REFERENCES

1. J.M. Bower, and D. Beeman, *The book of GENESIS: Exploring realistic neural modeling with the GEneral NEural SImulation System* (Springer-Verlag, New York) (1994).
2. M.L. Hines, and N.T. Carnevale, The NEURON simulation environment." *Neural Comput.*, **9**:6, (1997).
3. F. Chapeau-Blondeau, and N. Chambet, Synapse models for neural networks: From ion channel kinetics to multiplicative coefficient wij. *Neural Comput.* **7**, 713-734, (1995).
4. Kuipers, B. Qualitative simulations. *Artificial Intelligence*, **29**, 289-388, (1986).
5. J.L. Krichmar, *Intelligent Engineering Through Artificial Neural Networks. Volume 4* (ASME Press, New York) pp. 567-572 (1994).
6. J.L. Krichmar, G.A Ascoli, L. Hunter, and J.L. Olds, J.L. "A Model of Cerebellar Saccadic Motor Learning using Qualitative Reasoning", Lecture Notes in Computer Science, *Artificial and Natural Neural Networks*, G. Goos, J. Hartmants, J. van Leeuwen eds. Springer-Verlag NY (1997).
7. E. De Schutter, and J.M. Bower, An active membrane model of the cerebellar purkinje cell I. Simulation of current clamps in slice. *J. Neurophysiol.* **71**:1, 375-400, (1994).
8. E. De Schutter, E., and J.M. Bower, An active membrane model of the cerebellar purkinje cell II. Simulation of synaptic responses. *J. Neurophysiol.* **71**:1, 401-419, (1994).
9. M. Rapp, I. Segev, and Y. Yarom, Physiology, morphology and detailed passive models of cerebellar Purkinje cells. *J. Physiol. (London)* **471**, 87-99, (1994).
10. S.A. Edgley, and M. Lidierth, The discharges of cerebellar golgi cells during locomotion in the cat. *J. Physiol. (London)* **392**, 315-322, (1987).
11. J. Mitgaard, Membrane properties and synaptic responses of golgi cells and stellate cells in the turtle cerebellum *in vitro. J. Physiol. (London)* **457**, 329-354, (1992).
12. R.R. Llinas, and C. Nicholson, Reversal properties of climbing fiber potential in cat Purkinje cell: an example of a distributed synapse. *J. Neurophysiol.* **39**, 311-323, (1976).
13. R.R. Llinas, *Handbook of physiology. The nervous system. Motor control.* (American Physiology Society, Bethesda, MD), Sect. 1, Vol. II, pp. 831-876, (1981).

PERTURBATIVE M-SEQUENCES FOR AUDITORY SYSTEMS IDENTIFICATION

Mark Kvale,[1] and Christoph E. Schreiner[1]

[1]Sloan Center for Theoretical Neurobiology
Keck Center for Integrative Neuroscience, Box 0444
University of California, San Francisco
513 Parnassus Ave, San Francisco, CA 94143

INTRODUCTION

A popular approach to systems identification, i.e., identifying an accurate analytical model for the system behavior, is to use Volterra or Wiener expansions to model behavior via functional Taylor or orthogonal polynomial series, respectively. Both approaches model the response $r(t)$ as a linear combination of small powers of the stimulus $s(t)$. Although effective for mild nonlinearities, deriving the linear combinations becomes numerically unstable for highly nonlinear systems. A more serious problem is that many biological systems are adaptive, i.e., the system behavior is dependent on the stimulus ensemble. For instance, Rieke found that in the auditory nerve of the bullfrog linearity and information rates depended sensitively on whether a white noise or naturalistic ensemble is used.

One approach to handling these difficulties is to forgo the full expansion, and simply compute the linear response to small (perturbative) stimuli in the presence of various different ensembles, or operating points. By collecting linear responses from different operating points, one may fit nonlinear responses as one fits a nonlinear function with a piecewise linear approximation. For adaptive systems the same procedure would be applied, with different operating points corresponding to different points along the time axis. Perturbative stimuli have wide application in condensed-matter physics, where they are used to characterize linear responses such as resistance, elasticity and viscosity, and in engineering, perturbative analyses are used in circuit analysis (small signal models) and structural diagnostics (vibration analysis).

An effective stimulus for calculating the perturbative linear response of a system is the m-sequence. In physiology, m-sequences have been used primarily to compute system kernels, especially in the visual system. In this work, we use perturbative m-sequence stimuli to study the linear response of single units in the central nucleus of the inferior colliculus of a cat to amplitude-modulated (AM) stimuli. We add a small m-sequence signal to an AM carrier, which allows us to study the linear behavior of the system near a particular operating point in a non-destructive manner, i.e., without changing the operating point. Perturbative m-sequences allow one to calculate linear responses near the particular stimuli under study with only a little extra effort and work towards the goal of gaining as much information as possible about the system under study.

M-SEQUENCES

The binary m-sequence used here is a two-level pseudo-random sequence of $+1$'s and -1's. The sequence length is $L = 2^n - 1$, where n is the order of the sequence.

We model the IC response with a system F through which a scalar stimulus $s(t)$ is passed to give a response $r(t)$:

$$r(t) = F[s(t)]. \tag{1}$$

The functional F is taken to be a linear functional plus a DC component. The system can be written as the discrete convolution

$$r[t] = h_0 + \sum_{t_1=0}^{L-1} h[t_1]s[t - t_1] \tag{2}$$

with kernels h_0 and $h[t_1]$ to be determined. We assume that the system has a finite memory of M time steps (with perhaps a delay) so that at most M of the $h[t]$ coefficients are nonzero. To determine the kernels perturbatively, we add a small amount of m-sequence to a base stimulus s_0:

$$s[t] = s_0[t] + \alpha m[t]. \tag{3}$$

Cross-correlating the response with the original m-sequence yields

$$R_{rm}(\tau) = \alpha(L + 1)h[\tau] - h_0 - \alpha \sum_{t_1=0}^{L-1} h[t_1]$$

$$+ \sum_{t=0}^{L-1} \sum_{t_1=0}^{L-1} h[t_1]m[t]s_0[t + \tau - t_1] \tag{4}$$

A good bit of algebra yields the kernels

$$h(\tau) = \frac{1}{\alpha(L+1)} R_{rm}(\tau) - C_1 \frac{M}{L}$$

$$+ C_2 \frac{M^{1/2}}{\alpha N^{1/2} L^{1/2}}, \tag{5}$$

with the constants $C_1, C_2 \sim O(h[\tau])$ depending neural firing rate, statistics, etc., determined from experiment. If we take the kernel element $h(\tau)$ to be the first term in Eq. (5), then the last two terms in Eq. (5) contribute errors in determining the kernel and can be thought of as noise.

MODULATION M-SEQUENCES

Previous work of Moller and Langner has shown that many of the cells in the inferior colliculus are tuned not only to a characteristic frequency, but are also tuned

616

to a best frequency of modulation of the carrier. To measure modulation response, we create a modulation m-sequence:

$$s[t] = a\left(s_0[t] + b\,m[t]\right)\sin[\omega_c t], \tag{6}$$

where $|s_0[t]| \leq 1$ is the ambient signal, i.e., the operating point, $m[t] \in [-1, 1]$ is an m-sequence added with amplitude b, and ω_c is the carrier frequency. Demodulation gives the effective input stimulus

$$s_m[t] = a\left(s_0[t] + b\,m[t]\right). \tag{7}$$

TESTING THE METHOD

The m-sequences used in this experiment were of length $2^{15} - 1 = 32{,}767$. For each unit, 10 cycles of the m-sequence were presented back-to-back. After determining the proper characteristic frequencies, different stimuli were presented. The frequency of the stimulus never differed from the characteristic frequency by more than 500 Hz. The first stimulus was a carrier modulated with a pure sine wave with modulation depth 0.8: $s_m[t] = 1 + 0.8\sin[\omega_m t]$. The second stimulus, a full m-sequence modulation, was a carrier modulated with the m-sequence at full modulation depth: $s_m[t] = 1 + m[t]$. The third stimulus was a perturbative m-sequence modulation added to a sinusoidal operating point, $s_m[t] = 1 + 0.8\sin[\omega_m t] + 0.2m[t]$, so that the m-sequence signal is 6 dB down from the base signal. Figure depicts the three stimuli used. The stimuli were presented in random order so as to mitigate adaptation effects. The stimuli intensity were 20 dB above pure tone response threshold.

Performing the measurements, we found several results. First, in comparing the the spike rates for a pure sinusoid modulation vs. a combined sinusoid and m-sequence modulation, we found little difference between the two. This suggests that the m-sequence perturbation changes the operating point of the IC units only a little and that we are computing a valid linear response.

Second, we found that modulation transfer functions computed via the perturbative m-sequence with no sinusoid was essentially the same as that computed via conventional means. Thus the perturbative stimuli do indeed give adequate S/N ratios to compute kernels.

Third, given these checks, we compared the kernels generated by both a full m-sequence with no sinusoid and a perturbative m-sequence combined with an 80Hz sinusoid. The kernels are similar, except for a statistically significant dip at 13 m-sec in the perturbative kernel vs. the full kernel. This indicates a nonlinearity in the modulation transfer function. The change in kernel also shows that the kernel effectively changes in the presence of other stimuli, i.e., the sinusoid. Thus the IC does changes its operating point as a function of its environment.

ACKNOWLEDGEMENTS

This work was supported by The Sloan foundation and ONR grant number N00014-94-1-0547.

EXTRACELLULAR RECORDING FROM MULTIPLE NEIGHBORING CELLS: A MAXIMUM-LIKELIHOOD SOLUTION TO THE SPIKE-SEPARATION PROBLEM

Maneesh Sahani[1,2], John S. Pezaris[2], and Richard A. Andersen[1,2]

[1]Sloan Center for Theoretical Neurobiology
[2]Computation and Neural Systems Program
California Institute of Technology
Pasadena, CA 91125

INTRODUCTION

In recent years considerable attention among extracellular electrophysiologists has focused on the problem of simultaneously recording the activity of multiple neurons in behaving animals. Such recordings, it is hoped, will provide much-needed insight into the dynamics of neural ensemble computation and coding. Of particular interest are recordings from neighboring neurons, for example cells that lie within a single column of neocortex. Such cells are likely to share functional roles and to possess the anatomical interconnectivity needed for ensemble coding, making them plausible participants in local computational and signaling circuits.

Single cortical columns can be as little as 30μm in cross-section, and so it is difficult to introduce multiple independent electrodes into the same column *in vivo*. It is important, therefore, to distinguish the extracellular traces of action potentials from different cells gathered by a single electrode. The problem is made significantly easier by the use of a multi-tip electrode, for example the four-wire bundle commonly called a *tetrode*, that provides several slightly different electrical view-points on the same group of cells. We have recently adapted the tetrode technology, introduced by Recce and O'Keefe[1] for chronic recording in rat hippocampus, for use in behaving monkey experiments[2,3].

In the current paper we discuss a solution to the problem of separating waveforms from multiple cells in a tetrode recording. We adopt an explicitly probabilistic approach, constructing a parametric latent-variable model from which the data are presumed to be generated. We find estimates of the parameters of the model using maximum-likelihood techniques, and then, using these parameter values, infer the values of the latent variables, in particular the times of firing of the various cells. This two-stage maximum-likelihood process reflects a commonly made approximation to the full Bayesian posterior over the latent-variables[4,5]. The correct estimates of the firing times should be made by integrating over the possible parameter values; for strongly peaked posteriors, however, we can approximate this integral by evaluating the posterior at the most probable value of the parameters. Given a weak prior, this point is the same as the maximum-likelihood estimate.

DATA COLLECTION

Information about the construction of the electrode is available in an earlier paper.[3] Here we mention only those parameters of the construction and data collection that are relevant to the spike recognition problem.

The tetrode is a bundle of four individually insulated 13μm-diameter wires twisted together and cut so that the exposed ends lie close together. The potential at the tip of each electrode is amplified (custom electronics), low-pass filtered (9-pole Bessel filter, $f_c = 6.4$ kHz) to prevent aliasing and digitized ($f_s = 12.8$ to 20 kHz) (filters and A/D converter from Tucker Davis Technologies). This data stream is recorded to digital media; subsequent operations are currently performed off-line.

In preparation for inference, candidate events (where at least one cell fired) are identified in the data stream. The signal is digitally high-pass filtered ($f_c = 0.05f_s$) and the root-mean-square (RMS) amplitude on each channel is calculated. This value is an upper bound on the noise power, and approaches the actual value when the firing rates of resolvable cells are low. Epochs where the signal rises above three times the RMS amplitude for two consecutive samples are taken to be spike events. The signal is upsampled in the region of each such threshold crossing, and the time of the maximal subsequent peak across all channels is determined to within one-tenth of a sample. A 1 ms section is then extracted at the original f_s such that this *peak time* falls at a fixed position in the extracted segment. One such waveform is extracted for each threshold crossing.

GENERATIVE MODEL

Our basic model is as follows. The recorded potential trace $V(t)$ is the sum of influences that are due to resolvable *foreground* cells (which have a relatively large effect) and a *background* noise process. We write

$$V(t) = \sum_{\tau} \left(c_1^\tau S_1(t - \tau) + c_2^\tau S_2(t - \tau) + \cdots \right) + \eta(t) \tag{1}$$

Here, c_m^τ is an indicator variable that takes the value 1 if the mth cell fires at time τ and 0 otherwise. If cell m fires at τ it adds a deflection of shape $S_m(t - \tau)$ to the recorded potential. The effect of all background neural sources, and any electrical noise, is gathered into a single term $\eta(t)$. For a multichannel probe, such as a tetrode, all of $V(t)$, $\eta(t)$ and $S_m(t)$ are vector-valued.

In this paper we take the waveform shape $S_m(t)$ to be constant. In other work[6] we discuss the possibility of variation in the underlying waveforms independent of the common noise source $\eta(t)$. The c_m^τ are assumed to be independently Bernoulli distributed for each τ and m, with constant firing probabilities, except that we will enforce a refractory period between spikes from the same cell.

The distribution of the noise, $\eta(t)$, may intuitively be expected to be Gaussian; if the waveform of a spike on the cell membrane is constant (and the preparation well shielded), the noise in the recording is composed of thermal noise at the tip, noise in the

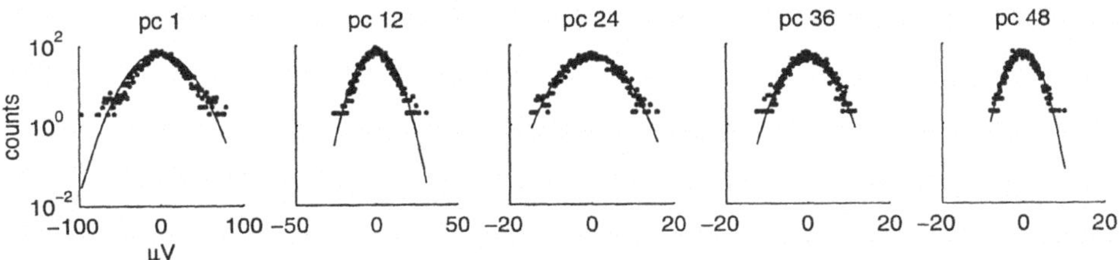

Figure 1. Distribution of background noise. We extract 1ms segments from a bandpassed recording sampled at 16 kHz from a four-channel electrode, avoiding the foreground spikes identified as described in the text. Each segment is thus a 64-dimensional object. We find the principal components of the ensemble of such vectors, and construct histograms of the projections of the vectors in these directions. A few of these histograms are shown on a log-scale (points), as well as a zero-mean Gaussian fit to the distribution projected along the same axes (lines).

signal amplification and conditioning electronics, and superimposed waveforms due to the electrical activity of many "background" cells too far away to be distinguished in the sorting process. The thermal and electronic noise are known to be Gaussian. Provided that the number of cells contributing to the background noise is large, and their firing relatively independent, the central limit theorem suggests that their influence too may be taken to be well approximated by a Gaussian, although not necessarily an isotropic one. The empirical distribution of the noise is shown in figure 1. It is clear that the Gaussian is a reasonable description, although a slight excess in kurtosis is visible in the higher principal components.

The distribution of $\eta(t)$, then, is described by a single correlation function. In practice the noise will be correlated over only a short time-lag and such a function will be equivalent to a Toeplitz covariance matrix, Σ_η. The parameters of the model to be discovered, then, are Σ_η, S_m and p_m. Once these are estimated we can proceed to find estimates for the spike times c_m^τ, which are the quantities we ultimately seek.

ESTIMATING NOISE PARAMETERS

The covariance of the background noise is measured directly from segments without foreground spikes. We can use the covariance to whiten the noise so that further inference proceeds in the context of decorrelated noise. To do this we fit an autoregressive (AR) process of order greater than the measured correlation length to the background, using the Yule-Walker equations. We then subtract the forward predictor given by this model. It is important to note that the resultant signal is not white: only the background is decorrelated. The complete signal is quite non-Gaussian and so cannot be decorrelated by second-order methods.

ESTIMATING WAVEFORMS

We now extract candidate spike waveforms from the whitened signal around the threshold crossings, as was described before. We will write V^i for the ith extracted event and assume that it represents a spike from a single cell (that is, only one of the c_m is non-zero). This is an unreasonable assumption; we can shore it up partially by eliminating from our collection of V^i segments that appear heuristically to contain overlaps (for example, double-peaked waveforms). Ultimately, however, we will need to make our inference procedure robust enough that the parameters describing the model are well estimated despite the errors in the data.

The benefit of this assumption is that it allows the waveforms to be modeled as arising from a mixture distribution, each waveform coming from a single multivariate Gaussian cluster with mean S_m (which is now viewed as a vector). We can achieve robustness by introducing additional mixture components, one centered at zero to account for false triggers of the event extraction heuristic and others with broad covariances to "mop up" any remaining overlaps. In what follows we will not write these clusters explicitly.

We consider the case of M cells, each of which generates events drawn from a multivariate Gaussian distribution with mean S_m and identity covariance (assuming that the noise has been whitened). For conciseness, we write the probability of a given observation V^i under the mth Gaussian as $G_m(V^i) = (2\pi)^{-2/d} \exp\left(-(V^i - S_m)^T(V^i - S_m)/2\right)$. The prior probability of a spike arising from the mth cell is written p_m, and the parameters are collected into a single vector $\boldsymbol{\theta}$. Neglecting, for the moment, the refractory constraint, the log-likelihood of the mixture model, given a sequence of observations $\{V^i\}$, is simply

$$l(\boldsymbol{\theta}; \{V^i\}) = \sum_i \log\left(\sum_m p_m G_m(V^i)\right). \tag{2}$$

The effect of the refractory period on this likelihood will be discussed below.

The EM approach to the fitting of such a Gaussian mixture model is well known[7, 8]. We introduce indicator variables c_m^i, similar to the c_m^τ. For each i exactly one of the corresponding c_m^i assumes the value 1 and the rest 0, thus indicating the cluster from which the given observation was drawn. If the c_m^i could be observed and thus used to augment the data V^i the (complete data) log-likelihood would be

$$l(\boldsymbol{\theta}; \{V^i, c_m^i\}) = \sum_i \sum_m c_m^i \left(\log p_m + \log G_m(V^i)\right). \tag{3}$$

The indicator variables allow us to bring the logarithm inside the summation, considerably simplifying the task of model fitting.

The EM procedure now proceeds in two steps, iterated to convergence. In the first (E) step we find the expected value of the log-likelihood (3) under the distribution $P(c_m^i \mid \{V^i\}, \boldsymbol{\theta})$. Since the log-likelihood is linear in the variables c_m^i, this simply involves replacing the c_m^i with the corresponding expected value. We write r_m^i for $E[c_m^i]$ (these are often called the *responsibilities*) and obtain

$$E[l(\boldsymbol{\theta}; \{V^i, c_m^i\})] = \sum_i \sum_m r_m^i \left(\log p_m + \log G_m(V^i)\right), \tag{4}$$

with, by Bayes' rule,

$$r_m^i = P(c_m^i \mid \{V^i\}, \boldsymbol{\theta}) = \frac{p_m G_m(V^i)}{\sum_{\tilde{m}} p_{\tilde{m}} G_{\tilde{m}}(V^i)}. \tag{5}$$

The second (M) step involves maximizing this expected log-likelihood over the model parameters $\boldsymbol{\theta}$. This is easily seen to reduce to independently fitting each of the Gaussians to the data, weighted by the corresponding responsibilities.

We now consider the alterations to this standard approach that are necessary to accommodate the refractory constraint. The likelihood $l(\boldsymbol{\theta}; \{V^i, c_m^i\})$ is identical to (3) in most cases, but for sequences of c_m^i that violate the constraint it diverges to $-\infty$. In taking the expected value of the log-likelihood, however, the probability of such divergence is 0 and so the E step results in a form similar to (4), except that the values of $E[c_m^i]$ are different. We will use s_m^i to denote these new responsibilities, reserving r_m^i for the values in (5).

To obtain the new responsibilities, consider first the simple case where only two spikes have been observed and the second appears within a refractory period of the first. We have a joint distribution over c_m^1 and c_n^2 with

$$P(c_m^1, c_n^2) = \begin{cases} 0 & \text{if } m = n \\ r_m^1 r_n^2 / Z & \text{otherwise} \end{cases} \tag{6}$$

where $Z = \sum_m \sum_{n \neq m} r_m^1 r_n^2$ is an appropriate normalizing constant. The expected values we seek are then just the marginals of this joint distribution, e.g.

$$s_m^1 = \sum_{n \neq m} r_m^1 r_n^2 / Z = r_m^1 (1 - r_m^2)/Z \tag{7}$$

where we have used the fact that $\sum r_n^i = 1$.

This result easily generalizes to the case of many spikes

$$s_m^i = \frac{r_m^i}{Z^i} \prod_{i,j \text{ refractory}} (1 - r_m^j) \tag{8}$$

where Z^i is the appropriate normalizer.

The M step is still a weighted Gaussian estimation, the weights now being the new responsibilities s_m^i.

SPIKE TIME INFERENCE

Three issues are left unresolved by this clustering process. First, the event identification heuristic could be improved upon once the true spike shapes have been determined. Second, if all events are to be clustered the sorting process must occur off-line, ruling out experiments in which rapid feedback about the cells' responses is needed. Third, superposed events have been discarded, rather than resolved into their constituent spike forms.

These issues are addressed by building matched filters for the identified spike waveforms. The filters are applied to the recorded signal to identify spike occurrences, thus improving event detection. The same filtering process can also be applied to further data from the same site as they are collected, facilitating real-time experimental monitoring of subsequent data. The algorithm is easily implemented on DSP hardware. Finally, as we shall see, the filtering process resolves spike superpositions.

Our filtering process arises as a direct solution to the maximum likelihood statement of the problem. For clarity, we will restrict ourselves here to the case of a single-channel continuous signal with spikes from two cells corrupted by white Gaussian noise. This simple formulation demonstrates the essence of the approach. The treatment is easily extended to four discrete-sampled channels with multiple cells.

We consider a recorded signal $V(t)$ which is composed of spikes of two shapes $S_1(t)$ and $S_2(t)$ occurring at times τ_1^i ($i = 1 \ldots N_1$) and τ_2^i ($i = 1 \ldots N_2$) respectively, and which is corrupted by stationary Gaussian noise with zero mean and standard deviation σ. The log-likelihood of the model given by the times $\{\tau_1^i\}$ and $\{\tau_2^i\}$ is proportional to

$$l(\{\tau_m^i, S_m\}; V(t)) \propto -\frac{1}{2\sigma^2} \int \left(V(t) - \sum_{i=1}^{N_1} S_1(t - \tau_1^i) - \sum_{i=1}^{N_2} S_2(t - \tau_2^i) \right)^2 dt. \tag{9}$$

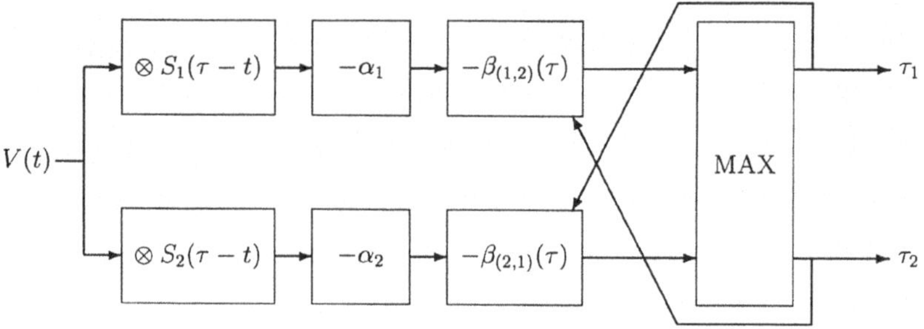

Figure 2. Block diagram illustrating the adaptive-threshold filter-based detection scheme for two spike shapes. The symbol $\otimes$ represents convolution. Consult text for an explanation of the other symbols.

We wish to maximize this likelihood with respect to the model.

We expand the square (bearing in mind that spikes from a single source cannot overlap), and, with some rearrangement and removal of terms independent of the times τ_m^i, we obtain the following expression:

$$
\begin{aligned}
l(\{\tau_m^i, S_m\}; V(t)) \propto \quad & \sum_{i=1}^{N_1} \left(2 \int V(t) S_1(t - \tau_1^i)\, dt - \int S_1(t)^2\, dt \right) \\
+ \quad & \sum_{i=1}^{N_2} \left(2 \int V(t) S_2(t - \tau_2^i)\, dt - \int S_2(t)^2\, dt \right) \\
- \quad & \sum_{\tau_1^i, \tau_2^j \text{ close}} \left(2 \int S_1(t - \tau_1^i) S_2(t - \tau_2^j)\, dt \right).
\end{aligned}
\tag{10}
$$

The notation "τ_1^i, τ_2^j close" means that the sum is taken over those pairs of times where the two associated spike shapes overlap.

The integrals within the first two summations dictate the matched filter form. The output of the mth matched filter (which has impulse response $S_m(-t)$) is compared to the threshold set by the squared power of the filter, α_m, to compute a time-dependent quantity $\mathcal{F}_m$.

$$
\alpha_m = \frac{1}{2} \int (S_m(t))^2\, dt
\tag{11}
$$

$$
\mathcal{F}_m(\tau) = \int V(t) S_m(t - \tau)\, dt - \alpha_m
\tag{12}
$$

Disregarding, for the moment, the final term in (10), this implies that we increase the likelihood of the model by choosing times τ_m^i which fall at the peaks of $\mathcal{F}_m$, provided that those peaks are positive and do not fall closer together than one spike-width.

The final term in (10) describes the interaction between the spikes. We can view it as a time-dependent modification of the threshold α_m caused by a spike in cell n. If a spike in the nth cell has been detected at τ_n, then the adjustment at time δ after the detection is $\beta_{(m,n)}(\tau_n + \delta)$, leading to output $\hat{\mathcal{F}}_m$.

$$
\beta_{(m,n)}(\tau_n + \delta) = \int S_m(t - \delta) S_n(t)\, dt
\tag{13}
$$

$$
\hat{\mathcal{F}}_m(\tau) = \int V(t) S_m(t - \tau)\, dt - \alpha_m - \sum_{j \neq i} \beta_{(m,n)}(\tau)
\tag{14}
$$

Thus, whenever a spike is detected according to (12), the thresholds for other spike shapes are transiently altered by the term β. This allows the correct resolution of superpositions. The process is illustrated in figure 2.

This adaptive-threshold process is thus seen to find the maximum likelihood estimate of the spike times, given the signal, the noise characteristics and the spike shapes.

DISCUSSION

The algorithms presented here are optimal, in the maximum-likelihood sense, given the simple Gaussian generative model described. They resolve overlaps correctly and can be run in real time. In this respect, they represent a significant advance over commonly used approaches to this problem.

The model examined may well prove to be too simple to apply to all experimental preparations[9,10]. It is, however, possible to extend this same framework to describe non-Gaussian variability as well as dynamic variation of spike shape during bursts[6].

ACKNOWLEDGMENTS

This work has benefited considerably from important discussions with both W. Bialek and S. Roweis. J. Hopfield has provided invaluable advice and mentoring to MS. Funding for various components of the work has been provided by the Keck Foundation, the Sloan Center for Theoretical Neuroscience at Caltech, the Center for Neuromorphic Systems Engineering at Caltech, the Office of Naval Research and the National Institutes of Health.

REFERENCES

1. M. L. Recce and J. O'Keefe, The tetrode: An improved technique for multi-unit extracellular recording, *Soc. Neurosci. Abs.* 15(2):1250 (1989).
2. J. S. Pezaris, M. Sahani, K. L. Grieve, and R. A. Andersen, Multiple single unit recording using tetrodes in macaque visual cortex: Electrode design and spike identification, *Soc. Neuroci. Abs.* 21:905 (1995).
3. J. S. Pezaris, M. Sahani, and R. A. Andersen, Tetrodes for monkeys, in: *Computational Neuroscience: Trends in Research, 1997,* J. M. Bower, ed., Plenum Press, New York (1997).
4. D. J. C. MacKay, Bayesian interpolation, *Neural Comp.* 4(3):415 (1992).
5. M. S. Lewicki, Bayesian modeling and classification of neural signals, *Neural Comp.* 6(5):1005 (1994).
6. M. Sahani, J. S. Pezaris, and R. A. Andersen, On the separation of signals from neighboring cells in tetrode recordings, in: *Advances in Neural Information Processing Systems 10,* M. I. Jordan, M. J. Kearns, and S. A. Solla, eds., MIT Press, Cambridge, MA (1998).
7. A. P. Dempster, N. M. Laird, and D. B. Rubin, Maximum likelihood from incomplete data via the EM algorithm (with discussion), *J. Roy. Stats. Soc. B* 39:1 (1977).
8. S. J. Nowlan, Maximum likelihood competitive learning, in: *Advances in Neural Information Processing Systems 2,* D. S. Touretzky, ed., Morgan Kaufmann, San Mateo, CA (1990).
9. M. S. Fee, P. P. Mitra, and D. Kleinfeld, Automatic sorting of multiple-unit neuronal signals in the presence of anisotropic and non-gaussian variability, *J. Neurosci. Meth.* 69:175 (1996).
10. M. S. Fee, P. P. Mitra, and D. Kleinfeld, Variability of extracellular spike waveforms of cortical neurons, *J. Neurophysiol.* 76(3):3823 (1996).

FROM CELLS TO SYSTEMS: Logos AND METALogos

Michael Stiber[1]* and Gwen A. Jacobs[1,2]

[1]Department of Molecular & Cell Biology
University of California
Berkeley, CA 94720-3200
stiber@u.washington.edu
[2]Center for Computational Biology
Montana State University
P.O. Box 173505
Bozeman, MT 59717-3505
gwen@nervana.montana.edu

INTRODUCTION

Neuroscientists study various anatomical, physiological, and functional components of nervous systems to better understand how the "low-level" activity of individual cells maps to behavior. In this research process, massive amounts of complex data are collected, but technology has not yet provided systems which integrate this information to help scientists analyze, visualize, and understand the data.

There are several aspects of this research process which are unusual when compared to most other scientific information processing activities. First of all, the base data gathered from experiments is diverse, including anatomical (2D and 3D images, 3D geometries), physiological (times series, point processes), and molecular (2D and 3D density distributions). Data is also gathered from individual animals, each exhibiting great variability in characteristics. Despite this apparent variability, each individual is assumed to be an instance of a particular kind of animal. Though data is gathered from individuals, questions to be answered apply to the idealized, "generic" animal. It is necessary, therefore, to generalize out an exemplar corresponding not to any one dataset, but rather to what all of the datasets have in common.

An additional complication is that data is captured from individual *cells*, yet the questions one wishes to answer typically are directed towards *systems*. So, not only is there a need to generalize from the individual dataset, one must also change derived information from one level of abstraction (the cell) to another (the system).

This paper describes the development of a prototype *researcher's associate*: a computer system which will perform a variety of routine and time-consuming tasks for neuroscience

---

*Current address: Computing & Software Systems, University of Washington, Bothell, 22011 26th Avenue SE, Bothell, WA 98021-4900.

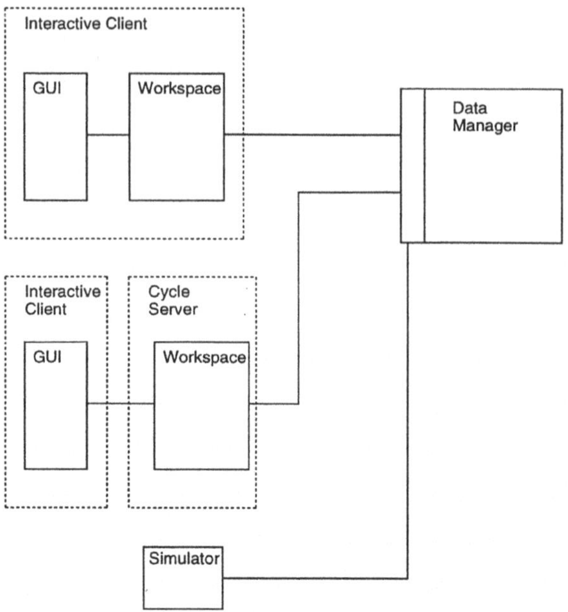

Figure 1. LOGOS system architecture.

researchers[1]. In its first phase, the system (dubbed LOGOS, and shown schematically in Figure 1) provides a distributed environment for data management and visualization. The second-generation system, METALOGOS, will incorporate knowledge-based techniques to enable researchers to work at the level of generalized systems, with METALOGOS translating commands and responses to the cellular dataset level.

LOGOS is unique in that, rather than being an encyclopedic archive of neurology-related information, it is an interactive system targeted toward neurobiology data that can help scientists visualize and understand neuronal structure-function relationships. It will allow the simultaneous visualization of large ensembles of neurons in their correct spatial relationships along with their individual functional properties, and consequently will also allow the prediction of emergent ensemble properties which arise from their organization within the nervous system.

The power of this simultaneous spatial visualization has already been demonstrated with smaller systems[2,3], and verifies the need for a complete infrastructure to dynamically and systematically bridge the gap from complex data to visualization of its myriad relationships.

SYSTEM ARCHITECTURE OVERVIEW

Figure 1 presents the basic LOGOS architecture, along with its potential interoperation with other systems. At the highest level, it is broken into two parts: an *interactive client* (top left) and a *data manager* (DM, right). The former runs on an investigator's desktop machine; the latter on some shared database server.

To allow investigators to use low-cost personal computers with LOGOS, the interactive client may be configured to host only a *graphical user interface* (GUI, middle left), which enables one to interact with the system and visualize data. All computation-intensive operations would then be performed on a shared, high-performance workstation (the *workspace*, or WS).

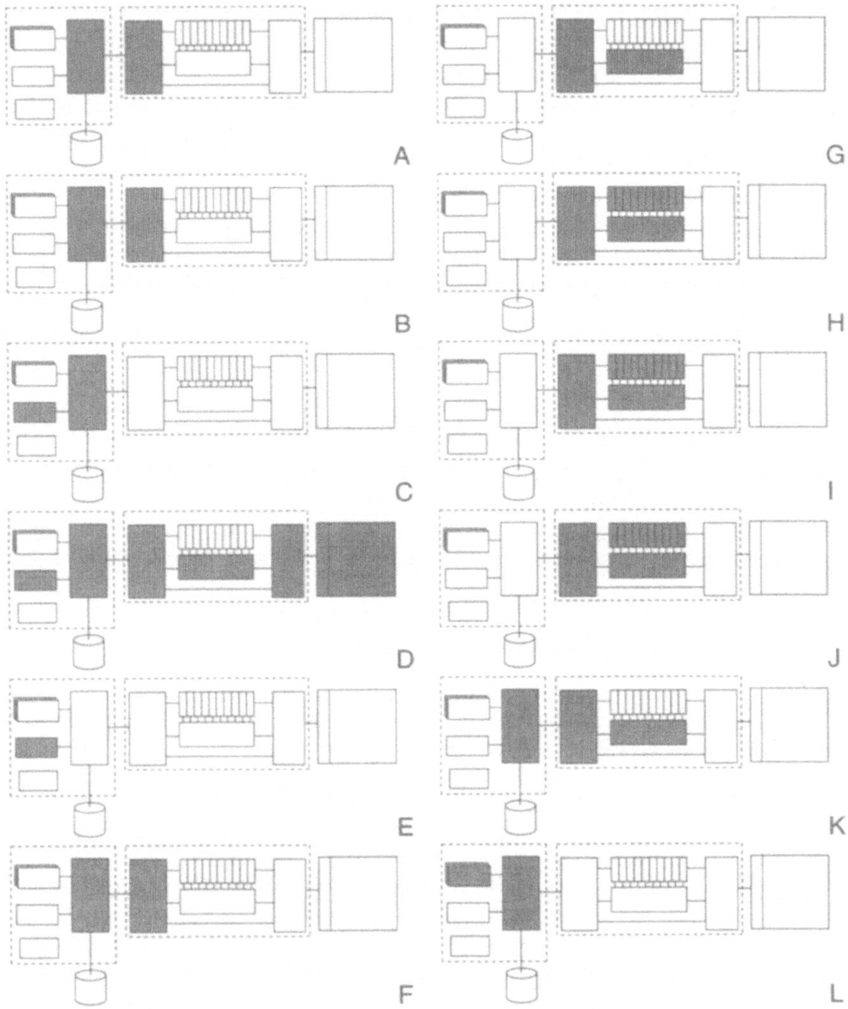

Figure 2. Example of LOGOS operation: computing a response map.

A third option exists for those interested in using their existing software (e.g., a simulation environment, bottom left): the data manager will provide a communication protocol for access from third-party applications. While this will not deliver the full LOGOS system, it will allow data sharing and possible extension of other neuroinformatics systems to include LOGOS functionality.

EXAMPLE OF SYSTEM OPERATION

Figure 2 presents an example of the flow of processing through the system to compute a *response map*[3]: a summary, combining anatomy and physiology, of how multi-afferent responses to a particular stimulus are mapped into the volume of a ganglion. A response map is a display, within the ganglion's volume, of the overall distribution of synaptic activity. In (A–L), the dashed lines surround the graphical user interface (GUI, left) and the workspace (WS, middle); the data manager (DM) is the rightmost block. Shaded blocks indicate subsystems that are active at each step.

First of all, the system will be started up (A–C). This involves the GUI connecting to the WS (A), the WS informing the GUI of the operations and data available to it (B), and the GUI presenting the user with the primary view of DM contents and system activity. This GUI/WS negotiation process is essential in a system where component capabilities may change (i.e., depending on the hardware on the user's desktop).

Before a query can be processed, the investigator must identify upon what data to operate (D, E). Based on either a single specification or a sequence of trial-and-error interactions, a set of cell data entries are transferred from the DM into the WS; the GUI provides feedback indicating WS contents (D). The user can then indicate some subset of WS contents as the object of further processing, and issue a request to compute the map for a particular stimulus (E).

This request is passed from the GUI to the WS (F), where it is broken down into a sequence of subtasks. First and foremost, one must consider that data was collected from a number of individuals with a normal range of anatomical variation. Thus, anatomical data for each cell must undergo a geometric transform to register all within some common coordinate system (G). LOGOS includes a number of methods for combining data from multiple cells into a group summary. The system can then perform the bulk of its processing within this idealized space (H–J), which involves:

H. computing a spatial density function for each cell based on its distribution of synaptic varicosities,

I. computing the response magnitude of each cell to the stimulus and applying that to the density function to produce its individual response map,

J. and summing the single-cell responses to produce the overall map.

The map, which is retained in the workspace (for further processing and/or eventual storage in the DM), is transmitted to the GUI (K), which in turn creates a viewer for visualization and graphical manipulation (L). Thus, from the investigator's point of view, one selects source data, issues a high-level command, and receives a graphical response which describes *system* operation; the mechanics of translating back and forth between the cell and system level is handled automatically by LOGOS and its library of operations for data transformation and integration.

CONCLUSION

This paper outlines the development of a neuroinformatics system as a:

- framework for neuroinformatics research,

- tool for neuroscience research,

- and platform for presentation of advanced scientific data management and analysis techniques to neuroscientists.

It is important to recognize the pivotal role of *both* neuroscience and computer science in this work; as such, any neuroinformatics project must include research areas of interest to both neuroscientists and computer scientists. The alternative would mean that new technologies wouldn't be as rapidly applied to neuroscience as they might be and that the unique aspects of neuroinformatics wouldn't receive the attention they deserve.

ACKNOWLEDGMENTS

This work was supported by NSF grant number BIR-9507314 to G.A.J.

REFERENCES

1. M. Stiber, G. A. Jacobs, and D. Swanberg, LOGOS: a computational framework for neuroinformatics research, *in*: "Proc. Ninth Int. Conf. on Scientific and Statistical Database Management," D. Hansen and Y. Ioannidis, eds., Olympia, WA (1997).
2. T. W. Troyer, J. E. Levin, and G. A. Jacobs, Construction and analysis of a database representing a neural map, *Microscopy Res. & Tech.* 29:329–43 (1994).
3. G. A. Jacobs and F. E. Theunissen, Functional organization of a neural map in the cricket cercal sensory system, *J. Neurosci.* (1996).

REGULARITY IN SPIKE FIRING WITH RANDOM INPUTS DETECTED BY METHOD EXTRACTING CONTRIBUTION OF TEMPORAL INTEGRATION OF A PAIR OF INCOMING SPIKES TO THE FIRING OF A NEURON

David C. Tam

Center for Network Neuroscience and
Dept. of Biological Sciences
University of North Texas
Denton, TX 76203
E-mail: dtam@unt.edu

ABSTRACT

A multi-unit spike train analysis method is introduced to detect the contribution of temporal integration of a doublet firing (a pair of spikes) in one neuron to the probability of spike generation in another neuron within the same network.   This is a conditional correlation method that estimates the conditional probability of firing of a spike in a neuron based on the probability of firing of not just a single spike but two sequential spikes (doublet) in another neuron prior to its firing.   The duration of temporal integration on the firing characteristics of spikes can be revealed by examining the relationship between the pre-interspike intervals (pre-ISIs) in one neuron and post-cross intervals (post-CIs) in another neuron statistically.   The "integration period" between two spikes are revealed by the appearance of a finite horizontal band of points (or lack of points) in the "pre-ISI vs post-CI scatter plot."   Furthermore, this analysis also reveals whether the coupled spike firing in one neuron is correlated with the first preceding spike in the other neuron.   Simulation results show that *regularity* in spike firing can be produced by a *randomly* firing driver neuron using this analysis.   The results also show that the firing characteristics of the driven neuron can be determined by the properties of the driven neuron (such as the integration period) relatively independent of the incoming rate of firing of the driver neuron.   The rhythmicity of synchronized firing in the driven neurons produced by entirely random inputs may provide insight in interpreting the significance of computational properties of a network of neurons.

1. INTRODUCTION

A multi-spike train analyses has been developed to decipher the significance and contribution of temporal integration to spike firings of a neuron in a network.   The relationship between the spike firing activities of two neurons detected by the multi-neuron recordings was usually examined traditionally by cross-correlation techniques. Yet these methods usually provide an estimate of probability between the spike firings in neurons, but they do not explicitly extract how many sequential spikes are correlated to the next spike firing. Thus, it is of interest to quantify the conditional probability of spike firing based on the temporal integration contributing to the generation of a spike. Although temporal integration is a well-known phenomenon, most spike train analyses were not designed specifically to extract how sequential spikes are correlate to the next spike firing (i.e., the contribution of temporal summation to the firing of a spike in a neuron). Conventional cross-correlation techniques (or other similar derived correlation statistics) primarily correlate the single-spike to single-spike firing relationship between neurons. Since temporal summation requires the integration of multiples spikes in time, such a contribution to the spike firing can only be obtained if the correlation spike train analysis method also includes the probability of firing based on not just one preceding spike, but *multiple* spikes. The input/output (I/O) relationship from multiple-spike firing to single-spike coupled-firing is yet to be addressed. The conditional cross-interval spike train analysis introduced here is designed specifically to address the contribution of temporal integration by correlating the firing of a "compared" neuron with the time between multiple firings in a "reference" neuron. The conditional probability of spike firing relative to the temporal integration period is obtained to reveal the contribution of multiple (two, in this case) spikes to the generation (or suppression) of spike firing in a neuron. We will also show by applying this analysis to simulated spike trains, *regularity* in firing can be produced by *randomly* driven inputs when the neuron has an extended, finite period of temporal integration. An extended description of this spike train analysis method will be found in [1].

2. METHODS

Let one of the spike train be called the reference spike train, A, with a total of N_A spikes. The other spike train will be called the compared spike train, B, with a total of N_B spikes. They are represented by

$$x_A(t) = \sum_{n=1}^{N_A} \delta(t - t_n) \tag{1}$$

$$x_B(t) = \sum_{m=1}^{N_B} \delta(t - t'_m) \tag{2}$$

$$\forall t_n, t'_m \text{ such that } t_n < t_{n+1} \text{ and } t'_m < t'_{m+1},$$

where t_n and t'_m are the times of occurrence of n-th and m-th spikes in spike trains A and B, respectively, and $\delta(t)$ is a delta function denoting the occurrence of a spike at time t. Note that we use the primed notation to denote the times of spike occurrence in spike train B and cross intervals between the two spike trains.

The pre-interspike interval (pre-ISI, τ_{n+1}) relative to the n-th reference spike in the reference spike train A is defined as

$$\tau_n = |t_{n-1} - t_n|. \tag{3}$$

The post-cross-interval (post-CI) between the compared and reference spike trains relative to the n-th reference spike in spike train A is defined as

$$\tau'_{n,m+1} = \left| t'_{m+1} - t_n \right|$$

(4)

such that $t'_m < t_n \leq t'_{m+1}$.

The probability of firing the next spike is given by the probability density function (pdf). The joint probability density function (joint pdf) of the next cross spike firing in the compared neuron at lag-time τ'_y (= post-CI) given that a preceding spike has fired in the reference spike train before the reference spike at lead-time τ_x (= pre-ISI) is given by

$$P(\tau_x \cap \tau'_y) = \frac{\displaystyle\sum_{n=2}^{N_A} \delta(\left| t_{n-1} - t_n \right| - \tau_x)\delta(\left| t'_{m+1} - t_n \right| - \tau'_y)}{N_A - 1}$$

(5)

$$\forall t_n, t'_m \text{ such that } t'_m < t_n \leq t'_{m+1} \text{ and } t_{n-1} < t_n$$

where τ_x is the "lead-time" and τ'_y the "lag-times" as defined in conventional correlation terminology, i.e., τ_x is similar to the lead-time in auto-correlation [2] and τ'_y is similar to the lag-time in cross-correlation [3]. The conditional firing probability density function (conditional pdf) is given by

$$P(\tau_x | \tau'_y) = \frac{\displaystyle\sum_{n=2}^{N_A} \delta(\left| t_{n-1} - t_n \right| - \tau_x)\delta(\left| t'_{m+1} - t_n \right| - \tau'_y)}{\displaystyle\sum_{n=2}^{N_A} \delta(\left| t'_{m+1} - t_n \right| - \tau'_y)}$$

(6)

Both the joint probability function and conditional probability function can be represented graphically by a two-dimensional function displayed in an xy-plot. Traditionally, the xy-plot representing the numerator of Eqs. 5 and 6 is used in most spike train analyses. This representation is similar to that represented by joint interspike interval (JISI) plots [4] or "return map" plots in nonlinear dynamics analyses [5, 6], pre-conditional cross interspike interval analyses [7] and post-conditional cross-interval analysis [8].

3. RESULTS

To illustrate how these analyses can be used to extract the temporal correlation, the spike trains of two synaptically connected neurons are simulated. Neuron A fires spontaneously with random firing patterns generated by a Poisson spike generating process, while neuron B is driven by neuron A with a 50% coupling probability strength.

Figure 1 shows the "pre-ISI/post-CI scatter plot" for neuron A (driver neuron) used as the reference neuron correlated with neuron B (driven neuron). A discrete well-defined 10 ms integration period is revealed as indicated by the horizontal dense cluster of points at 2.5 ms post-CI (between 4 and 10 ms pre-ISIs). (Note that this non-physiological sharp cut-off integration period of 10 ms is used for illustrative purpose only to show the phenomenon to be described below.) The points lying above this band at ~2.5 ms post-

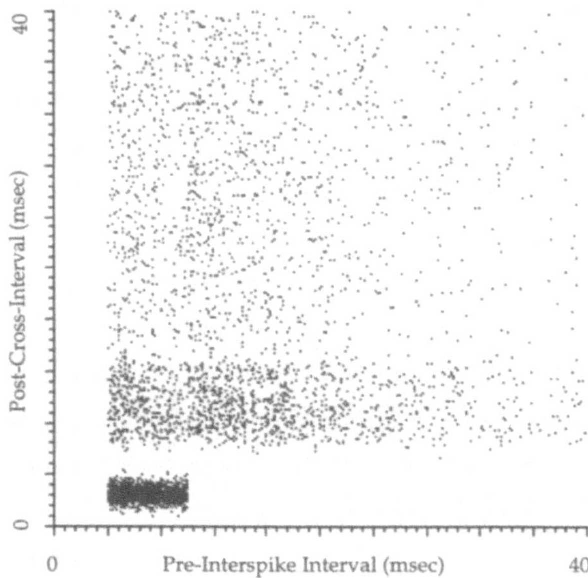

Figure 1. Pre-ISI vs Post-CI scatter plot revealing the temporal integration period by the horizontal band of points. (See text for explanation.)

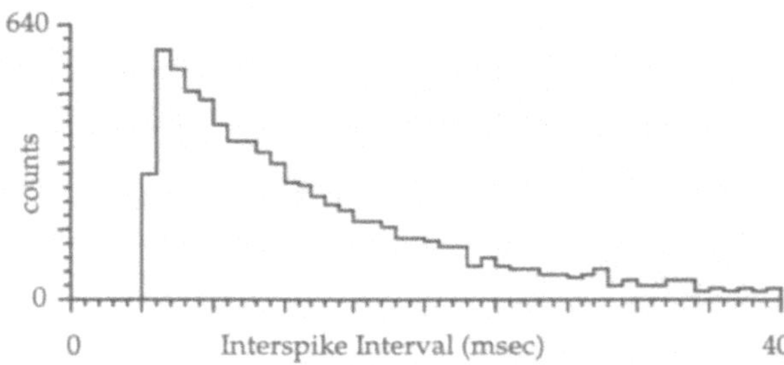

Figure 2. ISI plot for the driver neuron *A* revealing a Poisson distribution, congruent with the randomly firing characteristic of the neuron.

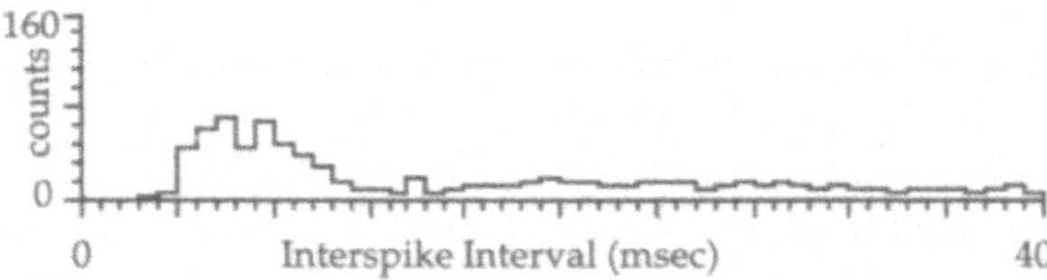

Figure 3. ISI plot for the driven neuron *B* revealing a peak distribution. This shows the regularity of firing even though it is driven by a randomly firing neuron *A*.

636

CI indicate that these two neurons are not tightly coupled in their firings. This is congruent with the fact that these neurons are coupled with a 50% connection strength. This conclusion can be made by observing the latency of firing between these two neurons along the post-CI axis. The latency during this 10 ms integration period (as indicated by the period within the 10 ms ISI of neuron A) is 2.5 ms when neuron B is driven by neuron A. When neuron B fails to follow the firing of neuron A, the latency varies with intervals much longer than the 2.5 ms (as indicated by the scattering of points above 2.5 ms post-CIs).

When the integration period is over (>10 ms), this 2.5 ms coupling disappears (i.e., no points are found along the 2.5 ms horizontal band beyond 10 ms pre-ISIs). Note that the integration period is revealed by the pre-ISI intervals (rather than the post-CI) because it takes into account the interval between the current (reference) spike and the previous spike (integration period) correlated to the next spike firing in the compared neuron. Thus, the relationship between the coupled firing due to temporal integration of two consecutive spikes is clearly quantified graphically.

The unexpected finding is that even though neuron B is driven directly by a randomly firing neuron A, the firing characteristics of neuron B is rather regular instead of random-like due to temporal integration. The ISI histograms illustrate the firing characteristics clearly for these neurons. A Poisson ISI distribution is found for neuron A, as expected for a randomly firing neuron (see Fig. 2), whereas the ISI distribution of neuron B is non-Poisson (see Fig. 3), with a preferred firing interval between 4 and 10 ms (indicated by the peak between 4 and 10 ms ISIs). Thus, this is an illustration of how temporal summation can be detected using this spike train analysis, and how temporal integration can affect the firing characteristics of a neuron.

4. CONCLUSIONS

We present a new spike train analysis technique to extract temporal summation of two consecutive spikes contributing to the firing of a neuron. The results show that not only temporal integration period can be extracted from the new method, but also the analysis has revealed that regularity of firing in a driven neuron can be produced by random inputs provided that the temporal integration period is extended over a finite period of time.

Acknowledgments. This research was supported by the Office of Naval Research (ONR grant numbers N00014-93-1-0135 and N00014-94-1-0686), and the Faculty Research Grant from the University of North Texas.

REFERENCES

[1] Tam, D. C. (1998) A cross-interval spike train analysis: the correlation between spike generation and temporal integration of doublets. Biol. Cybernetics (in press).
[2] Perkel, D.H., Gerstein, G.L. and Moore, G.P. (1967) Neuronal spike trains and stochastic point process. I. The single spike train. Biophys. J., 7: 391-418.
[3] Perkel, D. H., Gerstein, G. L. and Moore, G. P. (1967) Neuronal Spike Trains and Stochastic Point Process. II. Simultaneous Spike Trains. Biophys. J., 7: 419-440.

[4] Rodieck, R.W., Kiang, N.Y.-S. and Gerstein, G.L. (1962) Some quantitative methods for the study of spontaneous activity of single neurons. Biophys. J., 2: 351-368.

[5] Selz, K. A. and Mandell, A. J. (1992) Critical coherence and characteristic times in brain stem neuronal discharge patterns. In: *Single Neuron Computation.* (T. McKenna, J. Davis and S. Zornetzer, eds.) Academic Press, San Diego, CA. pp. 525-560.

[6] Smith (1992) A heuristic approach to stochastic models of single neurons. In: McKenna T, Davis J, Zornetzer S (eds) Single neuron Computation. Academic Press, San Diego, Calif., pp 561-588.

[7] Tam, D. C., Ebner, T. J., and Knox, C. K. (1988) Cross-interval histogram and cross-interspike interval histogram correlation analysis of simultaneously recorded multiple spike train data. J. of Neurosci. Methods, 23: 23-33.

[8] Tam DC, Gross GW (1994) Post-conditional correlation between neurons in cultured neuronal networks Proc World Congress on Neural Networks. San Diego, Calif., 2:792-797.

SUBJECT INDEX

The manufacturer's authorised representative in the EU is Springer
Nature Customer Service Centre GmbH, Europaplatz 3, 69115 Heidelberg,
Germany. If you have any concerns regarding our products, please
contact ProductSafety@springernature.com

Printed and bound by CPI Group (UK) Ltd, Croydon, CR0 4YY

23/06/2026

02146086-0003